Water Resources Monograph 19

MOUNTAIN RIVERS REVISITED

Ellen Wohl

American Geophysical Union
Washington, DC

Library of Congress Cataloging-in-Publication Data

Wohl, Ellen E., 1962-
  Mountain rivers revisited / Ellen Wohl.
       p. cm. — (Water resources monograph ; 19)
  Rev. ed. of: Mountain rivers, c2000.
  Includes bibliographical references and index.
  ISBN 978-0-87590-323-1
  1. Rivers. 2. Mountains. I. Wohl, Ellen E., 1962- Mountain rivers. II.
Title. III. Series: Water resources monograph ; 19.
  GB1203.2W64 2010
  551.48'309143—dc22
                                        2010044835

ISBN 978-0-87590-323-1
ISSN 0170-9600

Copyright 2010 by the American Geophysical Union
2000 Florida Avenue, N.W.
Washington, DC 20009

Cover photographs by Ellen Wohl. (front) Tributary of Tonahutu Creek, Rocky Mountain National Park, Colorado. (back) Rio Sardinal in Braulio Carillo National Park, Costa Rica.

CONTENTS

PREFACE

I wrote the first edition of this book, published in 2000, in response to a need expressed by one of my Ph.D. students at the time, David Merritt, who walked into my office one afternoon for a summary reference on mountain rivers. When I realized that such a reference did not exist, I set out to create one. The inclusion of topics reflected my own belief that rivers need to be examined not solely as physical systems but also as river ecosystems with chemical and biological components that exist in the context of pervasive and long duration human alteration of the environment. As research on topics related to mountain rivers grew dramatically during the past decade, I decided that it was time to write a second edition, and I reorganized the book to reflect my understanding of evolving knowledge.

As with the first edition, this second edition is aimed primarily at an audience already familiar with the basics of river process and form, although the reader with little knowledge of related topics, such as river chemistry, hyporheic zones, or riparian and aquatic ecology, can also gain a quick introductory overview of those topics from this volume. Advanced undergraduates, graduate students, and professional scientists and engineers who possess some general knowledge of river systems will find this volume of use, both for its own sake and to help them build on their existing knowledge of mountain rivers to better understand the unique aspects of these rivers. You can read the book straight through, because each section builds upon the sections that precede it, or use the book as a spot reference to provide a synthesis of current knowledge on specific topics.

The first edition benefited substantially from discussions with, and critical reviews by, Paul Carling (University of Southampton, England), Dan Cenderelli (U.S. Forest Service), Alan Covich (University of Georgia), Janet Curran (U.S. Geological Survey), Jim Finley (Telesto Solutions, Inc.), David Merritt (U.S. Forest Service), and LeRoy Poff (Colorado State University) and AGU reviews by John Costa (U.S. Geological Survey), Avijit Gupta (University of Leeds, England), and Malcolm Newson (University of Newcastle upon Tyne, England). Much of that material is still in this edition, and I thank each of these individuals for their efforts. The second edition has also benefited from discussions with Gordon Grant (U.S. Forest Service), Bob Hilton

Mountain Rivers Revisited
Water Resources Monograph 19
Copyright 2010 by the American Geophysical Union.
10.1029/2010WM001038

(Durham University), Neils Hovius (Cambridge University), Mark Macklin (University of Aberystwyth), and Grant Meyer (University of New Mexico) and reviews by Jim O'Connor (U.S. Geological Survey) and an anonymous reviewer, as well as the enhanced energy and concentration provided by Whole Foods' organic French roast coffee.

As with the first edition, I would like to dedicate this second edition to my graduate students. They continue to challenge, engage, and surprise me and to provide much of the pleasure that comes from working in fluvial geomorphology.

Ellen Wohl
Colorado State University

1. INTRODUCTION

Rivers shape many of the world's landscapes. In the process of transporting water, sediment, and dissolved chemicals from uplands, rivers redistribute mass across the Earth's surface. Rivers set the pace at which weathering and erosion lower landscapes, and control the gradient of adjacent hillslopes. Fundamentally, rivers organize terrestrial landscapes into drainage basins. As the rivers incise or aggrade in response to changes in baselevel, they create valleys that influence local climate; provide travel corridors for animals and humans; and support aquatic and riparian ecosystems that contain some of the Earth's highest levels of biodiversity.

Scientists have systematically studied rivers for more than two centuries. Among the questions asked have been: How do rivers interact with other variables such as climate, lithology and tectonics that influence landscapes? What governs the spatial distribution of river channels? What factors control the yield of water and sediment from hillslopes to rivers? How do interactions between water and sediment influence channel geometry through time and space?

This volume summarizes contemporary understanding of these and other aspects of rivers, in the context of rivers draining mountainous environments. Although the study of rivers is well-established, investigators typically focused on the lowland rivers along which most people live until the final decades of the 20th century. A substantial increase in the amount of research directed toward mountain rivers during the first decade of the 21st century supports the need for this second edition of *Mountain Rivers*, which was originally published in 2000. Increased attention to rivers in mountainous regions results from several trends within science and the greater society. Among these is the focus on numerically simulating landscape evolution over long timespans, which requires that modelers quantitatively parameterize rates of river incision and rates of crustal uplift in mountainous regions. Another factor driving increased investigation of mountain rivers is attempts to maintain or restore rivers as ecological refuges and as critical components of water supply in mountainous regions, which tend to be less densely populated than adjacent lowlands. Finally, mountain rivers with steep, coarse-grained, poorly-sorted beds, and limited sediment supply are typically poorly described by empirical equations for hydraulics and sediment dynamics developed for rivers with lower gradients, making the study of mountain rivers an intellectual and management challenge.

Mountain Rivers Revisited
Water Resources Monograph 19
Copyright 2010 by the American Geophysical Union.
10.1029/2010WM001039

1.1. Characteristics of Mountain Rivers

In this volume I define a mountain river as being located within a mountainous region and a mountainous region as having a mean elevation above sea level ≥ 1000 m [*Viviroli et al.*, 2003]. Each of the continents includes at least one major mountainous region (Figure 1.1). (Selected images appear in print. All images are available on the CD-ROM that accompanies the book.) Mountains cover 52% of Asia, 36% of North America, 25% of Europe, 22% of South America, 17% of Australia, and 3% of Africa, as well as substantial areas of islands including Japan, New Guinea, and New Zealand [*Bridges*, 1990]. Mountain rivers are thus widespread. Because of the steep topography of mountainous regions, mountain rivers typically have a gradient ≥ 0.002 m/m along the majority of the channel length [*Jarrett*, 1992], although substantial longitudinal variability of channel geometry is common in mountainous regions as a result of longitudinal variations in rock resistance, glacial history, and hillslope stability. Lower gradient reaches of channel typically occur upstream of glacial end moraines, massive landslide deposits, or beaver dams, for example, but these reaches create relatively short interruptions between the steeper channel segments up- and downstream.

As with lowland rivers, mountain rivers exhibit great variability in hydrologic regime; channel planform; channel gradient, grain size, and bedforms; sediment dynamics; and aquatic and riparian biota, both within individual mountain ranges and among diverse mountainous regions. Mountain rivers, as defined here, include first-order channels less than a meter wide fed by snowmelt draining an alpine meadow (Figure 1.2); wider rivers cutting steep-walled valleys that dense tropical rain forest vegetation cannot stabilize against periodic landslides (Figure 1.3); ephemeral channels incised into bedrock in arid mountains (Figure 1.4); boreal rivers with cutbanks exposing permafrost (Figure 1.5); and big, powerful rivers like the Indus that carry thousands of kilograms of sediment down to the adjacent lowlands each year (Figure 1.6). Perhaps the only consistent characteristic of mountain rivers is their typically steep gradients, although steep gradients tend to correlate with other characteristics, including

- erosionally resistant and hydraulically rough channel boundaries associated with bedrock and coarse clasts;
- highly turbulent flow with numerous longitudinal transitions between sub- and supercritical flow;
- limited supply of sediment of fine gravel and smaller size;
- bedload movement that is highly variable in space and time, with higher thresholds for initiation of motion than many lowland rivers;
- strongly seasonal discharge regime associated with glacial melt, snowmelt, or seasonal rainfall;
- substantial spatial variability in discharge as a result of spatial variability in precipitation and runoff caused by differences in elevation, basin orientation, and land cover;
- large longitudinal variations in channel geometry associated with variations in tectonics, lithology, glacial history, and sediment supply;

- in some cases, lesser temporal variations in channel geometry than lowland rivers because only infrequent floods or debris flows can exceed boundary resistance sufficiently to cause substantial channel change;
- relatively narrow valley bottoms with limited development of floodplains and lateral movements by rivers;
- in the absence of wide valley bottoms and the associated buffering of stream channels from hillslope processes, mountain rivers have the potential for orders-of-magnitude increase in water and sediment yield over a period of a few years following watershed-scale disturbances such as wildfire or timber harvest; and
- longitudinal zonation of aquatic and riparian biota influenced by river characteristics and by elevation as it relates to temperature and precipitation.

Mountain rivers tend exhibit high degrees of connectivity. *Landscape connectivity* [*Brierley et al.*, 2006] is high because individual landforms such as hillslopes and stream channels are closely coupled within a drainage basin. *Hydrological connectivity* [*Bracken and Croke*, 2007] is high because water moves rapidly from one landform to another and through the entire drainage basin relative to lowland watersheds with extensive groundwater storage. *Sediment connectivity* [*Fryirs et al.*, 2007] is high because limited storage means that sediment moves relatively rapidly from production sites on hillslopes through the drainage basin. Increasing research emphasis on different forms of connectivity reflects a desire to move beyond small spatial and short temporal scales of investigation in order to focus on emergent properties that evolve from the self-organization inherent in river catchments [*Phillips*, 2003; *McDonnell et al.*, 2007; *Reid et al.*, 2007b; *Ali and Roy*, 2009].

1.2. Advances Since the First Edition

Writing the second edition proved to be a much more time-consuming and expansive process than I had initially expected, but this reflects the dynamic nature of contemporary studies of geomorphology and mountain rivers. Many areas of investigation have expanded dramatically since the late 1990s and the volume of associated literature has grown correspondingly. Dramatic increases in the amount of research in topics such as: the interactions of tectonics, topography, and climate [*Willett et al.*, 2006]; hillslope hydrology and modeling [*Franks et al.*, 2005]; debris flows and associated hazards [*Jakob and Hungr*, 2005]; soil development and hillslope processes [*Heimsath et al.*, 2001; *Roering*, 2004]; hydraulics of steep channels [*Ferguson*, 2007]; braided river process and form [*Sambrook Smith et al.*, 2006]; diverse types of numerical models and associated predictions [*Wilcock and Iverson*, 2003; *Tucker and Hancock*, 2010]; geochronology [*Madsen and Murray*, 2009]; and instrumentation [*Jones et al.*, 2007] have made it challenging to keep track of and synthesize the literature. As a result, I have introduced several new sections to the second edition, substantially expanded other areas, and altered the organization of the volume to reflect changing research emphases within the community.

One broadly applicable change is the increasing emphasis on quantification, numerical modeling, and prediction in studies of the Earth's surface. This is exemplified by *Dietrich et al.*'s [2003] call for increased development and application of *geomorphic transport laws.* "A geomorphic transport law is a mathematical statement derived from a physical principle or mechanism, which expresses the mass flux or erosion caused by one or more processes in a manner that: 1) can be parameterized from field measurements, 2) can be tested in physical models, and 3) can be applied over geomorphically significant spatial and temporal scales" [*Dietrich et al.*, 2003, p. 103]. Geomorphic transport laws have been developed for some processes, including soil production from bedrock and river incision into bedrock, but do not yet exist for many geomorphic processes, including landslides, debris flows, and surface wash. Section 1.4 is designed to highlight the existing geomorphic transport laws relevant to mountain rivers and to provide an overarching conceptual framework for reading the succeeding, more detailed discussions of each of the processes and forms briefly mentioned in section 1.4.

1.3. Purpose and Organization of This Volume

This volume on mountain rivers is intended for the reader who already has a basic understanding of fluvial geomorphology, as developed in texts including *Leopold et al.* [1964], *Schumm* [1977], *Morisawa* [1985], *Richards* [1987], *Easterbrook* [1993], *Ritter et al.* [1995], *Bloom* [1998], *Knighton* [1998], *Bridge* [2003], or *Anderson and Anderson* [2010]. The emphasis of this volume is on channel processes and morphology, but the volume also includes brief reviews of other aspects of mountain rivers. The second chapter focuses on form and process at the scale of drainage basins (10^1-10^6 km^2), starting with interactions among tectonics, climate, and topography, and then reviewing hillslope processes, channel initiation and arrangement in a network, and valley geometry, including changes in process and form during the Quaternary. The third chapter covers process at the channel scale (10^{-2}-10^1 km^2), including hydrology, hydraulics, sediment dynamics, river chemistry, instream wood, and physical disturbances such as floods and debris flows. The fourth chapter examines types of channel morphology characteristic of mountain rivers and the fifth chapter discusses aquatic and riparian communities of mountain rivers. The sixth chapter explores human interactions with mountain rivers.

The diversity of topics addressed in this volume is designed to promote the realization that a mountain river is an integrated physical, chemical and biological system influenced by controls acting across various scales of time and space. The need to move beyond traditional disciplinary boundaries is reflected in the discussion of *Earth system science* starting in the late 20th century. A system is a collection of interdependent parts enclosed within a defined boundary; in this case, the interdependent parts within the boundary of the Earth are the lithosphere, hydrosphere, biosphere, and atmosphere. Emphasis on a systems approach reflects an increasing realization that we cannot effectively respond to global warming, contaminant dispersal, and other contemporary challenges unless we think about natural processes in ways that transcend disciplinary boundaries. The establishment of critical zone observatories in the United States (the *critical zone* is defined as the Earth's outer layer, from the lower atmosphere and vegetation canopy to the soil and groundwater, which sustains living organisms) is also designed to promote integrative study of surface processes and landforms. Rivers provide an obvious mechanism for integrative thinking because a seemingly simple, discrete channelized flow of water in fact reflects influences from high in the atmosphere to deep in the crust and across hemispheres.

This volume is primarily an integration and synthesis of existing knowledge of mountain rivers. Although it is not feasible to cite every published study on all aspects of mountain rivers, the list of references at the end of the volume is unusually long because I wanted to be as inclusive as possible. I have avoided citing abstracts or unpublished theses or dissertations unless these are the only published material relevant to a particular topic and I have mostly avoided citing references that are not in English. Because this volume focuses primarily on physical processes, the discussions and reference lists for river chemistry and for aquatic and riparian ecology are not as complete as those for other topics treated in this volume. Topics of which we have particularly limited knowledge are highlighted throughout this synthesis and the concluding summary emphasizes aspects on which further research is particularly needed.

1.4. A Mountain River Described and Enumerated

Headwater regions encompass substantial spatial and temporal variations in geomorphic processes. The upstream extent of the channel network represents the transition from hillslope to channel processes, and downstream portions of channel networks in steep terrain include the transition from debris flows to fluvial processes, as well as substrate transitions such as bedrock to gravel and gravel to sand [*Sklar and Dietrich*, 1998; *Montgomery*, 1999; *May*, 2007; *Stock and Dietrich*, 2003]. Figure 1.7 presents a schematic overview of the components of mountain rivers discussed in this volume and, where possible, examples of equations developed to quantify these components. These equations are discussed in detail in succeeding portions of the text. Some of the equations are developed from a theoretical basis such as a balance of forces; others are empirical equations that may be of limited usefulness when extrapolated beyond the data from which they were developed. Whether theoretically or empirically based, quantitative statements of geomorphic process and form help to guide and focus continuing research by identifying processes or forms that we cannot yet adequately parameterize or that deviate from existing observations.

Building on *Schumm*'s [1977] zonation of a fluvial system into three basic zones of production, transfer and deposition, Figure 1.7 organizes mountain rivers into three primarily spatial zones, each of which is dominated by a distinct suite of geomorphic processes and landforms. The *colluvial-fluvial transition* area occupies the uppermost portion of the drainage basin, where sediment produced from bedrock weathering is

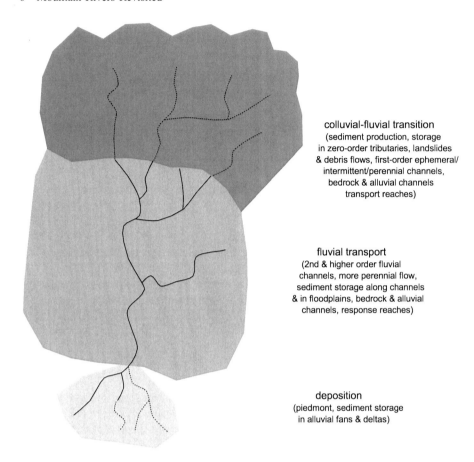

colluvial-fluvial transition
(sediment production, storage
in zero-order tributaries, landslides
& debris flows, first-order ephemeral/
intermittent/perennial channels,
bedrock & alluvial channels
transport reaches)

fluvial transport
(2nd & higher order fluvial
channels, more perennial flow,
sediment storage along channels
& in floodplains, bedrock & alluvial
channels, response reaches)

deposition
(piedmont, sediment storage
in alluvial fans & deltas)

Figure 1.7 Highly stylized illustration of the three primary zones of a mountain drainage basin, followed by some of the equations used to describe process and form in each of those three zones. Variables used in each equation are defined in subsequent portions of the text.

moved downslope into channels by mass movements such as debris flows and landslides, and where fluvial channels begin. Channels in the *fluvial transport zone* in the middle section of the basin typically have progressively less direct hillslope influences as wider valley bottoms and floodplains buffer materials coming from hillslopes by creating at least temporary storage zones. Lower gradients, less lateral confinement, and/or lower velocity and discharge facilitate deposition along channels in the *depositional zone*, which is typically beyond the mountain front but may also occur in locally wider valleys.

This downstream zonation of mountain drainage basins reflects progressive downstream trends in discharge, gradient, grain size and other stream characteristics that numerous investigators have documented across a range of mountain drainage basins. Other variables that do not show progressive downstream trends also characterize mountain drainage basins; hydraulic resistance and magnitude of bedload transport, for example, do not necessarily change progressively downstream. Most variables show both progressive downstream trends and dominantly local (10^1-10^3 m) variation, depending on the spatial scale under consideration: Gradient and grain size both decrease downstream at the scale of a larger mountain watershed, but can exhibit local reversals as a result of spatial and temporal variation in driving factors such as lithology, tectonic uplift or hillslope stability and associated sediment inputs (Figure 1.8).

The local, and potentially longitudinally discontinuous, values of some parameters support the concept of *geomorphic process domains*. Spatial variability in geomorphic processes governs temporal patterns of disturbances that influence ecosystem structure and dynamics [*Montgomery*, 1999]. Mass transfer in the uppermost portions of hillslopes might be dominated by avalanches and rockfall, for example, whereas debris flows exert a greater influence in the middle portions of the catchment, and fluvial processes dominate the lower portions.

One way to conceptualize mountain river form and process is within the framework of driving forces versus substrate resistance. Channel configuration at any point along the drainage network fundamentally reflects the ratio of hydraulic driving forces to substrate resistance. *Hydraulic driving forces* reflect the movement of a volume of water from higher to lower elevation and thus incorporate discharge and channel gradient. The potential energy converted to kinetic energy via the downstream flow of water can be expended on overcoming external frictional resistance, internal frictional resistance, and sediment transport; the expenditure of energy thus incorporates channel configuration, sediment supply, and the erodibility of the channel boundaries. The ratio of driving forces and *substrate resistance* varies temporally as tectonic uplift alters landscape relief or storms passing over the watershed or land use alter water and sediment yield to the channel. The ratio also varies spatially as progressively greater contributing area increases discharge in the channel or as the channel flows from glaciated to unglaciated portions of the catchment. Some forms of spatial variation, such as downstream increase in discharge, are well documented from a range of field settings and are best described as linear or exponential functions. Some forms of spatial variation, such as the magnitude of external frictional resistance, may show analogous downstream trends, but lack extensive field documentation. Other forms of spatial variation, such as bank resistance created by riparian vegetation, are not adequately described by linear or exponential functions and appear to predominantly reflect local controls that do not vary progressively downstream. Figure 1.9 lists channel forms and processes and what is known about their downstream trends in mountain rivers. Although limited work to date suggests that hydraulic driving force as reflected in stream power peaks in the upper third to middle part of the basin [*Knighton*, 1999], substrate resistance is so spatially variable in mountain drainage basins that it precludes generalizations. It may thus be more useful to apply the ratio of driving force to substrate resistance at the local scale rather than at the basin scale.

Parameter	Downstream Trend	Documentation
discharge (Q)	exponential increase	strong
gradient (S)	exponential decrease	strong
valley geometry	highly variable	limited
sediment supply	highly variable	limited
external resistance (f)	declines downstream	limited
total stream power	peaks at mid-basin	limited
suspended sediment	highly variable	limited
bedload transport	highly variable	moderate
bedforms	progressive change with S	strong
sinuosity	highly variable	limited
channel lateral mobility	highly variable	limited
bank resistance from		
riparian vegetation	highly variable	limited
instream wood	highly variable	limited

Figure 1.9 Downstream trends in selected parameters for mountain rivers and relative documentation (with progressively less documentation from strong through moderate to limited) of these trends based on field data from diverse settings.

Each of the very broad parameter categories outlined in Figure 1.9 is explored in greater detail in subsequent sections of this book, but Figure 1.9 provides a quick overview of our relative understanding of diverse patterns in mountain drainage basins. This figure also indicates how much work remains to be done.

1.4.1. North St. Vrain Creek, Colorado, USA

I use the specific example of North St. Vrain Creek in the Colorado Front Range, USA to further illustrate how individual parameters vary downstream or locally. I chose this watershed because it is one of the least altered by land uses in the region and because I have done much of my own research there. North St. Vrain Creek represents neither an exceptionally well-studied watershed nor a little known one; it falls somewhere between these extremes and in this respect represents many other mountainous drainages.

North St. Vrain Creek drains eastward from the Continental Divide (4050 m elevation) onto the Great Plains (1945 m elevation at the base of the mountains) and eventually joins the South Platte River (Figure 1.10A). The portion of the catchment within the mountains includes 250 km^2 of steep terrain underlain by Precambrian-age granites, gneiss, and schist [*Tweto*, 1979]. The Front Range has been relatively tectonically quiescent since the early Tertiary [*Crowley et al.*, 2002; *Anderson et al.*, 2006b]. Pleistocene valley glaciers extended down to approximately 2500 m elevation [*Madole et al.*, 1998]. Narrow, glaciated spines form the range crests at 4000 m elevation, below which lie widespread surfaces of low relief at 2300-3000 m elevation. Fluvial canyons are deeply incised into these low-relief surfaces [*Anderson et al.*,

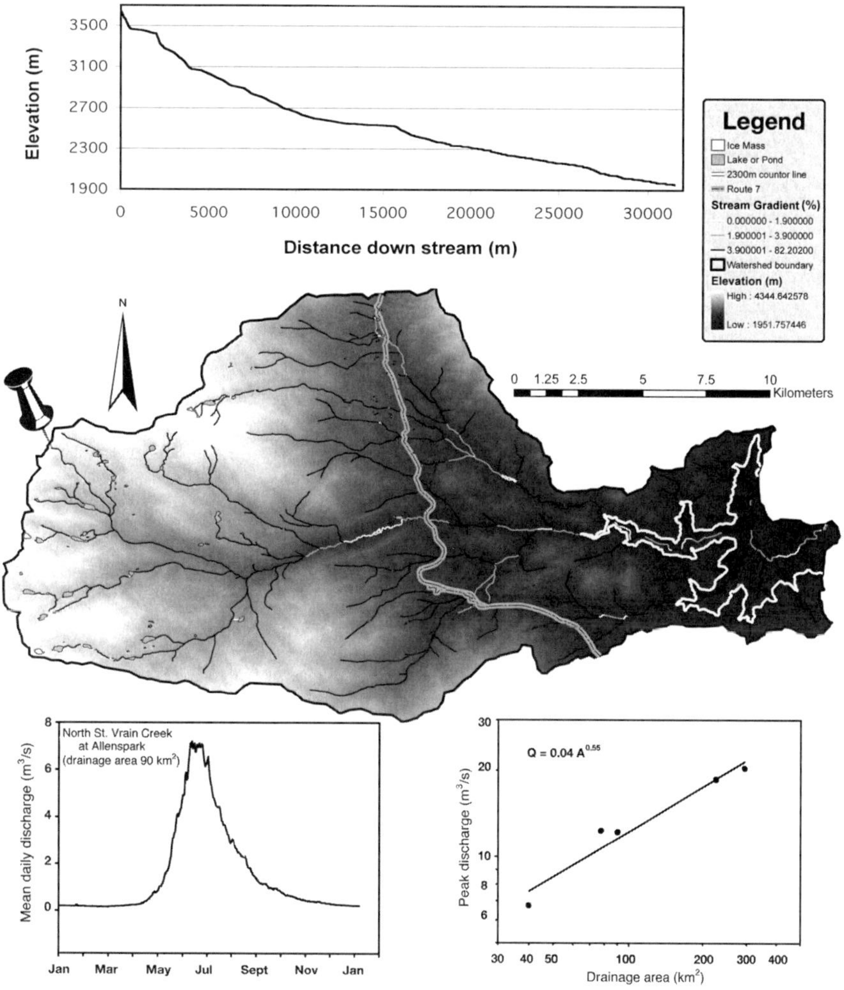

Figure 1.10 NSV map.

2006b]. Most bedrock outcrops in the region are densely jointed, and joint spacing and valley geometry correlate with the location of shear zones of Precambrian and Laramide age [*Abbott*, 1976]; wider, lower gradient portions of fluvial valleys typically correspond to more closely spaced joints and the location of shear zones [*Ehlen and Wohl*, 2002]. Variations in joint density, glacial history, and other large-scale controls create pronounced downstream variations in valley and channel geometry.

Snowmelt runoff dominates the annual hydrograph at all elevations within the catchment, producing a sustained May-June peak. On average, 85% of the annual flow

Parameter	Downstream Trend and Notes
discharge (Q)	exponential increase; as illustrated in Figure 1.10(A), peak annual discharge varies with drainage area with an exponent of 0.55, based on stream gage records covering multiple years at five sites with unregulated flow in and near the North St. Vrain catchment
gradient (S)	exponential decrease; as illustrated in Figure 1.10(A), longitudinal variation in stream gradient is readily obtained from 10-m DEM coverage of the catchment; spatial variation reflects primarily Pleistocene glacial history
valley geometry	highly variable; valley geometry can be directly estimated from 10-m DEMs via metrics such as connectedness (lateral distance between channel and base of valley wall) and entrenchment (ratio of channel width to valley width) or indirectly estimated from stream gradient on 10-m DEMs; Figure 1.10(A) illustrates spatial variation in stream gradient within the catchment, and the steepest gradient segments correspond to relatively deep, narrow valleys(< 50 m wide valley bottom), the moderate gradient segments to valleys of intermediate width and depth, and the lowest gradient segments to glacial troughs and broad valleys (> 50 m wide valley bottom) with meadows and wetlands
sediment supply	highly variable; little documentation in the North St. Vrain catchment, but volume and frequency likely vary with valley geometry, with coarser grained sediment episodically entering channels in mass movements along steep, narrow valley segments
external resistance (f)	declines downstream; limited documentation indicates that, as S decreases downstream, f also decreases (Wohl et al., 2004)
total stream power	peaks at mid-basin, as predicted by Knighton (1999), although values of stream power display a substantial amount of scatter rather than following smoothly ascending or descending trends (Wohl et al., 2004)
suspended sediment	highly variable; limited documentation indicates that suspended sediment increases during the annual snowmelt peak flow and following disturbances such as wildfire or debris flows
bedload transport	highly variable; limited documentation indicates increasing bedload transport in slightly finer grained channel segments downstream
bedforms	progressive change with S; spatial distribution of cascade, step-pool, plane-bed, and pool-riffle segments correlates well with S and can thus be predicted using 10-m DEM data (Wohl et al., 2004, 2007)
sinuosity	highly variable; like valley geometry, this correlates with S and can thus be indirectly estimated from 10-m DEM data, with high gradient corresponding to straight channels and lower stream gradients corresponding to greater sinuosity
channel lateral mobility	highly variable; as with sinuosity, this correlates with stream gradient and can be indirectly estimated from 10-m DEM data; steeper channels have lower lateral mobility than channels of lower gradient
bank resistance from riparian vegetation	highly variable; type of riparian vegetation varies with elevation and with valley geometry and can thus be indirectly estimated from 10-m DEM data; lower gradient channel segments flowing through relatively wide valleys are more likely to have dense herbaceous vegetation and willow (*Salix*) communities in relatively wide bands along the channel, whereas steep channel segments have limited riparian communities dominated by coniferous trees (Polvi, 2009)
instream wood	highly variable; limited documentation (e.g., Wohl and Jaeger, 2009; Wohl and Cadol, in press) indicates that higher wood loads and more frequent channel-spanning jams correspond to lower gradient stream segments

Figure 1.10 (continued)

occurs between May and September. Elevations below 2300 m also experience flash floods caused by summer convective storms. Rivers above this elevation have unit discharges of ~ 1 m^3/s/km^2, whereas rivers below 2300 m can have unit discharges of 40 m^3/s/km^2 [*Jarrett*, 1989]. Climate in the Front Range varies with elevation. Mean annual temperature varies from 1°C at the highest elevations to 11°C at the base of the range. Mean annual precipitation decreases from approximately 100 cm at the highest elevations to 36 cm at the mountain front, and the percentage of precipitation falling as snow also decreases with elevation.

Vegetation communities also vary with elevation, from alpine tundra above 3400 m, through subalpine spruce-fir forest, and montane pine forest below 2700 m [*Veblen and Donnegan*, 2005]. Wildfire and insect outbreaks are the most important forest disturbances in terms of extent, severity, and frequency. Three general types of historic fire regimes present in the catchment are: (i) infrequent, high-severity fires that kill all canopy trees over areas of hundreds to thousands of hectares and recur at intervals greater than 100 years in the subalpine zone; (ii) a complex pattern of low- and high-severity fires that burn areas of approximately 100 ha and recur at intervals of 40 to 100 years in the middle and upper montane zone; and (iii) frequent, low-severity firest that burn mainly the ground surface over areas of approximately 100 ha at intervals of 5-30 years in the lower montane zone [*Veblen and Donnegan*, 2005].

Beaver were trapped along the channels of the watershed starting in the early 19th century; the creek is named for French fur trapper Ceran St. Vrain. Although beaver have gradually recolonized the watershed, their populations are smaller than prior to trapping [*Wohl*, 2001]. The watershed is bisected by a two-lane highway; portions of the catchment upstream are largely in Rocky Mountain National Park and the mountainous portion downstream is largely in the Roosevelt National Forest. Flow in the creek is regulated starting at the base of the mountains. The information summarized in Figure 1.10 is drawn primarily from *Thompson et al.* [1996, 1999], *Wohl et al.* [2004], *Flores et al.* [2006], *Polvi* [2009], *David et al.* [2010], and *Wohl and Cadol* [in press]; with the exception of *David et al.* [2010], which is based on data collected in nearby drainages, these studies were conducted within the North St. Vrain catchment. Figure 1.10B reiterates Figure 1.9 with respect to the North St. Vrain catchment.

My research on North St. Vrain Creek and other mountainous catchments around the world has led me to conceptualize form and process in mountain rivers as illustrated in Figure 1.11. In this figure reach-scale gradient assumes primary importance. Gradient at channel lengths of 10^1-10^3 m can be a quasi-independent variable when the river does not have sufficient energy to create a smoothly concave longitudinal profile as a result of longitudinal variations in uplift rate, rock resistance, glacial history, sediment supply, or other parameters that influence gradient. Many other parameters correlate directly with reach-scale gradient (the solid arrows in Figure 1.11) and indirectly via intermediary parameters (the dashed arrows in Figure 1.11). Channel reaches of lower gradient, for example, correlate with wider valley bottoms or lower levels of connectedness (average distance from the channel edge to the valley edge) and higher values of entrenchment (ratio of valley width to channel width) [*Polvi*, 2009]. Wider valley

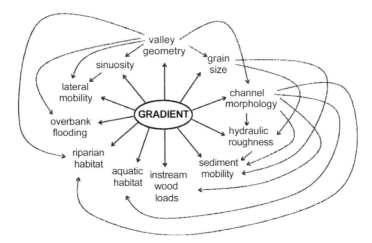

Figure 1.11 Schematic illustration of the correlations among variables along a mountain river. Reach-scale gradient assumes primary importance in this diagram because so many other variables directly or indirectly correlate with gradient, and because gradient is readily obtained from topographic data such as digital elevation models.

bottoms in turn correlate with greater sinuosity, lateral channel mobility and overbank flooding, riparian habitat associated with greater inundation and higher water tables, finer grain sizes in the streambed, and channel morphology such as pool-riffle or dune-ripple [*Montgomery and Buffington*, 1997; *Wohl et al.*, 2007; *Polvi*, 2009]. These channel morphologies associated with lower gradient have lower levels of hydraulic resistance [Darcy-Weisbach *f* or Manning's *n* coefficients; *Wohl et al.*, 2004; *David et al.*, 2010], greater sediment mobility, larger instream wood loads [*Morris et al.*, 2010; *Wohl and Cadol*, 2010], and greater pool volume than high-gradient channels.

Most mountainous catchments around the world now have some level of topographic data available, allowing reach-scale gradient to be quantified at varying degrees of spatial resolution. The correlations between reach-scale gradient and a wide variety of other parameters thus provides an entry point for understanding at least relative spatial variations in multiple parameters within a catchment. Field calibration of these relations can of course improve the ability to specify the degree of variation within variables such as grain size or instream wood load with respect to gradient.

2. MOUNTAIN DRAINAGE BASINS

This chapter begins with a brief overview of regional-scale interactions among climate, tectonics, and erosion as these influence the form and process of mountain drainage basins. Section 2 discusses hillslope process and form, including the production of sediment from bedrock and the distribution of that sediment along hillslopes through mass movements and diffusive sediment transport. Section 3 returns to climate in the context of understanding the types of precipitation that influence mountainous catchments, the spatial and temporal heterogeneities in the distribution of precipitation across high-relief terrain, and the surface and subsurface paths that water follows as it moves down hillslopes and into channels. An overview of the processes that influence how and where channels initiate (section 4) is followed by a review of different morphometric basin parameters and how these interact with the movement of water to influence hydrographs (section 5). Section 6 covers characteristics of valley morphology in mountain drainages. Sections 7 and 8 address longitudinal profiles, with an emphasis on processes and modeling of bedrock channel segments. Section 9 covers paleo-longitudinal profiles as preserved in terraces, and section 10 addresses alluvial fans, which can occur throughout mountainous drainages wherever wider valley bottoms and lower gradients facilitate persistent deposition. The common organizing framework of this chapter is that of the larger-scale variables and processes that influence the geomorphology of a drainage basin and how these interact with water falling as precipitation and then moving down hillslopes, with bedrock weathering and sediment production and transfer, and with tectonics. It is useful to begin with a brief review of how mountainous topography originates.

Mountainous regions are produced by four general types of deformation; folding, volcanism, fault block uplift, and vertical uplift [*Press and Siever*, 1986]. Folded mountains result from lateral compression, usually at the convergent boundary between two tectonic plates. Examples include the Appalachian Mountains of the eastern United States, the Alps of southern France, the Urals at the boundary between Europe and Asia, and the Transantarctic Mountains. The topography of folded mountains may be controlled by differential weathering of the lithologies exposed by uplift, with more resistant lithologies forming steeper slopes.

Volcanic mountains generally form at a divergent or convergent plate boundary, or at an intraplate hot spot such as the Hawaiian Islands. Examples of volcanic mountain

Mountain Rivers Revisited
Water Resources Monograph 19
Copyright 2010 by the American Geophysical Union.
10.1029/2010WM001040

ranges include the highlands of New Guinea, the North Island of New Zealand, the Japan Alps, and the Cascade Range of the northwestern United States. The geologic controls on volcanic islands are a function of the style of eruption and chemical composition of the lava.

Fault block uplift tends to produce mountain ranges with one very steep side parallel to the fault and a gentler side that does not have an active fault, as in the Teton Range of Wyoming, USA. Mountains produced by vertical uplift have faults parallel to both sides of the range, as in the Front Range of Colorado, USA.

Most of the world's major mountain belts include folded, faulted, and volcanic regions, as well as igneous plutons. The Himalaya mountain ranges, for example, include the folded and thrust-faulted zone of the Siwalik hills at the southern margin, and thrust-faulted and intruded rocks in the Middle Himalaya and the High Himalaya [*Bridges*, 1990]. The Andes Mountains of western South America include high volcanic peaks, folded belts, igneous intrusions, and extensive faults [*Bridges*, 1990].

2.1. Mountain Rivers and Tectonics

Early work on mountain rivers and tectonics emphasized the effect of mountainous topography and active uplift on drainage networks, noting that channels do not always follow existing slopes. Working in the western United States, *Powell* [1875, 1876] described both antecedent drainage networks in which pre-existing channels had maintained their spatial arrangement while the underlying landmass was deformed and uplifted, and superimposed channels which had incised downward to a buried structure. Either scenario could result in a river flowing through or across a mountain range (transverse drainage), rather than channels draining from the crest of the range downward to the neighboring lowlands (Figure 2.1). Thus, the drainage network was not a consequence of present topography. More recent studies discuss transverse drainage associated with the Coastal Range of Taiwan [*Lundberg and Dorsey*, 1990], isostatic uplift of the Apuseni Mountains, Hungary [*Thamo-Bozso and Kercsmar*, 2002], and the Betic Cordillera of southeastern Spain [*Stokes and Mather*, 2003]. *Humphrey and Konrad* [2000] argue that river sediment flux and tectonic uplift rate are the most important variables in determining whether a river will incise through or divert around an evolving bedrock uplift. Flume experiments testing the four general mechanisms proposed for transverse drainage (antecedence, superimposition, overflow, piracy) supported the ability of all of the mechanisms except superimposition to produce transverse drainage [*Douglass and Schmeeckle*, 2007]. Investigators continue to use drainage pattern, as well as terrace and channel geometry, as indicators of tectonic effects on rivers [*Schumm et al.*, 2000].

For cratonic or passive-margin settings, *Young* [1989] argues that although the alignment and form of individual valleys may reflect surface variations in lithology and structure, deeper crustal features control drainage patterns at the sub-continental scale. These deeper crustal features may only be discernible using remote-sensing technology to detect patterns such as Bouguer gravity anomalies [*Young*, 1989]. *Brookfield*

[1998] describes the importance of tectonic history in creating three regionally distinctive patterns among the major river systems of Asia. Differential compression and right-lateral shear produced highlands from which rivers of the Helmand-Farah system drain into arid depressions. Differential shear and clockwise rotation between the compressing Tibetan plateau and Southeast Asia produced large sigmoidal bends in widely separated rivers including the Chang Jiang, Mekong, and Salween. Southward thrusting and massive frontal erosion of the Himalaya caused progressive truncation of rivers including the Tsangpo, Indus, and Sutlej on the plateau [*Brookfield*, 1998]. Drawing on flume and field studies, *Schumm et al.* [2000] summarize how different alluvial channel morphologies respond to various types of tectonic deformation, with channel response partly governed by proximity to the axis of deformation. Bedrock rivers can show analogous spatially variable responses; following the 1999 Chi-Chi earthquake in Taiwan, river incision intensified near the fault scarp, whereas landslides induced by the earthquake mantle the river bed with sediment and impede bedrock incision in reaches distal to the fault [*Yanites et al.*, 2010a].

Recent work tends to emphasize more complex interactions in which rivers influence, as well as respond to, mountainous topography. Investigators have proposed for decades that arrangement and incision of valley networks can affect mountain relief and elevation. *Corbel* [1959] represents an early study indicating that rivers remove up to five times more sediment per unit area from mountain basins than from lowland basins. This type of comparison emphasizes the role of mountain rivers as conveyors of sediment from upland regions. Starting in the 1990s, investigators took this insight to the next level and proposed that the pattern of river incision can affect crustal structure in mountain belts by changing the distribution of stress in the crust [*Hoffman and Grotzinger*, 1993; *Beaumont and Quinlan*, 1994].

The effect of interactions among tectonic forces, climate and erosive processes in shaping mountainous topography has been the subject of much recent attention [*Koons*, 2009]. *Raymo et al.* [1988] and *Raymo and Ruddiman* [1992] propose that accelerated tectonic uplift increase weathering. Subsequent field studies support this assertion [*Carey et al.*, 2006]. Because chemical weathering is an important sink for CO_2, the removal of CO_2 from the atmosphere could have produced lower temperatures during the past 40 million years, facilitating glacial cycles. Glacial erosion may then have accelerated uplift and mountain building as removal of mass facilitated isostatic uplift [*Molnar and England*, 1990; *Hallet et al.*, 1996]. This is the so-called *glacial buzzsaw effect*; glacial erosion rapidly removes mass raised above the altitude of the local glacier equilibrium line [*Brozovic et al.*, 1997; *Whipple et al.*, 1999; *Montgomery et al.*, 2001; *Brocklehurst and Whipple*, 2002; *Mitchell and Montgomery*, 2006; *Naylor and Gabet*, 2007]. Evidence for the glacial buzzsaw effect comes primarily from field studies in areas with large glaciers. Investigations in regions with smaller alpine glaciers [*Foster et al.*, 2008], as well as tests using numerical models [*Tomkin and Braun*, 2002; *Tomkin*, 2007], suggest that the hypothesized effect is complicated by other factors such as whether the glacier base is frozen, how fracture density influences the pace of glacial erosion [*Dühnforth et al.*, 2010], and that there is a minimum magnitude of glacier erosion below which insufficient rock mass is removed to isostatically raise summit elevations [*Foster et al.*, 2010]. *Alley et al.* [2003] note that the long profiles of beds of highly erosive glaciers tend towards steady-state angles related to the overlying ice-air surface slopes, beyond which additional subglacial deepening depends on non-glacial processes. This suggests a limit to the erosion conceptualized in the glacial buzzsaw effect. Quantification of glacial erosion patterns using cosmogenic radionuclides [e.g., *Li et al.*, 2005] has substantially enhanced the ability to test hypotheses such as the glacial buzzsaw.

Hales and Roering [2009] propose a frost buzzsaw. Noting that rockfall controls erosion in the eastern Southern Alps of New Zealand, and that frost cracking is the primary rockfall mechanism, they correlate climate and elevational controls on frost cracking intensity with the elevation of the highest peaks and suggest that the height of these peaks is limited by a frost buzzsaw.

A further complication of the climate-tectonics-erosion interactions is that formation of large ice sheets results in substantial, albedo-induced cooling of the Earth's atmosphere. *Kuhle* [2007] describes a scenario in which development of a Tibetan ice sheet occurred as the Tibetan Plateau was lifted above snowline. Albedo-induced cooling from the ice sheet disrupted the summer monsoon circulation and facilitated the global depression of snowline and development of other ice sheets. Glacial-isostatic lowering of Tibet caused melting of the ice sheet during a period of positive radiation anomalies, which triggered an interglacial period. Glacial-isostatic rebound then lifted the plateau above snowline, triggering the next glacial period [*Kuhle*, 2007]. These alternating episodes of glacial advance and retreat influenced river dynamics by changing the supply of meltwater and sediment [*Rahaman et al.*, 2009] and in other ways: Steep fluvial knickpoints formed at the southeastern margin of the Tibetan plateau should erode rapidly back into the plateau, but *Korup and Montgomery* [2008] propose that the plateau edge has been preserved because numerous moraine dams on major rivers impede bedrock river incision.

Apart from large-scale glacial erosion, various field and modeling studies indicate that at smaller, regional scales, spatial gradients in the climate forcing that drives erosion can influence the development of geologic structures [*Hoffman and Grotzinger*, 1993; *Willett et al.*, 1993, 2001; *Avouac and Burov*, 1996; *Horton*, 1999; *Willett*, 1999]. This is expressed in the *tectonic aneurysm model* [*Zeitler et al.*, 2001] (Figure 2.2) in which local rheological variations arise in a deforming orogen as a result of deep and rapid incision. The crust weakens as the strong upper crust is locally stripped from above by erosion. The local geotherm is then steepened from below by a focused rapid uplift of hot rock. If efficient erosion continues, a positive feedback keeps material flowing into this weakened zone, which maintains local elevation and relief [*Koons et al.*, 2002; *A. L. Booth et al.*, 2009; *A. M. Booth et al.*, 2009]. These ideas led to numerous studies of the interactions among uplift, river incision, and climate fluctuations, and the extent and magnitude of glacial versus nonglacial erosion [*Harbor and Warburton*, 1993; *Burbank et al.*, 1996; *Hallet et al.*, 1996; *Whipple and*

Tucker, 1999; *Galy and France-Lanord*, 2001; *Lavé and Avouac*, 2001; *Dadson et al.*, 2003; *Snyder et al.*, 2003; *Korup et al.*, 2005; *Anderson et al.*, 2006a; *Barros et al.*, 2006; *Schaller and Ehlers*, 2006; *Smith*, 2006; *Anders et al.*, 2010; *Binnie et al.*, 2010; *Pelletier et al.*, 2010]. The result of this work is consensus that climate, erosion, and tectonics are strongly coupled through large-scale feedback systems [*Montgomery*, 2004a]. Gradients in climate [*Bookhagen and Burbank*, 2010] and tectonic forcing influence erosional intensity, which governs the development of topography, which in turn influences climate and tectonics. These interactions can be expressed in steady-state longitudinal river profiles along which different degrees of curvature reflect orographically-induced variations in precipitation [*Roe et al.*, 2002].

Several other studies also address the effect of valley incision on mountain topography. Modeling the effect of isostatically compensated valley incision on the elevation of mountain peaks, *Montgomery* [1994b] finds that this compensation could account for at most 5-10% of the present elevation of mountain peaks in the central Sierra Nevada of California, USA and the Tibetan Plateau and *Mitchell et al.* [2009] estimate that it adds <25% of height to peaks in the Cascade Range of Washington, USA. Such compensation could account for 20-30% of the present elevation of peaks in the Himalaya, however. *Montgomery and Stolar* [2006] propose that Himalayan river anticlines (major Himalayan rivers flow parallel to and down the axis of anticlines oriented transverse to the primary structural grain of the range) are the consequences of focused rock uplift in response to significantly larger net erosion along major rivers than in surrounding regions. Even in areas with less rapidly changing baselevel, such as the highlands of eastern Australia, the headward erosion of river gorges is the most important process denuding these highlands during the last 30 million years [*Nott et al.*, 1996]. Stream erosion of new drainage basins in extensional mountain ranges of the southwestern United States exceeds hillslope retreat, leading to elevation of summit plateaus [*Harbor*, 1997].

Recent developments in geochronology facilitate estimation of regional erosion rates in mountains. Techniques include use of cosmogenic ^{10}Be in sediment carried by streams (because minerals at depth are shielded from cosmic rays, ^{10}Be concentration when minerals reach the surface indirectly records their exhumation rate) [*Kirchner et al.*, 2001] and low-temperature thermochronologic data in which spatial patterns of mineral cooling ages are related to the rates at which buried rocks move toward the surface [*Safran*, 2003; *Schildgen et al.*, 2009]. Nuclides such as ^{10}Be or ^{26}Al are produced when secondary cosmic rays interact with the uppermost layer of the Earth's surface. The nuclides are produced within a characteristic depth scale of about 1 m, so that measured concentrations in rock exposures record erosion rates at that point and concentrations in sediments record an integrated denudation history while material passed through this depth interval [*Bierman and Nichols*, 2004]. Depending on the denudation rate, the resulting integration time scales are 10^3-10^5 years, providing a long-term estimate of denudation [*von Blanckenburg*, 2005]. Interpretation of erosional history from cosmogenic isotope ages requires some knowledge of geomorphic processes: Using a numerical simulation of cosmogenic nuclide production and distribution, *Niemi et al.* [2005] find that larger catchment areas must be sampled to accurately evaluate long-term erosion rates as the frequency of landsliding increases, and that sediment sampling is more appropriate than sampling bedrock surfaces in regions dominated by mass movement processes.

Luminescence dating is also applied to hillslope sediments. Luminescence techniques utilize the ability of some natural crystalline materials to store energy released by background radioactive decay over long periods. The stored energy can be released by stimulation by heat (thermoluminescence) or light (optically stimulated luminescence). *Fuchs and Lang* [2009] review luminescence dating of hillslope deposits and *Madsen and Murray* [2009] review optically stimulated luminescence dating of sediments <1,000 years in age.

Remote sensing imagery, digital elevation models, and geomorphometry – the quantitative description and analysis of geometric-topologic characteristics of the landscape – have been key to quantifying parameters such as relief, glacial and fluvial dissection, and hillslope and valley geometry [*Bishop et al.*, 2002, 2003; *Misukoshi and Aniya*, 2002; *Montgomery*, 2004a] and to modeling feedbacks among tectonic forcing, erosion, isostatic rebound, and rock exhumation [*Montgomery*, 2001b]. *Bishop et al.* [2004] review remote-sensing techniques and *Rasemann et al.* [2004] review geomorphometric variables and analysis in mountain environments using GIS. *Hengl and Reuter* [2009] provide a comprehensive overview of geomorphometry.

Physical experiments also provide insight into interactions among uplift and erosion at spatial scales from a single channel segment to entire watersheds [*Schumm et al.*, 1987; *Ouchi*, 2004]. *Lague et al.* [2003] use physical experiments to investigate landscape response to uplift and erosion and find that topography always reaches a steady state, with a mean elevation linearly dependent on uplift rate. Their steady-state surfaces exhibit a well-defined slope-area power law with a constant exponent of -0.12, a result consistent with a stream power erosion model (equation 2.31) that includes a non-negligible threshold for particle detachment.

Because bedrock channel incision can exert such an important control on hillslope stability and regional rates of uplift and erosion, many investigators have used river morphology to interpret the scale, magnitude, and timing of rock uplift, for which other evidence is often limited. Although measures such as mountain front sinuosity can be used [*Pérez-Peña et al.*, 2010], river morphology across drainage basins or regions is typically characterized in terms of gradient and longitudinal profile, which are readily obtained from digital elevation models (DEMs) [*Snyder et al.*, 2000; *Duvall et al.*, 2004; *Font et al.*, 2010]. Because longitudinal profile irregularities can reflect downstream variations in lithology and erodibility [*Valla et al.*, 2010] and glacial history [*Hobley et al.*, 2010], as well as rock uplift, profiles must be carefully interpreted within a geologic-geomorphic context. In addition to longitudinal profile, longitudinal variations in the width of bedrock channels indicate differential uplift [*Whittaker et al.*, 2007a, 2007b; *Attal et al.*, 2008; *Yanites et al.*, 2010b]. Numerical derivations of scaling relations for bedrock channel width, w, drainage area, A, gradient, S, and discharge, Q, have been derived from: flow resistance equations and mass conservation principles, producing $w \sim Q^{0.38}$ and $w \sim S^{-0.2}$ [*Finnegan et al.*, 2005]; assumptions that erosion rate scales with local shear stress, which results in $w \sim Q^{0.4}$ and $w \sim S^{-0.2}$ [*Wobus et al.*, 2006a]; and minimization of potential energy, for which $w \sim A^{0.5}$ [*Turowski et al.*, 2007]. Limited field investigations of downstream hydraulic geometry in bedrock channels tend to follow the lead of *Montgomery and Gran* [2001] in substituting A for Q, although this introduces uncertainties associated with hydroclimatic variation along a channel or drainage basin [*Flores et al.*, 2006]. The exponent in field-based $w \sim A^b$ relations has varied from 0.32 [*Montgomery and Gran*, 2001] to 0.55 [*van der Beek and Bishop*, 2003]. Compiling a large field data set from many regions, *Wohl and David* [2008] propose that scaling relations are relatively consistent among bedrock and alluvial channels, such that $w \sim A^{0.3}$ and $w \sim Q^{0.5}$, although bedrock channels tend to be consistently narrower than alluvial channels for a given drainage area. Because changes in rock erodibility or uplift rate can alter downstream scaling relations [*Wohl and Merritt*, 2001; *Cowie et al.*, 2006; *Jansen*, 2006], unexpected deviations in channel width from w-A or w-Q relations can be used to infer uplift.

Field data and numerical simulations indicate that adjustments to width and gradient as a result of increasing substrate resistance or uplift are typically tightly coupled [*Whipple*, 2004; *Stark*, 2006]. Holding substrate erodibility constant, gradient increases and width declines on rivers with higher uplift rates in southern California, USA [*Duvall et al.*, 2004]. Using physical experiments, *Turoswki et al.* [2006] demonstrate that, as uplift rate increases, channel width and cross-sectional area decrease and velocity increases approximately linearly. *Whittaker et al.* [2007] show that traditional hydraulic scaling laws break down along bedrock channels crossing an active fault in the central Italian Apennines; channel widths become decoupled from drainage area upstream of the fault and values of unit stream power are approximately four times those predicted by scaling relations. Similarly, *Amos and Burbank* [2007] find that small rivers crossing growing folds in New Zealand respond with channel narrowing up to some threshold of differential uplift, beyond which channel gradient also increases.

Regional rates of uplift can be compared to regional rates of denudation as an index of the efficiency of mountain hillslope and channel processes. Mountainous topography results from the imbalance between uplift caused by tectonics and denudation by tectonic (extensional faulting) or surface (glacial, hillslope, and fluvial) processes [*Burbank et al.*, 1996]. Early estimates of regional denudation rates came primarily from sediment yields averaged over decades or longer. *In-situ* produced cosmogenic nuclides are now widely used to infer denudation rates [*Vance et al.*, 2003; *Schaller et al.*, 2004].

Rates of both uplift and denudation can have substantial spatial and temporal variability. *Leopold et al.* [1964] use the 629,520 km^2 basin of the Colorado River in the southwestern U.S. as an example of spatial variability in denudation rate, as estimated from suspended sediment load expressed in centimeters derived from the drainage basin per unit of time. Denudation rates range from approximately 0.4 to 17 cm/ky and show fairly strong correlation with climate [*Leopold et al.*, 1964].

Oguchi [1996b] compares Holocene and contemporary denudation rates for a series of river basins in central Japan and finds that contemporary rates are up to three times higher than Holocene rates. Despite this variability, regional rates of the type listed in Table 2.1 may still be useful indicators of relative efficiency of weathering and erosion in various regions. (All tables are available on the CD-ROM that accompanies the book.) Both climate and relief strongly influence denudation rate.

Published rates of bedrock channel incision vary from 5 to 10,000 mm/ky, with the highest rates occurring in regions of tectonic uplift [*Wohl et al.*, 1994a; *Wohl*, 1998]. Most of these channel incision rates are long-term (Quaternary) averages for third-order or higher channels, but they indicate that tectonic uplift corresponds with increased transport ability and channel incision in mountainous regions, regardless of climate or lithology.

In summary, work within the past decade demonstrates that mountain rivers do not simply respond to tectonically controlled gradient; rather, spatially and temporally variable interactions among uplift, climate, and fluvial erosion allow rivers to both respond to and influence uplift, elevation, relief, and the distribution of mass across a landscape. These interactions are exemplified by the tectonic aneurysm model in which deep and rapid incision alters crustal properties such that a positive feedback develops and maintains local elevation and relief. Spatial variations in the gradient and width of bedrock rivers can reflect spatial variations in tectonics. Geochronological advances that facilitate quantification of uplift and denudation rates, mapping and modeling of river longitudinal profiles, and numerical and physical models of diverse landscape processes, all enhance our understanding of the interactions between mountain rivers and tectonics. However, as *Tucker* [2009] notes, there remains a pressing need to identify natural experiments in landscape evolution in which only one element varies significantly and for which the driving forces, initial conditions and/or boundary conditions are well constrained.

2.2. Hillslopes

Schumm [1977] divides the fluvial system into an upstream zone that serves as the primary sediment source for a drainage basin, a middle transfer zone, and a downstream zone that is primarily depositional, analogous to Figure 1.7. Mountain rivers occupy the upstream sediment-source zone of a drainage basin, and primarily reflect the controls of climate, geology, and land use as these influence water and sediment yield to the channel, and channel-boundary resistance. Geology is here taken to include lithology, structure, and tectonic regime. These characteristics will, in combination with climate, determine rate and manner of weathering, and thus slope morphology and processes of water and sediment movement.

This section explores form and process on hillslopes in some detail because hillslopes exert such strong influences on form and process in mountain rivers. The first subsection discusses how tectonics, lithology and climate influence weathering and erosion on slopes. This leads, in the second subsection, to the concepts of *steady-*

state hillslopes that exhibit no net change in surface elevation through time and *threshold hillslopes* that maintain a characteristic critical slope while being lowered. The third subsection addresses estimation of soil production in relation to bedrock properties and slope morphology. The fourth subsection builds on this information to examine downslope movement of sediment as aggregates in landslides and debris flows and the fifth subsection addresses diffusive sediment transport via creep, rain-splash, and overland flow, as well as models of diffusive transport. The final subsection draws on the preceding text to examine numerical and conceptual models of slope morphology and sediment movement.

2.2.1. Controls on Slope Morphology

Slope morphology is controlled by tectonics, lithology, climate and vegetation, which in turn control rates and processes of weathering and erosion and the downslope movement of water and sediment. Each of the primary controlling factors can vary regionally or locally and, because factors such as hillslope vegetation and landforms co-evolve, multiple possible outcomes can produce place-dependent results [*Marston*, 2010]. *Montgomery et al.* [2001] relate differences in morphology across the Andes Mountains to hemisphere-scale climate variations that in turn create differences in erosion rates and processes (e.g., relative dominance of glacial and fluvial processes). At much smaller scales, *Burnett et al.* [2008] demonstrate that south-facing slopes in Arizona are steeper and have significantly less weathered bedrock than north-facing slopes. The sandstone underlying both slopes weathers primarily by clay hydration, but the south-facing slopes are too dry for significant clay expansion. Similarly, *Birkeland et al.* [2003] document distinct differences in soil development on north- and south-facing slopes in the semiarid Colorado Front Range, USA, as do *Egli et al.* [2010b] for soils in the northern Italian Alps. As another example of local variations, *Roering et al.* [2005] find that deep-seated landsliding sensitive to the thickness and frequency of siltstone beds creates zones of lower gradient hillslopes and lower drainage density within the Oregon Coast Range, USA. In the Appalachian Mountains of North Carolina, USA, root reinforcement of unconsolidated slope materials varies because root tensile strength is greater in areas of divergent topography than in areas of convergent topography such as hollows [*Hales et al.*, 2009]. (*Schwarz et al.* [2010] provide a comprehensive review of different scales of root reinforcement on steep slopes and *Marston* [2010] provides a thorough review of interactions between geomorphic processes and vegetation on hillslopes.) Differences in joint density can create variations in rate of weathering, sediment production, and hillslope processes and morphology, as *Applegarth* [2004] documents for mountains in central Arizona, USA. Within a particular mountainous region, processes governing slope morphology are also likely to vary with elevation as climate and the resultant weathering, soil formation, vegetation, infiltration and other parameters vary [*Sarah*, 2004; *Dixon et al.*, 2009].

Controlling factors and resulting processes and morphology can also vary through time. Time periods of greater effective moisture created more intense chemical weathering in the arid southwestern United States during the Miocene-Pliocene, resulting in greater sediment production and delivery of colluvium from hillslopes [*Pederson et al.*, 2000, 2001]. At much shorter timescales, discrete disturbances such as wildfire can dramatically alter slope stability and morphology by changing land cover, infiltration and runoff [*Woods et al.*, 2007; *Pierson et al.*, 2008; *Moody et al.*, 2009].

Selby [1982] distinguishes among weathering-controlled slopes, transport-limited slopes, and transportational slopes. On *weathering-controlled slopes*, the rate at which regolith is produced is less than the potential rate at which it can be removed and the surface grain-size distribution may grow coarser over time [*Cohen et al.*, 2009]. The slope profile may reflect relative rock resistance [*Pederson et al.*, 2001]. The profile may also reflect the processes or structural controls that act to undercut or oversteepen the slope with respect to the profile angle at which the slope could theoretically be supported by rock resistance. On *transport-limited slopes*, the rate of regolith production is greater than the capacity of transport processes to remove it and the surface may grow finer grained over time [*D. Cohen et al.*, 2009; *S. Cohen et al.*, 2009]. On these slopes, regolith accumulates and the slope profile reflects the properties of, and processes acting on, the regolith. Under these conditions sediment production on slopes may decline as regolith thickness increases [*Heimsath et al.*, 2000]. *Transportational slopes* exhibit equilibrium between rate of weathering and rate of erosion.

Leopold et al. [1964] note that all types of slope morphology may be found in a given climate, although others generalize that slope morphology corresponds to climate [*Schumm*, 1956; *Abrahams and Parsons*, 1991; *Ritter et al.*, 1995]. Rounded hillslopes with convex upper portions and concave lower portions are more common in humid climatic zones with high weathering rates, thick regolith, and gradual downslope movement of sediment. Angular slopes with cliff faces and talus accumulations are more common in dry or cold regions with low weathering rates and greater physical weathering, extensive bedrock exposure, and mass movements.

The influence of climate on slope morphology and processes occurs first through the temperature and precipitation regime in which weathering occurs. Dry climates have predominantly physical weathering. Chemical weathering becomes more important as precipitation and temperature increase [*Leopold et al.*, 1964; *Ritter et al.*, 1995]. Polar and desert zones have very thin weathering profiles where the absence of water and plants produces low weathering rates. The rates increase through the temperate latitudes and reach a maximum in the humid tropics [*Selby*, 1982]. The rate of sediment removal from hillslopes does not follow the same pattern with respect to climate. The highest rates of sediment yield consistently come from semiarid and highly seasonal Mediterranean and tropical monsoonal climates [*Langbein and Schumm*, 1958; *Fournier*, 1960; *Douglas*, 1967; *Walling and Kleo*, 1979]. Rates are lower for arid regions because precipitation is not sufficient to mobilize sediment, and for humid regions because abundant vegetation effectively stabilizes hillslope sediment. These trends of weathering and sediment yield are reflected in the suspended sediment and solute loads of rivers. Rivers draining cold humid temperate regions carry the highest percentage of their total load in solution [e.g., *Dietrich and Dunne*, 1978], but rivers draining dry or seasonal tropical areas have both the highest suspended sediment concentrations and the greatest total sediment load [*Knighton*, 1984, Table 3.6].

The influence of geology on slope morphology and processes can be characterized by rating rock-mass strength as a function of (i) intact rock strength; (ii) weathering; (iii) spacing, orientation, width, and continuity of joints; and (iv) outflow of groundwater [*Selby*, 1980]. Higher values of rock-mass strength correlate strongly with steeper slopes [*Selby*, 1980, 1982] and parallel retreat occurs only when rock-mass strength is uniform into the slope [*Moon and Selby*, 1983].

Generalized models of hillslope evolution feature (i) *parallel retreat*, in which the slope profile remains constant while the whole landform erodes back; (ii) *slope replacement*, in which the steepest angle is progressively replaced by the upward expansion of a gentler slope near the base; and (iii) *slope decline*, in which the steep upper slope erodes more rapidly than the basal zone, causing a decrease of the overall slope angle [*Strahler*, 1950; *Schumm*, 1966; *Young*, 1972; *Ritter et al.*, 1995] (Figure 2.3). *Howard*'s [1994] simulation modeling of drainage basin evolution indicates that detachment-limited or weathering-controlled slopes undergo nearly parallel retreat and replacement with alluvial surfaces under fixed base level, whereas transport-limited slopes undergo gradual slope decline. *Kirkby* [2003] reviews numerical models of hillslope evolution and *Furbish* [2003] reviews coupled models of land-surface geometry and soil thickness applied to hillslope evolution.

The application of cosmogenic isotope [*Friend et al.*, 2000; *Parker and Perg*, 2005] and luminescence [*Thomas et al.*, 2007] techniques of geochronology is facilitating new insights into slope evolution over timescales of 10^3-10^6 years and the use of DEMs coupled with GIS software is improving our ability to quantitatively describe slope forms [*Lin and Oguchi*, 2009]. As illustrated in Figure 1.7 and explored in the next few sections of this chapter, we can parameterize some slope processes, such as diffusive sediment transport, although the broad applicability of these recently proposed equations and numerical models remains to be tested.

2.2.2. Steady-State Hillslopes

Steady-state conditions occur with no net change in surface elevation as a result of a balance between rock uplift and erosion [*Montgomery*, 2001b]. *Gilbert* [1877, 1909] and *Davis* [1898, 1899] recognize that the adjustment of hillslope evolution to river incision is essential to steady-state topography. *Penck* [1924, 1953] proposes that slope stability can control hillslope morphology by setting a limiting or threshold slope. This led to the concept of a *threshold hillslope* that maintains a characteristic critical slope while being lowered. Threshold hillslopes can be in bedrock or mantled with soil [*Montgomery*, 2001b]. These concepts were further established through *Mackin*'s [1948] graded river, *Strahler*'s [1950] graded drainage system, and *Hack*'s [1960] dynamic equilibrium in landscape evolution. *Willett and Brandon* [2002] identify four types of steady state that can characterize mountain systems: flux (accretionary flux equals erosional flux), topographic (elevation at all spatial points within a domain of interest does not change with time), thermal (a time-invariant subsurface temperature field within an orogen), and exhumational (time-invariant erosion rates within a specified spatial domain).

The concept of steady-state conditions is supported by work in the Nanga Parbat region of northern Pakistan, where slope angles are largely independent of age and rates of denudation and uplift [*Burbank et al.*, 1996], although *Ouimet et al.* [2009] document nonlinear behavior and a threshold in the relationship between erosion rate and mean hillslope gradient. In Nanga Parbat, mean slope angles are set by rock strength. The association of high river gradients with high uplift rates implies a dynamic equilibrium in which uplift and incision are balanced via adjustments in stream power. Landslides provide a mechanism by which hillslopes can adjust to changes in boundary condition resulting from river incision at their toes. If rapid river incision oversteepens the hillslopes, an increase in landslide activity increases sediment supply to the river (reducing the rate of channel incision) and returns the hillslope gradient to a stable level. Because the landslide threshold is influenced by fractured rock strength, the highest topography develops either where rocks are less fractured or where large rivers are most widely spaced. The spacing of large rivers thus controls mean relief [*Burbank et al.*, 1996]. *Arnett* [1971] demonstrates a similar influence of river incision on the gradients of, and sediment yield from, adjacent hillslopes in Australia. *Korup and Schlunegger* [2007] qualify these relations by distinguishing between hillslopes above an inner gorge, where adjustment of hillslopes through landsliding is mainly strength-limited and structurally controlled, such that mass movement processes are largely decoupled from channel incision, and hillslopes within the inner gorge with threshold conditions set by river incision.

Spatial differentiation among erosive processes and hillslope-channel coupling can create different rate laws for maintaining steady-state conditions. Examining two steady-state mountain ranges in Oregon, USA, *Montgomery* [2001b] delineates five zones of different drainage area-slope scaling relations corresponding to portions of the landscape characterized by different geomorphic processes (Figure 2.4A). *Stock and Dietrich* [2003] find that channel slope varies as an inverse power law of drainage area up to slopes of approximately 0.03-0.10 (Figure 2.4B). This is the zone of fluvial processes. At higher slopes, where debris flows dominate channel incision, the rate of increase in slope with decreasing drainage area declines. *Stock and Dietrich* [2003] interpret this to indicate that incision by debris flows limits the height of some mountains to substantially lower elevations than predicted by river incision laws.

2.2.3. Bedrock Weathering and Soils

Although bedrock-covered hillslopes occur in mountainous regions, soils are widespread in mountains and exert crucial influences on downslope movement of sediment and water, nutrient dynamics, and biota. In order to persist, soil must be replenished at a rate equal to or greater than that of erosion and *Lebedeva et al.* [2010] hypothesize that many soil profiles reach a steady-state thickness. Meteoric ^{10}Be can be used to date soils, although the technique is not straightforward and may have distinct limitations [*Egli et al.*, 2010a]. *Jungers et al.* [2009] used meteoric ^{10}Be to estimate virtual downslope soil velocities of 1.1-1.7 cm/yr in a well-mixed active transport layer ~ 60 cm thick in the Great Smoky Mountains of North Carolina, USA. The thickness of this transport layer is constant downslope and reflects the rooting depth and consequent root wad thickness of fallen trees, which in turn reflect depth to the soil/saprolite boundary.

Using cosmogenic ^{10}Be and ^{26}Al produced *in situ*, *Heimsath et al.* [1997] show that hillslope curvature (a surrogate for soil production) varies inversely with soil depth, and that soil production rates decline exponentially with increasing soil depth. Chemical weathering rates also decrease with increasing soil thickness [*Burke et al.*, 2007]. *Heimsath et al.* [1997] derive an empirical function for soil production, $-(\delta e/\delta t)$

$$\frac{\delta e}{\delta t} = -\left(\frac{\rho_s}{\rho_t} K \nabla^2 z\right) \qquad (2.1)$$

where ρ_s and ρ_t are soil and rock bulk densities, K is a diffusion coefficient with dimensions length^2/time, z is ground surface elevation, e is the elevation of the bedrock-soil interface, and t is time. Under these conditions, soil production should depend on hillslope curvature, $\nabla^2 z$. *Heimsath et al.* [1999, 2001] and *Almond et al.* [2007] demonstrate that where soil depth varies across an actively eroding landscape, rates of soil production also vary. Spatial variations in soil thickness can be simulated using parameterizations of simple soil creep (sediment flux equals a linear function of topographic slope) and depth-dependent soil creep, landsliding, and overland flow [*Braun et al.*, 2001]. Linear diffusion (simple creep) equations are only appropriate for shallow gradient, convex-upward regions of hillslopes, whereas depth-dependent creep equations are more broadly applicable [*Heimsath et al.*, 2005].

Integrating the work summarized above, *Yoo et al.* [2007] develop a process-oriented hillslope soil mass balance model based on the understanding that soil chemical weathering at any point on a hillslope depends on the flux of soil eroded from upslope as well as soil production from underlying bedrock at that point. Calculated soil transport is linearly related to the product of soil thickness and slope gradient, a conclusion reinforced by the numerical model of *Furbish et al.* [2009b].

As bedrock weathers, the resulting sediment moves downslope through mass movements such as landslides, debris flows, or rock falls, and through gradual diffusive processes such as rainsplash. The balance among these processes ultimately determines the thickness and type of soil present at any hillslope location. Although creep can involve aggregates of particles and is sometimes considered a type of slow mass movement, it is treated here as a gradual diffusive process.

2.2.4. Mass Movements

Mass movement involves downslope transport of aggregates rather than individual particles, and may occur as falls, slides, or flows [*Ritter et al.*, 1995]. Each type of movement may be strongly seasonal as a function of moisture availability and freeze-thaw processes [*Schumm and Lusby*, 1963; *Schumm*, 1964; *Caine*, 1976; *Yair and de Ploey*, 1979; *Miyabuchi and Nakamura*, 1991; *Sawada and Takahashi*, 1994].

Falls involve the free fall of rock or soil, usually in response to undercutting of the toe or face of a slope, weathering and enlargement of joints, or seismic vibrations. Rockwall recession rates correlate with geomechanical strength indices of bedrock [*Moore et al.*, 2009] and with climatically driven processes such as freeze-thaw weathering [*Hales and Roering*, 2009]. In densely jointed or fractured rocks, repeated falls may be a dominant source of coarse clasts for a river channel either directly or through maintenance of intermediary features such as talus slopes or rock glaciers [*Degenhardt*, 2009]. *Moya et al.* [2010] reconstruct rockfall frequency using dendro-morphological records.

A *slide* occurs when a mass of unconsolidated material moves without internal deformation along a discrete failure plane [*Carson and Kirkby*, 1972; *Ritter et al.*, 1995]. The slide may occur as a slump that has rotational movement of discrete blocks along a curved failure plane, or it may follow a fairly straight slide plane. This type of failure may be caused by a decrease in the shear strength of the regolith as a result of weathering, increase in water content, seismic vibrations, or freezing and thawing, or by an increase in shear stress caused by additions of mass or removal of lateral or underlying support [*Varnes*, 1958; *Brunsden and Jones*, 1976; *Selby*, 1982; *Ritter et al.*, 1995]. Slides typically change into flows as downslope movement continues.

A *flow* occurs when debris is sufficiently liquefied or vibrated that substantial internal deformation accompanies the movement. As with slides, flows may be primarily erosional at high gradients and depositional at low gradients.

In many of the world's mountainous regions, abrupt mass movements recur frequently (almost annually) within a drainage basin, and transport the majority of sediment to or along low-order stream channels [e.g., *Blackwelder*, 1928; *Pierson*, 1980; *Osterkamp et al.*, 1986; *Swanson et al.*, 1987; *Benda*, 1990; *Wohl and Pearthree*, 1991; *Jacobson et al.*, 1993; *Oguchi*, 1994c]. *Benda et al.* [2005] describe headwater streams as sediment reservoirs at periods of 10^1-10^2 years, which are episodically evacuated by debris flows or gully erosion. As stream order and drainage area increase, sediment yield per unit area or sediment delivery ratio decrease and sediment residence time increases because of increasing storage on hillslope or valley bottoms [*Schumm and Hadley*, 1961; *Boyce*, 1975; *Strand*, 1975; *Schumm*, 1977; *Dietrich and Dunne*, 1978]. Long-term average sediment yields for various mountainous regions are given in chapter 3 (Table 3.6). The inverse correlation between stream order and sediment yield reinforces *Schumm*'s [1977] conceptual model of the headwaters of a drainage basin as the primary sediment source for downstream portions of the basin.

2.2.4.1. Landslides. Landslides exert an important control on the early development of watersheds and valleys. Examining the early phase of mountain growth in the Finisterre Mountains of Papua New Guinea, *Hovius et al.* [1998] propose that isolated gorge incision initiates watersheds, which then expand by large-scale landsliding in a manner controlled by seepage. Further entrenchment occurs via fluvial incision of landslide scars and deposits [*Hovius et al.*, 1998]. Over longer timescales of continuing basin evolution, landslides create spatial and temporal variations in sediment delivery to mountain rivers [*Korup et al.*, 2004] (Figure 2.5). Landslides are a dominant source of sediment in mountain belts and they continue to influence river network evolution by determining basin area and the positions of drainage divides, as well as setting streamwise variations in sediment load and caliber [*Korup et al.*, 2010]. Moderate-sized landslides do the most work transporting material on hillslopes [*Guthrie and Evans*, 2007].

Landslides can be triggered by seismic activity [*Barnard et al.*, 2001; *Sato et al.*, 2005; *Morrissey et al.*, 2008; *Sanchez et al.*, 2010], freeze-thaw processes [*Arsenault and Meigs*, 2005], precipitation [*Guthrie and Evans*, 2004; *Cubito et al.*, 2005; *Crosta and Frattini*, 2008; *Saito et al.*, 2010], and land uses such as deforestation or removal of slope toes at road cuts [*Larsen and Torres-Sanchez*, 1998; *Barnard et al.*, 2001; *Remondo et al.*, 2005; *Imaizumi et al.*, 2008]. The sequencing of *landsliding episodes* triggered by events such as typhoons and earthquakes exerts an important influence on erosion and sediment yield [*Lin et al.*, 2008] and thus the geomorphic effects of landslides in mountain drainages. Over timespans of 10^3-10^6 years, landslides may be more common where progressive down-dropping of a basin or river incision destabilizes hillslopes [*Gelabert et al.*, 2003; *Lacoste et al.*, 2009]. The effects of changes in land use or land cover on landslide occurrence can be especially persistent. *Schmidt et al.* [2001] estimate that the loss of root reinforcement following timber harvest in the Oregon Coast Range, USA modifies hillslope stability for at least a century and *Roering et al.* [2003] find that landslides in the region tend to occur in areas of reduced root strength.

Landslides can also be the precursor of debris flows [*Blahut et al.*, 2010], as reviewed by *Sassa and Wang* [2005], although the transitional processes involved remain poorly understood [*Korup*, 2009]. *Brayshaw and Hassan* [2009] correlate the factors of channel gradient, angle of entry of landslide failure into the channel, initial failure volume, and the amount of in-channel stored sediment with whether a landslide will become a debris flow; steeper channels, low angles of entry, lower volumes of in-channel sediment, and larger initial failure are more likely to result in debris flows.

Korup [2005b] proposes a classification system for characterizing the physical contact between landslides and river channels and the resulting geomorphic consequences for drainage. He designates the geomorphic impact classes of landslides as being (i) buffered (no discernible contact between landslide toe and river channel), (ii) riparian (landslide slightly protruding or impinging on river channel, but clear dominance of fluvial erosion), (iii) occlusion (diversion of river channel around convex landslide toe), (iv) blockage (occurrence of a landslide-dammed lake), and (v) obliteration (complete burial of valley floor). *Hsu and Hsu* [2009] numerically simulate channel evolution and bed mobility in response to earthquake-induced landslide dams in China. Identifying the key controls on the geomorphic coupling between landslides and river channels remains an important need [*Korup*, 2009].

2.2.4.2. Debris flows. Debris flow is used here as a general term for a mass movement of sediment mixed with water and air that flows readily on low slopes [*Johnson and Rodine*, 1984]. This usage of debris flow thus subsumes debris slide, mud flow, mudslide, earth flow, debris torrent, lahar, and several other terms [*Hungr*, 2005]. Debris flows can be triggered by seismic activity [*Dong et al.*, 2009a]; intense or long duration rainfall or snowmelt [*Pike et al.*, 1998; *Springer et al.*, 2001; *Zimmermann*, 2004; *Larsen et al.*, 2006], particularly on slopes with an aspect that enhances exposure to wind-driven rainfall [*Pike and Sobieszczyk*, 2008]; rainfall following a forest fire [*Cannon*, 2001; *Cannon et al.*, 2001; *Gabet*, 2003]; timber harvest, or road construction [*May*, 2002]; volcanic eruption, landslides or earthflows [*Palacios et al.*, 2001; *Guadagno et al.*, 2005; *Malet et al.*, 2005; *Vallance*, 2005; *Imaizumi et al.*, 2008; *Saucedo et al.*, 2008]; rapid drainage of a glacial lake or a volcanic crater lake; or the impact of a high-speed stream of water from a cliff or channel knickpoint (the "firehose effect") [*Swanston and Swanson*, 1976; *Innes*, 1983; *Costa*, 1984; *Johnson and Rodine*, 1984; *Weirich*, 1987; *Wells et al.*, 1987; *Gavrilovic and Matovic*, 1991; *Jacobson et al.*, 1993; *Inbar et al.*, 1998]. *Wieczorek and Glade* [2005] review climatic factors that influence the occurrence of debris flows, and *Sidle* [2005] reviews the influence of timber harvest on debris flows. *Cannon et al.* [2001] distinguish between runoff-dominated processes of debris-flow generation that cause progressive sediment entrainment, and infiltration-dominated processes that trigger discrete failure of soil slips. The sediment content of overland flow is also important; high sediment concentration can limit infiltration and subsurface flow in debris-flow source areas [*Blijenberg*, 2007].

The presence of abundant unconsolidated material, steep slopes, a large but intermittent source of moisture, and sparse vegetation renders many of the world's mountainous regions particularly susceptible to debris flows [*Costa*, 1984]. Debris flows create a critical link between sediment production on hillslopes and transport in streams; *Eaton et al.* [2003] estimate that half of the long-term denudation at sites in the central Appalachian Mountains of the United States occurs episodically by debris-flow processes and *Mackey et al.* [2009] document denudation over 150 years at an active earthflow in northern California, USA that is more than 20 times faster than the regional erosion rate. Progressively smaller and steeper drainage basins have the potential to transport an increasingly larger percentage of sediment by debris flows because of their proportionally larger volume of rainfall or snow and steeper slopes [*Costa*, 1984], although lithological variations that influence clay production [*Sterling and Slaymaker*, 2007; *Moeyersons et al.*, 2008; *Webb et al.*, 2008] can create non-uniform regional spatial distribution of debris flows. Non-uniform temporal distributions of debris flows can reflect triggering processes or sediment supply [*Hartvich and Mentlík*, 2010]. *Bardou and Jaboyedoff* [2008] detect two types of debris flows in catchments of central Europe; large debris flows that occur early in basin evolution before a steady-state is reached, and smaller debris flows that are limited in size by sediment availability unless sediment storage is replenished by something such as a rock avalanche. Similarly, *Jakob and Friele* [2010] develop a magnitude-frequency model for a site in British Columbia, Canada that distinguishes relatively small events from larger events that result from transformation of rock avalanches.

Portions of drainage basins that serve as sediment storage sites, such as zero-order basins or hollows (small, unchanneled valleys) in humid mountain regions, can be particularly important for debris-flow initiation [*Dietrich and Dunne*, 1978; *Iida and Okunishi*, 1983; *Marron*, 1985; *Melelli and Taramelli*, 2004]. The underlying bedrock geometry of the hollows conforms fairly closely to the surface topography, with the thickest colluvium in the axis of the hollow [*Dengler et al.*, 1987]. The form of the hollow can also be a function of the size and durability of boulders supplied to slopes, rather than of the underlying bedrock [*Mills*, 1989]. In the central Appalachians, USA, deep v-shaped hollows are present where smaller boulders can be mobilized by surface water flow. Hollows are shallow and u-shaped where boulders are too large to be mobilized even by floods [*Mills*, 1989]. Colluvium is likely to be saturated in the upper to middle portions of colluvial hollows [*Dengler et al.*, 1987], facilitating failure during precipitation events. Instrumentation of a hollow in the Oregon Coast Range, USA indicates that debris-flow initiation occurs where local upward flow from bedrock penetrates overlying colluvium [*Montgomery et al.*, 2009].

Hollows are not necessary to generate debris flows, however, particularly under runoff-dominated processes. *Gabet and Bookter* [2008] describe progressively bulked debris flows developing from rill networks in which the locations of rill heads conform to slope-area thresholds characteristic of erosion by overland flow. Most studies of debris flows generated in recently burned areas also describe a progressive downstream increase in volume as rill and channel networks are eroded [*Santi et al.*, 2008]. *Savage and Baum* [2005] review methods of quantifying, mapping, and modeling the instability of steep slopes that can lead to landslides and debris flows.

Naef et al. [1990] describe an interesting situation in the Swiss Alps, where retreat of the Pleistocene glaciers left oversteepened and unsupported valley walls that led to "sackung," the downslope movement of bedrock by slow gravitational collapse. The fissures created during this collapse produce high infiltration and permeability, and large subsurface water storage capacity. These characteristics in turn prevent flooding along rivers draining the region, but may facilitate debris flows [*Naef et al.*, 1990]. Similarly, *Coltorti et al.* [2009] describe how deep-seated gravitational slope deformation, or sagging, in sandstones of Italy's Chianti Mountains may be the embryonic stage of a large landslide. The continuous sediment supply associated with these features may result in lower rainfall intensity values capable of triggering debris flows [*Turconi et al.*, 2010]. *Jomard et al.* [2010] use electrical resistivity tomography to recognize large landslide structures at depths and *Strozzi et al.* [2010] use remote sensing to monitor mountainous ground displacements associated with deep-seated rock mass movement.

Flow conditions can vary among debris flow, hyperconcentrated flows, and water floods both downstream and with time during a flow [*Takahashi*, 1991b; *Cenderelli and Wohl*, 1998; *O'Connor and Costa*, 1993; *Walder and O'Connor*, 1997; *Berti et al.*, 1999; *Cronin et al.*, 1999; *O'Connor et al.*, 2001; *Kataoka et al.*, 2008]. A water flood is a turbulent, Newtonian flow of water which carries relatively small amounts of sediment (1-40% by weight) and has a bulk density in the range of 1.01-1.3 g/cm^3 [*Costa*, 1984]. A hyperconcentrated flow is a stream flow enriched with a large amount of sediment (40-70% by weight), and a bulk density of 1.3-1.8 g/cm^3 [*Beverage and Culbertson*, 1964; *Costa*, 1984; *Pierson*, 2005; *Wang et al.*, 2009]. Hyperconcentrated flows are particularly common in association with volcanic debris flows, or lahars [*Pierson*, 2005]. A debris flow is a flow of water and sediment moving together as a single visco-plastic body that can be up to 90% sediment by weight, with a bulk density of 1.8-2.6 g/cm^3 [*Costa*, 1984]. *Iverson* [2005] describes debris flows as gravity-driven fluid-solid mixtures with abrupt surge fronts, free upper surfaces, variably erodible basal surfaces, and compositions that can change with position and time. He provides a thorough overview of debris-flow mechanics.

Bedload transport rate and maximum clast size can increase with increasing fluid density, if the flow around the grains is not laminar [*Rickenmann*, 1991a]. Debris flows in particular are notable for being able to transport enormous clasts that can exceed 100 m^3 [*Gavrilovic and Matovic*, 1991]. Material is entrained in debris flows via bed destabilization and erosion and via instability of stream banks undercut by bed erosion [*Hungr et al.*, 2005]. Bed material is destabilized as a result of drag forces at the base of the debris flow, which may be aided by strength loss resulting from rapid undrained loading, impact loading, and liquefaction of saturated channel fill [*Hungr et al.*, 2005]. In addition to transporting larger clasts, debris flows mobilize much greater volumes of sediment than stream flows of comparable recurrence interval [*Mao et al.*, 2009].

Differentiation of sediment deposits produced by each of the three types of flow is very difficult, but generally rests on clast fabric (orientation of clast long axis with respect to flow); deposit morphology; and grain size, sorting, and stratification. Of the three flow types, debris flows tend to produce deposits in which the long axes of cobbles and boulders show no preferred orientation, whereas hyperconcentrated and water flows produce deposits in which the clast long axes dip upstream [*Innes*, 1983; *Waythomas and Jarrett*, 1994; *Cenderelli and Kite*, 1998; *Cenderelli and Wohl*, 1998]. Debris flows typically have multiple steep-sided lobate boulder deposits representing the succession of surges that occur during a debris flow [*Zanuttigh and Lamberti*, 2007], as well as coarse grained, poorly sorted, sharp-crested levees, in contrast to the flat-topped, better sorted boulder berms produced by water floods and hyperconcentrated flows [*Sharp*, 1942; *Costa and Jarrett*, 1981]. Debris-flow deposits are likely to be matrix supported, with minimal sorting and stratification. Hyperconcentrated flows and water flows in mountainous regions may also produce relatively poorly sorted deposits, but these deposits are more likely to be clast supported and to have some stratification and grading [*Blackwelder*, 1928; *Hooke*, 1967; *Scott*, 1971; *Bull*, 1977; *Janda et al.*, 1981].

Valleys consistently subject to different process domains (e.g., debris flow, stream flow, snow avalanche) can have distinctly different morphometries that facilitate their identification, as well as hazard zoning [*De Scally et al.*, 2001]. The headwaters of systems dominated by fluvial and debris-flow erosion are highly incised and contain closely spaced channels, whereas those dominated by landslides are smoother and have lower drainage density, resulting in greater response times for sediment transport [*Schneider et al.*, 2008]. In regions with relatively frequent debris flows, such as the Oregon Coast Range of the USA, local valley slope decreases abruptly downstream from the portion of the network scoured by debris flows [*Stock and Dietrich*, 2006]. Strath terraces begin downstream from the furthest extent of debris flows, and drainage area-gradient (*A-S*) relations follow fluvial power laws such that *S* varies as an inverse power law of *A* [*Stock and Dietrich*, 2003]. This inverse power law does not extend to channel gradients greater than ~ 0.03 to 0.10. The *A-S* plot for steeper portions of the network curves in log-log space (Figure 2.4B), and landscape evolution models need a different incision law for these portions of the network [*Stock and Dietrich*, 2003]. Having observed lowering of bedrock channels by the impact of large particles entrained in debris flows, and the absence of fluvial erosion between debris flows [*Stock et al.*, 2005], *Stock and Dietrich* [2006] propose a debris-flow incision model in which lowering rate is proportional to the integral of solid inertial normal stresses from particle impacts along the flow and the number of up-valley debris-flow sources (equation 2.2):

$$-\frac{\partial z}{\partial t} = \frac{K_0 K_1}{\frac{T_0^2}{E_{eff}}\left(1-e^{-c_2 N_t}\right)^2\left(\frac{D_f}{D_f+D_e}\right)} \frac{N_t}{t_r} \frac{k_b v_0}{w(x)k_2 A^{a_2}} e^{a_1\left(\frac{A(x)}{A_{total}}\right)} \left[\cos(\theta)v_s\rho_p D_e^2\left(\frac{k_3 S^{a_3}}{k_2 A^{a_2}}\right)^w\right]^n$$

$$(2.2)$$

where $\delta z/\delta t$ is bedrock surface lowering rate, K_0 is a constant of proportionality relating excursions in solid inertial normal stress to bulk solid inertial normal stress, K_1 is a proportionality constant between rock strength and erosion rate, T_0 is rock tensile strength, E_{eff} is effective elastic modulus of fractured rock, c_2 is an exponent characterizing the decline in weathering with increasing trigger link magnitude, N_t is the number of upvalley trigger hollows or mobile debris-flow sources, D_f is characteristic bedrock fracture spacing, D_e is particle diameter characterizing the most effective size at eroding bedrock from a distribution, t_r is long-term recurrence interval of landsliding at a hollow, k_b is bulking coefficient, v_0 is landslide volume, w is debris-flow width at surge head, k_2 is proportionality constant between debris-flow depth and source area, A is drainage area, a_2 is a depth exponent, a_1 is a bulking exponent, A_{total} is drainage area at end of debris-flow long-profile, θ is valley slope angle, v_s is volumetric solids concentration, ρ_p is particle density, k_3 is a proportionality constant between streamwise debris-flow velocity and valley slope, S is valley slope, a_3 is velocity exponent, superscript w is an exponent relating shear strain rate to solid inertial normal stress, and n is an exponent characterizing the erosional efficiency of inertial normal stress.

The model predicts a spatially uniform lowering rate as down-valley increases in incision rate caused by increases in debris-flow frequency and length are balanced by reduced inertial normal stress at lower slopes. These predictions match actual topography, but *Stock and Dietrich* [2006] caution that the debris-flow dynamics are crudely parameterized and require further work.

Debris flows can erode steep or confined portions of headwater channels to bedrock [*Benda*, 1990; *Wohl and Pearthree*, 1991] (Figure 2.6), removing sediment that had been stored along the channel for decades to millennia, although drainage area alone is not an accurate predictor of the relative importance of erosion versus deposition during a debris flow [*Springer et al.*, 2001; *Morton et al.*, 2008]. Upper portions of a network with frequent debris flows have shorter transit times for sediment than downstream portions of the network in which fluvial transport dominates [*Lancaster and Casebeer*, 2007]. *Benda* [1990] calculates that debris flows remove an average of 5-10 m^3 of sediment per meter length of channel along headwater streams in western Oregon, USA. Erosion by debris flow in these channel networks continues a short distance into third-order channels. Working in Japan, *Imaizumi and Sidle* [2007] find that for catchments > 0.1 km^2, the percentage of channel network length affected by debris flows decreases with increasing drainage area as a result of reduced channel gradient. *Berger et al.* [2010] describe the use of sensors based on a resistance chain buried in the channel bed that can record the rate and timing of maximum bed erosion during debris flows and floods.

Many headwater channels may be primarily depositional sites between the occurrences of debris flows because the catchment area of these low-order channels is not large enough to generate discharge sufficient to mobilize the wood and coarse clasts introduced to the channel from the surrounding hillslopes (Figure 2.7). *Benda* [1990] estimates that 80% of the sediment delivered to first- and second-order channels in western Oregon accumulates during the time between successive scouring debris flows; fluvial transport removed only 20%. Working on mountain channels in Arizona, *Wohl and Pearthree* [1991] find that 10-year-old debris flow deposits in second-order channels are very little modified by subsequent water flows.

Along channels with more abundant sediment supply and more frequent debris flows, passage of a debris flow may disrupt the stable, coarse surface layer of a channel, initiating channel incision [*Zicheng and Jing*, 1987] or a complex response of aggradation and degradation [*Shimazu and Oguchi*, 1996]. Occurrence of a debris flow may also alter rainfall-runoff relations for a basin [*Agata*, 1994] and sediment supply for other processes of sediment transfer, including fluvial sediment transport [*Haas et al.*, 2004; *Johnson et al.*, 2008]. The effects of debris flows on fluvial sediment transport and channel morphology are mediated by complex interactions among riparian vegetation, instream wood, streambed armoring, and the degree to which hillslopes and channels are coupled [*Harvey*, 2001; *Montgomery et al.*, 2003; *Gomi et al.*, 2004].

The geomorphic effectiveness of a debris flow in terms of transporting sediment and altering channel morphology is a function of debris-flow recurrence interval (Table 2.2) in relation to sediment supply. As with floods, recurrence interval can be estimated using historical records [*Tropeano and Turconi*, 2004; *Marchi and Cavalli*, 2007]; botanical evidence in the form of damage to trees, age of trees that have regrown following debris-flow erosion or deposition, or lichenometry on debris-flow clasts [*Ballantyne*, 2002; *Wilkerson and Schmid*, 2003; *May and Gresswell*, 2004; *Stoffel et al.*, 2005; *Bollschweiler et al.*, 2008; *Arbellay et al.*, 2010]; or geological evidence [*Keefer et al.*, 2003], including cosmogenic exposure ages on large boulders [*Marchetti and Cerling*, 2005; *Dühnforth et al.*, 2007; *Marchetti et al.*, 2007]. As Holocene- to Quaternary-scale chronologies of debris flows are developed for different regions, it becomes apparent that, like floods, the occurrence of debris flows can cluster over periods of several centuries in response to climatic variation [*Keefer et al.*, 2003; *Pierce et al.*, 2004; *Sletten and Blikra*, 2007; *Stoffel et al.*, 2008].

Liu et al. [2008] derive magnitude-frequency relations for debris flows at valley and regional scales using statistical analysis of a large database from China such that

$$p(A) \sim A^{-n} \tag{2.3}$$

where $p(A)$ is the percentage of valleys with area A and n varies by region, and

$$V \sim \log A \tag{2.4}$$

where V is volume of debris flow. *Griswold and Iverson* [2008] similarly derive

$$A = \alpha_1 V^{2/3} \tag{2.5}$$

$$B = \alpha_2 V^{2/3} \tag{2.6}$$

where A is maximum valley cross-sectional area, B is total valley planimetric area likely to be inundated by a debris flow, V is volume, and the coefficients α_1 and α_2 are scale-invariant indices of the relative mobilities of rock avalanches ($\alpha_1 = 0.2$, $\alpha_2 = 20$), nonvolcanic debris flows ($\alpha_1 = 0.1$, $\alpha_2 = 20$), and lahars ($\alpha_1 = 0.05$, $\alpha_2 = 200$). Similar relations between A, B, and V are developed by *Berti and Simoni* [2007].

Debris flows can deposit substantial volumes of sediment in lower gradient (Figure 2.8) or less confined (Figure 2.9) reaches of a channel [*Benda*, 1990; *Wohl and Pearthree*, 1991]. They can also be a significant source of coarse clasts to the channel and floodplain [*Rutherfurd et al.*, 1994]. Debris-flow deposition has been predicted as a function of channel slope and tributary junction angle [*Benda and Cundy*, 1990]; past depositional sites; ground slope [*Conway et al.*, 2010]; boundary roughness [*Mizuyama et al.*, 1987]; grain-size distribution and hydrographs of debris flow and stream flow; and the flow hydrograph (yield strength, flow viscosity, discharge) [*Whipple*, 1992]. In forested regions, the wood entrained within debris flows can also significantly affect flow dynamics and deposition by reducing velocity and decreasing runout length, as numerically simulated for the Oregon Coast Range, USA by *Lancaster et al.* [2003]. Extremely large volumes of wood can also create dams that then give rise to dambreak floods [*Comiti et al.*, 2008] or facilitate deposition [*Rigon et al.*, 2008].

Debris-flow deposits take the form of debris fans at tributary channel junctions or mountain fronts (Figure 2.10), and levees and terraces along the channel [*Osterkamp et al.*, 1986; *Benda*, 1990; *Hewitt*, 1998; *Yanites et al.*, 2006]. The ability of subsequent water flows to re-work the debris-flow deposits is directly proportional to drainage area. Larger channels (third-order or higher) may not preserve distinct debris-flow deposits more than a few years after the debris flow, although they may have a non-uniform longitudinal distribution of coarse clasts as a result of tributary debris flow inputs [*Wohl and Pearthree*, 1991]. These larger channels receive increasingly greater proportions of sediment from debris flows; *Benda* [1990] estimates that 40% of the sediment delivered to second-order channels in the Oregon Coast Range, USA comes via debris flows (compared to 20% via stream transport), and debris flows deliver 70% of the sediment to third- through fifth-order channels. This debris-flow-delivered sediment causes aggradation at the tributary junctions of the larger channels, with the debris-flow deposit subsequently eroding to a boulder lag. Debris flows thus drive cycles of aggradation and degradation along mountain rivers of varying size, with the time span of the cycle varying at 10^2 to 10^3 years as a function of debris-flow recurrence interval and sediment supply [*Benda and Dunne*, 1987; *Bovis and Dagg*, 1987; *Okunishi et al.*, 1987; *Wieczorek et al.*, 1989]. These cycles of aggradation and degradation in turn control channel and valley-floor geometry, disturbance regime, structure and composition of riparian vegetation, and gradient of side slopes and channels [*Swanson et al.*, 1987; *Florsheim et al.*, 1991; *Hewitt*, 1998]. Geomorphically and ecologically important aspects of channel morphology, including spatial density of pools, substrate texture, channel width, and wood loads, are nonuniformly distributed through channel networks that include debris flows, with heterogeneity of channel morphology increasing in proximity to low-order channel confluences prone to debris flows [*Benda et al.*, 2003]. The probability of significant morphologic changes at tributary confluences increases with the size of the tributary relative to the main stem [*Benda et al.*, 2004]. Debris-flow disturbance can also facilitate colonization by invasive exotic plants [*Watterson and Jones*, 2006].

Channels with periodic debris-flow inputs may develop a debris fan or debris cone where the channel flows out from the mountain range (Figure 2.11). These fan-shaped accumulations of poorly sorted debris have gradients of 12-25° [*Brazier and Ballantyne*, 1989] and form a continuum with fans containing fluvial deposits, wholly alluvial fans [*Brazier et al.*, 1988; *Harvey*, 1992], and paraglacial fans created through the resedimentation of glacial deposits by fluvial and mass movement processes [*Ryder*, 1971a, 1971b; *Owen and Sharma*, 1998], as discussed further in section 2.10. Episodic deposition may dominate the morphology and sedimentology of these fans in both humid temperate [*Williams and Guy*, 1973; *Pierson*, 1980] and arid and semiarid regions [*Blackwelder*, 1928; *Hooke*, 1967; *Beaty*, 1990]. The frequency and extent of debris flows on the fan can be reconstructed using dendrochronologic records [*Mayer et al.*, 2010]. Fan size and morphology are directly related to the climate, lithology, tectonic setting, and mechanisms of sediment supply of the drainage basin [*Bull*, 1964; *Oguchi and Ohmori*, 1994; *Whipple and Trayler*, 1996; *Sorriso-Valvo et al.*, 1998]. Fans with a substantial input from debris flows with a high sediment concentration tend to have a steeper gradient and rougher surface than fans with flows of low sediment concentration [*Hooke and Rohrer*, 1979; *Beaty*, 1989; *Whipple and Dunne*, 1992; *Ikeda et al.*, 1993]. Differences in depositional processes create characteristic topographic differences that can be detected using remote sensing imagery [*Volker et al.*, 2007], as well as different stratigraphies that can be detected using ground-penetrating radar [*Sass*, 2006; *Sass and Krautblatter*, 2007]. Debris flows are likely to be primarily depositional on fans, whereas water floods may incise fanhead trenches [*Beaty*, 1990; *Scott and Erskine*, 1994].

The section on Hazards in chapter 6 discusses prediction, detection, and mitigation of debris flows, as well as numerical simulations of debris flows.

2.2.5. Diffusive Sediment Transport on Hillslopes

2.2.5.1. Creep. Creep occurs when particles displaced by bioturbation and in wetting-drying or freeze-thaw cycles move downslope under the influence of gravity [*Kirkby*, 1967]. Volumetrically significant examples of bioturbation include tree throw (Figure 2.12) and burrowing by soil-dwelling rodents (Figure 2.13) [*Heimsath et al.*, 1997; *Roering et al.*, 2002; *Yoo et al.*, 2005; *Gallaway et al.*, 2009]. Soils on forested slopes may experience higher transport rates than those on grass/shrub-dominated slopes because of the occurrence of tree throw in the former environment [*Roering et al.*, 2004]. *Heimsath et al.* [2002] use single-grain optical dating to determine the time elapsed since individual grains last reached the ground surface and demonstrate that grains throughout the soil profile repeatedly reach the surface where bioturbation is present.

Creep in cold climates is partly dependent on freeze-thaw processes. The depth and intensity of frost cracking in rock and soils depend primarily on mean annual temperature, and the effectiveness of these processes presumably varied through time with glacial-interglacial cycles [*Hales and Roering*, 2007]. Frost creep in northern Japan occurs only with diurnal freeze-thaw in nearly saturated soil [*Sato et al.*, 1997]. With increasing water or ice content, creep grades into solifluction or gelifluction, respectively; the very slow downslope flow of partially saturated regolith.

Movement associated with creep is greatest in the upper meter of the regolith, and occurs by deformation at grain boundaries and within clay mineral structures [*Selby*, 1982]. The movement is quasi-viscous and occurs at a shear stress large enough to produce permanent deformation, but not discrete failure [*Carson and Kirkby*, 1972]. Quantification of rates of soil production and downslope creep using cosmogenic ^{10}Be accumulations indicates that creep rate is proportional to surface gradient [*McKean et al.*, 1993], leading to soil flux proportional to a depth-slope product [*Furbish et al.*, 2009b]. Creep may be particularly important on steep mountain slopes, where it produces tree curvature (Figure 2.14), tilting of structures, turf rolls, soil cracks, regolith accumulations upslope of retaining structures, and terracettes [*Selby*, 1982]. Creep may also promote rapid mass movements by decreasing regolith shear strength [*Simon et al.*, 1990]. Where processes of creep are limited, hillslope erosion can be slower [*Kaste et al.*, 2007].

Roering [2004] formulates a model for granular creep by analogy to rate process theory used for chemical reactions. According to this theory, individual particles must be displaced to a height greater than adjacent particles in order to be displaced downslope. The height of neighboring particles varies inversely with slope angle such that barriers to movement decrease as slopes steepen until the slope approaches the friction-limited angle of repose, at which point the grains move in the absence of disturbance. Modeled granular behavior generally agrees with that observed in a physical experiment [*Roering*, 2004]. Creep can also occur in very coarse-grained materials such as talus [*De Blasio and Sæter*, 2009].

2.2.5.2. Rainsplash and overland flow. Precipitation falling on a surface may loosen or detach individual particles (*rainsplash*), making them more susceptible to entrainment by overland flow [*Mosley*, 1973; *Morris*, 1986; *Furbish et al.*, 2009a]. Using data from laboratory experiments and simulated rainfall in a field setting, *Dunne et al.* [2010] propose an equation for instantaneous rainsplash transport (q_{inst} in kg/m/cm):

$$q_{inst} = -\frac{aD^j e^{fC/D}}{\sqrt{2}\sqrt{1+2\cos\alpha_0}} \left[F \int_{\pi/2}^{3\pi/2} \Omega(\theta)\cos\theta d\theta + (1-F) \int_{-\pi/2}^{\pi/2} \Omega(\theta)\cos\theta d\theta \right]$$

(2.7)

where

$$\Omega(\theta) = \frac{(2\cos\theta\sec\beta\sin\alpha_0\tan\beta + \sqrt{2}\sqrt{1+\cos(2\alpha_0)\sec^2\beta + \cos(2\theta)\tan^2\beta)}}{(\cos^2\theta\sec^2\beta + \sin^2\theta)\cos\beta}$$

(2.8)

where a, j, and f are parameters related to the detachability of the soil, D is median drop diameter (mm), C is cover density (decimal fraction), α_0 is average particle takeoff angle, F is fraction of sediment mass splashed downslope by a single raindrop impact on a slope of angle β, and θ is travel direction ($\theta = \pi$ is directly downslope).

Where unvegetated, unfrozen slope surfaces are exposed during the summer months, overland flow may be capable of eroding measurable quantities of sediment and creating selective textural sorting of slope materials [*Dingwall*, 1972]. If the overland flow occurs as sheet flow that submerges individual roughness elements and forms a fairly continuous sheet of water across the hillslope, sediment may be stripped evenly from the slope crest and upper zone during sheet wash (Figure 2.15). *Sheet wash* is particularly effective on slopes where cohesion has been reduced by needle ice, trampling, or disturbance to vegetation [*Selby*, 1982]. In semiarid regions with sparse vegetation, sheet wash may account for up to 98% of all sediment moved from a hillslope [*Lustig*, 1965; *Leopold et al.*, 1966; *Emmett*, 1978]. In addition to larger vegetation cover, microbiotic soil crusts (Figure 2.16) can significantly reduce sediment detachment and erosion by processes such as rainsplash and sheet flow [*Uchida et al.*, 2000]. The presence of frozen soil during winter and spring can enhance overland flow and soil erosion during snowmelt [*Ollesch et al.*, 2005, 2006], and *Weigert and Schmidt* [2005] model this process with EROSION 3D. Correlations between point rates of soil loss derived from cosmogenic isotopes and slope morphometry indicate that slope-driven processes such as creep dominate on convex portions of hillslopes, whereas overland flow processes dominate in concave hollows [*O'Farrell et al.*, 2007].

The sediment transport capacity of overland flow can be expressed as a power function of slope and discharge

$$q_s = k_1 q^\beta S^\gamma \tag{2.9}$$

where q_s is sediment transport capacity per unit width of slope (q is discharge per unit width), S is local energy gradient, and k_1, β, and γ are derived empirically or theoretically [*Rustomji and Prosser*, 2001]. The relationship between discharge and contributing area can also be expressed as a power function. Published parameter values range from 0.7-2.1 for β and 0.4-2.1 for γ [*Prosser and Rustomji*, 2000]. Substantial spatial variations in runoff and sediment transport can result, however, from features such as vegetation patches and land use. In southeastern Spain, vegetation patches and agricultural terraces largely determine hydrological connectivity at the catchment scale, as reflected in values of runoff and sediment yield four and nine times higher, respectively, in scenarios without agricultural terraces [*Lesschen et al.*, 2009].

As overland flow progresses downslope and is concentrated by surface irregularities, the increasing water depth increases the shear stress at the base of the flow. Rills and gullies eroded into the hillslope act as conduits for sediment erosion down the slope [*Sutherland*, 1991] and rill-channel networks can exhibit equilibrium scaling characteristics similar to those of river networks [*Raff et al.*, 2004]. Rill geometry and sediment transport capacity can be affected by processes such as soil freeze-thaw cycling [*Gatto*, 2000] and headcut erosion [*Bennett et al.*, 2000a, 2000b; *Bennett and Casalí*, 2001; *Alonso et al.*, 2002]. Rills may deliver most of the water and sediment to channels even in arid regions with low infiltration capacities [*Yair and Lavee*, 1985]. On long semiarid hillslopes subject to Hortonian overland flow, slopewash becomes competent to transport sediment within a few meters of the drainage divide [*Dunne et al.*, 1995]. However, microtopography generated mostly by biotic processes forces the slopewash to develop a depth distribution that controls its transport capacity, and sediment is released from microtopographic mounds into the slopewash. As a result, the sediment supply is sufficient to prevent rill incision on the upper portion of the hillslope [*Dunne et al.*, 1995].

Although surface erosion dominates on hillslope profile convexities and deposition dominates on profile concavities [e.g., *Yamada*, 1999b], studies of sediment movement on two disturbed hillslopes in southeastern Australia suggest that sediment movement is extremely variable at a spatial scale of meters. Localized changes in gradient, ground cover, vegetation and microtopography facilitate alternating zones of erosion and deposition, with pulses of sediment moving between the zones over both short (individual events) and medium (40 years) timescales [*Saynor et al.*, 1994]. Other studies also indicate that, although a specific type of downslope movement of sediment may dominate a hillslope, most hillslopes are shaped through time by some combination of multiple processes, with spatial differentiation among processes reflecting variations in lithology [*Jimenez Sanchez*, 2002].

   2.2.5.3. Modeling diffusive transport. Diffusive sediment transport on slopes can be expressed as

$$q_s = -K\nabla z \qquad (2.10)$$

where q_s is sediment flux, ∇z is slope, and K is a diffusion coefficient with dimensions length^2/time [*Heimsath et al.*, 1997, 1999]. This is known as linear diffusion because it is primarily dependent on slope. *Roering et al.* [1999] propose that linear diffusion occurs at low gradients, but sediment flux increases rapidly as gradient approaches a critical value such that

$$q_s = \frac{K\nabla z}{1 - \left(\dfrac{|\nabla z|}{S_c}\right)^2} \qquad (2.11)$$

where S_c is the critical hillslope gradient and the other symbols are as in equation 2.10. Diffusivity K varies linearly with the power per unit area supplied by disturbance processes and varies inversely to the square of the effective coefficient of friction; in other words, sediment mobilization and transport vary with the shear strength of the soil so that sediments with more frictional resistance have lower diffusivities [*Roering et al.*, 1999]. A model developed by *Tucker and Bradley* [2010] also indicates that surface grains dislodged by random disturbance events move via diffusion at low slope angles, although characteristic grain motion length scale begins to approach the length of the slope as slope angle steepens, leading to planar equilibrium forms. Sediment flux at a point is not strictly a function of the gradient at that point only, but is an integral flux that reflects upslope topography. This gives rise to a nonlinear dependency on local slope in which variable upslope topography produces widely varying rates of sediment flux for a given local hillslope gradient [*Foufoula-Georgiou et al.*, 2010]. Physical modeling supports the existence of nonlinear hillslope sediment transport and suggests that initially steep slopes are lowered rapidly by landsliding before giving way to slower, creep-dominated transport [*Roering et al.*, 2001b]. Disturbances such as wildfire can lower the critical hillslope gradient by reducing slope roughness, which in turn greatly increases sediment flux [*Roering and Gerber*, 2005].

   Roering et al. [2001a] relate hillslope sediment flux and landscape lowering through the continuity equation

$$\rho_s \frac{\partial z}{\partial t} = -\rho_s \nabla \tilde{q} + \rho_r C_0(t) \qquad (2.12)$$

where $C_0(t)$ is the time-dependent rate of rock uplift, which they define as equivalent to base level lowering or channel incision rate along the hillslope margin (length/time), and ρ_r and ρ_s are the bulk densities of rock and sediment (mass/length^3), respectively. *Roering et al.* [2001a] introduce a dimensionless parameter, ψ_L (the ratio of nonlinear to linear components of sediment flux at a point), to express the relative importance of nonlinear transport. As the relative magnitude of nonlinear transport increases, the timescale for hillslope adjustment to base level lowering decreases; numerical modeling suggests that the adjustment timescale for steep, soil-mantled landscapes such as the Oregon Coast Range, USA is $\leq$ 50,000 years [*Roering et al.*, 2001a]. Comparing numerical simulations of landscape evolution using different transport models, *Roering et al.* [2007] and *Roering* [2008] find that nonlinear slope- and depth-dependent models better reproduce actual topography and soil thickness at sites in California and Oregon than do models in which flux varies proportionally with hillslope gradient.

2.2.6. Modeling Slope Morphology and Sediment Movement

Selby [1982] describes the feedback between slope morphology and downslope movement of water and sediment. Long straight slopes are dominated by subsurface flow that promotes solution and creep, and are unlikely to slide unless a severe storm or earthquake destabilizes the slope. Hillslope spurs shed water rapidly and are least affected by erosion. Convergence of subsurface and surface flow at slope concavities promotes solution and mass movements because of the high pore-water pressures generated at these locations [*Iverson and Reid*, 1992]. These areas thus have the most erosion. Sediment entrainment by sheetwash or rilling is concentrated at the base of slopes adjacent to channels, in concavities, and in areas of thin and impermeable regolith or disturbed vegetation. The concentration of flow in old mass movement scars increases weathering and promotes further mass movements [*Selby*, 1982]. These feedbacks are incorporated in computer simulations such as TOPOG [*O'Loughlin*, 1986] or SHE-TRAN [*Burton and Bathurst*, 1998; *Burton et al.*, 1998], which can be used to model shallow landslide initiation by coupling digital terrain data with near-surface through-flow and slope stability models [*Dietrich et al.*, 1993; *Montgomery and Dietrich*, 1994a]. *Lane et al.* [2007] investigate sediment generation from hillslopes and channel banks and its delivery to the channel network by coupling the models SHALSTAB and TOPMO-DEL, and *Hennrich and Crozier* [2004] use hydrologic models to infer slope stability. Landscape evolution models such as SIBERIA and CAESAR also of course include hillslope processes as part of their simulation of catchment-wide erosion and deposition over timescales from tens to thousands of years [*Hancock et al.*, 2010].

Slope morphology-process feedbacks are also incorporated in conceptual models such as the process-response model for hillslope-channel sediment transfer in devegetated mountainous terrain developed by *White and Wells* [1979]. They find that sediment yield from burned hillslopes is influenced by: (i) the amount of hillslope devegetation; (ii) seasonal variations in weathering and runoff; (iii) protective post-fire forest litter; and (iv) sediment production from burrowing animals. The multiple sediment pathways of the model are thus spatially and temporally variable. The model also differs significantly from the complex response model for alluvial valleys in that tributary channels adjust independently of trunk channels where a bedrock knickpoint separates tributary and trunk channel.

Benda and Dunne [1997b] describe sediment supply to channels as a process stochastically driven by rainstorms and other perturbations which occur on a landscape with its own spatial variability in topography, colluvium, and state of recovery from previous disturbances. Sediment supply then interacts with transport processes and with the topology of the channel network to create a sedimentation regime that varies systematically with drainage basin area.

Alpine glacier basins form a specialized subset in terms of water and sediment transfer to rivers. The mass balance of the glacier largely controls rate and magnitude of water and sediment transfer. Meltwater leaves the glacier via supraglacial, englacial, and subglacial paths, and may be temporarily stored in proglacial lakes before reaching a river channel [*Röthlisberger and Lang*, 1987]. Sediment may come directly from ice melt or via meltwater transport. Sediment carried via subglacial and proglacial melt-water, and by supraglacial and subglacial ice movement, constitutes the greatest volume in most alpine glacier basins [*Fenn*, 1987]. Rivers in these basins are not generally supply-limited with respect to sediment [*Gregory*, 1987], in contrast to mountain rivers in many regions.

In summary, slope morphology reflects interactions among tectonics, lithology climate, and vegetation – all of which vary spatially and temporally – that influence rates and processes of weathering and erosion. Slopes can be described as weathering-controlled, transport-limited, or transportational. Slope development through time can be modeled as parallel retreat, slope replacement, or slope decline. Although threshold and steady-state hillslopes can exist, spatial differentiation among erosive processes can create different rates of slope change. Mass movements such as landslides and debris flows strongly influence hillslope and channel processes in mountain drainages by altering regolith thickness, slope morphology, and the characteristics of sediment transfer down slopes and into channels. Diffusive sediment transport via processes such as creep, rainsplash and overland flow reflects not only physical controls such as slope gradient and depth of regolith, but also bioturbation and climatic fluctuations such as freeze-thaw cycles. Empirical functions now exist to describe soil production and processes of both mass sediment movement and diffusive sediment transport, but it remains challenging to parameterize these equations for diverse field settings and thus quantify differences in rates and processes acting on diverse hillslopes. Continuing advances in geochronology and remote sensing of slope morphology and processes facilitate testing and further model development.

2.3. Climate and Hydrology

Mountainous regions provide the major water sources for many parts of the world. The mountains provide seasonal water storage in the form of snow and longer storage in lakes, glaciers, and icefields. Human-built reservoirs can enhance this storage potential, and natural and human-built reservoirs buffer flood effects and provide energy for hydroelectric power [*Bandyopadhyay et al.*, 1997]. Although it has long been recognized that runoff from mountains can be the principal source of water for the surrounding lowlands in arid and semiarid regions [*Liniger*, 1992], recent quantitative studies demonstrate that the contributions from mountains to total runoff are about twice what would be expected based on their share of surface area and that more than half of the mountain areas in the world provide an important freshwater source for downstream regions [*Viviroli et al.*, 2007]. In the western United States, for example, 80% of the water used for agriculture, industry, and domestic purposes originates from mountain winter-spring snowpacks [*Price and Barry*, 1997]. Worldwide, mountains provide the water source for half of the world's population [*Price*, 2002]. Despite this abundance, mountainous headwaters are also potentially vulnerable to changes in water quality and quantity as a result of limited groundwater reservoirs, uneven temporal and spatial distribution of water, limited buffering from atmospheric pollution, and typically short, rapid hydrological connections between different components of the hydrologic system such as hillslopes and channels or riparian zones and channels [*Leibundgut*, 1998].

The hydrologic regime of a mountain river predominantly reflects climate as expressed directly through precipitation and indirectly through the influence of weathering, soils, and vegetation on runoff and infiltration. The hydrologic regimes of mountain rivers can be subdivided at the first level into those dominated by glacier melt, by snowmelt runoff, or by rainfall runoff (Table 2.3; Figure 2.17). Rainfall runoff regimes can be further subdivided based on the type of atmospheric circulation pattern producing the rainfall and the associated differences in rainfall intensity, duration, spatial extent, and frequency of occurrence. Although the specifics of precipitation distribution within a mountainous region commonly differ from the precipitation distribution of adjacent lowlands, regional atmospheric circulation patterns influence the mountainous region [*Barry*, 2008].

2.3.1. Generation of Precipitation

Hayden [1988] designates seventeen types of flood climate regions and each of these regions includes at least one mountain chain. At the first level, regions are divided into barotropic (gradients of pressure and temperature intersect) and baroclinic (weak and nearly parallel temperature and pressure gradients). Barotropic conditions occur predominantly in low-latitude tropical regions where convection through a deep layer of the atmosphere can generate high rainfall rates. Baroclinic conditions typify higher latitudes, which tend to have more modest rainfall rates [*Hayden*, 1988]. These different hydroclimatic regimes, in combination with substrate resistance and sediment supply as influenced by geology, produce a diverse array of water and sediment yields among mountainous regions.

Mountain meteorology typically includes greater spatial variability than adjacent lowlands because of complex interactions among global, mesoscale, and local land forcings [*Whiteman*, 2000; *Barry*, 2008]. Contemporary general circulation models can reasonably approximate global patterns of temperature, precipitation, and circulation, and thus provide boundary conditions for mountain climate systems [*Bush et al.*, 2004], but mesoscale variability in topography, land cover, albedo and other relevant landscape parameters strongly influence mesoscale to local mountain meteorology and are challenging to parameterize [*Craig*, 1998; *Hay and Clark*, 2003; *Vivoni et al.*, 2010]. *Bush et al.* [2004] review how GIS software can facilitate model downscaling, land-surface parameterization, land-atmosphere interactions, spatial data scaling and aggregation, and other techniques used to study mountain meteorology.

Precipitation in mountainous regions is strongly influenced by orography, aspect, and elevation [*Lang and Grebner*, 1998; *Anders et al.*, 2006; *Barry*, 2008]. Many mountain ranges have a pronounced rain shadow. As air masses rise in passing over the mountains, the drop in air temperature causes water vapor in the air masses to condense and fall as precipitation. The windward side of a mountain range is thus typically much wetter than the leeward side. This effect is most pronounced where the mountain range is close to a large body of water, as with the Cascade Range of Washington, USA, which parallels the coastline with the Pacific Ocean. Mean annual precipitation is 2500-3500 mm on the western side of the range, but drops to 400-1000 mm on the leeward eastern side [*Barry and Chorley*, 1987] (Figure 2.18). This effect may be reduced for mountain ranges that are far inland or that lie near a body of water with low sea surface temperatures. The Andes Mountains parallel the western coast of South America, but the western side of the mountains from central Peru south to Chile is a desert because the cold Humboldt Current minimizes evaporation from the Pacific Ocean immediately off the west coast.

Orographic effects may cause different types of precipitation on the windward and leeward sides of a mountain range, as well as differences in total precipitation [*Loukas et al.*, 2000]. Mountain ranges in northwestern-most Montana, USA, which are about 700 km from the Pacific Ocean, have snowmelt-dominated runoff with occasional mid-winter rain-on-snow floods [*Madsen*, 1995]. The next mountainous area to the east, which is only about 120 km further inland, does not have rain-on-snow floods because Pacific maritime fronts moving eastward no longer contain sufficient moisture to produce winter rains [*MacDonald and Hoffman*, 1995]. Studies elsewhere in the Rocky Mountains demonstrate that, although snow depth may be similar on windward and leeward sides of a mountain range, water equivalent decreases inland [*Rhea and Grant*, 1974]. Accurate estimation of snow input to a catchment and of the snow water equivalent remain difficult tasks [*Sommerfeld et al.*, 1990; *Fassnacht et al.*, 2003]. Precipitation intensity may also vary spatially in relation to average precipitation; storms in arid regions characteristically have very high intensity rainfall at the start of the storm [*Al-Rawas and Veleo*, 2009], for example.

In general, the effect of topography on spatial patterns of precipitation is strongly dependent on basic airflow characteristics and on synoptic or regional-scale features such as hurricanes. Both of these can vary significantly among storms [*Givone and Meignien*, 1990]. Wind-driven rainfall distributions can also show large variations over micro-scale topography, an effect that *Blocken et al.* [2006] successfully model using a two-dimensional computational fluid dynamics approach, although model verification is limited by available field data. Weather radar data improve model simulations of hydrologic processes by providing information on spatial distributions of precipitation [*Yang et al.*, 2003] and have improved precipitation estimates in mountainous regions during the past two decades [e.g., *Obled et al.*, 1991; *Jasper et al.*, 2002], although problems remain. Radar systems measure the reflectivity of radar beams by raindrops. Reflectivity is strongly dependent on drop size and precipitation type, and the applicability of rainfall radar depends on topography, distance from radar device, and characteristics of the radar and the precipitation [*Uhlenbrook and Tetzlaff*, 2005]. Ground station data are necessary to correctly transform radar reflectivities into rain intensities, but differences in data collection between radars and ground stations can be problematic [*Uhlenbrook and Tetzlaff*, 2005].

Aspect can also strongly influence moisture retention at regional to local scales. Rapid decrease of atmospheric moisture with altitude combines with rain-shadow effects in the leeside of mountain ridges to cause east-west oriented valleys within the inner portions of the European Alps to have considerably less mean annual precipitation than valleys along the border of the Alps [*Bacchi and Villi*, 2005]. Although north-facing slopes can retain snow longer than south-facing slopes in many regions [*Berg*, 1990; *Che et al.*, 2008; *Dornes et al.*, 2008], the difference in snow accumulation between northern and southern exposures is not always significant because of site-specific aspects of snow accumulation, as shown by *Löffler and Rößler* [2005] for Norwegian high mountain ecosystems. Modeling the spatial distribution of snow depth across a wind-dominated alpine basin in Colorado, USA, *Erickson et al.* [2005] find that an index of wind sheltering has the greatest effect on snow depth. Where present, differential snow accumulation indirectly influences hydrology through weathering, soil moisture, and vegetation.

The high relief of mountain environments also promotes more localized elevational effects on precipitation. Precipitation volume and type depend strongly on elevation [*Sevruk et al.*, 1998] and thus exhibit extreme spatial variability in mountains (Figure 2.19). Data from 25 rain gages in the foothills of South Africa indicate that differences in precipitation between ridges and valleys regularly exceed 200% during an individual storm [*de Villiers*, 1990]. Satellite radar data indicate five-fold differences in approximate annual precipitation between major valleys and their adjacent ridges in the Himalaya [*Anders et al.*, 2006]. *Price and Barry* [1997] suggest that the mosaic of climate on individual hill slopes consists of numerous topoclimates (100 m area) and site-specific microclimates (1-10 m scale) associated with patches of vegetation and irregular topography. This extreme spatial variability creates substantial challenges for spatial interpolation of precipitation data in mountainous regions [*Hay et al.*, 1998].

The Colorado Front Range illustrates some of the implications of the large spatial and temporal variability in mountain precipitation for flood intensity and estimates of flood recurrence interval. The Front Range lies in a semiarid region of the interior western United States. Precipitation increases with elevation from 380 to 1200 mm, as does the proportion of solid to liquid precipitation [*Ives*, 1980]. Seasonal snowmelt floods dominate river flow above approximately 2300 m elevation, creating a broad, less peaked flood hydrograph and unit discharges of approximately 1 $m^3/s/km^2$ [*Jarrett*, 1989]. Snowmelt peaks continue along rivers below 2300 m, but these elevations also have rainfall-generated floods that can reach unit discharge values of 40 $m^3/s/km^2$ when a rare, high intensity convective storm occurs over a catchment. Estimates of maximum flood magnitude and recurrence intervals based on snowmelt floods differ substantially from those based on combined snowmelt and rainfall floods (Figure 2.20) [*Jarrett*, 1990a]. Similar elevation limits for flash flooding are present throughout the Rocky Mountains, with latitude-dependent variations as a function of distance from the primary summer moisture source in the Gulf of Mexico [*Jarrett*, 1993; *Parrett and Holnbeck*, 1994].

Although precipitation increases with elevation in the case of the Colorado Front Range, which rises to 4300 m at the highest peaks, moisture-depleted air is more likely at elevations above approximately 3000-3500 m at higher latitudes. This creates high-elevation deserts and distinct zonation in geomorphic processes between the highest, arid elevations and the lower, more humid elevations of a mountain mass, as in the case of the Himalaya and the Tibetan Plateau [*Garner*, 1974; *Shroder and Bishop*, 2004].

Another example of elevational effects on hydrology comes from the Himalayan massif, where monsoonal flood peaks dominate rivers at the lower elevations. As much as 60-80% of discharge comes from monsoon rains for rivers in the lower parts of the southern slope. Glacier melt and snow melt increasingly control seasonal flow at higher elevations, contributing 50-70% of total discharge in the Pamirs and Tien Shan regions [*Gerrard*, 1990; *Wohl and Cenderelli*, 1998].

Precipitation can also exhibit long-period variability such that more zonal or meridional circulation patterns characterize periods up to several decades in length at regional to continental scales [*Hirschboeck*, 1987, 1988]. Periods of more meridional circulation tend to produce more severe flooding. Other investigators also detect fluctuations in flooding at timescales of decades [*Webb and Betancourt*, 1990] and of centuries to millennia [*Ely et al.*, 1993; *Wohl et al.*, 1994b; *Benito et al.*, 1996] across broad regions. Some of this variability is driven by global- to regional-scale fluctuations in oceanic and atmospheric circulation, including the El Niño-Southern Oscillation and the Pacific Decadal Oscillation [*Neal et al.*, 2002].

Although atmospheric models simulating generation of precipitation can be coupled to hydrological models simulating routing of precipitation through a drainage basin [*Toth et al.*, 2000; *Kim and Barros*, 2001], these models typically operate at discordant scales [*Leavesley and Hay*, 1998]. Collaboration between atmospheric scientists and hydrologists to resolve such issues is facilitated by international programs such as the Global Energy and Water Cycle Experiment (GEWEX) [*World Meteorological Organization*, 1987] and the Mesoscale Alpine Programme [*Binder and Schär*, 1996].

Perhaps most importantly, ongoing anthropogenic alteration of Earth's climate has largely negated the concept of *stationarity* – the idea that natural systems fluctuate within an unchanging range of variability – on which flood-frequency and hazard analyses, as well as regionalization of precipitation and discharge records, have long been based. In this situation, we need to identify nonstationary probabilistic models of relevant environmental variables and use those models to optimize water systems [*Milly et al.*, 2008].

2.3.2. Glacier and Snow Melt

Just as complex topography in mountainous regions affects precipitation distribution, it also affects retention of ice and snow. Short- and long-wave incoming radiation components can be influenced by small-scale topographic effects, creating complex spatial distributions of energy balance and patterns of snow retention and snow melt [*Fierz et al.*, 2003]. Snow melt also reflects spatial variations in snow water equivalent at the start of the melt season [*Anderton et al.*, 2002, 2004]. Vegetation cover and aspect influence snow accumulation and melt [*Andreadis et al.*, 2009]; *McCartney et al.* [2006] find the greatest snow water equivalent in portions of a subarctic alpine tundra catchment that have tall shrub vegetation, and earliest melt on southerly exposures and at lower elevations. Snow water equivalent then has a large impact on runoff volume [*McCartney et al.*, 2006].

The relative importance of glacier melt, snow melt, and rainfall typically vary by elevation within a mountainous region and by latitude and location with respect to atmospheric circulation patterns among mountainous regions [e.g., *Winiger et al.*, 2005]. Glacier and snow melt are more important at higher elevations and higher latitudes that may have especially sparse systematic records [*Gillan et al.*, 2010], and the seasonal melt contribution is delayed later into the summer with increasing elevation and latitude. As the basin area covered by ice increases, the ratio of summer to annual runoff increases, the occurrence of maximum monthly runoff is delayed, and interannual runoff variation is reduced [*Chen and Ohmura*, 1990; *Collins and Taylor*, 1990]. The relation between ice cover and interannual runoff variation is particularly important to human communities dependent on glaciated basins for water supply, because global warming and glacial retreat can substantially decrease the dependability of the water supply.

Non-linear relations exist between percentage of a basin covered by glacial ice and interannual variability in runoff; variability is less in moderately glaciated basins than in ice-free basins, but increases in basins with 40-66% glacial cover as the availability of energy for snow- and ice-melt becomes more important [*Collins*, 2006b]. Drainage area-discharge relations thus may not be linear in mountainous regions because of differences in the percentage of catchment area covered by permanent snow and ice. The Upper Hunza River, for example, drains 5,000 km^2 on low mountains north of the main Karakoram chain that do not have major glaciers. Discharge of the Upper Hunza River doubles at the confluence of a short river draining the 300 km^2 Batura glacier [*Gerrard*, 1990].

Runoff from glacier melt can be highly irregular during the melting season, depending on fluctuations in air temperature and storage or release of subglacially-stored water [*Hodgkins et al.*, 2009]. Rivers controlled by glacier melt or snow melt typically have a strongly diurnal discharge pattern during the melt season [*Slaymaker*, 1974; *Hodgkins*, 1997] (Figure 2.21). The timing of the daily peak is a function of percolation times from melt at the surface of the snowpack to the ground and distance downstream from the melt source. Travel times through the snowpack dominate streamflow in basins < 30 km^2, whereas snowpack heterogeneity causes the hour of peak flow to be highly consistent in basins > 200 km^2 [*Lundquist et al.*, 2005].

Glacier outflow hydrographs consist of a base flow component supplied by groundwater discharge, runoff from storage zones within the ice, runoff from the firn water aquifer, and regular drainage from lakes [*Gerrard*, 1990]. Supra- and subglacial meltwater feed a diurnally peaked component of flow, as does meltwater from the snow-free part of the glacier. Discharge from the glacier varies systematically during the melt season as a system of channels develops on and within the glacier [*Fenn*, 1987; *Nienow et al.*, 1998]. An initial peak flow may result from melt of seasonal snow cover on the glacial tongue and on non-glacial surfaces, followed by a second peak of melt from the glacial ice [*Aizen et al.*, 1995]. Interannual discharge fluctuations result from fluctuations in glacier mass balance, particularly as a function of summer weather [*Gerrard*, 1990]. Models of seasonal snow distribution [*Elder*, 1995; *Matsui and Ohta*, 2003] and runoff processes [*Ersi et al.*, 1995; *Flowers*, 2008] on glaciers and subalpine regions may be used to predict glacier- and snow-melt hydrograph characteristics. *Hock* [2003] reviews the differences between temperature-index and energy balance models of glacier-melt. *Pellicciotti et al.* [2008] provide an example of applying models of different complexity to simulate the energy balance of a glacier and glacier melt rates.

Catchments partly covered by glaciers also interact with rainfall in complex ways to produce runoff. Ice- and snow-covered portions of the basin can retain meltwater and rainfall, whereas snow-free portions of the basin rapidly return meltwater and rainfall to channels. The ability of the snowpack to retain or transmit rainfall depends in part on its structure; the presence of ice layers more than doubles the capacity of the snowpack to hold liquid water compared to a homogeneous snowpack [*Singh et al.*, 1998]. The proportions of a basin snow-covered and snow-free change through the seasons as the transient snow line shifts, creating seasonal variations in the magnitude of rainfall-induced flooding [*Collins*, 1998].

Rock glaciers form a unique subset of glaciers, although the discharge pattern of active rock glaciers can be similar to that of glaciers as a result of similar meltwater sources and flow paths [*Krainer and Mostler*, 2002]. Rock glaciers tend to have lower unit discharges, however, than ice glaciers.

2.3.3. Down Slope Pathways of Water

The processes by which water moves down slope exert an important control on the hydrology of mountain rivers. More than 95% of the water in stream flow passes over or through a hillside and its regolith before reaching the channel network [*Kirkby*, 1988]. Despite the critical importance of these processes to stream flow, our understanding of mechanics and spatial patterns in small watersheds is limited by our inability to obtain field measurements at relevant scales [*Robinson et al.*, 2008] and to explain existing observations of some phenomena [*Kirchner*, 2003, 2006].

Precipitation falling toward a hill slope can be intercepted by plants and evaporate or transpire back into the atmosphere. Interception losses can be 10-20% beneath grasses and crops and up to 50% beneath forests [*Selby*, 1982]. Evapotranspiration accounted for 9-12% of the annual water balance in an old-growth conifer forest in the Cascade Mountains of Washington, USA [*Link et al.*, 2005]. Evaporation of intercepted water during the rainfall season accounted for nearly a quarter of the water balance, in part because of abundant epiphytes in the canopy. Transpiration rates peaked in early summer, then declined with decreasing net radiation and soil moisture depletion [*Link et al.*, 2005]. Heterogeneity of terrain, local climate (radiation, temperature, wind speed), and plant cover also strongly influence evapotranspiration in mountain catchments [*Menzel and Lang*, 1998]. *De Jong et al.* [2005b], for example, do not find any steady or linear trends in evapotranspiration rate relative to altitude either within or between mountains in Switzerland and Poland.

Vegetation can also concentrate the movement of precipitation toward the ground surface via stemflow, or water flowing down the trunk or primary stems of a plant. Although typically a small percentage of total precipitation, in some environments this spatially concentrated flow can create a depression around the plant or initiate rills. Measured values of stemflow in diverse environments indicate values as low as 0.2% of total precipitation for a tropical montane forest in eastern Peru [*Gomez-Peralta et al.*, 2008], 2.2-7.2% for shrubs in semiarid loess regions of China [*Li et al.*, 2008], and 12.5% in a Mediterranean *Quercus* coppice [*Limousin et al.*, 2008], with a high of 30% in a seasonal cloud forest in Oman [*Hildebrandt et al.*, 2007].

A unique but locally important interaction between vegetation and fog or low clouds occurs in tropical montane cloud forests, in which cloud water is directly precipitated onto vegetation. Cloud forests, as the name implies, are frequently covered in cloud or mist. Vegetation in these forests intercepts quantities of cloud water equal to anywhere from 6% to 35% of mean annual precipitation [*Bruijnzeel*, 2005].

Soil moisture and infiltration rates can vary substantially through time and space in mountain environments [*Brandes et al.*, 1998; *Moritsuna et al.*, 1998; *Lin et al.*, 2006b]. Using a portable rainfall simulator and ring infiltrometer, *Harden and Scruggs* [2003] measure infiltration rates of 6-206 mm/h at four forested, equatorial sites on lower slopes of the Andes, 16-117 mm/h at six forested, temperate sites in the southern Appalachians, and 0-106 mm/h at four forested, tropical sites in the Luquillo Mountains of Puerto Rico. These rates exceed those of most natural rain events in these environments.

The soil cover is a dynamic medium that continuously changes in response to water movement and processes of weathering and sediment transport [*Brooks*, 2003; *Grant et al.*, 2004; *Kuhn and Yair*, 2004], and that exerts a fundamental control on slope processes and morphology in humid tropical and temperate regions through the regulation of water received at the soil-bedrock interface [*Brooks*, 2003]. Much effort is devoted to estimating the spatial distribution of soil depth [*Pelletier and Rasmussen*, 2009; *Tesfa et al.*, 2009] and soil moisture [e.g., *Chaplot and Walter*, 2003; *Ritsema et al.*, 2009] and to understanding how best to represent the distribution in models [*Chappell et al.*, 2005a]. Development of physically based models capable of including greater detail in soil profiles illustrates the linkages between soil hydrology and slope development [*Brooks*, 2003] and drives further field characterization of soils. *Smith* [2002] provides a thorough review of infiltration processes.

Precipitation reaching the ground surface can flow down slope at the surface as *Hortonian overland flow* (also known as infiltration excess overland flow) if the infiltration capacity is low relative to precipitation intensity. Hortonian overland flow occurs most commonly in the zones of sparse vegetation cover and thin regolith present in mountainous regions. In New Zealand catchments with low infiltration rates, moderate storms generate partial Hortonian overland flow and large storms generate widespread Hortonian overland flow [*Pearce and McKerchar*, 1979]. Infiltration into the debris-mantled belt that separates the steep upper slope from the channel can greatly reduce the overland flow contribution to storm runoff [*Yair and Lavee*, 1985]. Kinematic wave approaches are effective at routing Hortonian overland flow to channels [*Singh*, 2002]. The great majority of most watersheds does not produce Hortonian overland flow. *Lane et al.* [2009] describe the use of a topographic surface flow index to represent the hydrological connectivity of surface overland flow.

Precipitation can also move down slope as *saturation overland flow* (also known as saturation excess overland flow), which consists of direct precipitation onto saturated areas and return flow from the subsurface as saturation occurs. Saturation overland flow depends on the moisture content of the regolith before, during, and after precipitation. As prolonged precipitation allows deeper and less permeable regolith layers to become saturated, throughflow is deflected closer and closer to the surface as the level of saturation rises through the regolith [*Knighton*, 1984]. Saturation overland flow tends to move down slope more slowly than Hortonian overland flow. Patterns of antecedent soil moisture and vadose zone characteristics, rather than hillslope steepness, govern response times of runoff generation [*Montgomery and Dietrich*, 2002; *Fujimoto et al.*, 2008].

Saturation overland flow is rare outside of convergent flow zones [*Dietrich et al.*, 1992], but a study of 17 small catchments in New Zealand found that storm runoff during small rainfalls (1-100 day return period) always included saturation overland flow on small portions of the catchment [*Pearce and McKerchar*, 1979]. A catchment underlain by thick tephra deposits had a sharp initial hydrograph peak derived from saturation overland flow and a broad secondary peak generated by subsurface flow. Rapid subsurface flow from 40-90% of the catchment area produced the bulk of storm runoff during larger rainfalls in all of the 17 catchments [*Pearce and McKerchar*, 1979]. Fifty to seventy percent of discharge during the largest stream flow measured comes from surface runoff plus some shallow groundwater flow in small, forested headwater catchments in Finland [*Lepistö et al.*, 1994]. This contribution drops off markedly, to approximately 20-25% for lesser rainfalls (the remainder of discharge comes from shallow to intermediate groundwater flow not connected to the rainfall event) in these catchments with shallow soil depths and relatively high percentages (29-41%) of exposed bedrock. *Godsey et al.* [2004] also indicate that the formation of perched water tables and the generation of saturation overland flow can occur under forest canopies. *Bonell* [2005] provides a thorough review of runoff generation in tropical forests.

Infiltration depends on precipitation intensity and duration and antecedent moisture, as well as surface porosity and permeability, soil vertical conductivity, and water retention. Hillslope gradient, vegetation, and regolith grain size, compaction, depth, and areal extent control surface porosity and permeability. In alpine/subalpine basins of the Colorado Rocky Mountains, for example, subsurface flow correlates positively with the amount of surficial material in the basin and negatively with basin slope [*Sueker et al.*, 2000]. Reported infiltration capacities worldwide range from 0 to 2500 mm/hr [*Selby*, 1982]. Glaciated basins in particular can have large spatial variability in infiltration as a result of spatial differences in glacial and periglacial deposits [*Parriaux and Nicoud*, 1990]. Mountainous catchments can also have high spatial variability in regolith properties as a result of spatial variations in lithology and hillslope erosional and depositional features such as landslides, debris flows, and talus slopes [*Barontini et al.*, 2005]. *Godsey et al.* [2004] find distinct differences in runoff generation in two small (< 0.4 km^2) catchments near each other in Panama, for example, as a function of differences in lithology and soil thickness. Spatial variations in subsurface properties can be characterized with non-invasive methods such as ground penetrating radar, but this requires substantial investment of field time and labor [*Becker and McDonnell*, 1998].

The area of a drainage basin actually contributing water to a channel extends during precipitation and contracts after the precipitation ends, with contributing area ranging from 5% to 80% of the basin [*Dunne and Black*, 1970b; *Selby*, 1982]. Runoff models based on variable source areas typically use a simple function between groundwater storage and runoff such that runoff follows the simulated rise and fall of groundwater levels, although field evidence increasingly suggests that this is an oversimplification [*McDonnell et al.*, 2005] because, for example, the water table responds separately in riparian and hillslope zones [*Seibert et al.*, 2003].

Infiltrating water that remains in the subsurface can move downslope above the water table as *throughflow*, or below the water table as groundwater. Throughflow may depend on the general porosity and permeability of the unsaturated zone, and on the presence of preferential flow paths in the form of pipes or macropores [*Dunne*, 1980; *Jones*, 1981]. Pipes range from a few centimeters in length and diameter to hundreds of meters long and 2 m in diameter [*Selby*, 1982]. Pipes can form just above a zone of lower porosity and permeability or along a cavity created by a burrowing animal or by the decay of plant roots. Pipe networks can exist at more than one level in the regolith, with each level being activated by precipitation of different magnitude [*Gilman and*

Newson, 1980]. Piping is particularly common in arid and semiarid regions and is frequently found in association with badlands [*Graf*, 1988] (Figure 2.22), although piping can also be well developed in forested environments [*Chappell et al.*, 1998; *Hendrickx et al.*, 2005]. Collapse of the pipe roof, sapping, or concentration of flow downstream from the pipe outlet may enhance surface erosion [*Knighton*, 1984; *Mizuyama et al.*, 1994].

Van Schaik et al. [2008] describe the complex dynamics of preferential subsurface flow within a hillslope in a semiarid portion of Spain. They divide the subsurface into a fine-grained matrix domain and a macropore domain. Flow within macropores can contribute anywhere from 13% of total discharge during intense rainfall (when the relative importance of subsurface flow is suppressed by substantial surface runoff) to 80% during less intense rainfall. Flow within the macropores can drain to the finer matrix and fill up bedrock irregularities when the matrix is dry, but as the matrix grows wetter the loss of water from the macropores decreases and subsurface stormflow increases. *Kim et al.* [2004] also document shifting allocations of flow among different levels of the subsurface in the absence of strongly developed macropores, as well as throughflow as the dominant pathway for water delivery to the stream channel.

Subsurface flow dominates hillslopes with full vegetative cover and thick regolith [*Dunne and Black*, 1970a] and thicker soils increase the mean residence time of water on hillslopes and damp the temporal fluctuations of water movement in response to precipitation inputs [*Sayama and McDonnell*, 2009]. Water moving downslope via Hortonian overland flow generally has the most rapid rate of movement (50-500 mm/hr), whereas groundwater may move as slowly as 1×10^{-8} m/hr [*Selby*, 1982, Table 5.2] (Figure 2.23).

The distribution of water among different down slope pathways can alter in relation to precipitation magnitude, intensity or duration during a single storm, or on a regular annual basis in strongly seasonal climatic regimes. Published values of infiltration capacity in tropical soils are few and highly variable, for example, yet the fact that they tend to be very high and to exceed most rainfall intensities suggests that overland flow should be rare [*Dykes and Thornes*, 2000]. Observed changes related to seasonal rainfall, however, complicate this assessment. Soils in the seasonal tropics can develop water repellency during the dry season which, along with an extensive network of macropores and pipes, facilitates rapid down slope transmission of runoff early in the wet season. As the wet season continues, water repellency declines, infiltration into the soil matrix increases, soil cracks disappear, and runoff declines substantially. By the peak of the wet season, however, soil moisture conditions reach saturation throughout the soil profile, leading to saturation overland flow and increased runoff [*Calvo et al.*, 2005; *Hendrickx et al.*, 2005; *Niedzialek and Ogden*, 2005]. Other studies in tropical environments indicate that different soil types produce different hydrological responses through vertical variations in hydraulic conductivity [*Elsenbeer and Lack*, 1996; *Elsenbeer*, 2001].

McDonnell et al. [2005] note that the dominant runoff processes also change with spatial scale. The nature of the moisture release curve might condition runoff within a soil column, whereas partitioning between preferential and non-preferential flow as a function of soil structure and rain intensity become more important at the plot scale. Spatial variation in soil depth strongly influences laterally mobile flow in highly transient subsurface saturated areas at the hillslope scale. The relative amounts of water derived from different catchment geomorphic units such as hillslopes, riparian zones and bedrock outcrops strongly controls stream hydrograph behavior at the catchment scale [*McDonnell et al.*, 2005]. Diverse watersheds show a threshold-like behavior in terms of when hillslopes connect to riparian zones and the stream channel, and this may provide a method of reducing hillslope complexity into a simple measure of emergent behavior at the watershed scale [*McDonnell et al.*, 2005].

Elsenbeer [2001] proposes that hillslope flow paths occupy a spectrum from predominantly vertical to predominantly lateral. Lateral flow is highly non-linear, with threshold behavior to establish connectivity down the hillslope as small depressions in the underlying bedrock topography fill and spill or laterally discontinuous soil pipes and macropores self-organize into larger preferential flow systems as sites become wetter [*Sidle et al.*, 2001; *Hopp and McDonnell*, 2009]. Deep infiltration into bedrock can also be important where, despite thin soils and steep slopes, substantial losses of water into fractured bedrock occur. Evidence of such deep subsurface flow can take the form of slow or doubly peaked runoff response [*Onda et al.*, 2001].

Although groundwater storage in mountain regions tends to be much smaller than in adjacent lowlands, regional groundwater flow between watersheds is possible, as well as local groundwater flow into adjacent streams. Small groundwater reservoirs can respond to seasonal processes such as snowmelt runoff, as demonstrated by a study covering two months in Vermont, USA during which groundwater levels increased in association with increasingly shallow flow paths and flow through the organic horizon in the riparian and lower hillslope zones [*McDonnell et al.*, 1998]. Talus slopes are the primary groundwater reservoir in an alpine catchment in Colorado, USA, with rapid water transmittal through coarse debris at the talus surface and slower release of water from finer-grained sediments at depth [*Clow et al.*, 2003]. Ice stored in permafrost and rock glaciers forms the second largest groundwater reservoir at the site. Although snow- and glacier-melt contribute the majority of annual water flux to the basin, groundwater flowing from talus can account for ≥75% of stream-flow during storms and winter base flows [*Clow et al.*, 2003]. Similarly, *Yamazaki et al.* [2005] use Cl$^-$ tracers to estimate that 72% of the total runoff from a mountainous watershed in Japan comes from groundwater. Fractured bedrock can also serve as a groundwater reservoir in mountain catchments. In a catchment underlain by fractured granite in the Sierra Nevada of California, USA, snowmelt rates are near or below the bedrock permeability for most of the snowmelt season, facilitating direct infiltration from shallow soils until late in the melt season, when melt rates become double the bedrock permeability [*Flint et al.*, 2008]. During the few days of the latter conditions, snowmelt moves laterally at the soil-bedrock interface down-gradient to contribute directly to streamflow. In volcanic mountainous regions such as the Oregon Cascades of the USA, large, groundwater-fed rivers (e.g., McKenzie River, drains > 900 km^2)

exhibit much less variable hydrographs than adjacent, runoff-dominated rivers [*Jefferson et al.*, 2006].

Because streams with no groundwater flow typically go dry for a portion of the year [*Winter*, 2007], a simple indicator such as the percentage of headwater streams with perennial flow can be a good indicator of regional flow, as well as of the susceptibility of mountain aquifers to increased aridity [*Gleeson and Manning*, 2008]. *Newman et al.* [2006] review surface water-groundwater interactions in semi-arid drainages.

Representation of subsurface drainage is one of the core challenges of hydrologic modelling [*McDonnell et al.*, 2007]. Although the cost of implementing extensive instrumentation in mountain catchments limits the application of this method of studying down slope pathways of water, existing observational networks have provided important insights into mechanisms of down slope water movement, as well as spatial and temporal variability and issues of scaling [e.g., *Kirnbauer and Haas*, 1998]. *McGlynn et al.* [2002] review how perceptual models of hillslope hydrology have evolved through time at the Maimai catchment in New Zealand, which has been instrumented since the late 1970s. Multiple research approaches (dye tracers, isotopic tracers, integrated tensiometers and tracers, single throughflow pits, whole hillslope trenches) have facilitated different insights into the complexities of streamflow generation in this highly responsive catchment with steep slopes and a humid climate. Although soil water and groundwater comprise the majority of channel stormflow, flow varies widely across a slope section in association with bedrock topography. Perhaps the most important point that *McGlynn et al.* [2002] make is that our ability to conceptualize down slope pathways of water is strongly influenced by the types of measurements we make.

At the most basic level, 'black box' models use the amounts and/or composition of precipitation and base flow inputs and the stream outputs to quantify catchment response to storms without revealing much about processes of water movement down slope [*Kendall et al.*, 2001]. Chemical or isotope hydrograph separation using chemicals such as Cl or SiO_2 or naturally occurring conservative isotopes such as $\delta^{18}O$ or δ^2H can provide more insight into the time and geographic sources of stormflow [*Burns et al.*, 2001; *Kendall et al.*, 2001], although runoff sources and pathways remain poorly understood.

Investigators also study flowpaths and transit times for water in a catchment using physical experiments with sprinklers [*Lange et al.*, 2003], instrumentation such as microwave remote sensing [*Engman*, 1997], time domain reflectometry equipment [*Balin et al.*, 2006] and electrical resistivity tomography [*Uhlenbrook and Wenninger*, 2006] to measure subsurface moisture, as well as artificial tracers such as tritium added intentionally to the system, environmental tracers such as stable isotopes of oxygen and hydrogen, or temperature created by natural processes [*Jenkins et al.*, 1994; *Peters*, 1994; *Kendall et al.*, 1995; *Hoehn*, 1998; *Marc et al.*, 2001; *Uchida et al.*, 2003; *Uhlenbrook and Hoeg*, 2003; *Buttle and McDonnell*, 2005; *Tetzlaff et al.*, 2007b]. Ideal tracers interact least with the surroundings. Chemical tracers such as chloride, sulphate, silica or bromide can be applied artificially or occur naturally.

One of the unexplained paradoxes revealed by using water chemistry tracers to examine hillslope hydrology is the rapid mobilization of old water [*Kirchner*, 2003]. Although stream flow in small catchments can respond promptly to rainfall, fluctuations in passive tracers such as water isotopes and chlorides are strongly damped, indicating that storm flow consists of 'old' water that has been stored in the subsurface until it is pushed out by newly infiltrating water. Observations suggest that the mean residence time of water in hillslopes increases as the ratio of flow path length to flow path gradient increases, but we do not yet understand the mechanisms by which catchments store water for extended periods but then release it rapidly during storms.

Inability to explain such phenomena is particularly problematic because soil moisture is the major control on rainfall-runoff response at the watershed scale, especially where saturation overland processes dominate [*Robinson et al.*, 2008]. Measurements at a point using a range of *in situ* sensors and measurements at basin (2,500-25,000 km^2) and continental scales using remote sensing have left a gap at intermediate spatial scales of individual hillslopes and small catchments. At these scales we lack the type of data that might help to identify emergent behavior of small watersheds [*Kirchner*, 2006; *Robinson et al.*, 2008].

Infiltration capacity and overland flow in high-elevation or high-latitude catchments can also depend on the nature of soil ice. At 2000 m elevation in Nevada, USA, porous, concrete frost reduces infiltration rates and causes overland flow, whereas needle ice increases infiltration [*Haupt*, 1967]. Rapid fluctuations in the nature of soil ice and freeze-thaw cycles can thus create temporal and spatial variations in the relative importance of overland flow and throughflow, in turn causing variations in mountain-river hydrographs. Overland flow on slopes in British Columbia, Canada concentrates in a 30-40 m band below the snow line, creating a zone that retreats upslope with the snow line during the period of snowmelt [*Slaymaker*, 1974]. The contributing area of the basin thus changes rapidly in response to weather conditions.

Although local effects of soil ice on runoff and infiltration have been thoroughly investigated, we know much less about the importance of frozen soil on the hydrology of larger areas. Multi-year comparisons of snowmelt runoff from small catchments indicate no clear evidence for faster and larger runoff in winters with deep soil frost [*Shanley and Chalmers*, 1999], for example, in part because deep soil frost typically coincides with shallow snow cover [*Bayard and Stähli*, 2005]. Observations in the Swiss Alps indicate that a considerable portion of the meltwater generated at smaller scales does not show up as surface runoff at larger, catchment scales because it infiltrates unfrozen ground down slope before reaching a stream channel [*Bayard and Stähli*, 2005]. GIS provides a mechanism for combining data such as high-resolution multispectral images of snow and ice with additional layers such as topography and aspect for detailed investigation of the spatial pattern of snow and ice distribution [*Haeberli et al.*, 2004]. When combined with mechanistic studies such as those cited above in the Swiss Alps, and detailed models of heat and moisture transport in frozen soil [*Deangelis and Wood*, 1998], large-scale hydrologic models can be applied to surface and subsurface flow routing on hillslopes [*Koren*, 2006]. *Marcus et al.* [2004]

review applications of Geographic Information Science (GIS) to mountain hydrology. Combining GIS with digital elevation models and remotely sensed and ground-based measurements facilitates development of appropriate algorithms for characterizing hydrologic and geomorphic parameters such as basin area, flow paths, and drainage networks, as well as hydrologic modeling from the network to reach scales [*Marcus et al.*, 2004].

Permafrost forms a subset of soil ice, defined as regolith or soil with a temperature below 0°C continuously for more than one year (Figure 2.24). Approximately a quarter of the land surface in the Northern Hemisphere is underlain by permafrost [*Zhang et al.*, 2003a], of which mountain permafrost constitutes a substantial portion [*Hauck et al.*, 2005]. Permafrost can be spatially continuous or discontinuous, with patches of unfrozen ground separating larger patches of permafrost. Permafrost is also overlain by an active layer of varying thickness that thaws each summer. Investigators have already documented decreasing spatial extents and vertical thickness in both types of permafrost during the 20th century in response to warming air temperatures [*Jin et al.*, 2008; *Woo et al.*, 2008], and these trends are likely to continue and increasingly affect the hydrology of mountainous regions that include permafrost. Permafrost monitoring undertaken in the Swiss Alps in response to concerns over permafrost reduction indicates that soil resistivity, measured as a surrogate for soil temperature, can be accurately determined with a fixed electrode array that is accessible throughout the winter [*Hauck et al.*, 2005].

Monitoring of soil resistivity also illustrates the feedbacks among air temperature, snow cover, and soil temperature. Snow cover effectively decouples the ground from atmospheric influences during winter. *Hauck et al.* [2005] find that the freezing front gradually moves downward to a maximum of 6 m depth from December to May. As snowmelt starts again, soil temperatures gradually increase until the values of the previous September are reached at the end of August.

Modeling precipitation, melting of snow and ice, and down slope movement of water in mountainous catchments all remain significant challenges. Although simple degree-day models can predict melting of snow and ice relatively well in some cases [*Kayastha et al.*, 2005], for example, surface heterogeneities such as snow and ice pinnacles in cold, dry regions like the Andes complicate relations between air temperature and melting elsewhere [*Corripio and Purves*, 2005], forcing models of snowmelt to incorporate a wider variety of effects than just snow-covered area and air temperature [e.g., *Strasser and Etchevers*, 2005].

At a more basic level, assumptions regarding such fundamental physical processes as air turbulence and eddy size developed for flat terrain do not adequately describe steep terrain [*de Jong et al.*, 2005b]. Spatial and temporal heterogeneities in precipitation, infiltration, and runoff also limit the usefulness of averaged variables and obtaining spatially or temporally explicit data is difficult and expensive [*Terblanche et al.*, 2001]. As summarized by *Uhlenbrook and Tetzlaff* [2005], two decades of field investigations of how precipitation interacts with ground conditions to generate runoff have identified process and pattern at the plot scale [*Mosley*, 1982; *Hornberger et al.*,

1991; *Faeh et al.*, 1997], the headwater scale [*McDonnell*, 1990; *Uchida et al.*, 2002], and the catchment scale [*Merot et al.*, 1995; *Uhlenbrook et al.*, 2002; *Hrachowitz et al.*, 2009]. These field data inform physically based and conceptual applied hydrological models [*Uhlenbrook et al.*, 2003; *Fiori et al.*, 2009], but these models commonly have significant errors when applied to steep terrain [*Uhlenbrook and Tetzlaff*, 2005]. *De Jong et al.* [2005a] advocate geomorphological-geological zoning of process domains with respect to surface properties in mountainous watersheds as a means to improve the accuracy of coupling alpine meteorology and hydrology within models.

2.3.4. Modeling Hillslope Hydrology

Some hydrologic models include the entire catchment and the basic stages of precipitation, runoff, and stream flow routing during both storm flow and base flow conditions [*Downs and Priestnall*, 2003]. Top-down approaches are data-based and derive a model structure directly from the available data [*Post et al.*, 2005]. Bottom-up approaches develop complex models to describe small-scale processes operating in a catchment without considering hydrological processes operating at the catchment scale. The small-scale processes are then up-scaled to reproduce the hydrological response of the entire catchment [*Post et al.*, 2005]. Examples include MIKE SHE [*Refsgaard and Storm*, 1995], which simulates overland flow, infiltration, evapotranspiration, and groundwater; KINEROS2, which simulates overland flow on sediment of mixed grain sizes [*Smith et al.*, 1999]; or modifications of TOPMODEL [*Beven and Kirby*, 1984] that simulate snowmelt, infiltration, runoff, throughflow and groundwater flow [*Holzmann et al.*, 1998; *Schild et al.*, 1998]. The Representative Elementary Watershed (REW) approach represents a middle ground in which REWs, representative sub-watersheds organized around the river network, composed of an unsaturated zone, a saturated zone, a concentrated overland flow zone, a saturated overland flow zone, and a channel zone, are coupled [*Zehe et al.*, 2005]. *Barnes and Bonell* [2005] provide an overview of how to choose an appropriate model of catchment hydrology.

Beven [2001] thoroughly reviews rainfall-runoff modeling and uncertainties in ungaged basins, *Brooks* [2003] reviews models of hillslope hydrology, and *Götzinger et al.* [2008] discuss models that couple surface and groundwater systems. Many models require historical or contemporary data on the processes being simulated for calibrating at least a part of the model parameters [*Brath et al.*, 2004]. Model output can be strongly sensitive to multiple parameters [*Geza et al.*, 2009]. Satellite data on vegetation, soil moisture, precipitation, snow cover, water level, and other hydrologic variables can provide input to distributed hydrologic models [*Rango et al.*, 2003; *Shaban et al.*, 2004; *Lakshmi*, 2005; *Zuhal et al.*, 2007; *Brown et al.*, 2008; *Chaponniere et al.*, 2008].

McDonnell et al. [2005] discuss the need to move from models that rely heavily on calibration to models that are based on extraction of first-order process controls and better understanding of flow sources and pathways at the basin scale. The fundamental questions for measurement and modeling remain: Where does water go when it rains?

What flow path does it take to streams? How long does water reside in the catchment? Among the fundamental challenges to hillslope hydrology from a computational perspective are increasing uncertainty in parameter values when moving forward or backward in time to address larger scales, the difficulty of validating models when applied over longer time scales, and increasing structural uncertainty in models as process domains change [*Brooks*, 2003].

Rainfall-runoff models developed for lowland catchments can be adapted to mountain environments by embedding a snow module to simulate snow accumulation and melt [*Flerchinger and Cooley*, 2000; *Fontaine et al.*, 2002; *Das et al.*, 2006], and models specifically focused on snowmelt have been successfully used to simulate observed patterns [*Wang et al.*, 2001a; *Garen et al.*, 2005]. *Jonas et al.* [2009] describe methods to estimate snow water equivalent, which is time-consuming to measure, from snow depth. *Pomeroy et al.* [2007] describe the Cold Regions Hydrological Model platform, which was devised to incorporate algorithms for processes including snow redistribution by wind, snow interception, sublimation, snowmelt, infiltration into frozen soils, and hillslope water movement over permafrost. *Woo and Thorne* [2006] describe the use of macro-scale hydrological models to simulate snowmelt contribution from large mountainous catchments, coupled with catchment-scale models to simulate hydrographs for individual rivers.

Models have been developed specifically for glaciated catchments [*Klok et al.*, 2001; *Verbunt et al.*, 2003; *Konz et al.*, 2006; *Mernild et al.*, 2007; *Fox et al.*, 2008; *Koboltschnig et al.*, 2008], although they do not account for extreme events such as outbursts of glacially dammed water [*Willis et al.*, 2002; *Mernild et al.*, 2008]. The percentage of a catchment covered by glacial ice is particularly important; in catchments with >60% glacial ice, interannual variations in runoff mimic mean air temperature during the melt season, whereas in catchments with <2% glacial ice, the pattern of runoff is broadly inverse to air temperature and follows that of precipitation [*Collins*, 2006a, 2008]. Variations in meltwater storage between glaciers or within a single glacier through time can create lags of hours to days or longer for which models must account [*Singh et al.*, 2000]. Uneven rates of melting can also result from unusual meteorologic conditions. *Boon et al.* [2003] describe a three-day extreme melt event on a high Arctic glacier, during which 30% of the total summer melt occurred. Issues of scaling [*Hebeler and Purves*, 2008] and parameterization that are difficult to resolve in any hydrologic modeling effort also complicate models applied to glacier-melt. Models such as PERMEBAL (snow cover, permafrost, ground-surface temperatures; *Stocker-Mittaz et al.*, 2002), PERMACLIM (permafrost [*Guglielmin et al.*, 2003]) and ALPINE3D [*Lehning et al.*, 2006] are developed specifically for mountain catchments.

The level of process understanding and prediction typically sought in models developed for research purposes may not be necessary for more applied modeling efforts. Rapid and inexpensive methods to identify flow paths of water and to conceptualize hillslope hydrology can be applied to management questions such as hazard mitigation [*Dykes*, 2002] or conservation practices in agricultural lands [*Ticehurst et al.*, 2007]. Models can also be used to run virtual experiments for testing hypotheses of hillslope processes [*Weiler and McDonnell*, 2004].

2.3.5. Pressing Hydrologic Needs for Mountain Regions

Given the inherent spatial variability of precipitation and runoff generation in regions of high relief, as well as the rapid changes in mountain hydroclimatology associated with changing climate and land cover, several research needs emerge. First, we need to better understand the processes controlling the partitioning of energy and water fluxes within and out from mountain catchments [*Bales et al.*, 2006]. Processes such as snow accumulation and melt link energy fluxes and the water cycle in mountains, yet energy fluxes respond non-linearly to changes in climate and land cover. This is exemplified by the Sierra Nevada of California, USA, where snow-water equivalent has decreased at lower elevations over the past 50 years as average temperature has increased, but no consistent trends appear at higher elevations [*Howat and Tulaczyk*, 2005].

A second primary research need is to better understand feedbacks between hydrological fluxes and biogeochemical and ecological processes [*Bales et al.*, 2006], which are discussed in chapter 3. Carbon and nitrogen fluxes in mountain drainages exert an important influence locally and at regional scales. Limited existing data suggest that 60% of the carbon exchange in the western United States occurs at elevations between 700 m and 2000 m, because higher elevations are too cold and lower elevations are too dry [*Schimel et al.*, 2002; *Gordon et al.*, 2004], yet few data exist to evaluate contemporary flux rates or changes associated with changing climate and land cover.

Third, we need to enhance physical and empirical understanding of mountain hydrology using integrated measurement strategies and information systems [*Dunne*, 2001; *Bales et al.*, 2006]. Measurements of energy fluxes, precipitation, snow properties, soil moisture, stream discharge, and evapotranspiration all lack adequate spatial and temporal coverage in mountain regions and, given the inherent heterogeneity of mountains, this will always be the case to some degree. *McDonnell et al.* [2007] propose that we move beyond attempting to characterize landscape heterogeneity in existing models and instead explore organizing principles that might underlie the heterogeneity and complexity using approaches such as optimality principles and network theory. In other words, rather than focusing on what heterogeneity exists, we now need to ask why this heterogeneity exists. The connectivity of subsurface saturation provides an example of such an organizing principle for hillslope hydrology [*Hopp and McDonnell*, 2009].

Better understanding of sources of heterogeneity will allow us to address the primary research need of predicting the hydrological consequences of changes in land use, land cover, and climate [*Lettenmaier*, 2005]. Few mountain regions remain directly unaffected by land use [*Wohl*, 2006] and none are unaffected by global-scale changes such as climate warming. Although we can use predictions of changes in temperature and precipitation as inputs to existing hydrological models, we lack methods for estimating the equivalent of confidence bounds about the resulting predictions [*Lettenmaier*, 2005].

In summary, mountainous regions typically receive more precipitation and retain that precipitation longer in the form of ice and snow than adjacent lowlands, making the mountains crucial to the water supply of lowlands. Although global- to regional-scale atmospheric and oceanic circulation patterns strongly influence the distribution of precipitation across mountainous regions, local effects associated with elevation, aspect, and orography also exert significant influences on the volume and type of precipitation that make it very difficult to predict and model precipitation in mountainous regions. The presence of glacier and snow melt, in particular, commonly creates spatial and temporal nonlinearities in precipitation and runoff. The great majority of water in a stream passes through a hillslope and regolith before entering the stream. Between falling as precipitation and entering the stream, water can be intercepted and transpired by vegetation. Water moves downslope via Hortonian and saturation overland flow, throughflow (within a matrix or macropores), and groundwater flow. Because flow paths are influenced by diverse factors including hillslope gradient, vegetation, regolith grain size, compaction, depth, and areal extent, the presence of frozen ground, and precipitation magnitude, intensity or duration, downslope pathways of flow exhibit substantial spatial and temporal variability. Representation of subsurface drainage remains one of the core challenges of hydrologic modelling.

2.4. Channel Initiation and Development

2.4.1. Channel Initiation

Dietrich and Dunne [1993] argue that the channel head is a key landscape feature. The *channel head* is defined as the upstream boundary of concentrated water flow and sediment transport between definable banks [*Montgomery and Dietrich*, 1988, 1989; *Dietrich and Dunne*, 1993; *Anderson et al.*, 1997]. Banks can be identified by signs of sediment transport such as wash marks, small bedforms, and armored surfaces [*Dietrich and Dunne*, 1993]. The location of the channel head does not necessarily coincide with the location of the *stream head*, where perennial flow occurs. Channel segments of intermittent and ephemeral flow are commonly present above the stream head and below the channel head.

The distance from the drainage divide to the channel head controls drainage density and this in turn controls average hillslope length. The channel head is particularly sensitive to changes in external factors such as climate or land use, which affect runoff, surface erodibility, and sediment supply [*Montgomery and Dietrich*, 1992]. Shifts in the balance controlling channel-head locations may create erosion and deposition cycles which affect the whole drainage basin [*Dietrich and Dunne*, 1993].

Numerous studies indicate the difficulty in predicting the conditions under which channel initiation occurs. Multiple potential controls such as gradient, drainage area, infiltration capacity, and permeability/porosity, interact to influence processes such as overland flow, subsurface flow, and debris flow that affect channel-head location, and both controls and processes can have high spatial and temporal variability. Within this complex, variable system, channel initiation represents a threshold phenomenon in which surface or subsurface flow is sufficiently concentrated and persistent to produce a discrete channel (Figure 2.25).

Drainage networks may be controlled by surface or subsurface properties and processes. Surface-controlled networks develop through rilling, with the greatest surface irregularities concentrating flow in a master rill to which adjacent slopes are cross-graded via micropiracy [*Horton*, 1945]. Surface-controlled drainage networks may develop in at least three ways [*Dunne*, 1980]: (i) Rills may develop nearly simultaneously over the landscape and then integrate into a network. (ii) On a rising land surface, channels may extend downstream during slow warping or intermittent exposure of new land. (iii) An increase in slope or the lowering of base level may cause headward erosion of channels. "Erosional hot spots" occur where hydraulic forces are sufficiently amplified by channel constriction or locally steep gradients to overcome surface resistance and initiate headcuts [*Tucker et al.*, 2006]. Headcuts can then migrate upstream at rates that can be predicted by quantifying discharge, flow velocity, and other parameters [*Alonso et al.*, 2002; *Bennett and Alonso*, 2005]. (Rills may stabilize and fill through the process of gully gravure, in which coarse, erosion-resistant rock debris concentrates in rills and gradually entraps interstitial finer-grained erosional products [*Bryan*, 1940; *Osterkamp and Toy*, 1994].) The rate of headward growth of the master rill and the tributary rills will be partly a function of slope; steeper slopes will produce a more elongated network with less tributary development, other factors being equal [*Parker*, 1977; *Phillips and Schumm*, 1987]. Headward growth of rills does not necessarily produce a well-integrated channel network, however; arid or semiarid regions in which small convective storms generate the majority of surface runoff typically have short, actively eroding channel reaches separated by unchanneled or weakly channeled, vegetated, stable reaches [*Tucker et al.*, 2006].

Networks may also be controlled by spring sapping, in which subsurface flow returning to the ground surface enhances mechanical and chemical weathering and creates a pore-pressure gradient that exerts a drag on the weathered material [*Dunne*, 1980]. Heterogeneities in hydraulic conductivity and resistance to chemical weathering cause convergence of groundwater flow that leads to piping and sapping failure and a positive feedback with further flow convergence. As the channel erodes headward, this disrupts the flow pattern until water emerging along the valley sides is concentrated along a susceptible zone, initiating a tributary that also erodes headward and branches (Figure 2.26). The process continues until the increasing number of spring heads decreases the drainage area of each and limits water supply [*Dunne*, 1980]. Lithological differences in watersheds of the Ashio Mountains of Japan control subsurface flow paths and whether channels initiate at springs or through surface erosion [*Hattanji and Onda*, 2004].

Spring sapping may be an important mechanism of drainage development on various lithologies in humid regions [*Jones*, 1971], in unconsolidated materials [*Pornprommin and Izumi*, 2010], and in the sedimentary rocks of the semiarid Colorado Plateau, USA [*Laity and Malin*, 1985]. *Lamb et al.* [2006, 2008a], however, question whether seepage erosion alone is sufficient to cut canyons into resistant rock. In the Oregon Cascade Range, USA, drainage networks developed on Holocene and late Pleistocene basalt landscapes are dominated by spring- and groundwater-fed streams that transport little sediment and largely ineffective in incising river valleys [*Jefferson et al.*, 2010]. Landscape dissection occurs after the springs are replaced by shallow subsurface stormflow as the dominant mechanism of streamflow generation; landscape evolution is thus constrained by the time required for permeability to be sufficiently reduced for surface flow to replace groundwater flow [*Jefferson et al.*, 2010].

Drainage networks developed in karst terrains represent an important subset of networks controlled by subsurface processes. Karst features may develop in any mountainous region with carbonate rocks. Fluviokarst terrains contain transitions between surface and subsurface channels and valley networks, whereas in holokarst terrains the surface drainage network has been almost completely disrupted by subsurface piracy [*Ford and Williams*, 1989] (Figure 2.27). Within fluviokarst terrains, surface streams are typically diverted underground where incision of a clastic caprock exposes an underlying carbonate unit. The stream may be diverted at the clastic/carbonate contact [*White*, 1988] or may flow across the carbonate unit for some distance. Subsurface piracy routes may mimic the initial surface drainage pattern or may divert flow beneath surface drainage divides into adjacent catchments [*Ford and Williams*, 1989]. Within some mature fluviokarst basins, surface streams flow across the clastic-carbonate contact and onto low gradient floodplains within a topographically enclosed karst depression. Upon reaching the opposite wall of the depression, the stream flows directly into a cave. The distance that a stream flows before entering the subsurface is a function of contributing basin area [*Smart*, 1988; *Miller*, 1996]. The cave transmits stream flow beneath depression slopes and eventually discharges the water at a spring. Position of the spring may be controlled by lithology, structure, or base level [*White*, 1988; *Ford and Williams*, 1989]. In general, such karst terrains exhibit the profiles, gradients, and mouth elevations that one would predict for surface streams draining the same basin. Alpine karst is well developed in the Alps and Pyrenees, where glacial meltwater has strongly influenced the type and degree of karst features present [*Ford and Williams*, 1989].

The great majority of investigations focusing on the quantification of channel initiation have been conducted in humid regions. Early field-based studies of network development indicate that the source area above the channel head decreases with increasing local valley gradient in steep, humid landscapes (5-45° slopes) well-mantled with soil [*Montgomery and Dietrich*, 1988]. For the same gradient, drier regions tend to have larger source areas. *Montgomery and Dietrich* [1989] use field-mapped channel initiation points in northern California, USA to develop empirical equations relating source basin length (L), local valley slope (θ), and contributing drainage area (A):

$$L = \lambda \tan \theta^{-0.83}, \text{ where } \lambda = 67 \text{ m} \tag{2.13}$$

$$A = \lambda \tan \theta^{-1.65}, \text{ where } \lambda = 1978 \text{ m}^2 \tag{2.14}$$

$$A = 0.46 L^{1.99} \tag{2.15}$$

$$L = 1.48 A^{0.50} \tag{2.16}$$

Montgomery and Dietrich note an inverse relationship over a wide range of slopes between A and θ at channel heads. Subsequent work continues this trend of developing empirical, site-specific relations between topographic parameters, typically using bounding equations to quantify the range in channel head locations [*Montgomery and Dietrich*, 1992; *Dietrich et al.*, 1992; *Prosser and Abernethy*, 1996]. Some of these investigators also find an inverse relationship between A and θ [*Roth et al.*, 1996; *Roth and La Barbera*, 1997], although the nature of this relationship varies between low-gradient hollows with convergent topography and seepage erosion, and steeper topography where channel initiation is more likely to reflect saturation or Hortonian overland flow or landsliding [*Montgomery and Dietrich*, 1989; *Montgomery and Foufoula-Georgiou*, 1993; *Prosser and Abernathy*, 1996]. The inverse slope-area relationship does not always hold for diverse environments [*Bischetti et al.*, 1998; *Adams and Spotila*, 2005], partly as a result of differences in underlying bedrock that influence downslope pathways of water. *Hattanji and Matsushi* [2006] find that the S-A relation grows less consistent with larger relative groundwater contribution. New data sources and techniques such as LiDAR-derived digital terrain models facilitate the field delineation of hillslope and channel processes and channel initiation [*Tarolli and Dalla Fontana*, 2009].

Flow through fractured bedrock (Figure 2.28) can complicate relations between topographic metrics and channel initiation. Although topography is a reasonable surrogate for flow paths in steep upland topography because elevation potential largely dominates total potential, bedrock topography is likely to be more important than surface topography in controlling the routing of mobile water laterally downslope, and thus the initiation points of channel heads [*Anderson et al.*, 1997; *McDonnell*, 2003]. *Jaeger et al.* [2007] interpret the lack of systematic A-θ relations at sites in Washington, USA as reflecting the controlling influence of flow through fractured bedrock. The role of deeper subsurface flow in fractured bedrock may be particularly critical in arid or semiarid mountainous headwaters where peak runoff can originate from snowmelt over saturated or frozen ground or from rainfall runoff, but base flow comes from groundwater flow paths below the shallow, typically dry soils.

The smooth, long hills typical of low relief areas result from the large source area necessary to initiate a channel, whereas the discontinuous channels and dissected terrain of steeper areas suggest that channel-head locations are controlled by hillslope processes   such as landsliding that require much smaller source areas [*Montgomery and Dietrich*, 1989]. Drainage density in steep terrain thus depends at least in part on rate of weathering, regolith storage, and slope stability. *Zero-order basins*, also known as unchannelized valleys or *hollows*, may play an important role in regolith storage and slope stability. These features, which have been described for subhumid and humid mountains, store sediment that is periodically mobilized as a debris flow [*Hack and Goodlett*, 1960; *Dietrich and Dunne*, 1978; *Dietrich and Dorn*, 1984; *Mills*, 1989].

Dietrich et al. [1992, 1993] propose a digital terrain model for predicting the location of channel heads based on the assumption that they occur where saturation overland flow exerts a boundary shear stress that exceeds the critical value for substrate erosion. This is now generically expressed in a form similar to

$$AS^{\alpha} \geq C \tag{2.17}$$

where A is contributing catchment area, S is hillslope gradient, and C is channel initiation threshold [*Istanbulluoglu et al.*, 2002; *Dalla Fontana and Marchi*, 2003]. Subsequent field studies indicate that vegetation [*Prosser and Dietrich*, 1995; *Prosser et al.*, 1995], slope aspect [*Yetemen et al.*, 2010], and substrate grain size [*Istanbulluoglu et al.*, 2002; *Hattanji et al.*, 2006] influence the critical value, which helps to explain intraregional variability in values of *A*. *Rivenbark and Lane* [2004], for example, find that *A* varies from 4 to 13 ha in the southern Appalachian Mountains, USA. Limited, unpublished data from an arid region indicate that changes in critical shear stress (for channel initiation via overland flow) arising from differences in ground cover may dominate the proportionality constant between A and θ [*Montgomery and Foufoula-Georgiou*, 1993].

Numerical simulations can also be used to estimate the locations at which the threshold for channel initiation is exceeded. Some of these adapt the widely used TOPMODEL [*Sun and Deng*, 2003; *Kim and Lee*, 2004]; others develop new modeling approaches such as coupled hydraulics and sediment transport [*Simpson and Castelltort*, 2006]. *Lin et al.* [2006c] and *Kim and Kim* [2007] present methods for extracting channel initiation points from DEMs based on *S-A* relations. *Smith* [2010] develops a theoretical model of channel formation from mass balance equations for water and sediment, the St. Venant equations for water flowing down an energy surface gradient, and generalized representations of sediment transport.

Montgomery and Dietrich [1994b] use a plot of drainage area versus slope to delineate process thresholds for diffusive sediment transport, landslides, overland flow, and other processes. Valley maintenance then reflects spatial transitions in process dominance at timescales of 10^4-10^6 years, and temporal variance in the exceedance of a channel initiation threshold at shorter timescales of 10^2-10^3 years [*Montgomery and Dietrich*, 1994b].

Despite the scatter in the locations of channel heads potentially introduced by surface versus subsurface processes and spatial variations in climate and aspect,

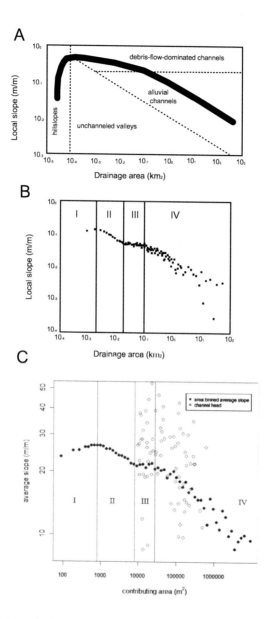

Figure 2.29 (A) Schematic illustration of relations between drainage area and local slope depicting transitions among hillslopes and various types of channels [after *Montgomery and Foufoula-Georgiou*, 1993, Figure 10]. (B) Diagram of local slope versus drainage area for a basin in Alabama, USA, illustrating the four regions of hillslope and channel process and form [after *Ijjasz-Vasquez and Bras*, 1995, Figure 1]. (C) Diagram of local slope versus drainage area for the Colorado Front Range, USA, with actual locations of channel heads indicated [after *Henkle*, 2010, Figure 13].

numerous investigators extract channel networks from digital elevation models (DEMs) by making assumptions about the locations of channel heads based on reversals or inflections in averaged hillslope profiles. These assumptions derive from the idea that channel heads correspond to the transition from convex to concave slope profiles [*Kirkby*, 1971, 1980; *Tarboton et al.*, 1991, 1992]. *Montgomery and Foufoula-Georgiou* [1993], however, note that this slope transition commonly coincides with the transition from divergent to convergent topography and thus more accurately reflects the start of valley development rather than channel heads. They proposed that the commonly observed reversal in slope gradient represents the transition from hillslopes to unchanneled valleys and channels dominated by debris flows, and that the next inflection downslope represents the start of alluvial channels (Figure 2.29A). *Ijjasz-Vasquez and Bras* [1995] designated four regions in slope profiles (Figure 2.29B) and interpreted the transition between regions I and II to represent the location of channel heads. Despite *Montgomery and Foufoula-Georgiou*'s [1993] warning that 'acquisition of even limited field data is recommended,' many investigators have located channel heads at the reversal in slope gradient in the absence of field data.

Tarolli and Dalla Fontana [2009] provide an exception in that they test this assumption using field data on the location of 30 channel heads in the eastern Italian Alps; they found that channel heads are mostly confined to region II. This supports the assumptions of *Ijjasz-Vasquez and Bras* [1995], although placing channel heads right at the transition between regions I and II would underestimate the minimum contributing area and overestimate the length of channel networks. Field-mapped channel head locations of 78 channel heads from the Colorado Front Range plot at the threshold between regions II and III, although some extend well into region IV [*Henkle*, 2010]. Most actual drainage areas for channel initiation in this semiarid region with snowmelt runoff are thus an order of magnitude larger, and plot in a significantly different portion of the slope-area graph, than would be assumed if the gradient reversal between regions I and II were used to locate channel heads. The actual location of channel heads in the semiarid study area is better represented by the first inflection point after the gradient reversal [*Montgomery and Foufoula-Georgiou*, 1993], which would be the transition between regions II and III, although a substantial portion of the channel heads are located in region IV (Figure 2.29C).

2.4.2. Channel Network Development

The development of an integrated drainage network on a hillslope has been studied with experimental apparatus; on surfaces of varying age for which all other variables are held constant (e.g., lava flows or glacial tills), and using computer simulations and landscape evolution models [*Collins and Bras*, 2010]. Experimental studies and field studies using surfaces of varying age indicate that drainage density tends to initially increase relatively rapidly, then change more slowly, and eventually decrease slightly as relief is reduced [*Glock*, 1931; *Ruhe*, 1952; *Leopold et al.*, 1964; *Flint*, 1973; *Parker*, 1977].

Computer simulations of drainage network development indicate that, in the absence of strong substrate controls such as orthogonal joints, the spatial arrangement of channels can be approximated fairly well with a random-walk model that predicts the most probable state under the constraints postulated [*Leopold and Langbein*, 1962; *Howard*, 1971]. Most models of river basin evolution also largely ignore substrate heterogeneity [e.g., *Willgoose et al.*, 1991]. *Rodriguez-Iturbe et al.* [1992a] propose that river networks follow power-law distributions in their mass and energy characteristics. These characteristics represent a balance among (i) the principle of minimum energy expenditure in any link of the network, (ii) the principle of equal energy expenditure per unit area of channel anywhere in the network, and (iii) the principle of minimum total energy expenditure in the network as a whole [*Rodriguez-Iturbe et al.*, 1992b]. It is not clear whether such principles apply to mountain river networks or other systems with strongly heterogeneous substrates.

Hovius et al. [1998] describe the development of watersheds during the early phase of mountain growth. As exemplified by observations from different parts of the Finisterre Mountains of Papua New Guinea, watersheds appear to initiate by isolated gorge incision. The watersheds then expand by large-scale landsliding that is controlled by groundwater seepage, and finally entrench by fluvial incision of landslide scars and deposits. Only infrequent, large landslides can modify the drainage pattern once a mountainous system of ridge and valleys is established.

In summary, channel initiation represents a threshold in which surface or subsurface flow is sufficiently concentrated and persistent to produce a discrete channel. The channel head is the upstream boundary of concentrated water flow and sediment transport between definable banks, whereas the stream head is the location of perennial flow; the two commonly do not coincide. Although the transition from hillslope to channel processes represents a crucial threshold for drainage basin form and process, our ability to quantitatively predict this transition remains limited by the multiplicity of potential controlling variables and processes. Gradient, drainage area, infiltration capacity, and permeability/porosity interact to influence processes such as overland flow, subsurface flow, and debris flow that affect channel-head location, and both controls and processes can have high spatial and temporal variability. Empirical equations relating source basin length, local valley slope, and contributing drainage area are most commonly used to predict the location of channel heads, but recent studies suggest that these equations do not transfer well between regions with diverse climates and slope processes. Consequently, we can delineate general patterns, such as larger source areas for channel initiation in drier regions, but more regionally specific delineation requires many more field data from diverse environments.

2.5. Basin Morphometry and Basin-Scale Patterns

Drainage basin morphometry can be characterized with linear, areal, and relief indices [*Strahler*, 1964; *Ritter et al.*, 1995; *Yamada*, 1999a]. As global coverage of DEMs and LiDAR data increases, automated methods are increasingly used to extract channels and channel networks from these data [e.g., *Passalacqua et al.*, 2010]. Morphometric indices are used to compare development of distinct drainage basins and to provide insight into hydrograph characteristics [e.g., *Strahler*, 1964; *Saxena and Prakash*, 1982; *Patton*, 1988a]. For example, *Luo et al.* [2010] infer hydraulic conductivity based on a concept of effective groundwater drainage length as reflected in surface drainage dissection patterns. Hypsometric curves of mass distribution within a drainage basin may be used to infer the history and processes of basin development because basin hypsometry reflects uplift rates and variations in erodibility of different lithological units [*Willgoose and Hancock*, 1998; *Walcott and Summerfield*, 2008]. *Pérez-Peña et al.* [2009], for example, correlate the distribution of active normal faults in the Granada basin of Spain with clusters of high or low values of the hypsometric integral. Spatial variation in morphometric parameters and basic drainage patterns can correspond to regional lithologic and tectonic influences [*Clark et al.*, 2004]. Geographical information systems and remote sensing techniques are increasingly applied to characterizing basin morphometry [e.g., *Kumar et al.*, 2001; *Vogt et al.*, 2003].

Figure 2.30 illustrates hypsometric curves for three small (4-7 km^2), steep (530-820 m relief) drainage basins in the USA. The catchment in the Colorado Front Range begins on a broad, low-relief upland and then descends steeply through a narrow

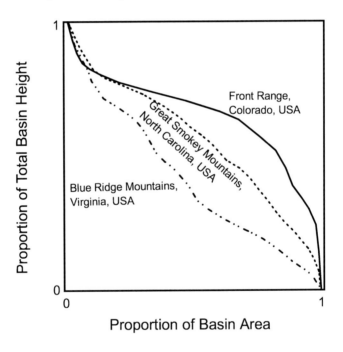

Figure 2.30 Hypsometric curves for three small drainage basins in mountainous regions of the Colorado Front Range, the Blue Ridge Mountains of Virginia, and the Great Smokey Mountains of North Carolina, all in the USA. Figure courtesy of Gregory S. Springer.

canyon into the Poudre River. The catchment in the Great Smoky Mountains of North Carolina is well dissected, with no alluvial fans, no floodplains, and minimal deposition in the valleys. The catchment in Virginia's Blue Ridge Mountains has alluvial fans at the base of the ridges, and extensive valley infill; more of the mass in this catchment is at lower elevations. The shapes of these hypsometric curves indicate the degree to which mass has been erosionally transferred from higher to lower elevations within each drainage basin.

Another example of using drainage basin morphometry to interpret basin development comes from *Oguchi* [1997b]. He finds a constant value of drainage density for 8 mountain river basins (A 12-78 km^2) formed on various lithologies and varying relief in central Japan. The uniform frequency or spacing of ridges and hollows in these rugged humid mountains suggests that erosion has continued without effecting a change in the spatial distribution of ridges and hollows, because valleys have undergone straight downcutting while maintaining the drainage density of antecedent drainage systems.

Drainage density and sediment yields for many mountain rivers changed with the retreat of large alpine glaciers and ice caps. Increased channelization of hillslope hollows in Japan resulted from landslides and gullying during a period of wetter climate at the Pleistocene-Holocene transition [*Oguchi*, 1997a, 1997c]. Sediment yields also increased during this period of rapid incision. *Schumm* [1997] summarizes studies indicating that lithologies which weather to produce a soil with a high permeability result in drainage networks that are less closely linked to climate than those formed on readily eroded materials with low permeability. Shale basins, for example, more readily reflect current climate conditions than granite basins. Drainage density response to climate change may ultimately depend on corresponding changes in the threshold of channelization in terms of minimum upslope area necessary to support a channel [*Montgomery and Dietrich*, 1992].

Comparison of basin morphometric indices for mountain and lowland rivers around the world indicates the expected differences in valley side slopes, relief, and relief ratios, but no significant difference in drainage density (Table 2.4). The development of basin morphometric indices dates to work by *Horton* [1945] and *Strahler* [1952b] and continues today as investigators search for scaling laws for river networks [*Rodriguez-Iturbe and Rinaldo*, 1997; *Vasquez et al.*, 2002]. *Kirchner* [1993], however, cautions that some of the inferences drawn from morphometric indices are unsupported because certain observed geometric properties are not specific to particular kinds of stream networks, but instead describe virtually all possible networks.

2.5.1. Basin Morphometry and Hydrology

Drainage basin morphometry strongly influences the movement of water from hillslopes to channels and along the channel network. Various investigators focus on the hydrophysical influence of specific drainage-basin characteristics as a means of developing predictive models for hydrographs [*Horton*, 1932, 1945; *Morisawa*, 1962;

Patton and Baker, 1976]. These studies are the precursors to more sophisticated models of runoff that incorporate geomorphic characteristics and the concepts of flood storage and flood routing to create geomorphic unit hydrographs [*Patton*, 1988a; *Rodriguez-Iturbe*, 1993].

The unit hydrograph or instantaneous unit hydrograph illustrates a basin's response to a unit average effective rainfall. *Nash* [1957] conceptualizes a basin as a series of linear reservoirs and proposes a gamma-law-type analytical expression of the unit hydrograph, with parameters that have to be calibrated. Much subsequent work has been based on this model. *Dooge* [1959] initiates the process of identifying a basin-scale transfer function using geomorphic characteristics to provide a physical basis for the unit hydrograph. *Rodriguez-Iturbe and Valdès* [1979] call this the *geomorphologic instantaneous unit hydrograph* and conceptualize it as a travel-time probability density function to the basin outlet in which each flow path is seen as an initial hillslope state and a set of *Strahler* [1957] stream-order states from the draining link to the outlet [*Cudennec et al.*, 2004]. *Saco and Kumar* [2002] explore the relative importance of kinematic dispersion (spatially varying celerities) and geomorphologic dispersion (heterogeneity of path lengths) on unit hydrograph outputs. *Moussa* [2003, 2009] proposes new catchment shape descriptors for use with geomorphic unit hydrographs. *D'Odorico and Rigon* [2003] explore how the differences in flow velocity between hillslope and channel components affect the probability distributions of travel times, and *Lee and Chang* [2005] add subsurface flow processes. *Wang and Tung* [2006] develop a framework for assessing the uncertainty features of a geomorphic unit hydrograph resulting from sampling errors.

Geomorphic unit hydrograph models have been successfully applied to mountain drainage basins in the former Czechoslovakia [*Pristachová*, 1990], the Canadian Rockies [*Ajward and Muzik*, 2000], Spain [*Agirre et al.*, 2005], and India [*Kumar and Kumar*, 2005]. Application of a geomorphic unit hydrograph in southern France suggests that the hydrograph is more sensitive to the channel topology and to the spatial distribution of rainfall when correlated with altitude than to the hydraulic properties of flow on hillslopes and in the channel [*Moussa*, 2008]. Geomorphically-based artificial neural networks also show promise for estimating direct runoff [*Zhang and Govindaraju*, 2003].

Lateral floodplain extent and its variability within a drainage basin exert a particularly important control on hydrology. Methods for delineating floodplain distribution include: hydraulic simulations of predicted water levels within a specified valley geometry [*Bates and Roo*, 2000; *Bates*, 2004], which are data- and computationally-intensive; terrain analyses based on DEMs and GIS-based tools that use simplifying assumptions about flood stage [*McGlynn and Seibert*, 2003; *Dodov and Foufoula-Georgiou*, 2005], including a constant flood level at all locations; and terrain analyses that are more hydrologically realistic in creating spatial variability in flood stage that reflects contributing area, channel-reach geometry, and geomorphic structure of the channel network [*Nardi et al.*, 2006]. *Nardi et al.* [2006] describe an automated technique for mapping floodplain distribution by linking a simplified inundation algorithm with river basin properties. They find power-law scaling of the floodplain width after averaging over stream orders in tests of the technique using catchments in Italy and in New Mexico, USA. This scaling, first proposed by *Bhowmik* [1984], takes the form of

$$w = q\,A^m \tag{2.18}$$

where w is floodplain width, A is contributing drainage area, and q and m are scaling parameters. Some of the difficulty in developing a model of a hydrograph that adequately simulates a range of drainage basins stems from our incomplete understanding of, first, how the actual contributing drainage net changes with rainfall intensity and duration and, second, the distribution of lag time as a function of stream order within a basin [*Patton*, 1988a].

Geomorphic characteristics that exert some control on the basin hydrograph include the following: (i) *Drainage area* is important as it controls the volume of runoff collected in a channel network. Discharge increases at a lesser rate than drainage area [*Patton*, 1988a]

$$Q_x = a\,A^b \tag{2.19}$$

where Q is discharge, x indicates recurrence interval, A is drainage area, and $b = 0.5$–0.9. (ii) *Drainage density* reflects the effectiveness of surface runoff and erosion, and influences concentration time of flow in a channel network. Drainage density reaches the highest values in semiarid and tropical environments, and in basins with non-resistant lithologies [*Gregory and Gardiner*, 1975]. However, drainage density is not sensitive to changes in the hydrologic response of a basin that occur during an individual storm [*Day*, 1978]. (iii) *Stream order and basin magnitude.* Strahler stream order correlates directly with discharge within a basin, presumably because of the correlation between drainage area and order for basins in similar climatic and geologic regions [*Leopold and Miller*, 1956; *Blyth and Rodda*, 1973; *Patton and Baker*, 1976]. (iv) *Drainage basin relief* is important in that higher relief is associated with steeper hillslopes, higher stream gradients, shorter times of runoff concentration, and larger flood peaks [*Patton*, 1988a]. (v) *Drainage basin shape* influences time of concentration and magnitude of peak discharge, with equant-shaped basins tending to produce larger, shorter peak discharges than linear basins [*Strahler*, 1964].

Drainage networks in a mountainous basin may be substantially modified if a severe or widespread storm produces extensive mass movements and associated erosion and deposition, or if seismic or volcanic activity alters basin relief. Localized convective storms strip colluvium and reduce hillslope-infiltration rates in the Santa Catalina Mountains of Arizona, USA [*Etheredge et al.*, 2004]. These changes "prime" the hillslopes and make them more responsive to areally extensive but lower intensity winter storms that create flooding in larger catchments. Channel networks may also respond to changes in water and sediment yield associated with land-use activities such as timber harvest or agriculture. Over longer timescales, drainage morphometry may respond to glaciation or climatic change.

2.5.2. Hydraulic Geometry

Hydraulic geometry [*Leopold and Maddock*, 1953] relates the response variables of flow width (*w*), depth (*d*), and velocity (*v*) to discharge (*Q*) via simple power functions:

$$w = aQ^b \qquad (2.20)$$

$$d = cQ^f \qquad (2.21)$$

$$v = kQ^m \qquad (2.22)$$

in which *a*, *c*, *k*, *b*, *f*, and *m* are numerical constants. At-a-station hydraulic geometry describes changes in hydraulic parameters as discharge fluctuates through time at a channel cross section, whereas downstream hydraulic geometry describes changes in hydraulic parameters as a discharge of the same frequency varies in magnitude downstream. Originally formulated primarily for lower gradient alluvial rivers, hydraulic geometry assumes that *Q* is the dominant independent variable of channel and flow geometry, and that dependent variables such as channel width, flow depth, slope, or velocity are related to discharge in the form of simple power functions. These basic assumptions may not hold for mountain river channels that are shaped by debris flows or glaciations [*Wohl*, 2004]. Mean annual or bankfull discharge is typically used for downstream hydraulic geometry analyses [*Park*, 1977b], and the assumption is that channel dimensions primarily reflect the forces exerted by discharges with this recurrence interval. The recurrence interval of the water discharge that shapes channel geometry, however, may vary greatly among basins (see discussion of bankfull discharge, chapter 3). *Navratil and Albert* [2010] suggest using two power functions to relate dependent variables to discharge; one for lower flows and one for higher flows, and they propose a model for detecting the breakpoint between lower and higher flows.

Hydraulic geometry grew out of *regime theory*, which is the concept that river channels adjust toward a dynamic equilibrium in which the channel is just able to convey the water and sediment supplied from upstream. Adjustments occur primarily through bed elevation, which affects transport capacity through channel gradient, but gradient can also be adjusted via sinuosity, and transport capacity also depends on width [*Ferguson*, 2003]. Consequently, regime theory focuses on the equilibrium slope and width for specified water and sediment discharge and channel boundary composition [*Gilbert*, 1914; *Lacey*, 1930; *Mackin*, 1948; *Blench*, 1951]. More recent work focuses on developing a 'rational' regime theory that is physically based [e.g., *Parker*, 1979; *Eaton et al.*, 2004].

At-a-station hydraulic geometry is useful for activities as diverse as monitoring stream flow (these relations form the basis for stage-discharge rating curves) and modeling minimum flow requirements for fish passage or boating [*Ferguson*, 1986]. Inflections in the rate of change in *w*, *d*, or *v* result from changes in roughness and channel confinement as *Q* varies [*Richards*, 1973; *Parker et al.*, 2007; *Wohl*, 2007]. The wide variation in exponents for at-a-station relations suggests that there is no unique hydraulic geometry that applies to streams in diverse regions [*Stewardson*, 2005; *Reid et al.*, 2010]. Because some of this variation can be explained by the location of a cross section within a channel reach (e.g., pool versus riffle), reach-averaged hydraulic parameters might be more strongly related to catchment- and reach-scale attributes than are cross-sectional parameters. *Stewardson* [2005] demonstrates that the five reach-scale hydraulic variables of mean and coefficient of variation of surface width and hydraulic depth, respectively, and the coefficient of variation of cross-sectional velocity, sufficiently characterize the gross hydraulics along a reach of river.

Downstream hydraulic geometry (DHG) is used for engineering stable channels, predicting the effects of flow regulation, and understanding the geomorphic role of floods or inferring past flood magnitudes from channel dimensions [*Ferguson*, 1986]. Typical exponent values for DHG are $b = 0.4$-0.5, $f = 0.3$-0.4, and $m = 0.1$-0.2 [*Park*, 1977b].

Numerous papers describe hydraulic geometry relations in specific field settings and a few of these papers describe mountain rivers [*Wohl*, 2005; *Golden and Springer*, 2006; *Pike et al.*, 2010]. *Osterkamp and Hedman* [1977] propose that the value of *b* developed for lower gradient channels also adequately describes higher gradient rivers. Using peak discharge information for 43 gaging stations during an average monsoon season in two drainage basins in the Middle Hills of Nepal, *Caine and Mool* [1981] find that the hydraulic geometry relations are generally similar to those reported from other regions of the world. The two Nepalese catchments are slightly longer than normal and have a quicker hydrologic response, both of which presumably reflect the steep terrain. *Fenn and Gurnell* [1987] find that the usual linear power functions describe the relation between discharge and the dependent variables for a proglacial braided channel in Switzerland. In contrast, *Ponton* [1972] finds that two tributaries of the Lillooet River of British Columbia, Canada do not follow the expected downstream hydraulic geometry trends because of gradient changes related to glaciation. *Phillips and Harlin* [1984] describe a mountain river in Colorado, USA that does not follow predictable downstream changes in hydraulic geometry because of substantial changes in alluvial substrate. Working on a proglacial river in British Columbia, Canada, *Ashworth and Sauks* [2006] note that although the *Q-w* relation for a given location and channel configuration is transferable to adjacent reaches with different morphology, the prospect of a more universal relation for braided rivers is remote because of differences in hydraulics and channel geometry between rivers. *Arp et al.* [2007] find weak DHG relations within single catchments and regionally in the Sawtooth Mountains of Idaho, USA. Sediment sources from hillslopes and tributaries and sediment sinks in the form of lakes create such strong heterogeneity in grain size that channel width and depth vary substantially in the absence of variation in discharge as bed and bank resistance change in passing from coarse sediment (wide, shallow channels) to finer sediment (deep, narrow channels).

Wohl [2004] proposes that DHG relations should have some spatial limits within a drainage basin because non-fluvial processes such as debris flows can influence the geometry of channels draining small areas of steep terrain. She proposes a ratio of stream power to sediment size above which DHG is well developed, as defined by coefficients of determination > 0.5 for the DHG relations (Figure 2.31). Channel networks with strong colluvial inputs or glacial inheritance and relatively low stream power have poorly developed DHG relations [Wohl et al., 2004]. Higher values of stream power can result in well-developed DHG relations even in channels with strong colluvial or bedrock influences [Wohl, 2005; Wohl and Wilcox, 2005; Pike et al., 2010]. The presence of coarse sediments and lateral constrictions can also limit adjustments in channel width to increasing discharge more in some types of channel units (i.e., steps, cascades) and channel morphologies than in others, creating scatter in DHG relations based on cross-sectional measurements [Vianello and D'Agostino, 2007].

Substituting drainage area for Q, Montgomery and Gran [2001] find that $b = 0.3$-0.5 characterizes bedrock as well as alluvial channels, although bedrock channels tend to be narrower and deeper than alluvial channels with equal drainage areas. Wohl and David [2008] also find consistent DHG exponents among bedrock and alluvial channels, suggesting that erosional resistance is not the primary control on scaling relations for channel geometry.

DHG relations have also been analyzed in the context of optimal energy expenditure, as defined by minimum energy expenditure within a river link, constant energy expenditure per unit channel bed area, and minimum energy expenditure in the whole network [Ibbitt, 1997]. Molnar and Ramirez [2002] use data from two mountain catchments in New Zealand to find the optimal combination of DHG exponents w and f, which are not far removed from data-derived values of w and f for these catchments. They interpret the discrepancies to reflect heterogeneity in geology and other watershed factors. Once the hydraulic geometry of natural rivers is calibrated for properties such as relative bank strength, DHG relations are well described by optimality theory [Millar, 2005].

An approach related to hydraulic geometry is to consider downstream trends in flow energy expenditure. Knighton [1999] proposes a model of downstream variation in stream power. Based on the assumption that the longitudinal profile of the river has an exponential form, the model predicts that total stream power peaks at an intermediate distance between the drainage divide and the mouth, the position of which depends on the ratio of downstream rate of change of discharge and downstream rate of change in slope (Figure 2.32). Unit stream power, which is more sensitive to rate of change in slope, is predicted to peak closer to the headwaters, about half-way between the source and the location of the total stream power maximum, although this will depend on how channel width varies downstream. Tests of the model in lowland alluvial channels indicate that site-specific variations in discharge and slope create discrepancies between the observed and predicted locations of the maxima. These discrepancies might be more pronounced in high-relief drainage basins because of substantial downstream variability in slope.

As might be expected, the degree to which a mountain river approximates alluvial hydraulic geometry relations and energy expenditure models will reflect the relative importance of substrate and sediment supply in controlling channel geometry. Because mountain rivers commonly do not follow the trends of channel and valley gradients predicted by downstream hydraulic geometry, it is appropriate to consider in more detail the controls on valley morphology and gradient when working in mountain rivers.

2.5.3. Downstream Fining

Most coarse-grained rivers show a downstream decrease in average grain size at scales of tens to hundreds of kilometers, although this downstream trend can be interrupted by coarse sediment from hillslope or tributary inputs (Figure 2.33) [*Wohl and Pearthree*, 1991; *Grimm et al.*, 1995; *Malarz*, 2005; *Attal and Lavé*, 2006] or from knickzone erosion [*Deroanne and Petit*, 1999]. *Rice* [1998, 1999] develops a procedure for identifying significant lateral sediment sources using drainage basin area, network magnitude, and the basin area-slope product to define individual channel links within which downstream fining occurs. *Sklar et al.* [2006] develop a predictive equation for the travel distance required to abrade coarser tributary sediments to the size of the mainstem inputs from upstream

$$ L_{\Delta D}^{*} = \frac{1}{\alpha} \ln \left(\frac{D_t}{D_m} \right) \tag{2.23} $$

where $L_{\Delta D}^{*}$ is the distance over which the grain size perturbation decays, D_t is mean grain size from the tributary, and D_m is mean grain size in the mainstem.

Along mountainous catchments in which hillslopes are closely coupled to channels, grain-size distribution in channels may closely mirror the size distribution of hillslope sediment supply because local resupply from hillslopes offsets the influence of channel processes such as selective entrainment or abrasion on grain size [*Sklar et al.*, 2006]. Mountain rivers can have downstream coarsening to a grain size maximum that corresponds to maximum unit stream power and an inflection in the drainage area-slope relation that represents the transition from debris flow-dominated channels to fluvially-dominated channels, with downstream fining below this transition [*Brummer and Montgomery*, 2003; *Pike et al.*, 2010].

Differences in lithology and weathering of coarse material introduced to the channel can also strongly influence patterns of downstream grain size in mountain rivers because of the close coupling between hillslope sediment sources and channels [*Miller*, 1958; *McPherson*, 1971; *Attal and Lavé*, 2006]. In the Marsyandi River of the Nepalese Himalaya, clasts from the more rapidly uplifted and eroded Higher Himalayan lithologic units are over-represented on downstream gravel bars because of their major contribution to sediment influx, whereas easily erodible lithologies are under-represented relative to resistant rock types [*Attal and Lavé*, 2006]. Subsequent flume experiments indicate that abrasion rates differ by more than two orders of magnitude for differing lithologies, so that the most resistant lithologies control bed load flux and fining ratio [*Attal and Lavé*, 2009].

Changes in channel cross-sectional geometry at tens to hundreds of meters, and corresponding changes in transport capacity, can create substantial variation in patterns of downstream fining [*Constantine et al.*, 2003; *Rengers and Wohl*, 2007]. Channel engineering that alters cross-sectional geometry or bank stability can also influence patterns of grain size [*Surian*, 2002].

Downstream decrease in grain size has been attributed primarily to (i) selective sorting (entrainment, transport, and deposition) [*Brierley and Hickin*, 1985; *Paola et al.*, 1992; *Werritty*, 1992; *Ferguson and Wathen*, 1998; *Seal et al.*, 1998; *Gomez et al.*, 2001], (ii) abrasion in place [*Schumm and Stevens*, 1973] or during transport, or (iii) some combination of these processes [*Mikos*, 1994; *Moussavi-Harami et al.*, 2004], as well as fracturing or comminution [*Chatanantavet et al.*, 2010]. Experimental simulations suggest that the relative importance of selective sorting is partly a function of clast resistance, with abrasion roughly equal to selective sorting in producing downstream fining for less resistant lithologies [*Parker*, 1991]. Lateral grain sorting through differential mobility of sediment patches with varying mean size produces observed downstream fining patterns in aggrading and laterally unconfined rivers, whereas vertical sorting through incorporating part of the coarse surface layer in the deposit produces downstream fining in narrowly confined channels [*Seal et al.*, 1998]. *Powell*'s [1998] review of sediment sorting in coarse-grained alluvial rivers notes that multiple scales of bed-material sorting exist, with relationships between pattern and process likely to vary as a function of scale. *Chatanantavet et al.* [2010] develop a physically based model that suggests selective sorting by differential transport influences downstream fining only in streams with relatively fine gravel sizes and lower slopes.

Downstream reduction in grain size as a result of abrasion can be quantified using some form of *Sternberg*'s [1875] law:

$$D = D_0 e^{-\alpha x} \tag{2.24}$$

where initial grain size D_0 wears down to D at distance x from the origin at a rate given by rock durability parameter α, which sets the abrasion length scale, $1/\alpha$ [*Sklar et al.*, 2006]. Values of α can vary by at least two orders of magnitude (10^{-3}/m to 10^{-5}/m). *Chatanantavet et al.* [2010] suggest that the effects of abrasion and fracturing should be considered separately, rather than being lumped into α.

Predictive models provide a means to test our understanding of the many feedback linkages which govern these patterns and processes. *Hoey and Ferguson* [1997] simulate downstream fining by selective transport and deposition in the absence of lateral water or sediment inputs. *Gasparini et al.* [2004], using a numerical landscape model to explore relations between streambed grain sizes and river basin concavity, reproduce downstream fining through selective particle erosion. They also reproduce a transitional regime where the mobility of sand and gravel changes with streambed grain-size distribution and a sand-dominated region where the mobility of sand and gravel is constant. Several investigators document the gravel-to-sand transition in mountain drainage networks [*Ohmori*, 1991; *Knighton*, 1999; *Singh et al.*, 2007; *Parker et al.*, 2008b], which reflects decreasing competence associated with changes in sediment supply, bed gradient or channel geometry. *Sambrook Smith and Nicholas* [2005] invoke the interactions between flow and sediment supply; as sand deposition decreases the effective bed roughness, near-bed velocities increase, shear stress and turbulent kinetic energy decrease, and burst and sweep events become less frequent around topographic high points. These responses to changing bed conditions are most pronounced where effective roughness height k approaches 10 mm or less ($k/d_{50} = 0.4$), which is typically found immediately upstream of gravel-sand transitions in natural rivers. Reduced bed shear stress and turbulent bursts and sweeps around gravel clasts result in greater mobility of the sand fraction relative to coarser particles, creating a predominantly sand-bed channel downstream [*Sambrook Smith and Nicholas*, 2005].

As noted by *Sklar et al.* [2006], further work is needed to understand the interactions between bedrock lithology, climate, erosion rate, and hillslope sediment production and transport as these influence the grain-size distribution of sediment supplied by hillslopes to the channel network. Further work is also required to quantify the abrasion rate of bed material in rivers, constrain α, and distinguish in the field particle size reduction due to abrasion from other sources of spatial variability in grain size [*Sklar et al.*, 2006]. Improved models of downstream fining will incorporate grain splitting as well as abrasion (Figure 2.34).

In summary, the influences of drainage basin morphology on the hydrograph are reflected in the geomorphologic instantaneous unit hydrograph, a travel-time probability density function to the basin outlet. Drainage area, drainage density, stream order and basin magnitude, and drainage basin relief and shape all influence the basin hydrograph. Downstream hydraulic geometry is well described by power functions between discharge and the dependent variables of width, depth, and velocity for mountain drainages in which the ratio of hydraulic driving force to substrate resistance exceeds a critical threshold. Downstream fining patterns can be strongly influenced by spatial variations in geomorphic process (debris flows vs water flows), sediment inputs from hillslopes and tributaries, lithology and weathering, channel geometry, and knickzone erosion. Downstream fining reflects some combination of selective sorting, abrasion, and comminution.

2.6. Valley Morphology

Valley morphology provides the template in which channel form and process adjust to changes in water and sediment yield and boundary resistance. Because valley morphology typically changes over longer timespans than channel morphology, longitudinal variation in characteristics such as valley-bottom width and gradient can constrain channel morphology and adjustments to perturbations [*Snyder and Kammer*, 2008; *Fryirs and Brierley*, 2010].

Büdel [1982] attributes consistent valley geometries to each of the ten major morphoclimatic zones that he designates. Similarities in valley cross-profile, longitudinal profile, drainage pattern, and rate of incision among valleys in each zone reflect the similar relief-forming mechanisms operating in that zone. Büdel distinguishes between valleys in the periglacial region, which have smooth longitudinal profiles, broad, gravel-floored valleys, and rapid rates of incision, for example, and the stepped, narrow, bare bedrock valleys of tropical regions. Other investigators conclude that tectonic regime or lithology, structure, and rock strength [e.g., *Harden*, 1990; *Ehlen and Wohl*, 2002; *Lifton et al.*, 2009] exert the dominant control on reach-scale valley cross-sectional and longitudinal morphology. If one among these various controlling factors dominates valley morphology, that dominance varies among individual drainage basins.

A common feature of mountain drainage basins is an asymmetry of valley slope profiles on opposite sides of a given valley. This asymmetry occurs in all climatic regions and is caused by differences in local climate and weathering regime on opposite sides of the valley. The asymmetric profiles thus illustrate the effect of subtle differences in climate on similar lithologies [*Leopold et al.*, 1964]. *Hack and Goodlett* [1960] find that drier slopes in the humid Appalachian Mountains, USA tend to be gentler and longer, with coarser regolith, higher drainage density, and a predominance of slope wash and channel erosion, whereas wetter slopes are dominated by creep. *Melton* [1960] uses data from valleys in arid to subhumid temperate climates across the United States to develop the hypotheses that (i) low channel gradients favor the development of valley asymmetry in east- or west-trending valleys, with the north-facing slopes becoming steeper as the rivers are moved against their toes by debris from the south-facing slopes and (ii) steep channel gradients in v-shaped valleys favor more symmetric development of valley sides because erosional debris is carried downstream rather than accumulating as asymmetrical valley fill that shifts the channel laterally. Observations in semiarid regions, however, indicate that moister north-facing slopes on which snow accumulates are gentler than drier south-facing slopes [*Leopold et al.*, 1964; *Wohl and Pearthree*, 1991] (Figure 2.35).

Valley asymmetry can also take the form of longitudinal offsets where smaller rivers are deflected by strike-slip faults, or of asymmetrical widening of larger rivers downstream from strike-slip faults [*Zhang et al.*, 2004]. *Lin and Oguchi* [2006] extract longitudinal and transverse profiles of watersheds from DEMs and develop a method to identify points where morphometric characteristics change abruptly.

Gangodagamage et al. [2007] extract river corridor width (lateral distance from the center line of the river to the left and right valley walls at a fixed height above the water surface) from 1-m LIDAR topography of the South Fork Eel River basin, a mountainous catchment in northern California, USA. They find distinct variations in corridor width associated with changes in substrate and suggest that such an analysis can be usefully combined with field observations and numerical modeling. *Gallant and Dowling* [2003] develop an algorithm for using digital elevation models to identify valley bottoms based on their topographic signatures as flat, low-lying areas.

Cirque and valley glaciers may strongly control valley morphology and rates of valley incision in mountainous regions glaciated during the Pleistocene and/or the Holocene. In these regions, many rivers are slightly modifying valleys most recently largely shaped by glaciers, although glacial erosion and deposition are extremely variable along a valley [*MacGregor et al.*, 1998]. Deeper glacial erosion may occur, for example, where valley width decreases, at a change in valley gradient, or where a tributary glacier enters and reduces the effective width of the main glacier. Comparison of the dimensions of fluvial and glaciated valleys in central Idaho, USA reveals that glaciated valleys are wider and deeper than fluvial valleys for a given drainage area, although the power-law exponents for the two valley types are the same [*Amerson et al.*, 2008]. Because the glacially-shaped longitudinal valley profile can persist for hundreds of thousands of years [*Fabel et al.*, 1998], glacial history sets the framework for the river valley, controlling such characteristics as depth of valley alluvium and bedrock valley gradient, and therefore the spatial organization of channel types [*Brardinoni and Hassan*, 2007]. *Anderson et al.* [2006a] review the effects of different glacial scenarios on valley profiles.

Steeper valley walls in glaciated basins also increase the potential for hillslope instability such as bedrock landsliding [*Meigs*, 1998; *Amerson et al.*, 2008]. Large ice sheets such as the Cordilleran Ice Sheet of western North America can cause drainage divides to migrate by tens of kilometers, cause enhanced canyon erosion downstream from breached divides, and even eliminate drainage divides when large proglacial lakes drain [*Riedel et al.*, 2007]. And as long as a glacier or icefield is present, it may create large seasonal and shorter period fluctuations in meltwater and sediment discharge to rivers, which in turn influence river incision or aggradation and thus indirectly influence valley morphology.

Hanging tributary valleys are widespread in catchments that experienced alpine glaciation. The smaller tributary glaciers do not erode the valley as effectively as larger glaciers, so that the tributary junction remains perched well above the main valley bottom when the ice recedes (Figure 2.36). Hanging valleys can also form in tectonically active and incising landscapes with no history of glaciation [*Crosby and Whipple*, 2006; *Wobus et al.*, 2006b]. *Crosby et al.* [2007] use relations for sediment flux-dependent bedrock incision [*Sklar and Dietrich*, 1998, 2004] to explain the formation and persistence of hanging tributaries at threshold drainage areas. The distribution of temporary and permanent hanging valleys within a catchment results from interactions among: the magnitude of base level fall; the upstream attenuation of incision; the lag time of the response in sediment flux; and the non-systematic variation in tributary drainage areas within the channel network [*Crosby et al.*, 2007]. Rapid mainstem incision can over-steepen tributary junctions beyond a threshold slope or low tributary sediment flux during mainstem incision can limit the tributary's ability to incise at a rate similar to that of the main channel [*Crosby et al.*, 2007]. As might be expected, the amount of over-steepening needed to form a hanging valley increases as tributary drainage area increases up to some maximum value, above which large tributaries can keep pace with base level fall and mainstem incision. *Crosby et al.* [2007] predict S_{hang}, the channel gradient at which incision rate falls below the rate of base level fall and permanent hanging valleys form, as

$$S_{hang} = \left[2\sqrt{\frac{1}{3S_t}} \cdot \cos\left(\frac{1}{3}\cos^{-1}\left(\frac{\frac{-1}{2S_t K' \beta A^{1-bc-1/4}}}{\sqrt{\frac{1}{27S_t^3}}}\right) + \frac{4}{3}\pi\right) \right]^{-2} \tag{2.25}$$

where S_t is steady-state transport-limited gradient; $K' = K_{SA}/(k_w k_q^b)$, with K_{SA} as a dimensional coefficient expressing the channel's erodibility, k_w as a coefficient for the relation between channel width and discharge, and k_q as a coefficient for the relation between discharge and drainage area; β is the percentage of eroded material that is transported as bedload; A is drainage area; and S_t is the steady-state transport-limited gradient. *Crosby et al.* [2007] also predict the maximum drainage area at which a temporary hanging valley can form as

$$A_{temp} = \left(\frac{k_w k_q^b \ I_{max}}{K_{GA} \beta \ U_{initial}}\right)^{\frac{1}{1-bc}} \tag{2.26}$$

where K_{GA} is a dimensional constant equal to (r/L_s), with r as the fraction of the volume detached off the bed with each collision and L_s as saltation hop length; I_{max} is maximum incision rate; and $U_{initial}$ is background rate of base level fall. Although equations such as these require substantial data for accurate parameterization at a site, they facilitate quantitative testing of predicted relations in diverse field settings.

2.7. Longitudinal Profiles and Bedrock Channel Incision

Longitudinal profiles of rivers have been studied for more than a century [e.g., *Powell*, 1875, 1876; *Gilbert*, 1877; *Davis*, 1902; *Mackin*, 1948; *Hack*, 1957; *Pazzaglia et al.*, 1998; *Weissel and Seidl*, 1998; *Hodges et al.*, 2004], although there was a hiatus between circa the 1950s and the 1990s. Interest in processes and rates of bedrock channel incision, and in linking bedrock channel incision to hillslope processes and to landscape evolution, increased dramatically after the 1990s, driven in part by parallel advances in geochronology and numerical simulations of landscape processes [*Bishop*, 2007].

Because overall stream gradient decreases downstream, longitudinal profiles of large drainage basins generally have a concave upward shape, which *Hack* [1957] expresses as

$$S = k \, L^n \tag{2.27}$$

where S is slope, k incorporates mean bed particle size, L is distance, and the exponent n is an index of profile concavity. Mountain rivers may have concave profiles, but they are also very likely to have straight or convex profiles with knickpoints (Figure 2.37). These characteristics reflect the relative inability of smaller, headwater channels to incise rapidly enough to keep pace with tectonic uplift, base level fall, glacial effects, climate change, or sediment supply, partly because of greater bedrock exposure and consequently less erodible channel substrates than downstream channel reaches [*Merritts and Vincent*, 1989; *Howard et al.*, 1994]. Mountain rivers can also have segmented longitudinal profiles reflecting different historical process domains. *Anderson et al.* [2006b] describe two-part longitudinal profiles for rivers in the Colorado Front Range, USA; profiles above the glacial limit are significantly flattened from their original fluvial slopes, and fluvial profiles downstream from the glacial limit display a prominent convexity associated with a transient response to lowering of base level. In contrast, rivers in the northwest Indian Himalaya display increased channel concavity downstream from glacially modified reaches, and require $> 500,000$ years to recover [*Hobley et al.*, 2010]. Disequilibrium in the form of deviation from a semi-log straight line longitudinal profile may persist for millions of years along headwater bedrock channels [*Goldrick and Bishop*, 1995].

Bedrock longitudinal profiles and terraces have typically been used as an index of rock uplift rate in steady-state landscapes and in landscapes experiencing transient increases in erosion rate [*Snyder et al.*, 2000; *Rădoane et al.*, 2003; *Hodges et al.*, 2004; *Mather and Hartley*, 2006; *Mugnier et al.*, 2006; *Wobus et al.*, 2006c, 2006d; *Kirby et al.*, 2007; *Whittaker et al.*, 2008; *Pritchard et al.*, 2009; *Font et al.*, 2010; *Roberts and White*, 2010]. *Anderson* [1994] uses the term *incision wave* to describe a scenario in which wave-like patterns of stream incision propagate through numerically modeled bedrock channels. A crest where available stream power is greatest and vertical incision is most rapid is flanked by limbs with less available stream power and slower incision rates. Successive wave trains of knickpoint migration after a base level fall can be linked to removal of the alluvial cover, followed by bedrock incision [*Loget and Van Den Driessche*, 2009]. Field evidence from southern Spain supports this model [*García et al.*, 2004].

As *Goldrick and Bishop* [2007] note, there is no inherent mechanism in Hack's *SL* index (equation 2.27) for differentiating disequilibrium steepening from that which is an equilibrium response of the river to more resistant lithology. Consequently, they proposed the DS model

$$H = H_0 - k \left(\frac{L^{1-\lambda}}{1-\lambda} \right) \tag{2.28}$$

where H is elevation of the bed, H_0 is an estimate of the theoretical elevation of the divide if hydraulic processes were active right to the drainage head, and λ is the exponent of the relationship between discharge and L. This can also be expressed as

$$S = -k \, L^{-(am/n)} \tag{2.29}$$

where a is the exponent of the length/area relationship and the m/n ratio is the profile concavity index. The DS model allows equilibrium steepening caused by an increase in channel substrate resistance (parallel shift in the DS plot, or ln slope versus ln downstream distance) to be distinguished from disequilibrium steepening caused by long profile rejuvenation (disordered outliers on the DS plot) [*Goldrick and Bishop*, 2007].

One of the key components of explaining mountain-river longitudinal profiles is understanding the processes by which channels incise [*Hancock et al.*, 1998; *Wohl*, 1998]. Bedrock channels are particularly important in this respect because the rate of bedrock incision may limit the rate at which base level change is transmitted along a drainage basin [*Tinkler and Wohl*, 1998]. Natural discontinuities or thresholds occur along a channel between substrates of bedrock, coarse-grained alluvium, and fine-grained alluvium [*Yatsu*, 1955; *Schumm*, 1956; *Howard*, 1980]. The gradient of alluvial channels is determined by hydraulic regime [*Moshe et al.*, 2008], whereas the gradient of bedrock channels may be an independent variable in that weathering must precede erosion [*Howard*, 1980, 1998].

Alluvial channel segments will develop if the local or ultimate base level remains constant or lowers very slowly, or where the gradient of the underlying bedrock is lower than that required for transport of sediment supplied from upstream [*Howard*, 1980]. Local base levels may be resistant lithologies, differential rock uplift [*Kobor and Roering*, 2004], large logjams [*Montgomery et al.*, 1996a], or mass movements from the hillslope that overwhelm local transport capacity and create alluvial reaches upstream [*Paul et al.*, 2000; *Korup*, 2004; *Lancaster and Grant*, 2006; *Schuerch et al.*, 2006]. *Pratt et al.* [2002] discuss how enhanced monsoonal precipitation in the Himalaya during the early Holocene increased landsliding and generated a pulse of sediment that temporarily overwhelmed the Marsyandi River's transport capacity, creating an alluvial reach until the river incised through the aggraded sediment to bedrock. In general, channels are expected to have fine-grained alluvium where sediment concentrations are high but grain size range is narrow. Coarse-grained alluvial channels are favored by low sediment loads and relatively large proportions of coarse sediment [*Howard*, 1987].

Fine-grained alluvial channels may be referred to as *live-bed* or *regime* channels because sediment transport occurs at all but the lowest flows. Coarse-grained alluvial channels are sometimes called *threshold* or *stable* channels because sediment moves only near bankfull discharge [*Howard*, 1980] or during extreme floods or debris flows. Mountain rivers tend to have substrates of coarse-grained alluvium or bedrock, so that channel incision occurs only episodically. The conditions under which sediment transport and channel incision occur along a channel in coarse-grained alluvium may be predicted in terms of the hydraulic conditions necessary to mobilize most of the sediment (see chapter 3). These channels typically have gradients near the threshold of motion [*Howard et al.*, 1994]. Two key unresolved questions about the long-term evolution of mixed bedrock-alluvial channels are (1) how and when the bedrock is eroded and (2) whether channel gradient is determined by the necessity to transport the alluvium or to erode the bed [*Howard*, 1998].

The differing effects of coarse sediment are particularly noticeable along mixed bedrock-alluvial rivers in which lower gradient reaches tend to be sediment-mantled and higher gradient reaches are bedrock-floored [*Tooth and McCarthy*, 2004]. *Jansen* [2006] discusses one example from southeastern Australia where modulation of bedrock exposure at the reach scale, coupled with adjustments to channel width and gradient, facilitates uniform incision across rocks of different erodibility. *Tooth et al.* [2007] demonstrate how differences in the ratio of sediment supply to transport capacity not only create alluvial river segments, but dictate channel planform and dynamic parameters such as avulsion frequency and style. Geometric and grain-size parameters can be similar between alternating alluvial and bedrock channel reaches when local processes create limited bedrock exposure along a largely alluvial channel [*Keen-Zebert and Curran*, 2009]. Subsurface imaging such as two-dimensional electrical resistivity [*Hsu et al.*, 2010] can be used to detect the contact between sediment and bedrock and thus determine thickness of alluvial fill and the depth of sediment that must be mobilized before bedrock boundaries directly influence hydraulics during a flow.

2.7.1. Processes of Bedrock Channel Erosion

Bedrock erosion along mountain rivers influences the stability and erosion rate of adjacent hillslopes [*Burbank et al.*, 1996] and thus the rate of sediment supply to channels [*Korup et al.*, 2004]. Increased sediment supply can overwhelm local transport capacity and create an alluvial cover that mantles the bedrock and limits incision [*Sklar and Dietrich*, 1998] until the sediment is flushed downstream. Because of these feedbacks between hillslope and channel processes, and because bedrock incision can limit longitudinal profile adjustment, understanding the processes and rates of bedrock channel erosion is central to understanding the dynamics of mountain rivers.

Erosive mechanisms operating along cohesive channel boundaries include *abrasion, corrosion, cavitation, plucking/quarrying* (Figure 2.38) associated with hydraulic lift forces, and what *Chatanantavet and Parker* [2009] call *macroabrasion* (a process of fracturing the bedrock into pluckable sizes mediated by particle impacts). Most investigators report erosion rates resulting from the combined effects of all of these processes, although they may infer the relative importance of specific processes. The relative importance of each process in any channel depends on flow and substrate characteristics [*Springer et al.*, 2003; *Wohl and Springer*, 2005]. Lithologic variability along the Ukak River, Alaska, USA, for example, results in strong plucking and rapid erosion in well-weathered and fractured rocks and inner-channel incision and slower erosion in solid, cohesive bedrock [*Whipple et al.*, 2000]. Similarly, densely jointed phyllite along the bed of the Ocoee River, Tennessee, USA is rapidly eroded by quarrying, whereas massive quartzite is more slowly abraded through pothole erosion [*Goode*, 2009]. Interpreting field data from the Indus River in Pakistan, *Hancock et al.* [1998] suggest that block quarrying is the most efficient bedrock erosive process when joints and bedding planes are sufficiently close, whereas abrasion is most effective in regions of separated flow and more massive substrate. Where more than one mechanism is present, plucking is faster than abrasion [*Hartshorn et al.*, 2002].

Sculpted forms such as potholes, grooves, or undulating walls are abrasive features that are particularly well developed in some bedrock channels [*Richardson and Carling*, 2005] (Figure 2.39) and can be used to infer hydraulics during formative flows [*Zen and Prestegaard*, 1994]. *Springer et al.* [2006] note that high erosion efficiencies, expressed as rate of deepening, are necessary if small potholes are to persist as a channel incises. Efficiencies decline as potholes deepen because comparatively deep potholes must erode orders of magnitude larger volumes of substrate than shallow potholes as the bed incises. Pothole dimensions of radius and depth are strongly correlated using a simple power law, and suspended sediment likely plays a larger role than grinders in pothole erosion [*Springer et al.*, 2006]. In channel reaches with heterogeneous bed topography, the likelihood of pothole formation tends to be greatest at intermediate relative bed elevations, suggesting that local hydraulics and tools versus cover relationships govern pothole formation and maintenance [*Goode*, 2009]. *Johnson and Whipple* [2007] use flume experiments to explore the interactions among hydraulics, abrasion, and channel form. They note the feedbacks inherent in a sequence of abrasion from sediment preferentially drawn to topographic lows in the channel, further erosion of these low points, creation of tortuous flow paths and sculpted forms as a result of spatial focusing of erosion, followed by energy dissipation and inhibition of further incision.

Channel gradient and flow width/depth ratios influence flow structure and distribution of shear stress and abrasion by suspended sediment in bedrock channels, and flow structure and erosive forces exert feedbacks that strongly influence channel morphology, the type and distribution of sculpted features, and the relative rate of bed erosion [*Wohl and Ikeda*, 1997; *Wohl et al.*, 1999; *Carter and Anderson*, 2006; *Johnson and Whipple*, 2010; *Goode and Wohl*, 2010b]. Correlations among bedrock channel morphology, hydraulics and substrate variables reflect these interactions [*Wohl and Merritt*, 2001; *Goode and Wohl*, 2010b]. Where longitudinal variability in uplift rates or rock erodibility creates spatial differences in erosion rates and mechanisms, flow hydraulics and channel geometry adjust to minimize variation [*Springer et al.*, 2003].

Bedrock erosion also depends on the thresholds for removal of material and upon the frequency and duration at which these thresholds are exceeded. Although many investigators have assumed that extreme flows are necessary to overcome substantial erosional thresholds [*Baker and Pickup*, 1987], *Hartshorn et al.* [2002] find that river incision at a rate of 2-6 mm/yr along the LiWu River, Taiwan is driven by relatively frequent flows of moderate intensity. *Johnson et al.* [2010] find that sustained floods with smaller peak discharges can be more erosive in an ephemeral sandstone channel than flash floods with higher peak discharges as a result of changes in bed alluviation. *Stark et al.* [2010] infer a climatic signature for incised river meanders in the western North Pacific, where river sinuosity peaks in the typhoon-dominated subtropics, characterized by extreme rainfall and floods, with rock erodibility exerting a secondary control on river sinuosity. *Annandale* [2002] proposes an erodibility index, K, to quantify a substrate's ability to resist scour

$$K = M_s K_b K_d J_s \qquad (2.30)$$

where M_s is intact mass strength, K_b is block size, K_d is discontinuity shear strength, and J_s is joint density. *Stock et al.* [2005] propose that incision rates are proportional to the square of rock tensile strength, although the threshold of motion of a thin sediment mantle, rather than bedrock hardness or rock-uplift rate, controls channel slope in weak bedrock lithologies with tensile strengths below about 3-4 MPa. This latter point reflects the fact that, in bedrock channels with at least partial alluvial cover, erosional thresholds depend strongly on entrainment of sediment and the abrasive effects of sediment in transport.

Using an experimental channel, *Finnegan et al.* [2007] demonstrate that when the channel is free of alluvial cover, incision is focused over a fraction of the bed width that varies strongly with bedload supply and transport capacity. Increases in sediment supply widen the area of incision. Simultaneously, spatial variation of incision enhances bed roughness, which eventually inhibits efficient transport of bedload. *Goode and Wohl* [2010a] document reduced distances of bedload transport in portions of a bedrock channel with greater topographic complexity of the bed. In Finnegan et al.'s flume experiments, reduced transport leads to deposition of alluvial cover and the suppression of incision at high rates of bedload supply. *Johnson and Whipple* [2010] demonstrate a linear increase in erosion rate with sediment flux in another experimental channel, and a linear decrease in erosion rate with the extent of alluvial bed cover. Similarly, *Turowski et al.* [2008] find that cover and tool effects of mobile sediment dominate the partitioning of lateral and vertical erosion along the Liwu River, Taiwan such that temporary alluviation of the channel thalweg during large floods promotes preferential bank erosion. *Lague* [2010] develops a numerical model that simulates sediment transport and bedrock incision at daily time scales but can be run for timespans of thousands of years.

Erosive processes in cohesive but non-indurated materials, such as silt and clay, display many similarities to bedrock channels, including abrasion and the development of sculpted forms. Processes unique to these substrates include upward-directed seepage, which creates an additional driving force, and matric suction as an additional source of resistance to bed erosion [*Simon and Collison*, 2001].

Although most investigations of erosion in bedrock channels focus on long-term average rates inferred from terraces or cosmogenic nuclide dating of bedrock surfaces, within the past few years some investigators have directly measured rates and processes of channel erosion over shorter timescales. Using high-precision repeat surveys, *Hartshorn et al.* [2002] document up to 182 mm of erosion into quartzite during a 10-month period along the LiWu River in Taiwan. Using erosion pins, *Stock et al.* [2005] measure rates up to 16 cm/yr for relatively erodible volcanic and sedimentary rocks in the northwestern U.S. and Taiwan. These rates greatly exceed long-term erosion-rate estimates for the sites. *Reusser and Berman* [2007] describe an innovative approach in which they use LiDAR-generated detailed DEMs of bedrock straths dated using [10]Be to calculate volumes of rock removed and corresponding erosional fluxes during incision.

2.7.2. Models of Bedrock Channel Incision

Howard [1980] expresses the erosion of bedrock channels as a rate law

$$E = k\, A^m\, S^n \tag{2.31}$$

where E is average erosion rate, which increases in proportion to drainage area A and gradient S; m and n are constants, and k includes the inherent bed erodibility and the magnitude and frequency characteristics of the flow. If erosion rates were directly proportional to bed shear stress, m would equal 0.38 and n would equal 0.81 [*Howard*, 1980]. Because substrate erodibility can vary so greatly along bedrock streams, these streams do not necessarily have a simple downstream hydraulic geometry and gradient is a semi-independent variable [*Howard*, 1980].

Equation 2.31 and variants of it are known as *stream power incision laws* because erosion is proportional to S and to A as a surrogate for Q. Channels described by this equation are known as *detachment-limited* because river erosion is controlled by the erodibility of the bedrock. Rivers can also display *transport-limited* conditions in which volumetric transport capacity (Q_{eq}) is a function of stream power, sediment flux is equal to transport capacity, and incision or deposition rate equals the downstream divergence of sediment flux [*Willgoose et al.*, 1991]

$$Q_{eq} = K_t A^m S^n \tag{2.32}$$

where K_t is a sediment transport coefficient, m and n are area and slope exponents as in equation 2.31. *Brocard et al.* [2006] describe transitions from detachment-limited to transport-limited conditions as Q or A increase and incision rate decreases in rivers of the western Alps. The location of the transition also shifts upstream with increasing bedrock erodibility. *Tomkin et al.* [2003] review other models for incision of bedrock rivers.

Seidl and Dietrich [1992] apply the stream power erosion rate law to mountain channels in the western United States and find that a single rate law cannot approximate incision across an entire channel network because of differences in mechanisms of erosion across the network. Steep tributaries incise primarily by scour through periodic debris flows, whereas other channels erode by knickpoint propagation or by vertical wear via abrasion and dissolution [*Seidl and Dietrich*, 1992; *Montgomery and Foufoula-Georgiou*, 1993]. Some applications of the erosion rate law to mountainous streams in regions of moderate tectonic activity and semiarid climate indicate that the law does not accurately predict vertical fluvial incision rates because downstream trends in *A-Q* vary significantly, and because stream power is apportioned among processes such as lateral incision and bedload transport, as well as vertical bedrock incision [*Mitchell and Pazzaglia*, 1999]. A detachment-limited stream power law works well in describing the incision of the Colorado River and its tributaries in Glen Canyon, USA, however, where incision rates are constrained by *in situ* [10]Be concentrations on gravel-covered strath terraces [*Cook et al.*, 2009]. Several other studies also indicate that bedrock channel incision is linearly related to stream power [*Young and McDougall*, 1993; *Rosenbloom and Anderson*, 1994; *Seidl et al.*, 1994; *van der Beek and Bishop*, 2003].

Stock and Montgomery [1999] use mapped paleoprofiles and modern river profiles to simulate the lowering of ancient river profiles under scenarios of varying m, n, and k. They find that along rivers with a relatively stable base level, incision occurs concurrently along the entire profile and there is a strong area dependence on incision rate. For rivers subject to abrupt base level change, long-term lowering rate is a function of the frequency and magnitude of knickpoint erosion and there is a small area dependence on incision rate. For rivers with a stable base level, k may vary over five orders of magnitude as a function of lithology [*Stock and Montgomery*, 1999].

Calculating bedrock channel erosion based on S and A implies that incision rate is primarily proportional to stream power. *Sklar and Dietrich* [1998, 2001] note that incision also depends non-linearly on coarse sediment supply such that sediment promotes erosion at low supply rates by providing tools for abrasion, but inhibits erosion at high supply rates by burying the bedrock streambed and protecting it from erosive processes. This implies that grain-size distribution and volume of hillslope sediment supply exert an important control on bedrock channel incision and geometry. *Sklar and Dietrich* [2004] model bedrock abrasion, E, by saltating bedload as

$$E = k_1 \left[\frac{q_s}{\left(\frac{\tau *}{\tau_c *-1} \right)^{0.5}} \right] - k_2 \left[\frac{q_s^2}{\left(D_s^{3/2} \left(\frac{\tau *}{\tau_c *-1} \right) \right)^2} \right] \tag{2.33}$$

where the tools and cover effects are represented by the first and second terms, respectively, and

$$k_1 = \frac{0.08 Y R_b g}{k_v \sigma_T} \tag{2.34}$$

$$k_2 = \frac{0.014 Y (R_b g)^{0.5}}{(\tau *_c)^{1.5} k_v \sigma_T^2 \rho_s} \tag{2.35}$$

valid for $q_s \le q_t$. The relevant symbols are q_s sediment supply per unit width, $\tau *$ nondimensional shear stress, $\tau_c *$ nondimensional critical shear stress, D_s grain diameter, Y Young's modulus of elasticity, R_b nondimensional buoyant density, σ_T rock tensile strength, ρ_s sediment density, and q_t sediment transport capacity per unit width. Measurements of bedrock wear in a laboratory abrasion mill agree well with predictions from this model [*Sklar and Dietrich*, 2004]. *Sklar and Dietrich* [2006] discuss how to apply the model represented by equation 2.33 at a landscape scale and note that the threshold of motion is the most important effect that should be represented explicitly. *Lamb et al.* [2008b] update this model to include incision by impacting suspended and bedload sediment, relating incision (E^*) to the three dimensionless quantities of normalized sediment supply, normalized effective impact velocity cubed, and relative supply of bed load:

$$E^* = \frac{E\sigma_T^2}{\rho_s Y (gD)^{3/2}} = \frac{A_1}{k_v} \left[\frac{q}{(UH\chi + U_b H_b)} \right] \left[\frac{w_{i,eff}}{(gD)^{1/2}} \right]^3 \left[1 - \frac{q_b}{q_{bc}} \right] \qquad (2.36)$$

where E is rate of vertical erosion, σ_T is rock tensile strength, ρ_s is density of sediment, Y is Young's modulus of elasticity, D is sediment diameter, A is cross sectional area of a sediment particle, k_v is an empirical rock erodibility coefficient, q is volumetric sediment supply per unit channel width, U is depth-averaged streamwise flow velocity, U_b is depth-averaged streamwise bedload velocity, H is depth of flow, χ is an integral relating the flux of suspended sediment to near-bed volumetric sediment concentration, flow depth, and flow velocity, H_b is thickness of the bed load layer, $w_{i,eff}$ is effective impact velocity, q_b is volumetric bed load flux per unit channel width, and q_{bc} is volumetric bed load transport capacity per unit channel width.

The importance of sediment supply and tool effects is supported by the failure of detachment-limited shear stress/stream power models to describe well-constrained field examples of channels thought to be in steady state [*Turowski et al.*, 2009a]. *Johnson et al.* [2009] explore how coarse alluvial cover increases bed roughness and sets thresholds of motion, creating a situation in which channel gradient is set mainly by the sediment load rather than bedrock properties, despite long-term incision into bedrock. They note that alluvial cover can inhibit incision even when only moderate amounts of cover are present.

Davy and Lague [2009] incorporate some of these processes in their mesoscale erosion/deposition model, which differs from earlier models by explicitly taking into account a mass balance equation for stream flow. The model couples the dynamics of stream flow and topography through a sediment transport length function ξ, which is equal to the ratio between the sediment river load and the deposition flux. This function represents the average distance covered by a particle in the stream flow before being trapped on topography. If ξ is small, the model reduces to a transport-limited equation. If ξ is large, sediments are never re-deposited after being eroded and the model simulates detachment-limited behavior.

Threshold of motion is governed by critical shear stress and by the frequency and duration of flows exceeding critical shear stress. The simplest models implicitly assume uniform flows through time. *Snyder et al.* [2003] use a stochastic distribution of floods to model exceedence of erosion thresholds and *Lague et al.* [2005] and *Molnar et al.* [2006a] investigate the sensitivity of longitudinal profiles and incision rates to changes in flow regime. *Stark et al.* [2009] propose a simple model of self-buffered bedrock channel erosion that incorporates downstream spreading of bed sediment in a manner that allows for a broad-tailed, power law probability distribution of transport velocities of bed sediment over the long-term.

Chatanantavet and Parker [2009] develop a physically based model of bedrock channel incision that incorporates abrasion, plucking and macroabrasion:

$$E = E_a + E_p = \left\{ \left[\beta\alpha\left(\frac{m_b+1}{\chi^{m_b}}\int_0^{\chi} E\chi_1^{m_b}d\chi_1\right)\right] + q_pp_{pb}\right\} * \left[1 - \frac{\left(\frac{m_b+1}{\chi^{m_b}}\int_0^{\chi} E\chi_1^{m_b}d\chi_1\right)}{q_{ac}}\right]^{n_o}$$

$$(2.37)$$

where E is total bedrock incision rate, E_a is vertical rate of bed incision due to abrasion, E_p is rate of bedrock incision due to plucking and macroabrasion, β is the abrasion coefficient, α is the fraction of the load that consists of particles coarse enough to accomplish wear, m_b is an exponent defined as $n_b/(1-n_b)$, where n_b is an exponent in width-area relation, χ is a surrogate for distance x defined as drainage area/channel width, q_p is volume rate of entrainment of pluckable chunks into bedload transport per unit bed area per unit time, p_{pb} is volume fraction of material in the battering layer that consists of pluckable particles, and q_{ac} is capacity transport rate of wear particles. This model implies that plucking-macroabrasion processes can operate more efficiently than abrasion, although lower tensile rock strength facilitates abrasion.

Continued advances in modeling of the profile evolution of bedrock channels will require the development of rate laws for dissolution; abrasion [*Foley*, 1980a, 1980b; *Chatanantavet and Parker*, 2009]; knickpoint evolution (see next section); and boulder production, breakdown, and removal [*Howard et al.*, 1994; *Seidl et al.*, 1994, 1997; *Hancock et al.*, 1998; *Carling et al.*, 2002]. Models will need to incorporate varying erosional thresholds and magnitude, frequency, and duration of exceedence, as well as upstream-downstream coupling [*Gasparini et al.*, 2006]. The ability to model profile evolution as a function of these processes will also facilitate our understanding of changes in rates and types of hillslope and channel processes that have occurred during the past two million years [*Pazzaglia et al.*, 1998], as well as our ability to predict future patterns associated with changing climate and runoff regime.

Continuing development of geochronologic methods and high-resolution topographic data [*Anders et al.*, 2009], combined with increasing interest in rates of bedrock channel incision for incorporation into landscape evolution models, has substantially increased the number of field studies that document rates of bedrock channel incision (Table 2.5).

2.8. Knickpoints and Gorges

A knickpoint is a step-like discontinuity in the longitudinal profile of a river channel. A knickpoint may be distinguished from a bed-step associated with a step-pool sequence in that knickpoints generally occur singly or in relatively small groups, and erode headward with time in contrast to the stationary bed-steps. Knickpoints can occur in unconsolidated or weakly consolidated alluvium (Figure 2.40), but they are best developed in bedrock or cohesive alluvium. Knickpoints can be stepped, buttressed, or undercut, and the headward erosion can take the form of parallel retreat or of rotation such that the angle of the knickpoint face with the vertical increases with time. The world's greatest waterfalls – such as the Sutherlan (580 m tall) in New Zealand, the Kjelfossen (792 m) in Norway, Angel Falls (979 m) in Venezuela, and Yosemite Falls (739 m) in USA – exemplify the vertical drop that can be associated with a knickpoint. *Harbor et al.* [2005] describe identification of knickpoints using deviation from a predicted slope-area relationship for a catchment or landscape region and *Hayakawa and Oguchi* [2009] develop a semi-automated method of knickzone extraction using DEMs and GIS.

Knickpoints can be created when a base level fall or uplift of the drainage basin increases channel gradient and the stream's incisional capability so that a knickpoint originating at the channel mouth erodes upstream [*Crosby and Whipple*, 2006; *Kober et al.*, 2006]. Knickpoints can also migrate upstream along tributaries when incision in response to base level change or uplift occurs more rapidly along the main channel than along the tributaries [*Seidl and Dietrich*, 1992]. *Chatanantavet and Parker*'s [2009] model of bedrock channel incision suggests that knickpoints can also be autogenic rather than externally driven. The steep face of the knickpoint can be maintained during headward erosion in bedded substrates where a more resistant upper unit serves as a caprock that is continually undermined by erosion of the weaker units beneath [*Holland and Pickup*, 1976; *Wohl et al.*, 1994a] (Figure 2.41). A steep face can also be maintained in jointed substrates where plucking of blocks from the knickpoint face occurs [*Clemence*, 1988; *Bishop and Goldrick*, 1992; *Pyrce*, 1995]. *Lamb and Dietrich* [2009] mathematically model persistence of a vertical face in jointed substrate. A knickpoint eroding upstream in massive, homogeneous substrate is likely to become less pronounced with time as the knickpoint face decreases in slope and the incising reach above the knickpoint lip increases in slope [*Gardner*, 1983; *Stein and Julien*, 1993; *Frankel et al.*, 2007]. As an illustration of this, a study comparing two nearby drainage basins in the mountains of Japan that are very similar except for lithology demonstrates that the river flowing over jointed granite have numerous knickpoints, whereas the river flowing over homogeneous sedimentary rocks has a smooth, straight profile [*Tanaka et al.*, 1993]. *Hayakawa and Oguchi* [2006,2009], however, infer only a relatively weak lithologic influence on the abundance of knickzones along mountain streams in central Japan. Comparison of numerous stepped- and straight-profile channels in Hawaii, USA suggests that even channels on bedded and jointed rocks can have a straight longitudinal profile if they have a stable base level history [*Seidl et al.*, 1994].

Knickpoints can also form as a result of an increase in the ratio of water/sediment discharge. This can occur during a single flood (Figure 2.42) or over a period of time as a result of land use or other controls operating within the drainage basin. A knickpoint can form where a particularly resistant material is exposed in the channel bed [*Miller*, 1991; *Phillips et al.*, 2010]. A knickpoint of this type will either disappear or become a more gradually steepened knickzone (Figure 2.43) once the river has incised an inner channel through the resistant unit [*Biedenharn*, 1989]. Examining the long profiles of numerous rivers in France, however, *Larue* [2008] finds that knickpoints of lithological origin maintain a strong vertical stability during river incision, whereas those of tectonic origin or base level lowering migrate upstream at a rate controlled by river discharge. Finally, a knickpoint or knickzone can also form where a large point source of sediment such as a rock slide overwhelms fluvial transport capacity [*Korup et al.*, 2006; *Phillips et al.*, 2010]. (Upstream knickpoint migration can also cause stream incision and removal of the toe of adjacent hillslopes, destabilizing the slope and increasing failures such as rock slides [*Bigi et al.*, 2006].)

Although some of the world's largest waterfalls occur in low-relief continental shield provinces, knickpoints are also common in mountainous regions because of tectonic activity, the greater likelihood of finding resistant bedrock close to the surface, and the continuing fluvial adjustment to Pleistocene glacial erosion (Figure 2.44). *Hayakawa and Oguchi* [2009], for example, document a characteristic frequency of knickzone distribution common to steep bedrock reaches in Japanese mountain watersheds. Knickpoints are some of the sites of the greatest concentration of energy dissipation along a river [*Young*, 1985] and rates of knickpoint retreat can be two orders of magnitude greater than erosion rates elsewhere along the channel [*Seidl et al.*, 1997]. Reported rates of knickpoint migration range from 2 to 100 cm/kyr in bedrock [*Tinkler and Wohl*, 1998; *Wohl*, 1999], although *Simon and Thomas* [2002] report rates of 0.7-12 m/yr in resistant, cohesive clays in Mississippi, USA and *Lamb et al.* [2007] report a rate of 6 cm/yr for knickpoints formed in basalt in Hawaii, USA.

The prevalence of knickpoints along mountain rivers formed in bedrock may indicate that other forms of channel incision cannot keep pace with uplift, resulting in steepening of the channel longitudinal profile until a knickpoint is formed and incision occurs fairly rapidly. The formation and headward retreat of a knickpoint can thus determine the extent to which a bedrock channel reach is serving as a local base level limiting incision along upstream alluvial reaches. Although knickpoints are commonly regarded as local base levels that limit upstream transmission of base level fall, *Berlin and Anderson* [2009] describe a scenario in which stream channels with drainage area sufficient to propagate knickpoints upstream are capable of steepening upstream of the knickpoints. As *Stock and Montgomery* [1999] note, two critical and unresolved questions with respect to the role of knickpoints in bedrock channel incision are: (i) what sets the frequency, amplitude, and decay rate of knickpoints?, and (ii) how do rivers with knickpoints in tectonically active regions evolve concave profiles?

Another question that has arisen for rivers with knickpoints is whether the rate of knickpoint retreat correlates with drainage area or stream power and thus falls under the stream power erosion law. *Bishop et al.* [2005] find a statistically significant power relationship between catchment area and distance of headward recession for knickpoints and conclude that knickpoint retreat rates are consistent with the stream power law. *Crosby and Whipple* [2006] find that retreat distances correlate well with drainage area for knickpoints on tributaries at sites < 1 km upstream of tributary-mainstem junctions in the Waipaoa drainage of New Zealand, but conclude that knickpoint location primarily reflects thresholds of channel incision at small drainage areas.

Developing a numerical simulation of basins with constant and temporally varying uplift rates, *Niemann et al.* [2001] find that knickpoints travel through the basins with constant vertical velocity when uplift rate varies temporally, which means that all knickpoints from a given perturbation in uplift rate should be found at the same contour level. A model developed by *Wobus et al.* [2006b] has similar implications.

Hayakawa and Matsukura [2003] develop an empirical equation for rate of knickpoint retreat in rivers of the Boso Peninsula, Japan, and apply the equation to other regions. Their approach is based on the idea that knickpoint retreat reflects the ratio of erosive force, F, to bedrock resistance, R, which they model as

$$\frac{F}{R} = \left(\frac{AP}{WH}\right)\sqrt{\frac{\rho}{S_c}} \qquad (2.38)$$

where A is drainage area upstream from the knickpoint, P is annual precipitation, W is width and H is height of knickpoint face, ρ is density of water, and S_c is unconfined compressive strength of the bedrock, measured with a Schmidt hammer. Equation 2.38 implies that knickpoint recession rate can alter in response to changes in waterfall lip length as well as changes in flow volume [*Hayakawa and Matsukura*, 2009]. The empirical equation for knickpoint retreat, D/T in m/yr is

$$\frac{D}{T} = 99.7 \left[\frac{A}{P}\sqrt{\left(\frac{\rho}{S_c}\right)}\right]^{0.73} \qquad (2.39)$$

Berlin and Anderson [2007] also model knickpoint location and recession in Colorado, USA using drainage area and rock erodibility. *Harbor et al.* [2005] document differences in knickpoint recession in the Appalachian Mountains of the eastern United States and attribute these differences to spatial variation in rock erodibility.

Bishop et al. [2005] develop an empirical relationship between distance of knickpoint retreat and catchment area for uplifted rivers in eastern Scotland and conclude that lithology exerts only a secondary control on knickpoint recession. The strong correlation between rate of retreat and area suggests that, like other forms of erosion in bedrock channels, knickpoint retreat can follow the stream power law (equation 2.31). *Bishop et al.* [2005] speculate that there is some minimum value of total power for which knickpoint retreat is able to keep pace with uplift; below that value, knickpoints presumably grow progressively taller with time.

Recent studies focus on the mechanics of erosion at the knickpoint (Figure 2.45) and in the plunge pool. *Haviv et al.* [2006] propose that erosion amplification at a knickpoint lip is

$$\frac{E_{lip}}{E_{(x=L_a)}} = \left(1 + \frac{0.4}{Fr^2(x=La)}\right)^{3n} \tag{2.40}$$

where $E_{(x=La)}$ is the erosion rate at the upstream end of the flow acceleration zone above a knickpoint, Fr is the Froude number at this site, and n is between 0.5 and 1.7. Flow acceleration and erosion amplification at a knickpoint lip result from low pressures at the lip cross section and are more important where the flow cross section at the lip is in contact with air around its entire perimeter (i.e., where the falling jet is not confined by sidewalls and where knickpoints consist of a resistant caprock and erodible footrock). Equation 2.40 suggests that erosion at the lip could be up to 2-5 times higher relative to other sites along the channel with the same hydraulic geometry. *Haviv et al.* [2006] use this in numerical simulations to infer that amplified erosion at the knickpoint lip forms an oversteepened reach with a length longer than the flow acceleration zone, as long as incision wave velocity (the rate at which a channel point at a given, fixed elevation is translated horizontally) at the upstream end of the flow acceleration zone is higher than the retreat rate of the knickpoint face. The oversteepened reach is most pronounced at greater channel gradients.

Lamb et al. [2007] describe valley-head erosion dominated by abrasion from impacting sediment particles in plunge pools at the base of knickpoints in Hawaii, USA. They model upstream knickpoint propagation with a nonlinear dependence on the sediment flux passing over the knickpoint and a linear dependence on the ratio of kinetic versus potential energy of sediment impacts.

Deploying a six-component force/torque sensor at the base of a jet over a horseshoe-shaped knickpoint in a flume, *Pasternack et al.* [2007] find that jet impact causes moderate near-bed stress, but the stress vector shifts to a much more forceful upthrust downstream of the impact point, with peak stress downstream from the boil crest. Instantaneous fluctuations in near-bed lift exceed drag by a factor of 5-10, providing a mechanism for plunge-pool scour at the base of knickpoints. Other flume experiments indicate that aeration of the jet reduces plunge pool scour [*Minor et al.*, 2002].

The headward retreat of a knickpoint through a resistant substrate unit can leave behind strath terraces and a deep, narrow canyon with walls that are only slightly modified by slope processes [*Harbor et al.*, 2005; *Ward et al.*, 2005; *Frankel et al.*, 2007]. Such gorges form in erosional or tectonic escarpments, including the Ontario Escarpment of eastern Canada [*Gilbert*, 1896; *Tinkler et al.*, 1994], and in mountain belts such as the Lachlan Fold Belt of southeastern Australia, where gorge extension and incision are a major mechanism of highland denudation [*Nott et al.*, 1996]. These gorges are commonly characterized by alternating pools and rapids. The pools have been interpreted as plunge pools formed during periods of slower upstream migration of the knickpoint, whereas the rapids represent either the plunge pool bars or periods of faster recession that resulted in downstream deposition of eroded materials [*Philbrick*, 1970]. Rapids can also result from localized inputs of coarse sediment associated with tributary junctions or fracture zones (Figure 2.46). *Scheidegger* [1995; *Scheidegger and Hantke*, 1994] notes that many gorges along mountain rivers are oriented parallel to regional joint patterns and shear zones, suggesting that channel incision occurs more effectively along these zones of weakened rock. The maintenance of a deep, narrow channel cross section following knickpoint recession (Figure 2.47) both reflects the erosional resistance of the canyon walls and maximizes the shear stress and stream power per unit area for a given discharge and channel gradient.

2.9. Terraces

The best indicators of changes in river longitudinal profile through time are terraces. Terraces represent channel and floodplain surfaces no longer subject to active fluvial modification. Terraces can be created by crossing the threshold of critical power as a result of change in climate, land use, base level, or tectonic regime, or through intrinsic processes associated with sediment transport that is discontinuous in time and space. The *threshold of critical power* is defined as the ratio of power available (stream power) to power required (resisting power) for entrainment and transport of bedload [*Bull*, 1979]. When the threshold is exceeded, degradation occurs. As expressed by *Gilbert* [1877], *Mackin* [1948], and *Leopold and Bull* [1979], a *graded stream* is one in which the slope is adjusted to prevailing water and sediment discharges, such that the channel is neither aggrading nor degrading and slope remains constant through time. A change in base level or water or sediment discharge causes a response in the form of aggradation or degradation that can lead to terrace formation. This raises questions of how, and how far, changes in controlling factors are transmitted along a channel [*Merritts et al.*, 1994].

Bull [1991] focuses on how interactions between hillslopes and channels during episodes of climatic change result in aggradation, degradation, or stability along mountain channels. Changing base level may also cause aggradation or degradation independent of climatic change (Figure 2.48), although fluctuating sealevel during the Quaternary is inextricably tied to climate change. The post-glacial isostatic rebound of regions covered by ice sheets also altered base level and caused channel incision [*Krzyszkowski and Stachura*, 1998].

Bull [1991] compares geomorphic responses to climate change of a semiarid to subhumid mountain range in southern California, USA, a humid mountain range in central New Zealand, and a hyperarid mountain range in southern Israel. In each case the ratio of water discharge and sediment discharge is critical in determining whether or not the threshold of critical power is exceeded, but a single climatic change can produce different results in channel behavior depending on whether the change is from an arid to semiarid climate, for example (increase in sediment yield), or from a semiarid to a subhumid climate (decrease in sediment yield). In the Charwell River basin of central New Zealand, channel response varies along the basin as a function of elevational controls on climate (Figure 2.49). Thus, although mountain rivers in each of the three regions that Bull studied underwent Quaternary episodes of aggradation and degradation, these episodes were not synchronous between regions or even within single large drainage basins that spanned a range of elevations. During full-glacial and subsequent transitional climates, hillslopes in hot deserts accumulated colluvium that was stripped during the Holocene to aggrade valley floors. Hillslopes in the humid mesic mountain ranges, in contrast, had maximum sediment yields during full-glacial climates, resulting in channel aggradation. The humid mountain ranges also had more frequent episodes of aggradation, suggesting a more rapid rate of hillslope and channel adjustment to climate change. In arid regions where sources of bedload are limited, sufficient accumulation of hillslope regolith must precede channel aggradation [Bull, 1991].

Studies of mountain-river terrace chronologies and slope incision generally support Bull's work. Schick [1974] notes that specific aspects of climate change, such as the frequency of large floods, may be particularly important in controlling terrace formation along arid-region mountain rivers, a conclusion also reached by Bull and Knuepfer [1987] in the Charwell basin of New Zealand. Patton and Boison [1986] find Holocene channel aggradation in semiarid regions. Aggradation occurred during glacial intervals in the subhumid Rocky Mountains of Wyoming, USA [Moss and Bonini, 1961; Moss, 1974]; during the Pleistocene-Holocene transition in the humid Coast Range of western Oregon, USA [Personius et al., 1993]; during the late glacial in the humid mountains of central Japan [Oguchi, 1996a]; and during the interglacial in the extremely humid Melinau drainage basin of Borneo [Rose, 1984]. The ability to resolve Quaternary changes in geomorphic processes has been substantially enhanced by continuing advances in geochronologic techniques, including cosmogenic isotopes and luminescence dating [Klose, 2006].

Working in the active orogenic belt of northern California, USA, Merritts and Vincent [1989] demonstrate that a river incising into bedrock in response to an effective base level fall can maintain uniform incision and a steady longitudinal profile along its length if the river is able to transmit all of the base level fall from its mouth upstream along its length. For a given rate of uplift, smaller rivers will be steeper than larger rivers in order to maintain the stream power required for incision. If stream power along a channel reach falls below what is required for incision to match uplift rate, then the gradient will increase along that reach relative to the next downstream reach and a terrace may be formed [Merritts and Vincent, 1989]. Boll et al.'s [1988] numerical simulation of terrace formation indicate that, although the inputs of climate and tectonics may change continuously, the output of the simulation is discontinuous in the formation of terrace steps.

Because changes in the causal factors of terrace formation can occur simultaneously, one of the most straightforward means of classifying terraces is to focus on terrace composition rather than genesis. Strath terraces have low-relief bedrock treads mantled with a thin veneer of alluvium (Figure 2.50). Straths have been predicted to form in response to periods of balanced sediment supply [*Formento-Trigilio et al.*, 2003], altered sediment supply [*Pazzaglia and Brandon*, 2001; *Wegmann and Pazzaglia*, 2002; *Poisson and Avouac*, 2004; *Fuller et al.*, 2009], glacial-interglacial transitions [*Pan et al.*, 2003], tectonically induced changes in rock uplift [*Rockwell et al.*, 1984; *Molnar et al.*, 1994; *Cunha et al.*, 2005], falling local base level [*Reneau*, 2000], eustatic base level fall [*Pazzaglia and Gardner*, 1993; *Merritts et al.*, 1994], or autocyclic oscillations in erosion rate in laterally migrating channels [*Hasbargen and Paola*, 2000]. Numerical modeling suggests that formation of strath terraces requires input variability that produces a changing ratio of vertical to lateral erosion rates [*Hancock and Anderson*, 2002]. Regardless of causal mechanism, the existence of strath terraces implies a period of vertical stability during which a relatively planar bedrock valley bottom is formed by lateral stream erosion, followed by a period of incision as transport capacity increases beyond sediment supply. Strath terraces tend to be more extensive in rivers flowing over weak rock and more poorly developed and/or preserved over less erodible rock [*Montgomery*, 2004b]. Spatial variation in joint density can also influence strath formation in resistant rocks [*Wohl*, 2008]. *García* [2006] proposes that a minimum catchment area is required for strath formation because stream power must be greater than needed to keep pace with rock uplift rates. As uplift rate and rate of river incision increase, strath terraces are less likely to be formed [*Merritts et al.*, 1994].

Fill terraces are alluvial sequences too thick to be mobilized throughout their entire depth by the river. Like strath terraces, multiple causal scenarios can produce fill terraces, including: fluctuating water and sediment discharge during glacial cycles [*Sugai*, 1993; *Candy et al.*, 2004; *Kamp et al.*, 2004; *Hanson et al.*, 2006; *Arboleya et al.*, 2008], volcanic eruptions [*Yokoyama*, 1999], climatic change [*Reneau*, 2000; *Eden et al.*, 2001; *Hsieh and Knuepfer*, 2001; *Monecke et al.*, 2001; *Starkel*, 2003; *Anderson*, 2005; *Litchfield and Berryman*, 2005; *Bookhagen et al.*, 2006; *Leigh and Webb*, 2006], and changing land use [*Keesstra et al.*, 2005] (Figure 2.51); repetitive lateral shifting and stillstands during continuous downcutting [*Mizutani*, 1998]; fluctuating base level [*Brown et al.*, 2003; *Litchfield and Berryman*, 2005]; complex response to base level change [*Schumm and Parker*, 1973; *Akiyama and Muto*, 2006]; and sediment waves that migrate down basins over periods of 10^1-10^2 years in response to hillslope mass movements or land use [*James*, 2006]. Fill terraces can also be event-based terraces where a flood, debris flow, or landslide overwhelms sediment transport capacity [*Miller and Benda*, 2000; *Strecker et al.*, 2003; *Wohl et al.*, 2009] (Figure 2.52). Fill terraces are more likely than strath terraces to be reduced in planform size via lateral channel erosion after the terrace is formed [*Moody and Meade*, 2008].

Each type of terrace can record a prior longitudinal profile of the river, although formation of strath terraces can lag by several thousand years the input changes that cause their formation [*Merritts and Vincent*, 1989] and both types of terraces can be tectonically deformed after formation. *Germanoski and Harvey* [1993] and *Merritts et al.* [1994] demonstrate that physically continuous terrace treads may not be time-equivalent surfaces. Tectonically altered longitudinal profiles derived from terraces can be used to constrain timing and location of deformation [*Keller et al.*, 2000; *Blisniuk and Sharp*, 2003; *Tsai and Sung*, 2003; *Necea et al.*, 2005; *Mason et al.*, 2006; *Scharer et al.*, 2006; *Shyu et al.*, 2006].

Increasing spatial resolution and availability of LiDAR data and DEMs facilitate designation and mapping of terraces and other valley-bottom features [*Jones et al.*, 2007], even as the subsurface geometry of fill terraces and other valley-bottom deposits is increasingly being imaged using shallow geophysical techniques such as ground penetrating radar and electrical resistivity ground imaging [*Froese et al.*, 2005]. *Demoulin et al.* [2007] describe an automated method to extract fluvial terraces from DEMs.

Terraces are readily formed along many mountain rivers, in part because the relatively low sediment storage along hillslopes and valley bottoms facilitates aggradation of a new, high surface whenever sediment supply increases and incision into bedrock whenever sediment supply decreases. Terrace formation is also enhanced by the relatively high rates of uplift and associated channel incision in many mountainous regions; by the substantial fluctuations in water and sediment discharge associated with advances and retreats of upstream glaciers; and by the occurrence of landslides that cause upstream aggradation and subsequent degradation when the landslide toe is eroded [*Ryder and Church*, 1986]. Many mountainous areas also experienced incision or aggradation leading to terrace formation in association with more frequent or larger outburst floods as alpine glaciers retreated at the end of the Pleistocene [e.g., *Herget*, 2004]. Terraces formed by these various mechanisms are typically preserved only as discontinuous fragments because of subsequent slope instability and river incision along narrow mountain valleys. Discontinuous terrace remnants along ephemeral channels in arid regions can also result from periodic incision during major floods [*Schick*, 1974; *Schick and Magid*, 1978] and from alternating episodes of aggradation-degradation that appear to be inherent in the operation of these channels [*Womack and Schumm*, 1977].

The longitudinal continuity of terraces depends on processes of both formation and preservation. Studies of terraces along rivers in coastal mountain ranges, for example, note distinct longitudinal differences [*Rose*, 1984]. Working in the Coast Ranges of northern California, USA, *Merritts et al.* [1994] find that the lower reaches of rivers are dominated by the effects of oscillating sea level. Aggradation and formation of fill terraces extends tens of kilometers upstream during sea level highstands, with a depositional gradient about half that of the original channel. The middle and upper reaches of rivers are dominated by the effects of long-term uplift. Strath terraces with gradients steeper than the modern channel bed form along middle reaches that are upstream from the aggradation associated with sea level rise, but far enough downstream from the drainage divide that stream power exceeds what is necessary to transport the prevailing sediment load [*Merritts et al.*, 1994].

Longitudinally discontinuous terraces can form where a river crosses an active tectonic structure such as a fold or fault, with planform changes rather than incision upstream and downstream of the crossing point [*Pearce et al.*, 2004]. Longitudinally discontinuous terrace remnants can also form along rivers draining forested catchments where persistent logjams create local bed aggradation and subsequent terrace remnants when the logjam finally disperses [*Montgomery and Abbe*, 2006; *Wohl et al.*, 2009].

Rates of fluvial incision [*Mills*, 2000; *Schaller et al.*, 2005; *Litchfield and Berryman*, 2006] and knickpoint propagation [*Brocard et al.*, 2003], timing of effective base level rise or fall [*Frankel and Pazzaglia*, 2006], and timing and magnitude of changes in water and sediment yield induced by climatic or land-use changes [*Kasai et al.*, 2001] have been determined from terrace chronologies. Terrace chronologies are also used for field tests of conceptual models such as that of an incision wave [*García et al.*, 2004] or exhumation via knickzone propagation [*Zaprowski et al.*, 2001]. Ages of terraces have been estimated using: ^{14}C dating [*Personius et al.*, 1993; *Merritts et al.*, 1994]; weathering rinds [*Colman and Pierce*, 1981; *Birkeland*, 1982; *Knuepfer*, 1988; *Adams et al.*, 1992; *Pazzaglia and Brandon*, 2001]; dendrochronology of terrace vegetation [*Pierson*, 2007]; soils [*Knuepfer*, 1988; *Jakab*, 2007; *Tsai et al.*, 2007]; lichenometry [*Baumgart-Kotarba et al.*, 2003]; optically stimulated luminescence [*Hanson et al.*, 2006; *Pederson et al.*, 2006]; U-series dating on pedogenic carbonate in terrace sediments [*Sharp et al.*, 2003; *Candy et al.*, 2004]; cosmogenic isotopes [*Anderson et al.*, 1996; *Hancock et al.*, 1999; *Hetzel et al.*, 2002; *Schildgen et al.*, 2002; *Wolkowinsky and Granger*, 2004; *Reusser et al.*, 2004, 2006; *Riihimaki et al.*, 2006]; paleomagnetic dating [*Pan et al.*, 2003]; tephrachronology [*Dethier*, 2001]; and ages of bounding lithologic units such as basalt flows [*Westaway et al.*, 2004; *Maddy et al.*, 2005]. *Harkins and Kirby* [2008] describe a method of estimating when hillslope processes replace fluvial processes in modifying terrace morphology based on degradation of terrace risers, although care must be taken in selecting the model of hillslope evolution to be applied to terraces with different sediment compositions [*Nash and Beaujon*, 2006]. Over longer time periods during which substantial changes may have occurred in lateral channel location, drainage network pattern can be reconstructed from relict fluvial gravels [*Larsen et al.*, 1975; *Scott*, 1975; *Bartholomew and Mills*, 1991; *Anthony and Granger*, 2007], drainage reversal from river terrace remnants and clast imbrication patterns in fill terraces [*Craw and Waters*, 2007], and river capture from terraces [*Maher et al.*, 2007].

Early numerical simulations of terrace formation such as that by *Boll et al.* [1988] have subsequently diversified as investigators use cellular automaton models of landscape evolution [*van de Wiel et al.*, 2007] and other approaches [*Veldkamp and Tebbens*, 2001] to investigate causes and processes of terrace creation.

2.10. Alluvial Fans

Although some mountain rivers end in deltas where they enter a lake or the ocean, various types of alluvial fans are a more typical downstream geomorphic punctuation point along mountain rivers (Figure 2.53). Alluvial fans are a characteristic feature of mountainous drainages in which channels exit laterally confined canyons and enter a glacial trough or other wider valley segment, an intermontane basin, or a piedmont environment in which the discharge conditions promote frequent avulsion [*North and Warwick*, 2007]. Debris and water flows tend to deposit sediment immediately downstream from the constriction, creating depositional features that can be dominated by mass movements (debris fans) [*Benda et al.*, 2003], hyperconcentrated flows [*Pope and Wilkinson*, 2005], snow avalanches [*de Scally and Owens*, 2005], fluvial processes and particularly floods [*Harvey*, 2007], or some combination of depositional processes [*Gomez-Villar and Garcia-Ruiz*, 2000; *Garfi et al.*, 2007]. All of these depositional features are here described as *alluvial fans*, although most investigators distinguish different types of fans based on depositional processes and morphology [e.g., *Saito and Oguchi*, 2005; *de Scally et al.*, 2010]. Coalescence of adjacent alluvial fans forms an *alluvial apron* or *bajada* [*Bull*, 1977]. Unusually large fan-shaped bodies are referred to as *fluvial megafans* [*Arzani*, 2005; *Chakraborty and Ghosh*, 2010]; most of these occur between 15° and 35° north and south of the equator where rivers that undergo moderate to extreme seasonal fluctuations in discharge enter actively aggrading basins [*Leier et al.*, 2005]. Also related to alluvial fans are the *floodouts* described by Australian investigators where a marked reduction in channel capacity compared with upstream reaches results in an increasing proportion of overbank flows and associated depositional features [*Pickup*, 1991; *Tooth*, 1999].

Although fan stratigraphy is quite heterogeneous, with patchy distribution of soil properties and vegetation [*Bedford and Small*, 2008], fan sediments tend to be relatively porous and permeable. The limited alluvium beneath channels in mountainous terrain can limit infiltration losses, but nearly half of the flood volume in arid or semiarid regions can be lost to underlying shallow aquifers once the channel passes beyond the mountain front [*Vivoni et al.*, 2006]. *Herron and Wilscon* [2001] find that a small alluvial fan in the temperate, subhumid climate of southeastern Australia slows and/or stores between 20% and 100% of surface runoff delivered to the fan from a 26-ha catchment, depending on antecedent conditions. Alluvial fans can thus hydrologically buffer a catchment. *Niswonger et al.* [2005] use a 1d unsteady stream flow model to evaluate seepage losses along mountain-front streams that discharge intermittently onto alluvial fans in arid regions. Wetlands can occur at the base of fans in humid environments where lower gradients promote deposition of finer, less permeable sediments [*Woods et al.*, 2006].

Tributary fan geometry can also create sediment buffers for larger, mainstem rivers by temporarily storing sediment introduced by floods and debris flows on the tributaries. Tributary fans influence the gradient and width of laterally confined receiving mainstem valleys or channels, creating longitudinal variation in geometry, energy expenditure, and channel morphology [*Kieffer*, 1989; *Hammack and Wohl*, 1996; *Grams and Schmidt*, 1999; *Webb et al.*, 1999; *Miller et al.*, 2001; *Larsen et al.*, 2004; *Yanites et al.*, 2006; *Dubinski and Wohl*, 2007].

Bull [1964] proposes that, within a particular climatic and tectonic setting, lithologies that weather rapidly and produce very fine sediments typically result in steeper fans of larger volume than lithologies that weather to sand-sized sediment, although lithologies that weather to very coarse sediment deposited in debris flows or rockfalls

[*Beaty*, 1989, 1990; *Blair and McPherson*, 1998; *Blair*, 1999] can have the steepest slopes [*Sorriso-Valvo et al.*, 1998]. *Bull* [1962] proposes that source area (A_d) dominates fan size (A_f) and slope (S_f):

$$A_f = aA_d^n \qquad S_f = cA_d^d \tag{2.41}$$

in which the slope of the regression line (n) varies, but averages 0.9 for fans in the United States, and the coefficient a varies by more than an order of magnitude as a result of lithology, climate, and the space available for fan deposition [*Viseras et al.*, 2003]. *Oguchi and Ohmori* [1994] find that higher rates of weathering and erosion in Japan produce larger values of a and smaller values of n. *Giles* [2010] correlates fan area with fan volume.

Whipple and Trayler [1996], however, propose that relative fan size is independent of climate and source lithology in steady-state landscapes, and much more dependent on tectonic setting and particularly subsidence rates. They propose that fan size varies with climate or lithology only during times of significant departure from steady state caused by a sudden increase in uplift, a change in climate, or exposure of lithologies with greatly differing erodibilities. *Whipple and Trayler* [1996] propose that the volumetric rate of fan deposition, V_f, in a steady-state landscape is

$$V_f = E_{to} \left(\frac{1-\lambda_r}{1-\lambda_s} \right) U_r A_d^{1-r} \tag{2.42}$$

where E_{to} is the trap efficiency of a 1-km^2 catchment, λ_r is rock porosity, λ_s is the porosity of the fan sediments, U_r is the rock uplift rate, and r is a constant which can vary with process (e.g., debris flow versus fluvial transport). For landscapes with transient uplift rates, climate, or lithological erodibility, the expression becomes

$$V_f = \beta(t)E_{to}(t)\left(\frac{1-\lambda_r}{1-\lambda_s} \right) U_r(t)A_d^{1-r'} \tag{2.43}$$

where $\beta(t)$ is the catchment response function and $E_{to}(t)$ and r' may be different than their steady-state values. Analogously, rapid uplift and erosion in Taiwan create constant-slope alluvial fans across source areas of 80-2900 km^2 [*Lin et al.*, 2009].

Chakrabarti and Abhinaba [2007] propose that tectonic environment and catchment size control the sedimentary processes acting on alluvial fans. *Densmore et al.* [2007] use a 1d model to explore the sensitivity of simple catchment-fan sediment routing systems to tectonic and climatic perturbations. In many cases the interactions between climate and tectonics make it difficult to separate their effects and to infer the relative importance of one factor over the other [*Harvey*, 2005; *Dade and Verdeyen*, 2007]. Lateral erosion by stable or incising axial river channels can limit the size of alluvial fans [*Leeder and Mack*, 2001].

Debris flows of high sediment concentration can produce relatively steep and rugged topography on the upper fan, whereas low sediment concentration can smooth the lower fan surface by filling channels and other depressions [*Whipple and Dunne*,

1992]. *Webb and Fielding* [1999] describe small alluvial fans in Antarctica produced by debris flow and sheetflood processes generated by melting of the active layer during summer (rainfall is absent). Various morphometric indices can be used to distinguish spatial differences in depositional processes [*de Scally and Owens*, 2004] and to distinguish dominant depositional process on individual fans [*Al-Farraj and Harvey*, 2005]. Airborne laser swath mapping digital topographic data that can be used to calculate surface roughness at a scale of meters facilitates these distinctions. Using such data, *Frankel and Dolan* [2007] find that alluvial fans in arid regions grow smoother with time until a threshold is crossed and roughness increases at a larger wavelength as a result of surface runoff and development of new channel networks. Subsurface imaging such as ground-penetrating radar can also be used to understand fan stratigraphy and infer depositional history [*Hornung et al.*, 2010].

Fan profiles within and away from the channel tend to be segmented, with the steepest slopes along the flanks [*Hooke and Rohrer*, 1979]. Radial fan profiles tend be concave, whereas cross-fan profiles are convex [*Bull*, 1977]. Intermittent regional uplift can create segmented longitudinal profiles of channels on fans, as well as channel incision (Figure 2.54) and the formation of terraces [*Bull*, 1964; *Arboleya et al.*, 2008]. Local changes in base level associated with features such as temporary lake development can produce similar changes in the channels on a fan [*Colombo*, 2005]. Fanhead incision can also result from prolonged or pulsed changes in the ratio of water and sediment supply [*Poisson and Avouac*, 2004; *Davies and Korup*, 2007]. Numerical modeling of fan evolution, however, suggests that fan entrenchment can occur in the absence of external forcing as a result of autogenic feedbacks among flow width, sediment transport and the rate of fan aggradation [*Nicholas and Quine*, 2007b]. Base level fall can also produce fan progradation rather than incision [*Harvey*, 2002b]. Differential uplift or subsidence in which the source area rises relative to the depositional zone typically produces thick alluvial fans [*Bull*, 1977; *Whipple and Trayler*, 1996].

Stock et al. [2007] note that channel gradient typically decreases downstream from values of 0.10-0.04 to 0.01 on fluvially-dominated fans. This trend has been explained via (i) a transport hypothesis in which slope is adjusted to transport the supplied sediment load, which decreases down-fan as deposition occurs, and (ii) a threshold hypothesis in which higher threshold slopes are required to entrain coarser particles in the channels of the upper fan and lower slopes are required to entrain finer grains down-fan. *Stock et al.* [2007] attribute the observed changes in gradient to downstream fining associated with overbank deposition of coarse particles down-fan.

Numerical simulations of fan processes and morphology have rapidly grown progressively more complex during the past decade [e.g., *Salcher et al.*, 2010]. *De Chant et al.* [1999] describe a diffusivity model to estimate the scale-independent surface profile of alluvial fans across which most of the surface is depositionally active and the profile of fans is characterized by channeling. *Coulthard et al.* [2002] and *Nicholas and Quine* [2007a] simulate coupled upland river basin and fan evolution using a cellular automaton model. *Clevis et al.* [2003] use a forward numerical model to simulate the effects of pulsed tectonic activity on fan dynamics. *Karssenberg and Bridge* [2008] use a 3d numerical model to simulate sediment dynamics and resulting morphology on floodplains, deltas, and fans.

Fans can be geomorphically complex environments in which channelized water, hyperconcentrated, and debris flows alternate spatially and temporally with unconfined sheet flow or with rapid infiltration that produces sieve deposits and subsurface flow [*Griffiths et al.*, 2006], and within-channel aggradation alternates with deposition [*Mather and Hartley*, 2005]. Physical experiments indicate that sheetflow can give way to channelized flow in the absence of any change in external variables as experimental fans evolve [*Clarke et al.*, 2010]. Episodic aggradation at the fanhead leads to oversteepening and initiation of incision [*Scott and Erskine*, 1994]. Channels avulse and are abandoned through stream capture [*Field*, 2001]. Secondary channel networks form on temporarily inactive portions of the fan surface through processes such as rilling and piping, while active distributary channel networks occupy other portions of the fan. The resulting stratigraphy records paleoenvironmental conditions, but is challenging to interpret.

Investigators have used calculations of changes in the volume of fan deposition [*Nanninga and Wasson*, 1985] through time to infer: climatically driven changes in sediment supply [*Ono*, 1990; *Oguchi*, 1994b, 1994c; *Kochel et al.*, 1997; *Ritter et al.*, 2000; *Nott et al.*, 2001; *Jennings et al.*, 2003; *Thomas et al.*, 2001; *Brown et al.*, 2003; *Klinger et al.*, 2003; *Oguchi and Oguchi*, 2004; *Hartley et al.*, 2005; *Weismann et al.*, 2005; *Waters et al.*, 2010], including alpine glacier retreat [*Hornung et al.*, 2010]; human-driven changes in sediment supply [*Gabris and Nagy*, 2005; *Gómez-Villar et al.*, 2006; *Kasai*, 2006; *Chiverrell et al.*, 2007] or deposition [*Davies and McSaveney*, 2001; *Davies et al.*, 2003; *Spaliviero*, 2003]; tectonically driven changes in supply and deposition [*Mather et al.*, 2000; *Ben-David et al.*, 2002; *Zisu et al.*, 2003; *Collins et al.*, 2008]; the magnitude-frequency of event-based sedimentation associated with storms or wildfires [*MacArthur et al.*, 1990; *Meyer et al.*, 1992, 1995, 2001; *Pierce et al.*, 2004; *Jull and Geertsema*, 2006; *Pierce and Meyer*, 2008]; and to construct short- and long-term sediment budgets for the upstream catchment [*Oguchi*, 1997c; *Johnson and Warburton*, 2002a; *McEwen et al.*, 2002]. Paraglacial fans are alluvial fans composed predominantly of reworked glacial deposits [*Ryder*, 1971a, 1971b]. Depositional volumes through time on these fans can be used to infer the rates at which mass movements and fluvial processes modify landscapes during deglaciation [*Owen and Sharma*, 1998; *Weissmann et al.*, 2002; *Barnard et al.*, 2004; *Dühnforth et al.*, 2008]. Ground penetrating radar [*Ekes and Friele*, 2003] and seismic refraction [*Schrott et al.*, 2003] can be used to image subsurface fan architecture. Uplift, tilting, and incision of fan deposits can also be used to infer uplift rates and timing of deformation [*Cunningham et al.*, 2003; *Fu et al.*, 2005; *Harkins et al.*, 2005; *Yeats and Thakur*, 2008].

Chronologies of fan processes can be developed using historical records [*Crosta and Frattini*, 2004; *Kotarba*, 2005], dendrochronology [*May and Gresswell*, 2004; *Wilford et al.*, 2005; *Turk et al.*, 2008], luminescence techniques [*Robinson et al.*, 2005; *Yeats and Thakur*, 2008], clast weathering [*Kotarba*, 2005], soil development

[*Kochel et al.*, 1997], cosmogenic isotopes [*Barnard et al.*, 2004, 2006], rock varnish [*Hooke and Dorn*, 1992], bomb isotopes such as ^{7}Be and ^{210}Pb [*Wallbrink et al.*, 2005], and radiocarbon dating [*Lewis and Birnie*, 2001; *Sanborn et al.*, 2006]. *Pelletier et al.* [2007] provide a cautionary note with regard to using rock varnish and desert pavements in arid regions; they find that although pavements may require tens of thousands of years to develop from a newly abandoned alluvial surface, disturbed pavements can redevelop over relatively short timescales of decades if a mature eolian epipedon is present.

Alluvial fans can also result from a single extreme event [*Blair*, 2001], such as the Roaring River alluvial fan in Colorado, USA, which resulted from a dambreak flood in 1982 (Figure 2.55). The flood entrained abundant glacial sediment along a steep, laterally confined portion of the Roaring River, then deposited an estimated 280,000 m^3 of sediment [*Jarrett and Costa*, 1986] at the junction with the lower gradient, wider Fall River. Successive phases of sedimentation resulted from a noncohesive sediment-gravity flow, sheetflood, and channelized water flows, and produced three distinct depositional lobes [*Blair*, 1987]. Investigators have also documented fans dominated by a specific and sometimes repetitive type of disturbance, including tributary glaciation [*Bali et al.*, 2004], volcanic eruptions [*Torres et al.*, 2004], wildfire-induced mass movements [*Jull and Geertsema*, 2006], and landslides [*Korup et al.*, 2004]. In these situations, alluvial fans can buffer stream systems from hillslope disturbances [*Lancaster and Casebeer*, 2007], although on other fans in-channel sediment sources dominate fan deposition and the coupling of channels and fans is particularly important [*Johnson and Warburton*, 2006].

The geomorphically complex and dynamic environment of alluvial fans also presents special challenges for hazard zoning and mitigation. Because fans are typically the only gently sloping surfaces available in high-relief environments, and because hazards such as debris flows [*Berti and Simoni*, 2007], sheetfloods [*Mukhopadhyay et al.*, 2003; *Mather and Hartley*, 2005], and channel avulsions [*Mack et al.*, 2008] are intermittent in time [*Grodek et al.*, 2000], urbanization and other development concentrates on fans [*Zorn et al.*, 2006]. Among the approaches used to mitigate hazards on fans are: channelization; bank stabilization; debris interception barriers and detention basins [*French et al.*, 2001; *Prochaska et al.*, 2008]; numerical simulations of the spatial distribution, type, and intensity of geomorphic processes [*Takahashi et al.*, 2001; *Shao et al.*, 2002; *Berti and Simoni*, 2007]; and warning devices [*Kellerhals and Church*, 1990]. *Pelletier et al.* [2005] describe an integrated approach to assessing hazards on fans and creating a probabilistic flood hazard map using hydraulic modeling, satellite-image change detection, field-based mapping of recent flood inundation, and surficial geologic mapping. As with other geomorphic features, the increasing availability of high-resolution LiDAR [*Jones et al.*, 2007] and airborne laser swath mapping [*Staley et al.*, 2006] data facilitate differentiation of geomorphic units on alluvial fans. The increasing availability of frequently repeated satellite imagery also facilitates change detection on fans [*Torres et al.*, 2004], as well as calibration of parameters for numerical simulations of flow processes on fans [*Catani et al.*, 2003].

2.11. Summary

Hillslope processes controlling water and sediment yield are linked to channel processes in all drainage basins. In mountain drainage basins this link is particularly strong because of the proximity of hillslopes and channels. Many mountain rivers have relatively narrow valleys with limited floodplains. Hillslopes serve as a primary and direct source of water and sediment, and changes in hillslope morphology or processes are not buffered by valley-bottom storage. Mountain drainage basins may thus respond more readily than lowland basins to fluctuations in climate, geology, or land use. In high-relief mountain regions, climate changes translate into elevational shifts in vegetation and weathering regime, and the typically thin regolith and low slope stability ensure that these shifts will alter water and sediment yield. Changes in tectonic regime or exposed lithology are also more likely to alter water and sediment yield in mountain drainage basins than in lowland basins because of the absence of filtering mechanisms associated with thick, stable regolith. Although the world's mountain regions are diverse in climate and geology, drainage basins formed in these regions share a strong structural influence on drainage development, and spatially and temporally heterogeneous inputs of precipitation and hillslope paths for the down slope movement of water into channels.

Some of the equations for bedrock weathering and soil production, incision by debris flows, and diffusive sediment transport outlined in this chapter represent important steps toward parameterizing hillslope processes in landscape evolution models [*Hancock et al.*, 2010; *Tucker and Hancock*, 2010] and toward coupling hillslope and channel numerical models. Ultimately these equations will help to construct process-based relief-denudation relationships [*Roering et al.*, 2007], but we need more tests of the applicability of these equations across diverse field settings, as well as additional parameterizations of specific hillslope processes.

3. CHANNEL PROCESSES

This chapter focuses on the movement of water and sediment within the stream channel. The chapter begins with an overview of discharge in mountain rivers, including methods of estimating flow and the importance of different magnitudes and frequencies of flow in shaping channels (section 3.1), then moves on to exchanges between stream flow and the underlying hyporheic zone (section 3.2), and the chemical reactions that influence the solute content of stream waters (section 3.3). These sections are followed by the mechanics of water flowing down steep mountain rivers (section 3.4), and how the water interacts with the channel boundaries and sediment supplied from other sources to influence sediment dynamics (section 3.5) and erosion of channel boundaries (section 3.6). The chapter then discusses the physical functions of wood in mountain rivers (section 3.7), followed by an integrative section on the downstream trends in channel dynamics resulting from interactions among all of these factors (section 3.8).

3.1. Hydrology

3.1.1. Discharge Estimation and Flow State

One of the challenges to understanding the spatial and temporal variability of mountain-river hydrology is the sparsity of direct, systematic discharge records [e.g., *Braun et al.*, 1998]. The percent of mountainous area in each of the world's major terrestrial regions ranges from 39% (Asia) to 16% (Africa), but the percent of water level observations in mountains is generally less than 13% (except for Asia, with 34%) [*Bandyopadhyay et al.*, 1997]. The most thoroughly instrumented mountain watersheds are those maintained for research purposes, and these commonly include discharge gages. Examples of these networks in the United States include experimental forests maintained by the Forest Service [*Gottfried et al.*, 2002] and Long-Term Ecological Research sites funded through the National Science Foundation. These highly instrumented sites with records that cover at least a decade remain the exception for mountain catchments. Although the World Meteorological Organization

Mountain Rivers Revisited
Water Resources Monograph 19
Copyright 2010 by the American Geophysical Union.
10.1029/2010WM001041

recommends the highest densities of instrument networks in mountains, these regions typically have low-density networks.

Different recurrence intervals of flow can be estimated from regional discharge-drainage area relations calibrated using gage records [*Sanborn and Bledsoe*, 2006] and, in some cases, other parameters such as catchment geology [*Tague and Grant*, 2004]. *Surian and Andrews* [1999] demonstrate that regional flow duration equations can accurately define relatively large, infrequent flows in some mountainous regions. Systematic hydrologic records are inadequate for a detailed understanding of flood hydrology in many mountainous regions because of the combined sparsity of precipitation and gaging stations and the spatial and temporal variability in precipitation and flow [e.g., *Klemes*, 1990; *Kostka and Holko*, 1994]. Even at sites with systematic precipitation and streamflow records, these data commonly span less than a hundred years, making it difficult to accurately estimate the magnitude and frequency of extreme floods using conventional flood-frequency analysis. The design of dams and other structures, land-use management, and the siting of critical installations require an evaluation of flood risk, and large flows can be particularly important for understanding long-term sediment transport, channel adjustment, and landscape evolution. To fill the gap in systematic records, flood-frequency curves developed from short-term systematic records can be extrapolated to longer time spans [*Jakob and Jordan*, 2001]. Systematic records from lower elevation areas near the mountains can also be used to approximate mountain-river hydrology via storm transposition, or scenarios of probable maximum precipitation/probable maximum flood based on worst-case rainfall-runoff predictions can be used [*Jarrett and Costa*, 1988; *Schulze et al.*, 1994]. Numerous studies indicate that these techniques tend to overestimate maximum floods at higher elevations [*Baker et al.*, 1987; *Jarrett and Costa*, 1988; *Carling*, 1994a; *Carling and Grodek*, 1994; *Pruess et al.*, 1998]. This overestimation can have serious economic consequences for land-use zoning and the design of flood-control or hydraulic structures.

Sieben [1997] characterizes these and other approaches as either statistical models or deterministic models. A statistical model is based on statistical analyses of empirical data; represented trends are assumed to be consistent and continuous and are extrapolated to extreme events. A deterministic model is based on a mechanistic concept suitable for deterministic simulation of the processes of interest, with imposed boundary conditions that correspond to an extreme event scenario.

Existing discharge records can also be supplemented with new instrumentation or remotely sensed data. For smaller channels, weirs (Figure 3.1) can be established [*Hudson et al.*, 1990; *Moschen*, 1990] or pressure sensors can be installed at relatively stable cross sections for discharge gaging. *Fonstad and Marcus* [2005] and *Legleiter et al.* [2009] describe hydraulically assisted bathymetry, which uses stream gage information on discharge, image brightness data, and Manning-based estimates of stream roughness to calculate water depth from aerial photographs or hyperspectral imagery without ground-truth information at the time of image acquisition. The increasing availability of airborne laser-scanner data also facilitates the use of high-resolution topographic representations in hydrologic modeling [*Hollaus et al.*, 2005]. These types of approaches can be particularly useful in regions difficult to physically access [*Pike*, 2006].

Indirect methods can also be used to estimate discharge at the reach scale. Most of these methods were developed for lower gradient rivers, and the assumptions that are typically made regarding flow state when indirectly estimating discharge have led to much discussion of the appropriateness of these methods for mountain rivers. Most indirect methods of discharge estimation, such as slope-area computations based on the Manning equation [*Dalrymple and Benson*, 1967], step-backwater methods [*O'Connor and Webb*, 1988], or a simplified slope-area method that does not require an estimate of Manning's roughness coefficient [*Sauer et al.*, 1985], all assume steady, uniform flow. Floods along steep rivers, however, may be unsteady, rapidly varying, and debris-charged [*Glancy and Williams*, 1994]. Indirect discharge estimates can therefore be in error because of scour and fill, rapid changes in flow, substantial sediment transport, and flow transitions between subcritical and supercritical [*Jarrett*, 1987; *Sieben*, 1997]. *Bathurst* [1990] estimates that the accuracy of the slope-area method when applied to mountain rivers may typically be as poor as 30%.

The greatest deviations from steady, uniform flow may occur in specific climatic environments with relatively short duration, high magnitude flows. Small, ephemeral desert channels can have especially unpredictable and brief flows. After five years of research in the mountainous 0.6 km^2 Nahal Yael drainage of southern Israel, *Schick* [1970] reported that typical runoff events are produced by a period of effective rainfall lasting 15 minutes, with a median flow rise lasting 2.5–5 minutes, and peak discharges of 4-8 minutes. Such flows may nevertheless yield up to 152 tons/km^2 of sediment. For these types of channels, *Glancy and Williams* [1994] suggest the use of flow-triggered video cameras to assess water-surface profiles; recording streamflow gages; direct velocity measurements to verify roughness coefficients; and scour chains to record depth of scour and fill. These types of data can be used to assess how well the flow approximates steady, uniform conditions.

Flows in small channels of the seasonal tropics can also have very short duration and high magnitudes [*Dykes and Thornes*, 2000; *Wohl et al.*, 2009b], although these flows occur much more frequently than in arid regions [*Wohl and Jaeger*, 2009]. River discharge in tropical regions typically remains poorly documented relative to other climatic regions.

The presence of seasonal ice along mountain rivers also creates unique flow conditions and channel processes that make it difficult to estimate flow magnitude using indirect methods. Flat-bottomed valleys underlain by permafrost at high latitudes are typically covered by aufeis over most of their length at the end of winter. Aufeis is a thick shelf of ice well above the level of base flow. The largest quantity of meltwater runoff passes through a channel on top of the aufeis, as does much of the sediment transported during the year [*Priesnitz and Schunke*, 2002]. As the aufeis gradually melts, these supra-aufeis channels disappear, leaving some unusual sediment deposits (Figure 3.2).

Tinkler [1997b] proposes a method for indirect estimation of discharge during high flows. He finds that the *Kennedy* [1963] equation for the mean velocity of a critical wave train,

$$v = 1.2495\sqrt{\lambda} \tag{3.1}$$

where λ = wavelength of the standing waves, can be used to estimate the mean velocity in rigid-boundary channels with little sediment transport at wavelengths of 0.5-7 m. This provides an accurate, economical, and fairly simple means to measure mean velocity during high discharges and is not hampered by the requirement of steady flow. It does, however, require direct observation or photography of standing waves during peak flow.

Much of the debate regarding the appropriateness of applying traditional indirect discharge estimations to mountain rivers arises from the observation of hydraulic jumps along these rivers. Hydraulic jumps form at the rapid change in depth associated with a transition from a supercritical to a subcritical flow regime [*Elevatorski*, 1959; *Dingman*, 1991]. Hydraulic jumps create intense turbulence and large kinetic energy losses, with the amount of loss depending on the change in depth across the jump [*Roberson and Crowe*, 1993]. Spatial and temporal variations in air content of water going through a jump help to induce pressure fluctuations that correlate with energy dissipation. *Vallé and Pasternack* [2002] use time domain reflectometry to measure air content at natural jumps in field settings. Hydraulic jumps in mountain rivers occur (1) over substantial obstacles on the bed, (2) at converging or irregular banks or at a strong eddy that acts as an effective bank, (3) at a contraction or expansion of flow, or (4) below vertical drops [*Kieffer*, 1985; *Zgheib*, 1990; *Carling*, 1995; *Tinkler*, 1997b]. Flow velocities decrease substantially across hydraulic jumps and bedload can be deposited downstream, stabilizing the position of the jump [*Whittaker and Jaeggi*, 1982; *Kieffer*, 1985; *Carling*, 1995]. Because hydraulic jumps indicate at least locally occurring supercritical flow, their presence along mountain rivers suggests that supercritical flow may be more prevalent along these rivers than along low-gradient channels. If supercritical flow dominates along steep channels during floods, then most indirect discharge estimation methods would provide a very inadequate approximation of actual discharge.

The consensus regarding flow state along mountain rivers is that supercritical flow across an entire cross section does not persist for lengths of more than approximately 8 m along most channels because of extreme energy dissipation in the form of hydraulic jumps, turbulence, and obstructions [*Trieste and Jarrett*, 1987; *Trieste*, 1994]. *Jarrett and Costa* [1986], for example, note the occurrence of predominantly subcritical flow for a dambreak flood in a channel reach with an average slope of 0.10. Measuring flow velocities in rapids on the Colorado River with an acoustic Doppler current profiler, *Magirl et al.* [2009] demonstrate that flow is subcritical throughout all the measured rapids, despite local slopes of 0.03 m/m and the presence of standing waves. *Tinkler* [1997a], however, discusses the persistence of central zones of critical or supercritical flow for lengths greater than 8 m within a range of channel types. *Grant*

[1997] hypothesizes that channel hydraulics and bed configuration interact to produce Froude numbers close to 1 that are observed along many steep, competent streams. Supporting evidence comes from *Magirl et al.* [2009], who indicate that supercritical flow in the rapids of the Colorado River is rare even at large discharge.

Total energy loss must increase as slope and depth increase if flow is to remain critical. The increased roughness required for flow to remain critical comes principally from increased shearing and vorticity between central critical/supercritical flow and adjacent subcritical lower-velocity water [*Tinkler*, 1997a] and from bedload transport where possible. *Tinkler* [1997a] suggests the following boundary to define the lower limit of slope for channels in which central flow is critical or supercritical:

$$S = g \ n^2 \ h^{-0.33} \tag{3.2}$$

where S is channel gradient, g is gravity, n is Manning's roughness coefficient, and h is flow depth. Careful application of such equations can help to define probable flow conditions and to constrain errors associated with indirect discharge estimation, but our understanding of mountain-river hydraulics remains insufficient for confident reconstruction of discharge in many cases.

In summary, spatial and temporal variability in hydraulics make it difficult to use traditional indirect methods of discharge estimation, such as the Manning equation, to supplement the typically sparse systematic discharge records of mountainous regions. Long-term instrumented watersheds thus become particularly valuable for understanding temporal and spatial variability in discharge and for linking hydrology and hydraulics to sediment and channel dynamics.

3.1.2. Paleoflood Indicators

Historical, botanical, and geologic records of past flow have been used successfully in mountainous regions to extend short-term or non-existent systematic discharge records [*Jarrett*, 1990b; *Enzel et al.*, 1993; *Rico et al.*, 2001; *Johnson and Warburton*, 2002b; *Harden et al.*, 2010]. *Historical records* that can be used to estimate the magnitude and frequency of past flow include diaries and journals, damage or insurance reports, and marks of peak flow stage that people create on buildings or along channel boundaries [*Hooke and Redmond*, 1989; *Fanok and Wohl*, 1997; *Thorndycraft et al.*, 2002; *Gurnell et al.*, 2003; *Gob et al.*, 2008]. These types of records are limited by the length of human occupation of a region and are typically qualitative. They are also filtered through human perceptions in the sense that the lower discharge level defining a "flood" can become smaller as population density along a channel increases. An example of historical records of past flow comes from Buddhist monasteries in Nepal, where monks preserve the memory of floods that occurred prior to systematic discharge records [*Bjonness*, 1986]. Historical records can include repeat photography that provides insight into various forms of hillslope and channel change related to floods or more persistent hydrologic conditions [*Webb*, 1996; *Webb et al.*, 2004]. Indirect indicators of runoff can also be developed by using remote sensing imagery to estimate changes in water storage in small irrigation reservoirs, which are widespread in regions such as semiarid Africa [*Liebe et al.*, 2009].

Botanical records take the form of damage to riparian trees caused by high or low flows, or the structuring of riparian vegetation by age and type in response to stability and inundation of riparian habitat [*Sigafoos*, 1964; *Everitt*, 1968; *Yanosky*, 1982a; *Hupp*, 1988; *Hupp and Osterkamp*, 1996; *Hupp and Bornette*, 2003; *Ruiz-Villanueva et al.*, 2010]. Flood damage to riparian trees includes impact scars [*Yanosky and Jarrett*, 2002; Figure 3.3] and adventitious or split-base sprouts caused by flood-borne debris [*Hupp*, 1988], adventitious roots or changes in ring characteristics associated with overbank sedimentation [*Hupp and Simon*, 1991; *Friedman et al.*, 2005a], and anatomical changes in tree roots exposed by bank erosion [*Sigafoos*, 1964; *Hupp and Osterkamp*, 1996; *McAuliffe et al.*, 2006; *Malik and Matyja*, 2008]. Tree rings can provide a chronology for each of these morphological responses to flow conditions [*Yanosky*, 1982b; *Phipps*, 1985]. Variations in tree-ring width and symmetry can also indicate past episodes of high and low flow [*Yanosky*, 1983, 1984; *Jones et al.*, 1984; *Meko*, 1990; *Earle*, 1993; *Yuan et al.*, 2007; *Axelson et al.*, 2009]. Stands of individual riparian species of differing ages reflect frequency of inundation and lateral stability of a channel [*Sigafoos*, 1961; *Everitt*, 1968; *Helley and LaMarche*, 1973; *Gottesfeld and Gottesfeld*, 1990; *Hupp*, 1990; *Hupp and Simon*, 1991; *Scott et al.*, 1996, 1997]. Botanical records provide chronologically precise information for a wide range of discharges, but these records are limited by the presence and age of woody riparian vegetation [*Wohl and Enzel*, 1995].

Geologic records of past flow can be used for regime-based reconstructions, competence estimates, paleohydraulic estimates, or stage estimates [*Jacobson et al.*, 2003]. Most of these techniques for estimating past discharge can be related to flood occurrence using chronologic methods involving: tree-ring analysis [*Sigafoos*, 1964; *Yanosky*, 1984]; radiocarbon dating of organic materials associated with the flood deposits [*House and Baker*, 2001; *Yu et al.*, 2003; *Sheffer et al.*, 2008]; cosmogenic isotope analyses of fluvially-abraded bedrock or very large clasts [*Fenton et al.*, 2002]; luminescence dating of fine-grained fluvial or eolian sediment [*Nador et al.*, 2007]; lichenometry or weathering rinds on bedrock and boulders in the active channel or on terraces [*Knuepfer*, 1988; *Macklin et al.*, 1992; *Gob et al.*, 2005, 2008]; soils analysis; or other relative methods [*Costa*, 1978].

Regime-based reconstructions of relatively frequent floods rely on alluvial channel geometry as preserved in relict channels [*Harden et al.*, 2010]. This approach is poorly suited to the confined valleys and partially bedrock-controlled rivers present in most mountainous regions, but is applicable in broader mountain valleys such as glacial troughs. Using this approach in the Uinta Mountains of Utah, USA, *Carson et al.* [2007] quantify long-term (10^3 yr) fluctuations in bankfull discharge and relate these to glacial and climatic fluctuations. A related technique is the use of geomorphic and botanical indicators to estimate bankfull discharge (see section 3.1.4) in ungaged catchments. Typically, this is done by finding some type of feature that consistently correlates with a flow magnitude that recurs on average every 1-2 years along gaged channels and then using the occurrence of this feature on ungaged channels to estimate bankfull. Commonly used features include the top of the channel banks and changes in vegetation [*Pike and Scatena*, 2010].

Competence estimates involve using either the average clast size, bedforms, or bedding structure to estimate past mean flow [*Baker*, 1974; *Church*, 1978; *Maizels*, 1983, 1989; *Nott and Price*, 1994], or using the largest clasts likely to have been transported by fluvial processes to estimate the shear stress, unit stream power, or velocity of the associated peak flow (Figure 3.4) [*Birkeland*, 1968; *Inbar and Schick*, 1979; *Bradley and Mears*, 1980]. The underlying assumption is that the channel is transport-limited (rather than supply-limited) with respect to competence. Care must be taken to avoid debris-flow deposits or rockfall into the channel that may have been rounded in place. Competence estimates of past flow are generally imprecise in comparison to other indirect methods of estimating discharge and hydraulics [*Costa*, 1983; *Williams*, 1983; *Komar*, 1987a,b; *Wohl*, 1992a] and are limited to information about the single largest flow down the channel, generally with minimal information as to when such a flow occurred. The same stochastic and site-specific controls that make it difficult to predict sediment entrainment and deposition along coarse-grained con-temporary channels make it difficult to retrodict the conditions of past flow as represented by coarse-clast deposits. Most equations developed from contemporary process observations focus on entrainment, whereas paleohydrology is of necessity focused on sites of deposition [*O'Connor et al.*, 1993]. An exception comes from the work of *Bordas and Silvestrini* [1992], who propose thresholds for coarse sand deposition, as a function of unit stream power, for isolated grain, bulk, and mass transport. *Komar and Carling* [1991] and *Wilcock* [1992b] question how representa-tive one or a few of the largest particles are of the transported sediments and of flood hydraulics. They note that flow competence relations differ from one stream to another, depending on the pattern of grain sorting, which is a function of the bed-material grain-size distribution. Nevertheless, competence estimates may be the best option available along some mountain channels.

Paleohydraulic estimates rely on depositional or erosional indicators of flow hydraulics. *Carling* [1995] describes lateral gravel berms deposited adjacent to hy-draulic jumps in bedrock channels and notes that the angle subtended by the berm crestline with respect to a regular bankline can be used to estimate Froude number of flow. *Carling and Grodek* [1994] and *Nott and Price* [1994] use the bars at the downstream end of plunge pools to estimate discharge and hydraulics through the pool. *Zen and Prestegaard* [1994] find that the geometry of flow separation zones and thus the location of lateral potholes are scaled by flow Reynolds number. These potholes can thus be used to reconstruct paleoflow depths and velocities if parts of the channel bed adjacent to the flow obstruction are preserved.

Stage estimates use erosional or depositional records of maximum stage, in combination with channel geometry, to infer past discharges. Erosional records take the form of scour lines (Figure 3.5) [*Wohl*, 1995], lichen limits [*Gregory*, 1976], or the truncation of landforms such as alluvial fans that impinge on the channel [*Shroba et al.*, 1979]. Depositional evidence of peak stage includes silt lines [*O'Connor et al.*, 1986], organic debris [*Carling and Grodek*, 1994], boulder bars [*Elfström*, 1987; *Jarrett et al.*, 1996; *Cenderelli and Cluer*, 1998], and fine-grained slackwater sediments (Figure 3.6) [*Baker*, 1987; *O'Connor et al.*, 1994; *Kite et al.*, 2002). Some of these features are more accurate as stage indicators than are others [*Jarrett and England*, 2002]. Fine-grained organic debris or silt lines formed by the adhesion of fine sediment and organic materials along the valley walls can be highly accurate but ephemeral features unless they are protected from subsequent erosion, as at the back of an alcove or cave [*Springer*, 2002] formed in a bedrock valley wall. Boulder bars and slackwater sediments, on the other hand, are likely to provide a minimum estimate of peak stage, although the degree of underestimation produced by using these indicators can vary, depending on the geometry of the depositional site and the sediment concentration of the flow [*Baker and Kochel*, 1988].

The use of *paleostage indicators (PSIs)* relies on the assumptions that channel geometry has not changed substantially since the time of the flood that created the PSI, and that scour and fill were minimal during the flood. PSIs are thus most commonly used along stable boundary channels formed in bedrock or very coarse alluvium. Channels with a fairly deep, narrow cross-sectional geometry that maximizes stage change as a function of discharge make it easier to differentiate PSIs from individual floods. Erosional landforms or boulder bars can facilitate estimation of only the largest flood to have occurred along a channel, but slackwater sediments can record numerous floods. Slackwater sediments settle from suspension at sites of flow separation such as channel expansions, tributary channel mouths, or channel-margin alcoves and caves [*Baker*, 1983, 1987; *Springer and Kite*, 1997]. At these sites, each depositional unit records a flood and a chronology of depositional units can be established with radiocarbon, luminescence, or ^{137}Cs dating [*Ely et al.*, 1992].

Historical, botanical, and geologic records of past floods can be used to understand variations in flood magnitude and frequency over centuries to millennia [*Greenbaum et al.*, 2001] and to infer climatic drivers of these variations [*Ely et al.*, 1993, 1996; *Ely*, 1997; *Redmond et al.*, 2002; *Thorndycraft et al.*, 2002; *Harden et al.*, 2010]. Paleo-flood data can be used to determine non-exceedance levels of flood magnitude [*Levish*, 2002] and to refine the estimated probable maximum flood for a basin because these records indicate the largest discharge to have occurred within a period that may be several hundred years or, in some cases, ten thousand years long [*Webb and Rathburn*, 1988]. These records can also be used to improve flood-frequency estimates for more frequent flows if the non-systematic data represent the population of all floods above a fixed discharge threshold that have occurred during a known timespan. For non-systematic data, the discharge threshold may have varied through time as a function of changes in depositional geometry at a site. As a result, different statistical distributions must be applied to each type of data [*Benson*, 1950; *Stedinger and Cohn*, 1986; *Stedinger and Baker*, 1987; *Salas et al.*, 1994; *Blainey et al.*, 2002; *England et al.*, 2003a, 2003b].

In summary, numerous types of information can be used to indirectly estimate the magnitude and chronology of floods occurring prior to systematic stream gaging. The accuracy, type of flood (extreme floods, annual floods), and length of record vary substantially between sites, but in many cases paleoflood indicators add valuable information to sparse systematic gage records from mountain drainages. The increasing use of paleoflood indicators in regions such as the U.S. Rocky Mountains [*Jarrett*, 1989, 1990b; *Grimm et al.*, 1993; *Pruess et al.*, 1998], for example, is helping to define the magnitude and frequency of flood occurrence and the hydraulic conditions during floods.

Field and lab studies of flow hydraulics along mountain rivers are complementary to the techniques discussed above in that they permit the refinement of indirect methods of discharge estimation and the use of paleoflood data to define the extreme conditions present along a given channel. The non-uniform character of slope, width, sediment size and composition, and discharge and sediment transport along mountain rivers can create various types of wave phenomena, such as propagation of hydrographs or aggradation fronts, that can develop into discontinuous profiles [*Sieben*, 1997]. Although these discontinuities in flow and sediment movement make it very difficult to model flood hydraulics and morphology along mountain rivers, continuing development of numerical simulation methods is steadily improving model applicability to mountain rivers.

3.1.3. Modeling Stream Discharge

Rather than using field-derived estimates of discharge, data on the spatial distribution of basic hydrological variables such as rainfall and runoff can be combined with digital terrain analysis to estimate discharge at varying points in a channel network and to create spatially explicit water budgets [*Montgomery et al.*, 1998]. Various studies use a 2d finite element model of flow with hillslope hydrology models, remote sensing data, or digital elevation models to simulate the extent of flood inundation, as well as flow depth and velocity in lowland environments [*Bates et al.*, 1997, 1998; *Charlton*, 1999; *Stewart et al.*, 1999; *Bates and De Roo*, 2000; *Marks and Bates*, 2000]. These methods may also be applicable to some segments of mountain rivers. Other approaches include combining gaged and ungaged stream flows with physical and statistical data to develop probability models for all uncertain parameters [*Campbell*, 2005]. The Bayesian methods used in this approach can be applied to rainfall-runoff models [*Sharma et al.*, 2005] or to models of stream flow. Both types of models can also be run using a Generalized Likelihood Uncertainty Estimation scheme [*Beven and Binley*, 1992], which conditions the parameter distributions and generates prediction uncertainty envelopes that incorporate parameter uncertainties [*Wyatt and Franks*, 2006]. A great deal of attention is given to spatial variability in parameters and processes and to uncertainty in parameter estimation, as reflected in the papers collected in *Sivapalan et al.* [2006]. Where glacier-melt influences the hydrology of a mountain basin, hydrological models can be improved by incorporating even a single annual glacier mass balance [*Konz and Seibert*, 2010].

Physically based models can also be combined with Artificial Neural Networks to forecast flooding in mountainous regions [*Cullmann et al.*, 2006]. Coupled surface-subsurface models can be used to evaluate river-aquifer exchange [*Frei et al.*, 2009]. Hydrological models can be coupled with models of water chemistry to study contaminant transport [*Bronstert and Plate*, 1997] or with models of nutrients such as carbon and nitrogen [*Band et al.*, 1993]. *Nelson et al.* [2003] review modeling of flow and sediment transport.

Various types of hydrologic models are increasingly being applied to predictions of the hydrologic effects of climate change in mountainous regions. These models tend to show a progressive increase in mean flows within glacial basins as the glaciers melt and recede more rapidly [*Kuhn and Batlogg*, 1998; *Rees and Collins*, 2006a] and more complex shifts in the timing and magnitude of snowmelt-dominated runoff systems as the elevational range of snow and rainfall shift. One of the greatest challenges of mountain water resources in the coming decades may be that of dramatically reduced water supply in catchments heavily dependent on glacial meltwater once the glaciers disappear.

In summary, indirect discharge estimation at the catchment scale typically relies on some type of precipitation-runoff and hydrologic routing model based on known or inferred characteristics of precipitation, topography, and drainage network. Rapid improvement of these models during the past decade has enhanced their accuracy when applied to mountainous drainage networks.

3.1.4. Bankfull Discharge

Discharge is gaged, indirectly estimated, and modeled in mountain rivers for several reasons, including estimation of water supply and quantification of forces acting on channel boundaries and transporting sediment. In a geomorphic context, knowledge of the magnitude, frequency and duration of floods is necessary to understand the geomorphic effects of differing flow levels; do the channel cross-sectional and planform geometry predominantly reflect annual peak flows, for example, or larger floods with a longer recurrence interval? This question is particularly important in stream management and restoration, and is commonly discussed in the context of bankfull discharge.

Bankfull discharge is one of the most widely used yet most controversial concepts in fluvial geomorphology. Originally defined morphologically as the discharge that produces a stage just before the flow begins to overtop the stream banks [*Wolman and Leopold*, 1957], bankfull discharge has gradually assumed many more implications because it has been equated with a specific recurrence interval and geomorphic function. Because numerous investigators find that a discharge which nearly overtops the banks recurs approximately every 1-2 years on many channels [*Leopold et al.*, 1964; *Harman et al.*, 1999; *Castro and Jackson*, 2001], bankfull discharge has been defined based on recurrence interval rather than channel morphology. And because this discharge transports the majority of suspended sediment in some rivers, bankfull discharge has been interpreted as the single most important flow for controlling channel process and form [*Wolman and Miller*, 1960; *Dunne and Leopold*, 1978]. There are numerous problems with each of these assumptions.

Defining bankfull discharge creates the first problem. Many streams have inset channels, uneven banks, or multiple side-slope convexities reflecting different flow magnitudes, which makes it extremely difficult to consistently estimate bankfull from channel morphology. *Radecki-Pawlik* [2002] and *Navratil et al.* [2006] compare several different field-based, morphological definitions of bankfull, using top of bank, bank inflection, ratio of channel width to mean depth [*Wolman*, 1955], level of significant change in the relation between wetted area and top channel width [*Williams*, 1978a], and first maximum local bank slope [*Riley*, 1972]. They find that using the different methods results, on average, in discharge estimates that vary by a factor of three at a given site and they recommend that bankfull discharge be reported as a range of discharges rather than a single number.

The second problem is that bankfull discharge defined from a specified component of channel morphology may have very different recurrence intervals among different sites [*Pickup and Warner*, 1976; *Williams*, 1978a; *Gomez et al.*, 2007]. Morphologically defined bankfull discharge has a recurrence interval of 40-90 days, for example, along mountain rivers in Puerto Rico [*Pike and Scatena*, 2010]. Thus, even a consistently defined morphologic feature is not likely to represent the same flow recurrence interval on diverse channels. Duration of bankfull discharge can also vary between sites. Studying snowmelt-dominated mountain rivers in Colorado and Idaho, USA, for example, *Segura and Pitlick* [2010] find that the frequency of flows greater than bankfull decreases downstream from ~ 15 days/year in headwater reaches to ~ 6 days/year in downstream reaches, implying that basin response to precipitation and runoff is nonlinear.

The third problem is that, whether defined from channel morphology or recurrence interval, bankfull discharge does not necessarily transport the majority of suspended sediment or dominate channel form [*Emmett and Wolman*, 2001; *Powell et al.*, 2001; *Thompson and Croke*, 2008]. Using suspended sediment records from hundreds of rivers across the United States, *Simon et al.* [2004] conclude that discharge with a 1.5-year recurrence interval does transport the largest proportion of the annual suspended sediment load on most rivers. And discharges at or below bankfull do transport the majority of bedload in some mountain rivers [*Ryan et al.*, 2002; *Torizzo and Pitlick*, 2004]. There is thus a reasonable basis for the generalizations underlying bankfull discharge as a sort of 'workhorse' of fluvial geomorphology that transports much of the sediment moved in a channel. This does not necessarily extend to channel form, however, particularly along channels with erosionally resistant boundaries. Several studies demonstrate that channel morphology typically reflects multiple recurrent discharge magnitudes [*Pickup and Warner*, 1976; *O'Connor et al.*, 1986; *Phillips*, 2002; *Surian et al.*, 2009; *Turowski and Rickenmann*, 2009].

The phrase *dominant discharge*, which is sometimes equated with bankfull discharge, reflects the assumption that there is a single magnitude of flow which, if constantly maintained within a stream, will produce the same average channel dimensions and morphology as those produced by a stable stream's entire hydrologic regime [*Crowder and Knapp*, 2005]. *Effective discharge* (the discharge that transports the largest amount of sediment; *Schmidt and Morche*, 2006), bankfull discharge, mean annual flood, and the 1.5-year flow have all been proposed as constituting dominant discharge [*Rosgen and Silvey*, 1996; *Griffiths and Carson*, 2000]. The idea of a single, particularly important discharge is an oversimplification of the complexity of interactions among flow and channel process and form through time. This oversimplification becomes dangerous when taken to extremes in river management. River restoration based on the concept of bankfull discharge, for example, can neglect the geomorphic and ecological importance of less frequent, larger flows, as discussed in chapter 6. Effective discharge is also commonly used as an index value for scaling channel dimensions and for designing stable channels during restoration, yet predictions of effective discharge depend on the accuracy of the bedload transport equation across a range of flows [*Barry et al.*, 2008] In general, as hydrologic variability and boundary resistance increase, infrequent floods exert a progressively more important influence on sediment transport and channel morphology. *Migon et al.* [2002], however, sound a cautionary note for the Sudetes Mountains of central Europe that now applies to many mountainous regions around the world; features newly created by extreme events are commonly quickly obliterated by humans, thus reducing the long-term geomorphic effects of these events.

In summary, syntheses across numerous field sites suggest that a relatively frequent discharge just below that which creates overbank flow transports the majority of sediment and strongly influences channel geometry in many lower gradient alluvial rivers. Bankfull discharge can be difficult to define morphologically in mountain rivers that experience debris flows or occasional very large floods, however, and it may be these larger, less frequent flows that transport the majority of sediment or dominate channel geometry along rivers formed in erosionally resistant substrate such as bedrock or boulders or along rivers with a highly variable flow regime. More work is needed to evaluate how well conceptual models of bankfull and dominant discharge apply to mountain rivers and to document the characteristics of flows that transport the majority of sediment and shape channel parameters in mountainous regions.

3.1.5. Floods

Floods along mountain rivers can differ from those in lowland systems because of the close coupling between hillslopes and channels. *Gallino and Pierson* [1985] describe a landslide on the flank of Mount Hood, Oregon, USA that transformed rapidly into a debris flow. The debris flow eroded and incorporated large volumes of channel fill as it surged down Polallie Creek and created a large debris fan that temporarily dammed the East Fork Hood River at the confluence of the two channels. Within 12 minutes, a lake of 104,600 m^3 formed behind and then breached the dam, sending a flood wave down the East Fork Hood River [*Gallino and Pierson*, 1985]. Similar cases of hillslope mass movement temporarily damming a channel and facilitating an outburst flood have been described from British Columbia, Canada [*Russell*, 1972] and from Japan [*Ouchi and Mizuyama*, 1989].

The sediment supplied from hillslope mass movements can also strongly influence total sediment load along a channel during a flood [*Williams and Guy*, 1973; *Mizuyama*, 1991; *DiSilvio*, 1994; *Cheng et al.*, 2005; *Lin et al.*, 2008], as well as the spatial patterns of erosion and deposition along the channel [*Jacobson et al.*, 1993; *Miller*, 1994; *Cenderelli and Kite*, 1998]. *Nolan and Marron* [1985] describe contrasting channel responses to major storms in two mountainous areas of western California, USA. Channels in northwestern California have pervasive, long-lasting widening and filling of intermediate- and high-order channels during regional storms with recurrence intervals in excess of 50 years. Sediment delivered to these channels by landslides from the structurally weak rock units of the Mesozoic Franciscan assemblage overwhelms transport capacities throughout much of the channel network. Streamside hillslope failures are less common in the Tertiary sedimentary and volcanic units of the Santa Cruz Mountains of west-central California and channel sediment transport capacity is exceeded only locally during large storms. Widespread bedrock control along the channel banks limits channel widening. As a result, there is only moderate scour in steep low-order channels and moderate fill in high-order channels in the Santa Cruz Mountains [*Nolan et al.*, 1984].

Debris flows and landslides can introduce large amounts of wood to a flooding channel, as well as sediment. The wood can facilitate formation of temporary debris dams (Figure 3.7) that create secondary pulses when they burst, or floating wood can enhance erosion of channel banks and riparian areas, as well as structures in and along the channel [*Comiti et al.*, 2008a].

Drainage area and channel gradient, as well as magnitude and frequency of hillslope failures, also strongly influence how hillslope sediment affects channel morphology. Debris flows in the central Appalachians, USA during Hurricane Camille in 1969 caused scour in drainage areas less than 1 km^2 and channel gradients steeper than 0.1; mixed erosion and deposition with continuous reworking of the valley flood along streams draining up to 65 km^2; and localized, discontinuous valley-floor reworking along basins larger than 100 km^2 [*Miller*, 1990a]. *Fuller* [2007] describes analogous spatial differences in flood geomorphic effectiveness along rivers in New Zealand as a function of channel and valley geometry.

Floods along mountain rivers can be generated by various types of rainfall, rain-on-snow, snowmelt, or the failure of either natural or artificial dams [*Weingartner et al.*, 2003]. In high-latitude mountain regions, ice jams can also exacerbate flooding during the spring [*Buzin*, 2000; *Smith and Pearce*, 2002]. Regions with the highest discharge per unit drainage area in the United States tend to combine regional atmospheric conditions that produce large precipitation volumes and steep topography that enhances precipitation by convective and orographic processes and quickly concentrates flow in channels [*O'Connor and Costa*, 2004]. Numerous studies demonstrate that flood magnitude and frequency vary over decadal or longer timespans in relation to large-scale atmospheric circulation patterns such as the North Atlantic

Oscillation [e.g., *Maas and Macklin*, 2002]. Disturbances such as wildfire [*Conedera et al.*, 2003] or human-induced changes in land cover can also greatly increase runoff and flood discharge for a given precipitation input (see chapter 6).

The magnitude-frequency differences between the largest floods and the more frequent floods are a function of the mechanisms that produce the floods. In semiarid mountainous regions of the western USA where floods are produced by intense thunderstorms, the 100-year flood can be more than ten times the mean annual flood, whereas in snowmelt regions the 100-year flood may be less than twice the size of the mean annual flood (Figure 3.8) [*Pitlick*, 1994b]. Variability in precipitation amount and intensity is more important than basin physiography and drainage area in controlling variation in flood frequency distribution in both temperate and tropical regions [*Pitlick*, 1994b; *Smith et al.*, 2005]. Examining extreme floods produced by dissipating tropical storms in the Appalachian Mountains of the eastern United States, *Sturdevant-Rees et al.* [2001] find that the relationship between drainage network structure and storm motion significantly influences regional flood hydrology; the south-to-north motion of Hurricane Fran in September 1996 amplified peak discharge at the outlet of south-to-north flowing river catchments. In glaciated basins, extreme floods can occur when a late afternoon thunderstorm coincides with maximum meltwater from the lower part of a glacier [*Gerrard*, 1990].

3.1.5.1. Outburst floods. Dam failure can produce a flood peak discharge more than an order of magnitude larger than precipitation-generated floods along a channel. Such failure can be associated with an artificial dam [*Jarrett and Costa*, 1986], or failure of a caldera lake [*Houghton et al.*, 1987; *Waythomas et al.*, 1996; *Procter et al.*, 2010] or a landslide dam [*Reneau and Dethier*, 1996], but many mountain outburst floods originate from glacier lakes. Thirty-five outburst floods have occurred during the past 200 years, for example, in the Karakoram Himalaya [*Hewitt*, 1982]. Approximately 230 supra- or proglacial lakes are present in the Cordillera Blanca of Peru and a widespread glacier recession beginning in the 1920s initiated a series of damaging floods [*Lliboutry et al.*, 1977]. At least eleven outburst floods occurred in the Nepalese and Tibetan Himalaya between 1935 and 1991 [*Cenderelli*, 1998]. Space and airborne remote sensing imagery and GIS platforms facilitate analysis of changes in the size and spatial distribution of glacial lakes and hazard evaluation [*Haeberli et al.*, 2004], although monitoring and modeling are complicated by the potential for different drainage pathways and mechanisms [*Huss et al.*, 2007].

The meltwater that supplies a glacier-outburst flood can be trapped where a glacier occupies the main valley and the tributaries are free of ice, or where a recessional moraine across an alpine valley dams meltwater released as the glacier continues to retreat. Meltwater can also be trapped in the angle between two confluent glaciers; where small lakes develop on the collapsing margin of a retreating glacier; or where water pockets develop within or beneath a glacier [*Gerrard*, 1990]. When the pressure of accumulated water becomes sufficient, the water is released rapidly through or under the ice dam. This is most likely to occur in late summer and autumn.

The Icelandic term *jökulhlaup* (literally translated as 'glacier leap') is used to describe the sudden and rapid release of water impounded by, within or under, or on the surface of, a glacier [*Sturm et al.*, 1987]. This water can drain through subglacial tunnels, by overtopping the dam at the contact of the valley margin and glacial ice, or by rupture of the ice dam [*Costa and Schuster*, 1988; *Driedger and Fountain*, 1989; *Walder and Costa*, 1996], although subglacial drainage is most common [*Walder and Costa*, 1996]. Once the ponded water begins to drain, energy dissipated by the flowing water melts the ice and enlarges the subglacial channels until the water is drained [*Clarke*, 1982]. The subglacial channels are then blocked by roof collapse or plastic flow of the ice [*Björnsson*, 1992], allowing the water to accumulate once more. This cycle repeats on an annual or irregular basis as a function of glacier dynamics [*Evans and Clague*, 1994; *Cenderelli*, 2000].

Roberts et al. [2000, 2001] describe the complex processes that can occur within the glacier during a jökulhlaup. Rapid increases in jökulhlaup discharge generate basal hydraulic pressures in excess of ice overburden so that flood waters can be forced through the surface of the glacier, leading to the development of a range of supraglacial outlets [*Roberts et al.*, 2000]. Flood water flowing subglacially or through upglacier-dipping fractures is supercooled as it is raised to the surface faster than its pressure-melting point can increase as glaciostatic pressure decreases. Sediment entrained subglacially and carried by this flood water is deposited englacially within the fractures, and the presence of mixed fines and boulder-sized clasts suggests that englacial fracture discharge has a high transport capacity [*Roberts et al.*, 2001].

Volcanic and meteorologic events can trigger jökulhlaups by causing rapid increases in the volume of water ponded by the ice, or outbursts can result from continually changing meltwater levels within a glacier in the absence of discrete external triggers [*Tweed and Russell*, 1999; *Russell et al.*, 2000]. As glaciers recede, ice dams grow thinner and progressively less water is required to exceed the threshold for creating jökulhlaups, which thus become more frequent but smaller in magnitude [*Tweed and Russell*, 1999]. The warmer summers and shorter, warmer winters likely to occur under global warming thus may increase the incidence of jökulhlaups as glacial ice melts and thins more rapidly [*Vuichard and Zimmermann*, 1987]. The magnitude and frequency of jökulhlaups can be predicted numerically with reasonable accuracy for sites with data available on flood history or on reservoir volume and the characteristics of the ice dam and flood-water route [*Maizels and Russell*, 1992].

Ice-dammed lakes are present in every glaciated region of the world [*Tweed and Russell*, 1999] and are likely to have been even more common during the last stages of the Pleistocene [*Baker and Bunker*, 1985]. Because jökulhlaup discharges are typically much larger than rainfall- and snowmelt-generated floods occurring in a given drainage basin, jökulhlaups can perform substantial geomorphic work [*Post and Mayo*, 1971; *Hewitt*, 1982; *Haeberli*, 1983]. A 1984 jökulhlaup from Ape Lake in British Columbia, Canada widened the alluvial reaches of the upper Noeick River (562 km^2 drainage area) from a pre-flood average of 75 m to almost 200 m and caused up to 0.75 m of aggradation along the floodplain [*Desloges and Church*, 1992]. At least nine large, well-documented jökulhlaups occurred along mountain channels in western Canada from 1850 to 1997 [*Clague and Evans*, 1997].

Most studies focus on the geomorphic effects of jökulhlaups on sparsely forested, high-mountain valleys [*Carling and Glaister*, 1987; *Konrad*, 1998; *Shroder et al.*, 1998; *Cenderelli and Wohl*, 2001, 2003; *O'Connor et al.*, 2001] or on Icelandic-type, unforested glacial outwash plains [*Maizels*, 1991, 1997; *Fay*, 2002; *Magilligan et al.*, 2002; *Carrivick et al.*, 2004]. In the latter setting, jökulhlaups with a rapid rising limb result in high rates of sediment entrainment and deposition, but do not produce well-defined bedforms, whereas jökulhlaups with more prolonged rising and falling limbs allow time for large-scale bedforms to develop, as well as processes such as winnowing and bed armoring to occur [*Rushmer et al.*, 2002; *Rushmer*, 2007]. Jökulhlaups and other types of outburst floods can also occur in forested mountainous environments [*Clague and Evans*, 1997; *Kershaw et al.*, 2005], where wood entrained by the flood can mediate erosional and depositional effects [*Oswald and Wohl*, 2008] (Figure 3.9).

The most common pattern described for the geomorphic effects of outburst floods in mountainous environments is that of erosion in narrower, steeper valley segments, deposition in wider, lower gradient sections of the valley, and downstream decreases in the magnitude and extent of geomorphic effects as the outburst flood attenuates and approaches the magnitude of more commonly occurring peak flows [*Cenderelli and Wohl*, 2003]. Sediment transport rates can be substantially increasing during the years immediately following an outburst flood as sediment deposited along the channel during the waning stages of the flood is re-mobilized by subsequent lower flows and as channel banks eroded during the flood continue to supply sediment after the flood [*Bathurst et al.*, 1990; *Bathurst and Ashiq*, 1998].

Water ponded behind a moraine dam typically is catastrophically released only once, in contrast to jökulhlaup floods. Numerous moraine-dammed lakes are present in glaciated mountainous regions because of the retreat of glaciers since the Little Ice Age circa 1500 to 1850 AD [*O'Connor and Costa*, 1993; *Cenderelli*, 2000]. These moraine dams are prone to failure because their steep, unvegetated slopes are unstable; the dams consist of poorly sorted sediment that may be uncompacted and noncohesive; and the moraine may have only a thin veneer of sediment over a melting ice core [*Costa and Schuster*, 1988]. The dam can overtop and breach, the ice core can melt and collapse, or the dam can be weakened by seepage and piping erosion [*Yesenov and Degovets*, 1979; *Fushimi et al.*, 1985; *Costa and Schuster*, 1988; *Kershaw et al.*, 2005]. As with jökulhlaups, catastrophic drainage of the ponded water produces a rapid rise in discharge and a flood peak that commonly exceeds those associated with other types of floods in the drainage basin. This in turn can cause substantial geomorphic change along downstream channel segments [*Blown and Church*, 1985; *Vuichard and Zimmerman*, 1987; *Cenderelli and Wohl*, 2001, 2003].

Outburst floods in mountainous regions commonly also occur from the failure of landslide dams blocking the main valley drainage. If these dams fail, it is usually within one year of formation [*Costa and Schuster*, 1988]. Although seepage may be present, most dams fail from overtopping of ponded water [*Costa and Schuster*, 1988].

The hydrograph of the resulting flood depends on the erosion rate of the breach and the water volume released [*Walder and O'Connor*, 1997] and is typically steep and geomorphically effective. Outburst floods from landslide dams have been reported from mountains in many regions, including western Canada [*Evans*, 1986], south-central China [*Tianche et al.*, 1986], southwestern China [*Dai et al.*, 2005], the Karakoram Himalaya [*Code and Sirhindi*, 1986], the western USA [*Gallino and Pierson*, 1985; *Costa and Schuster*, 1988], and the Andes [*Snow*, 1964]. *Waythomas* [2001] documents the complicated interactions within a volcanically active region of Alaska, USA in which volcanic debris dams repeatedly formed and failed during the Holocene, in one case impounding water that destabilized an upstream glacier-dammed lake. Explosive volcanic eruptions can create spatially complex disturbances that result in destruction of vegetation and hillslope mass movements and associated changes in water and sediment yields to channels [*Major and Mark*, 2006].

3.1.5.2. Geomorphic effects of floods. The erosional and depositional features produced by floods along mountain rivers vary widely as a function of flood hydraulics, channel boundary resistance, and sediment supply. Erosional features in cohesive substrates include potholes, longitudinal grooves, pool scour, inner channels, knickpoints, plucking of jointed rock, and flute marks [*Baker*, 1973, 1988a; *Baker and Pickup*, 1987; *Miller and Parkinson*, 1993; *Tinkler*, 1993; *Wohl*, 1993, 1998]. Erosional features formed in unconsolidated materials take the form of longitudinal grooves, channel widening and incision, stripped floodplains, anastomosing erosion channels, cutoff chutes, and erosion of impinging tributary fans [*Stewart and La-Marche*, 1968; *Baker*, 1988a; *Miller and Parkinson*, 1993]. Changes during a single flood can be dramatic, as when the active bed width of the armored Reuss River, Switzerland more than doubled (35 to 75 m) during an August 1987 flood with an estimated recurrence interval between 250 and 700 years [*Naef and Bezzola*, 1990]. Another example comes from the 1996 failure of an artificial dam on the Ha!Ha! River in Quebec, Canada [*Lapointe et al.*, 1998]. The resulting peak discharge of 1100 m³/s was eight times the 100-year flood. More than 9 million m³ of sediment were eroded from the river valley and portions of the channel incised up to 16 m and widened by up to 250 m. Flood-induced erosion can also expose the stratigraphy of older fluvial and mass-movement deposits, facilitating interpretation of longer-term valley evolution [*Eaton et al.*, 2003a].

Depositional features associated with floods along mountain rivers include channel gravel bars, gravel splays, gravel and sand sheets, wake deposits, slackwater deposits (Figure 3.10), terrace-like boulder berms, in-channel aggradation, extensive overbank gravels, and a change from a single channel to braiding or bifurcating [*Scott and Gravlee*, 1968; *Stewart and LaMarche*, 1968; *Baker and Kochel*, 1988; *Miller and Parkinson*, 1993; *Hasegawa and Mizugaki*, 1994; *Warburton*, 1994; *Cenderelli and Cluer*, 1998; *Zielinski*, 2003; *Ortega and Heydt*, 2009]. These features enhance habitat diversity for riparian vegetation and the ages of woody vegetation record flood history at a site [*Meyer*, 2001].

Floods with 50-100 year recurrence intervals in the Italian Alps are invariably characterized by high sediment transport and extreme channel aggradation [*DiSilvio*, 1994]. This aggradation decreases downstream; a July 1987 flood on the Adda River produced up to 4.5 m of bed aggradation in the upper reaches ($S = 1.2\%$, $w = 30$ m) and 1.5 m of aggradation in the lower reaches ($S = 0.25\%$, $w = 90$ m). In-channel flood aggradation is most serious in channels with bed slopes between 0.2% and 2% [*DiSilvio*, 1994]. Use of remote sensing imagery can supplement field-based studies of the extent and nature of flood erosional and depositional features [*Dhakal et al.*, 2002].

In many of the world's mountainous regions, high-magnitude low-frequency floods may be the dominant geomorphic events because only these floods generate the large hydraulic driving forces necessary to alter the resistant channel boundaries of mountain rivers. *Wolman and Miller* [1960] define *dominant discharge* as the flow which transports the most sediment. Using suspended-load gaging station records from rivers throughout the USA (A 26-358,500 km²), they interpret the dominant discharge to be a relatively frequent (approximately 1-2 year recurrence interval) flow of moderate magnitude for most streams. *Wolman and Gerson* [1978] expand the concept of geomorphic work and dominant discharge to *geomorphic effectiveness*, defined as the modification of channel morphology. They note that a flood can significantly alter channel and floodplain morphology without transporting extremely large quantities of suspended sediment and they emphasize that effectiveness also depends on the rate of recovery of channel morphology to the form that prevails between successive floods, versus flood recurrence interval.

Subsequent studies in a wide variety of geomorphic settings delineate the conditions under which either extreme or relatively frequent floods dominate channel morphology. *Kochel* [1988] summarizes the drainage basin factors (basin morphometry, climate, lithology) and channel factors (sediment load, channel boundary material) that control channel and floodplain response to large-magnitude floods. Many mountain channels have the characteristics of high relief, thin soils, sparse vegetation, hillslope failures, coarse bedload, high channel gradient, flashy hydrograph, and resistant channel boundaries that lead to major geomorphic response to large floods [*Kochel*, 1988]. (*Batalla and Vericat* [2009] propose an index of flashiness based on the rate of discharge increase per unit time, $\Delta Q/t$ in m³/s/h.) Support for the idea that extreme floods are geomorphically more important in highland than in lowland channels comes from several studies comparing highland and lowland channel response to floods in the Darjeeling Himalaya [*Froehlich and Starkel*, 1987], southern New England, USA [*Patton*, 1988b], the humid tropics [*Gupta*, 1988], central Appalachian Mountains, USA [*Miller*, 1990a; *Eaton et al.*, 2003b], and western Oregon, USA [*Grant and Swanson*, 1995]. In a review of the effects of floods in resistant-boundary channels, *Wohl* [2008b] concludes that the aggregate population of large floods creates the greatest geomorphic effects in headwater channels, whereas individual large floods are likely to be most geomorphically important in the middle portion of drainage basins (Figure 3.11).

De Jong's [1994] study of flood features in the Schmiedlaine catchment of the German Alps illustrates the controls on the localized geomorphic effectiveness of floods. Extreme floods (150 year recurrence interval) along the Schmiedlaine reorganize the channel and deposit the largest scale features, such as megaclusters, step-pool sequences, and log jams. The persistence of these features is controlled by position along the channel. Those at the apex of a bend, where stream power is reduced, are likely to remain unaltered, whereas those at the entrance or exit to a bend are more likely to be re-worked by normal flood flows [*de Jong*, 1994].

What becomes clear when reviewing these studies is that the geomorphic influence of a given flood on a mountain channel will be largely a function of the balance between flood driving forces and channel boundary resistance [*Baker*, 1977, 1988a; *Pitlick*, 1988; *Miller*, 1990b]. This balance can vary rapidly with time and along the channel because of variations in (i) flood hydraulics resulting from the flood hydrograph at a site, downstream changes in flood discharge, or channel morphology in terms of width/depth ratio and gradient [*Miller*, 1994; *Procter et al.*, 2010] and (ii) channel-boundary resistance as a function of bedrock characteristics and exposure, and sediment characteristics, storage, and supply [*Scott and Gravlee*, 1968; *Froehlich and Starkel*, 1987; *Harvey et al.*, 1987; *Warburton*, 1994; *Miller*, 1995]. *Bull* [1979, 1988] emphasizes the importance of thresholds between driving and resisting forces in controlling channel aggradation versus degradation at a site.

An example of spatial variations in the geomorphic effects of an extreme flood comes from a study of *glacier-lake outburst floods (GLOFs)* along mountain rivers in the Khumbu Himal region of Nepal [*Cenderelli*, 1998; *Cenderelli and Wohl*, 1998; *Wohl et al.*, 2001; *Cenderelli and Wohl*, 2001, 2003]. Wider, lower gradient channel reaches (w = 150-200 m, S = 0.046) had abundant deposition during a 1985 GLOF, whereas narrow, steeper channel reaches (w = 15-60 m, S = 0.074) with channel boundaries formed in glaciofluvial outwash had extensive erosion. Narrow, steep channel reaches formed in bedrock had minimal channel change during the flood because of the much greater channel-boundary resistance (Figure 3.12). Similar patterns of channel change on steep rivers during large floods have been described from the Colorado Front Range, USA [*Shroba et al.*, 1979; *Jarrett and Costa*, 1986], the southern Appalachian Mountains, USA [*Miller and Parkinson*, 1993], the Klamath Mountains of northern California, USA [*Stewart and LaMarche*, 1968], the Alps of southern Germany [*Schmidt*, 1994], and volcanic Mount Ruapehu in New Zealand [*Procter et al.*, 2010].

Several authors propose thresholds quantified via stream power per unit area for substantial channel modification during floods. *Magilligan* [1992] uses data from five floods on low gradient, alluvial channels to identify a minimum threshold of 300 W/m^2 for substantial erosional modification. Working on a steeper, confined channel, *Lapointe et al.* [1998] also identify a 300 W/m^2 threshold for major scouring of the alluvial valley bottom. Summarizing flood data from six canyon rivers, *Wohl et al.* [2001] propose a lower threshold value for substantial flood modification (erosion or deposition) of channel boundaries in the form

$$\omega = 21A^{0.36} \qquad (3.3)$$

where ω is stream power per unit area (W/m^2) and A is drainage area (km^2). Other investigators identify distinct thresholds associated with different types of erosion (e.g., boulder transport and bedrock erosion) at specific sites [*Costa and O'Connor*, 1995; *Kale*, 2005].

The specific channel morphology produced during flooding will also depend on antecedent conditions and on the magnitude of the flood relative to earlier floods [*Eaton and Lapointe*, 2001]. *Thompson* [1987] describes an upland channel in the United Kingdom that had alternated during the preceding 150 years between meandering and braided as a function of the magnitude and frequency of flooding. A series of moderate floods produced gradual change as processes of bar dissection and channel division led to braiding. Periods of more frequent and severe flooding produced a meandering planform. *Harvey* [1987] categorizes upland channels in northwestern England as stable, sinuous channels in areas of low sediment supply, and unstable, braided channels in catchments of high sediment input. In contrast to the example from *Thompson* [1987], a 1982 storm with a return period in excess of 100 years produced massive erosion and sediment input that caused two of the stable channels to become braided [*Harvey*, 1987]. Returning to the 1985 GLOF in the Khumbu Himal, the geomorphic effects of the flood decreased dramatically downstream. This resulted from a combination of (i) peak discharge attenuation with distance from the damburst source (from approximately 2300 m^3/s at 7 km downstream, to 1400 m^3/s at 27 km downstream), (ii) a decreasing ratio of outburst-flood peak discharge to seasonal snowmelt peak discharge (ratio of 60 at 7 km, to 6 at 7 km) (Figure 3.13), and (iii) the passage of a comparably sized GLOF down the lower channel reaches in 1977 [*Cenderelli*, 1998; *Cenderelli and Wohl*, 2001]. The immense outburst floods of meltwater from the continental ice sheets and large icecaps that occurred at the end of the Pleistocene created large, extensive and persistent geomorphic features that continue to influence valley and channel morphology in many high-latitude and high-altitude regions [*Baker et al.*, 1993; *Clayton and Knox*, 2008; *O'Connor and Burns*, 2009; *Wiedmer et al.*, 2010].

As land use within mountain catchments and urbanization along mountain rivers increase, human-induced changes in land cover and channel geometry also increasingly influence flood erosional and depositional processes and forms. Structures such as check dams that impede sediment movement typically create enhanced erosion downstream [*Hooke and Mant*, 2000]. Changes in land cover that increase runoff can enhance flood peaks and channel erosion [*Hicks et al.*, 2005], whereas those that increase sediment yield and trigger channel aggradation can decrease flood conveyance and increase flooding without an increase in peak discharges [*Stover and Montgomery*, 2001].

In addition to the long tradition of field-based characterization of flood erosional and depositional features, and inference of flood processes, various types of flood simulations are now employed, ranging from the hydraulic and sediment transport models mentioned in section 3.1.3 to cellular automata developed specifically for floods in dryland environments [*Bunch et al.*, 2004].

In summary, floods along mountain rivers can result from rainfall, rain-on-snow, snowmelt, or the failure of natural or artificial jams, and can be exacerbated by ice jams. Outburst floods are particularly characteristic of mountainous regions and can produce peak discharges more than order of magnitude greater than precipitation-induced flooding. Sediment introduced to rivers via debris flows and landslides from adjacent slopes can strongly influence patterns of erosion and deposition during floods. These patterns typically include erosion in narrower, steeper valley segments, deposition in wider, lower gradient sections of the valley and, for outburst floods, downstream decreases in the magnitude and extent of geomorphic effects as the flood attenuates and approaches the magnitude of more commonly occurring peak flow. Many mountain channels have the characteristics of high relief, thin soils, sparse vegetation, hillslope failures, coarse bedload, high channel gradient, flashy hydrograph, and resistant channel boundaries that lead to major geomorphic response to large floods.

3.2. The Hyporheic Zone

The hyporheic zone is the portion of unconfined, near-stream aquifers where stream water is present. Biologists define this zone as the subsurface inhabited by stream macroinvertebrates, hydrologists define the zone as a flow-through subsurface region containing flowpaths that originate and terminate at the stream, and geochemists define it as a mixing zone between surface water and deep-sourced groundwater; *Gooseff* [2010] proposes that the hyporheic zone be defined based on the timescale of flow, analogous to definitions of floodplains based on inundation frequency, such that there are 2-hour or 24-hour hyporheic zones.Exchanges between groundwater and stream water associated with this zone may substantially impact stream water chemistry, temperature, and the availability of nutrients. Downwelling flows transfer solutes and surface water that is rich in dissolved oxygen advectively into the sediment, for example, altering subsurface habitat for organisms, whereas upwelling flow can transfer nutrients and algal cells to streams [*Tonina and Buffington*, 2009]. Along rivers with broad, gravel floodplains, the hyporheic zone can extend up to 2 km laterally from the active channel [*Stanford and Ward*, 1988]. Along mountain channels more closely constrained by bedrock, the hyporheic zone can extend less than 30 m into the floodplain [*Wondzell and Swanson*, 1999] or even less than 1 m [*Wroblicky et al.*, 1998].

Water exchange between the stream channel and adjacent aquifer is governed by vertical and horizontal hydraulic gradients, although pressure and velocity fluctuations in surface flow decrease exponentially in the subsurface, mainly within the first two layers of grains [*Detert et al.*, 2008]. Numerical simulations suggest that pore-water velocity is faster when the shear velocity on the bed surface and/or the permeability of the surface increase, although the penetration depth of pore-water velocity fluctuations into the bed decreases when the shear velocity is larger [*Higashino et al.*, 2009, 2010].

The upwelling or downwelling hyporheic flux, e, can be expressed as

$$e = -KA\frac{d^2h}{dl^2} - K\frac{dA}{dl}\frac{dh}{dl} - A\frac{dK}{dl}\frac{dh}{dl} \qquad (3.4)$$

where K is the sediment hydraulic conductivity, A is the cross-sectional area of a sediment volume within the riverbed, and dh/dl is the spatial gradient of the energy head h, which is total streambed pressure, expressed in meters of water and defined as

$$h = z + h_p + C\frac{U^2}{2g} \qquad (3.5)$$

where z is the elevation head (changes in bed elevation), h_p is the static pressure head (changes in flow depth), C $(U^2/2g)$ is the dynamic pressure head (changes in flow velocity and momentum), C is a generic loss representing changes in momentum resulting from form drag or channel contraction/expansion, and U is mean flow velocity [*Tonina and Buffington*, 2009]. The turnover length, L_m, or length scale for complete mixing between surface and hyporheic waters, depends on river discharge (Q), wetted channel perimeter (P), and the rate of exchange (q_h, the average downwelling flux per unit streambed area) [*Buffington and Tonina*, 2009]

$$L_m = \frac{Q_r}{q_h P} \qquad (3.6)$$

Local sites of upwelling and downwelling are determined by geomorphic features such as streambed topography [*Harvey and Bencala*, 1993; *Kasahara and Wondzell*, 2003; *Cardenas et al.*, 2004; *Gooseff et al.*, 2006; *Marzadri et al.*, 2010; *Reidenbach et al.*, 2010] and mobile bedforms [*Packman and Brooks*, 2001], structures including individual pieces of instream wood [*Wondzell et al.*, 2009] and logjams [*Lautz et al.*, 2006; *Hester and Doyle*, 2008], meanders [*Wroblicky et al.*, 1998], and associated variation in channel-bed grain-size distribution, permeability heterogeneity, and depth to bedrock [*Salehin et al.*, 2004; *Sawyer and Cardenas*, 2009] (Figure 3.14). Bedform-induced advection is modulated by discharge and the degree of topographic submergence [*Tonina and Buffington*, 2007]. Recharge to the hyporheic zone occurs where stream water slope increases, such as at the transition from pools to steeper channel segments [*Harvey and Bencala*, 1993], or where permeability or bed depth increase downstream [*Anderson et al.*, 2005]. Return occurs where stream water slope decreases, such as at the transition from steeper channel segments to pools [*Harvey and Bencala*, 1993], or where permeability or bed depth decrease downstream [*Anderson et al.*, 2005]. The complexity of flow exchanges can result in channel segments that concurrently lose and gain water [*Payn et al.*, 2009]. Hyporheic flow paths recharged by stream water are generally short (1-10 m) [*Harvey and Bencala*, 1993], although the spacing between upwelling and downwelling is closely related to the spacing of channel units such as pools and riffles [*Marion et al.*, 2002; *Gooseff et al.*, 2006], so that average hyporheic flow path lengths increase from headwater to mid-order mountain streams [*Anderson et al.*, 2005]. As might be expected, stratified beds favor development of horizontal flow paths within the bed and change the rate of solute transfer across the stream-subsurface interface relative to homogeneous beds [*Marion et al.*, 2008a; *Sawyer and Cardenas*, 2009].

The characteristics of the hyporheic zone can also vary as a function of channel type. A gorge stream flowing in a single straight channel along a deep valley may have high transport capacity, unstable bed sediments, and limited alluvial fill, so that lateral and vertical exchange processes are of minor importance [*Brunke and Gonser*, 1997]. A braided channel with rapid lateral migration across coarse alluvium may maximize lateral and vertical exchange, whereas the finer sediment loads of meandering channels may reduce such exchange [*Brunke and Gonser*, 1997].

Buffington and Tonina [2009] propose that mechanisms driving hyporheic exchange vary systematically with different channel morphologies and associated fluvial processes in mountain drainage basins. Variation in alluvial volume is likely more important in unconfined braided or pool-riffle channels, for example, than in confined cascade and step-pool channels. Rate and spatial extent of hyporheic exchange also differ by channel type (Figure 3.15).

Large-scale hyporheic exchange processes can be controlled mainly by catchment geology [*Brunke and Gonser*, 1997]. Comparing headwater streams in New Mexico, USA, for example, *Morrice et al.* [1997] find that a sandstone-siltstone catchment with a fine-grained alluvium of low hydraulic conductivity has much more limited groundwater-surface water exchange than volcanic catchments with alluvium of intermediate grain size and hydraulic conductivity, or a granite/gneiss catchment of coarse, poorly sorted alluvium with high hydraulic conductivity.

The characteristics of the hyporheic zone tend to vary widely in space and time. Temporal variations may be a function of discharge. The size of the hyporheic zone lateral to two mountain streams in New Mexico, USA decreases by approximately 50% during high flows [*Wroblicky et al.*, 1998]. Although the overall pattern of hyporheic flow changes little over the course of an average year along fourth- and fifth-order mountain streams in Oregon, USA [*Wondzell and Swanson*, 1996a], floods with recurrence intervals of 50 to 100 years dramatically change the hyporheic zones [*Wondzell and Swanson*, 1999]. In some unconstrained reaches, channel incision and lowering of the water table during these floods decreases the extent of the hyporheic zone. In other reaches, the extent of the hyporheic zone increases where channel incision steepens head gradients, or shifts in location where lateral channel jumps alter exchange flow paths. In constrained reaches with less depth and area of sediment available to be reworked by the flood, less change occurs in the hyporheic zone [*Wondzell and Swanson*, 1999].

In cold climates, seasonal changes associated with ice cover can affect hyporheic exchange. *Cardenas and Gooseff* [2008] find that, for equivalent depth and discharge, covered channels have shallower hyporheic zones and larger fluxes than uncovered channels. Hyporheic exchange decreases as cover roughness increases, perhaps partly as a result of decreasing bed roughness [*Cardenas and Gooseff*, 2008]. Snowmelt runoff can cause a daily pumping of the hyporheic zone [*Loheide and Lundquist*, 2009].

Although conceptual models of surface-hyporheic flow paths and exchanges go back to *Vaux* [1968], quantification of these exchanges required robust, automated streambed instrumentation to map spatial and temporal flow patterns [*Constantz*, 2008]. Water temperature is frequently used as an indicator of water sources and exchanges because it requires no active addition of a tracer to the flow and is relatively easy and inexpensive to monitor [*Stonestrom and Constantz*, 2003; *Seydell et al.*, 2008; *Lautz*, 2010]. The key to using temperature as an indicator is to measure instantaneous temperature differences between instream and hyporheic water [*Arrigoni et al.*, 2008], although temperatures measured inside a pipe buried in streambed sediment can lag diurnal temperature signals originating in the river by up to 1.5 hours [*Cardenas*, 2010]. Figure 3.16 illustrates the conceptual basis for using temperature as an indicator of surface-subsurface exchange in a stream. Near-surface electrical resistivity imaging coupled with an electrically conductive stream tracer (dissolved NaCl) can also provide *in situ* imaging of spatial and temporal dynamics of hyporheic exchange [*Ward et al.*, 2010a,b].

Temperature and introduced tracers such as sodium chloride are used to quantify hyporheic exchange in field and flume settings [*Elliott and Brooks*, 1997; *Thibodeaux and Boyle*, 1987] and to calibrate models of hyporheic exchange. Models vary from one- to three-dimensional [*Storey et al.*, 2003; *Cardenas et al.*, 2004; *Gooseff et al.*, 2006]. Some investigators use an approach based in groundwater flow or hydraulic heads, but a more popular approach is to use a modified form of the stream solute advection dispersion model such as transient storage models by *Bencala and Walters* [1983], *Jackman et al.* [1984], and *Hart* [1995]. These models include the stream transport mechanisms of advection and dispersion, as well as lateral inflow and outflow, and transient storage [*Runkel et al.*, 2003]. The net mass transfer from the stream to the retention domains is assumed to be proportional to the difference of concentration between the stream water and a storage zone of constant cross-sectional area [*Marion et al.*, 2008b]. The residence time distribution of water and solute in storage is assumed to be exponential, which limits the models to a relatively narrow range of short timescales [*Gooseff et al.*, 2003]. Subsequent models such as STAMMT-L (solute transport and multirate mass transfer-linear coordinates; *Haggerty and Reeves*, 2002) attempt to overcome the strong simplifications inherent in transient storage models. STAMMT-L uses an advection-dispersion mass transfer equation with an additional source/sink term that represents mass exchanges with the storage zones through a convolution integral of the instream solute concentration and a residence time distribution [*Marion et al.*, 2008b]. Comparative tests indicate that the transient storage model OTIS (one-dimensional transport with inflow and storage [*Runkel and Chapra*, 1993]) simulates concentrations well over short time scales (circa 30 minutes), whereas STAMMT-L more accurately simulates late-time stream concentrations of tracers over circa 24 hours [*Gooseff et al.*, 2003]. The STIR (solute transport in rivers) model uses a stochastic approach to derive a relation between instream solute concentration and the resident time distributions in different retention domains (surface dead zones, hyporheic exchange) [*Marion et al.*, 2008b].

One of the more difficult tasks of modeling and measurement is to separate surface storage in eddies and pools from retention associated with hyporheic exchange [*Runkel et al.*, 2003; *Gooseff et al.*, 2005]. *Phanikumar et al.* [2007] use acoustic Doppler current profiler data and wavelet decomposition to separate stream flow into regions of slow and fast velocity and to estimate the relative sizes of main channel and storage zones. Transport modeling based on tracer data provides estimates of the sizes of surface and subsurface storage zones. Coupling these estimates with the velocity data facilitates assessment of the relative importance of surface and subsurface storage in different stream reaches. As might be expected, a straight reach with limited alluvium is dominated by surface storage, whereas surface storage is only a fraction of total storage along a gravel-bed meandering reach with wide floodplains [*Phanikumar et al.*, 2007]. *Briggs et al.* [2009] use tracer breakthrough curves from the main channel, cross-sectional stream velocity distributions, and stream tracer concentration time series data from several locations in the main channel and adjacent surface transient storage zones to inform a transient storage model with both surface and subsurface storage zones.

Hyporheic flow appears to constitute a very small portion of stream discharge in some mountain river segments; 0.02% during winter baseflow discharge, 0.8% of discharge during summer low flow, and less than 0.1% of storm discharge along 14 km^2 McRae Creek in western Oregon, USA [*Wondzell and Swanson*, 1996a]. However, *Payn et al.* [2009] document mass loss of surface discharge of over 10% along a mountain headwater stream in Montana, USA via deeper, slower flowpaths. Along larger, lowland alluvial rivers with deeper and broader valley fill, such as the Willamette River of Oregon, hyporheic discharge may constitute 15% of surface discharge [*Laenen and Risley*, 1997]. Hyporheic flow is likely to be important along lower-gradient alluvial segments of mountain rivers such as those present in alpine meadows or glacial troughs [*Loheide and Lundquist*, 2009].

Engineering modifications which involve straightening channels, eliminating secondary channels, reducing channel-floodplain connectivity, removing instream wood, altering flow regime, or changing channel grain-size distribution may have substantial impacts on hyporheic zones [*Brunke and Gonser*, 1997; *Wondzell and Swanson*, 1999; *Wondzell et al.*, 2009]. Conversely, stream restoration can utilize steps, pools, log dams and other instream geomorphic structures to enhance hyporheic exchange and nutrient processing [*Doll et al.*, 2003; *Lautz and Fanelli*, 2008; *Knust and Warwick*, 2009]. Structures that span the channel are more effective at driving hyporheic flow than structures that partially span the channel, and background ground-water discharge and sediment hydraulic conductivity also exert an important control on hyporheic exchange [*Hester and Doyle*, 2008]. The success of such structures at enhancing localized hyporheic exchange can be assessed with a technique such as that described by *Carling et al.* [2006], who use activated carbon granules to capture the tracer Rhodamine WT dye as a reflection of the spatial variation in interstitial flow speeds in experimental fine gravel beds.

In summary, although the proportion of total discharge involved in hyporheic exchange may be quite small in mountain rivers, it may exert an important control on river chemistry and aquatic ecology. Bed topography, variations in bed grain size and depth to bedrock, and structures such as logjams govern hyporheic exchange. Rapid advances in modeling hyporheic exchange can be used to predict the effects of channel engineering on this exchange.

3.3. River Chemistry

The first comprehensive study of surface water chemical composition in the United States was conducted during the first decade of the 20th century. Approximately 155 sites were sampled initially, but the program was then greatly curtailed until the 1950s [*Hem et al.*, 1990]. The U.S. presently maintains hundreds of stream gaging stations where water quality measurements are made, via programs such as the National Streamflow Quality Accounting Network (NASQAN) and the 59 study sites of the National Water Quality Assessment (NAWQA) program. Analogous programs elsewhere include the European Union's Water Information System for Europe (WISE) and Water Framework Directive.

The chemical contents of precipitation falling over a landmass vary with distance from the ocean, with pollution inputs, and through time. The precipitation then reacts with plants, soil, regolith, and bedrock, so that the chemistry of water entering a river depends more on the hillslope flow paths followed by the water than on the chemistry of the original precipitation. The mixing of at least three sources of water (groundwater, soil water, and overland flow) produces the chemistry of most rivers [*McDonnell et al.*, 1991]. Relatively little natural change occurs in the water chemistry once the water enters a river because of the short residence time of water within the channel [*Drever*, 1988]. Any chemical change that does occur is usually associated with biological processes and trace elements.

Alpine streams draining catchments with thin regoliths and poorly developed soils may be more influenced by precipitation chemistry than are other types of rivers [*Meixner et al.*, 2000]. Studies in mountain drainages indicate that subsurface flow contribution increases as the areal extent and thickness of regolith and soils in the basin increase [*Sueker*, 1995]. An alpine catchment of bedrock and talus may be dominated by surface and shallow subsurface flow [*Finley et al.*, 1995], whereas a forested mountain catchment with well-developed soils may be dominated by groundwater discharge even during snowmelt [*Shanley et al.*, 1995]. Discharge from unchanneled valley heads, or zero-order hollows, and the mixing of water from various hillslope sources can influence river chemistry in soil-mantled, forested mountains [*Asano et al.*, 2009]. Oxygen isotopes tracers in two small catchments in the Colorado Front Range, USA indicate that subsurface flow from a talus field contributes ≥ 60% of streamflow even during the early snowmelt season in a 225-ha catchment, whereas new water dominates (82%) discharge in an 8-ha catchment lacking significant talus deposits [*Liu et al.*, 2004a]. The talus area can serve as a source of nitrogen during snowmelt, whereas nitrogen may be limited in analogous catchments without nitrogen [*Meixner et al.*, 2000]. Typical 24-hour storms displace only a fraction of the stored water in a steep headwater catchment in the Oregon Coast Range, USA, so that 'old water' dominates runoff chemistry [*Anderson and Dietrich*, 2001].

Field studies such as these provide a strong caveat to the idea that the river chemistry of hydrologically flashy mountain catchments reflects primarily precipitation chemistry, as long as soil or hillslope regolith is present. Water chemistry in mountain channels can also be strongly influenced by subsurface flow through fractured bedrock, particularly during low-flow periods [*Anderson and Dietrich*, 2001; *Tsujimura et al.*, 2001], or by flow through a well-developed riparian zone [*Hooper et al.*, 1998]. Studying a group of mountainous rivers with similar geology and climate in Argentina, *Lecomte et al.* [2009] correlate: basin size with major ions, conductivity, and pH, which tend to increase downstream; channel gradient with the total concentration of heavy metals (high concentrations occur in lower gradient areas that promote desorption); and high drainage density with low Cl⁻ concentration, which reflects maintenance of the chemical signature of atmospheric precipitation. Understanding the influence of water sources and flow paths on river chemistry remains a key issue in the hydrologic sciences [*National Research Council*, 1991]. Examples of models designed for this purpose are provided in *Soulsby et al.* [1999], *Hooper* [2003], and *Xie et al.* [2005].

The importance of precipitation input varies for different elements [*Nedeltcheva et al.*, 2006]. A study of sulfur isotopes of sulfate in the Bear Brook watershed in Maine, USA, for example, indicates that sulfur isotopes in this first-order intermittent stream are controlled by the relative contribution of marine versus non-marine sulfate in precipitation; fractionation of stable sulfur isotopes within the watershed has only a minor influence [*Stam et al.*, 1992]. However, *Aulenbach et al.* [1996] find that a relation is not evident between trends in precipitation and surface water chemistry in 15 small watersheds in the USA either for individual inorganic solutes or for solute combinations, despite the small watersheds and generally unreactive bedrock.

The primary constituents of river chemistry are the dissolved ions HCO_3^-, Ca^{2+}, SO_4^{2-}, H_4SiO_4, Cl^-, Na^+, Mg^{2+}, and K^+; dissolved nutrients N and P; dissolved organic matter; dissolved gases N_2, CO_2, and O_2; and trace metals [*Berner and Berner*, 1987; *Allan*, 1995]. The sum of the concentrations of the dissolved major ions, known as the total dissolved solids (TDS), is highly temporally and spatially variable in response to precipitation input, discharge, lithology in the drainage basin, and the growth cycles of terrestrial vegetation [*Berner and Berner*, 1987]. An average natural value for rivers is 100 mg/l (20 times the concentration in rain) with, on average, another 10 mg/l contributed by pollution. Ca^{2+} and HCO_3^- from limestone weathering tend to dominate in general [*Berner and Berner*, 1987] and can be locally important even if carbonate minerals constitute only a minor phase of the bedrock [*Anderson and Dietrich*, 2001]. Because dissolved fluxes in rivers are related to rates of continental weathering and rates of weathering are dependent on uplift, the largest dissolved fluxes today occur in rivers draining the Himalayan and Andean mountains and the Tibetan Plateau [*Raymo et al.*, 1988]. The factors influencing TDS are summarized in Table 3.1.

Gibbs [1970] classifies rivers on the basis of their chemistry as being dominated by (1) precipitation (low TDS, high Na/(Na + Ca) ratio), (2) rock weathering (intermediate TDS, low Na/(Na + Ca) ratio), or (3) evaporation and fractional crystallization of $CaCO_3$ (high TDS, high Na/(Na + Ca) ratio). Subsequent investigators emphasize the role of geology and erosional regime as the major controls on river chemistry [*Garrels and Mackenzie*, 1971; *Drever*, 1982; *Berner and Berner*, 1987]. *Stallard and Edmond* [1983], for example, describe: (1) rivers draining intensely weathered materials (cation-poor, siliceous rocks, deeply weathered soils and saprolites), with TDS less than 20 mg/l, that correspond to Gibbs' precipitation-controlled rivers, (2) rivers draining siliceous terrains of cation-rich igneous and metamorphic rocks and terrestrial shales, with TDS of 20-40 mg/l, that fall between Gibbs' precipitation- and rock-dominated rivers, (3) rivers draining marine sedimentary rocks, with TDS of 40-250 mg/l, that correspond to Gibbs' rock-dominated rivers, and (4) rivers draining evaporates, with TDS greater than 250 mg/l, that correspond to Gibbs' evaporation-crystallization rivers. In general, rivers draining sedimentary rocks tend to have at least two times the TDS of rivers draining igneous and metamorphic rocks [*Holland*, 1978]. Because crystalline rocks compose many of the world's mountain ranges, mountain rivers tend to have low concentrations of TDS.

Topographic relief may exert an important influence on river chemistry. Greater relief usually corresponds with greater erosion and faster exposure of fresh bedrock for chemical weathering [*Clow and Sueker*, 2000]. Chemical weathering in the soil may be incomplete if the rocks are resistant to weathering, so that spatial variations in lithology can significantly influence river chemistry [*Berner and Berner*, 1987]. Climate exerts a strong influence on river chemistry by governing the rate and intensity of weathering. As an example, a simple weathering law for basaltic lithologies, validated on small and large catchments, relates the consumption of atmospheric CO_2 to regional runoff and mean annual temperature [*Dupré et al.*, 2003]. *West et al.* [2005] demonstrate similar variations in silicate weathering rates in relation to physical erosion rates, rainfall, and temperature.

Glaciation may also influence river chemistry. More reactive minerals may contribute to a river's dissolved load in proportion to their abundance in the local rock, a process known as *selective weathering*. Selective weathering is more likely to occur in previously glaciated catchments than in non-glaciated catchments, as demonstrated for mountains as diverse as the Himalaya [*Reynolds et al.*, 1995a; *Gazis et al.*, 1998] and the U.S. Rocky Mountains, where selective weathering of calcite and biotite occurs in talus relative to tundra environments [*Williams and Platts-Mills*, 1998], and these minerals can be the dominant source of solutes despite constituting a very small fraction of the bedrock [*Drever and Hurcomb*, 1986; *Drever*, 1988]. Chemical denudation rates may also be higher for glaciated areas. A study of cationic denudation for a drainage basin in the Cascade Mountains of Washington, USA found that subglacial waters have a significantly higher yield (800 to 2390 meq/m^2/yr) than extra-glacial waters (580 meq/m^2/yr), although both yields are higher than the world average of 190 meq/m^2/yr [*Axtmann and Stallard*, 1995].

Studies of Swiss alpine glaciers indicate that the chemistry of glacial meltwater may be highly variable in time and space [*Brown and Fuge*, 1998]. Solute concentrations vary inversely in phase with diurnal variations in meltwater discharge, because the average contact time of meltwater with the sediment beneath the glacier varies with discharge [*Collins*, 1995a]. Also, meltwaters initially flow slowly through small channels leading from the bases of moulin shafts. Flow accelerates downstream as water volume increases where confluents join to produce larger conduits. The dissolution of suspended sediments raises solute concentrations considerably more per unit length of channel in the smaller conduits with low velocity flow than in larger conduits with faster flow [*Collins*, 1995b]. Borehole sampling on alpine glaciers indicates that subglacial waters may have distinctly different chemical compositions within a glacier as a result of chemical weathering of bedrock or glacial flour that contains different quantities of reactive trace minerals such as carbonates and sulfides [*Lamb et al.*, 1995]. Glaciers in the High Arctic may also have highly variable solute concentrations controlled by meltwater source, rate of melting, and subaerial chemical weathering and flow pathways [*Hodgkins et al.*, 1997, 1998; *Anderson et al.*, 2003].

Mountain rivers that do not have headwater glaciers also exhibit strong temporal variations in water chemistry in association with seasonal snowmelt. New snow has greater variability in oxygen isotopes than the snowpack, which in turn has greater isotopic variability than snowmelt. These changes reflect isotropic redistribution during snow metamorphism and melting [*Taylor et al.*, 2001]. A study of seven streams in high-elevation catchments of the Sierra Nevada in California identifies three phases in the chemical composition of streams during snowmelt [*Melack and Sickman*, 1995]: (1) from the onset of snowmelt to peak discharge, solute concentrations decrease; (2) at or near peak discharge, concentrations are at a minimum; and (3) with declining discharge, solute concentrations increase. Although the details vary, this pattern seems to be consistent for alpine catchments around the world [*Miller and Drever*, 1977; *Denning et al.*, 1991; *Baron*, 1992; *Baron et al.*, 1995; *Campbell et al.*, 1995a, 1995b; *Clow and Mast*, 1995; *Mast et al.*, 1995; *Stottlemyer*, 2001; *Peterson et al.*, 2005]. A study of the headwaters of the Urumqi River basin, Tian Shan, China, for example, finds that solute concentrations in streamwater are highest at the initiation of snowmelt, decline through the melt season and into the summer, and then increase as the contribution of baseflow increases [*Fengjing et al.*, 1995]. In this catchment, as in others, the *ionic pulse* (the release of solutes from the snowpack and the flushing of weathering products from the soil) and the dissolution of eolian particles are important as meltwater percolates through the snowpack [e.g., *Tranter et al.*, 1986]. In catchments with contaminants, the ionic pulse can be toxic to aquatic organisms. Despite long-term decreases in acidic deposition in the Hubbard Brook Experimental Forest of New Hampshire, USA, episodic acidification occurs during snowmelt and mobilizes aluminum to concentrations toxic to fish [*Demers et al.*, 2010]. Dissolved organic carbon mobilized from shallow soils during snowmelt in this catchment also results in the mobilization of mercury from the same sources. This represents a large portion of the annual mercury export from the catchment [*Demers et al.*, 2010]. *Lilbaek and Pomeroy* [2007] develop a model in which the cumulative load of an ion infiltrating into frozen unsaturated soil can be estimated as a function of meltwater ion concentration and infiltration rate.

Seasonal fluctuations in mountain river chemistry can reflect seasonal changes in the primary hydrologic flow path from the snowpack early in the melt season, to the shallow soil zone, and finally the bedrock aquifer [*Finley et al.*, 1995; *Miller et al.*, 1995; *Rice and Bricker*, 1995). Tropical catchments with distinct weathering environments as a function of depth below the surface can produce different solute chemistries as rainfalls of differing intensity and duration access different flow paths [*White et al.*, 1998]. Even relatively small mountain catchments can have substantially different subsurface flow paths that vary in relative importance through time, producing detectable differences in river chemistry [*Aubert et al.*, 2002; *Mul et al.*, 2007].

In addition to regular seasonal changes in stream chemistry, mountain rivers may be affected by episodic events such as timber harvest (see chapter 6), mass movements, or volcanic eruptions. Immediately following the 1980 eruption of Mount St. Helens in Washington, USA, for example, some streams northeast of the volcano showed sulfate and chloride increases [*Klein and Taylor*, 1980]. Organic compounds produced when the hot, eruptive debris buried or destroyed forests on the slope of the volcano also influenced water chemistry.

The presence of tarns or other types of lakes and wetlands in mountain basins influences water chemistry. *Robinson and Matthaei* [2007] document how lake position influences seasonal dynamics in stream temperature and nutrients in a mountain catchment in Switzerland. *Matyjasik and Keate* [2003] describe differences in aquatic chemistry among the distinct types of wetlands that occur in mountains, including peatlands, fens and marshes.

River chemistry is of course very important to aquatic organisms, as discussed by *Patrick* [1995]. Dissolved organic matter is an important energy-nutrient source for many aquatic species. Ca, Mg, oxidized S, N (as nitrates and ammonia), and phosphates, along with small amounts of Si, Mg, and Fe, are also desirable for many species. Aquatic species in rivers vary greatly in their tolerance for acidity [*Horecky et al.*, 2006] and trace metals; in some forms these trace metals stimulate growth, in other cases they are toxic. The pH of the water affects the solubility of various elements and thus their availability to aquatic organisms. Numerous studies show that the survival of juvenile salmonid fish in headwater streams, for example, is influenced by pH, the ionic forms of Al, and by Ca concentration [*Baker and Schofield*, 1982; *Brown*, 1982; *Neal et al.*, 1997]. Colloids such as iron oxyhydroxide minerals may absorb metals onto their surfaces and thus make the metals less available to organisms.

Because of the presence of features such as glaciers, rock glaciers, and permafrost, climatic warming will likely alter the aquatic chemistry of many mountain watersheds by altering subsurface flow paths and biogeochemical reactions. Working on arctic streams, *Keller et al.* [2007] find that increasing thaw depth of permafrost leads to increasing carbonate and Ca supply to soils and streams, as well as spatially variable increases in P and K supply.

In summary, the chemistry of mountain rivers is influenced by precipitation inputs and by hydrologic flowpaths to the river. Hydrologic flowpaths can include flow through fractured bedrock, soil, hillslope regolith, and riparian zones. Topographic relief and glacial history can also influence river chemistry, which can exhibit strong seasonal variations related to glacier- and snow-melt. Understanding the influence of water sources and flow paths on river chemistry remains a key issue in the hydrologic sciences.

3.3.1. Dissolved Nutrients

The primary nutrients in river water are nitrogen and phosphorus. Nitrogen gas, N_2, must be fixed, or combined with hydrogen, oxygen and carbon, in order to be used by terrestrial organisms. The three major land inputs of fixed nitrogen, in forms such as NO_3^- and NH_4^+, include biological fixation (approximately 60%), precipitation and dry deposition of previously fixed nitrogen (24%), and the application of industrially fixed nitrogen in fertilizer (16%) [Berner and Berner, 1987]. Rivers are important conduits for N transport, although they can also remove and transform dissolved N [Hall and Tank, 2003]. River output of nitrogen is about 18-20% of the total nitrogen loss from the land [Van Breemen et al., 2002]. Of this, 85% is organic nitrogen, and most of the rest (dissolved inorganic nitrogen and erosion of minerals containing nitrogen) is derived from organic matter decomposition. In general, biological recycling of nitrogen is very efficient [Ranalli and Macalady, 2010]; the total river output of organic nitrogen is only 8% of the nitrogen assimilated annually by the terrestrial biosphere [Berner and Berner, 1987]. Rates of dissolved nitrogen uptake from the water column are especially high in shallow streams in which algae and microbes in attached biofilms are present [Hall and Tank, 2003].

Riverine processing of carbon and nitrogen is heterogeneous in time and space as a result of *biogeochemical hot spots* (patches that show disproportionately high reaction rates relative to the surrounding matrix) and *hot moments* (short periods of time that exhibit disproportionately high reaction rates relative to longer intervening time periods) [McClain et al., 2003]. Riverine hot spots include anoxic zones beneath riparian environments [Lowrance et al., 1984] or the convergence of ground and surface waters in hyporheic zones [Harvey and Fuller, 1998]. Riverine hot moments occur when episodic hydrological flow paths reactivate and/or mobilize accumulated reactants, such as rare flows occupying secondary channels in an arid environment [McGinness et al., 2002] or snowmelt enhancing leaching of dissolved organic carbon in high-elevation watersheds [Boyer et al., 2000]. Maintenance of these environments and hydrologic variability thus becomes crucial to maintaining nutrient uptake and processing. Although spatial variation in nitrogen export from catchments within a relatively small region can be substantial, distributed topographic indicators designed to represent hydrologic flushing mechanisms for nitrogen export (e.g., proportion of wetlands) can be used to predict this export [*Creed and Beall*, 2009].

Denitrification in stream sediments is a significant mechanism of within-stream nitrogen loss [*Nihlgard et al.*, 1994], although the magnitude of this denitrification varies through time and space as residence time of water in sediment varies with hydraulic conductivity and groundwater discharge [*Gu et al.*, 2008a, 2008b]. Studies in the Sonoran desert of the southwestern United States, for example, indicate that the subsurface may be a source of nitrate to the nitrogen-limited streams [*Holmes et al.*, 1994]. Nitrification in downwelling regions along desert channels is several times that in upwelling regions [*Jones et al.*, 1995a], illustrating the importance of chemical linkages between the surface stream and the hyporheic zone [*Stanley and Boulton*, 1995].

Nitrogen input to streams also varies with the hydrologic pathway of water entering from adjacent slopes [*Band et al.*, 2001], including factors such as soil thickness and hillslope gradient [*Scanlon et al.*, 2010], and with land use on those slopes [*Wang et al.*, 2004]. *Welsch et al.* [2001] find that *Beven and Kirkby*'s [1979] topographic index correlates positively with nitrate concentrations in shallow subsurface stormflow in a small catchment in the Catskill Mountains of New York, USA, indicating a topographically driven flushing of high nitrate in shallow soil during storms. Headwater streams with substantial populations of anadromous fish also receive marine nitrogen via upstream migration, spawning, and death of the fish [*Bilby et al.*, 1996].

Riparian zone biogeochemical interactions may also influence nitrogen dynamics. In temperate and wet tropical areas, nitrate fluxes can be relatively constant because denitrification and plant uptake remove allochthonous nitrate within a few meters of travel along shallow riparian flowpaths [*McClain et al.*, 1999]. The effectiveness of nitrogen removal by a riparian zone in Switzerland varies seasonally, however, from up to 95% of the total nitrogen entering the riparian area in late summer to only 27-38% during winter [*Maitre et al.*, 2003]. In contrast, relatively little nitrogen processing occurs during transport from upland through riparian zones in arid areas because precipitation moves rapidly across the riparian zone as surface runoff [*McClain et al.*, 1999]. The presence of water ponded upstream from logjams or beaver dams can significantly increase nitrogen uptake by increasing sites for processing of organic matter [*Naiman and Melillo*, 1984; *Naiman et al.*, 1986; *Fanelli and Lautz*, 2008]. Recently developed hydrological-biogeochemical models that simulate nutrient exchange at the sediment-water interface of rivers suggest that riparian denitrification exceeds benthic denitrification [*Thouvenot-Korppoo et al.*, 2009].

Headwater streams are disproportionately important in riverine processing of some forms of nitrogen because they have the most rapid uptake and transformation of inorganic nitrogen [*Peterson et al.*, 2001]. During seasons of high biological activity, first- through fourth-order streams typically export downstream less than half of the input of dissolved inorganic nitrogen from their watersheds [*Peterson et al.*,

2001]. These streams can also account for the majority of total stream length within a watershed [*Seitzinger et al.*, 2002]. Spatial variations in nitrogen export from headwater catchments reflect differences in the thickness and spatial extent of the surface organic horizon as controlled by vegetation type and climate-controlled rates of organic decomposition [*Lawrence et al.*, 2000; *Sickman et al.*, 2002]. Total annual stream discharge is a positive predictor of annual dissolved organic nitrogen output, with peak concentrations occurring after the onset of seasonal rains but before the peak in the hydrograph, probably due to flushing of products of decomposition that build up during the dry summer [*Vanderbilt et al.*, 2003].

Nutrient uptake lengths in headwater streams are particularly short because of autotrophic (algae and moss) and heterotrophic (bacteria and fungi) uptake and release of nutrients [*Nihlgard et al.*, 1994]. Because the heavy shade over many forested mountain rivers limits primary production, leaves and small woody material from the surrounding environment are the major sources of energy to many forested headwater streams [*Cummings*, 1974]. Most of this leaf litter input is ingested by aquatic macroinvertebrates, which in turn excrete substantial amounts of potassium, calcium, and dissolved organic carbon to the stream water [*Nihlgard et al.*, 1994]. The complexities of surface flow and surface-subsurface exchange in mountain rivers also strongly influence nutrient uptake lengths. Wood and logjams, for example, promote storage and processing of fine and coarse particulate organic matter [*Bilby*, 1981] (Figure 3.17). Interactions between flow velocity and obstacles which promote retention of particulate organic matter exert a stronger control on nutrient uptake lengths than do biological processes [*Minshall et al.*, 1983]. The magnitude and frequency of flow pulses, which involve discharge fluctuations that drive expansion-contraction cycles in areas inundated below the bankfull level [*Tockner et al.*, 2000b], also influence nutrient uptake lengths. Interbiome comparisons among headwater streams suggest that those with greater hydraulic complexity and fine particulate retention, as well as less hydrologic variability, have shorter nutrient uptake lengths [*Mulholland et al.*, 2001; *Webster et al.*, 2003].

It is particularly important to understand nitrogen dynamics now that human activities have markedly increased nitrogen inputs to watersheds and in some cases decreased the ability of river ecosystems to process nitrogen, resulting in excess nitrogen loads and eutrophication in coastal areas [*Boyer et al.*, 2002, 2006; *Seitzinger et al.*, 2002; *Galloway et al.*, 2004]. Much of the increased nitrogen comes from inorganic fertilizer use, but emissions from fossil-fuel combustion also contributed substantially to the doubling of nitrogen inputs to the U.S. from human activity during 1961 to 1997 [*Howarth et al.*, 2002]. More traditional farming practices and poor management of animal waste can also severely contaminate local, shallow aquifers and streams in mountain environments [*Hajdu and Füleky*, 2007], as can increased nitrogen delivery from ski-resort septic systems or golf course fertilization [*Gardner and McGlynn*, 2009]. Evidence that many large river systems have limited capacity to process excess nitrogen comes from comparisons of pre- and post-industrial nitrogen riverine fluxes; pre-industrial fluxes were greatest from the largest rivers, such as the

Amazon, whereas post-industrial fluxes are greatest from rivers in the industrialized zones of North America, Europe and southern Asia [*Green et al.*, 2004].

Even seemingly pristine mountain catchment without direct nitrogen sources within the watershed can receive excess nitrogen through atmospheric transport and deposition. High-elevation watersheds in Rocky Mountain National Park, Colorado, USA, have been changed by the effects of nitrogen deposition [*Baron et al.*, 2000; *Burns*, 2004], with nitrogen coming from vehicle and industrial emissions and agricultural sources such as crop fertilizer and animal waste [*Heuer et al.*, 2000], all of which occur primarily in the lowlands beyond the mountain range (Figure 3.18). The granitic bedrock and shallow soils in mountain watersheds do not provide much chemical buffering [*Baron*, 1992; *Clow and Sueker*, 2000] and the short growing season limits the amount of nitrogen that plants can take up [*Baron et al.*, 1994]. By 2004, soils, waters, and plants in the mountain watersheds of the national park showed evidence of changes caused by nitrogen deposition [*Baron et al.*, 2000; *Fenn et al.*, 2003].

In contrast to nitrogen, most phosphorus that is lost from land moves via river runoff, although phosphorus tends to be efficiently utilized by biological systems. Although the ultimate source of phosphorus is weathering of geologic materials, the main source in many river catchments is precipitation and dry deposition [*Graham and Duce*, 1979]. Because phosphorus is relatively insoluble, it is often a limiting nutrient in biological systems [*Berner and Berner*, 1987].

In summary, nitrogen is a primary nutrient in rivers, and rivers are important conduits for N transport, although they can also remove and transform dissolved N. Denitrification in stream sediments is a significant mechanism of within-stream nitrogen loss that varies temporally and spatially. Nitrogen input to streams varies with the hydrologic pathway of water entering from adjacent slopes. Headwater streams are disproportionately important in riverine processing of some forms of nitrogen because they have the most rapid uptake and transformation of inorganic nitrogen, but excess nitrogen inputs from anthropogenic sources can overwhelm the uptake ability of headwater catchments.

3.3.2. Organic Matter and Gases

Organic matter can be present in river water both in particulate and dissolved forms [*Berner and Berner*, 1987]. Dissolved organic matter in general is composed of different classes of organic compounds with differing reactivity and ecological roles. The quantity of different compounds varies with time and space, responding to seasonal variation and individual rainfalls. *Jaffé et al.* [2008] analyze variations in dissolved organic matter composition in stream waters using UV-visible absorbance and fluorescence spectra.

Dissolved organic matter is usually expressed as dissolved organic carbon (DOC), which is typically between 2 and 15 mg/l, but can reach 60 mg/l in rivers draining wetlands. By comparison, precipitation has DOC values of 0.5-1.5 mg/l and soil water has up to 260 mg/l [*Drever*, 1988]. DOC varies with the size of the river, the climate, and vegetation [*Thurman*, 1985]. Mountain rivers in alpine regions tend to have the lowest concentrations (> 1 mg/l) [*Berner and Berner*, 1987]. About half of the DOC present in a river is fulvic acids, the fraction of humic substances soluble at all values of pH [*Drever*, 1988]. The average river ratio of DOC to TDS is low (1:18) [*Berner and Berner*, 1987].

In small, turbulent, unpolluted mountain streams, diffusion maintains O_2 and CO_2 near saturation, although concentrations change seasonally and daily with temperature [*Berner and Berner*, 1987]. Flow paths through, and residence times of water in, catchments are among the primary controls on DOC variation in headwater streams of the Colorado Rocky Mountains, USA [*Boyer et al.*, 1995]. DOC concentrations in the streams increased during the rising limb, peaked before maximum discharge, and then decreased rapidly during the period of snowmelt. Soils are the primary source of DOC to these streams, and the variations in concentration during the snowmelt season reflect both the initial flushing of carbon produced prior to snowmelt and the changes in flow depth and rapidity of meltwater as the season progresses [*Boyer et al.*, 1995]. *Köhler et al.* [2009] model the observed dynamics of total organic carbon (TOC) using runoff as a proxy for soil wetness conditions and changing flow pathways and air temperature. Using the model to simulate different scenarios for future climate in a headwater stream in the boreal zone of northern Sweden, they find that TOC concentrations increase under all scenarios.

As for nitrogen, processes in headwater mountain streams may exert a disproportionate influence on carbon transfer. High-standing oceanic islands of the southwest Pacific, for example, constitute only ~ 3% of the global landmass, but streams draining these islands contribute 17-35% of the estimated particulate organic carbon entering the world's oceans [*Lyons et al.*, 2002].

Understanding the dynamics of organic matter in streams is particularly important because DOC is an important intermediate in the global carbon cycle and is highly reactive, influencing riverine ecosystem function by controlling microbial food webs [*Kaiser et al.*, 2004]. Inland waters of various types have a significant role in carbon dynamics and, because the water cycle is extremely sensitive to climate change, waterborne carbon fluxes will respond to climate change [*Dessert et al.*, 2003; *Galy et al.*, 2008a; *Battin et al.*, 2009]. Although the efficiency with which rivers retain and oxidize organic carbon depends on the evolution of microbial communities in response to *geophysical opportunities* (riverine features that increase the residence time of organic molecules in transport) [*Battin et al.*, 2008] and substantial differences in carbon dynamics occur between rivers with well-developed floodplains and small mountainous rivers [*Galy et al.*, 2008b], most studies of fluvial transfer of terrestrial organic carbon to the oceans treat the river as a black box, rather than focusing on the specific aspects of the river that facilitate retention or transport of carbon. Carbon transfer to the river depends on factors such as landslides triggered by tropical cyclones, the proportions of fossil carbon (from bedrock) and carbon from soils and terrestrial plants, and land use [*Lyons et al.*, 2002; *Hilton et al.*, 2008a]. Carbon transport to the ocean also reflects river discharge, with greater transport during floods [*Hilton et al.*, 2008b]. *Madej*'s [2010] first-order estimate of a carbon budget for Redwood Creek, California, USA provides an exception to the statement above in that she explicitly considers the influence of stream structure on carbon dynamics by accounting not only for carbon inputs via landslides, but also instream carbon storage in terms of wood and sediment stored because of the presence of wood.

In summary, transfer of carbon to mountain rivers reflects delivery processes such as landslides and catchment sources such as bedrock geology, soils, and plants. Dissolved organic carbon in a mountain river varies in relation to climate, terrestrial vegetation, flowpaths and residence times of water, and the valley-bottom and channel features that influence retention and processing of fine-grained organic material by microbial communities. Few studies of carbon transfer explicitly consider the influence of stream structure on carbon dynamics.

3.3.3. Trace Metals and Pollutants

Trace metals generally occur at concentrations less than 1 mg/l in river waters. These can be derived from rock weathering or from human activities such as mining, burning fuels, smelting ores, or disposing of waste products [*Drever*, 1988]. Typically, trace metals are of most concern as a source of river pollution. A trace metal may be a contaminant or a hazard. A *contaminant* is any element or substance that occurs in the environment at concentrations above background levels, where background refers to the natural concentration of an element in natural materials at a given location [*Gough*, 1993]. A contaminant that occurs at a level potentially harmful to organisms constitutes a *hazard*. Natural sources of contaminants include mineralized areas and soils enriched in elements such as selenium. A contaminant may enter a river from a point source, which is a single source with a small area (e.g., a mine) or from a non-point source such as agricultural fields or a marine shale or sandstone enriched in uranium [*Gough*, 1993]. During the 1980s and 1990s, Ronald Eisler of the U.S. Geological Survey put together a series of very useful summaries of the hazards to living organisms associated with different types of contaminants [e.g., *Eisler*, 1989, 1993].

In many of the world's mountains, mine drainage of metal-rich water released during reactions between water and rocks containing sulfide minerals is typically associated with acidic water and may be a source of contaminants and hazard. There are between 100,000 and 500,000 abandoned or inactive mine sites in the United States, most of which are in the mountains of the western United States [*King*, 1995]. To date, these mines have created hazards to which the national government has responded by funding more than fifty restoration sites related to non-fuel mining activity [*King*, 1995]. In the Colorado Rocky Mountains, for example, abandoned mines in the Summitville District have seeps throughout the mine workings, and water draining through mine adits and heap-leach sites. This water eventually reaches the Alamosa River, which can have concentrations of Al, Cu, Fe, and Mo greater than 1 mg/l after rain storms [*King*, 1995]. Attempts to reduce this acid-mine drainage include grouting fractures in abandoned mines to prevent the inflow of oxygenated groundwater, and hindering the bacteria that enhance the chemical reactions between water and sulfide minerals [*Gough*, 1993]. The downstream chemical impacts of acid mine drainage can be extremely complicated as a result of seasonal dilution by snowmelt, spatial dilution or concentration via tributary inflows, and biogeochemical processes that can cause daily variations in solute concentrations of up to 40% [*Sullivan and Drever*, 2001]. Climate change may exacerbate the biological effects of acid-mine drainage as longer dry periods alternate with less frequent, more intensive precipitation in the western United States [*Nordstrom*, 2009]; under these conditions, higher average concentrations of acids and metals will occur, as well as larger sudden increases in concentrations during rising limb of discharges following dry periods. *Runkel et al.* [2007] describe a reactive stream transport model that can be used to simulate constituent concentrations during instream transport. *Pelletier et al.* [2008] describe a three-dimensional contaminant transport numerical model of dilution and mixing in channel networks.

Another potentially substantial source of pollution in many mountain rivers is acid rain. Fossil fuel combustion and the smelting of nonferrous metals produce sulfuric and nitric acid that can be dissolved in precipitation, deposited as particles, or form acid precursor gases such as SO_2 and NO_x that are absorbed directly by plants and other surfaces [*Cortecci and Longinelli*, 1970; *Drever*, 1988; *Moldan and Cerny*, 1994]. Although ion exchange along streambeds can buffer against pH change [*Norton et al.*, 2000], some rivers may become acidified in areas receiving long-term acid deposition, such that the carbonate alkalinity is zero or negative, and pH is usually less than 5. The input of acid anions may vary seasonally, with an increase at the time of snowmelt as acid anions, sulfate, and nitrate in the snow are flushed into surface waters [*Christophersen and Wright*, 1981]. The time lag between the input of acid anions from the atmosphere and acidification of the river will depend on soil thickness and cation-exchange capacity in the drainage basin, which will in turn depend on the rate of chemical weathering and the age and stability of surfaces in the basin [*Drever*, 1988; *Velbel*, 1993]. Because retention in high-elevation watersheds can be limited, atmospheric deposition can quickly alter stream chemistry [*Michel et al.*, 2000]. Lithology is an excellent predictor of the relative buffering capacity of a watershed for mid-Atlantic watersheds in the USA [*O'Brien et al.*, 1997]. Several quantitative models have been developed to relate the chemistry of runoff at a site to the input of acid anions from the atmosphere [e.g., *Henriksen*, 1980; *Chen et al.*, 1984; *Cosby et al.*, 1985; *Ball and Trudgill*, 1995; *Neal et al.*, 1995; *O'Brien et al.*, 1997; *Sullivan et al.*, 2004].

Mountain rivers in eastern North America and northern and central Europe were most affected by acid rain [*Drever*, 1988] during the 20th century, but China's rapid economic and industrial growth during the first decade of the 21st century has also caused acid rain over large areas of Asia [*Zhao et al.*, 2008]. Contemporary worldwide anthropogenic sulfur emissions into the atmosphere total between 70 and 100 million metric tonnes per year, with approximately 60 million metric tonnes of natural sulfur emissions. In northern Europe, 90% of atmospheric sulfur is of anthropogenic origin

[*Hultberg et al.*, 1994]. The U.S. Geological Survey established the 200 sites of the National Atmospheric Deposition Program for monitoring wet atmospheric deposition in the United States. The program, begun in 1978, is designed to determine whether on-going and future regulatory actions to reduce air pollution are resulting in an improvement in the quality of precipitation chemistry in the United States. Europe has analogous programs such as ADIOS (Atmospheric Deposition and Impact on the Open Mediterranean Sea), begun in 2001. Although acid rain has been reduced in regions such as the northeastern United States, many acid streams and lakes have not recovered because the cation-exchange capacity in watershed soils has been stripped [*Wathne et al.*, 1990]. The type, amount, and distribution of stored sulfur pools in the ecosystem strongly influence how quickly a catchment recovers once sulfur deposition decreases [*Armbruster et al.*, 2003].

Atmospheric deposition can also be an important source of mercury to mountain river catchments. Mercury is highly toxic to living organisms and is the leading cause of impairment in estuaries and lakes within the United States [*Brigham et al.*, 2003]. In addition to natural sources such as volcanoes and geothermal springs, mercury enters the atmosphere from combustion of coal (87% of inorganic mercury transmission to the atmosphere), waste incineration, industrial uses, and mining [*Brigham et al.*, 2003]. Wildfires can also be an important source of mercury release to the environment, although the amount of mercury emitted depends on tree species composition, which affects pre-fire mercury accumulation as well as fire severity. Wildfires and prescribed burns in the United States release between 13% and 42% of the U.S. anthropogenic mercury flux [*Biswas et al.*, 2007]. Human activities have more than doubled the amounts of mercury in the atmosphere during the past 150 years [*Schuster et al.*, 2002] and the effects of this increase now appear in the most remote mountain environments. The Western Airborne Contaminants Assessment Project investigated mercury concentrations in eight U.S. national parks, including 3 in Alaska. Results indicate bioaccumulation of mercury in salmonids to levels that impair the health of individual organisms and ecosystem function [*Schwindt et al.*, 2008]. Inorganic mercury emitted to the environment is transformed into highly toxic methylmercury via microbial processes, and methylmercury bioaccumulates and biomagnifies in aquatic food webs [*Wiener et al.*, 2002]. In addition to mercury, dust – silt and clay sized particles – can be transported around the world and can affect everything from climate to nutrient and contaminant loading [*Goudie*, 2009].

In summary, trace metals that create contaminants and hazards in mountain rivers can come from sources as diverse as acid-mine drainage, acid rain, and atmospheric deposition of mercury. Mountain rivers in apparently remote, pristine areas can be severely affected by atmospheric deposition.

As a general summary, the chemistry of mountain rivers tends to be characterized by low concentrations of TDS primarily because of the predominance of crystalline rocks in mountain drainage basins. The chemistry of hillslope inputs to the river is highly spatially variable because of rapid exposure of bedrock and relatively little mediation by chemical weathering in soils. Subtle differences in bedrock lithology may be discernible in the dissolved loads of mountain streams even when the drainage is formed on deeply weathered rocks, as in the Coweeta watershed of North Carolina USA, where saprolite developed on metasedimentary schists and gneisses averages 6 m in thickness [*Velbel*, 1992]. Mountain river chemistry is characterized by selective weathering and a contribution of solutes by reactive minerals disproportional to the abundance of minerals in the local rock. Low concentrations of DOC, and O_2 and CO_2 levels near saturation, are common in mountain rivers. These characteristics may vary as a function of lithology, climate, vegetation cover, relief, and flowpath for water reaching the river channel. Flowpaths, in particular, may vary as a function of regolith accumulation and valley morphology.

3.4. Hydraulics

Despite rapid growth in the technical literature during the past decade, the hydraulics of mountain channels remain less understood than those of lower gradient alluvial channels. Standard hydraulic equations developed for lower gradient sand-bed channels do not adequately describe mountain rivers along which steep gradients and large grain and form roughness promote non-logarithmic velocity profiles, localized critical and supercritical flow, and well-developed, localized lateral and vertical flow separation. The broad category of 'mountain rivers' also encompasses tremendous variability, from cascade channels in which individual grains protrude above the water surface of even the largest discharges, to pool-riffle channels with high levels of relative grain submergence but well-developed downstream alterna-tions between converging and diverging flow associated with individual pools and riffles. The complex hydraulics of mountain rivers are perhaps best described as stochastic, but recent work suggests that mean hydraulic conditions can be quantified if the effects of slope, grain and form roughness on velocity distribution are quan-tified as a function of flow stage. Much of the work done on hydraulics of mountain rivers focuses on developing equations that (1) adequately predict resistance coeffi-cient in relation to gradient, relative submergence, flow depth, particle size distribu-tion, and bedform types, (2) quantify the contribution of the different components of roughness to total flow resistance, or (3) characterize cross-stream and vertical velocity distributions and the associated forces of lift or shear stress exerted on the channel boundaries.

3.4.1. Resistance Coefficient

Although energy must be conserved within a system, potential energy can be transformed to kinetic energy, and viscous forces can transform kinetic energy of river flow into heat energy that represents a loss to the river system [*Roberson and Crowe*, 1993]. Energy equations for rivers thus quantify the transfer of energy between potential and kinetic energy, and thermal energy losses resulting from resistance to flow [*Julien*, 1995]. A commonly used version is that of the Bernoulli equation modified with a head loss function:

$$\frac{\alpha v_1^2}{2g} + \frac{p_1}{\gamma} + z_1 = \frac{\alpha v_2^2}{2g} + \frac{p_2}{\gamma} + z_2 + h_L \tag{3.7}$$

where α is the energy coefficient, a velocity head correction factor that varies from 1.03 to 1.36 for fairly straight, prismatic channels. This coefficient is generally higher for small channels, where it may exceed 2. The largest known value for laboratory measurements is 7.4 [*Chow*, 1959]. For the other variables, v is mean velocity (m/s), g is gravity (m^2/s), p is the system pressure (N/m^2), γ is the specific weight of the fluid (N/m^3), z is the elevation of the fluid element (m), h_L is the head loss resulting from resistance (m), and the subscripts denote values at upstream (1) and downstream (2) locations. The first term ($\alpha\, v^2/2g$) represents kinetic energy, and the second (p/γ) and third (z) terms represent potential energy.

The head loss over a given length of channel (the energy slope) represents the loss of mechanical and kinetic energy caused by resistance along the channel, including skin friction and form resistance (Figure 3.19). *Form resistance* is further subdivided into grain, form, bank, obstruction, and sediment-transport components of resistance [*Roberson and Crowe*, 1993]. Because flow velocities vary from a maximum near the free-surface of the flow to zero at the wall, shear forces are created and produce viscous energy dissipation, known as *skin friction* [*Tritton*, 1988]. Form drag occurs because localized flow separation can create a high pressure upstream from an object and low pressure in the object's wake. The resulting pressure-gradient force opposes flow and creates viscous energy losses downstream from the object [*Tritton*, 1988; *Roberson and Crowe*, 1993]. All of these forms of energy dissipation are subsumed into a single resistance coefficient such as Manning's n, but accurate estimation of such coefficients has proved difficult in mountain rivers.

The three most commonly used resistance coefficients (n, f, c) relate to velocity as follows:

$$\textit{Manning} \qquad v = (R^{0.667} S^{0.5})/n \tag{3.8}$$

$$\textit{Darcy–Weisbach} \qquad v = (8gRS/f)^{0.5} \tag{3.9}$$

$$\textit{Chezy} \qquad v = c\,(RS)^{0.5} \tag{3.10}$$

where R is hydraulic radius (cross-sectional area/wetted perimeter), S is channel gradient, and g is gravity. (Based on a dataset with 1037 discharge measurements from the USA and New Zealand, an exponent on S of 0.33 rather than the traditional value of 0.5 reduces the variance associated with estimating flow resistance [*Bjerklie et al.*, 2005].)

Inaccurate estimation of a flow resistance coefficient may produce inaccurate indirect discharge measurements. A discharge value computed by the slope-area method, for example, is inversely proportional to Manning's n value, so that selection of an n value is very important [*Eddins and Zembrzuski*, 1994]. The default method for quantifying n, f, or c is to measure the other variables in equations 3.8-3.10 and then calculate the resistance coefficient. This can be time- and labor-intensive, dangerous, or impractical in remote sites not easily reached during the discharge of interest. Obtaining directly measured values of v, R and S is also problematic in channels with large spatial variability. Consequently, investigators typically try to develop empirical relations for predicting resistance coefficient from more readily measured variables. Several methods have been proposed for estimating n values, including visual comparison to various rivers for which n-values have been empirically determined [*Barnes*, 1967; *Hicks and Mason*, 1991]. Another method involves a subdivision of n into a base value and additive values [*Cowan*, 1956; *Arcement and Schneider*, 1989]:

$$n = (n_b + n_1 + n_2 + n_3 + n_4) \quad m \qquad (3.11)$$

Equation 3.11 applies to reach-scale n values. Roughness coefficients are determined for subsections of each cross section in a reach, and then composited. For equation 3.11,

n_b = base value for a straight, smooth, uniform channel, in natural materials (0.028-0.70) for gravel to boulder bed

n_1 = a correction factor for bank-surface irregularities (0.0-0.02)

n_2 = a value for variations in shape and size of the channel cross section (0.0-0.015)

n_3 = a value for obstructions (0.0-0.05)

n_4 = a value for vegetation and flow conditions (0.002-0.1), and

m = a correction factor for meandering of the channel (1.0-1.30).

The various components of equation 3.11 can be visually estimated using the descriptions in *Arcement and Schneider* [1989].

Limerinos [1970] relates n to hydraulic radius and particle size based on data from primarily lower gradient (0.00068-0.024) channels with small gravel to medium-sized boulders (d_{84} of 2-75 cm; R/d_{84} of 0.9-47.2) for bed material at discharges of 5.6-427 m^3/s:

$$n = \frac{0.1129 R^{0.167}}{1.16 + 2.0\log\left(\dfrac{R}{d_{84}}\right)} \qquad (3.12)$$

where d_{84} is particle intermediate diameter (m) that equals or exceeds 84% of the particles, and R is hydraulic radius (m).

Bray [1979] uses a dataset of gravel-bed rivers (S of 0.00022-0.015, Q of 5.5-8140 m^3/s, R/d_{84} of 11-85) and modifies equation 3.12 to:

$$n = \frac{0.113R^{0.167}}{1.09 + 2.2\log\left(\dfrac{R}{d_{84}}\right)} \tag{3.13}$$

Griffiths [1981] uses a gravel-bed river dataset from New Zealand (Q of 0.05-1540 m^3/s, S of 0.000085-0.011, d_{50} of 0.013-0.301 m; R/d_{50} of 3-53) to relate R and d_{50} to resistance:

$$\frac{1}{\sqrt{f}} = 0.76 + 1.98\log\left(\frac{R}{d_{50}}\right) \tag{3.14}$$

where f is the Darcy-Weisbach friction factor (see equation 3.9). This can be re-stated as

$$n = \frac{0.1129R^{0.167}}{0.76 + 1.98\log\left(\dfrac{R}{d_{50}}\right)} \tag{3.15}$$

Hey [1979] develops an equation for gravel-bed rivers using data from rivers in the United Kingdom (Q of 0.995-190 m^3/s, S of 0.0090-0.031, d_{84} of 0.046-0.25 m, R/d_{84} of 0.97-17.24):

$$\frac{1}{\sqrt{f}} = 2.03\log\left(\frac{aR}{3.5d_{84}}\right) \tag{3.16}$$

where a varies between 11.1 and 13.46 as a function of channel cross-sectional shape. This can be reformulated as

$$n = \frac{0.1129R^{0.167}}{2.03\log\left(\dfrac{aR}{3.5d_{84}}\right)} \tag{3.17}$$

Noting that the Darcy-Weisbach friction factor tends to underestimate the rate of change in resistance at a site (as discharge varies) at higher gradients, *Bathurst* [1985] develops an empirical equation for the friction factor in gravel-bed rivers (Q of 0.14-195 m^3/s, S of 0.004-0.04, d_{84} of 0.113-0.74 m, $R/d_{84} < 10$):

$$\frac{8}{f} = 5.62\log\left(\frac{R}{d_{84}}\right) + 4 \tag{3.18}$$

which can be reformulated as

$$n = \frac{0.3193R^{0.167}}{56.2\log\left(\dfrac{R}{d_{84}}\right) + 4}$$ (3.19)

Numerous investigators have tried to develop a single index of grain roughness that may be incorporated into calculations of the flow resistance coefficient. This index most commonly takes the form of roughness height, k_s. In the absence of strong secondary currents, total flow resistance can be expressed by the Darcy-Weisbach friction factor, f:

$$f = \left[2.03\log\left(\frac{12.2R}{k_s}\right)\right]^{-2}$$ (3.20)

Several investigators have fit field or flume data to the equation above to calculate k_s and then expressed k_s as equal to $C\,d_x$ where C is a constant and d_x is the grain diameter for which x% is finer. These values range from $k_s = 1.25d_{35}$ to $k_s = 3.5d_{90}$ [Millar and Quick, 1994, but those from high-gradient channels commonly use a large d_x value. Hey [1979, 1988] uses $3.5d_{84}$, for example; Parker and Peterson [1980] use $2d_{90}$, and Lawless and Robert [2001a] recommend $0.73d_{50}$ for plane-bed channels. The use of $k_s = 3.5d_{84}$ originated with experiments using quasi-homogeneous sand roughness and assuming a logarithmic velocity distribution [Nikora et al., 1998]. Wiberg and Smith [1991] demonstrate theoretically that the log formula for hydraulic resistance with $k_s = 3d_{84}$ is valid even for flows with large relative roughness despite velocity profile deviations from the log law in these flows. The value of k_s also depends significantly on the concentration of bed roughness elements [Wiberg and Smith, 1991], suggesting that k_s is a function of several characteristic scales, rather than just d_{84} [Gomez, 1993].

Because the same grain-size distribution can offer greater or lesser resistance to flow depending on packing [Ferguson, 2007] (Figure 3.20), statistics and spectral analysis of the bed microtopography may better characterize roughness height [Nikora et al., 1997; Aberle et al., 2010]. The standard deviation of bed elevation (σ_z) in a digital elevation model or dense longitudinal profile works well, but this can be very laborious to obtain in field settings [Aberle and Smart, 2003]. Nikora and Walsh [2004] propose the use of high-order structure functions to describe bed topography at scales comparable to d_{50}, and Aberle and Nikora [2006] use the probability distribution of bed elevations and its moments and structure functions to derive roughness parameters. The assumption implicit in equation 3.20 is that grain roughness dominates total resistance. This may or may not be appropriate for specific channel segments, partic-ularly those with well-developed bedforms such as step-pool sequences that create substantial form resistance. It is probably more accurate to designate f_{grain} on the left side of equation 3.20, rather than simply f.

Nikora et al. [1998] propose using a random field approach for characterizing the roughness of gravel-bed channels. They demonstrate that the bed elevation distribution is close to Gaussian and apply the second-order structure function $D\ (\varDelta x,\ \varDelta y)$ of bed elevation $z\ (x,\ y)$ to describing gravel-bed roughness:

$$D(\varDelta x, \varDelta y) = \frac{1}{(N-n)(M-m)} \sum_{i=1}^{N-n} \sum_{j=1}^{M-m} \{Z(x_i + n\delta x, y_j + m\delta y) - Z(x_i, y_j)\}^2 \qquad (3.21)$$

where $\varDelta x = n\delta x$, $\varDelta y = m\delta y$, δx and δy are the sampling intervals and N and M are the total numbers of measuring points of bed elevations in directions x and y, respectively. This model describes the gravel-bed roughness only at the level of individual particles; hydraulically important particle clusters are ignored, restricting the model's use in flow resistance and bedload investigations. However, the model may be able to improve the ability to predict bulk flow velocity by providing an estimate of roughness height k_s (here, $k_s = f(\sigma_z, \varDelta x, \varDelta y)$, where σ_z is degree of particle sorting in the z direction).

Sometimes flow depth is comparable to or smaller than k_s, in which case the zone of flow that approximates a log-law profile does not exist. *Katul et al.* [2002] develop a model for flow resistance in these shallow flows that draws on work in vegetation canopy turbulence. The key model parameter is the characteristic length scale describing the depth of the Kelvin-Helmholtz wave instability, which is proportional to the mean height of roughness elements on the stream bed, or k_s. *Katul et al.* [2002] predict roughness in these very shallow flows using

$$n = \frac{h^{0.17}}{\sqrt{g}4.5f\left(\dfrac{h}{D}\right)} \qquad (3.22)$$

where h is flow depth, f is the Darcy-Weisbach coefficient, and D is the mean height of roughness elements.

Most of these methods of estimating a flow resistance coefficient are not designed specifically for steep channels. Field data indicate that n values are much greater on high gradient cobble- and boulder-bed streams than on low-gradient streams having similar relative grain submergence (R/d_{50}) values [*Jarrett*, 1987]. As gradient increases, energy losses increase as a result of wake turbulence and the formation of localized hydraulic jumps downstream from boulders [*Jarrett*, 1992]. Using empirical data from numerous channels with slopes greater than 0.002 (d_{84} of 0.1-0.8 m, S of 0.002-0.052, Q of 0.34-127 m³/s, R of 0.15-2.2 m), *Jarrett* [1984, 1985, 1987] develops an equation that uses energy gradient and hydraulic radius to predict an n value:

$$n = 0.32 \quad S_e^{0.38} R^{-0.16} \qquad (3.23)$$

where S_e is energy gradient. *Jarrett* [1994] notes that n values computed in this manner are on average 53% larger than field-estimated n values, and that discharges computed using equation 3.23 vary accordingly from indirect peak discharge measurements using field-selected n values.

Working on experimental channels with gradients up to 20% and high sediment transport rates, *Smart and Jaeggi* [1983] develop the empirical equation:

$$\left(\frac{8}{f}\right)^{0.5} = 5.75\left[1-\exp\left(-0.05\left(\frac{R}{d_{90}}\right)\left(\frac{1}{S^{0.5}}\right)\right)\right]^{0.5}\log\left(8.2\left(\frac{R}{d_{90}}\right)\right) \tag{3.24}$$

which can be reformulated as

$$n = \frac{0.3193R^{0.167}}{5.75}\left[1-\exp\left(-0.05\left(\frac{R}{d_{90}}\right)\left(\frac{1}{S^{0.5}}\right)\right)\right]^{0.5}\log\left(8.2\left(\frac{R}{d_{90}}\right)\right) \tag{3.25}$$

Bathurst [2002] derives empirical equations distinguished by gradient. For $S > 0.8\%$,

$$\left(\frac{8}{f}\right)^{1/2} = 3.10\left(\frac{d}{D_{84}}\right)^{0.93} \tag{3.26}$$

where d is flow depth and D_{84} is grain size.

Wahl [1994] tests several of these equations for n using a composite data set. He finds that all of the equations significantly underestimate n for relatively low discharges (< 0.2 median annual flood discharge). Moderate to large discharges (> 0.2 median annual flood discharge) are approximated fairly well by the equations. Extrapolation of any of the equations to discharges greater than about 1.5 times the median annual flood discharge is not yet warranted because of the lack of data for large discharges [*Wahl*, 1994], especially as roughness coefficient is commonly inversely related to hydraulic radius and discharge [*Coon*, 1994].

Working specifically on a small mountain stream ($R < 0.25$ m, S of 0.02-0.16, R/d_{84} of 0.5-2.8), *Marcus et al.* [1992] evaluate the Limerinos, Jarrett, Bathurst, and Cowan methods of estimating n, as well as four other techniques. Observed roughness values are significantly under-predicted, often by an order of magnitude, by all the techniques except Jarrett's, which over-estimates roughness by an average of 32%. Under-estimation results from inadequately addressing the effects on flow resistance of large sediment sizes, low ratios of flow depth to hydraulic radius, steep slopes, and severe turbulence, as well as the tendency for observers to use lower visual estimates of n because of experience in lower gradient channels. *Marcus et al.* [1992] conclude that discharge should be measured directly whenever possible, particularly during low flows when R/d_{84} values are low.

More recent attempts to develop empirical equations for predicting resistance in steep channels tend to employ f rather than n because f is non-dimensional and is physically interpretable as a drag coefficient if resistance is equated with the gravitational driving force per unit bed area and assumed proportional to the square of velocity [*Ferguson*, 2007]. Step-pool channels exhibit a strong correlation between f

and dimensionless unit discharge ($q*$) and the ratio of step height to step length [*Comiti et al.*, 2007], where

$$q* = \frac{q}{\sqrt{gD_{84}^3}} \qquad (3.27)$$

David et al. [2010] find that f correlates strongly with bed gradient, flow, channel type (cascade or step-pool), and $q*$. Compiling data from numerous sources including those used in *Comiti et al.* [2007], *Ferguson* [2007] combines the Manning equation for deep flows with a linear resistance relation for roughness layers in shallow flows to develop a pair of non-dimensional hydraulic geometry relations and a variable-power resistance equation

$$\left(\frac{8}{f}\right)^{0.5} = \frac{a_1 a_2 \left(\dfrac{d}{D}\right)}{\left[a_1^2 + a_2^2 \left(\dfrac{d}{D}\right)^{1.67}\right]^{0.5}} \qquad (3.28)$$

where a_1 is a constant, typically 6.7 if D_{50} is used as a roughness scale, or 8.2 if D_{84} or D_{90} is used; a_2 is a constant, typically ~ 1 to 4; d is mean flow depth; and D is a representative grain diameter. Equation 3.28 plots as a smooth curve, asymptotic to a 1/6-power relation at $d/D \gg 1$, but to a linear relation at $d/D < 1$. This provides a single resistance equation applicable to shallow and deep flows over coarse beds and, although not explicitly stated, assumes linearity of effects. In comparing reach-averaged velocities calculated from equation 3.28 to measured values from several sites, considerable predictive uncertainty remains, likely as a result of measurement errors and theoretical limitations [*Ferguson*, 2007].

Values for either f or n tend to be very high in mountain rivers relative to those reported for lower gradient systems, and to vary substantially in relation to stage and to channel morphology [*Lee and Ferguson*, 2002]. Reported values of f for small step-pool channels during low flow range from 5 to 380 [*Curran and Wohl*, 2003; *MacFarlane and Wohl*, 2003]. *Reid and Hickin* [2008] measure resistance coefficients spanning three to six orders of magnitude (f, 0.29-12,700; n, 0.047-7.95) on plane-bed channels in British Columbia, Canada.

Given the spatial and temporal heterogeneity in hydraulic variables and the difficulty of measuring relevant parameters such as grain size, channel gradient, and flow depth directly, some investigators have proposed reach-scale predictive methods based on indirect measurements, such as *Tinkler's* [1997b] approach equation (3.1). *Hodel et al.* [1998] estimate roughness by applying photogrammetry to stereo photographs of the water surface along a reach in which velocity is simultaneously measured using a salt tracer. *Cooper et al.* [2006] estimate hydraulic resistance using a rapid and inexpensive acoustic technique known as grazing angle sound propagation, which measures the temporal dynamics of turbulent water surfaces. Flume experiments suggest that the standard deviation in acoustic pressure decreases as hydraulic resistance increases over a range of relative submergence and bed slopes typical of gravel-bed rivers. Although such measures provide no predictive value, they can be very useful for quantifying resistance under specific flow conditions. Other approaches focus on network-scale variations in flow resistance. *Orlandini* [2002] uses data from two mountainous catchments in New Zealand to approximate spatial variation in resistance to flow by combining a hydraulic equation of the Manning type, a relationship for discharge and upstream drainage area, and relationships for mean flow velocity, resistance coefficient, hydraulic depth, and friction slope (Figure 3.21). This approach will be useful in understanding the relative magnitude and spatial distribution of resistance and energy expenditure in mountainous catchments, even if the equations do not provide highly accurate representations of resistance at all points in the network. If recently demonstrated correlations among gradient, channel morphologic type, and flow resistance [*David et al.*, 2010] can be applied more broadly, it will be possible to predict at least the relative magnitude of resistance throughout a network by extracting reach-scale gradient from digital elevation models or other topographic data.

In summary, there is not at present a well-tested, consistently accurate equation for calculating the resistance coefficients of mountain rivers. Steep gradients, poorly sorted beds, coarse particles with median diameters that may approach the maximum flow depths, and localized flow transitions all complicate the estimation of total resistance. Recent increases in published hydraulic data from experimental and natural channels have contributed substantially to advances such as equation 3.28, which shows promise of being more broadly applicable than earlier equations, but the fact that resistance tends to correlate strongly with features specific to certain channel types, such as step and pool dimensions [*Comiti et al.*, 2007; *David et al.*, 2010], suggests that it may not be feasible to develop resistance equations that are accurate across a range of steep channel types.

3.4.2. Resistance Partitioning

Under conditions of steady, quasi-uniform, two-dimensional, fully developed turbulent flow over a deformable channel bed, flow resistance is caused by (1) viscous and pressure drag on grains of the bed surface (*grain roughness*), (2) pressure drag on bed undulations (*form roughness*), and (3) pressure and viscous drag on sediment in transport above the bed surface [*Griffiths*, 1987]. Steep, coarse-grained channels can also have *spill resistance* (Figure 3.22) from locally supercritical flow and wave drag on elements protruding above the water surface. Grain resistance represents the channel-bed roughness that induces energy losses resulting from skin friction and form drag from individual bed particles. As depth increases, the individual particles influence a lower proportion of the flow and the effect of grain resistance is diminished. However, the coarse, poorly sorted clasts and relatively shallow flow of mountain rivers tend to make grain resistance more important in these channels than in most lower gradient rivers, even during high flows. Consequently, grain roughness can dominate at high submergence, with form roughness more important at low submergence [*Ferguson* 2007].

The relative importance of grain and form roughness in relation to flow depth varies among individual channel reaches. Working in a steep, small channel (flow depth 0.2-0.5 m, width 2.9 m, gradient 0.034), *Thorne* [1997] finds that a stable bed element that is small relative to the flow depth (≤ 1 m in streamwise dimension) does not represent a significant obstacle to flow. Streamwise acceleration over the feature and a minor component of steering occur. In contrast, large particles or clusters ≥ 2 m in the streamwise dimension substantially obstruct flow. The steering of flow around the obstacle is more significant than acceleration over it, and the creation of an extensive backwater zone significantly influences the local evolution of the channel bed. Along the pool-riffle, gravel-bed Virgin River of Utah, USA, the relative magnitude of form drag increases with discharge, reducing the shear stress on the bed surface by about 40 percent at bankfull discharge [*Andrews*, 2000].

Spatial density of individual obstacles also exerts an important influence on flow resistance. Isolated roughness elements have no wake interaction and resistance is proportional to the number of elements. As elements become more closely spaced, the wake behind each element overlaps with the next element to create wake interference, until the elements are so closely spaced that they form a more or less smooth surface with skimming flow [*Nowell and Church*, 1979]. Minimum resistance is related to the transition from isolated roughness to wake interference flow [*Canovaro et al.*, 2007], which corresponds to a length/height ratio of roughness elements of approximately 9-10 [*Davies*, 1980; *Wohl and Ikeda*, 1998]. In step-pool channels, the ratio of flow depth to roughness height also governs the transition from nappe (flow alternates downstream between super- and subcritical conditions in passing over steps) to skimming (entire flow is supercritical) flow [*Chanson*, 1994] (Figure 3.23). The partitioning of flow resistance changes significantly when flow transitions from nappe to skimming; reach-averaged flow velocities increase substantially with the onset of skimming flow as spill resistance declines and grain resistance increases [*Comiti et al.*, 2009a].

Although energy dissipation might be expected to be inversely proportional to velocity, most energy dissipation at steps comes from potential energy losses rather than velocity head losses [*Pasternack et al.*, 2006]. The tailwater depth controls the jump regime of the step and thus how much energy conversion occurs. Simulating horseshoe waterfalls in flume experiments, *Pasternack et al.* [2006] find that hydraulic jumps below steps efficiently convert kinetic energy to potential energy; maximum energy dissipation occurs when no jump is present and downstream tail depth is exactly critical.

The varying relative contributions of grain, form, and spill roughness as a function of submergence is one of the challenges to creating a single, widely applicable resistance equation for mountain rivers. Resistance equations developed specifically for shallow flows [e.g., *Bathurst*, 1978, 2002; *Jarrett*, 1984; *Rickenmann*, 1991b; *Katul et al.*, 2002] tend to substantially over- or under-predict resistance when extrapolated to deeper flows [*Ferguson*, 2007].

Downstream variations exist in the variables that influence water-surface slope and resistance, such that individual grains are most important in the headwaters, bedforms dominate the middle reaches, and channel bends become increasingly important at lower gradients [*Prestegaard*, 1983b]. Similarly, flow resistance changes throughout a channel network, with transient bedforms dominating flow resistance in sand-bed channels, bed material and pool-riffle sequences creating most of the resistance in gravel-bed rivers, boulder form drag dominating boulder-bed channels, and ponding dominating step-pool channels [*Bathurst*, 1993].

Sediment movement can also effectively alter grain resistance by altering near-bed velocity flow field in the proximity of entrained or deposited particles. *Bottacin-Busolin et al.* [2008] derive a probability distribution of individual grain resistance from statistics of the near-bed velocity field and of the entrainment risk. *Recking et al.* [2008] propose flow resistance equations for three flow domains: (i) no bedload and a constant bed roughness $k_s = D$, where D is a representative grain diameter; (ii) a transitional domain in which the bed roughness evolves from D to $2.6D$ with increasing flow; and (iii) high bedload transport over a flat bed with a constant bed roughness $k_s = 2.6D$.

Clifford et al. [1992b] suggest that C, the multiplier of characteristic grain size, is attributable to the effect of small-scale form resistance, and that there are two discrete scales of bed roughness, associated with grain and microtopographic roughness elements. Grain resistance is quantified using a modified form of Keulegan's relation [*Parker and Peterson*, 1980]:

$$C_G^{-1/2} = \frac{1}{\kappa}\left[\ln\left(\frac{D}{D_{90}}\right) + A\right] \tag{3.29}$$

where κ is von Karman's constant (0.40 for clear water), and

$$A = \ln(11/m) \tag{3.30}$$

where $m = k_s/d_{90}$. *Prestegaard* [1983a] uses this equation to determine the component of total resistance attributable to grain resistance at bankfull stage along 12 gravel-bed channel reaches (Q of 0.8-33.0 m^3/s, S of 0.0012-0.036, d_{84} of 0.053-0.315 m). She then assumes that the remaining resistance is associated with bars. Bar resistance thus accounts for 50-75% of the total resistance in these gravel-bed channels. Similarly, form resistance accounts for up to 90% of total resistance in pool-riffle channels during bankfull and near-bankfull flows [*Millar*, 1999]. In additive approaches to resistance partitioning, however, partitioning estimates are highly sensitive to the order in which components are calculated and inflate the values of difficult-to-measure components that are calculated by subtraction from measured components [*Wilcox et al.*, 2007].

The percentage of total resistance caused by bars increases as stage declines [*Parker and Peterson*, 1980]. In a study of 62 sites in the United Kingdom (S of 0.0012-0.0215), bar resistance ranges from 7% to 86% of the total resistance during high flows [*Hey*, 1988]. Using the flume data from *Meyer-Peter and Muller* [1948] and *Smart and Jaeggi* [1983], *Griffiths* [1987] finds that dimensionless form shear stress exceeds dimensionless particle shear stress for values of approximately three times critical Shields stress in subcritical flow, and five times in supercritical flow. For given

Shields stress and increasing relative roughness, relative form shear stress increases in subcritical flow because of bedform influences but decreases in supercritical flow because of transport rate effects [*Griffiths*, 1987].

The factor R/k_s (equation 3.20), often expressed as R/d_{84}, is *relative grain submergence*. Flow can be subdivided into large-scale roughness ($0 < R/d_{84} < 1$), intermediate-scale roughness ($1 < R/d_{84} < 4$), and small-scale roughness ($R/d_{84} > 4$) [*Bathurst*, 1985; *Ugarte and Madrid*, 1994]. Empirical formulas for n are then calculated for each roughness criteria. For example, *Ugarte and Madrid* [1994] propose:

$$n = 0.183 + \ln\left[\frac{1.3014 S^{0.0785}\left(\dfrac{R}{d_{84}}\right)^{0.0211}}{F^{0.2054}}\right]\frac{d_{84}^{1/6}}{g^{0.5}} \qquad (3.31)$$

where F is Froude number, for $1 < R/d_{84} < 12.5$ and S of 0.2-0.4%. To some extent, this incorporates the changes in resistance as a function of flow depth discussed earlier. More field data including detailed hydraulic and geometric measurements across a range of channel types and flow magnitude, are necessary before we can accurately estimate n or f for varying channel configurations and flow conditions.

Studies of grain and form resistance were mostly conducted on pool-riffle gravel-bed rivers until the first decade of the 21st century. Because riffles serve as hydraulic controls of flow along such channels, the flow resistance depends on variations in flow geometry between pools and riffles, and reflects local accelerations and decelerations. Several recent studies of resistance and partitioning focus on step-pool channels, which have substantial spill resistance associated with steps at lower flows [*Comiti et al.*, 2009a]. *Lee and Ferguson* [2002] find that k_s cannot be reliably estimated as a fixed multiple of d_{84} in the steps of step-pool channels, although they are able to develop a good relation between f and k_s by optimizing separate values of k_s for each of their multiple study reaches. They interpret this to mean that field-derived values of d_{84} do not adequately represent the integrated effects of various types of flow resistance, but none of the alternative measures that they evaluate (e.g., k_s/d_{84}, k_s/d_{50}, k_s/d_{max}, measures of bed topography) gives consistent estimates of k_s, indicating the "need for further work on methods to relate the roughness length scale to measurable properties of steep stream beds" [*Lee and Ferguson*, 2002, p. 71]. Subsequent studies suggest that parameters such as step dimensions and relative bedform submergence correlate more strongly with f in step-pool channels [*Comiti et al.*, 2009a; *Yochum*, 2010].

An additional source of roughness in some mountain rivers comes from instream wood. Individual pieces of wood can create grain or form resistance, depending on their size relative to adjacent bed topographic elements and flow depth [*Wilcox and Wohl*, 2006]. Logjams, or wood incorporated into bed steps, can produce form roughness. Step-pool channels with wood tend to have deeper flows, taller and more widely spaced steps, and greater flow resistance [*MacFarlane and Wohl*, 2003]. Relatively few investigators have quantified these effects. An equation for the drag coefficient of wood in low-gradient, sand-bed channels [*Shields and Gippel*, 1995] does not adequately describe the roughness produced by wood in steep channels with non-uniform bed gradients [*Curran*, 1999]. Step-pool channels without wood have shorter, more closely spaced steps and lower values of *f* than analogous channels with wood [*MacFarlane and Wohl*, 2003]. In flume experiments and natural channels, wood located near step lips or forming steps increases step height and creates greater flow resistance than wood located farther upstream on step treads [*Wilcox and Wohl*, 2006; *Yochum*, 2010].

Bank roughness in mountain streams has been neglected relative to bed roughness (Figure 3.24), but a few studies have dealt explicitly with this source of boundary resistance. Based on the *Houjou et al.* [1990] approach, *Buffington and Montgomery* [1999c] develop an empirical equation relating the ratio of wall (bank) and bed shear to the width/depth ratio of a channel and the ratio of bed to bank roughness height:

$$\frac{\tau_w}{\tau'} \approx 2.55 \left(\frac{w}{h}\right)^{-1.1} \left(\frac{z_{ob}}{z_{ow}}\right)^{-0.21} \tag{3.32}$$

where τ' = the bed shear stress, τ_w = the bank shear stress, w = channel width, h = channel depth, z_{ob} = roughness height of the bed, and z_{ow} = roughness height of the banks. The preliminary analysis of *Buffington and Montgomery* [1999c] suggests that this approach is useful in partitioning flow resistance in steep mountain channels, although statistical analyses of field data that include bed and bank roughness measures from bouldery step-pool and cascade channels suggest that the contributions of bank roughness are less effective in creating resistance than bed roughness [*Yochum*, 2010]. Field techniques for quantifying the ratio representing bed versus bank roughness length scales as affected by width variability and vegetation require further investigation.

Kean and Smith [2006a, b] use spatially dense surveys of bank geometry from a sand-bed stream in spectral analysis to detect regular and irregular sequences of small-scale topographic features. They define features in terms of protrusion height, stream-wise length scale, and spacing between crests. They determine form drag on individual elements by using the drag coefficient and a reference velocity that includes the effects of roughness elements further upstream. They then develop a model to calculate the drag on each element, as well as spatially averaged total stress, skin friction stress, and roughness height of the boundary. The model indicates that drag on small-scale topographic features substantially alters the near-bank flow field; neglecting the effects of lateral stresses results in a 56% over-estimate in discharge [*Kean and Smith*, 2006a].

Even less attention has been given to roughness at the top of the channel than to bank roughness, but ice cover in cold climates can effectively increase boundary roughness and flow depth and decrease flow velocity [*Muste et al.*, 2000]. Although rooted, submerged aquatic plants are less common in mountain rivers than in lower gradient systems, these plants can be present and can also influence boundary roughness [*Nepf*, 1999; *Ghisalberti and Nepf*, 2004; *Shucksmith et al.*, 2010] and deposition of sediment [*Cotton et al.*, 2006; *Zong and Nepf*, 2010].

In summary, the relative importance of grain, form and spill resistance varies throughout a mountain river network in relation to factors such as gradient, grain size and channel morphology, and temporally with changing stage. Much research has focused on relating total resistance to a characteristic roughness height, k_s, which is most commonly equated to a representative grain diameter. Recent work suggests that, for the strongly vertical bedforms of step-pool channels, k_s may be better described in terms of bedform rather than grain parameters. The relative contributions of bank and bed roughness to resistance in various steep channels remain largely unknown.

3.4.3. Velocity and Turbulence

Flow velocity and the associated shear stress and lift forces are strongly related to flow resistance because velocity distribution is also affected by bed material size distribution, relative submergence, and channel gradient, and ultimately reflects the balance between energy available and energy expended in overcoming resistance. The ratio of lift and shear force depends upon surface roughness, for example; the rougher the surface, the greater the shear stress compared to dynamic lift [*Dittrich et al.*, 1996]. Empirical mean velocity equations for mountain rivers recognize these inter-relations between velocity and flow resistance, as for example *Rickenmann* [1994a]:

$$v = 0.37 g^{0.33} Q^{0.34} S^{0.20} / d_{90}^{0.35}, \quad \text{for} \quad S > 0.6\% \tag{3.33}$$

$$v = 0.96 g^{0.36} Q^{0.29} S^{0.35} / d_{90}^{0.23}, \quad \text{for} \quad S > 1.0\% \tag{3.34}$$

or *Ferro and Pecoraro* [2000]:

$$\frac{V_m}{u^*} = r_0 + r_1 \log \left(\frac{h}{d_{84}} \right) \tag{3.35}$$

where V_m is mean velocity, u^* is shear velocity, r_0 and r_1 are numerical constants, h is flow depth, and d_{84} is a grain size.

The empirical data from which mean velocity is estimated can be obtained using at least two methods, point measurements and dilution tracers. Point measurements utilize various types of current meters, including mechanical impellor (cup and vane), electromagnetic and ultrasonic current meters, and laser velocimeters. As reviewed by *Clifford and French* [1993b], mechanical impellor current meters are low cost, durable field instruments, but have a poor frequency response (< 1 Hz), provide only one-dimensional information, and require maintenance and re-calibration. Electromagnetic current meters are robust, of intermediate cost, have a good frequency response (5-20 Hz), tolerate particle and other contamination in the flow, and provide one-, two-, or three-dimensional measurements, but frequency response is affected by head design (Figure 3.25). Ultrasonic current meters are expensive, fragile, and sensitive to particle and air bubble contamination, but provide two- [*Ferro*, 2003] and three-dimensional [*Thompson*, 2007; *Wilcox and Wohl*, 2007] measurements with an excellent frequency response (up to 30 Hz). Laser velocimeters have the highest cost and are sensitive to suspended sediment, but produce very high frequency response and three-dimensional measurement without perturbing the flow [*Clifford and French*, 1993b]. *Buffin-Bélanlanger and Roy* [2005] review how sampling frequency and record length govern which turbulence parameters can be extracted from velocity time series.

A primary difficulty with any type of point measurement used to estimate mean velocity along mountain rivers is that numerous measurements are necessary to estimate a mean value which accounts for the high levels of vertical, lateral, temporal, and downstream variability in velocity characteristic of mountain rivers. The development of acoustic Doppler profilers (ADPs) has made it feasible to obtain spatially dense point-velocity measurements in a very short period of time [*Whiting*, 2003; *Ellis and Church*, 2005; *Kostaschuk et al.*, 2005; *Magirl et al.*, 2009]. Spatially dense point-velocity measurements can then be analyzed using spatial statistics to detect patterns at the cross-sectional or reach scale [*Legleiter et al.*, 2007].

Although point measurements are most appropriate for characterizing the spatial and temporal variability of velocity along mountain channels, dilution tracers may more accurately characterize the mean velocity of a channel reach. Dilution tracer techniques involve introducing a fluorescent dye [*Gees*, 1990; *Graf*, 1995] or a chemical such as NaCl [*Day*, 1977] into the flow. The tracer can be introduced steadily over a finite time period or instantaneously (slug injection). The tracers must be miscible and not alter the density or velocity of the fluid flow. Some minimum channel length (the mixing length) is required before the cross-sectional distribution of the tracer concentration is nearly uniform for constant flow injection, or the amount of dilution is constant for slug injection. In turbulent flow, complete mixing may occur within 15 times the mean channel width [*Elder et al.*, 1990]. Beyond this mixing length, the properties of the tracer reflect flow velocity rather than injection procedure. For a slug injection of NaCl, for example, tracer concentration is measured with a conductivity meter and times of peak concentration at two points separated by a known distance are used to calculate mean flow velocity [*Planchon et al.*, 2005] (Figure 3.26). The attenuation of the tracer concentration can also be used to quantify temporary storage in zones of flow separation [*Richardson and Carling*, 2006], although it can be difficult to separate surface storage from delays induced by hyporheic exchange. Both salt tracers and current meters can yield adequate results within a 2% average deviation, but salt tracers are most appropriate for channels with turbulent flow and irregular geometry [*Benischke and Harum*, 1990].

Spatial and temporal distributions of velocity and other hydraulic variables may also be examined using flow models. One-dimensional step-backwater models such as HEC-RAS [*Hydrologic Engineering Center*, 1997] are the models most widely applied to resistant-boundary channels [*O'Connor and Webb*, 1988], but these models do not adequately approximate the highly turbulent flow along high-gradient channels. Two-dimensional models such as RMA2 [*Donnell et al.*, 1997; *Bates et al.*, 1997], HIVEL2D [*Berger and Stockstill*, 1995], TELEMAC [*Daubert et al.*, 1989; *Bates and Hervouet*, 1999; *Hervouet*, 2000], or FLUENT [*Nicholas*, 2001] require considerably more expertise to use, as well as more spatially detailed channel geometry and hydraulic calibration data to adequately simulate a range of flow conditions [*Miller and Cluer*, 1998], but have become more accessible and widely used within the past decade. Three-dimensional models, including computational fluid dynamics approaches, have a limited predictive ability because of problems of specifying topographic complexity, but they do provide more reliable estimates of bed shear stress and the three-dimensional flow field that is important for mixing processes [*Lane et al.*, 1999; *Nicholas and Sambrook Smith*, 1999; *Booker et al.*, 2001; *Casas et al.*, 2010]. Most of these models assume static boundaries. The 1d model HEC-6 predicts scour and deposition within rivers [*US Army Corps of Engineers*, 1998], although the model simulates only uniform changes in river-bed elevation over the entire width of the channel, and neglects lateral changes [*Rathbun and Wohl*, 2001]. Other models such as GSTARS provide quasi-2d simulations using a stream-tube approach to accommodate differential scour and deposition over the width of a cross section [*Yang et al.*, 1998]. *Lane* [1998] reviews hydraulic models.

Flow along channels with low gradients and fine bed material is commonly approximated by a semilogarithmic velocity profile in which velocity varies with the logarithm of distance from the bed [*Leopold et al.*, 1964] such that the profile includes a laminar sublayer, a turbulent boundary layer with logarithmic profile, and an outer layer that deviates slightly from logarithmic [*Ferguson*, 2007]. In the region immediately above a rough surface, the mean profile of turbulent shear flow is described via Prandtl's law of the wall

$$u/u_* = (1/\kappa)\ln(30y/k_s) \quad \text{or} \quad u/u_* = (1/\kappa)\ln(y/y_0) \qquad (3.36)$$

where $u_* = (\tau/\rho)^{0.5}$ is a velocity scale (the shear velocity), $\kappa \sim 0.41$ is von Kármán's constant, k_s is a roughness length scale, y is distance from the bed, and y_0 is the value of y where the flow velocity is zero [*Bridge*, 2003; *Church*, 2008]. Prandtl's law of the wall indicates that velocity is proportional to roughness-scaled distance from the wall. When D/d approaches 1, where D is a characteristic bed material grain diameter, the equation breaks down because the flow becomes a series of jets between large roughness elements [*Church*, 2008]. When the flow becomes deep, an outer layer develops above the wall layer.

The law of the wall for smooth boundaries outside the viscous sublayer and buffer layer applies where the boundary Reynolds number is less than 5. In this case, the velocity depends only on the dimensionless distance y_d from the bottom:

$$u/u_* = 2.5 \ln (\rho u_* y/\mu) + B' \qquad (3.37)$$

where 2.5 is the inverse of the von Kármán constant (0.4 for clear water), μ is the dynamic viscosity of the fluid and B' has a value between 5 and 6. For rough boundaries where the boundary Reynolds number exceeds 65, bottom roughness (k or D_{50}) must be taken into account:

$$u/u_* = 2.5 \ \ln \ (y/k) + B' \qquad (3.38)$$

where B' has a value of 8.5. *Le Roux* [2004] proposes an integrated law of the wall for the transitional regime:

$$u/u_* = 2.5 \ \ln \ (Re_* y/k) + 5.3 - 0.1206(Re_* - 5) \qquad (3.39)$$

where Re_* is the boundary or roughness Reynolds number.

Even fairly well-sorted beds can have a well-developed roughness sublayer near the bed, where the velocity profile becomes more uniform than the log-law profile, under conditions of low relative submergence [*Nakagawa et al.*, 1991; *Tsujimoto*, 1991]. Flow along mountain rivers includes a substantial portion of sublayer flow between larger boulders or below the surface that defines the general bed-water boundary. This low-velocity flow gives way fairly abruptly to high-velocity flow above the boulders. *McLean and Nikora* [2006], for example, find that for a range of rough-bed flows, the vertical distribution of double-averaged velocity (in time and in a volume occupying a thin slab parallel to the mean bed) consists of a linear region below the roughness tops and a logarithmic region above them. This two-part velocity profile has been described as s-shaped [*Marchand et al.*, 1984; *Bathurst*, 1988; *Jarrett*, 1991), although the profile varies widely as a function of grain and bed roughness (Figure 3.27) [*Robert et al.*, 1993; *Bergeron*, 1994; *Wohl and Ikeda*, 1998]. The two parts consist of an inner layer that reflects grain resistance and an outer layer that reflects total resistance [*Lawless and Robert*, 2001a]. *Byrd* [1997], for example, finds that 40% of the local profiles of streamwise velocity measured in a steep (0.034 m/m), bouldery channel in Colorado, USA approach a linear form, whereas 10% are nearly logarithmic, and the remaining 50% have various forms. *Le Roux and Brodalka* [2004] develop an Excel™ VBA program for calculating shear velocity using several input parameters including velocity profile data.

Use of the semilog velocity profile results in a significant overestimation of flow resistance along mountain rivers at high flows [*Bathurst*, 1994], with the degree of overestimation depending on k_s and R/k_s. *Ferguson* [2007] develops a non-dimensional hydraulic geometry approach to velocity:

$$V^* = a^{1-m} S^{(1-m)/2} q_*^{\ m} \qquad (3.40)$$

where $V^* = V/(gD_{84})^{0.5}$, m is the downstream hydraulic geometry exponent for velocity, and $q^* = q/(gD_{84}^{\ 3})^{0.5}$.

In addition to variation in mean velocity as a function of height above the bed, turbulent fluctuations decrease at greater distances from the bed, both in terms of amplitude and relative to the mean velocity at the depth being considered [*Smart*, 1994]. When the bed material is in motion, turbulence increases relative to the mean velocity, especially in the lower third of the profile [*Smart*, 1994]. Turbulence in the roughness sublayer is commonly measured using laser Doppler anemometry [*Hammann and Dittrich*, 1994]. *Singh et al.* [2010] use spectral analysis of measured velocity fluctuations to detect distinct power law scaling regimes that reflect evolving multiscale bed topography.

Turbulence in coarse-grained channels can be conceptually subdivided into macro-turbulence, which occurs at scales between flow width and depth (e.g., eddies), meso-turbulence, which occurs at scales between flow depth and the dissipative scale (v^3/ε, where v is kinematic viscosity and ε is energy flux), and micro-turbulence in the form of dissipative eddies [*Nikora*, 2008]. The energy of the mean flow is transferred into turbulent energy through velocity shear and through flow separation behind variously scaled roughness elements [*Nikora*, 2008].

Turbulence in high-gradient, coarse-grained channels can differ in significant ways from turbulence in other channels. As effective roughness increases, coherent flow structures become more defined throughout the flow depth [*Hardy et al.*, 2010]. Long velocity records from the gravel-bed North Fork Toutle River, Washington, USA, sampled at 2 Hz, indicate low-frequency fluctuations over a range of coherent wave periods [*Dinehart*, 1999]. If these represent discrete fluid structures, the structures would be 2-3 times longer than predicted by empirical relations for mean boil periods in lowland streams and those predicted by published relations for eddy lengths.

Furbish [1993] mathematically describes longitudinal flow structures in coarse-grained mountain channels using depth-averaged equations of momentum and continuity that are linearized and solved in the wavenumber domain. The velocity field and water-surface topography can exhibit a systematic structure over a distance of tens of channel widths and longer, although the structure is partially obscured by noise caused by local variations in channel width and bed roughness. The structure takes the form of a filament of high streamwise velocity that exhibits a near-oscillatory structure as it threads back and forth across the channel [*Furbish*, 1993]. Subsequent work indicates that bed topography significantly affects the water surface, whereas velocity structure is affected equally by width variations [*Cudney*, 1995].

All scales of turbulence exert important influences on momentum transfer [*Tennekes and Lumley*, 1994] and sediment transport [*Clifford*, 1993a; *Carling and Tinkler*, 1998]. Turbulence generation and amplification require (1) Reynolds numbers high enough to induce instability, (2) shear flow with the introduction of vorticity, and (3) a perturbation to the flow [*Mollo-Christensen*, 1971; *Tennekes and Lumley*, 1994]. A vortex is a periodic, whirlpool-like turbulence feature [*Lugt*, 1983] that is associated with flow separation. Flow separation occurs when flow along a physical boundary develops an adverse pressure gradient (decelerating flow in the upper part of the profile), becomes unstable and detaches from the wall, and forms reverse flow adjacent to the physical boundary [*Schlichting*, 1968; *Tritton*, 1988]. Because fluid inertia can be important, flow separation is more common at high Reynolds numbers [*Tritton*,

1988]. The separated boundary layer functions as a zone of high shear between the downstream and recirculating flow [*Tennekes and Lumley*, 1994]. This zone of high shear produces vorticity.

Flow separation may occur at channel bends [*Leeder and Bridges*, 1975], channel expansions [*Middleton and Southard*, 1984; *Carling*, 1989b], pools [*Kieffer*, 1985; *Thompson et al.*, 1996, 1998, 2007], bedforms, and large grains (Figure 3.28). Flume experiments over poorly sorted fluvial gravels (d_{84} of 22.1 mm, S of 0.002, h of 15 cm) reveal a near-bed zone dominated by obstacle-derived vortices (and an outer zone of mean unidirectional flow [*Nakagawa et al.*, 1991; *Kirkbride*, 1993]. Interaction between the zones results in the intermittent shedding of vortices from the lee of obstacle clasts into the outer zone. If vortices form at the zone of flow separation, vortex shedding can create a series of paired eddies with opposite senses of rotation, known as a von Karman vortex street [*Tritton*, 1988]. As the vortices disintegrate downstream or detach from the bed they create areas of strong upwelling known as boils. *Matthes* [1947] suggested the term kolk for the combined vortex-boil system. A kolk represents the most powerful concentration of energy in natural rivers and strongly influences bed scour and sediment transport.

Kolks are related to the bursting phenomenon, which includes four main momentum transfer mechanisms [*Nelson et al.*, 1995]: (1) *Sweeps* – high velocity pulses originating in the main flow that move toward the bed. Sweep-like motions can dominate shear stress production in the near-bed region [*Roy et al.*, 1996], or in the outer flow region [*Ferguson et al.*, 1996] of gravel-bed rivers. (2) *Bursts* – slow-moving parcels of water originating in the boundary layer that move toward the main flow. (3) *Outward interactions* – high velocity pulses from the boundary layer toward the main flow. (4) *Inward interactions* – low velocity pulses from the main flow toward the bed. The bursting phenomenon represents the exchange of eddy inertia and momentum between the boundary layer and the free-stream zone [*Thompson*, 1997].

Sweep impacts can be grouped and can create patches of entrainment on a mobile bed [*Best*, 1992]. Visualization experiments suggest that larger clasts on a bed are entrained by downstream rushes, whereas the smaller clasts are more likely to be entrained by chaotic transient vertical flows associated with the obstacle-derived vortices [*Kirkbride*, 1993]. Motion pictures taken through the clear water of Duck Creek in Wyoming, USA indicate that the collective motion of bed gravels (d_{50} of 4 mm) is characterized by frequent, brief, localized, random sweep-transport events that in the aggregate transport approximately 70% of the total load moved [*Drake et al.*, 1988].

Field data from gravel-bed rivers with pools and riffles indicate systematic differences in turbulence characteristics in association with changing bedforms [*Clifford and French*, 1993a; *Clifford*, 1996]. Under the high relative roughness of riffles, vortex shedding processes dominate turbulence, although the distribution of individual roughness elements does not strongly influence the spatial variability of turbulent flow properties [*Lamarre and Roy*, 2005]. Slowly moving fluid emanating from the near-bed region impinges upon higher levels with greater frequency and across larger areas as flow increases [*Buffin-Bélanger et al.*, 2000, 2006]. Under the low relative rough-ness of pools, the outer zone flow structure is dominated by inner-outer zone roll-up structures more akin to burst-sweep features: The distortion of streamwise subparallel vortices and the localized collapse of the sublayer organization cause bursts [*Robinson*, 1990]. The ejection of low momentum fluid into the outer flow results in a compen-sating inrush of outer flow fluid towards the bed, which probably initiates the next generation of vortices and distorts the sublayer to initiate further bursting [*Grass*, 1971; *Kirkbride*, 1993]. High speed sweeps occur as wall-directed inrushes of higher than average downstream velocity fluid [*Best*, 1993]. Once the sweeps contact the channel boundary, they spread laterally and lose momentum downstream [*Grass*, 1971].

Robert's [1998] measurements of velocity profiles in pools and riffles illustrate the results of these differences in turbulence-generating mechanisms. At low flows, riffles are characterized by higher near-bed velocity gradients than pools, and have a greater resistance and bed shear stress. The differences in near-bed velocity gradients decrease as discharge increases, although velocities higher in the water column remain significantly different between pools and riffles. Within a riffle, boundary roughness can generate complex secondary currents and flow convergence, although the spatial structure of the flow field becomes smoother as stage increases [*Legleiter et al.*, 2007].

Several investigators have documented the distinctive velocity and turbulence features associated with forced pools created by lateral constrictions. The constriction creates a backwater in the upstream portion of the pool, which causes flow acceleration and steeper water-surface slopes past the constriction [*Kieffer*, 1985; *Schmidt et al.*, 1993; *Thompson et al.*, 1999]. The jet of accelerated flow is bounded laterally by recirculating eddies. Vortices along the shearing zone between the jet and the eddies [*Clifford*, 1993b; *Thompson*, 2004], and the high near-bed velocities and turbulence associated with the jet, promote bed scouring during high flows [*Thompson*, 2007] (Figure 3.29). Many of these flow patterns can be simulated with carefully calibrated two-dimensional flow models [*Thompson et al.*, 1998; *Harrison and Keller*, 2007].

Flow is steered by the channel boundary, as well as shaping the channel boundary by erosion, transfer, and deposition of sediment [*Church*, 2003]. This creates "pinned," or persistent, secondary circulations that scale with the channel dimensions. Channel width is scaled by the flow that the channel must transmit and by the bank materials. Channel geometry then scales turbulence because the flow and channel gradient establish the rate at which potential energy must be transformed into kinetic energy and dissipated in turbulence [*Church*, 2003].

Analogous differences in turbulence related to bedforms exist along step-pool channels [*Wohl and Thompson*, 2000; *Wilcox and Wohl*, 2007]. Bed-generated turbu-lence dominates locations immediately upstream from bed-steps and at step lips. Locations immediately downstream from bed-steps are dominated by wake turbulence from mid-profile shear layers associated with roller eddies where the flow from the step plunges into the pool below. Adverse pressure gradients (decelerating flow in the upper profile) up- and downstream from steps may be enhancing turbulence generation, whereas favorable pressure gradients (accelerating flow in the upper profile) at steps suppress turbulence [*Wohl and Thompson*, 2000] (Figure 3.30). Three-dimensional measurements of point velocity in step-pool channels indicate large contributions to turbulent kinetic energy from the vertical component of velocity [*Wilcox and Wohl*, 2007]. Discharge and morphologic position significantly affect turbulence intensity, with the greatest turbulence intensities occurring in pools and at high discharges.

Each step in a step-pool sequence has the potential to create a *hydraulic jump*. The classical hydraulic jump is a steady unsubmerged jump with a horizontal supercritical jet, a rapidly varied aerated roller, a gradually varied outflow, and the downstream tailwater (Figure 3.31A). Other jump types form as a function of the ratio of upstream hydraulic head and bed step height to the tailwater depth. Submerged jumps occur when the tailwater elevation exceeds jump height (Figure 3.31B). Sloping jumps occur when the jump forms on or below a sloping bed that impinges on a horizontal bed (Figure 3.31C) and are typically unsubmerged. Form roughness elements creating submerged and unsubmerged jumps are highly irregular in step-pool channels, but inclination angle and projection length of the jet, as well as submergence and steepness of the jump, control the underlying bed shear that drives morphologic change near jumps [*Vallé and Pasternack*, 2006].

Investigations of the effect of channel confluences on velocity and turbulence have focused primarily on lower-gradient rivers, but confluences also occur in mountain stream networks. Characteristic patterns of flow occur in the vicinity of channel confluences: lower velocities in the area of mixing and accelerated flow downstream from the confluence; a stagnation zone at the apex of the channel junction; a zone of flow deviation and strong fluid upwelling close to the avalanche face and at the margin of the tributary mouth bar; and lower velocities over the bar at the downstream corner of the junction [*De Serres et al.*, 1999]. The water surface is super-elevated in the zones of stagnation and mixing, and tilted at the edge of the mixing layer [*Biron et al.*, 2002]. These patterns in water surface reflect turbulence; a zone of quasi-two-dimenstional turbulence energy at low frequencies is present in the mixing zone, for example [*Rhoads and Sukhodolov*, 2004]. Although the shear layer at the mixing zone occupies a limited portion of the cross-sectional area of flow, turbulence kinetic energy in the shear layer is two to three times greater than in the rest of the flow [*Sukhodolov and Rhoads*, 2001]. Helical motion enhances patterns of mixing at confluences [*Rhoads and Sukhodolov*, 2001]. Similar investigations using arrays of three-dimensional current meters document the turbulent mixing occurring at the interface between flow in the channel and across the floodplain [e.g., *Carling et al.*, 2002].

Downstream passage of mobile bedforms such as sand dunes or gravel bedload sheets may also have coincident fluctuations in velocity and flow resistance [*Dinehart*, 1999; *Prent and Hickin*, 2001]. Streamwise momentum is reduced when grains collide with one another on the bed or in the flow, causing a decrease in streamwise flow velocity and an increase in flow resistance that *Gao and Abrahams* [2004] denote as f_{bt}.

At present, the details of sand bedforms are better understood than are those of gravel bedforms. Flow over sand ripples and dunes is dominated by shear layer instability associated with separation zones in the lee of the bedform [*Bennett and Best*, 1996]. Amalgamation into larger ripples generates larger-scale turbulence that begins to advect through the entire flow depth. Turbulent coherent structures penetrating into the outer flow induce return flows of greater magnitude that exert greater shear stress as they impact the bed, which increases sediment transport and produces higher bedforms that evolve into dunes [*Bennett and Best*, 1996; *Fernandez et al.*, 2006]. Analogous patterns may be less consistently developed above coarser-grained features such as pebble clusters because of interference from surrounding grain roughness, but *Lawless and Robert* [2001b] identify six regions with distinct vertical flow characteristics around pebble clusters: flow acceleration up the stoss side; recirculation behind the cluster; vortex shedding from the crest and shear layer; flow reattachment downstream from the cluster; upwelling of flow downstream from the point of reattachment; and recovery of flow (Figure 3.32).

Measurements of coherent flow structures have relied primarily on experimental or field studies using flow visualization or single-point measurement techniques. Computational fluid dynamics has recently been applied to studies of the generation, evolution, and destruction of flow structures [*Hardy et al.*, 2007] and, combined with the ability to rapidly obtain spatially dense 3d point-velocity measurements, suggests the potential to dramatically improve numerical simulations of turbulence. *Church* [2003] discusses the scales of turbulent flow.

Understanding of flow and turbulence characteristics is increasingly applied to understanding the mechanics of bedload entrainment and transport along mountain rivers [*Hunt and Papanicoloau*, 2003]. Flume experiments indicate the complexities of bedload-flow interrelations, however; although bedload affects flow velocity by modifying the rate of dissipation of turbulent kinetic energy, bedload can increase or decrease flow velocity depending on bed roughness and the relative magnitude of flow and sediment transport variables [*Carbonneau and Bergeron*, 2000]. Specifically, bedload can increase flow turbulence and decrease mean velocity over smooth beds, but reduce turbulence and increase velocity over rough beds. The conditions under which sediment movement occurs are commonly defined in terms of bed shear stress.

In summary, the spatially and temporally variable velocity of mountain rivers can be characterized using point measurements or dilution tracers. Point measurements, particularly if obtained at high spatial densities using an instrument such as an acoustic Doppler current profiler, are more useful for examining spatial variations and turbulent patterns. Dilution tracers better represent spatially averaged velocity. Velocity profiles along steep, coarse-grained mountain rivers typically have a low-velocity zone between larger boulders and a high-velocity zone above the characteristic roughness height. Mountain rivers exhibit various scales of turbulence. Although momentum transfer via bursts and sweeps occur in diverse settings, systematic differences in turbulence characteristics occur in relation to spatial differences in bedforms and grain-size distributions.

3.4.4. Bed Shear Stress

Much of our understanding of bed shear stress and the conditions under which sediment begins to move is based on the work of Albert Shields during the 1930s [*Buffington*, 1999]. Shields conducted flume experiments using a variety of sediment types with narrow grain-size ranges and initially planar beds. He then described initial motion as discharge increased, although it is not clear whether Shields specified initial motion by extrapolating paired measurements of shear stress and bedload transport rate to a zero level of transport (reference technique) or through visual observation of the flume bed surface (visual technique) [*Buffington*, 1999]. Shields expressed incipient grain motion as a dimensionless ratio of the critical bed-shear stress (τ_c') to submerged grain weight per unit area

$$\tau_c' = \tau_c^*(\rho_s - \rho)gD \tag{3.41}$$

where τ_c^* is dimensionless critical shear stress, ρ_s is sediment density, ρ is water density, and D is characteristic grain size. Most subsequently developed equations for incipient grain motion take this basic form.

Attempts to measure or calculate bed shear stress values in mountain rivers are complicated by the channel-bed roughness and the associated turbulence and velocity fluctuations. Because of these conditions, the Shields' entrainment function commonly used for lower gradient channels with uniformly sized grains does not adequately predict initiation of motion [*Reid et al.*, 1985; *Graf*, 1991]. *Dietrich and Whiting* [1989] review nine equations available for estimating local bed shear stress from field data. Many of these equations use terms such as the fluctuating component of downstream or vertical fluid velocity, which can be difficult to define accurately in field studies.

The only method for directly measuring shear stress may be expressed as

$$\tau_b = \rho \, \overline{u'w'} \tag{3.42}$$

where τ_b is boundary shear stress, u', w' are the fluctuating component of downstream and vertical fluid velocity, respectively, and ρ is density of fluid. *Cooper and Tait* [2010] suggest that flow velocity data should be evaluated by spatial averaging for the Reynolds equations to produce time- and space-averaged (double-averaged) momentum equations in order to better understand the physical mechanisms that generate boundary shear stress.

Spatially detailed velocity and turbulence measurements have proven difficult to obtain in natural rivers with extremely rough or mobile beds [*McLean and Smith*, 1979], leading to the use of relations derived from theoretical models. The two most commonly employed relations are:

$$\tau_b = \rho \, g \, h \, S \tag{3.43}$$

where h is depth of flow and S is downstream water-surface slope, and

$$\tau_b = \frac{\rho(u\kappa)^2}{\left(\ln\left(\dfrac{z}{z_0}\right)\right)^2} \tag{3.44}$$

where κ is von Karman constant, u is velocity, z is near-vertical coordinate, perpendicular to bed, and z_0 is roughness parameter including effect of saltating grains.

In general, z_0 is hypothesized to be proportional to the saltation height of the moving grains, or controlled by a representative coarser fraction of the moving or static bed surface. Equation 3.43 is only approximately correct over short channel reaches because of convective accelerations [*Dietrich and Whiting*, 1989]. When using equation 3.44, it is difficult to make reliable velocity profile measurements over mobile beds and beds with large grain sizes. It becomes necessary to make profile measurements very close to the bed to avoid the form drag of bedforms [*Dietrich and Whiting*, 1989].

Bed shear velocity u_* ($u_* = (\tau_0/\rho)^{0.5}$, where τ_0 is local shear stress and ρ is fluid density) can be estimated within 3% using the depth-averaged velocity in the vertically averaged logarithmic velocity profile, if the channel has a relatively simple flow geometry that approximates a log profile [*Wilcock*, 1996]. Estimates of u_* made from a single near-bed velocity measurement are less precise by a factor of 3 (for Wilcock's sites; d_{90} of 85-120 mm, Q of 23-80.5 m^3/s, h of 0.75-2.6 m) because of the larger uncertainty associated with a single measurement [*Wilcock*, 1996]. Estimates of u_* from the slope of the near-bed velocity profile are the least precise, but can be made without independent knowledge of bed roughness.

Part of the difficulty in obtaining useful estimates of bed shear stress arises from the relatively high temporal and spatial variability of this factor along mountain rivers. The turbulence described previously can lead to substantial variability in velocity and shear stress at a point during constant discharge. Heterogeneities of the channel bed caused by grains and bedforms can also create substantial velocity and shear stress variations across a cross section or downstream during a constant discharge. Measuring shear stress across alternate bars on a gravel-bed channel (h of 0.4 m, S of 0.0010, d_{84} of 16.1 mm), *Whiting and Dietrich* [1991] find that large cross-sectional area changes resulting from variation in depth force large streamwise accelerations and cross-stream flow off the central bar. These topographically driven downstream and cross-stream accelerations produce a pattern of boundary shear stress that decreases out of the upstream bend, increases over the bar top, and then decreases in deeper flow [*Whiting and Dietrich*, 1991]. The convective accelerations across the channel become progressively less important as stage increases [*Whiting*, 1997]. The recent development of measurement methods such as acoustic Doppler current profiling facilitates rapid acquisition of spatially detailed velocity and depth measurements that can be used to calculate bed shear stress [*Sime et al.*, 2007].

The primary reason for estimating bed shear stress is that of calculating the bedload sediment entrainment and transport, and the associated erosion, deposition, and channel change. Because rates of sediment entrainment and transport can increase in a rapid and nonlinear manner with increasing bed shear stress, the accuracy of estimation of bed shear stress becomes critical, particularly at conditions near the entrainment threshold [*Wilcock*, 1996]. Spatial and temporal variability in bed shear stress are sufficient that the sum of local transport rates across a cross section can be substantially different from the total load calculated using the section-averaged shear stress [*Carson and Griffiths*, 1987; *Wilcock et al.*, 1994; *Wilcock*, 1996]. Because of these complications, estimates of bed shear stress used for estimates of sediment movement commonly focus on temporally and spatially averaged conditions or probability distributions [*Powell*, 1998]. Ability to quantify bed shear stress and initiation of sediment motion also controls ability to predict rates of bedload transport under differing flow regimes.

In summary, Shields' original experimental work relating incipient grain motion to bed shear stress was based on initially planar beds and a narrower grain-size distribution than typically occurs in mountain rivers. Most equations for incipient grain motion nonetheless utilize an equation with a format similar to Shields' by relating grain motion to dimensionless critical shear stress and a characteristic grain size. It has proven difficult to measure or calculate bed shear stress for use in these equations because of temporal and spatial variations in velocity, turbulence, and bed roughness.

3.4.5. Stream Power

Stream power is typically expressed as (i) total stream power, Ω

$$\Omega = \gamma Q S \qquad (3.45)$$

where γ is the specific weight of water (assumed to be 9800 N/m^3), Q is discharge and S is channel gradient, (ii) stream power per unit area, ω

$$\omega = \tau v \qquad (3.46)$$

where τ is boundary shear stress ($= \gamma R S$, where R is hydraulic radius), and v is velocity, or (iii) specific stream power, ω

$$\omega = \Omega / w \qquad (3.47)$$

where w is channel width. Each of these measures of stream power quantifies the energy available to perform geomorphic work against the channel boundaries, which can be related to processes such as bedrock channel incision (section 2.7.1) and sediment entrainment and transport.

Knighton [1999] develops a model of downstream variations in total and specific stream power, assuming that the longitudinal profile of the river has an exponential form. The model predicts that total stream power peaks at an intermediate location, depending on the ratio of the downstream rates of change of discharge and slope, respectively. Specific stream power is more sensitive to rate of change in slope and peaks closer to the headwaters. The model accurately predicts patterns along the River Trent in the UK [*Knighton*, 1999]. Limited field tests of the model in mountain catchments indicate substantial deviations from hypothetical patterns as a result of local influences on fluvial process and form [*Fonstad*, 2003]. *Jain et al.* [2006] discuss different methods of analysing stream long profiles and the resulting differences in longitudinal distribution of stream power. They attribute variations in stream power within headwater reaches to discharge variability, whereas variability in channel gradient becomes more important in mid-basin and downstream reaches. Rates of longitudinal change in stream power have also been related to relative channel response to high-magnitude floods; the greatest responses correspond to rapid downstream decreases in stream power [*Reinfelds et al.*, 2004].

Costa and O'Connor [1995] illustrate conceptually how differences in stream power magnitude and duration relative to erosional thresholds in a channel govern the geomorphic effectiveness of different kinds of floods (Figure 3.33). *Kale and Hire* [2007] have published one of the few studies to date that attempts to quantify these relations by integrating stream power through time in relation to discharge variations. They use continuous daily discharge data from 1978-1990, hydraulic geometry equations and the relationship between discharge and water-surface slope to compute the daily specific stream power for a site on the lower Tapi River in central India. The total energy generated is then estimated by integrating the data under the stream power graph. Annual peak floods contribute anywhere from 3 to 34% of the total energy expended during a monsoon season and suspended sediment load is strongly related to energy [*Kale and Hire*, 2007].

In summary, our ability to quantify the magnitude of different sources of resistance in steep, coarse-grained mountain rivers remains limited by the lack of broadly applicable equations for each component of resistance. Techniques such as ground-based LiDAR are dramatically improving physical characterization of channel boundary roughness, however, and in combination with improved hydraulic measurements and modeling, will likely facilitate substantial advances in the next few years. Improved ability to quantify bed roughness will in turn improve the ability to predict near-bed hydraulics and thus the magnitude of forces associated with velocity, turbulence, shear stress, and stream power exerted against the channel boundaries. As we empirically characterize bed configuration and hydraulics over a wider range of channel conditions (in terms of gradient, grain-size distribution, bedforms, and relative submergence), we will better be able to determine whether we need multiple equations to describe parameters such as shear stress under different channel configurations or whether it is possible to develop a single, broadly applicable equation.

3.5. Sediment Processes

3.5.1. Bed Sediment Characterization

3.5.1.1. Sampling and measurement. Critical shear stress for sediment movement will be a function of local hydraulics, as well as grain size, shape, sorting, and packing

[*Barta et al.*, 1994; *Bartnik and Michalik*, 1994; *Moore and Diplas*, 1994]. Shear forces dominate clast entrainment only when a clast protrudes into the flow, for example; lift force dominates when the clast is level with neighboring particles [*Ergenzinger and de Jong*, 1994]. One of the first challenges to accurately predicting bedload transport is therefore that of characterizing the bed surface sediments that affect local hydraulics and may provide the source material for bedload transport.

The grain-size distribution of channel-bed sediments can be characterized via bulk (volumetric) sampling or clast measurements *in situ*. In either case, the spatial variability in grain-size distribution (e.g., between pools and riffles, or pools and steps) along the channel must be addressed. A representative sample can be obtained by combining subsamples of equal volumes [*Wolcott and Church*, 1991]. *Singer* [2008] describes a sampler that can be deployed from a boat, extract samples up to 16 kg in weight, and penetrate approximately 5 cm below the bed surface.

Bulk sampling can be extremely difficult along mountain rivers with sediment coarser than gravel. In a study of a braided gravel-bed river (d_{50} of 16 mm), *Mosley and Tinsdale* [1985] conclude that accurate determination of mean grain size requires a sample of approximately 100 kg, although samples in which the weight of the largest clast is less than 5% of the total weight have unbiased estimates of mean grain size. Bias is likely in small samples of river bedload and good precision requires very large samples of poorly sorted gravel deposits [*Ferguson and Paola*, 1997], although matrix-supported sediments and unimodal distributions can be characterized more easily than framework-supported sediments and bimodal distributions [*Haschenburger et al.*, 2007]. As the b-axis of the largest clast at a site approaches 64 mm, required sample sizes become impractical (> 400 kg) for field sieving by hand [*Church et al.*, 1987]. Consequently, some type of *in situ* clast measurement is more commonly used, although such methods do not consistently sample clasts smaller than 15 mm [*Fripp and Diplas*, 1993] and tend to sample most precisely grains between the 66th and 91st percentiles [*Green*, 2003].

In situ clast measurements may employ a grid [*Wolman*, 1954]; a random walk [*Leopold*, 1970]; a visual comparator [*Billi*, 1994]; areal sampling with clay or with wax [*Diplas and Fripp*, 1992]; Fourier or other spectral analysis of particle outlines on digitized photographs [*Diepenbroek and de Jong*, 1994; *Buscombe et al.*, 2010]; a systematic unaligned method in which the study area is divided into a number of arbitrary cells within each of which a sampling position is chosen [*Wolcott and Church*, 1991]; or an area count of all clasts exposed at the surface within a given unit area [*Kellerhals and Bray*, 1971; *Wohl et al.*, 1996]. *Petrie and Diplas* [2000] develop techniques to calculate the confidence intervals for grid, aerial, and volume samples. Recent advances in methodology tend toward analysis of airborne imagery obtained from airplanes, helicopters, balloons, and other platforms [*Carbonneau*, 2005; *Carbonneau et al.*, 2005; *Graham et al.*, 2005; *Verdú et al.*, 2005; *Dugdale et al.*, 2010] or toward ground-based photographic or laser scanning imagery [*Heritage and Milan*, 2009; *Hodge et al.*, 2009; *Warrick et al.*, 2009], and statistical descriptions of surface organization using probability density distributions of the bed-surface elevation [*Marion et al.*, 2003]. Use of any type of imagery requires resolving the minimum area required to obtain a representative sample (between 50 and 200 times that of the largest grain), the effect of lower-end truncation on grain-size percentiles, the effect of river-bed structure such as imbrication and hiding, and the use of individual particle measurements versus size classes [*Graham et al.*, 2010].

Each grain on a surface projects, on average, an area proportional to the square of its sieve diameter, facilitating the comparison of bulk and surface samples [*Diplas and Sutherland*, 1988]. If the results of surface sampling are adequately described by the binomial distribution [*Fripp and Diplas*, 1993], then surface samples require only about a third of the material that a bulk sample requires to meet the low accuracy levels described by *DeVries* [1971] [*Crowder and Diplas*, 1994]. This still requires 200-400 clasts for surface samples [*Fripp and Diplas*, 1993]. *Fripp and Diplas* [1993] propose a method to determine the required sample size based on the percentile value (size fraction) of interest and the level of accuracy required in the study.

Different methods of surface sampling may not produce equivalent results. Approximately 500 random sites are equivalent to 100 systematic unaligned or grid sites [*Wolcott and Church*, 1991]. Random walk and grid methods produce statistically indistinguishable values of d_{50} and d_{84} when performed by a single operator, although multiple operators can generate statistically different population measures [*Wohl et al.*, 1996]. Differences between operators occur because of differences in clast selection and visualization of mutually perpendicular axes, and become statistically significant as sample size increases past 100 clasts [*Hey and Thorne*, 1983]. The standard deviation about replicated means increases linearly with sediment size, with larger standard deviations for samples collected by different individuals than for sampling by one person [*Marcus et al.*, 1995].

Wolman [1954] advocates a surface sample size of 100 clasts and this remains the most commonly used sample size. Each clast is commonly characterized by the diameter of the intermediate, or *b*, axis. Clast measurements can be used to determine (i) frequency by weight, where particle volume is computed and a constant specific weight is assumed, so that the frequency of each size interval can be expressed, or (ii) frequency by number, where the frequency of each size interval is expressed as the percentage by number of the total number of particles in the original sample. *Leopold* [1970], *Kellerhals and Bray* [1971], and *Diplas and Fripp* [1992] propose methods for converting frequency by number to frequency by weight data, although the Kellerhals and Bray method may bias the finer particles [*Ettema*, 1984]. *Kondolf et al.* [2003a] review different methods of characterizing surface and subsurface bed sediments.

Buffington and Montgomery [1999a] propose a two-tier system of ternary diagrams for classifying textural patches (grain-size facies) in gravel-bed channels. The procedure identifies the relative abundance of major size classes (boulder, cobble, gravel, sand, silt) and subcategories of the dominant size (very coarse, coarse, medium, etc). The procedure can be used to create facies maps which provide stratification for sampling physical and biological conditions.

In summary, quantifying the grain-size distribution of bed sediments remains an important and difficult task in poorly sorted, coarse-grained mountain rivers. Acquisition of air- or ground-based imagery that can be digitally processed is being increasingly used to obtain spatially extensive and accurate data more quickly and less expensively than traditional sampling methods.

3.5.1.2. Coarse surface layers. The surface of the channel bed can also be described in terms of selective size distribution by depth. *Carling and Reader* [1982] distinguish among three types of coarse surface layers: (1) *pavement* – a coarse surface layer that is rarely disrupted (Figure 3.34); (2) *armor* – a coarse lag layer developed at waning flows that is regularly disrupted; and (3) a *censored layer* that forms as matrix material is removed from around the surface framework particles as stage increases. The bed of a channel can have a coarse-grained surface that is characteristically one grain diameter thick and is both coarser and better sorted than the underlying material [*Moss*, 1963, 1972; *Gessler*, 1965]. *Bray and Church* [1980] propose the term 'armor' for such a surface when particle motion is a relatively frequent occurrence over a period of years. 'Coarse surface layer' will be used here as a generic term that does not imply any specific mechanism of formation or frequency of mobility or disruption.

Theoretical models for formation of coarse surface layers vary from one-step models that predict equilibrium compositions and scouring depths, to multiple-step models that describe different stages of coarsening processes [*Sutherland*, 1987]. No consensus has yet been reached as to how the coarse surface layer forms, however, at least in part because the mechanism of formation undoubtedly differs among channels (Table 3.2). *Carling and Reader* [1982], for example, describe a scenario for upland streams in the United Kingdom in which fine sediments are winnowed from the interstices of the coarse surface layer by intermediate flows, forming what *Sutherland* [1987] describes as a static armor layer. In contrast, a mobile armor layer is a coarse surface layer that is maintained in the presence of an upstream sediment supply and during flows capable of moving all grain sizes [*Powell*, 1998; *Muskatirovic*, 2008]. *Dunkerley* [1990] describes coarse grains carried in traction to the sites which they armor, and deposited over preexisting bodies of sediment. Armor layers can thus develop through kinematic sorting [*Wilcock*, 2001] in zones where reduced flow velocities cause deposition of large clasts, and where particle clustering or traction clogging occurs. *Whiting and King* [2003] distinguish supply-limited components of an armor layer (particles for which the supply is less than the ability of the stream to transport the sizes) and hydraulically-limited components (particles which the stream cannot transport). *Garde et al.* [2006] develop a simple mathematical model to predict the particle size distribution of the armor layer.

Coarse surface layers typically are well developed when local bedload supply from upstream is less than the ability of the flow to transport that load, as occurs downstream from dams [*Dietrich et al.*, 1989]. *Gran and Montgomery* [2005] document progressive changes in sediment supply, bed roughness, and bedload transport following the 1991 eruption of Mount Pinatubo in the Philippines. Finer-grained sediment is initially mobilized preferentially through selective transport. As the bed coarsens, gravel-size clasts interact to form structures that increase bed roughness and critical shear stress and inhibit clast mobility. With progressive declines in sediment inputs, the channel armors through winnowing. Other investigators also document the formation of stable, coarse surface layers as sediment supply declines [*Liébault and Piégay*, 2001; *Liébault et al.*, 2002; *Grams et al.*, 2007].

Ephemeral channels typically have less developed armored surfaces than perennial channels [*Almedeij and Diplas*, 2005; *Hassan*, 2005], although *Hassan et al.* [2006] infer that sediment supply is a first-order control on bed-surface armoring, whereas hydrograph characteristics play a secondary role. Dryland rivers in Australia, which have low sediment supply and infrequent mobilization of bedload, exhibit less organization of bed-material surface facies than fully self-adjusting rivers with frequent reworking of bedload [*Hoyle et al.*, 2008].

Using a one-dimensional model in which discharge fluctuates repeatedly, *Parker et al.* [2008a] find that the upstream end of the modeled reach repeatedly aggrades and degrades in response to changes in sediment transport, whereas the bed adjusts a short distance downstream so that bed elevation and surface grain-size distribution become invariant in time even though bedload transport continues to fluctuate. *Parker et al.* [2008a] refer to the length of adjusting bed as the hydrograph boundary layer.

Some of the variability in mechanisms of forming and maintaining the coarse surface layer may reflect position within the drainage basin, as this influences channel gradient, grain-size distribution, and channel morphology. Based on data from 13 gravel-bed drainage basins (drainage area 1.5 to 28,000 km^2) from around the world, significant volumes of fine bedload are transported during discharges less than bankfull in channels with high ratios of bed-material d_{50} to bedload d_{50}, while the coarser bed-material substrate remains stable [*Lisle*, 1995]. Moving downstream, the bed as a whole is accessed for bedload by deeper annual scour and the difference between transport of finer and coarser portions of the grain-size distribution decreases [*Lisle*, 1995]. The coarse surface layer may also be spatially variable at the reach scale as a result of variations in shear stress and sediment transport associated with alternate bar topography [*Lisle and Madej*, 1992]; as a result of hydraulic roughness caused by large instream wood [*Buffington and Montgomery*, 1999b]; or as a result of variations in rate of sediment supply [*Buffington and Montomery*, 1999c; *Vericat et al.*, 2006]. The fine sediments winnowed from a stable armor layer can be temporarily stored in the separation flow downstream from large bed elements. Alternatively, the fine sediments can be transported as an under-capacity bedload over a stable cobble-bed. The resulting open-work surface gravel has the same grain-size distribution as the subsurface layer and is not really an armor layer [*Sutherland*, 1987].

Bedload is characterized by grain-size distributions which are finer than, and can be completely distinct from, the grain-size distributions of the bed material [*DiSilvio and Brunelli*, 1991]. Bedload and subpavement size distributions are similar, which *Parker and Klingeman* [1982] interpret as indicating that the coarse half of a subpavement moves through a reach at nearly the same rate as the fine half. Because coarser grains are intrinsically less mobile than finer grains, some mechanism must act to nearly equalize mobility. They hypothesize that the pavement seen in gravel-bed streams at low flows is in place during typical flows capable of moving all available grain sizes. This pavement provides the equalizing mechanism by exposing proportionally more coarse grains to the flow. Pavement is thus a mobile bed phenomenon present under a range of flows [*Andrews and Erman*, 1986; *Suzuki and Kato*, 1991; *Wilcock and DeTemple*, 2005; *Clayton and Pitlick*, 2008]. *Andrews and Erman* [1986] find that a significant quantity of sediment representing a majority of the particle sizes present at the channel surface is transported during a period of sustained, large discharge. The relatively coarser surface layer that is present during smaller discharges is also in place and unchanged. Only a few clasts, though representing nearly all available sizes, are entrained at any instant by even the peak flows. Differential entrainment is thus not a significant process and the coarse bed surface is constructed and maintained when all sizes of bed material are moving.

In summary, perhaps the only safe generalization regarding coarse surface layers is that they are typical of coarse-grained, poorly sorted mountain rivers that have not recently experienced a complete disruption of the bed such as that caused by an unusual flood or a debris flow. Evidence suggests that coarse surface layers can form and persist via a variety of mechanisms that partly reflect sediment supply, flow regime, channel geometry including sources of roughness, and position within the drainage basin.

3.5.2. Particle Clusters

Even the most well developed and stable coarse surface layers typically have small-scale ($< 10^{-1}$ m) spatial variability in roughness associated with individual grains or particle clusters. Particle clusters, also known as pebble clusters, imbricate clusters, cluster bedforms, and microforms, were first described by *Dal Cin* [1968]. Closely nested groups of clasts aligned parallel to flow constitute particle clusters, which are typically 0.1-1.2 m in length in the streamwise direction [*Brayshaw*, 1984]. An obstacle clast anchors a stoss-side accumulation of imbricated particles and a wake tail [*Papanicolaou et al.*, 2003]. Particle clusters have been observed in gravel-bed channels with low bed-material transport rates [*Church et al.*, 1998], high bed armoring [*Biggs et al.*, 1997], and bimodal sediment size distribution [*Hendrick et al.*, 2010], and typically occur in riffle, alternate bar, or plane-bed sections of the channel. Clusters form during the recessional limb of floods as particles are deposited around exceptionally large clasts [*Brayshaw*, 1984].

Particle clusters influence hydraulics and sediment processes in gravel-bed rivers. Localized flow acceleration and deceleration, recirculation, increased turbulence intensity, vortex shedding, flow reattachement, fluid upwelling, and flow recovering are all present downstream from clusters [*Buffin-Bélanger and Roy*, 1998; *Lacey et al.*, 2007]. Clusters increase boundary resistance up to a maximum value related to longitudinal spacing of the clusters [*Hassan and Reid*, 1990], analogous to the effects described for other roughness elements in section 3.4.2. The flow resistance induced by clusters is also inversely related to bedload flux [*Hassan and Reid*, 1990]. Clusters delay incipient motion and limit the availability of bed material for transport [*Brayshaw*, 1984], although the wake particles are entrained first, followed by the stoss particles and finally the obstacle clast [*Billi*, 1988; *Wittenberg and Newson*, 2005]. Clusters can also disperse without movement of the obstacle clast [*De Jong*, 1991]. Creation of a significant component of form drag via the construction of clusters may precede disruption of the bed during a transport event [*Clifford et al.*, 1992a], although clusters are resistant to entrainment during floods and can thus provide refuge for benthic organisms [*Biggs et al.*, 1997]. Experimental removal of clusters resulted in increased sediment yield in a small headwater stream in British Columbia, Canada [*Oldmeadow and Church*, 2006].

Tribe and Church [1999] experimentally produce clusters during a two-dimensional kinematic simulation of a gravel streambed. These structures develop simultaneously with the armor layer by particles moving from less stable positions into more stable configurations against each other. The clusters substantially reduce sediment transport. The timescale of particle cluster development suggests that the bed structure of gravel-bed channels may reflect the history of dominant flows rather than more recent flows [*Church et al.*, 1998].

Flume experiments reveal that clusters develop in uniform sediment at 1.25 to 2 times the Shields parameter of an individual particle and start disintegrating at about 2.25 times the Shields parameter [*Strom et al.*, 2004]. The clusters follow an evolutionary sequence of individual particles with no cluster, followed sequentially by a two-particle cluster, a comet-shaped cluster of loosely packed in-line clasts, a triangular cluster of well-packed particles, a stable rhomboid or diamond shape with a low drag coefficient and, as shear stress increases, disintegration and a return to no clusters [*Papanicolaou et al.*, 2003]. Sediment availability also affects the architecture and size of clusters [*Papanicolaou and Schuyler*, 2003], with clusters forming in mixed sand-gravel beds as sand supply decreases [*Gran et al.*, 2006]. The presence of clusters can be predicted based on values of the non-dimensional parameter wQ^2/gd_{84}^6, where w is bankfull channel width and d_{84} is bed sediment [*Strom and Papanicolaou*, 2009].

In summary, particle clusters – closely aligned groups of clasts – increase boundary roughness and influence hydraulics, sediment dynamics, and bed stability. Clusters are stable at lower values of bed shear stress and may form during the recessional limb of floods. Clusters have been described primarily in riffle, alternate bar, or plane-bed channel segments.

3.5.3. Sediment Entrainment

Although the amount of sediment moving as bedload may be less than a quarter of that moving in suspension along many rivers [*Williams and Rosgen*, 1989], studies of high-gradient rivers indicate that bedload composes a much higher proportion of the total sediment load along these channels than along low-gradient channels [*Hayward and Sutherland*, 1974; *Nanson*, 1974; *Mosley*, 1978; *Bradley and Mears*, 1980; *Harvey*, 1980; *Schick and Lekach*, 1993]. Bedload is also generally much more important than suspended load in forming and changing the channel of a mountain river [*Pitlick and Thorne*, 1987; *Leopold*, 1992]. As a result, more attention has been devoted to bedload processes along mountain rivers than to suspended sediment processes.

Emphasis during the 1980s on gravel-bed rivers led to increased attention on bedload entrainment and transport from poorly-sorted channel beds. Earlier attempts to quantify conditions of entrainment balanced drag and body forces acting on a grain by specifying critical shear stress, packing, pivoting angle, and grain diameter. This approach ignores lift force and describes instantaneous conditions which may deviate substantially from the measured mean shear stress [*Richards*, 1990]. Consequently, this deterministic approach has been largely replaced by physically based stochastic models that treat the two distinct phenomena of velocity fluctuations and variations in grain size and pivoting angle [*Richards*, 1990].

Pivoting angle represents the particle's resistance to movement. This is dependent on (i) the ratio of grain diameter d to the underlying grain size, as this affects relative protrusion of the grain into the flow, (ii) particle shape, and (iii) packing (imbrication) [*Komar and Li*, 1986; *Li and Komar*, 1986]. Some of these variables, such as particle shape, appear to become more important as the size of entrained and transported particles increases relative to the size of the bed roughness elements [*Demir and Walsh*, 2005]. Although Shields' critical shear stress predictions work well for uniformly sized sediment and for the median grain size in a mixed distribution [*Shvidchenko et al.*, 2001], critical shear stress is lower than predicted when d is larger than the length scale of the bed roughness, k_s ($d/k_s > 1$), and higher when $d/k_s < 1$ [*Wiberg and Smith*, 1987]. *Armanini and Gregoretti* [2005] quantify this effect, which they term the relative degree of exposure. Values of critical shear stress for mobilizing sediment from the armor layer increase systematically with average channel gradient [*Mueller et al.*, 2005]. A pivoting angle expression may be developed for entrainment of a single grain [*Wiberg and Smith*, 1987], such as:

$$\zeta = \cos^{-1}\left[\left(\frac{d}{k_s}+z\right)\left(\frac{d}{k_s}+1\right)\right] \tag{3.48}$$

where z = grain protrusion and ζ = pivoting angle. There have been few attempts to estimate representative pivoting angles and exposure parameters, however, for entire regions of a mixed grain size bed so that general transport rates can be predicted [*Richards*, 1990]. As a result, many assumptions are made when applying an equation such as 3.48 to an actual field setting [*Chase*, 1994].

An exception to the lack of field studies of pivoting is that of *Johnston et al.* [1998]. They use a digital load cell to directly measure the force required to pivot or slide a particle out of its resting place for 8000 particles in five rivers. These measurements indicate that relative grain size (d_i/k_s) is the only statistically significant variable for predicting median ζ within a site. *Johnston et al.* [1998] also conclude that particle resistance to motion should be measured in the field. Location-general empirical relations that attempt to predict pivoting angle at a given site without direct *in situ* measurements will have uncertainty on the order of $\pm$ 10 degrees.

Instantaneous fluctuations in the flow field around particles exert a strong control on entrainment. Bed irregularities can substantially increase these fluctuations; flume experiments demonstrate peak deviations of about 30% of the mean horizontal force for a sphere in a smooth bed, but deviations increase to twice the mean downstream from a step and four times the mean when the sphere protrudes roughly half its diameter above the bed [*Schmeeckle et al.*, 2007]. The magnitude and duration of the instantaneous turbulent forces applied to a sediment grain, as well as their product, or impulse, determine the grain's threshold of motion [*Diplas et al.*, 2008; *Valyrakis et al.*, 2010].

Experimental flume studies of initial motion in mixed grain size beds variously focus on vibration *in situ*, single grain movement [*Andrews and Smith*, 1992], or general motion of the bed [*Richards*, 1990]. These studies indicate that the grain shear stress effective in entrainment and transport is systematically less than the total shear stress estimated from the law of the wall or the depth-slope product because of form drag effects from pebble clusters and other bedforms [*Petit*, 1989]. Other flume studies suggest that discharge or stream power may be more effective predictors of entrainment and transport for steep, poorly sorted beds [*Bagnold*, 1977; *Bathurst*, 1987b; *Bathurst et al.*, 1987; *Ashmore*, 1988; *Ferguson*, 1994]. A cautionary note for all flume studies lies in the fact that the stability of a graded sediment bed depends not only on the applied shear stress of the antecedent flow, but also on its duration [*Monteith and Pender*, 2005].

Field studies of entrainment are difficult to compare because of a variety of measurement techniques and sampling procedures [*Komar*, 1987b]. Initial entrainment may be defined by the largest grain size caught in a bedload sampler, or by extrapolating size-fractional transport rates to a low value at which shear stress is identified and related to grain-size characteristics [*Parker et al.*, 1982]. Shear stress may be estimated using various time- and space-averaging procedures, as explained earlier. Entrainment and deposition can also be inferred from measured changes in bed elevation [*Trayler and Wohl*, 2000; *Konrad et al.*, 2002; *Brasington et al.*, 2003].

Field studies of entrainment led to the concept of *Parker et al.*, 1982; *Parker and Klingeman*, 1982; *Andrews*, 1983]. According to equal mobility, the exposure of large grains compensates sufficiently for their greater submerged weight so that all grain sizes are entrained at approximately the same shear stress. Critical shear stress is thus independent of a grain's size, but dependent on the substrate median size d_{s50}. This relation is expressed by using a hiding factor d_i/d_{s50} [*Andrews*, 1983]:

$$\theta_{ci} = 0.0834 \left(\frac{d_i}{d_{s50}} \right)^{-0.872} \tag{3.49}$$

At shear stresses exceeding this critical value, all grain sizes are transported at rates in proportion to their presence in the bed material [*Powell*, 1998], although the entrainment of bed particles is sporadic and only a small portion of the available sediment is in motion at any given time [*Andrews and Erman*, 1986]. Lower transport rates and lower excess shear stresses require a greater degree of surface coarsening in order to equalize the mobility of different grain-size fractions [*Parker*, 1990]. Some subsequent studies of bedload transport in mountain rivers support the concept of equal mobility [*Marion and Weirich*, 2003], but results are mixed.

Parker and Toro-Escobar [2002] distinguish weak and strong forms of the hypothesis of equal mobility. The weak form states that, in order to transport the coarse and fine portions of the bedload distribution through a river reach at equal rates, the coarser sediment must be over-represented on the bed surface, giving rise to mobile-bed armor. The strong form of the hypothesis states that the grain-size distribution of the coarse portion of the bedload should be similar to that of the substrate and finer than that of the surface layer. Experimental tests confirm the weak form of equal mobility in that the surface layer is significantly coarsened compared with the bedload, even when sand is excluded from the distribution [*Parker and Toro-Escobar*, 2002].

On the other hand, some work supports *selective entrainment* by grain size [*Komar*, 1987a, 1989; *Ashworth and Ferguson*, 1989; *Komar and Shih*, 1992; *Shih and Komar*, 1990a,b; *Habersack and Laronne*, 2001]. Entrainment experiments conducted with variously shaped particles indicate that the likelihood of grains of differing size being entrained at the same incident flow depends on the packing and grain-size distribution of the bed material [*Carling et al.*, 1992]. Equal mobility may apply to unimodal and weakly bimodal sediment distributions, whereas critical shear stress increases with grain size for strongly bimodal sediments [*Wilcock*, 1993]. Studying mountain rivers with high sediment loading resulting from the 1991 eruption at Mount Pinatubo, *Montgomery et al.* [1999] suggest that equal mobility and selective entrainment represent end-member concepts that apply to channels with low (or intermittent) and high (or continuous) sediment supply, respectively.

Field studies indicate variability in the mode of sediment entrainment as a function of discharge, time, and grain-size distribution (Figure 3.35). Sand becomes mobile along pool-riffle channels at lower values of shear stress than does gravel, but equal mobility is approximated at the higher measured values of shear stress and when bedload transport is integrated over longer timespans of an entire flow season [*Church et al.*, 1991; *Kuhnle*, 1992a]. Similarly, selective transport best describes ordinary snowmelt high flows (1-2 year recurrence interval) along Squaw Creek, Montana, USA (step-pool, pool-riffle, and pool-alternate bar), although equal mobility may occur during extreme flows that also drastically change channel morphology [*Bunte*, 1996]. Analyzing 20 years of bedload data from the steep, coarse-grained Rio Cordon in Italy, *Mao and Lenzi* [2007] conclude that equal mobility occurs only during the largest floods recorded (recurrence interval > 50 yr), when the levels of excess shear stress are approximately 1.45 times the critical shear stress. The step-pool sequences present in the Cordon and analogous streams promote selective entrainment and transport by effectively reducing bed shear stress and causing an apparent increase in critical shear stress [*Mao et al.*, 2008c]. *Wathen et al.* [1995] find that gravel transport in a pool-riffle channel is slightly size-selective, whereas sand transport is close to equal mobility. Similar size-specific differentiation between equal mobility and selective mobility has been observed by *Mao and Surian* [2010]. Sand and gravel transport at discharges of 0.1-1.0 m^3/s along a stable, boulder-bed step-pool channel in Colorado approximates equal mobility [*Blizard and Wohl*, 1998]. Following a large release of fine sediment from a reservoir into a cobble-bed, pool-riffle channel, sediment transport became supply-limited with respect to clay to medium sand first at upstream sites and then at downstream sites, and eventually became supply limited with respect to coarse sand to pebbles [*Wohl and Cenderelli*, 2000].

Kuhnle [1992b] makes an important distinction between equal entrainment mobility, where all sediment sizes in the bed begin to move at the same strength of flow, and equal transport mobility, during which all sediment sizes are transported according to their relative abundance in the bed material. One does not imply the other. *Kuhnle* [1992b] does not detect equal entrainment mobility, but equal transport is approached at high values of shear stress (at low shear stress, fines are over-represented). The tendency to either equal mobility or selective entrainment appears to depend on local conditions and may be influenced by the manner in which entrainment is defined in the field [*Richards*, 1990].

Part of the variability in entrainment and transport may also be associated with supply-limited versus capacity-limited conditions. In two-phase bedload transport the initial phase is supply-limited, with finer fractions of the bed material in transport over an armored surface, whereas the bed is disrupted and all grain sizes are in transport during the later phase [*Jackson and Beschta*, 1982; *Bathurst*, 1987a]. Marginal bedload transport as flow approaches the entrainment threshold during the first phase may result from mobilization of sediment from fine-grained patches [*Vericat et al.*, 2008]. During the initial phase, fine sediment may infiltrate into the bed, but during the later phase the bed may become a source of suspended sediment. Sand infiltration into a coarse-grained bed decreases as the volume of sand in motion increases, and significant infiltration occurs only to a depth equivalent to a few median grain diameters of the bed material [*Wooster et al.*, 2008]. Filling of interstitial pockets by sand or fine gravel passing over a coarser gravel bed can enhance entrainment of coarser particles and cause a shift from selective mobility to equal mobility as reduced turbulence in the near-bed region causes fluid acceleration and increased drag on coarse surface particles [*Venditti et al.*, 2010]. Fine sediment may also be incorporated into the flow by a gradual winnowing process to a depth equivalent to the d_{50} of the armor layer [*Beschta and Jackson*, 1979; *Carling*, 1984b]. Successive flows that disrupt the bed may eventually deplete the supply of fine sediments and change the suspended sediment rating curving [*Milhous*, 1982].

Based on data from several mountain streams in Colorado and Wyoming, USA, the transition from phase I sand transport to phase II sand and gravel transport typically occurs at about 80 percent of the bankfull (1.5-year recurrence interval) discharge [*Ryan et al.*, 2005]. During phase I transport across stable beds that are relatively hydraulically smooth, supply-limited sand can move as flow-parallel sand ribbons or flow-transverse barchan dunes [*Kleinhans et al.*, 2002]. Sand in transport across hydraulically rougher beds is less likely to move in bedforms.

Bunte [1990] hypothesizes a similar alteration in bedload transport during a high discharge to explain the variable transport rates observed at Squaw Creek, Montana, USA. During rising discharge, selective entrainment and transport of the fine sediments create a bedload distribution that increases regularly with discharge. When the removal of fines sufficiently exposes the cobbles on the bed, a threshold is crossed and the entire bed becomes mobile. This is a temporary condition, however, because of interactions between grain sizes (e.g., hiding, clustering) and a decrease in shear stress associated with local deposition, which promote pulses of bedload transport [*Bunte*, 1990].

Empirical observations of magnetically tagged gravels and exposure of bed tags indicate that some areas of gravel-bed channels have little, if any, bed activity even during floods [*Konrad et al.*, 2002; *Haschenburger and Wilcock*, 2003; *Hassan and Woodsmith*, 2004; *Haschenburger*, 2006]. *Haschenburger* [1999] uses the exponential density function to describe frequency distributions of scour and fill depths along gravel-bed channels. She finds that this function provides a plausible model of these channels, in which a limited area of the bed scours or fills relatively deeply (for a 15-m wide channel, up to 1 m vertically of scour or fill during the highest discharges observed). In general, there is an increasing depth of activity over an increasing proportion of the channel bed as peak discharge increases [*Haschenburger*, 1999]. *Kaufmann et al.* [2008] characterize these types of spatial unevenness in bed mobility using an index of relative bed stability. This index is the ratio of bed surface geometric mean particle diameter to estimated critical diameter at bankfull flow, based on a modified Shields' criterion for incipient motion.

Both field and flume studies indicate the importance of form drag as an influence on sediment entrainment in gravel-bed rivers [*Brayshaw*, 1985; *Petit*, 1989, 1990; *Hassan and Reid*, 1990; *Reid et al.*, 1992; *Best*, 1993]. For a given particle size, the greater the bed roughness, the greater will be the shear stress required to initiate motion. This may be incorporated into shear stress equations using the roughness length y_0, which is the height of zero velocity obtained by extrapolation of the law of the wall fitted to field velocity profile data [*Richards*, 1990]:

$$\theta_{ci} = 0.069 (d_i/y_0)^{-0.699} \qquad (3.50)$$

Most sediment transport equations predict increasing transport rates with increasing stage. The effect of form roughness, however, is such that changes in relative protrusion of bed particles may result in a decline in transport rates with increasing depth [*Bagnold*, 1977; *Reid et al.*, 1985]. Because of the many stochastic variables influencing particle critical shear stress, this parameter may be characterized by a probability distribution rather than a single value [*Powell*, 1998].

As noted earlier, one of the problems with applying the Shields equation for critical shear stress to entrainment in poorly sorted channels with highly turbulent flow is that the equation neglects lift forces. Direct measurements indicate that lift does not scale with the velocity difference across a grain [*Nelson et al.*, 2001; *Schmeeckle et al.*, 2007], contrary to expectations based on the Bernoulli equation (3.7), making it difficult to incorporate lift into a force balance. *Vollmer and Kleinhans* [2007] incorporate lift force into formulation of the dimensionless Shields parameter:

$$\theta_{crit} = \frac{\left(\dfrac{\cos\alpha - \sin\alpha}{\tan\varphi}\right) k_{dens}\tan\varphi'}{k_{eff}[(c_D k_{std} + (k_{std}c_L + k_{turb})\tan\varphi')]} \tag{3.51}$$

where α is angle of bed inclination (degrees), φ is pivot angle (degrees), φ' is angle of repose (degrees), k_{dens} is packing density parameter (m^2), k_{eff} is effective critical shear stress scaled by the time-averaged critical shear stress, c_D is drag coefficient, k_{std} is a factor representing the hydraulic of the approaching flow, c_L is lift coefficient, and k_{turb} is turbulent lift force scaled by the time-averaged boundary shear stress. *Vollmer and Kleinhans* [2007] provide the supplementary equations for calculating individual parameters in equation 3.51.

Using data from flumes and natural channels, *Lamb et al.* [2008c] note that the critical Shields stress for initial sediment motion increases with channel slope such that particles of the same size are more stable on steeper slopes, presumably as a result of changes in particle emergence, local flow velocity and turbulence. They formulate a version of the critical Shields stress τ_{*cT} to reflect this:

$$\tau_{*cT} = \frac{hS}{rD} = \frac{2}{C_D}\frac{u_*^2}{\langle u^2 \rangle}\left(\frac{\tau_T}{\tau_T - \tau_m - \tau_w}\right)\left(\frac{\tan\phi_0 - \tan\beta}{1 + (F_L/F_D)\tan\phi_0}\right)\left[\frac{V_p}{A_{xs}D}\frac{1}{r}\left(\frac{\rho_s}{\rho} - \frac{V_{ps}}{V_p}\right)\right] \tag{3.52}$$

where h is flow depth, S is channel slope, r is the submerged specific density of the sediment, D is the diameter of a particle, C_D is the drag coefficient, u_* is the shear velocity, u is the local velocity, τ_T is the total driving stress at the bed, τ_m is the stress spent on the bed morphology, τ_w is the stress spent on the channel walls, φ_0 is the friction angle between grains, β is the bed-slope angle, F_L is lift force, F_D is drag force, V_p is the total volume of the particle, A_{xs} is the cross-sectional area of the particle that is perpendicular to and exposed to the flow, V_{ps} is the submerged volume of the particle, ρ_s is sediment density, and ρ is fluid density.

A final important point with respect to shear stress and sediment entrainment comes from studies on Turkey Brook in the United Kingdom [*Reid et al.*, 1985]: the bed shear stress at entrainment averages three times the shear at the cessation of motion, an effect analogous to that described by *Hjulström* [1935] with respect to velocity. This complicates effects to understand hydraulic conditions creating entrainment based on field measurements during flows when entrainment is not occurring. Also, although studies of entrainment from poorly sorted, coarse-grained beds typically ignore the effects of cohesion, cohesion can be present and can substantially alter processes of detachment and entrainment [*Jain and Kothyari*, 2009].

In summary, bedload entrainment is typically estimated using physically based stochastic models that treat the two distinct phenomena of velocity fluctuations and variations in grain size and pivoting angle. Critical shear stress for entrainment is lower than predicted by the Shields equation when grain size is larger than the length scale of the bed roughness and higher when grain size is smaller than the bed roughness. Instantaneous fluctuations in the flow field around particles also exert a strong control on entrainment, and values of critical shear stress for mobilizing sediment from the armor layer increase systematically with average channel gradient. Field studies of entrainment are difficult to compare because of a variety of measurement techniques and sampling procedures, but have been interpreted to indicate either equal mobility or selective entrainment. The mode of sediment entrainment can vary as a function of discharge, time, and grain-size distribution.

3.5.4. Measurement of Bedload Transport

Once the bed material grain-size distribution has been characterized and the conditions under which entrainment occurs have been specified, another basic requirement for predicting bedload transport along mountain rivers is empirical data. It is important to distinguish between bedload discharge, which refers to instantaneous transport rates (e.g., kg/s, m³/s), and bedload yield, which refers to amounts of sediment (e.g., tons, m³) moved over longer time periods of floods or a seasonal cycle [*Carson and Griffiths*, 1987]. In the following discussion, bedload is equivalent to bed material load and is that part of the sediment load composed of grain sizes represented in the bed and moving by rolling, sliding, or saltating. In contrast, dissolved load moves in solution, and wash load is that part of the sediment load moving in suspension and composed of grain sizes finer than those of the bed.

Bedload transport along steep, gravel-bed rivers is notoriously difficult to directly measure [*Ryan and Troendle*, 1997; *Hicks and Gomez*, 2003]. Any sampler placed in the flow may perturb local hydraulics sufficiently to create anomalously high or low transport conditions. The Helley-Smith sampler is commonly used either hand-held on a rod or by lowering from a boat or bridge (Figure 3.36). Efficiency ratings for the Helley-Smith sampler have been calculated as sampler efficiency and hydraulic efficiency. Sampler efficiency is a ratio of sediment collected to sediment that would have passed the nozzle area without the sampler present [*Glysson*, 1993]. A field calibration of the Helley-Smith sampler demonstrated sampler efficiencies of 90-100% for particles 0.5-16 mm in diameter [*Emmett*, 1980]. Sampler efficiency dropped below 70% for particles greater than 16 mm, although the reduction may have been caused by the small number of particles in motion. Hydraulic efficiency is the ratio of water discharge through the sampler orifice to discharge through the same area without the sampler [*Kuhnle*, 1992b]. Calculated hydraulic efficiencies range from 1.43 to 1.53 [*Hubbell*, 1987; *Kuhnle*, 1992b]. Because this does not create a sampler efficiency greater than 100%, it is not regarded as a problem. Helley-Smith samplers commonly have either 76- or 152-mm intakes. Using a sampler with an intake size much larger than bed grain size (~5D) increases the accuracy of sampling [*Vericat et al.*, 2006]. U.S. Geological Survey protocol recommends a sampling time of 30-60 seconds so as to account for instantaneous transport variations, but not clog the sampler bag [*Glysson*, 1993]. Sampler efficiencies drop off after the sampler is 40% full [*Emmett*, 1980], but sampling times of up to 10 minutes may be appropriate under conditions of low sediment discharge [*Ashworth and Ferguson*, 1989].

Bunte et al. [2004, 2007] develop an alternative to the Helley-Smith in the form of portable bedload traps with a 0.3 × 0.2 m opening and a 0.9-m-long trailing net (Figure 3.37). Traps are positioned on ground plates anchored in the streambed to minimize disturbance during sampling and to facilitate sampling times up to 1 hour. Bedload rating and flow competence curves obtained with the traps are steeper than those obtained from Helley-Smith samplers, but the portable traps are not practicable where a significant portion of the total load consists of fines [*Bunte et al.*, 2004]. Transport rates measured with both types of samplers approach similar results near or above bankfull flow [*Bunte et al.*, 2008]. Under the relatively low bedload transport common in coarse-grained channels, short sampling times (2 minute) using the portable traps under-predict transport rates obtained from longer (10 and 60 minute) sampling by factors of 2-3 at moderate flows and a factor of 5 at flows near bankfull [*Bunte and Abt*, 2005].

One of the challenges of using a point sampler such as the Helley-Smith or a portable trap is that bedload movement commonly varies by an order of magnitude or more across a channel cross section [*Carling*, 1994b]. Because of this, sampling errors decrease as the number of samples collected increases and the number of traverses of the channel over which samples are collected increases [*Gomez and Troutman*, 1997]. It is best if sampling is conducted at a pace that allows a number of bedforms to pass through the sampling cross section.

An alternative to point sampling of bedload is to use some type of sediment trap and to measure filling rates [*Reid et al.*, 1980; *Garcia et al.*, 2000; *Habersack et al.*, 2001; *Sear*, 2003]. In 1973 a portion of the East Fork River in Wyoming, USA was temporarily diverted so that a concrete trench could be emplaced across the bed. Hydraulically operated horizontal gates, opening to 15-cm wide and flush with the bed, allowed bedload to be trapped in individual sections or across the whole channel. A conveyor belt transported trapped sediment to the bank for measurement [*Leopold and Emmett*, 1976; *Bagnold*, 1977]. *Laronne et al.* [1992, 2003] and *Ergenzinger et al.* [1994b] describe slot samplers set at intervals across other channels. A Birkbeck-type pressure pillow (a water-filled pressure pillow with pressure transmitters) is located in the slot and the system is fully automated to sample at 0.25-sec intervals. The slots have a capacity of 0.25-0.4 m³, and usually fill before the cessation of bedload transport. *Klingeman and Milhous* [1970] adapt a vortex-tube, which utilizes vortices that develop in slots perpendicular to flow to trap sediment, to Oak Creek in Oregon, USA, and *Hayward and Sutherland* [1974] use a similar design on the Torlesse Stream catchment in New Zealand. When *Sterling and Church* [2002] compare results from pit traps and Helley-Smith samples in Harris Creek, Canada, they find that the traps yield nearly 100% efficiency for material larger than 2.8 mm, whereas the Helley-Smith sampler is more variable in catch and trapping efficiency, with the highest efficiency for finer material. The main limitation to the use of pit traps is the width of the channel (at 15 m, the East Fork River is the widest channel so instrumented to date) and the capacity of the sampler.

A sediment trap can also take the form of a storage area such as a large pit excavated at the downstream end of a reach, which can be monitored for volume of sediment fill following a single flood or an entire flow season [Lenzi et al., 1990]. Examples of these systems exist on the Rio Cordon in Italy (5 km^2 drainage area) (Figure 3.38), the Erlenbach in Switzerland (0.7 km^2; Figure 3.39), the 284 ha Arnás catchment in the Spanish Pyrenees [Lana-Renault et al., 2006], East St. Louis Creek (8 km^2) in the Fraser Experimental Forest of Colorado, USA and Nahal Yael, Israel (0.6 km^2 drainage area; Schick, 1970; Schick et al., 1987b).

Using a sediment trap as the reference, a bedload rating curve constructed from a series of Helley-Smith samples overpredicts yield by 36% on a small, steep (0.07 m/m) boulder channel in Switzerland [Warburton, 1990]. Loads calculated with the Schoklitsch formula overpredict yield by 111%. Warburton [1990] concludes that the Helley-Smith and Schoklitsch methods do not account for temporal and spatial variation in transport rates. In contrast, comparison of more than 30 years of weir pond data and more than 1500 Helley-Smith samples from mountain channels in Colorado, USA indicates that the values of total bedload from each method are comparable [Troendle et al., 1996].

Bedload movement can also be estimated from a subset of marked tracer particles that are assumed to represent some proportion of total movement [Hassan and Ergenzinger, 2003]. Markers can be miniature radios, naturally occurring magnetic minerals [Gottesfeld and Tunnicliffe, 2003], or artificially emplaced magnets, paint, or radioactive injections [Hassan et al., 1984; Chacho et al., 1994; Ergenzinger and de Jong, 1994; Michalik and Bartnik, 1994; Thompson et al., 1996; Habersack, 2003; Wong et al., 2007]. The mobility of tracer particles can be reduced with time as they are mixed to less active locations in the subsurface or transported downstream to reaches with different shear stress or bed grain size [Ferguson and Hoey, 2002; Ferguson et al., 2002]. Because recovery rates can range as widely as 33% [Hassan et al., 1984] to 69% [Thompson et al., 1996], the number of tracers placed in a stream is typically in the tens to hundreds.

Rickenmann [1994b] describes the use of hydrophones, sensors installed in the channel bed that measure the acoustic signals resulting from the impact of bedload grains transported over the measuring cross section. Early attempts to indirectly measure bedload using acoustic collision meters [e.g., Richards and Milne, 1979; Bänziger and Burch, 1990] had problems with adequately calibrating the meters, particularly for high flows. Newer technology has alleviated many of these problems and various types of acoustic devices and impact sensors are now used to measure bedload in diverse mountain streams [Bogen and Møen, 2003; Downing et al., 2003; Froehlich, 2003; Mizuyama et al., 2003; Raven et al., 2010]. A pressure pillow placed on the channel bed can be used to calculate the thickness of overlying sediment as a function of the difference between the internal pressure of the pillow and the hydrostatic pressure [*Kurashige*, 1999]. Continuously logging impact sensors consisting of a steel plate fixed to bedrock or a large boulder can also be used [*Richardson et al.*, 2003], although these provide conservative estimates of bedload movement at high flows because the sensors can only record up to three impacts per second [*Reid et al.*, 2007a]. *Tunnicliffe et al.* [2000] describe a magnetic induction system for measuring bedload movement based on a sensor that produces signals when its magnetic field is enhanced by a passing particle, although calibration of sediment flux requires independent knowledge of particle magnetic susceptibility [*Hassan et al.*, 2009]. *Gomez* [1987] and *Ergenzinger and de Jong* [2003] review bedload measurement techniques.

Bedload mobility can also be inferred by determining residence time of bed sediments. *Thompson et al.* [2007] determine residence times of fine sediments in step-pool and plane-bed channels of southeastern Australia using Optically Stimulated Luminescence dating and find that these estimates generally agree with estimates based on competence equations.

Bedload transport during flume studies is typically measured by weighing sediment that exits the flume at set intervals, which can be time- and labor-intensive. *Zimmerman et al.* [2008] describe the use of high-resolution video-based transport measurements over a table lighted from beneath.

In summary, the temporal and spatial variability of bedload movement in mountain rivers makes it challenging to collect or observe representative samples. Commonly used methods include Helley-Smith or portable trap samplers, fixed sediment traps, marked tracer particles, and impact or pressure sensors.

3.5.5. Mechanics of Bedload Transport

Numerous studies using each of these types of bedload measurement demonstrate that bedload movement is episodic along a given channel reach [*Schick et al.*, 1982; *Lekach and Schick*, 1983; *Reid et al.*, 1985; *Kuhnle and Southard*, 1988; *Whiting et al.*, 1988; *Hassan et al.*, 1991; *Bunte*, 1992; *Hoey*, 1992; *Gottesfeld and Tunnicliffe*, 2003; *McNamara and Borden*, 2004; *Singh et al.*, 2009]. This episodicity is explained in terms of number of channels and sediment storage along braided rivers [*Hoey and Sutherland*, 1991; *Young and Davies*, 1991; *Warburton and Davies*, 1994]. For single flow-path channels, the formation and migration of bedforms [*Reid and Frostick*, 1986; *Gomez et al.*, 1989; *Young and Davies*, 1991; *Cudden and Hoey*, 2003], longitudinal sediment sorting [*Iseya and Ikeda*, 1987], and the lateral shifting of bedload streets [*Leopold and Emmett*, 1976; *Ergenzinger et al.*, 1994a] are invoked to explain episodic bedload transport. Different stage-bedload discharge rating curves for adjacent channel reaches causes downstream changes in sediment transport that may develop by a feedback mechanism into bed waves [*Griffiths*, 1989]. The passage of bedload sheets through a reach is accompanied by selective deposition of fine bedload in the coarse surface layer [*Carling*, 1994b]. This causes physical and hydraulic smoothing of the bed and associated adjustment of hydraulic parameters. As the sheet passes, reentrainment causes further readjustment to hydraulically rougher conditions. Changes in the sorting of mixed-size sediment may also affect bedform dimensions and dynamics [*Wilcock*, 1992a], as well as the bedload transport rate [*Chen and Stone*, 2008] and relative mobility of different grain-size fractions [*Clayton*, 2010]. Features such as longitudinal slope discontinuity and alternating repetition of scour and fill are attributed to processes associated with heterogeneous sediment transport.

Bedload sheets are migrating slugs of bedload one to two grain-diameters thick that alternate between fine and coarse particles [*Whiting and Dietrich*, 1985; *Iseya and Ikeda*, 1987]. The sheets form when bedload segregates into alternating mobile zones of low grain roughness and high grain roughness. The larger grains move rapidly across the smooth areas of finer grains and then accumulate downstream where other large grains create high grain-to-grain friction [*Lisle*, 1987]. When the coarse grains trap sufficient fine particles in their interstices, they become mobile again [*Whiting et al.*, 1988]. This process, which results from very efficient vertical and longitudinal grain sorting [*Recking et al.*, 2009], has been described as an alternation between a smooth bed in which sand controls the mobility of the sediment mixture, and a congested bed in which gravel controls mobility [*Ikeda*, 1984; *Ikeda and Iseya*, 1986, 1988]. Flume experiments also reveal that during large floods when the surface framework particles are entrained, the bed dilates and material falls through the pore spaces [*Frostick et al.*, 2006].

Field investigations suggest that substrate disturbance depth during passage of a bedload sheet or traction carpet is similar in magnitude to particle exchange depth and moving layer thickness and is a small multiple of the bed surface d_{90} [*De Vries*, 2002]. Increased bedload transport under these conditions results from increased mobile fraction of the bed area and grain velocity, rather than increased thickness of the mobile layer [*De Vries*, 2002].

Cross-sectional to reach-scale heterogeneity of bed structure and relief can be responsible for some of the spatial and temporal heterogeneity of bedload movement [*Laronne and Carson*, 1976; *Measures and Tait*, 2008; *Mueller and Pitlick*, 2005]. Clusters of bed particles [*Brayshaw et al.*, 1983] can impede particle entrainment because larger particles shield smaller particles. Large particles can also protect fine-grained lee deposits (Figure 3.40). Bedload transport rates increase as discharge fluctuates, independent of changes in sediment supply, because lee sediments form a temporarily stable deposit behind each flow obstruction for a given discharge [*Thompson*, 2008]. Increases or decreases in discharge disrupt this temporary stability and increase sediment supply to the main flow. Larger features such as transverse ribs formed by lines of large clasts across the channel [*Koster*, 1978] and bed steps with an associated plunge pool [*Whittaker and Jaeggi*, 1982] typically form in steep channels and create hydraulic jumps and other flow heterogeneities. Bars and pools are associated with diverging and converging flow that promotes deposition and scour, respectively [*Ashmore*, 1982; *Thompson*, 1986]. These bars and pools can be fixed by bedrock outcrops, channel bends, or instream wood [*Lisle*, 1986; *O'Connor et al.*, 1986]. Wood can temporarily store a wedge of sediment upstream and, when the wood breaks or is mobilized, the sediment is released suddenly [*Beschta*, 1987; *Adenlof and Wohl*, 1994; *Douglas and Guyot*, 2005].

Heterogeneous bedload movement along mountain rivers can also be associated with differing sediment supply developing in response to progressive bed armoring [*Gomez*, 1983; *Lenzi et al.*, 2004b], changes in land use or channel configuration [*Young et al.*, 2001; *Gomi and Sidle*, 2003; *Surian and Cisotto*, 2007], or associated with volcanic eruptions, hillslope mass movements and/or debris flows along the channel [*Hack and Goodlett*, 1960; *Lisle*, 1987; *Benda*, 1990; *Wohl and Pearthree*, 1991; *Cenderelli and Kite*, 1994; *Gran and Montgomery*, 2005].

Because of the close coupling between hillslope and channel processes and the difficulty in directly measuring downstream sediment movement over longer time periods, sediment transport and routing for mountain rivers are sometimes estimated using mathematical models that incorporate topography, lithology, climate, and sediment transfer and storage through a series of reservoirs at the drainage-basin scale [*Pickup et al.*, 1983; *Kelsey et al.*, 1986; *Lisle*, 1987; *Mizutani*, 1987]. Filled channel segments, or reservoirs, respond to variations in sediment supply primarily by changes in stored sediment volume, with little change in transport rate [*Lisle and Church*, 2002]. Channel segments that are closer to supply-limited conditions respond to changes in sediment supply through armoring and form roughness. The types and degrees of sediment connectivity between channel segments vary with sediment sources and stream competence [*Hooke*, 2003]. Airborne remote sensing, digital photogrammetry, differential GPS, and LiDAR technologies that facilitate rapid acquisition of high resolution and high precision topographic data sets over various spatial scales greatly facilitate the measurement and modeling of reach-scale sediment movement and morphological change [*Rumsby et al.*, 2008].

Bedload movement can also vary in a manner analogous to the hysteresis curve commonly used to describe suspended sediment transport during a flood [*Schöberl*, 1991; *Rickenmann*, 1994b; *Trayler and Wohl*, 2000; *McNamara and Borden*, 2004]. *Kuhnle* [1992a] describes greater bedload transport rates during rising stage and hypothesizes that they are caused by a lag in the formation and destruction of bedforms and/or the bed pavement relative to flow. *Moog and Whiting* [1998], in a ten-year study of six gravel-bed mountain rivers in Idaho, USA, also find a hysteresis effect in bedload transport. At a given flow rate, more bedload is carried by discharges preceding the first annual occurrence of a "threshold" rate because readily moved sediment supplies that accumulate in the channels during low-flow periods from late summer to early spring are removed by rising spring-time discharges up to the threshold. *Hassan and Church* [2001] describe a similar seasonal hysteresis effect. The threshold discharge is greater than mean annual discharge and about one-half bankfull discharge. *Church and Hassan* [2002] describe fractional transport rates of partial mobility, full mobility, and overpassing/suspended along Harris Creek in British Columbia, Canada as a function of increasing flow.

Hysteresis in bedload transport can also be related to an effect described by *Schick et al.* [1987a,b]. They note that ephemeral desert channels have less well developed coarse surface layers than perennial channels, perhaps because the brief and violent desert floods cause disruption of the coarse surface layer at a rate faster than it can form during the relatively short and less geomorphically effective periods of hydrograph recession. Desert channels with a cobble-pebble-sand size distribution have coarse clasts present in both the surface and subsurface. Tracer particles indicate that these coarse clasts are intermittently buried and re-exposed from one flood to the next in an equilibrium vertical exchange. Clasts that have an inter-event buried phase are transported significantly farther in their subsequent move than those that do not, suggesting that rising limb bed-scour may be an important control on bedload transport rate along some channels.

Sand can significantly affect the mobility of coarser particles even in the absence of burial of coarser particles. Gravel transport rates can increase by orders of magnitude as sand content increases in the range of 15-27%, for example, even though the proportion of gravel in the bed decreases [*Wilcock et al.*, 2001]. Sand can be preferentially transported at low flows that move little or no coarse sediment. As the amount of sand present increases, local sand patches and stripes can decrease the protrusion of coarse grains and the hiding effect that limits transport of finer grain sizes in mixed grain distributions [*Ferguson et al.*, 1989; *Wilcock*, 1993; *Seal and Paola*, 1995], as well as decreasing the reverse flow typically found in the lee of protruding clasts [*Sambrook Smith and Nicholas*, 2005]. Moderate amounts of sand can thus decrease the entrainment of coarser grains. Once a coarse particle is entrained, however, it can move faster over the relatively smooth bed and may move farther because sand filling pore spaces between large grains reduces resting places. Sand thus has a non-linear effect on the transport of coarser particles [*Wilcock and Kenworthy*, 2002].

Bedload transport can also be related to suspended sediment concentrations. Studies of sand moving in suspension and as bedload over an open-work gravel bed indicate that the depositional rate for the sand infilling the gravel is strongly linearly correlated with the suspended sediment concentration, although turbulent resuspension of sediment prevents deposition in a surface layer of gravel of thickness approximately equal to the mean grain size of the gravel [*Carling*, 1984b].

In summary, spatial and temporal heterogeneity in bedload transport has been related to sediment supply, channel morphology, bed roughness, hydraulics, and interactions between flow and sediment supply. Bedload transport in diverse mountain rivers exhibits hysteresis.

3.5.6. Downstream Bedload Transport Patterns, Rates, and Frequency

At the simplest level, bedload transport in alluvial channels tends to follow an exponential distribution with a mean that varies with grain size [*Hill et al.*, 2010]. This distribution can be obscured, however, by spatial and temporal variations. In addition to the episodic pulses of bedload movement that have been measured along many different types of mountain rivers, bedload transport can vary downstream because of differential erosion and deposition associated with bedform sequences. *Carling and Glaister* [1987] and *Carling* [1990] describe segregation of gravel from sand in the flow separation zone downstream from a negative step or bar-front. Painted tracer particles in pool-riffle channels indicate higher sediment-transport competence in pools than in riffles at high flow [*Petit*, 1987; *Thompson et al.*, 1996]. Riffles contain significantly smaller deposited tracer particles than pool centers, pool exit-slopes, and runs [*Thompson et al.*, 1996]. Along North St. Vrain Creek, Colorado, USA, riffles provide the most stable high-flow depositional sites for particles 16-90 mm in size; larger particles are usually deposited on runs or pool exit-slopes [*Thompson et al.*, 1996]. These grain-size trends reflect hydraulic controls. A central jet of high velocity flow present in the pool during high discharge scours the pool thalweg, leaving only very coarse particles. This jet dissipates in upwelling and boils at the pool exit-slope, promoting deposition of sediment transported through the pool center. As flow continues beyond the pool into the next downstream riffle, shallower flow depths and high bed roughness further decrease flow competence [*Thompson et al.*, 1996]. Riffles and alternate bars are thus sites of sediment deposition during high flows [*Andrews and Nelson*, 1989; *Harvey et al.*, 1993]. During low flows, the steepened water-surface gradients over riffles and bars may promote bar dissection and the removal of finer particles [*Harvey et al.*, 1993], which are then stored in pools until the next high flow [*Lisle and Hilton*, 1992, 1999]. Pool-riffle channels can thus have downstream sorting in sediment size in association with bedforms, causing trends of sediment entrainment and deposition to reverse between high and low flow conditions. Path length distributions become more symmetrical as flow approaches channel-forming conditions, with a mean equal to the pool-bar spacing [*Pyrce and Ashmore*, 2003a, b]. Using magnetic tracer particles along ephemeral channels with pools and alternate bars, *Hassan et al.* [1991] find that when mean particle travel distance approaches the scale of bar spacing, trapping in the bars interrupts particle movement.

Step-pool channels show similar trends. Clasts in the pools have the highest probability of entrainment during the rising limb of floods and pools are also favored locations for deposition during the falling limb [*Whittaker*, 1987b; *Schmidt and Gintz*, 1995]. The large, step-forming clasts remain stable except during infrequent (approximately 50 yr recurrence interval) large flow [*Grant et al.*, 1990]. The form and spill roughness associated with steps and pools reduces bed shear stress and elevates the critical shear stress. Particle entrainment tends to be size selective [*Mao et al.*, 2008], except when rare high flows create discharges approximately four times the critical threshold for most particles and equal mobility occurs [*Mao and Lenzi*, 2007]. In the mountains of Colorado, USA, pool-riffle channels are partially mobile as the more readily entrained particles (d_{50} = 4-40 mm) move across a static bed [*Ryan*, 1994a,b]. In contrast, transport is minimal in step-pool channels, where sands and small gravels (d_{50} = 1 mm) move in pulses during the falling limb of the hydrograph.

Ergenzinger et al. [1994a] describe similar bedload-channel bed interactions along a channel with pools and alternate bars. Bedload pulses are preferentially generated during smaller floods when sediment that has accumulated in local deposits (bars) is entrained again. Roughness elements that protrude into the flow cause locally increasing vertical velocities, which in turn drive vortex cells that are important to initiating entrainment from a bar. As the flow cells are enlarged and displaced locally, differential erosion, deposition and transport occur across the cross section, and bedload pulses are created and magnified. The probability of a bedload pulse occurring thus depends on the magnitude and duration of discharge, and the organization of secondary flow. As the pulse moves downstream, it is unlikely to maintain a constant volume because of continuing interactions with the flow and channel-bed topography. During large bedload pulses, the vortex cells may be replaced by a more chaotic two-layered flow [*Ergenzinger et al.*, 1994a].

Fine and coarse fractions of bedload can also be differentially routed across features such as point bars, with fine grains being swept inward over the point bar and coarse grains routed outward toward the pool [*Clayton and Pitlick*, 2007]. *Hassan et al.* [2008] classify bedforms into four categories to facilitate examining their effects on sediment transport; microforms such as particle clusters, mesoforms such as step-pool or riffle-pool sequences, macroforms such as bars and wood jams, and megaforms such as sedimentation zones and floodplains.

In addition to cross-sectional or reach-scale pulses of bedload movement, large volumes of coarse sediment introduced to a channel from a point source or during a relatively short time interval can move downstream along the drainage basin in discrete waves of bedload. Examples of this phenomenon come from studies of dispersal of mine tailings [*Knighton*, 1989; *James*, 1993]; sediments contaminated by plutonium released from the nuclear facility at Los Alamos, New Mexico, USA [*Graf*, 1996]; sediments from debris flows or landslides [*Wohl et al.*, 1993; *Madej and Ozaki*, 1996; *Sutherland et al.*, 2002; *Hoffman and Gabet*, 2007]; gravel augmentation downstream from dams [*Sklar et al.*, 2009]; and dam removal [*Major et al.*, 2008]. The sediment wave can be dispersive or transgressive [*Cui et al.*, 2003a], and may move passively along a channel or interact with the channel boundaries such that the boundaries are eroded locally [*Wathen and Hoey*, 1998]. Although dispersion strongly influences sediment waves in coarse-grained channels, significant translation can occur where the input sediment is much finer than the ambient bed material [*Lisle*, 2008; *Sklar et al.*, 2009]. Flume experiments indicate that variations in longitudinal slope will initially be the dominant influence on wave evolution and the sediment wave will decay in amplitude and increase in wavelength without significant migration along the channel [*Lisle et al.*, 1997]. After the sediment wave has decayed sufficiently, it may migrate downstream as a persistent and coherent feature. *Needham and Hey* [1992] describe a one-dimensional, physically-based model developed for predicting the initiation and propagation of bed waves and the associated morphological response of the river. Figure 3.41 illustrates *Madej*'s [2001] conceptual model of the development of channel-bed structure and organization after a large sediment introduction. *Cui et al.* [2003b] describe a numerical model that successfully reproduces the dispersive behavior of sediment waves observed in flumes.

A slightly different situation arises when widespread, low-intensity excess sediment delivery produces a longitudinal distribution of gravel bars with a broad-scale wave-like form as the sediment moves through a channel network over 10^1-10^2 years [*Jacobson and Gran*, 1999]. Although such an increase in gravel bars may be less noticeable than a discrete sediment wave, it likely occurs in many channel networks where changes in climate or land use increase sediment delivery over a period of many years across substantial portions of a watershed.

Bedload transport rates and the distances that bedload clasts are transported vary widely among mountain rivers as a function of channel morphology, grain-size distribution, magnitude and frequency of discharge, and other factors. The bedload transport rates and distances listed in Tables 3.3 and 3.4 typically include variation at a site of 1-2 orders of magnitude. This variability undoubtedly reflects the complex interactions among the primary variables which control bedload movement; sediment supply, channel-bed sedimentology (texture, packing, armoring, bedforms), hydraulics, and grain characteristics (size, shape, density).

Einstein [1937] initiates the use of statistical distributions to describe particle movements along a channel. He regards particle displacement as a random process and fits the distribution of step lengths with a gamma-related probability distribution. Subsequent studies use exponential or gamma distributed step lengths [e.g., *McNamara and Borden*, 2004], either of which leads to a compound Poisson distribution of total path length during a flood, when each particle may move repeatedly [*Hassan et al.*, 1991; *Hassan and Church*, 1992]. *Carling and Hurley* [1987] find a good match between a Poisson distribution and bedload transport during short, discrete flows on two upland streams in the United Kingdom. A Poisson distribution also best describes the travel length of cobbles in a step-pool channel in Bavaria [*Ergenzinger and Schmidt*, 1990]. Using several hundred magnetically tagged clasts in two gravel-bed streams in Israel, *Hassan et al.* [1991] find that the compound Poisson model fits the data relatively well for movement during low flows, when the shear stress is only slightly above the threshold of particle movement and most particles move a relatively small number (1-2) of steps. During high flows, most particles move in several distinct phases that are not adequately described by a homogeneous random process, and the measured data differ significantly from the fitted curves [*Hassan et al.*, 1991]. Grains smaller than D_{50} tend to show only moderate variation in travel distance, but travel distance of larger grains correlates inversely with grain size [*Church and Hassan*, 1992]. This likely reflects the trapping of smaller grains by larger grains, whereas the travel distance of larger grains is limited by the energy required for their transport [*Church*, 2003]. *Júnior* [2005] finds that a one-dimensional gamma functions model fits transport of sand-sized grains in a laboratory flume. Under weak transport, the distribution of particle displacement is local and stochastic, but under stronger transport the grains cluster and the characteristic distance of particle displacement scales with the channel dimensions of transverse oscillation (pools and riffles or alternate bars) [*Pyrce and Ashmore*, 2003a,b].

The frequency at which clasts of varying size are entrained and the total bedload transport during flows of varying magnitude are of importance when attempting to determine the dominant discharge. Working on five ephemeral gravel streams with gradients of 0.014-0.030, *Begin and Inbar* [1984] estimate that frequent flows with a probability of 0.9-0.7 of being equaled or exceeded produce the median grain size (0.1-46 mm) along the channels. Flows about 3.5 times the mean annual discharge initiate motion of cobble and gravel bed material along the upper Colorado River, which is close to a threshold between braided and meandering ($S = 0.0015$) [*Pitlick*, 1994a]. Flows equivalent to at least a 5-year flood are necessary to create significant movement of the bed material in this system [*Pitlick*, 1994a]. Similarly, a study of 24 non-braided gravel-bed rivers ($S = 0.0009$-0.026; $d_{50} = 23$-120 mm) in the Colorado Rocky Mountains, USA finds that transport of bed-material particles is a relatively frequent occurrence that began at flows slightly less than bankfull that are equaled or exceeded on average several days each year [*Andrews*, 1984]. A later study on 23 headwater gravel-bed streams in snowmelt-dominated portions of Idaho, USA indicates that flows between mean annual discharge and bankfull discharge move 57% of the bed-load, whereas flows above bankfull discharge move only 37% of the bedload [*Whiting et al.*, 1999]. (Bankfull discharge on these rivers has a 2-year recurrence interval.) The *bedload effective discharge* on these rivers, defined as the discharge that transports more sediment than any other flow over the long term, averages 80% of the bankfull discharge, and has an average recurrence interval of 1.4 years. In contrast, the effective discharge on the steep (13.6% mean gradient) Rio Cordon, Italy is 2.45-2.65 m^3/s, which has a recurrence interval of 1.7 years and slightly exceeds the estimated bankfull discharge of 2.3 m^3/s [*Mao et al.*, 2005]. The ratio of effective discharge to bankfull discharge is independent of basin size, grain size, and gradient, but does increase with the relative magnitude of large, infrequent flows [*Whiting et al.*, 1999]. Based on six years of data from two small upland streams in the United Kingdom, 90% of the total bedload is transported by 29% of the floods in one stream (drainage area 11.7 km^2) and 38% of the floods in the other stream (drainage area 2.2 km^2) [*Carling and Hurley*, 1987].

In contrast to these findings, long-term research in the small, high-relief catchment of Nahal Yael, Israel demonstrates that bedload transport along this ephemeral channel occurs primarily during high-magnitude, low-frequency flows [*Schick et al.*, 1982]. Studies on a bouldery step-pool subalpine channel with snowmelt runoff also indicate that the channel bed remains stable during the peak flow of average years, with only much finer sediment (sand and gravel) mobilized [*Adenlof and Wohl*, 1994; *Blizard and Wohl*, 1998; *Trayler and Wohl*, 2000]. The duration of near-bankfull flows (approximately 1.5 year recurrence interval) correlates well with sand and cobble transport, however, along these snowmelt channels [*Troendle*, 1993].

As might be expected, the frequency of bedload movement is a function of the ratio between hydraulic driving forces and channel-substrate resistance. Very coarse-grained or poorly sorted mountain rivers may require extreme flows that recur infrequently before substantial mobilization of the coarsest sediment occurs. A portion of the finer (sand to gravel/cobble) fraction, however, may be mobilized annually. Clasts in mountain rivers with more uniform and finer grains (cobble or smaller) may be mobilized annually or every few years. Some investigators find that duration of flow above a threshold such as bankfull discharge correlates best with total volume of bedload discharge [*Rickenmann et al.*, 1998; *Troendle*, 1993], whereas others find the strongest correlation between peak flow and bedload volume [*Lana-Renault et al.*, 2006]. These differences may reflect hydroclimatic regime; *Rickenmann et al.* [1998] worked in perennial snowmelt and rainfall systems and Troendle worked in a perennial, snowmelt-dominated system, whereas *Lana-Renault et al.* [2006] worked in an ephemeral, rainfall-dominated system.

Issues of bedload transport magnitude and frequency are important in regions such as the Colorado Rocky Mountains, USA, where increasing flow diversions from mountain rivers and flow regulation below dams necessitate definition of flows necessary to maintain channel conveyance or desirable aspects of aquatic and riparian habitat [*Jowett*, 1997]. On the Gunnison River (drainage area 20,534 km^2, S of 0.0019-0.0084) in Colorado, USA monthly mean discharge during the snowmelt season has decreased 63% since the construction of upstream reservoirs in the mid-1960s; bankfull discharge and the 10-year and 25-year floods have decreased to approximately half of pre-reservoir levels [*Elliott and Parker*, 1992]. As a result, bed-material entrainment has decreased. Channel reaches that formerly had periodically entrained coarse beds are now paved and stable, particularly downstream from tributary confluences with periodic supplies of coarse debris-flow sediments [*Elliot and Parker*, 1992; *Dubinski and Wohl*, 2007]. This has led to encroachment of riparian vegetation along the channel margins [*Auble et al.*, 1994], which further reduces sediment mobility. Similarly, concerns about maintaining populations of the endangered Colorado pikeminnow (*Ptychocheilus lucius*) along the Yampa River, Colorado led to minimum streamflow hydrograph recommendations for baseflow, rising and recessional limbs, and peak discharge based on field observations and experimental and computer simulations of sediment entrainment and transport [*O'Brien*, 1987; *Stewart et al.*, 2005]. In this case, the issues of concern were (1) ensuring a relatively sand-free cobble substrate during the summer spawning period, (2) providing backflooded slackwater habitat for juvenile fish, and (3) maintaining the pool-riffle sequence.

As water diversions increase from mountain rivers, there will be growing pressure to precisely specify the minimum flow regime required to maintain characteristics such as spawning habitat for pikeminnow. Such specifications in turn rely on mathematical descriptions of the hydraulic conditions under which thresholds of bedload transport, lateral channel movement, or overbank inundation are crossed [e.g., *Pitlick and Van Steeter*, 1998; *Rathburn et al.*, 2009]

A related concern is that of predicting the likely impacts of climate change on mountain rivers. The EROSLOPE project, for example, evaluates the effects of varying scenarios of climate change by determining probabilities of hillslope and channel-bed erosion in Alpine drainage basins [*Ergenzinger*, 1994]. *Pizzuto et al.*

[2008] combine hydrological and sediment transport models to simulate scenarios of climate and land use change and their effects on a gravel-bed river in Maryland, USA.

In summary, cross-sectional or reach-scale pulses of bedload movement can reflect differential erosion and deposition associated with bedform sequences. Network-scale pulses may reflect downstream movement of a point source of sediment. Compound Poisson distributions are commonly used to describe particle movements along a channel. The ability to describe and predict the frequency of bedload movement is particularly important in the context of quantifying dominant discharge, as well as the effects of flow regulation and climate change.

3.5.7. Bedload Transport Equations

Numerous bedload transport equations have been developed from empirical data collected on steep, coarse-grained channels. Because these equations are empirical, their applicability is typically limited to the specific conditions of sediment, hydraulics, and channel morphology under which the data were collected. The data from which the equations were developed can also be too limited to adequately characterize the range of bedload transport conditions present at a specific site, or the transport equations may over-simplify the variability present in bedload transport. Transport equations developed from flume studies may be even more compromised with respect to these considerations, although *Dogan et al.* [2009] recommend quantification of similarity of predictor variables between field and laboratory data as a means of assessing the transferability of lab data. As noted by *Church and Hassan* [2005], classical sediment transport formulae represent high flux conditions, for example, whereas bedload transport in coarse-grained channels typically remains at very low rates. In other words, most bedload transport equations were developed for, and are really only applicable to, capacity-limited systems where the amount of bedload in transport is limited by the available flow energy, whereas many coarse-grained channels are supply-limited systems where the amount of bedload in transport is limited by the available sediment. After reviewing several bedload transport equations developed for gravel-bed channels, *Carson and Griffiths* [1987] conclude that there may be no simple, readily predictable relationship between instantaneous coarse sediment transport rates and channel flow parameters. *Church and Hassan* [2005] reinforce that conclusion nearly two decades later. Bed roughness, hydraulics, bed-sediment packing and sorting, and limits on sediment supply may be so stochastic and site-specific that generalized bedload transport equations never adequately approximate actual bedload movement [*Hunziker and Jaeggi*, 2002].

Bedload transport equations tend to focus on (i) grain tractive stress [e.g., *Einstein*, 1937, 1942; *Brown*, 1950; *Meyer-Peter and Muller*, 1948; *Bagnold*, 1956; *Baker and Ritter*, 1975; *Parker et al.*, 1982; *Wilcock and Crowe*, 2003], (ii) stream power per unit area [e.g., *Bagnold*, 1977, 1980], (iii) stream discharge [e.g., *Schoklitsch*, 1962; *Bathurst*, 1987a; *Ashiq et al.*, 2006], or (iv) stochastic functions for sediment movement. The Einstein-Brown formula is based on uniform grain size and plane-bed conditions, which are seldom present along mountain rivers. Although the Meyer-Peter and Muller equation does take into account particle roughness, it either grossly under-predicts (if d_{50} for the surface layer is used) or overpredicts (d_{50} for the subsurface layer) relative to actual transport in gravel-bed channels [*Parker et al.*, 1982; *Carson and Griffiths*, 1987]. The Bagnold hypothesis assumes that a necessary constraint for the maintenance of equilibrium bedload transport is that fluid shear stress at the bed must be reduced to a critical value associated with the incipient motion of grains. *Seminara et al.* [2002] demonstrate, however, that under equilibrium bedload transport no areal concentration of bedload is sufficient to reduce fluid shear stress at the bed to the critical value.

The Parker formula, developed from the low unit transport rates of the Oak Creek data set, may not apply well to channels with higher transport rates. This equation is based on the assumption that, in gravel-bed streams with pavement, all bedload sizes experience equal mobility beyond a threshold. Bedload relations developed empirically for each of 10 grain-size ranges in the Oak Creek data are collapsed into a single curve described by:

$$W^* = 0.0025^{[14.2(\varphi-1)-9.28(\varphi-1)]} \tag{3.53}$$

where W^* is the dimensionless total bedload, defined by

$$W^* = \frac{rq_B}{[g^{0.5}(HS)^{1.5}]} \tag{3.54}$$

where r is the submerged specific gravity of the sediment, q_B is the volumetric total bedload per unit width of gravel bed, S is the downstream slope of the energy grade line, and

$$\varphi_{50} = \frac{\tau^*_{50}}{\tau^*_{r50}} \tag{3.55}$$

τ^*_{50} is the Shields stress for the median diameter of the subpavement, defined by

$$\tau^*_{50} = \frac{\tau}{\rho r g d_{50}} \tag{3.56}$$

where d_{50} is subpavement median grain size and $\tau^*_{r50} = 0.0876$. This equation is valid only for φ_{50} less than 1.65. An extension equation beyond $\varphi_{50} = 1.65$ is

$$W^* = 11.2[1-(0.822/\varphi_{50})]^{4.5} \tag{3.57}$$

Dawdy and Wang [1993] note the need for a protocol governing determination of surface and subsurface grain-size distributions, local energy slope, cross section location, and data collection when using this formula.

The *Wilcock and Crowe* [2003] equations for predicting fractional bedload transport in gravel-bed streams are among the more widely used approaches:

$$\frac{\tau_{ri}}{\tau_{rs50}} = \left(\frac{D_i}{D_{s50}}\right)^b \tag{3.58}$$

$$\tau_{rm}^* = 0.021 + 0.015exp[-20F_s] \tag{3.59}$$

$$W_i^* = 0.002\phi^{7.5} \text{ for } \phi < 1.35$$
$$W_i^* = 14\left(1 - \frac{0.894}{\phi^{0.5}}\right)^{4.5} \text{ for } \phi \geq 1.35 \tag{3.60}$$

where τ_{ri} is the reference shear stress of size fraction i, defined as the value of τ at which W_i^* is equal to a small reference value $W_r^* = 0.002$, D_i is grain size of fraction i, D_{s50} is median grain size of bed surface, $b = 0.12$ for $D_i/D_{sm} < 1$ and $b = 0.67$ for $D_i/D_{sm} > 1$, τ_{rm}^* is the reference dimensionless Shields stress for mean size of bed surface, F_s is proportion of sand in surface size distribution, W_i^* is dimensionless transport rate of size fraction i, and φ is τ/τ_{ri}. The model incorporates a hiding function (τ_{ri}/τ_{rs50}) that resolves discrepancies observed among earlier hiding functions; uses the full size distribution of the bed surface, including sand; and incorporates a nonlinear effect of sand content on gravel transport rate that was not included in previous models [*Wilcock and Crowe*, 2003]. Applying these equations to the regulated, gravel-bed Trinity River in California, USA, *Gaeuman et al.* [2009] achieve the best fit to observed bedload transport with modifications to equation 3.59 and the hiding function used to scale τ_{rm}^* to other grain size fractions. *Recking* [2010] also proposes a new hiding function that is a power law of the D_{84}/D_{50} ratio. This function corrects the ratio between the Shields number and its critical value and contributes to a surface-based bed load transport formula able to match observed values from large field and flume data sets within an order of magnitude.

Bagnold [1977, 1980] proposes that bedload transport rate increases non-linearly with stream power above a critical value, ω_c, such that

$$\omega_c = c_1 D^{1.5}\log(c_2 d/D) \tag{3.61}$$

where c_1 and c_2 are numerical constants, D is the diameter of mobilized particles, and d is mean flow depth. Although *Carson and Griffiths* [1987] conclude that the Bagnold equation tends to overpredict transport, *Gomez and Church* [1989] find that the Bagnold formula is the most successful of 12 sediment transport formulae that they test against gravel-bed channel data (Table 3.5). The success rate for this formula is 25%, however, and it does contribute a "spectacular failure" [*Gomez and Church*, 1989]. *Ferguson* [2005] notes shortcomings of the original equation and develops two modified versions by adding terms for channel gradient and relative grain size. Among the shortcomings that Ferguson notes is the failure to account for hiding and protrusion effects on the mobility of mixed-size stream beds. Using tracer clasts in differently sized rivers of Belgium, *Petit et al.* [2005] find that the largest river has the lowest values of ω_c, and the value increases as grain and bedform resistance increase in smaller rivers.

Studies prior to that of *Gomez and Church* [1989] compared theoretical and actual results by either (1) plotting calculated bedload discharge versus actual water discharge in a sediment rating curve and then comparing the rating curve from actual bedload discharge, or (2) using a discrepancy ratio of observed/calculated bedload discharge. *Gomez and Church* [1989] combine both methods as

$$Y_j = X_j + E + E_j + e_j \qquad (3.62)$$

where Y = observed bedload, X = calculated bedload, j = a set of values, E = mean bias of the computational procedure ($E = Y(\text{mean}) - X(\text{mean})$), E_j = the local bias of the procedure for the jth case, and e_j = the random error contributing to empirical variance.

Using this procedure, most of the formulae (Meyer-Peter, Schoklitsch, Meyer-Peter and Muller, Parker, duBoys-Straub, Einstein, Yalin, Ackers, Bagnold) characteristically overpredict, perhaps because they do not account for surface coarsening or variations in sediment supply rate [*Gomez and Church*, 1989], as well as failing to account for the stress borne by rarely mobile grains [*Yager et al.*, 2007]. *Yager et al.* [2007] modify the *Fernandez Luque and Van Beek* [1976] equation to account for these properties of steep, coarse-grained channels:

$$q_{sm}* = 5.7(\tau_m* - \tau_{cm}*)^{1.5}\left(\frac{A_m}{A_t}\right) \qquad (3.63)$$

where $q_{sm}*$ is dimensionless transport rate of mobile sediment, τ_m* is the dimensionless stress borne by the mobile sediment, $\tau_{cm}*$ is the dimensionless critical shear stress of the mobile sediment, A_m is the bed-parallel area of mobile sediment, and A_t is the total bed area.

Other tests of bedload formulae against data from mountain rivers favor different equations: *Georgiev* [1990] favors equations by Schamov and by *Schoklitsch* [1962]; *Bathurst et al.* [1987] favor Schoklitsch; *Carling* [1983] favors *Yalin* [1963]; *Hayward* [1979] favors *Bagnold's* [1966] concept of the proportion of stream power utilized in bedload transport; *Barry et al.* [2004] favor *Parker et al.* [1982]; and *Carling* [1989a], *Martin* [2003], and *Martin and Ham* [2005] favor Bagnold. The Schoklitsch formula is

$$q_{sb} = [2.5(\rho_s/\rho)]S^{1.5}(q - q_c) \qquad (3.64)$$

where q_c is the critical discharge at which bedload transport begins:

$$q_c = 0.26\left(\frac{\rho_s}{\rho} - 1\right)^{1.67}\frac{d_{40}^{1.5}}{S^{1.17}} \qquad (3.65)$$

For steep channels, *Bathurst et al.* [1987] modify q_c:

$$q*_c = q_c/(g^{0.5}d_{16}^{1.5}) = 0.21 \ S^{-1.12} \qquad (3.66)$$

for S = channel slope (0.25-10%). *Rickenmann* [2001] also notes that bedload transport is a function of the product of effective discharge and channel gradient, and defines the relative efficiency of the stream in transporting bedload in terms of deviation from this transport function. He interprets the large variations and strong decreases in efficiency in smaller streams to result from substantial form resistance for relative flow depths (h/d_{90}) smaller than $\sim$ 4-6

A simple power function of total discharge, Q, best describes bedload transport for a large dataset from Idaho, USA [*Barry et al.*, 2004]:

$$q_b = \alpha Q^\beta \tag{3.67}$$

where q_b is bedload transport per unit width, and α and β are empirical values.

The great majority of bedload transport equations, even those developed specifically for gravel-bed or mountain rivers, are not based on data from step-pool channels. Exceptions include equations by *Suszka* [1991], and the equation of *Smart* [1984], developed for channels with gradients of 0.04-20%:

$$\phi = 4\left[\left(\frac{d_{90}}{d_{30}}\right)^{0.2} S^{0.6} C\theta^{0.5}(\theta - \theta_{cr})\right] \tag{3.68}$$

where ϕ; *is dimensionless sediment transport:*

$$\phi = \frac{q_b}{[g(s-1)d^3]^{0.5}} \tag{3.69}$$

q_b is volumetric sediment discharge per unit channel width, s is ratio of sediment density to water density, and d is mean grain diameter. For equation 3.68, S is channel slope, C is flow resistance factor [$= V/(gHS)^{0.5}$], θ is dimensionless shear stress, and θ_{cr} is critical Shield's parameter.

Equation 3.68 best describes sediment transport in supply-limited step-pool systems where sediment moves along the channel in waves not controlled by bed and hydraulic variables [*Whittaker*, 1987a, b]. By contrast, *Blizard*'s [1994] comparison of 7 bedload transport formulae (Einstein-Brown, Ackers-White, Smart-Jaeggi, Bagnold, 2 forms of Schoklitsch equation) to data from a subalpine step-pool channel demonstrates that both forms of the Schoklitsch formula and the Ackers-White equation are most successful, although even these tend to overpredict bedload discharge by up to three orders of magnitude. Another comparison of four equations (Yang, Meyer-Peter-Muller, Laursen, Einstein) to empirical data from step-pool channels in Idaho, USA used an "effective" slope, which removes all vertical drops from the profile and more accurately models the energy slope in the pools [*Johnejack and Megahan*, 1991]. Under these conditions, the Laursen and Einstein equations predict transport within an order of magnitude, but do not adequately predict the gradation of the transported sediment or the percentages of suspended and bedload. The threshold of incipient motion is not exceeded with the Yang and Meyer-Peter-Muller equations [*Johnejack and Megahan*, 1991].

Another shortcoming of most calculations of bedload transport in rivers, including those in numerical models of aggradation and degradation, is that they are one-dimensional

[Ferguson, 2003]. Hydraulic variables and calculations of bedload transport rate are averaged over the channel width, although large spatial variations in these properties, as well as strongly transverse and vertical components of hydraulic forces, are typically present in mountain rivers. Using statistical models to represent spatial variability in shear stress, *Ferguson* [2003] finds that bedload flux increases with the variance in shear stress, especially in coarse-grained rivers.

The results summarized here still support *Carson and Griffith's* [1987] conclusion that a simple, readily predictable, and widely applicable equation for bedload transport in steep, coarse-grained channels probably cannot be developed, a conclusion reinforced by *Diplas and Shaheen's* [2008] overview of the topic. What may be the most efficient approach is to develop a series of equations, with each equation having a narrow range of applicability that is defined by the most important controlling parameters, including channel gradient, relative roughness, grain-size distribution [e.g., *Armanini*, 1992], bedforms present, and velocity distribution. In the absence of such predictive equations, bedload transport is commonly extrapolated from records of multi-year sediment yield or sediment budgets.

These difficulties in finding broadly applicable mathematical expressions for bedload transport have not prevented investigators from developing numerical simulations of bedload transport processes [e.g., *Nikora et al.*, 2002]. The earliest models date to the late 1980s [e.g., *Naden*, 1987] and examine the interactions among flow, sediment transport, and bedforms in gravel-bed rivers. The applicability of these models is limited by their complexity and the lack of empirical data for parameterizing and testing the models. Subsequent one-dimensional [*Hassan and Church*, 1994; *Ferguson and Church*, 2009] and two-dimensional [*Tribe and Church*, 1999; *Malmaeus and Hassan*, 2002] kinematic models simulate vertical mixing and surface structures. An important component of such models is the rules governing particle interactions such as collision, entrapment, and the presence of a resistance field around stationary obstacles that creates indirect particle interaction [*Malmaeus and Hassan*, 2002]. *Schmeeckle et al.* [2001] use physical experiments to develop a collision Stokes number as a measure of the momentum of an interparticle collision versus the viscous pressure force in the interstitial gap between colliding particles, and a model of particle motion after rebound for particles of arbitrary shape.

Papanicolaou et al. [2004] describe the use of a 1d, coupled hydrodynamic-sediment transport model developed for steep mountain streams. Two-dimensional simulations based on hydraulic models coupled with sediment transport rules are also used to simulate bedload transport and channel morphology at larger spatial scales in order to predict changes in bed elevation [*Li et al.*, 2008], bed-surface texture, or sediment concentration [*Hairsine et al.*, 2002; *Sander et al.*, 2002] as sediment supply and discharge vary. *Nelson et al.* [2003] review 1d, 2d, and 3d sediment transport models.

In summary, measurement and modeling of bedload entrainment and transport remain significant challenges in mountain rivers because of substantial and highly variable entrainment thresholds, highly stochastic bedload movement, and limited field datasets that characterize relevant hydraulic, grain, and channel geometry variables. Physical experiments and numerical models provide useful means to simplify the complexity of real channels in order to explore the role and relative importance of individual variables, and will facilitate continued improvements in the understanding of bedload dynamics. Equations such as those developed by *Wilcock and Crowe* [2003] and *Yager et al.* [2007] that incorporate hiding functions and the component of dimensionless critical shear stress borne by rarely mobile bed sediments, respectively, are particularly appropriate for coarse-grained and supply-limited mountain rivers.

3.5.8. Bedload Yield and Sediment Budgets

Bedload yield can have large seasonal or interannual variations as a result of changes in sediment supply or sediment storage. Sediment supply along mountain rivers is strongly influenced by drainage basin and channel processes. Drainage basin sediment inputs include hillslope, valley bottom, glacial, and tributary channel sources that can be gradual (e.g., slope wash, average tributary inflows, soil creep) or abrupt (e.g., debris flows, rockfalls, tributary flash floods). These inputs can also be seasonally driven [e.g., *Wetzel*, 1994; *Beylich and Gintz*, 2004], aperiodic [*Schiefer et al.*, 2006], or variable over timescales of 10^3 yr because of large-scale variations in climate and base level [e.g., *Oguchi*, 1994a; *Coulthard et al.*, 2008; *Marden et al.*, 2008; *García and Mahan*, 2009]. Temporal variations in sediment input and output make it problematic to extrapolate long-term sediment yield from shorter-term monitoring records [*Schiefer et al.*, 2010]. The common characteristic seems to be that most of the sediment delivered to a channel moves during a very small fraction of the total time under consideration, whether that is a single year or multiple years. More than 75% of the long-term sediment flux from mountain rivers in Taiwan, for example, occurs during typhoon-generated floods that occur during less than 1% of the time [*Kao and Milliman*, 2008]. Half of the historical sediment load from a small mountainous watershed over a period of 70 years was delivered in less than 5 weeks [*Farnsworth and Milliman*, 2003]; a striking example of the importance of infrequent events in small catchments. In the wet tropics of Australia, the 1% of the total rainfall with the highest intensity contributes 9% of the transported sediment; for the 30% of the total rainfall with the highest intensity, the figure rises to 87% of the transported sediment [*Yu*, 1995]. Decadal variations in precipitation characteristics and resultant floods and hillslope instability [*Erskine and Saynor*, 1996; *Trustrum et al.*, 1999] can therefore greatly influence short-term sediment yields [*Johnson and Warburton*, 2002a; *Schiefer et al.*, 2006]. Following initial mobilization from sites of weathering, sediment can then be stored for varying lengths of time in depositional features associated with glacial, mass movement, eolian and fluvial processes (Figure 3.42).

Because sediment generated at erosion sites can be stored within a basin, sediment delivery ratio is used to represent the difference between volume of sediment generated and volume of sediment stored or transported from the basin. Sediment delivery ratio has been expressed via empirical regionalization relationships with drainage area, A, as

$$y = bA^\theta \tag{3.70}$$

where y is areal average sediment yield and θ is an empirical parameter that varies from -0.52 to 0.12 [*Milliman and Meade*, 1983; *Lu et al.*, 2005]. Sediment delivery ratio varies widely through time and space, typically between 0 and 1 [*Walling*, 1983; *Wasson et al.*, 1998; *Marutani et al.*, 1999]. In addition to A, factors such as topography, land cover, land use, lithology, process domains, and cyclic channel processes (e.g., arroyo cutting and filling) create spatial and temporal variability in sediment delivery ratio [*Fryirs and Brierley*, 2001; *Kasai et al.*, 2001; *Miller et al.*, 2001; *Springer et al.*, 2001; *Zierholz et al.*, 2001; *Verstraeten et al.*, 2003; *Molina et al.*, 2009; *Nyssen et al.*, 2009; *Warrick and Mertes*, 2009]. *Lu et al.* [2005] model variations in sediment delivery ratio with a simple conceptual model using a hillslope store that addresses transport to the nearest channels and a channel store that addresses sediment routing in the channel network.

The coupling between different components of the drainage basin exerts an important control on length of time that sediment is stored. Within-hillslope coupling, hillslope-channel coupling, and within-channel coupling (tributary-mainstem and reach-to-reach) all influence sediment storage and transfer [*Jones*, 2000; *Harvey*, 2002a; *Wichmann et al.*, 2009]. *Fryirs et al.* [2007] conceptualize sediment yield from the upper Hunter River catchment in Australia as being controlled by (dis)connectivity within and between landscape compartments such as terraces, floodplains, and hillslopes, and *Taylor and Kite* [2006] examine how valley-width morphometry and style of sediment delivery from hillslopes influence the efficiency of sediment transport and yield. As might be expected, timescales of sediment storage increase with spatial scales, and relate to magnitude-frequency characteristics of transport events [*Harvey*, 2002a]. One of the primary difficulties when attempting to generalize with respect to processes of erosion and sediment transport in mountain areas is the dominant influence and spatial variability of the bedrock geology and the thickness and nature of the regolith [*Bogen*, 1995]. A typical mountain river is essentially a number of local erosion and sedimentation subsystems (see also chapter 4), and downstream sediment yield records may not accurately characterize the system as a whole [*Bogen*, 1995]

Studies along the Fall River, Colorado, USA and the Colorado River, Grand Canyon, USA illustrate tributary-mainstem coupling. In July 1982 the earthen-fill Lawn Lake Dam in Rocky Mountain National Park, Colorado failed catastrophically. Drainage of 740,000 m³ of water from Lawn Lake generated a flood that peaked at 500 m³/s along the steep ($S = 0.10$), narrow Roaring River that drains the lake [*Jarrett and Costa*, 1986]. The floodwaters eroded a large quantity of glacial till from the channel margins of the Roaring River. A fan 17 ha in area and containing 280,000 m³ of sediment [*Blair*, 1987] was deposited where the Roaring River joins the meandering Fall River ($S = 0.002$) in a broad glacial valley. Within a year of the dam failure, large snowmelt floods moved much of the resulting sand and gravel downstream from the junction of the two rivers to an undisturbed, previously cobble-bedded, highly sinuous reach about 2 km long on the Fall River [*Pitlick and Thorne*, 1987]. Within this zone, the channel aggraded approximately 0.7 m, to the level of the floodplain. Subsequent lower snowmelt discharges then gradually transported sediment downstream such that the upper end of the storage zone was eroded to the pre-flood bed while the downstream end still had abundant flood sediment, producing large spatial variations in bedload transport along the channel [*Pitlick and Thorne*, 1987].

Debris flows and flash floods along channels tributary to the Colorado River in Grand Canyon repeatedly introduce large volumes of sediment (peak discharge averages 100-300 m³/s) to the main river in the form of debris fans [*Melis et al.*, 1995]. These fans constrict the main channel, creating steep rapids. During large discharges on the Colorado River (exceeding approximately 1400 m³/s), the constriction creates supercritical flow that facilitates erosion of the debris constriction until it is approximately half the upstream width, at which point erosion decreases and the fan becomes fairly stable [*Kieffer*, 1989]. Coarse clasts eroded from the constriction may be deposited 2-3 channel widths downstream, where the high velocity jet through the constriction dissipates [*Kieffer*, 1987; *Kieffer et al.*, 1989]. Analogous processes have been described from other canyon rivers with large point sources of coarse sediment from the valley walls or tributaries [*Hammack and Wohl*, 1996; *Elliott and Hammack*, 1999; *Larsen et al.*, 2002].

The rate of sediment supply from hillslopes to mountain rivers is a function of geology (lithology, structure, tectonics) and climate, as these control rates of weathering and erosion. Based on sediment yields from 21 sites (drainage area 3-75 km²) in the mountains of central Japan, Quaternary volcanic rocks yield twice as much sediment as intrusive rocks and ten times the sediment from mixed Paleozoic lithologies [*Mizutani*, 1987]. Similarly, areas with weakly consolidated bedrock represent only ~10% of total drainage area in the Western Transverse Ranges of California, USA, but contribute ~50% of the suspended sediment discharge [*Warrick and Mertes*, 2009]. Differences in weathering characteristics between the granodiorite in the upper basin and granite in the lower basin of semiarid Ash Creek in Arizona, USA, combined with a strong elevation-precipitation gradient, result in the upper basin producing most of the runoff and the lower basin most of the sediment [*Harvey et al.*, 1987]. A study of 20 mountain streams in central and southern California, USA found that a period of dry climate (1944–1968) had consistently low annual river sediment flux, whereas a subsequent period of wet climate (1969–1999) had a mean annual suspended sediment flux approximately five times greater [*Inman and Jenkins*, 1999]. Coupled changes in climate and base level can also enhance hillslope sediment supply, as in coastal catchments of New Zealand, where ~25% of postglacial sediment yield resultz from channel incision following base level fall and ~75% of the sediment derives from hillslope erosion processes [*Marden et al.*, 2008].

Sediment supply to mountain rivers from drainage basin sources differs from this supply to low-gradient alluvial rivers in that the proximity of steep hillslopes to many mountain rivers commonly results in large volumes of coarse-grained sediment being introduced from the hillslope directly into the channel, and may result in faster rates of sediment production (Figure 3.43) [*Nichols et al.*, 2005a]. *Ackroyd and Blakely* [1984] describe the catastrophic removal of a 600-m^3 sieve deposit along a mountain channel in New Zealand. The deposit had accumulated in the channel from landslides on perched scree fields and was stable for 30 years until a flood mobilized it, causing bedload transport rates at least four times average transport rates. *Mizutani*'s [1987] study of mountain rivers in central Japan documents an inverse correlation between sediment yield and drainage density, suggesting the more efficient transfer of sediment from hillslopes to headwater channels.

Mountain rivers typically have a lower ratio of valley-bottom to channel width than lower-gradient rivers. As a result, floodplains along mountain rivers tend to be poorly developed and to store minimal sediment relative to low-gradient systems. *Nakamura et al.* [1987, 1995] compare four drainage basins (A = 11-1345 km^2, S = 0.004-0.1) in a mountainous area of northern Japan. Using the ages of riparian trees to determine the time of sediment deposition, they find that sediment residence time and average transport distance increase downstream. The upper portions of the catchments have more frequent slope failures and landslides which produce sediment that is quickly transported downstream from the steep, narrow portions of the channels. Very large point inputs of sediment, however, can so overwhelm local transport capacity that they result in substantial storage along the valley bottom. *Morche et al.* [2006] describe a large rockslide that occurred in the Wetterstein Mountains of Germany circa 1800 A.D. The rockslide created a lake that then progressively filled with sediment over the next 200 years, decoupling sediment transport between the upper and lower portions of the catchment until the lake completely filled.

In-channel sediment inputs can also contribute to sediment yields. Along mountain rivers these inputs come from bed and bank erosion of storage sites such as riffles and bars (erosion during low flows), and pools (erosion during high flows). The strongest correlation in the *Mizutani* [1987] study was between sediment yield and channel mean gradient. Because steep channel gradients tend to be associated with direct hillslope sediment inputs, minimal valley-bottom storage, and high values of shear stress and stream power per unit area, it is difficult to know which factor(s) best explains this correlation. Ultimately, bedload yield reflects the balance between hydraulic driving forces and sediment availability. *Wathen et al.* [1997] attempt to quantify within-reach sediment storage using tracer clasts, but find that exhaustion of tracer supply over time and difficulty of recovering tracers limits the usefulness of this approach.

Glacial dynamics can strongly influence sediment supply to mountain rivers in alpine glacier basins, as reviewed by *Gurnell* [1995]. In glaciated basins of the Karakoram Himalaya, 40-50% of the annual sediment yield derives from glaciers [*Collins*, 1996]. The presence of bedrock under a glacier can result in sediment yields that are an order of magnitude lower than for glaciers of similar size underlain by till. The size of the glacier and the volume of discharge strongly control total annual sediment yield for glaciers underlain by sediment, and the level of activity of the glacier influences the amount and timing of sediment yield. Major conduits within the glacier are more stable in their position and are likely to be associated with lower and less variable production of suspended sediment than are smaller conduits. Transitions among small, linked conduits and major conduits during the ablation season are likely to result in substantial changes in sediment availability to the changing drainage network. On rivers draining glaciers in the Norwegian mountains, the relationship between water discharge and sediment transport is subject to continuous change through the season and from year to year as the subglacial drainage system changes [*Bogen*, 1995], making it difficult to develop predictive relations between discharge and sediment yield [*Pearce et al.*, 2003; *Tomkins and Lamoureux*, 2005]. The weak correlations between water discharge and sediment transport during rising discharges may reflect the erosion of spatially variable sediment sources as subglacial tunnels expand and sediments are entrained from the glacier bed or released by the melting of debris-loaded ice. The stronger correlations during falling discharges may reflect a stronger control by hydraulic variables on the amount of sediment remaining in transport.

Sediment yields from drainage basins with only relict (Pleistocene) glacial deposits can also be influenced by the glacial history. Catchments larger than 1 km^2 (up to 3 x 10^4 km^2) have higher unit sediment yields than smaller catchments [*Owens and Slaymaker*, 1992]. Sediment yields from the larger catchments are dominated by secondary remobilization of Pleistocene sediments stored along valley bottoms, whereas smaller catchments can still be storing sediment eroded from hillslopes. Eroding streamside glacial deposits dominate sediment yields in many watersheds in the Catskills Mountains of New York, USA [*Nagle et al.*, 2007].

Volcanic activity can also strongly influence sediment supply to rivers in some mountainous regions [*Sasahara et al.*, 2005; *Tagata et al.*, 2005]. Twenty years after the 1980 eruption of Mount St. Helens in the Cascade Range of Washington, USA, annual sediment yield below the massive avalanche deposit created by the eruption remained 1-2 orders of magnitude (10^3-10^4 Mg/km^2) greater than pre-eruption yield (~ 10^2 Mg/km^2) [*Major et al.*, 1999, 2000]. Recovery time and magnitude of increase varied in relation to eruption effect on each drainage basin (e.g., lahar deposits versus deforestation from the blast). Wildfires can also substantially influence sediment yields in the short term [*MacDonald et al.*, 2000; *Blake et al.*, 2005; *Reneau et al.*, 2007].

Climate change can influence sediment supply and delivery over varying time spans. Using a cellular river basin evolution model, *Coulthard et al.* [2008] find that quasi-linear relationships between water and sediment discharge can exist during stable climate periods, but an increase in flood magnitude and frequency over a sustained period (> 10 yr) can create a very large increase in sediment delivered by identically sized floods. Consequently, different climate periods can have significantly different relationships between water and sediment discharge as a function of sediment supply and storage.

Sediment yields reported from mountainous drainage basins around the world vary by four orders of magnitude (Table 3.6) as a result of interactions among the various factors controlling sediment yield. Sediment yield can be highly variable even within a specific region. Specific sediment yields (m^3/km^2) vary by three orders of magnitude for a given catchment size in Switzerland [*Spreafico and Lehmann*, 1994], with similar magnitudes of variation in Taiwan [*Kao and Milliman*, 2008]. Consistent characteristics, however, are that all rivers with large sediment loads originate in mountains and that the majority of the sediment load within a river basin will come from the mountainous portion [*Milliman and Syvitski*, 1992]. More than 80% of the sediment load in the Amazon River basin, for example, comes from the Andes, which constitute only about 10% of the river basin area [*Meade et al.*, 1985]. Although *Dedkov and Moszherin* [1992] report the highest suspended sediment yields from glacial and subnival zones in a comparison of 1,872 mountain rivers, climate does not appear to exert an important influence on global sediment yield patterns outside of glaciated regions; equatorial rivers and higher-latitude mountain rivers of similar size do not have significantly different sediment yields [*Milliman and Syvitski*, 1992]. Land use does appear to be important; Milliman and Syvitski attribute high erosion rates in mountainous regions throughout much of southern Asia to poor soil conservation, deforestation, and over-farming. In general, global sediment budgets consistently underestimate sediment fluxes from small mountainous rivers, many of which discharge directly onto active margins [*Milliman and Syvitski*, 1992].

Van de Wiel and Coulthard [2010] sound an important cautionary note for studies attempting to correlate changes in sediment yield to external factors such as climate or land use. They identify and describe a mechanism of self-organized criticality in the bedload sediment output from the drainage basin evolution model CAESAR. Their results suggest that sedimentary records used to infer past environmental conditions could reflect primarily internal system drivers rather than external drivers.

One of the most commonly used methods of calculating sediment yield for a drainage basin is that of measuring sediment accumulation in natural or artificial reservoirs [e.g., *Khosrowshahi*, 1990; *Fujita et al.*, 1991; *Schick and Lekach*, 1993; *Valero-Garcés et al.*, 1999; *Slaymaker et al.*, 2003; *Pelpola and Hickin*, 2004; *Hart and Schurger*, 2005; *van Rompaey et al.*, 2005; *Gellis et al.*, 2006; *Schiefer et al.*, 2006]. *Campbell and Church* [2003] describe a reconnaissance sediment budget for a site in British Columbia, Canada, developed from aerial photograph analysis and field investigation of volumes of erosional and depositional landforms, an approach also applied to other field sites [*Brasington et al.*, 2003; *Gaeuman et al.*, 2003; *Aalto and Dietrich*, 2005; *Hetherington et al.*, 2005; *Veyrat-Charvillon and Memier*, 2006; *Wichmann et al.*, 2009] and facilitated by the increasing availability of aerial and ground-based LIDAR [*Heritage and Hetherington*, 2005] and boat-based laser scanners [*Alho et al.*, 2009]. As repeat topographic surveys with high spatial resolution become more affordable, digital elevation models (DEMs) built from such surveys can be used to produce DEM of difference maps for estimating the net change in sediment storage [*Wheaton et al.*, 2010]. *Schrott et al.* [2003] and *Otto et al.* [2009] describe a similar analysis using geophysical methods to quantify thickness of different types of valley-fill deposits, and *Straumann and Korup* [2009] propose a region-growing algorithm based on a slope-gradient criterion to automatically extract areas of post-glacial fluvial and lacustrine valley fills from digital topography at the mountain-belt scale. A more localized, reach-based approach uses V^* (the fraction of scoured residual pool volume that is filled with sand and fine gravel) as an index of the supply of mobile sediment in a stream channel [*Lisle and Hilton*, 1992]. V^* at the reach scale reflects stream power and wood load, as these influence sediment storage [*Sable and Wohl*, 2006], as well as sediment supply.

Other methods of estimating sediment yield rely on equations to predict (1) upland sediment supply (e.g., various modifications of the Universal Soil Loss Equation, or USLE [*Wischmeier and Smith*, 1978; *Renard et al.*, 1991; *Yu*, 2005; *Smith and Dragovich*, 2008]; the USDA Forest Service's WATSED model [*USFS*, 1992]; EPIC, designed by *Klaghofer and Summer* [1990] for alpine catchments; KINEROS2 [*Martinez-Carreras et al.*, 2007]; WATEM/SEDEM [*Keesstra et al.*, 2009]); (2) suspended or bedload transport along the channel (*Cornwell et al.*, 2003; *Sear et al.*, 2003; or HEC-6 model [*Hydrologic Engineering Center*, 1977]); or (3) integrated hillslope-channel processes (e.g., the Stanford Model [*Crawford and Lindsley*, 1962], the Colorado State Model [*Simons et al.*, 1975; *Simons and Li*, 1976; *Li*, 1979], SHETRAN/SHESED [*Wicks and Bathurst*, 1996], *Benda and Dunne*, 1997a, *Hooke*, 2000, or DHSVM [*Doten et al.*, 2006]). *Becht et al.* [2005] describe the use of a spatial model calibrated by field measurements as part of the interdisciplinary SEDAG project on sediment cascades in Alpine systems. Models of upland erosion typically include factors such as slope, rainfall, soil texture, and land cover or land use [*Sharma and Sharma*, 2003] and may focus on sheet erosion [*Aksoy et al.*, 2003] or include rill erosion or mass movements. Long-term estimates of sediment yield can also be obtained using cosmogenic isotopes [*Matmon et al.*, 2003].

More than one study has demonstrated that contemporary sediment yields based on relatively short-term monitoring of suspended sediment yield tend to exceed long-term erosion rates determined from methods such as cosmogenic nuclides [*Clapp et al.*, 2000; *Bierman et al.*, 2005], suggesting either contemporary increases in sediment yield or unrepresentative sampling of long-term fluctuations in sediment yield. In contrast, *Ferrier et al.* [2005] find that long-term erosion rates inferred from cosmogenic ^{10}Be are faster than those inferred from several decades of sediment flux measurements at catchments of 0.2-720 km^2 in the northern California Coast Ranges, USA. Changes in land use and land cover can create disparities between short- and long-term sediment yields by altering processes of sediment production, delivery, and storage [*Wasson et al.*, 1998; *Gellis et al.*, 2004; *Constatine et al.*, 2005; *Leithold et al.*, 2005; *Page et al.*, 2008].

Sediment budgets can be developed by identifying and quantifying the sources, storage, and transport of sediment, typically on an average annual basis. This identification and quantification can use field mapping; aerial photographs; measurements of hillslope erosion using sediment fences (Figure 3.44) or erosion pins; mapping and chronology of sediment storage sites [*Stokes and Walling*, 2003]; soil isotopic ratios influenced by nuclear weapons tests [*Stokes and Walling*, 2003]; naturally occurring radionuclides [*Reusser and Bierman*, 2010]; mineralogy and petrology to identify sediment source areas [*Vezzoli*, 2004; *Hatfield and Maher*, 2009]; or reservoir sedimentation rates [*LaMarche*, 1968; *Caine*, 1974; *Dietrich and Dunne*, 1978; *Griffiths*, 1980; *Lance et al.*, 1986; *Roberts and Church*, 1986; *Sutherland*, 1991; *Froehlich and Walling*, 1992; *Bradley and Williams*, 1993; *Richards*, 1993; *Foster et al.*, 1994; *Olley and Murray*, 1994; *Saynor et al.*, 1994; *Wallbrink et al.*, 1994; *Loughran and Campbell*, 1995; *Oguchi*, 1997b; *Hill et al.*, 1998; *Valero-Garcés et al.*, 1999; *Johnson and Warburton*, 2002a; *Walling et al.*, 2003; *Malmon et al.*, 2005; *Nichols et al.*, 2006; *Smith and Dragovich*, 2008]. For rapid, recent deposition, low-cost geochronologic alternatives include using manufacture dates of garbage buried in alluvial sediments [*Kurashige et al.*, 2003]. Spatial distribution of different components of, and processes within, the sediment budget can be modeled using GIS platforms [*Dalla Fontana and Marchi*, 1998; *Ali and De Boer*, 2003; *Curtis et al.*, 2005; *Wilkinson et al.*, 2006]. Long-term budgets can also use geophysical methods to quantify volumes of sediment in different types of deposits (e.g., alluvial fans, floodplains) [*Schrott et al.*, 2003].

Slaymaker [2006] notes that, although many hydrologic parameters are scale invariant, geomorphic parameters are subject to scale distortion, which makes it improbable that sediment budgets are scale invariant. He reviews five ways of identifying scaling relations in sediment budgets, but notes that there is no *a priori* basis for selection of any particular method.

Spatially distributed sediment budgets can be used to target specific areas for erosion control [*Wilkinson et al.*, 2005] or for field instrumentation [*Molnar et al.*, 2006b], and to understand how changes in land use, land cover, or climate move through the drainage basin to cause changes in channel morphology [*Holliday et al.*, 2003; *Descroix et al.*, 2005]. *Reid and Dunne* [2003] provide a comprehensive overview of sediment budgets.

One final point to bear in mind with respect to sediment yield is that more than half the sediment eroded from many catchments is carried in solution [*Dietrich and Dunne*, 1978; *Selby*, 1982; *Summerfield and Hulton*, 1994]. The rate is higher in catchments with a greater input of subsurface water [*Hohberger and Einsele*, 1979] and with carbonate or other soluble lithologies [*Waylen*, 1979; *Selby*, 1982].

In summary, bedload yield and sediment budgets for mountain rivers can exhibit substantial spatial and temporal variability as a result of diverse sediment inputs, transport mechanisms, and storage sites within a catchment. Sediment can enter channels from hillslope sources, valley-bottom landforms such as terraces and floodplains, tributaries, and in-channel sources. The coupling or connectivity between different components of a drainage basin strongly influences sediment transfer and storage time, with mountain rivers having greater connectivity than most lowland systems. The rate of sediment production and supply reflect geology and climate, as well as glacial history and contemporary glacier dynamics, volcanic activity and disturbances associated with land use. Most of the sediment delivered to a channel moves during a very small fraction of the total time under consideration, largely during floods or debris flows. The sediment delivery ratio, which relates average sediment yield to drainage area, is used to represent the difference between volume of sediment generated and volume of sediment stored or transported from the basin. Commonly used methods of calculating sediment yield for a drainage basin rely on measuring sediment accumulation in natural or artificial reservoirs, as well as erosional and depositional landforms. These approaches are facilitated by the increasing availability of aerial and ground-based LIDAR and boat-based laser scanners. As repeat topographic surveys with high spatial resolution become more affordable, digital elevation models (DEMs) built from such surveys can be used to produce DEM of difference maps for estimating the net change in sediment storage. Other methods of estimating sediment yield rely on equations to predict upland sediment supply, suspended or bedload transport along the channel, or integrated hillslope-channel processes. Sediment budgets can be developed by identifying and quantifying the sources, storage, and transport of sediment, typically on an average annual basis. This identification and quantification can use field mapping; aerial photographs; measurements of hillslope erosion using sediment fences or erosion pins; mapping and chronology of sediment storage sites; soil isotopic ratios influenced by nuclear weapons tests or naturally occurring radionuclides; mineralogy and petrology to identify sediment source areas; or reservoir sedimentation rates.

3.5.9. Processes of Deposition

As mentioned in the previous sections, bedload in transport along a mountain river typically moves in episodic steps when considered at the timescale of an individual flow, an annual hydrograph, or a longer period of time. Coarse-sediment deposits along mountain rivers can take many forms as a function of sediment supply, channel morphology, and hydraulics. Examples include ripples, dunes, alternate bars, lateral gravel berms, transverse bars, point bars, steps, boulder clusters (Figure 3.45), and lobate deposits across the channel [*Baker*, 1978; *Cenderelli and Cluer*, 1998; *Bridge*, 2003]. Once a bar has been formed, it functions as a sediment storage area and influences local hydraulics and sediment transport [*Jaeggi*, 1987]. The clasts creating surface roughness for the bar also influence local depositional patterns, as does in-stream wood [*Buffington and Montgomery*, 1999b]. The scale and intensity of turbulence create a turbulence template under which only clasts large enough to tolerate the local turbulence are deposited [*Clifford et al.*, 1993].

Studies of coarse-sediment deposition along high-gradient channels have focused on hydraulic thresholds for deposition [*Bordas and Silvestrini*, 1992]. Attention has also been given to the classification of textural patches [*Buffington and Montgomery*, 1999a] and to the use of depositional patterns to infer flow hydraulics [*Elfström*, 1987]. The locations of coarse-grained deposits can reflect sites of declining or minimum stream power per unit area during large discharges [*O'Connor et al.*, 1986; *Wohl*, 1992a; *O'Connor*, 1993]. Differences in morphologic, sedimentologic, and fabric characteristics of deposits can also reflect changes in flow processes along a flood route [*Cenderelli*, 1998; *Cenderelli and Wohl*, 1998]. Studies focusing on the mechanics of deposition [*Ashworth et al.*, 1992; *Ashworth*, 1996] use both field observations and flume experiments, with flume experiments facilitating detailed process observations. In one of the early studies, *Lewin* [1976] examines re-adjustment of the gravel-bed River Ystwyth, Wales ($S = 0.004$) following artificial straightening in 1969. Primary mid-channel transverse bars formed rapidly at a fairly regular downstream spacing during infrequent high flows. This in turn led to bank erosion and the primary bars became the cores of point bars, with additional lateral and tail accretion and chute formation, until the channel became sinuous.

Baumgart-Kotarba's [1986] work on the braided Bialka River in the Carpathian Mountains of Poland indicates that channel bars form only when flow velocity is sufficient to transport cobbles and simultaneously scour the pools. Large discharges (recurrence interval > 4 yr) significantly widen the channel and form new bars, with the major sediment supply for the bars coming from bank undercutting. Coarse clasts deposited at zones of flow divergence form the nucleus of the bar, which then accretes downstream or laterally with smaller clasts. The dip angles of the clasts on the bars increase with current velocity and are highest on the distal portions of the bars [*Baumgart-Kotarba*, 1986]. In general, processes of deposition and channel adjustment in braided rivers are among the most difficult to measure and to model because of the huge quantity of field data necessary to describe the distributions of depth, slope, velocity, and sediment in these intrinsically variable systems [*Davies*, 1987] and because of the continual feedbacks and adjustments between channel morphology and sediment movement [*Ferguson and Ashworth*, 1992; *Laronne and Duncan*, 1992; *Ashworth*, 1996].

Flume studies largely focus on the conditions that initiate and maintain bar formation. Carling's flume studies indicate that clast deposition can be initiated at the sites of energy loss associated with flow transitions such as hydraulic jumps [*Carling*, 1995] or the shear zone of flow separation [*Carling*, 1989b]. The resultant accretionary gravel and boulder berms take the form of (1) simple berms that consist of a well-defined single curvilinear ridge parallel or oblique to flow; (2) complex berms of one curvilinear ridge with an integral series of lobes or re-curved ridges; and (3) compound berms of two or more independent subparallel berms [*Carling*, 1989b]. These berms minimize energy losses and ensure that high flows are transmitted efficiently through channel transitions.

Iseya et al. [1990] observe lobate coarse clast deposits along the Higashi-Gochi ($S = 0.09\text{-}0.11$) tributary of the Oi River, Japan following a flood with a recurrence interval of approximately 30 years (Figure 3.46). These lobate deposits have a unit of massive coarse clasts overlain by a stratified unit that includes layers of open-work, matrix-fill and matrix-supported clasts. Flume experiments suggest that a high frequency of coarse clasts develops at the downstream end of bedload pulses. These clasts rapidly aggrade along portions of the channel with high grain and form roughness. The later stages of the bedload pulse then aggrade more slowly, allowing stratification and a finer matrix to develop [*Iseya et al.*, 1990]. A similar process produces stationary alternate bars that divert flow across the experimental channel and cause pool scour, leading to alternate sequences of bars and pools [*Lisle et al.*, 1991].

A transverse negative step in the channel bed, such as that present downstream from a bed step, large clast, instream wood, or bar front, creates flow separation in the form of a transverse roller eddy. At the point of flow separation, sand can be carried downstream in suspension trajectories in the main flow, while gravel traveling by rolling or sliding is deposited immediately below the step as the flow separates. The initial gravel deposit has an open-work structure with poorly developed fabric, but increasing volume of deposition creates a depositional slope with grading and a sand matrix. Distinct facies can thus be produced while flow remains steady [*Carling and Glaister*, 1987; *Carling*, 1990]. *Wende* [1999] describes very coarse-grained deposits (intermediate clast axis > 8 m) downstream from negative bedrock steps created by hydraulic quarrying of jointed bedrock. These deposits include single clasts leaning on the steps, clusters of imbricated boulders formed only on the stoss side of the steps, and clusters extending up- and downstream beyond the step [*Wende*, 1999].

Instream wood can exert strong localized controls on processes of deposition along forested rivers. Whether individual pieces of wood, logjams extending across small channels [*Douglas and Guyot*, 2005], or accumulations that cover only a small portion of the bed of larger channels but initiate bar formation [*O'Connor et al.*, 2003], the presence of wood facilitates flow separation and reduced transport capacity, leading to deposition both up- and downstream from the wood (Figure 3.47).

Different methods of change detection can be employed to monitor form and volume of depositional sites in rivers, including aerial digital photogrammetry, terrestrial oblique digital imagery and automated digital photogrammetry [*Chandler et al.*, 2002], and high-resolution ground surveys [*Brasington et al.*, 2003; *Raven et al.*, 2009].

In summary, the diverse forms of coarse-sediment deposits in mountain rivers reflect hydraulic thresholds for deposition, and the deposits can thus be used to infer flow hydraulics. Typically, coarse deposits reflect sites of declining or minimum values of hydraulic variables such as shear stress or unit stream power. Channel geometry and erosional resistance of the channel boundaries, flow hydraulics, sediment supply, and instream wood all influence the deposition of coarse sediments.

3.5.10. Suspended Sediment

Although more attention has been devoted to bedload processes in mountain rivers, suspended sediment can also exert an important influence on channel processes and morphology, water quality, and the availability of aquatic habitat. The sources and transport of suspended sediment have received some attention, with more work being done on the effects of fine sediment on aquatic habitat or the quantification of fine sediment stored in or on the bed as an indirect indicator of anthropogenic effects.

Mountain rivers are typically regarded as having low suspended loads relative to lower gradient channels, although suspended load may exceed bedload along some mountain channels [*Barsch et al.*, 1994a; *Alexandrov et al.*, 2009]. The high-gradient channels in *Williams and Rosgen*'s [1989] compilation of sediment data for 93 streams in the USA have suspended sediment concentrations of 1-2,840 mg/L, although channels affected by the 1980 eruption of Mount St. Helens had concentrations up to 29,100 mg/L for several years after the blast. By comparison, low-gradient alluvial channels listed in this compilation have concentrations of 14-15,700 mg/L [*Williams and Rosgen*, 1989]. Worldwide, however, mountain catchments plot above the mean line on *Walling and Kleo*'s [1979] suspended sediment yield-catchment area graph [*Gurnell*, 1987b].

Data on relative proportions of suspended and bed load from sites around the world indicate that suspended load dominates in rivers of the equatorial zone, constituting 85-99% of particulate transport, whereas bedload increases to nearly half the load in rivers draining portions of central Europe influenced by Pleistocene glaciation, with values as high as 87% bedload in some basins [*Babiński*, 2005]. *Bogen and Bønsnes* [2003], however, measure suspended:bedload ratios of 1.5:1 for glaciated catchments in northern Norway. Synthesizing across 90 glacier basins, *Gurnell et al.* [1996] find that suspended sediment yield per unit area exceeds the global average defined by *Walling and Kleo* [1979], with basins containing predominantly warm-based glaciers producing higher yields than those with predominantly cold-based glaciers. Volcanically active warm-based glaciers in Iceland produce some of the highest unit suspended sediment yields [*Gurnell et al.*, 1996]. Published data of glacially derived sediment indicate a large range, from < 30% to > 75%, in the proportion of total load transported as bedload.

Mountain rivers draining alpine glaciers can have seasonally high suspended sediment concentrations that change in grain size and volume as the glacial meltwater pathways evolve during the course of the melt season [*Haritashya et al.*, 2010]. Norwegian proglacial rivers have low concentrations (< 40 mg/L) during early snow-melt, higher concentrations (50-100 mg/L) as glacier melting progresses, and the highest concentrations (> 500 mg/L) during floods generated by both rain and glacier melting [*Bogen*, 1995]. For the braided channel network downstream from the glacier, suspended sediment concentration depends both on sediment supply from upstream and on the amount of sediment deposited in, or eroded from, the braided river system. The highest concentrations are reached during a rapid increase in water discharge and when discharge during the preceding period is of intermediate size [*Bogen*, 1980]. Swift flow during high discharge restricts sedimentation, as does low sediment supply during low discharge. However, peak suspended sediment occurs prior to peak melt-water discharge from the 286 km^2 Gangotri Glacier basin in the Garhwal Himalaya of India [*Singh*, 2006]

In contrast to subglacial suspended sediment supply to basal channels in temperate glaciers, suspended sediment can be acquired principally via subaerial sediment supply to ice-marginal channels at non-temperate glaciers [*Hodgkins*, 1996]. Arctic glaciers can have seasonal trends in diurnal hysteresis between suspended sediment concentration and discharge, with progressive changes from a suspended-sediment-concentration-lead to a suspended-sediment-concentration-lag. The ice-marginal channels have a continuing sediment supply from heavily debris-covered ice-cored moraines and the hysteresis probably results from a delaying effect because of sediment circulation in low-velocity water adjoining the channel, which varies with discharge [*Hodgkins*, 1996].

Studies documenting higher ratios of suspended load in mountainous settings include: ~25% bedload from the 5 km^2 Rio Cordon in the Italian Alps [*Lenzi et al.*, 2003b]; <30% bedload from the 0.33 km2 Izas catchment in the central Spanish Pyrenees [*Alvera and García-Ruiz*, 2000]; ~35% bedload from the 1620 km^2 upper Marsyandi River of central Nepal [*Pratt-Sitaula et al.*, 2007]; and ~5% bedload from the 112 km^2 Eshtemoa catchment in Israel [*Alexandrov et al.*, 2009]. In some cases, such as the Izas catchment, the relatively high proportion of suspended load likely reflects the readily weathered bedrock underlying the catchment.

Organic matter typically constitutes a minor portion of suspended sediment, but can constitute more than half of suspended sediment in old-growth temperate forests [*Madej*, 2005]. It is too simplistic to think of suspended sediment as simply discrete sediment particles; much suspended material consists of flocs that are complex micro-ecosystems with an active biological community and microprocesses that control the structure and transport of the material [*Droppo*, 2003].

Mountain rivers in arid and semiarid regions can carry appreciable suspended sediment [e.g., *Barsch et al.*, 1994b; *Powell et al.*, 1996; *Alexandrov et al.*, 2009]. Comparing suspended sediment transport from hundreds of sites across the United States, *Simon et al.* [2004] find that semiarid environments have the highest median suspended sediment concentrations, whereas humid regions with erodible soils and steep slopes have the highest sediment yields. *Lekach and Schick* [1982] report concentrations up to 285,000 ppm along Nahal Yael, Israel. The silt-clay fractions of this load are derived by wash from underneath the stony surface layer on the basin slopes, and reach a maximum of 30,000 ppm, beyond which the additional suspended sediment is all sand supplied by the channel.

The limited work done on tropical mountain rivers indicates that they can also have high suspended sediment loads during storms. *Gellis et al.* [2001] measure concentrations ranging from 6,900 ppm to 140,000 ppm on two Puerto Rican catchments between which erosion and sediment yields differed markedly as a result of differing geology and land use. The magnitude of peak flow and maximum rate of hydrograph rise correlate most strongly with suspended sediment concentration in these small, steep, hydrologically flashy catchments [*Gellis*, 2001].

Mountain river basins subject to recent disturbances can also have a substantial increase in suspended sediment load. During a four-year period following the 1988 forest fire in the Yellowstone region of Wyoming, USA, a headwater catchment experienced suspended sediment increases of up to 473%, whereas lower channel reaches had only 60% increases [*Ewing*, 1996]. These types of increases can be moderated by accumulations of wood on hillslopes and in small channels that form logjams and trap sediment [*Gellis*, 1993]. Wildfire substantially increases suspended sediment loads during the rising limb of the snowmelt hydrograph and several summer thunderstorms, whereas bedload transport rates do not increase following the fire [*Ryan and Dixon*, 2008]. Anthropogenic disturbances including timber harvest and road construction also typically cause substantial increases in suspended sediment loads [*Douglas*, 1996, 2003; *Lai et al.*, 2005].

As might be expected, periods of greatly increased precipitation, whether single events or seasonal or longer-term climatic fluctuations, also create variability in suspended sediment loads [*Ross and Gilbert*, 1999; *Hudson*, 2003; *Beylich et al.*, 2006; *Wulf et al.*, 2010]. Fluctuations in sources and sinks of suspended sediment associated with channel dynamics (e.g., arroyo cutting and filling) also create decadal or longer-term variations in suspended sediment loads [*Amin and Jacobs*, 2007].

Work on headwater channels in Japan indicates that suspended sediment concentration can peak before or during peak water discharge, depending on whether the suspended sediment is coming primarily from the channel bed or from the hillslopes [*Kurashige*, 1994, 1996]. The amount of fine sediment entrained from the channel bed will depend on whether the coarse surface layer remains stable, thus limiting the supply of fines, or the whole bed surface is mobilized so that the greater supply of fine sediment in the subsurface can also be accessed by flow [*Diplas and Parker*, 1992].

Because the sources of suspended sediment may not be uniformly distributed throughout the watershed, predicting suspended sediment in relation to precipitation can be difficult [*Banasik and Bley*, 1994], although some studies indicate that precipitation intensity exerts an important control on suspended sediment concentration [*Lana-Renault et al.*, 2007]. The contribution of suspended sediment from specific source areas can be directly measured using traditional field monitoring techniques, or indirectly estimated using models or physical reasoning [*He and Owens*, 1995]. Source areas can also be identified via "fingerprinting" one or more of the diagnostic properties of suspended sediment and potential source materials. Diagnostic properties include sediment mineralogy, color, or chemistry, mineral magnetics, heavy-metal content, nitrogen stable isotopes, or radionuclide activity (^{137}Cs, ^{210}Pb, ^{7}Be, ^{226}Ra) [*He and Owens*, 1995; *Walling*, 2003; *Froehlich and Walling*, 2005, 2006; *Fox and Papanicolaou*, 2008]. Using ^{7}Be as a tracer, *Bonniwell et al.* [1999] find that new suspended sediment (derived from the adjacent landscape rather than the streambed) varies from 12% to 96% of the total, and that suspended sediment in a mountain stream draining 389 km^2 in Idaho, USA moves in steps approximately 60 km long near the peak of the hydrograph and declines to about 12 km near baseflow. *Whiting et al.* [2005] use radionuclides to demonstrate that the contribution to suspended load from soils declines from 50% in the upper part of a watershed in Montana, USA (31,690 km^2) to 11-26% downstream, and transport distances increase from a few kilometers in the headwaters to hundreds of kilometers downstream.

Suspended sediment has been observed to move in clouds or waves. On a proglacial stream in Switzerland, these waves are associated with the progressive encroachment of the stream across its floodplain during glacier melt [*Gurnell and Warburton*, 1990]. On the Torlesse Stream, New Zealand, waves are of 1-3 hr durations and are related to a sudden influx of sediment into the channel as a result of bank collapse [*Hayward*, 1979]. When suspended sediments are being transported along the Torlesse they can account for up to 90% of the sediment yield, but during most stormflows there are prolonged periods of no suspended sediment transport. As a result, suspended sediment contributes less than 10% of the annual sediment yield from the catchment [*Hayward*, 1979]. Measurements of suspended sediment transported in meltwaters from a Swiss glacier indicate that considerable variations in seasonal and interannual sediment transport result from drainage basin instabilities associated with initiation of the drainage network in spring, from temporary flow constrictions in subglacial channels, and from precipitation-enhanced high discharges [*Collins*, 1990]. Studies on the River Exe in the United Kingdom also support the idea that suspended sediment transport may be more strongly influenced by sediment supply than by transport energy [*Walling and Webb*, 1987].

As with bedload, actual sampling of suspended load can be problematic [*Hicks and Gomez*, 2003]. A standard procedure in the USA is to use a DH-48 sampler to obtain samples of the flow that are point-integrated (sampling at a specific flow depth until the sampler is full) or depth-integrated (lowering and raising the sampler at a constant rate from the flow surface to just above the channel bed) [*Leopold et al.*, 1964] (Figure 3.48). The data for Nahal Yael, Israel come from Hayim 7 multiple stage automatic samplers, which abstract several 1-L samples during the rising stage of flow [*Lekach and Schick*, 1982]. The filling of a particular container begins when the stage rises to approximately 3 cm above the container's intake, and lasts approximately 1 minute. By arranging several samplers at varying heights above the channel bed, point-integrated samples can be obtained for successive stages of the rising limb. Other types of automatic samplers triggered by changes in flow stage or set for specific time intervals include those made by Teledyne ISCO.

Another approach is to use continuously recording turbidimeters that can be connected to dataloggers and deployed in arrays once they are calibrated to the suspended sediment loads in a particular field setting [*Orwin and Smart*, 2005]. Because turbidity is easier to measure than suspended sediment concentration, several studies develop and validate models relating these two parameters [*Chikita et al.*, 2002; *Lewis*, 2003; *Langlois et al.*, 2005; *Teixeira and Caliari*, 2005].

Schindl et al. [2005] review other methods of measuring suspended sediment, including acoustic techniques, focused beam reflectance, laser diffraction, nuclear, optical [*Schoellhamer and Wright*, 2003], and remote spectral reflectance. *In situ* laser sensors that measure grain-size distribution, settling velocities, and suspended sediment concentrations are particularly useful in remote locations where direct sampling is difficult, costly, and episodic [*Melis et al.*, 2003; *Thonon and Van Der Perk*, 2003].

The degree of vertical mixing is an important consideration in designing a sampling strategy for suspended sediment. Comparing samples collected from the water surface with those collected from close to the bed using a vacuum pump, *Carling* [1984a] finds that the suspended sediment load in shallow rough-bedded streams is fully mixed, with an insignificant depth-dependent concentration gradient. In contrast, narrow and deep streams carry proportionally more suspended sediment close to the bed as discharge increases because the efficiency of turbulent transfer to the surface is reduced.

De Boer [2005] describes the use of a cellular model of erosion and sediment transport to derive rating curves for suspended sediment concentration in relation to discharge. Developing rating curves from empirical data can be difficult as a result of variation in methods of sampling suspended sediment or insufficient samples covering the range of flow conditions [Smith and Croke, 2005]. Variance in the curves typically increases with drainage area [Cashman and Potter, 2006].

Other models such as SedNet [Post and Hartcher, 2006] use an approach similar to the USLE (section 3.5.8) to estimate suspended sediment loads by constructing budgets that account for the primary sources and stores of sediment. The relationship between suspended sediment concentration and discharge can be accurately simulated over millennial time spans with models such as HydroTrend [Kettner et al., 2007], a climate-driven hydrologic-transport model that creates time series of water and suspended sediment discharges as a function of climate and local catchment characteristics that influence the hydrology of the contributing rivers [Mulder and Syvitski, 1996]. Chikita et al. [2007] describe a simulation of discharge and suspended sediment load from a glacier basin in Alaska.

The effect of high suspended sediment concentrations on flow processes is now better understood than when the first edition of this book was written. At that time, vertical velocity and sediment profiles, frictional resistance and sediment transport could not be accurately predicted for high-concentration flows [Bradley and McCutcheon, 1987]. When sediment exceeds 20% by volume, standard descriptions such as the log velocity profile for the law of the wall or the Manning resistance equation typically do not adequately describe flow conditions [Bradley and McCutcheon, 1987]. More recent work demonstrates that suspended sediment can attenuate turbulence when the ratio of particle size to turbulence length scale is small and enhance turbulence when the ratio is large. Cao et al. [2003] develop an empirical turbulent eddy viscosity-based closure model for the mean velocity of suspended-sediment-laden flows.

Although suspended sediment delivery to the world's oceans is declining because of the storage effects of reservoirs built on larger rivers [Walling and Fang, 2003], suspended sediment delivery to many mountain rivers has increased historically as a result of changing land uses. The effects of suspended sediment on aquatic habitat typically become of concern when the supply of suspended sediment to a channel increases to the level that fine sediment deposition alters the grain-size distribution of the channel-bed material. Land uses such as timber harvest [Corn and Bury, 1989; Madej and Ozaki, 1996], mining [Wagener and LaPerriere, 1985; Van Haveren, 1991], construction [Wolman, 1967; Boon, 1988], agriculture [De Boer, 1997], or grazing [Trimble and Mendel, 1995; Myers and Swanson, 1996a] can result in increased fine sediment entering a mountain river. This can in turn cause infilling of pools [Lisle and Hilton, 1992; Rathburn and Wohl, 2003] or of spawning gravel frameworks [Harvey et al., 1993; Kondolf and Wilcock, 1996; Wilcock et al., 1996b] and consequent loss of aquatic habitat for macroinvertebrates and fish. It can also cause reduction of instream photosynthesis [*Cuker et al.*, 1990], change in nutrient avail-ability [*Farnworth et al.*, 1979], and affect sight feeding, respiration, and orientation of fish [*Karr and Schlosser*, 1977; *ASCE Task Committee*, 1992]. Most stream organisms can withstand short-term exposure to elevated levels of suspended sediment, but chronic exposure is more detrimental [*ASCE Task Committee*, 1992].

In summary, although mountain rivers are typically regarded as having relatively low suspended sediment concentrations, numerous studies indicate the potential for substantial suspended sediment in rivers with: glaciated catchments; arid, semiarid or tropical climates; or recent disturbances such as wildfire or timber harvest. Field monitoring or fingerprinting can be used to identify the sources of suspended sediment, which can be directly measured using handheld or automated samplers, or inferred from turbidity or remote sensing using optical, acoustic, and other types of signals.

In general, the entrainment, transport, and deposition of all grain sizes present along a mountain river are strongly influenced by hillslope processes as these govern sediment supply and by in-channel interactions between flow and substrate. Parallel developments in methods of mapping and quantifying sediment volumes, identifying sediment sources, and modeling sediment transport and deposition are substantially improving estimates of sediment yield and construction of sediment budgets. Although some mountain rivers have erosionally resistant banks formed of extremely coarse grains or of bedrock, other channels can obtain a substantial component of the sediment moving in suspension and along the bed from erosion of stream banks. Bank stability and processes of erosion therefore influence sediment supply and transport, as well as channel stability and geometry.

3.6. Bank Stability

A stream's ability to erode its banks, and the processes and rates of bank erosion, exert a fundamental control on sediment supply and channel geometry. Bank erosion also presents a management problem in many contexts. Bank erosion can result in damage to infrastructure along channels, such as buildings and bridges, as well as property loss. Bank erosion can alter riparian structure and function and thus affect riparian and aquatic communities. Altered width/depth ratios, channel substrate, sed-iment supply, and water quality resulting from bank erosion can influence physical processes in the channel and aquatic communities. And bank erosion is commonly regarded as an indicator of channel instability and the need for management to mitigate this instability. For all of these reasons, as well as to improve our understanding of fundamental channel dynamics, much attention is given to analyzing and modeling bank stability.

The strength of channel banks reflects the frictional properties of the bank material and the effective cohesion [*Simon et al.*, 1999]. Effective cohesion results from matric suction within the unsaturated part of the bank and/or root reinforcement from vege-tation [*Eaton*, 2006]. Spatial and temporal heterogeneities in these properties result from bank stratigraphy, moisture content, and riparian vegetation [*Rinaldi and*

Casagli, 1999; *Parker et al.*, 2008]. Erosion of stream banks can occur via: mass failure through slumping or toppling of a slab [*Osman and Thorne*, 1988]; entrainment of individual particles by shearing along the bank face [*Rinaldi and Darby*, 2008]; detachment via shrink-swell or freeze-thaw cycles [*Lawler*, 1993b; *Prosser et al.*, 2000; *Yumoto et al.*, 2006; *Wynn et al.*, 2008]; or groundwater seepage [*Fox et al.*, 2007]. The relative effectiveness of these processes typically varies downstream, with mass failure dominating lower portions of drainage basins and processes such as freeze-thaw and fluvial entrainment likely to be more important along mountain rivers [*Lawler*, 1992]. *Fonstad and Marcus* [2003] demonstrate the existence of power-law relationships between the number of bank failures of a given size throughout a watershed and the magnitude of those bank failures such that lower-gradient, alluvial streams are more susceptible to large bank failures.

Stream banks along mountain rivers may have a linear profile, but more typically display a vertical upper section composed of overbank deposits that are strongly influenced by cohesion of fine-grained sediment and the roots of vegetation, and a lower section of coarser, less cohesive sediment originally deposited in the channel (Figure 3.49). Many investigators assume that the vertical upper bank has an average height proportional to the effective cohesive strength and that the failure mode of the upper bank is adequately described by a simple slab failure stability analysis [*Darby and Thorne*, 1996; *Darby et al.* 2000; *Istanbulluoglu et al.*, 2005]. This works better in lower-gradient rivers than in gravel-bed channels, where stability of the lower bank needs to be modeled separately using a modified friction angle approach [*Millar and Quick*, 1993]. *Eaton* [2006] combines the two approaches in a composite bank stability analysis based on effective bank cohesion for the upper bank and friction angle for the lower bank.

Recognizing the effect that stratigraphy can have on bank stability, *Simon et al.* [2000] develop a bank stability algorithm and then the BSTEM model for layered cohesive stream banks by combining the Coulomb equation for saturated banks with the *Fredlund et al.* [1978] equation for unsaturated banks. The heterogeneity of bank properties is most effectively represented with a probabilistic representation of effective bank material strength parameters [*Parker et al.*, 2008]. *Rinaldi and Darby* [2008] summarize methods of stability analysis applied to river banks. Parameter uncertainties in bank stability models are typically high enough that the likelihood of generating unreliable predictions is typically very high (>80%) [*Samadi et al.*, 2009]. These uncertrainties result primarily from the natural variability of the parameters.

Models have also been developed that focus specifically on the effects of riparian vegetation on bank stability. These effects include increased mass [*Thorne*, 1990; *Abernethy and Rutherfurd*, 2001; *Simon and Collison*, 2002], increased shear strength derived from root reinforcement of the sediment [*Wu et al.*, 1979; *Pollen and Simon*, 2005; *Pollen*, 2006], and variations in bank pore-water pressure caused by vegetative influences on infiltration, evaporation, and transpiration [*Thorne*, 1990; *Simon and Collison*, 2002; *Pollen-Bankhead and Simon*, 2010]. Meadow vegetation (Figure 3.50) can be very effective in reducing bank erosion along mountain rivers

[*Micheli and Kirchner*, 2002a, b], as can woody vegetation, although the magnitude of the effect varies between plant species [*Pollen and Simon*, 2005; *Docker and Hubble*, 2008]. *Pizzuto et al.* [2010] describe cyclic bank erosion on decadal time-scales along channels in the eastern United States. Small volumes of sediment are removed between large trees on the bank, creating a scalloped bank morphology buttressed by large trees that gradually become undercut and lean into the channel until they topple, which results in a larger volume of sediment removal from the bank and restarts the cycle.

Existing models suggest that riparian vegetation exerts the greatest influence on bank stability when growing on low, shallow banks of weakly cohesive sediments and when growing at the ends of the incipient failure plane at the bank toe or at the intersection of the failure plane with the floodplain, although the change in simulated factors of safety is less than 5% [*Van De Wiel and Darby*, 2007]. Following *Lawler's* [1992] downstream zonation of bank erosion processes, *Abernethy and Rutherfurd* [1998] note that flow resistance created by vegetation is most important where fluvial entrainment dominates bank erosion, and increased bank shear strength caused by root reinforcement is most important in downstream reaches where mass failure dominates bank erosion. The effect of vegetation on bank strength declines as the absolute scale of the channel increases [*Eaton and Millar*, 2004].

Rates of bank erosion can be directly measured using repeat cross-sectional surveys [*Lane et al.*, 1994], analysis of channel planform through time from aerial photographs, or use of pins [*Wolman*, 1959] or photo-electronic erosion pins [*Lawler*, 1991, 2005, 2008]. Rates of bank erosion can also be inferred from indicators such as anatomical changes in tree roots exposed by bank erosion [*Malik and Matyja*, 2008]. *Lawler* [1993a] reviews methods of assessing bank erosion. *Pizzuto* [2003] reviews models of bank erosion and associated channel migration.

In summary, bank strength reflects the frictional properties of the bank material and the effective cohesion resulting from matric suction in unsaturated materials and root reinforcement from vegetation. Spatial and temporal variability in these properties reflect bank stratigraphy, moisture content, and riparian vegetation. Banks along mountain rivers typically have a vertical upper section of fine-grained overbank deposits influenced by cohesion and vegetation, and a lower section of coarser, less cohesive channel sediment.

3.7. Instream Wood

Large woody debris (LWD) is composed of pieces of wood typically defined as being 10 cm or greater in diameter and 1 m or greater in length, although definition of minimum size varies between studies [*Hassan et al.*, 2005]. *Abbe and Montgomery* [2003] define effective piece size based on geomorphic effects and *Young et al.* [2006] suggest defining minimum sizes based on the prevalence of instream piece sizes relative to those in riparian zones. Because debris can carry a negative connotation, LWD will be described in the remainder of this section simply as wood. Wood exerts an important control on channel processes in many mountain rivers and is one of the topics summarized in this book which has undergone a dramatic expansion in research since the first edition in 2000. At the time of the first edition, the great majority of wood research had been conducted in the Pacific Northwest of the United States and Canada, where wood was recognized to play a critical role in salmonid habitat. Since that time, investigators have documented the geomorphic and ecological roles of wood from boreal to tropical environments, with increasing emphasis on topics such as wood recruitment, transport and decay, patterns of wood distribution at reach to network scales, and quantifying the hydraulic and morphologic effects of wood. As a result of this ongoing research, several conceptual models of wood dynamics at network scales now exist.

Wood in stream channels increases boundary roughness and bank resistance [*Keller and Tally*, 1979; *Curran and Wohl*, 2003; *Wilcox and Wohl*, 2006; *Wilcox et al.*, 2006; *Manners et al.*, 2007; *Comiti et al.*, 2008b] and deflects flow toward the channel bed and banks, thus accentuating local boundary erosion [*Montgomery et al.*, 1995a; *Wallerstein*, 2003; *Hassan and Woodsmith*, 2004]. As noted by *Mutz* [2003], however, detailed information on the distribution of hydraulic variables in the vicinity of single logs or jams close to a natural streambed is largely missing (*Manga and Kirchner* [2000], *Hygelund and Manga* [2003], and *Daniels and Rhoads* [2004] are exceptions) and both field and flume experiments are needed to fill this gap.

Wood facilitates localized storage of sediment and organic material by creating obstacles to flow and associated zones of flow separation [*Swanson et al.*, 1976; *Keller and Swanson*, 1979; *Keller and Tally*, 1979; *Marston*, 1982; *Buffington and Montgomery*, 1999b; *Faustini and Jones*, 2003; *Haschenburger and Rice*, 2004; *Douglas and Guyot*, 2005; *Andreoli et al.*, 2007] (Figure 3.51). In one study using sediment tracer particles on 1st- and 2nd-order step-pool channels in Idaho, approximately 70% of the tracer material remaining in the study reaches was deposited behind logjams [*Ketcheson and Megahan*, 1991]. *Piégay and Gurnell* [1997] note similar patterns in headwater channels in Europe. Use of radionuclides (^{14}C, ^{210}Pb, ^{137}Cs, ^{7}Be) to date sediment stored in association with wood indicates greater sediment storage volumes and times in river segments with wood than in segments without wood [*Fisher et al.*, 2010; *Skalak and Pizzuto*, 2010].

Although stable wood increases the residence time of sediment, wood that breaks or is mobilized can release a pulse of sediment that moves downstream to the next depositional site, resulting in bed degradation at the site of wood removal [*Bugosh and Custer*, 1989; *Adenlof and Wohl*, 1994]. Stored sediment and wood may also be periodically flushed from the channel by a debris flow that scours the channel to bedrock [*Swanston and Swanson*, 1976] and creates a massive debris jam at a lower gradient reach downstream [*Kochel et al.*, 1987]. Such jams can effectively trap sediment for decades until the debris decomposes [*Keller and Swanson*, 1979; *Abbe and Montgomery*, 1996]. By increasing hydraulic roughness at the reach scale, wood can also promote deposition of finer bed material throughout the reach [*Buffington and Montgomery*, 1999b].

By trapping fine sediment and organic matter, wood facilitates processing of nutrients by stream organisms [*Naiman and Sedell*, 1979; *Webster and Swank*, 1985; *Bilby*, 2003]. A study of mountain rivers in New Hampshire, USA indicates that 75% of the standing stock of organic matter in 1st-order streams is contained in wood dams; this drops to 20% in 3rd-order streams [*Bilby and Likens*, 1980]. Experiments with marked wood along a 2nd-order channel in Puerto Rico indicate that wood is typically associated with debris dams and large amounts of organic litter [*Covich and Crowl*, 1990]. Wood further influences aquatic organisms by enhancing substrate and habitat diversity [*Bilby and Likens*, 1980; *Harmon et al.*, 1986; *Bisson et al.*, 1987; *ASCE Task Committee*, 1992; *Fausch and Northcote*, 1992].

At the reach-scale, wood alters channel morphology by creating forced alluvial reaches [*Montgomery et al.*, 1996a, 2003] or by changing the dimensions of bedforms such as steps [*MacFarlane and Wohl*, 2003] and pools [*Robison and Beschta*, 1990; *Richmond and Fausch*, 1995; *Beechie and Sibley*, 1997; *Mao et al.*, 2008a] (Figure 3.52). High-gradient rivers in forested regions commonly have sediment stored upstream from wood, with plunging flow over the wood step and a scour pool at the base [*Bilby and Ward*, 1989]. In this situation, 30-80% of the vertical drop of the stream can be influenced by wood [*Keller and Swanson*, 1979] and much of the stream's energy is dissipated at wood-created falls and cascades that occupy a fairly small proportion of total channel length [*Heede*, 1972; *Thompson*, 1995].

The pools associated with wood receive particular attention because of their importance to salmonid fish [*Dill et al.*, 1981; *Fausch*, 1984; *Sullivan et al.*, 1987]. Plunge pools, backwater pools, dammed pools, and lateral scour pools associated with either root wads or large wood are among the types of pools enhanced by wood [*Bisson et al.*, 1982]. Up to 75% of all pools along mountain rivers may be associated with wood [*Robison and Beschta*, 1990; *Young et al.*, 1990]. Pool volume is inversely related to stream gradient and directly related to amount of wood [*Carlson et al.*, 1990]. Flume experiments [*Beschta*, 1983] indicate that, for wood suspended above the streambed (as opposed to buried in the bed), larger diameter pieces create longer and deeper pools. Maximum depth and area of scour occur when wood is perpendicular to flow [*Cherry and Beschta*, 1989]. Pool area also correlates positively with the wood volume at the pool [*Bilby*, 1985]. *Linstead* [2001] provides an example of quantifying the effects of wood on fish habitat with the widely used Physical Habitat Simulation Model (PHABSIM).

Logjams in unconfined mountain rivers can increase channel width [*Nakamura and Swanson*, 1993], enhance overbank flooding [*Jeffries et al.*, 2003] and floodplain geomorphic and ecological heterogeneity [*Montgomery and Abbe*, 2006; *Pettit and Naiman*, 2006], promote bar growth and consequent channel shifting [*O'Connor et al.*, 2003], and initiate lateral channel migration and anastomosing morphology [*Brummer et al.*, 2006; *Montgomery and Abbe*, 2006]. Logjams can be aggregated in portions of the network where valley and stream morphology reduce transport capacity, such as lower gradient, wider stream segments [*Kraft and Warren*, 2003; *Morris et al.*, 2010; *Wohl and Cadol*, 2010].

Wood is recruited into channels via mass movements from adjacent valley slopes [*May and Gresswell*, 2003a, 2003b; *Wohl et al.*, 2009] (Figure 3.53), bank erosion (Figure 3.54) [*Martin and Benda*, 2001], individual tree mortality [*Lienkaemper and Swanson*, 1987; *Cadol et al.*, 2009], tree canopy damage during wildfire, tropical storms and ice storms [*Young et al.*, 1994; *Kraft et al.*, 2002; *Phillips and Park*, 2009; *Bendix and Cowell*, 2010], and transport from upstream [*Benda et al.*, 2003]. Valley geometry and hillslope processes, channel dynamics, and forest characteristics (age, species, disturbance regime) thus all influence processes of wood recruitment.

Benda and Sias [2003] propose two linear equations that provide a useful conceptual framework for understanding the influences on wood recruitment and retention within a given channel segment:

$$\Delta S_c = \left[L_i - L_0 + \frac{Q_i}{\Delta x} - \frac{Q_0}{\Delta x} - D \right] \Delta t \qquad (3.71)$$

where ΔS_c is change in storage within a reach of length Δx over time interval Δt, L_i is lateral wood recruitment, L_o is loss of wood to overbank deposition during floods and abandonment of jams, Q_i is fluvial transport of wood into the reach and Q_o is fluvial transport out of the reach, and D is *in situ* decay. Lateral wood recruitment reflects several types of supply:

$$L_i = I_m + I_f + I_{be} + I_S + I_e \qquad (3.72)$$

where I_m is chronic forest mortality, I_f is toppling of trees following fire and windstorms, I_{be} is inputs from bank erosion, I_s is wood from landslides, debris flows, and snow avalanches, and I_e is exhumation of buried wood. Each of the variables in equations 3.71 and 3.72 can be parameterized using further equations; for example,

$$I_m = [B_L M H P_m] N \qquad (3.73)$$

where B_L is volume of standing live biomass per unit area, M is the rate of mortality, H is the average stand height, P_m is the average fraction of stem length that becomes in-channel wood when a riparian tree falls, and N is 1 or 2, depending on whether one or both sides of the channel are forested.

In practice, many of the fundamental variables can be difficult to quantify for a specific segment of mountain stream, but several regional models now exist that estimate wood recruitment and retention. These include *Meleason et al.* [1999], *Beechie et al.* [2000], *Welty et al.* [2002], and *Benda et al.* [2003] for the U.S. Pacific Northwest and *Bragg et al.* [2000] for the central Rocky Mountains in the USA. Both *Beechie et al.* [2000] and *Welty et al.* [2002] use linked models that simulate forest stand development and wood recruitment and depletion, respectively. *Hassan et al.* [2005] provide a more thorough review of existing models.

Any model is only as effective as the calibration data used to parameterize individual variables and rates of change, which requires detailed and extensive field studies to capture the range of natural and human-induced variability in wood loads

[*Fox and Bolton*, 2007] and in the many factors that influence instream wood dynamics. Significant regional differences also exist in most of the variables in equations 3.71 to 3.73 [e.g., *Gurnell*, 2003]. Wood in the Pacific Northwest tends to have longer residence times than the limited estimates available for other regions, for example, although estimates within the Pacific Northwest range from a mean residence time as low as 12 years to 100 years [*Hyatt and Naiman*, 2001; *Hassan et al.*, 2005] as a function of differences in piece size, channel dynamics, wood load, and decay by species [*Bilby et al.*, 1999]. Maximum residence time has been estimated at up to 1400 yr on the Queets River, Washington, USA [*Hyatt and Naiman*, 2001], and long residence times have been reported from central Sweden [mean > 200 yr, *Dahlström et al.*, 2005] and from southeastern Australia [max 240 yr, *Webb and Erskine*, 2003]. On the other hand, residence times in the range of 40-100 years characterize wood in the southern Appalachian Mountains of the USA [*Hart*, 2002], residence times as low as 3.4 yr have been reported for the Colorado Rocky Mountains [*Wohl and Goode*, 2008], and preliminary data from the wet and seasonal tropics suggest residence times < 3 yr [*Wohl et al.*, 2009]. Low residence times in the Rocky Mountain streams likely reflect low wood loads and enhanced transport, whereas those in the tropics reflect both high transport capacity [*Cadol et al.*, 2009] and rapid decay. Tables 3.7a and 3.7b summarize wood loads for diverse regions.

The likelihood and importance of transport from upstream increase for higher-order channels [*Martin and Benda*, 2001], in part because pieces that are shorter than the bankfull channel width are more likely to move [*Lienkaemper and Swanson*, 1987]. Drag and buoyancy are the main mobilizing forces [*Alonso*, 2004]. *Bocchiola et al.* [2006a] estimate the floating threshold $h*$ for a piece of wood as

$$h* = \frac{\rho_w d_w}{\rho_{\log} D_{\log}} \tag{3.74}$$

where ρ_w is water density, d_w is flow depth, $\rho_{\log}$ is wood density, and $D_{\log}$ is log diameter. *Bocchiola et al.* [2008] note that a floating threshold is well approximated by setting $h* = 1.26$. Entrainment is primarily influenced by piece angle relative to flow direction, the presence of a rootwad, wood density, and piece diameter [*Braudrick and Grant*, 2000]. Field experiments indicate that transport distance exponentially increases with water depth at peak flows and that the sequence of flows influences transport; flow magnitude greater than the previous flows is necessary for retransport of most logs [*Haga et al.*, 2002]. Flume experiments indicate that wood is deposited where flow depth is less than buoyant depth, typically at the head of mid-channel bars, in shallow zones of flow expansion, and on the outside of bends [*Braudrick and Grant*, 2001]. The presence of obstacles such as large boulders and other wood also influences transport distance and location of deposition [*Bocchiola et al.*, 2006b, 2008] (Figure 3.55).

The likelihood that wood will move as individual pieces rather than as debris jams increases for higher-order channels [*Braudrick et al.*, 1997]. The higher proportion of wood oriented perpendicular to flow in small streams reflects the relative immobility of wood in these streams [*Bilby and Ward*, 1989; *Richmond and Fausch*, 1995]. Much of the movement that does take place in small channels occurs during years of high discharge [*Lawrence*, 1991; *Berg et al.*, 1998]. Spatial and temporal patterns of wood retention and transport in 1st- to 7th-order streams in Oregon indicate that less than 30% of debris volume occurs within the active channel; the majority of wood is retained along channel margins and floodplains [*Gregory*, 1991] (Figure 3.56). No more than 10% of the wood was redistributed in any year and debris longer than the active channel width rarely moved [*Gregory*, 1991]. Wood is preferentially deposited and stored in wider, lower gradient segments of rivers with longitudinal variability in channel geometry [*Wyzga and Zawiejska*, 2005]. Because transport capacity is limited in very small watersheds and wood is effectively stored in very large watersheds, wood export per unit watershed area is highest in intermediate-size watersheds [*Seo and Nakamura*, 2009]. Volumes of annual wood transport into reservoirs across a variety of basin sizes in Japan imply the existence of a transport threshold (~ 10-20 km^2) below which little wood is exported because the river is supply-limited with respect to wood, and an export threshold (~75 km^2) above which wood exports no longer increase significantly with watershed size because the river becomes supply-limited with respect to wood [*Fremier et al.*, 2010].

Differences in wood retention throughout the channel network also influence organic carbon budgets at watershed scales. Comparing fluvial export of wood to 131 reservoirs throughout Japan, *Seo et al.* [2008] find that wood-related export of organic carbon per unit area is relatively high in small watersheds, where a large proportion of wood is retained on narrow valley floors until it fragments or decays and is exported in forms other than large wood. Organic carbon export is highest in intermediate-sized watersheds, in which wood supplied from upstream and recruited by bank erosion is consistently transported downstream. Organic carbon export declines in large watersheds because of limited recruitment and transport of wood supplied from upstream on large floodplains [*Seo et al.*, 2008].

Total wood load along a channel reach reflects geology as this controls valley-side slope and the occurrence of mass movements, as well as channel width, discharge, and drainage area [*Keller and Tally*, 1979; *Robison and Beschta*, 1990; *Rosenfeld*, 1998]. Small streams tend to have high wood loads because they have narrow valleys, steep side slopes, numerous mass movements, narrow channels, and small upstream drainage areas [*Keller and Tally*, 1979]. Working on a headwater channel in Vermont, *Thompson* [1995] interprets the relations among volume of standing timber adjacent to the channel, wood, and sediment storage behind wood to indicate that wood is involved in a negative feedback such that channel degradation leads to increased standing timber recruitment and wood sediment-storage sites until the channel once again becomes stable.

The relative importance of wood in affecting flow energy dissipation and sediment storage decreases as channel gradient decreases [*Keller and Swanson*, 1979]. Wood is deposited along channel margins in lower gradient, wider streams [*Bisson et al.*, 1987], and associated sediment storage and boundary scour occupy a smaller proportion of the channel [*Bilby and Ward*, 1989]. At the upstream end of a basin, steep, bedrock-confined streams seldom contain appreciable wood because large pieces remain above the flow and small pieces are rapidly transported downstream [*Nakamura and Swanson*, 1993].

Timber harvest in a mountain drainage basin can impact hillslope water and sediment yield and the supply of wood to channels [*Heede*, 1991b; *Gomi et al.*, 2003]. Not only does timber harvest directly remove wood and reduce recruitment potential to streams, but ancillary activities such as construction of roads can alter fluvial transport of wood through a channel network [*Czarnomski et al.*, 2008]. Conversely, logging can increase hillslope instability and the mass movements that introduce large amounts of wood to channel networks [*Bunn and Montgomery*, 2004] or result in greater wood recruitment by leaving remaining trees more vulnerable to windthrow [*Gomi et al.*, 2006]. Comparison of logged and unlogged basins in western Washington, USA reveals that although the number of wood pieces is similar, logged basins have smaller pieces [*Ralph et al.*, 1994]. Other studies show fewer wood pieces in logged than in old-growth basins [*Murphy and Hall*, 1981; *Silsbee and Larson*, 1983; *Richmond and Fausch*, 1995; *Nowakowski and Wohl*, 2008]. Numerous studies comparing logged and old-growth basins consistently demonstrate that channels in logged basins have reduced pool area and depth and increased riffle area [*Murphy and Hall*, 1981; *Hogan*, 1986; *Bisson et al.*, 1987; *Carlson et al.*, 1990; *Ralph et al.*, 1994; *Richmond and Fausch*, 1995].

Disturbance in the form of forest fire can also affect wood characteristics through changes in water and sediment yield, as well as changes in wood recruitment over decades as trees establish and grow [*Nakamura and Swanson*, 2003; *Jones and Daniels*, 2008]. Comparisons of adjacent burned and unburned watersheds 2-3 years [*Young*, 1994] and 11 years [*Zelt and Wohl*, 2004] after the 1988 Yellowstone, USA fire indicate larger and more mobile wood in the burned creek initially, leading to more widely spaced, large jams with associated thick accumulations of fine sediment in the burned creek a decade later. Standing dead trees that fall in the decades following a riparian fire can be recruited to the stream in pulses associated with flooding [*Bendix and Cowell*, 2010].

Prior to the 21st century, wood was removed from stream channels to improve flood conveyance, enhance fish passage, or for timber salvage, and wood continues to be removed in some mountain streams [*McIlroy et al.*, 2008]. Removal of wood increases bedload movement [*Heede*, 1985]. Rivers from which wood has been removed are less sinuous, wider, shallower, and have less pool volume and overhead cover than comparable, undisturbed rivers [*Bilby*, 1984; *Klein et al.*, 1987; *Schmal and Wesche*, 1989; *Fausch and Northcote*, 1992]. *Francis et al.* [2008] suggest that because so many contemporary rivers lack wood relative to their condition prior to intensive human disturbance, investigators have overlooked the many geomorphic effects of wood when interpreting the post-glacial history of river valleys across the northern temperate zone.

In addition to reducing wood loads within mountain rivers, historical land uses can alter the non-linear processes created by wood, such as the formation of logjams [*Warren et al.*, 2007]. Absence of very large logs that can act as key pieces which initiate jams, for example, leads to fewer and smaller jams, which in turn increases the mobility of other wood pieces, reduces bar formation and pool volume, and can reduce channel complexity created by anastomosing [*Collins et al.*, 2002]. Conversely, many of these functions can be restored if wood and logjams are artificially reintroduced to rivers for restoration [*Keim et al.*, 2000] (Figure 3.57). *Brooks et al.* [2006] describe the geomorphic effects of wood reintroduction along the Williams River in southeastern Australia, where 436 logs within 20 engineered log jams were introduced over an 1100-m-long reach. Within five years, channel erosion was reversed and sediment storage increased, as did pool and bar area and morphologic diversity [*Brooks et al.*, 2006].

Although the idea of actively reintroducing wood to rivers is gradually becoming more popular, there remain perceptual challenges to retaining and reintroducing wood, as reflected in the results of surveys of university students in nine countries [*Piégay et al.*, 2005] and in eight US states [*Chin et al.*, 2008]. Participants in the survey viewed 20 photographs, 10 of rivers with wood and 10 of rivers without wood. Participants rated the photographs according to how esthetically pleasing the photographs appeared, how natural the scene looked, how dangerous they felt the river to be, and the extent to which they perceived a need for improvement within the channels. With the exceptions of Sweden, Germany, and Oregon (USA), students perceived rivers with wood to be less esthetically pleasing, more dangerous, and needing more improvement than rivers without wood. Survey administrators interpreted the exceptions to reflect greater environmental education and awareness (Sweden and Germany versus France, India, Italy, Poland, Russia, and Spain) and awareness of the particular wood-related issue of salmonid conservation (Oregon versus Colorado, Connecticut, Georgia, Illinois, Iowa, Missouri, and Texas).

Network-scale trends that appear consistently in diverse field areas include an inverse relation between wood load per unit area of channel and drainage area [*Lienkaemper and Swanson*, 1987] and increasing fluvial transport of wood downstream, as reflected in differences in wood load and in the presence of jams, particularly downstream increases in transport or allochthonous jams, as opposed to *in situ* or autochthonous jams [*Gurnell et al.* 2002; *Abbe and Montgomery*, 2003; *Warren et al.*, 2007]. Working in the Yellowstone region of the USA, *Marcus et al.* [2002] propose that 1st and 2nd order stream segments are transport-limited with respect to wood, 3rd and 4th order streams can effectively redistribute wood into jams without changing the total volume of wood in the channel, and 6th order and larger channels are supply-limited with respect to wood. *Wohl and Jaeger* [2009b] document similar patterns in mountain stream networks in Colorado, USA. Figure 3.58 illustrates conceptual models of wood characteristics at the network scale. *Lancaster et al.* [2001] is one the few studies to date to develop a numerical model that integrates hydrologic and geomorphic processes at the catchment scale with forest dynamics to quantitatively simulate sediment and wood storage in small mountain watersheds over timescales of 10^3 years.

Hassan et al. [2005] identify several knowledge gaps with respect to wood. These include the role of mass wasting on wood recruitment and storage, the relative importance of different wood depletion processes, the effects of changes in ecosystem variables across a watershed or landscape, the effects of management practices, and long-term (greater than circa 10^1 yr) variations in wood dynamics.

In summary, instream wood: increases boundary roughness; enhances sediment storage, including fine sediment and organic matter; can alter channel geometry by creating forced alluvial reaches or enhancing overbank flows and formation of anastomosing channels; and increases the diversity and stability of aquatic habitat for diverse organisms in mountain rivers. Wood is recruited to rivers via fluvial transport from upstream and from riparian and hillslope areas via individual tree mortality, hillslope mass movements, and mass tree mortality. Wood is removed from channels through downstream fluvial transport and transport into overbank areas. The residence time of wood in a stream segment varies greatly throughout a basin and between regions, partly as a reflection of differences in recruitment and transport, as well as differences in wood decay. Wood load (volume of wood per unit area of channel) typically decreases downstream, as wood mobility increases, and the geomorphic and ecologic functions of wood also vary longitudinally within a catchment.

3.8. Channel Stability and Downstream Trends

As with beauty, channel stability may be in the eye of the beholder. The question of whether a channel is stable has important implications, however, because perceived instability can trigger management actions designed to stabilize or restore the channel. As first explained by *Schumm and Lichty* [1965], physical factors that influence channel morphology, such as hillslope gradient, can be independent of channel processes at timespans of less than a hundred years, but dependent on channel processes at longer timescales. And a channel that appears to be rapidly changing and unstable during a two-year period that includes a large debris flow, may demonstrate quasi-equilibrium when considered over a century

Numerous definitions could be proposed for channel stability as a result of considering differing timespans or differing magnitudes of fluctuation in such channel characteristics as width/depth ratio or bedform configuration. *Brunsden and Thornes* [1979] define a transient form ratio

$$TF = \frac{\text{mean relaxation time}}{\text{mean recurrence time of events}}, \tag{3.75}$$

such that $TF > 1$ for unstable systems with transient forms (that is, because the recurrence interval of events capable of producing change is shorter than the time necessary for the channel to equilibrate, the channel is constantly changing) and $TF < 1$ for stable channels with characteristic forms (the channel remains fairly constant because the system reaches equilibrium long before the next event capable of producing change; because only very infrequent events are capable of changing channel form; or because frequent, moderate events produce little change). The mean recurrence time of events capable of producing substantial change along mountain channels varies widely as a function of geology, tectonic regime, climate, and land use. *Hey* [1987] characterizes headwater channel reaches, particularly in tectonically active mountain belts, as being inherently unstable relative to lowland rivers or to rivers in tectonically stable or unglaciated regions. In general, debris flows and floods are the primary natural agents of change along mountain rivers.

The concept of channel stability, as applied to mountain rivers, is largely a function of the magnitude of channel change caused by debris flows and large floods, and the frequency of these events, in relation to channel change that occurs between successive events. A channel can be largely unchanging or stable between disturbances when considered at graded or steady time spans, with either a steady-state or very gradual trend of aggradation or degradation (TF < 1). The higher the channel boundary resistance, the more extensive the bedrock or very coarse substrate controls on width/depth ratio and gradient, the more likely the channel is to have stable patterns of erosion, deposition, and channel morphology and to be substantially altered only by infrequent events. Alternatively, a channel can continually alternate between differing morphologies and rates of sediment transport because of frequent disturbances (TF > 1) (Figure 3.59A). This situation can also be described as one in which a threshold of channel operation is frequently crossed.

Bull [1979] expresses the threshold of critical stream power as a ratio between driving and resisting forces that governs whether a channel will aggrade or degrade. For the scenario of TF > 1 in Figure 3.59A, the first debris flow strips the channel to bedrock, creating a very efficient conveyance system that continues to transport sediment out of the channel reach, increasing degradation. In this case the erosion accompanying the debris flow causes the channel to cross a threshold from aggradation to degradation. The second debris flow deposits sufficient sediment along the channel to increase boundary roughness and decrease channel conveyance, and causes the channel to cross another threshold from progressive degradation before the debris flow to progressive aggradation after the debris flow. This is an example of *Schumm*'s [1973] extrinsic threshold, or a channel response driven by an external influence.

Schumm [1973] also describes intrinsic thresholds inherent in the operation of a system. Intrinsic thresholds have been described for the operation of steep arid or semiarid drainages where localized deposition over a period of time can result from downstream loss of transport capacity because of evaporation and infiltration. Eventually the valley floor becomes over-steepened and the channel incises, even though the water and sediment yields from the basin side slopes have not changed [*Schumm and Parker*, 1973; *Patton and Schumm*, 1975; *Womack and Schumm*, 1977].

Depending on the time interval of observation, a channel undergoing complex response can be described as stable or unstable. The channel steadily erodes when considered over the longer timespan of cyclic time (circa $10^3 m$ yr). At this timescale, a

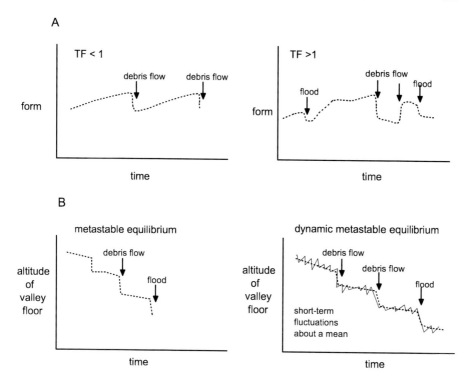

Figure 3.59 (A) Transient form ratio applied to channel-bed elevation and considered during graded or steady time. For TF < 1, sediment gradually accumulates along the channel and the bed aggrades to constant level until infrequent debris flows strip the channel to bedrock. For TF > 1, channel-bed elevation and material are continually changing as floods and debris flows alternately erode and deposit sediment along the channel. (After *Brunsden and Thornes*, 1979) (B) When considered over cyclic time, the channel-bed has net degradation, with most of the work being accomplished during infrequent debris flows and floods. (After *Chorley and Kennedy*, 1971; *Schumm*, 1977)

mountain river dominated by infrequent floods or debris flows may be best represented as having metastable or dynamic metastable equilibrium (*Chorley and Kennedy*, 1971; *Schumm*, 1977; Figure 3.13b). The crossing of either an extrinsic or intrinsic threshold can also trigger the type of complex response whereby channel behavior alternates between aggradation and degradation both downstream and with time. Finally, lag times up to tens of thousands of years can produce long-term trends in channel stability. As a result, any attempt to describe channel stability must be strictly defined with respect to temporal and spatial scales and the magnitude of channel change or variability [e.g., *Ritter et al.*, 1999]. Within these constraints, individual mountain channels may unstably alternate between aggradation and degradation at a timescale of decades [e.g., *Froehlich et al.*, 1990], or have a transient form ratio less than 1 at a similar timescale [e.g., *Harvey et al.*, 1979; *Nakamura and Swanson*, 1993].

The balance among driving forces, substrate resistance, and sediment supply, and the consequent geomorphic importance of differing flow magnitudes and stability of a channel, has predictable downstream trends [*Wohl*, 2008b]. Channel morphology and sediment transport along steep, headwater channels are more likely to be largely controlled by debris flows and floods [*Seidl and Dietrich*, 1992; *Shimazu*, 1994]. As channel gradient decreases and width/depth ratio and valley-bottom width increase downstream, the magnitude of change caused by extreme events relative to changes induced by moderate events decreases and frequently recurring fluvial processes become more effective at modifying debris flow and extreme flood features [*Heede*, 1981; *Patton*, 1988b; *Wohl and Pearthree*, 1991; *Miller and Parkinson*, 1993; *Greenbaum and Bergman*, 2006], although this modification can require decades for extremely large floods with recurrence intervals greater than a century [*Sloan et al.*, 2001]. These changes are typically associated with a decrease in the portion of bed roughness caused by individual clasts or wood; a decrease in very localized turbulence and an increase in larger-scale persistent secondary flow such as eddies; a decrease in stream power per unit area and an increase in total stream power; an increase in bed sediment mobility and consequently channel responsiveness to varying flow conditions; an increase in width/depth ratio; and a decrease in channel and valley-bottom gradient. These downstream trends in physical channel characteristics partly control downstream trends in channel morphology (chapter 4) and in aquatic and riparian biota (chapter 5). Temporal trends in channel stability can also be at least broadly predicted using conceptual models such as complex response [*Schumm and Parker*, 1973] or some version of the numerous channel evolution models [*Schumm et al.*, 1984; *Simon and Hupp*, 1986; *Leyland and Darby*, 2008].

3.9. Summary

The hydrologic regime of a mountain catchment can be dominated by runoff from glacier melt, snowmelt, rain-on-snow, rainfall, or some combination of these processes. Runoff-generating mechanisms commonly change with elevation, so that drainage area-discharge relations are unlikely to be linear in mountain regions. Runoff is generally strongly seasonal and spatially variable at scales from a single hillslope to an entire drainage basin.

Discharge records tend to be sparse for mountain rivers. Floods along these systems can be brief, unsteady, and debris-charged, and critical and supercritical flow are more common than along low gradient channels. These conditions can make indirect discharge estimation difficult and relatively inaccurate. Despite these limitations, historical, botanical, paleocompetence, paleohydraulic, and paleostage records have all proven useful in supplementing systematic discharge records in mountain regions.

Mountain channels characteristically have steep gradients, large grain and form roughness, and highly turbulent flow with non-logarithmic velocity profiles. Research efforts have focused on: (i) trying to predict resistance coefficients as a function of gradient, relative submergence, flow depth, particle-size distribution, or some other parameter, (ii) quantifying the contributions of grain and form roughness; and (iii) characterizing the distributions of velocity, lift force, bed shear, and unit stream power. Several methods have been developed for estimating the resistance coefficient, although these are mostly not designed specifically for steep channels. At present, no consensus exists on the method most applicable to mountain rivers. The relative importance of different sources of roughness changes throughout a drainage network (e.g., grain roughness may be more important in headwater channels, form roughness becomes more important downstream) and with channel type (e.g., grain roughness may dominate boulder-bed channels, whereas form roughness dominates gravel-bed channels). Although vertical velocity distributions commonly do not approximate a logarithmic curve, both velocity profiles and mechanisms of turbulence generation vary widely as a function of boundary roughness. Bed shear stress has proven difficult to measure and there is no consensus on the best method of measurement. Most investigators use temporally and spatially averaged conditions or probability distributions to characterize bed shear.

The grain-size distribution of the channel bed can be characterized by bulk sampling or, most commonly, by *in situ* clast measurement. Several methods have been used for the latter, but these do not necessarily produce consistent results. The channel bed of mountain rivers commonly has a coarse surface layer which may be either mobile or static. There is no consensus at present on how this layer develops, at least in part because mechanisms of formation and maintenance probably vary throughout a channel network and with time.

Clast entrainment is commonly characterized by physically based stochastic models that treat velocity fluctuations and variations in grain size and pivoting angle. The alternative models of equal mobility and selective entrainment both seem to apply as a function of differing conditions of discharge and sediment supply, and as a function of differing locations within a catchment.

Relatively few good datasets of bedload transport exist for mountain rivers. Bedload transport consistently demonstrates high spatial and temporal variability, and this has been attributed to progressive bed armoring, heterogeneity of bed structure and associated hydraulics, the sediment storage associated with wood, the migration of bedforms, and other factors. Bedload transport equations focus on grain tractive stress, unit stream power, discharge, or stochastic functions for sediment movement. Although many equations have been developed, no single equation has been found to consistently outperform the others when applied to mountain rivers. It will probably be most effective to develop several different types of equations, each of which applies to a specific channel type and sediment supply. At present, a compound Poisson distribution of total path length seems to best approximate bedload transport along coarse-grained channels.

Coarse sediment can enter mountain rivers directly from adjacent hillslopes, and there is relatively little storage of sediment along valley bottoms. Sediment yields vary by four orders of magnitude among mountain drainages, but are consistently higher than yields from lowland drainages. Sediment production and yield in a mountain catchment can be strongly influenced by climate, tectonics, lithology, contemporary glacial processes, or the legacy of Pleistocene glaciation. Suspended sediment loads from many mountain rivers tend to be relatively low except in arid or glaciated catchments, or following a disturbance such as a volcanic eruption or forest fire.

Wood exerts a significant influence on many mountain rivers. Wood alters boundary roughness and flow resistance; promotes localized scour; creates bed steps in higher gradient channels, islands or bars in lower gradient channels, or forced alluvial reaches that would otherwise be bedrock channel segments; stores sediment and organic material; and increases substrate and habitat diversity. Single, largely immobile pieces of wood in headwater channels give way to greater wood mobility and the creation of logjams downstream.

Channel stability along mountain rivers is largely a function of the magnitude and frequency of debris flows and floods. Because these extreme flows can more effectively overcome the high boundary resistance of mountain channels, erosional and depositional features created by these flows tend to dominate channel and valley morphology. Debris-flow recurrence intervals vary widely in mountain catchments, from less than one year to > 5,000 years, as a function of climate, lithology, and tectonics. Floods can result from precipitation or from dam failure. The geomorphic importance of any debris flow or flood will be influenced by the ratio of hydraulic driving forces to substrate resisting forces, by antecedent conditions, and by the magnitude of the flow relative to the magnitude of earlier flows.

4

4. CHANNEL MORPHOLOGY

4.1. Spatial and Temporal Variability in Channel Morphology

The morphology of an ideal alluvial channel flowing across a homogeneous, weakly resistant substrate reflects only the hydraulic and sediment transport processes occurring within the channel. This is what *Leopold and Langbein* [1962] described as a channel that is the author of its own geometry. In practice, most alluvial channels are subject to some morphologic constraint as a result of differing substrate resistance or tectonic movement [*Ouchi*, 1985; *Schumm*, 1986]. Alluvial channels do tend, however, toward the smoothly concave-upward longitudinal profile characteristic of a graded stream [*Mackin*, 1948]. Mountain rivers, in contrast, are likely to have strongly segmented longitudinal profiles that can be straight or convex upward, as well as concave (see also chapter 3). These profile characteristics can reflect one or more of the following controls.

1.   The stream power in the headwater portion of a drainage is low relative to substrate resistance and the channel is unable to incise at a rate equal to tectonic uplift [*Merritts and Vincent*, 1989; *García*, 2006].

2.   In a bedrock-dominated system, lithologic and structural variations downstream can strongly influence substrate resistance and sediment supply. The Poudre River, Colorado, USA provides an example of structural controls. Where the channel follows the course of shear zones active during the Precambrian and the Laramide orogeny (circa 70-35 million years ago), the river valley is broad (300-500 m) and lower in gradient (0.014 m/m). The valley is much narrower (50-100 m) and steeper (0.028 m/m) where the channel does not follow the shear zone (Figure 4.1). Greater joint density in the wide, lower gradient segments of the channel [*Ehlen and Wohl*, 2002] reduces rock-mass strength and contributes to the formation of straths and strath terraces [*Wohl*, 2008a]. Another example comes from incised meanders on the Colorado Plateau, USA, where channel gradient, channel cross-sectional shape, and bend symmetry correlate strongly with bedrock type [*Harden*, 1990].

3.   The glacial or climatic history of a region may have produced downstream variability in sediment supply to the channel, thus affecting channel morphology differently along a downstream gradient. Glaciated drainages, in particular, can have

Mountain Rivers Revisited
Water Resources Monograph 19
Copyright 2010 by the American Geophysical Union.
10.1029/2010WM001042

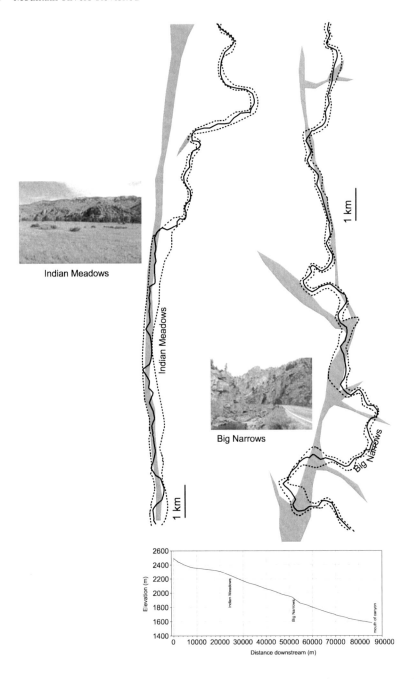

Figure 4.1 Plan view of two segments of the Poudre River in the Colorado Front Range, USA. Solid line denotes the path of the river, dashed lines are the edges of the valley bottom, shaded areas represent a Precambrian-age shear zone. The river valley is most narrow and sinuous where it diverges from the shear zone.

channel gradients that are independent of discharge and drainage area [*Day*, 1972; *Ferguson and Ashworth*, 1991]. The terminal moraine of late-Pleistocene glaciation in the Poudre River valley, for example, effectively creates a local base level. The 10 km length of valley upstream from the moraine is filled with finer-grained alluvium, creating a 0.2% gradient along which the Poudre Rievr meanders. The valley segments immediately up- and downstream have steep, narrow, coarse-grained channels (Figure 4.2). *Phillips and Harlin* [1984] describe appreciable variations in downstream hydraulic geometry exponents for a subalpine stream where it crosses a meadow occupying a former lake bed in the Sangre de Cristo Mountains of southern Colorado, USA. Another example of downstream variability and the effects of climate change on mountain rivers comes from *Bull*'s [1991] work in the New Zealand Alps. An increase in temperature and precipitation result in a decrease of hillslope sediment yield for portions of the drainage basin below 1200 m in elevation, because of an increase in soil thickness and vegetation density. Above 1200 m elevation (above present treeline), the same climatic change results in an increase in erosion, decreased soil thickness, and an increase in sediment yield.

4.   Other factors such as land use, forest fires, floods, and mass movements can also affect sediment supply and thus channel morphology. *Wohl and Pearthree* [1991] describe a progression in channel morphology following forest fires and debris flows in the Huachuca Mountains, Arizona, USA. Debris flows strip the steeper portions of ephemeral tributary channels to bedrock, creating a trapezoidal channel that then gradually fills with sediment during the next few decades until it becomes a vegetated swale. The tributary debris flows cause local aggradation and non-uniform distribution of coarse sediments in the perennial main channels, with the effects becoming less pronounced with distance downstream from the tributary and with time. The details of channel morphology along any given portion of these mountain channels thus partly reflect the recent history of forest fires and debris flows. *Ryan and Grant* [1991] find a 30-fold increase in open riparian canopies on low-order tributaries of the Elk River basin, Oregon, USA between 1956 and 1979. These openings are caused by hillslope mass movements associated with clearcuts and roads. Channel response to the increased sediment supply includes formation of gravel bars, bed and bank erosion, and loss of wood. *Hewitt* [1998] describes "naturally fragmented river systems" in the Karakoram Himalaya where rockslide barriers create alternating gorges and aggraded valley reaches at a scale of tens of kilometers along each river, and similar descriptions come from Japan [*Shimazu and Oguchi*, 1996], New Zealand [*Korup*, 2004] the northwestern United States [*Benda et al.*, 2004], and many other mountainous regions.

One of the primary characteristics of mountain channel morphology is thus pronounced and abrupt longitudinal variations in gradient, valley width, channel pattern, and grain size that reflect substrate resistance and climatic and tectonic history. Wider valley segments are likely to have a lower gradient channel formed in coarse alluvium, whereas narrower valley segments are likely to have a steep bedrock channel (Figure 4.2). These steep, narrow channel segments limit the response of intervening alluvial segments to base level lowering or decreased sediment supply. The erosional resistance of the bedrock segments allows them to act as local base levels that limit incision of upstream alluvial segments.

In contrast to this spatial variability, mountain channels in some regions of the world maintain relatively stable channel morphology through time. Because of the high channel-boundary resistance, only extreme and relatively infrequent events substantially alter the channel boundaries. Following these extreme events, the channel can again be stable for decades to centuries (see also chapter 3). A 1985 glacier-lake outburst flood along the Langmoche Khola in the Nepalese Himalaya, for example, eroded approximately 2,600,000 m^3 of sediment, including boulders up to 2.7 m in diameter, from 26 km of a valley that varied from 10 m to 200 m wide [*Vuichard and Zimmermann*, 1987]. Doing field work in the region 10 years later, *Cenderelli and Wohl* [1998, 2000, 2003] observed that the longitudinal boulder bars created by the 1985 flood remained in pristine condition, with no re-working of clasts by subsequent flows (Figure 4.3). Flows following the 1985 flood had incised a channel approximately 10 m wide and 2 m deep into the outburst-flood deposits, which spread across a valley > 100 m at some locations. Another example comes from *Kite and Linton*'s [1993] study of the 1985 flood on the Cheat River in West Virginia, USA. This flood was the only event during at least a century capable of transporting the large boulders that strongly influence channel morphology along the Cheat Narrows and Cheat Canyon.

Proglacial and other braided mountain channels, in contrast to the relatively stable channels described above, typically change substantially during the summer melt season or between years in response to fluctuations in glacier mass balance [*Fenn and Gurnell*, 1987; *Gregory*, 1987], or to fluctuations in sediment yield from adjoining hillslopes. Small drainage basins in wet climates can also experience relatively frequent (recurrence interval < 10 yr) debris flows and floods that substantially alter channel and valley configuration [*Sawada et al.*, 1983].

The temporal variability of mountain channel morphology is a function of disturbance frequency and magnitude. In some regions, channel morphology may be largely relict from Pleistocene deglaciation. Some investigators argue, for example, that the coarse clasts forming step-pool sequences along alpine and subalpine channels of the Colorado Rocky Mountains, USA can no longer be mobilized by lower, post-glacial discharges [*Gordon*, 1995]. Larger channels at lower elevations in this same region may have much shorter recurrence intervals for substantial disturbance. The 1976 flood along the Big Thompson River in the Colorado Rocky Mountains created dramatic erosional and depositional features in the coarse-grained alluvium overlying the resistant bedrock substrate (Figure 4.4). Channel segments steeper than 2% gradient generally were scoured, especially on the outside of bends and where the channel was constricted. Deposition occurred along portions of channel with a gradient less than 2%; boulders as large as 3.6 m in diameter were moved [*Shroba et al.*, 1979]. Floods of similar magnitude occurred in this region in 1864, 1935, and 1965. *Starkel*'s [1972] description of the geomorphic effects from a flood produced by intense rainfall in the

Darjeeling Himalaya provides another example of frequently disturbed channels. Widespread slope failures contributed massive amounts of coarse sediment to the rivers, and even small channels carried boulders 3-5 m in diameter. Headwater channels incised up to 2-3 m into bedrock and channel width increased greatly. *Starkel* [1972] estimates a recurrence interval of 20-25 years for such events in the Darjeeling Himalaya.

The nature of channel adjustment to an abrupt disturbance or to a prolonged change in water and sediment yield will depend on the magnitude and duration of the external change, and on the channel boundary resistance. Very rapid adjustments can occur in the channel width/depth ratio or bed configuration, whereas changes in channel planform and reach or basin gradient occur over decades or longer [*Knighton*, 1984, 1998]. As the grain size of the channel substrate decreases downstream, boundary resistance decreases and channel width/depth ratio and bedforms become more responsive to short-term changes in controlling variables. This is well illustrated by comparing the response of step-pool versus pool-riffle channels to flow diversion, for example [*Ryan*, 1994b]. Step-pool channels from which most flow has been diverted for decades show little change in geometry or grain-size distribution, whereas pool-riffle channels tend to fill with sediment and grow narrower as riparian vegetation encroaches along the margins.

Channel adjustment is sometimes explored in the context of extremal hypotheses. Extremal hypotheses are models based on the assumption that the equilibrium channel morphology corresponds to the morphology that maximizes or minimizes the value of a specific parameter [*Darby and Van De Wiel*, 2003]. Examples include the minimization of stream power [*Chang*, 1980, 1988] or unit stream power [*Yang and Song*, 1979], minimization of energy dissipation rate [*Yang*, 1976; *Yang et al.*, 1981], and maximization of friction factor [*Davies and Sutherland*, 1983; *Abrahams et al.*, 1995] or sediment transport rate [*White et al.*, 1982]. Although the use of extremal hypotheses has been criticized as an "essentially metaphysical method of predicting steady-state channel dimensions, which offers no explanatory power" [*Ferguson*, 1986] and as reflecting "common (but far from inevitable) side effects rather than self-regulation" [*Phillips*, 2010], verification exercises indicate that "model predictions based on extremal hypotheses provide global, if not exacting, agreement with a wide range of observations" [*Darby and Van De Wiel*, 2003]. The underlying assumption in extremal hypotheses is typically that a high rate of energy expenditure at a specified channel site will eventually result in boundary deformation until the rate of energy expenditure declines, such that numerous site-specific adjustments minimize spatial variation in energy expenditure along a channel [e.g., *Grant*, 1997; *Wohl et al.*, 1999].

As the spatial coverage and resolution, temporal resolution, and availability of remote sensing imagery increase, and the cost decreases, various types of images are increasingly used to characterize channel morphology at an instant in time or through time with repeat imagery [*Fonstad and Marcus*, 2005; *Snyder*, 2009; *Bailly et al.*, 2010; *Bird et al.*, 2010; *Bowman et al.*, 2010; *Lane et al.*, 2010]. The existence of basin-scale, high resolution datasets facilitates the evaluation of traditional conceptual models and extremal hypotheses [*Fonstad and Marcus*, 2010].

In summary, mountain rivers characteristically display pronounced longitudinal variability in channel morphology as a result of longitudinal variation in glacial history, lithology and structure, response to relative base level change, and other factors. Temporal variability differs between regions and between segments of a basin in relation to the magnitude and frequency of events such as floods and debris flows that are capable of substantial sediment transport. Various extremal hypotheses have been used to characterize longer term trends and adjustment of channel morphology.

4.2. Channel Classification Systems

Channel classification can be particularly challenging for mountain rivers because of the pronounced spatial variability in morphology described earlier. There are numerous channel classification systems that may be at least partially applicable to mountain channels [*Church*, 1992; *Kondolf et al.*, 2003b]. At the drainage basin scale, channels can be classified in terms of (i) sequence with respect to structure (consequent, subsequent, resequent, etc); (ii) sequence with respect to time (superimposed, antecedent [see *Powell*, 1875, 1876]) and evolution (youthful, mature, old [see *Davis*, 1899]); (iii) spatial pattern (dendritic, radial, rectangular, etc. [see *Howard*, 1967]); (iv) channel planform (straight, meandering, braided, anastomosing [see *Schumm*, 1985, 2005; *Lewin and Brewer*, 2001; *Eaton et al.*, 2010b]); (v) flow regime (perennial, intermittent, ephemeral, or glacier-melt, snowmelt, rainfall, rain-on-snow, combined runoff sources, or other subdivisions [*Hannah et al.*, 2005]); or (vi) hierarchy within a stream ordering network [*Strahler*, 1952a]. Although several versions of stream order exist, the system proposed by Strahler is the most widely used; a first-order channel has no tributaries, and two equal-order channels must join to create the next highest order. Channels are typically interpreted from dashed lines or solid blue lines on topographic maps, although they may also be defined in terms of the degree of contour crenulation.

Channel classifications can be applied from the regional scale down through watersheds, channel, and reaches, where a reach is a length of channel typically at least several channel widths in length and over which relevant characteristics of a river remain essentially the same [*Ferguson*, 2008]. The most useful scale for classification depends on the intended purposes of the classification and the scale of data on which the classification is based. Comparing large-scale environmental river classifications for invertebrate assemblages, water chemistry, and hydrology, *Snelder et al.* [2008] find that differences in classification performance between rivers defined from geographic regionalization and numerical classification of individual river valley segments are small relative to unexplained variation.

Mountain channels are also typically classified at the reach scale. Reach classifications can be based on hydrology, channel planform, bedforms, or some combination of these characteristics (Table 4.1), although *Simon et al.* [2007] note the limitations to using form-based systems for rivers in which form can adjust to altered water and sediment inputs through time. Mountain channels typically fall into the cascade, step-pool, plane-bed, or pool-riffle categories of *Montgomery and Buffington* [1997] (Figure 4.5), or into one of the bedrock categories of *Wohl* [1998, 1999]. *Montgomery et al.* [1996a] note that the occurrence of bedrock and alluvial channels can be described by a threshold model relating local sediment transport capacity to sediment supply. Valley-spanning logjams can create alluvial channels in what would otherwise be bedrock reaches. Variation among observers in defining and categorizing the parameters on which a classification system is based can create some uncertainty [*Roper et al.*, 2008], and some channel reaches will always appear to be transitional given the boundaries that classification systems impose on the continuity of form characteristic of natural systems.

The next three sections focus on three types of reach-scale channel morphology that typically occur in a downstream progression, although longitudinal variations in valley morphology and gradient can disrupt this downstream sequence. The bedforms most characteristic of these channel morphologies are part of a continuum from disorganized large clasts to regularly spaced pools and riffles. This continuum reflects downstream changes in channel gradient, grain size/substrate, and the expenditure of flow energy. The presence of specific bedforms and channel morphologies can thus be used to infer reach-scale stability and response to disturbance [*Montgomery and Buffington*, 1997, 1998].

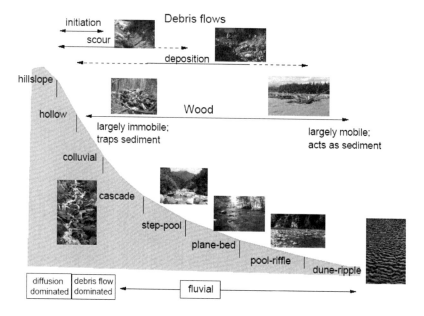

Figure 4.5 Schematic illustration of the channel classification from *Montgomery and Buffington* [1997, Figure 4].

4.3. Channel Morphologic Types

4.3.1. Step-Pool Channels

Step-pool channels are characterized by downstream alternations between steps composed of clasts, wood, and/or bedrock, and plunge pools that form at the base of each step (Figure 4.6) [*Chin*, 1989]. *Hayward* [1980] distinguishes among boulder steps composed of a group of boulders arranged in a straight or curved line across the channel; riffle steps formed by a collection of larger-than-average clasts that steepen the channel, and occur at slopes less than 0.05 m/m; and rock steps where the channel is confined by bedrock. To these may be added wood steps of pieces longer than 1 m and wider than 10 cm. Step geometry is typically defined in terms of step height and step spacing, although differences in the manner of defining step height between investigators can enhance between-site variations in data [*Nickolotsky and Pavlowsky*, 2007]. Step-pool channels are typically interpreted to reflect a supply-limited system [*Whittaker*, 1987b] and are most common along reaches where relatively immobile coarse clasts or wood can trap sediment in a wedge that tapers upstream. Flow plunging over the immobile obstacle scours a pool at the obstacle's base, creating a step-pool sequence.

Step-pool sequences have been described for various lithologies, in climatic regimes from cold temperate [*Grant et al.*, 1990] to hyperarid [*Wohl and Grodek*, 1994] and tropical [*Wohl and Jaeger*, 2009], and at step spacings from tens of centimeters [*Abrahams et al.*, 1995] to tens of meters [*Bowman*, 1977]. *Wohl* [1992b] proposes that alternating steep and gentle gradient channel reaches at a scale of hundreds of meters may be large-scale step-pool sequences. *Montgomery and Buffington* [1997] note that step-pool sequences are most common at reach gradients of 0.03 to 0.10, although others report these forms at much higher gradients [*Grant et al.*, 1990; *Wohl and Grodek*, 1994]. Comparisons of different analytical techniques for identifying individual step-pool bedforms indicate that visual identification is most effective [*Wooldridge and Hickin*, 2002].

Laboratory experiments with a mobile gravel bed demonstrate that steps can originate as antidunes [*Whittaker and Jaeggi*, 1982; *Ashida et al.*, 1984; *Grant and Mizuyama*, 1991; *Grant*, 1994; *Hasegawa and Kanbayashi*, 1996; *Parker*, 1996; *Wang et al.*, 2004]. In these experiments, larger particles that come to rest under the crests of standing waves trap other, smaller particles and create a step of imbricated grains. Subsequent flume studies, however, document steps forming without antidunes but under hydraulic conditions within the region of antidune formation [*Weichert et al.*, 2008]. In addition, studies of steps formed by coarse clasts along ephemeral tributaries with flows insufficient to submerge the clasts [*Wohl and Grodek*, 1994], and steps formed in bedrock [*Duckson and Duckson*, 1995; *Wohl*, 2000b], suggest that other mechanisms can also operate in step formation. *Chin* [1999a] suggests that, although field data on Froude number, flow depth, and step wavelength from the Santa Monica Mountains, California, USA are consistent with the antidune origin of step-pool formation, true antidunes may be difficult to maintain in steep headwaters where large, less mobile roughness elements can disturb the regularity of flow and sediment transport. Studying the 60 km^2 catchment of Shatford Creek in British Columbia, Canada, *Zimmermann and Church* [2001] conclude that steps are randomly located, stable structures resulting from the interlocking of larger clasts, which they call *key-stones*. *Milzow et al.* [2006], however, find that step lengths along the Vogelbach in Switzerland are statistically different from randomly generated sequences. *Curran and Wilcock* [2005] find that steps formed in a flume follow a Poisson distribution, which they interpret as indicating that the mechanisms by which steps form include those without any interaction between the bed and the water surface.

It thus appears that antidunes may create steps and pools in flume experiments and field settings with a relatively narrow range of grain sizes and at the lower end of the gradient spectrum under which step-pools form, whereas steeper channels with poorly sorted grain-size distributions are more likely to have step-pool sequences originating from the relative immobility of wood or keystones, which *Zimmermann et al.* [2010] refer to as the jammed state. Erosion of material supporting keystones and their subsequent mobilization can destroy steps [*Lenzi*, 2001].

Flows of a magnitude sufficient to form steps have a recurrence interval on the order of 50 years in the western Cascade Mountains of Oregon, USA [*Grant et al.*, 1990]. Other investigators also infer relatively long recurrence intervals (> 10 years) for flows capable of fully mobilizing the grains that form step-pool sequences [*Lenzi*, 2001; *Thompson and Croke*, 2008]. In contrast, studies of a small (6.5 km^2) basin in the Japanese Alps indicate that debris flows that occur approximately once each year thoroughly destroy the existing channel and deposit extensive sediment to create a uniform channel-bed gradient [*Sawada et al.*, 1983]. Subsequent water flows re-create a step-pool sequence nearly identical to that existing prior to the debris flow and, as the bed once again becomes armored, sediment transport decreases substantially until passage of the next debris flow. On the Lainbach River, Germany, minor floods, with a discharge less than 10 m^3/s (average Q ~ 2 m^3/s) maintain the step-pool system [*Ergenzinger*, 1992]. During an extreme flood of 200 m^3/s in 1990 the channel rapidly changed from step-pool to braided, but subsequently resumed a step-pool morphology. Similarly, *Lenzi* [2001] documents how unlimited sediment supply during extraordinary floods with recurrence intervals of 30-50 years on the 5 km^2 Rio Cordon, Italy causes sediment deposition in pools and lengthening of step spacing, effectively reducing bed steepness. Step-pool sequences and maximum roughness are gradually restored during succeeding ordinary floods with recurrence intervals of 1-5 years, but conditions remain capacity-limited for several years following an extraordinary flood [*Lenzi et al.*, 2003b; *Turowski et al.*, 2009b]. In general, the mobility of steps is a function of particle size [*Chin*, 1998], as well as hydrologic regime in the channel. Steps tend to be stable when considered at small temporal and spatial scales, so that the steps function as independent variables that dissipate stream energy and regulate channel hydraulics [*Chin*, 1998]. At larger scales, steps become dependent variables that reflect changes in water and sediment discharge. Flume experiments suggest that the stability of the bed increases as the jamming ratio (channel width/D_{84step}) decreases for jamming ratios less than six [*Zimmermann et al.*, 2010]. Rough banks can also significantly increase the stability of the bed, although bank roughness is traditionally ignored in field studies.

Most studies of step-pool channels have sought correlations between step characteristics and the potential control variables of reach-scale gradient, discharge, and sediment supply and grain size [*Chartrand and Whiting*, 2000; *Maxwell and Papanicolaou*, 2001]. Several studies demonstrate a consistent correlation between step spacing and height and channel gradient [*Heede*, 1981; *Whittaker*, 1987b; *Wohl and Grodek*, 1994] (Figure 4.7), although others find stronger correlations between step height and length and hydraulics [*Chartrand and Whiting*, 2000; *Milzow et al.*, 2006]. Channel gradient is interpreted as an indicator of rate of flow energy expenditure. Step characteristics in relation to gradient do not differ significantly between substrate types along channels with mixed alluvial and wood steps [*Wohl et al.*, 1997], mixed alluvial and bedrock steps [*Grodek et al.*, 1994; *Wohl and Grodek*, 1994], and mixed lithology bedrock steps [*Duckson and Duckson*, 1995], suggesting that flow energy expenditure strongly controls step characteristics. In alluvial step-pool channels, particle size appears to dominate step height and discharge dominates step wavelength [*Chin*, 1999b; *Chartrand and Whiting*, 2000]. Spectral analysis [*Chin*, 2002] reveals multiple significant peaks that reflect variance in step-pool spacing resulting from external factors, although periodicity in the occurrence of steps suggests that these features behave analogously to other bedforms in reflecting mutual adjustment between flow, sediment, and energy expenditure [*Chin*, 1999b].

Step characteristics represent a means of adjusting boundary roughness and step-pool bedforms appear to evolve toward a condition of maximum flow resistance because maximum resistance implies maximum stability [*Abrahams et al.*, 1995]. In a series of flume experiments, *Abrahams et al.* [1995] demonstrate that maximum flow resistance occurs when steps are regularly spaced and have a ratio of mean step height: mean step length:channel slope between 1 and 2, a condition observed along many natural step-pool channels [*Duckson and Duckson*, 2001; *Lenzi*, 2001; *MacFarlane and Wohl*, 2003; *Wohl and Wilcox*, 2005]. These results are inconsistent with the antidune model of step formation because the Froude numbers at which flow resistance is maximized are well below those values usually associated with antidunes. *Canovaro and Solari* [2007] take a slightly different approach, equating steps to macro-roughness stripes over which flow resistance is maximized when the spacing between stripes is about 10 times the average macro-roughness height, a condition which they note as being present in many naturally occurring step-pool sequences. Mean values of frictional resistance do not vary between step-pool and pool-riffle channels despite different sources of resistance and greater relative grain submergence in pool-riffle channels, which suggests that channel morphology along steep streams adjusts to minimize spatial variation in resistance among channel types [*Wohl and Merritt*, 2008].

Chin and Phillips [2007] explore step-pool development as an autogenic self-organization process in which entropy increases as initially undifferentiated planar channels develop steps and pools, then decreases when step-pool sequences with consistent geometry develop. During this organization process, flow resistance increases and slope decreases as vertical sinuosity develops, which suggests a minimization of stream power. Studying bedrock step-pool sequences in Thailand, *Carling et al.* [2005] propose that the observed strongly periodic spacing of steps reflects a situation in which the energy generated at an upstream step is effectively dissipated before the next downstream step, so that a step-pool sequence sustains an equilibrium reach-scale energy slope.

Relatively little is known of the hydraulics and sediment transport of step-pool channels compared to pool-riffle channels, although several studies enhance our understanding [*Chin and Wohl*, 2005; *Church and Zimmermann*, 2007]. Working on a step-pool channel in Bavaria, *Stüve* [1990] finds that energy loss and roughness are extremely variable between successive cross sections at low discharges. As discharge increases, there is a tendency toward more uniform energy loss between cross sections. *Chanson* [1994, 1995] describes nappe flow and skimming flow in rigid, concrete stepped chutes and spillways. *Nappe flow* consists of a succession of free falls at low discharges, with the free-falling jet followed by a hydraulic jump at the downstream end of the step tread. With increasing flow rate or decreasing step length, the hydraulic jump disappears and a skimming flow regime develops. In *skimming flow* the water flows as a coherent stream, with recirculating vortices occurring at the base of each step riser [*Chanson*, 1996]. Strong aeration at the leading edge of flood waves and through hydraulic jumps reduces buoyancy, locally intensifies bed erosion, and decreases the friction factor of a flow [*Vallé and Pasternack*, 2002; *Chanson*, 2004]. The rate of energy dissipation in this separation zone downstream from the step crest plays an important role in flow resistance [*Egashira and Ashida*, 1991]. In natural channels the occurrence of nappe versus skimming flow is presumably influenced by the more irregular step geometry associated with plunge pool erosion [*Abrahams et al.*, 1995]. It may be that step geometry is partly controlled by the presence of step-forming clasts large enough to withstand the erosional energy of hydraulic jumps present during nappe flow except during the largest discharges. This is supported by flume experiments. *Comiti et al.* [2009a], for example, document a transition from nappe to skimming flow in self-formed step-pool sequences in an experimental channel. Flow resistance decreases significantly when flow passes from nappe to skimming conditions.

One-dimensional velocity profiles and velocity fluctuations over steps and pools throughout the snowmelt hydrograph indicate that flow becomes more turbulent (as judged by coefficient of variation of velocity) as stage increases, particularly at lower-gradient reaches with less variable bed roughness [*Wohl and Thompson*, 2000]. Flow also becomes more turbulent as gradient increases and as bed roughness increases. Velocity profiles suggest that pools immediately downstream from steps have wake turbulence from mid-profile shear layers. Locations immediately upstream from steps, at step lips, and in runs are dominated by bed-generated turbulence. Adverse pressure gradients upstream and downstream from steps may be enhancing turbulence generation, whereas favorable pressure gradients at steps are suppressing turbulence (see also chapter 3). The wake-generated turbulence leads to higher energy dissipation in step-pool reaches relative to more uniform-gradient reaches [*Wohl and Thompson*, 2000].

Using a 3d acoustic Doppler velocimeter in the same channel, *Wilcox and Wohl* [2007] find that cross-stream and vertical components of velocity contribute averages of 20% and 15%, respectively, to overall velocity vector magnitudes. The vertical component of velocity contributes substantially to turbulent kinetic energy. Discharge and morphologic position significantly affect mean streamwise velocities and turbulence intensities for all flow components. Substantially higher velocities occur upstream from steps than occur in pools and the greatest turbulence intensities occur in pools and at high discharges.

Step-pool sequences can produce high values of the Darcy-Weisbach friction factor relative to lower-gradient gravel-bed rivers [*Curran and Wohl*, 2003]. Resistance partitioning suggests that grain resistance creates a relatively minor contribution to total resistance and that total resistance increases with increasing step height [*MacFarlane and Wohl*, 2003; *Wilcox and Wohl*, 2006; *Wilcox et al.*, 2006; *David et al.*, 2010]. *Lee and Ferguson* [2002] find that velocity and resistance vary strongly with discharge at field sites and in flume experiments and conclude that a single resistance law cannot describe the full range of flows experienced by natural step-pool channels. They evaluate various parameters that might be used to quantify effective roughness height (k_s), including step D_{84}, and standard deviation in bed elevation along stream and cross-stream, without finding any consistent indicator of k_s. Using an analogous field dataset, *Comiti et al.* [2007] conclude that dimensionless unit discharge (q^*) is the most important independent variable in explaining observed variations in velocity and resistance, followed by bed gradient and the ratio of step height to step length. *David et al.* [2010] find similar correlations for another field dataset of step-pool and cascade channels.

Although less attention has been given to pools in step-pool sequences, these features can be described similarly to pools occurring in pool-riffle sequences, using residual pool volume, pool infill ratio, and simple length and depth measurements (Figure 4.8). Recent investigations of scouring processes in pools associated with naturally formed step-pool sequences, experimental channels, and check dam sequences [*Lenzi and Comiti*, 2003; *Lenzi et al.*, 2004a], increase our understanding of processes contributing to step-pool morphology. Working in an experimental channel, *Lenzi et al.* [2003c] find that when the ratio of the critical water depth h_c and the sill spacing L rises above a characteristic value (in their experiments, 0.06-0.07), interference by the downstream sill renders scouring less effective. *Lenzi et al.* [2003d] use surveys of check dams to propose the following semi-empirical equation for maximum scour depth with respect to crest elevation, y_s:

$$0.6 \leq \frac{y_s}{z + H_s} \leq 1.4 \qquad (4.1)$$

where z is measured drop height (m) and H_s is critical specific flow energy (m). *Comiti et al.* [2005] develop an upstream-forced cascade model for step-pool formation, in which the energy of the falling jet controls the geometry of the pools:

$$l_{step}l_{berm} \sim 3E \sim 3 \quad (z + H_s) \tag{4.2}$$

such that l_{step} is step spacing, l_{berm} is berm distance between step lip and downstream boulder berm that forms the start of the next step tread, and E is jet energy, which is given by the specific critical flow energy H_s plus the potential drop energy z.

Field studies using tracer particles indicate that particle shape, size, and position on the channel bed influence transport frequency and length for bedload along step-pool channels. Elongated pebbles have longer transport distances than platy (disc) shaped particles [*Schmidt and Ergenzinger*, 1992; *Moore and Diplas*, 1994]. Shear forces dominate erosion when the particle is exposed, but lift force becomes much more important when the particle is level with neighboring particles [*Ergenzinger and de Jong*, 1994]. Particles located in pools are much more likely to be entrained and will be transported longer distances than those on steps or riffles [*Schmidt and Ergenzinger*, 1992]. Total displacement length is independent of particle size for smaller, fully mobile grain sizes and decreases rapidly with grain size for larger fractions in a state of partial transport [*Lenzi*, 2004]. Mean travel distance of marked clasts can be estimated from excess stream power [*Lamarre and Roy*, 2008] Sediment transport is episodic during a flood, with alternating transport steps and non-movement intervals [*Schmidt and Ergenzinger*, 1992]. Attempts to predict sediment transport along step-pool channels using some of the standard sediment relations developed for coarse sediments and steep gradients (e.g., Ackers-White, Bagnold, Einstein-Brown, Schoklitsch, Smart-Jaeggi) indicate that these relations tend to overpredict actual measured bedload transport by more than three orders of magnitude [*Blizard*, 1994; see discussion, chapter 3], in part because the additional form drag exerted by steps greatly alters the distribution of effective shear stress [*Lenzi et al.*, 2006]. Actual bedload transport is highly temporally and spatially variable. Although bedload discharge may correlate with hydraulic variables in these channels, bedload movement is strongly influenced by such factors as instream wood location and stability, and by local bed gradient and step height and spacing [*Adenlof and Wohl*, 1994; *Busskamp*, 1994; *Trayler*, 1997; *Blizard and Wohl*, 1998; *Trayler and Wohl*, 2000]. *Lenzi et al.* [2006] propose the use of dimensionless critical unit discharge q^*_{ci} as an effective parameter for quantifying grain entrainment in step-pool channels.

Montgomery and Buffington [1997] characterize step-pool sequences as *transport reaches* that are relatively insensitive to changes in water and sediment yield caused by human activities. Examining the effect of flow diversion on mountain streams in Colorado, USA, *Ryan* [1994b, 1997] finds no change in the width of step-pool channels after 20-100 years of flow diversion, in contrast to the 30-50% decrease in width of similar pool-riffle channels. *Madsen's* [1995] assessment of channel responses to increased water and sediment yield associated with timber harvest in northwestern Montana, USA indicates that step-pool channels are much less sensitive than pool-riffle channels. The only response of the step-pool channels is a fining of pool particle sizes with increased sediment yield. This is in contrast to the pool-riffle channels in the study, which serve as *response reaches* [*Montgomery and Buffington*, 1997] in which substantial channel change is observed. Even relatively unresponsive step-pool channels can be pushed toward detectable morphologic change by extreme changes in flow or sediment, however; *David et al.* [2009] document significant differences in bank stability, substrate grain-size distribution, wood load, and pool residual depth between control channels and analogous channels with ski slope development (and associated increases in water and sediment yield).

Stream restoration now includes check dams that mimic step-pool morphology [*Lenzi*, 2002; *Comiti et al.*, 2009b] and recreation of step-pool sequences as urban development and other forms of land use encroach into steep terrain (Figure 4.9). The principle of maximum flow resistance has proved particularly useful in this context, as has allowing the location of steps to adjust through time so that the steps behave as dependent fluvial variables rather than artificially fixed forms [*Chin et al.*, 2009].

In summary, vertical steps alternate downstream with plunge pools in step-pool channels. Step-pool sequences may result from antidunes or from less mobile clasts or wood that created a jammed state. Relatively high-magnitude, low-frequency floods mobilize step-forming clasts in most step-pool channels, although more commonly occurring smaller flows can maintain step-pool sequences. Flow in these high-gradient, typically coarse-grained channels is strongly three-dimensional and appears to transition from nappe flow at lower discharges to skimming flow at higher discharges, with an associated decrease in flow resistance and increase in average velocity. Bedload transport is spatiall and temporally discontinuous, with greater mobility for particles in pools than those in steps. Step-pool channels are transport reaches that are less insensitive to changes in water and sediment supply than are pool-riffle channels.

4.3.2. Plane-Bed Channels

Plane-bed channels lack the well-defined, rhythmically occurring bedforms that characterize both step-pool and pool-riffle channels [*Montgomery and Buffington*, 1997]. It is important to distinguish plane-bed channel form in gravel- and boulder-bed channels, which can persist through conditions of bed stability and mobility, from upper- or lower-regime plane-bed bedload transport of finer non-cohesive sediment [e.g., *Abrahams and Gao*, 2006; *Wong et al.*, 2007], when well-sorted grains move without aggregating into ripples, dunes, or antidunes. This section refers to the former channel type.

Plane-bed channels, which are most common at channel gradients of 0.01-00.03, can have a surface layer of coarse clasts or can be formed on bedrock. Long stretches of relatively planar bed may be punctuated by occasional channel-spanning rapids (Figure 4.10) or pools. Plane-bed channels tend to be intermediate between step-pool and pool-riffle channels with respect to gradient, D_{84}, and channel width [*Wohl and*

Merritt, 2005, 2008]. *Reid and Hickin* [2008] document well-developed at-a-station hydraulic geometry relations and power relations between Darcy-Weisbach *f* and *Q* for five plane-bed channels in British Columbia, Canada. Resistance typically varies by at least two orders of magnitude over the range of flow sampled and varies over six orders of magnitude among the channels. Even in these fairly planar beds, form roughness associated with particularly large grains and other irregularities such as single steps accounts for up to 90% of the total resistance [*Reid and Hickin*, 2008].

Much of the field and flume research directed at plane-bed channel segments involves measuring distributions of velocity and shear stress in relation to boundary roughness created by individual grains and pebble clusters (see section 3.5.2). Vertical velocity profiles for plane-bed segments have an inner layer, which reflects grain resistance and increases with increased bed roughness, and an outer layer that reflects total resistance [*Lawless and Robert*, 2001a].

Plane-bed channels lie at the boundary of *Montgomery and Buffington*'s [1997] transport and response reaches, and may function as either sources or storage sites for sediment as water/sediment ratios change. Although the characteristics of plane-bed channels indicate supply-limited conditions during most discharges, correlation of bedload transport rate and discharge during higher flows that mobilize the bed [*Jackson and Beschta*, 1982; *Sidle*, 1988] suggests that during high discharge these channels are transport-limited. The presence of sufficient wood may increase flow convergence and divergence to the level that pool scour and bar deposition transform the channel to a forced pool-riffle morphology [*Montgomery and Buffington*, 1997].

4.3.3. Pool-Riffle Channels

Keller and Melhorn [1978] provide one of the early systematic examinations of pool-riffle channel morphology and the processes which create and maintain this morphology. They describe pools and riffles as meandering in the vertical dimension, produced by the occurrence of regularly spaced deeps (pools) with intervening shallows (riffles) and note that these features are found in both bedrock and alluvial channels. The regular spacing of pools has been related to channel width [*Keller and Melhorn*, 1978] and to channel gradient [*Wohl et al.*, 1993], although variability of channel-boundary resistance associated with bedrock [*Roy and Abrahams*, 1980; *Wohl and Legleiter*, 2003], wood [*Montgomery et al.*, 1995a; *Buffington et al.*, 2002], or other obstructions [*Lisle*, 1986], and longitudinal variations in valley width and associated flow convergence [*White et al.*, 2010] may also influence pool spacing. The regularity of pool spacing along channels has led several investigators to interpret the undulating bed topography of pools and riffles as a means of regulating the dissipation of flow energy and specifically of maintaining quasi-equilibrium [*Dolling*, 1968], minimizing potential energy loss per unit mass of water [*Yang*, 1971], or minimizing power expenditure [*Cherkauer*, 1973]. It is important, however, to distinguish between pool-riffle sequences formed in relatively well-sorted alluvium with few largely immobile obstructions and pool-riffle sequences formed in poorly-sorted, coarse-grained alluvium with abundant obstructions. Pool-riffle sequences created under the former conditions are more likely to be influenced primarily by feedbacks between hydraulics and sediment in transport [*Lofthouse and Robert*, 2008], whereas channel boundary configuration likely plays a critical role in creating pool-riffle sequences in the latter channels, which are typical of mountain rivers.

Gilbert [1914] and *Keller* [1971] propose the *velocity-reversal hypothesis* to explain the maintenance of pool-riffle sequences. Using field data, Keller notes that as discharge increases, the near-bed velocity in a pool increases more rapidly than the velocity in an adjacent riffle. Keller hypothesizes that the flow in pools might be more competent at high stage than the flow over riffles, thus explaining the common observation that pools scour at high flow and fill at low flow, whereas riffles are depositional sites at high flow and scour at low flow [e.g., *Jackson and Beschta*, 1982; *Campbell and Sidle*, 1985; *Thompson*, 1994; *Sear*, 1996; *Thompson et al.*, 1996]. *Keller*'s [1971] data, however, do not actually demonstrate a velocity reversal. Subsequent investigators have found it difficult to demonstrate velocity reversal through pools and riffles with changing discharge, in part because Keller's original hypothesis emphasizes bed velocity, which is difficult to measure during high flows in many streams. Indirect methods such as hydraulic models can be used to simulate bed velocity [*Booker et al.*, 2001; *MacWilliams et al.*, 2006] and many of these simulations, whether 2d or 3d, suggest that near-bed velocity reversal does occur, although the field data to verify these results remain limited. As an example of this approach, *Cao et al.* [2003] simulate bed shear stress and velocity with a 2d model. They find that channel constriction can, but may not necessarily, create competence reversal, depending on channel geometry, flow discharge, and sediment properties.

Subsequent investigators modify Keller's original hypothesis by proposing that velocity might instead converge or become nearly equal at high flows [*Richards*, 1976; *Robert*, 1998; *Carling*, 1991; *Wohl*, 2007]; that shear stresses might vary differentially with discharge although a velocity reversal is unlikely [*Teleki*, 1972; *Bhowmik and Demissie*, 1982; *Wilkinson et al.*, 2004]; that shear stress [*Dolling*, 1968] or unit stream power [*O'Connor et al.*, 1986] in pools might exceed that over riffles because of greater depth; that flow convergence resulting from an upstream constriction and the routing of flow through the system might exert a more important influence on pool-riffle morphology than velocity reversal [*MacWilliams et al.*, 2006; *Sawyer et al.*, 2010]; or that a competence reversal might occur in pools that are hydraulically rougher than riffles [*Carling and Wood*, 1994]. Others [*Andrews*, 1979; *Lisle*, 1979; *Keller and Florsheim*, 1993; *Milan et al.*, 2001] argue that velocity reversal does indeed occur. *Caamaño et al.* [2009] propose that velocity reversal depends critically on the ratio of riffle width to pool width, residual pool depth, and flow depth over the riffle, and they develop a simple one-dimensional criterion for a reversal threshold

$$\frac{B_R}{B_P} - 1 = \frac{Dz}{h_{Rt}} \qquad (4.3)$$

where B_R is riffle water surface width, B_p is pool water surface width, Dz is residual pool depth, and h_{Rt} is riffle thalweg depth. Reversal occurs when $(B_R/B_P) - 1 > Dz/h_{Rt}$.

Much of the uncertainty over the existence of a pool-riffle velocity reversal has been generated by the assumption that, to convey the same discharge at high flows, a lower velocity over a riffle must be compensated by a change in cross-sectional area, presumably by an increase in width [*Richards*, 1978]. This assumption relies on cross-sectionally averaged velocity. The presence of strong eddy flow along the margins of pools during high discharges permits the formation of a central jet of high velocity flow that does in fact exhibit velocity reversal with respect to riffle velocity [*Thompson*, 1994, 1997; *Thompson et al.*, 1998, 1999] (Figure 4.11). Using a high-resolution 2d flow model in a bouldery pool-riffle channel, *Harrison and Keller* [2007] find that maximum shear stress and velocity occur over riffles during low flow. As discharge approaches bankfull, these maxima shift to the pool head because of strong flow convergence created by large roughness elements. This effect declines at higher flows as large roughness elements are submerged, suggesting that pool-riffle sequences are maintained by flows at or near bankfull in bouldery channels.

Clifford [1993a] proposes that a velocity reversal is not necessary to explain pool and riffle formation and maintenance if riffles and pools are undulating components of a single coarse-grained bedform unit. In this case, pool-riffle sequences may result from kinematic waves [*Langbein and Leopold*, 1968], which in return require systematic spatial variation in clast arrival and departure probabilities capable of maintaining an undulating bed. *Clifford* [1993a] uses field data to support the idea that such variations arise from (i) spatial differences in the structural arrangement of surface sediments (specifically, more microtopographic structuring on riffles than in pools) and (ii) differences in the behavior of the turbulent velocity/stress field at riffles and at pools (greater turbulent kinetic energy over riffles for most of the flow range). The whole pool-riffle sequence could be initiated with the generation of roller eddies upstream and downstream from a major flow obstacle such as a very large clast, with the obstacle then being removed as part of the process of bed scour [*Clifford*, 1993b]. Local scour of the pool creates deposition downstream, which then generates the next-downstream flow irregularity.

Other investigators propose similar conceptual models for the initiation of pool scour, but emphasize the role of lateral constrictions rather than bed obstacles [*Lisle*, 1986; *Wohl et al.*, 1993; *Thompson et al.*, 1996; *Wilkinson et al.*, 2008]. Pool depth appears to be particularly sensitive to the characteristics of the constriction that create the pool [*Thompson*, 2002a; *Wohl and Legleiter*, 2003]. *Lisle* [1986] presents the following equation for predicting the minimum width of an obstruction that will form a pool:

$$B = W_b - W_s \sec\beta - L_s \sin\beta \qquad (4.4)$$

where B is width of obstruction (m), W_b is bed width of approach channel (m), W_s is width of scour hole measured perpendicular to the axis of the scour hole at the widest point of obstruction (m), L_s is length of scour hole (m), and β is the deflection angle of flow (degrees). Constrictions affecting 33-50% of the approaching flow tend to form pools, whereas smaller constrictions form only local scour holes [*Lisle*, 1986]. Experimental reintroduction of wood in the form of engineered logjams that laterally constricted the channel created and enhanced pool-riffle sequences along a gravel-bed river in Australia [*Brooks et al.*, 2004].

Several investigators [*Lisle*, 1986; *Clifford*, 1993b; *Smith and Beschta*, 1994; *Thompson and Hoffman*, 2001; *Thompson*, 2002a, 2006a, c; *Thompson and Wohl*, 2009] propose that the vortex shedding associated with the channel constriction present near the head of a laterally constricted pool is capable of substantially altering sediment-transport patterns and pool geometry. Pool formation has been attributed to vortex scour [*Matthes*, 1947; *Lisle*, 1986; *Clifford*, 1993b; *Smith and Beschta*, 1994] and the deepest parts of pools are commonly found immediately downstream from constrictions where vortex flow is strongest [*Lisle*, 1986; *Clifford*, 1993b; *Thompson*, 1997]. Pool elongation reduces rate of energy expenditure by influencing patterns of vortex formation and dissipation [*Thompson*, 2004]. Sediment deposition at the downstream end of pools has also been related to boil formation [*Matthes*, 1947; *Lisle*, 1986; *Lisle and Hilton*, 1992] and to deposition in recirculating-eddy systems [*Kieffer*, 1987; *Schmidt*, 1990; *O'Connor*, 1993; *Carling et al.*, 1994; *Thompson*, 1994; *Cluer*, 1995]. *Lofthouse and Robert* [2008] focus on process of riffle formation and maintenance as these influence the length of pool-riffle sequences. They propose that transport capacity is compromised as the path length of a pool extends beyond a critical threshold for a given bed-material size. This can promote initial shortening of the downstream riffle and formation of a new riffle slightly upstream.

DeVries et al. [2001] describe a relatively inexpensive scour monitor constructed of plastic golf balls fitted internally with ring magnets and strung on a two-conductor cable enclosing a reed switch. They use this device to record changing depth of scour in a pool during a high flow that mobilizes bed material. *Jansen and Brierley* [2004] use cut-and-fill assemblages of poorly sorted pool sediments to infer late Holocene flood history on a bedrock channel.

Sites of coarse sediment input have also been used to explain riffle locations [*Webb et al.*, 1987; *Wohl et al.*, 1993] and pools have been correlated with sites of lower boundary resistance along bedrock channels [*Dolan et al.*, 1978]. *Wohl and Legleiter* [2003], on the other hand, correlate pool location with the downstream spacing of resistant bedrock outcrops in the valley walls that create lateral constrictions, in part because they find high variability in pool spacing. *Thompson*'s [2001] simulation model for pool and riffle formation indicates that, although the location of individual laterally-forced pools can be primarily random in character, a semi-rhythmic spacing of pools averaging four to eight times bankfull width can still result. Because of minimum pool- and riffle-length criteria imposed by hydraulic-feedback mechanisms, Thompson's results suggest that whether obstacles, variability in boundary resistance, or coarse sediment input will initiate formation of a pool or riffle is partly dependent on flow energy and on rates of energy expenditure.

Leopold et al. [1964] proposed that pools are spaced at five to seven times the average channel width. Averaging across diverse substrate types, *Keller and Melhorn* [1978] found an average pool spacing of six times the channel width, although *Roy and Abrahams* [1980] used the same data to demonstrate a greater mean pool spacing for bedrock streams than for alluvial streams. *Wohl and Legleiter* [2003] also demonstrate larger and more variable pool spacing for bedrock channels with laterally constricted pools. Pool spacing in forested mountain drainage basins in Alaska and Washington, USA depends on wood load and channel type, slope, and width [*Montgomery et al.*, 1995a]. Mean pool spacing in pool-riffle channels with low wood loads averages 2-4 channel widths, implying that channel morphology is very sensitive to the presence of wood and other obstructions [*Smith et al.*, 1993]. Plane-bed channels with similar wood loads have pools at 9-13 channel widths. Forced pool-riffle channels, which have higher wood loads, have pools spaced less than 2 channel widths and gradients that overlap pool-riffle and plane-bed channel types. In all channel types, however, less than 40% of wood causes the formation of a pool. *Montgomery et al.* [1995a] conclude that channel width strongly influences pool spacing in streams with similar wood loads, but high wood load is likely to decrease pool spacing.

Pools may be subdivided on the basis of the primary erosional and depositional processes that create them [*Bisson et al.*, 1982]. Plunge pools form by scour associated with flow plunging vertically over an obstacle (Figure 4.12). Dammed pools form when debris spanning a channel ponds water upstream. A root wad protruding into the flow or wood that directs the current along its length may form a lateral scour pool. A backwater or eddy pool can be formed by an eddy behind debris and other structures located at the channel margin [*Bisson et al.*, 1982]. Each of these types of pools has different patterns of flow, as well as sediment and nutrient retention.

Attention has also been devoted to differentiating pools and riffles. Criteria that have been proposed for distinguishing between pools and riffles include bed material size [*Leopold et al.*, 1964]; water-surface slope [*Yang*, 1971]; and the index v^2/d, where v is mean flow velocity and d is mean flow depth [*Wolman*, 1955]. Because these criteria may depend on discharge or the history of previous discharge, other investigators [*Richards*, 1976; *O'Neill and Abrahams*, 1984; *Takahashi*, 1990; *Carling and Orr*, 2000] propose using bed topography and bedform differencing techniques [*Hanrahan*, 2007].

Although more attention has been devoted to pools than to riffles, *Legleiter et al.* [2007] use a 3d acoustic Doppler velocimeter to characterize spatial and temporal variations in velocity along a coarse-grained, high gradient (0.009 m/m) riffle as discharge varies through a snowmelt season (Figure 4.13). They document the development of secondary currents and flow convergence at higher stages, and the effective diminution of grain-scale boundary roughness, which results in the spatial structure of the flow field becoming smoother and more continuous as stage increases. *Montgomery and Buffington* [1997] characterize pool-riffle sequences as response reaches that exhibit channel change in response to changes in water and sediment discharges. Increases in water yield associated with timber harvest in northwestern Montana, USA cause bank erosion, sediment deposition, and substrate fining in pool-riffle channels [*Madsen*, 1995], in contrast to the lack of response in step-pool channels in the region. In-channel scour triggered in part by increased peak flows appears to be an important source of sediment in these channels. Flow diversions result in channel narrowing by 30-50% along pool-riffle channels in the mountains of Colorado, USA, again in contrast to the lack of response in step-pool channels [*Ryan*, 1994b, 1997]. Land uses that increase sediment yield to laterally-forced pools in bedrock and boulder-bed streams in the Colorado Front Range, USA correlate with reduced pool volume, after adjustment for the effects of discharge and gradient [*Goode and Wohl*, 2007].

Several studies [*Lisle*, 1982; *Wohl et al.*, 1993; *Madej and Ozaki*, 1996; *Wohl and Cenderelli*, 2000; *Kasai et al.*, 2004] demonstrate that a large increase in sediment load along a pool-riffle channel will cause preferential filling of the pools, effectively creating a more uniform reach gradient and flow depth. *Lisle* [1982] explains this response as leading to an increase in the effectiveness of moderate discharges (< 2 yr recurrence interval) to transport bedload and shape the bed because of a reduction in channel form roughness. As the excess sediment is removed from the channel reach, the pool-riffle sequence gradually re-forms. The proportion of residual pool volume filled with fine bed material, V^*, can be used to monitor and evaluate the supply of excess fine sediment in gravel-bed channels [*Lisle and Hilton*, 1992, 1999], although this metric can also be influenced by lithology and wood or other features that modify local hydraulics and fine sediment deposition [*Sable and Wohl*, 2006]. One-dimensional sediment transport models can accurately model more than half of the pool scour observed following a large increase in sediment and associated pool filling, although 2d models more accurately represent pool hydraulics [*Rathburn and Wohl*, 2001, 2003]. *Hassan and Woodsmith* [2004] document preferential localized pool scour of coarse sediment caused by wood.

Because salmonids tend to use pool-riffle channel segments in preference to plane-bed or step-pool reaches [*Moir et al.*, 2004], as well as preferentially using specific portions of the pool-riffle bedform sequence for spawning [*Hanrahan*, 2007], techniques that can be used to map the spatial distribution of pool-riffle channel segments within a channel network provide valuable management tools.

In summary, pool-riffle channels meander in the vertical dimension, with regularly spaced downstream alternations between deep pools and shallow riffles. Pool-riffle sequences can form in relatively well-sorted alluvium with few largely immobile obstructions and in poorly-sorted, coarse-grained alluvium with abundant obstructions that act to localize pool scour or riffle deposition. Numerous variations on the theme of a velocity reversal – in which the near-bed velocity in a pool increases more rapidly than the velocity in an adjacent riffle as discharge increases – have been proposed to explain the formation and maintenance of pool-riffle sequences. Processes resulting in pools and riffles undoubtedly vary among diverse channels, and pool-riffle sequences forced by lateral obstructions are particularly likely to have pool geometry reflecting the complex secondary hydraulics that develop around the obstruction. Downstream pool spacing can vary widely around the commonly assumed value of 5-7 times the channel width in relation to factors such as erosional resistance of the channel boundaries and instream wood loads. Pool-riffle sequences tend to be response reaches that are relatively sensitive to changes in water and sediment supply, with increased sediment resulting in preferential filling of pools.

Schumm's [1977] division of an entire drainage basin into source, transport, and depositional zones may also be applied to the mountainous portion of a drainage basin in that the steepest, low-order channels are likely to receive sediment directly from the hillslopes; the intermediate step-pool channels may mostly transport the sediment downstream; and the lower pool-riffle channels will respond to the sediment with increased storage and changes in channel configuration. Superimposed on this general downstream progression along a mountain river may be the presence of bedrock reaches with knickpoints and gorges (chapter 2), incised alluvial channel segments, or reaches with braided or anastomosing channels.

4.4. Incised Alluvial Channels

Although incised alluvial channels may be more common in regions of lower relief than in mountainous terrain, low-gradient segments of mountain rivers can incise through the same processes that drive incision in other types of channels [*Jaeggi and Zarn*, 1999]. Alluvial channels in diverse environments that incise as a result of processes such as base level fall or change in water or sediment discharge [*Schumm*, 1999] pass through a sequence of channel morphology with time that is summarized in *channel-evolution models* [*Schumm et al.*, 1984; *Simon and Hupp*, 1986]. The most widely used model includes six stages (Figure 4.14) that correspond to shifts in dominant adjustment processes and associated rates of sediment transport, bank stability, sediment accretion, and ecologic recovery [*Simon and Castro*, 2003]. The timing associated with each stage of channel evolution and the relative magnitudes of vertical and lateral erosion vary widely between different streams and stream segments [*Simon and Rinaldi*, 2006], but the entire sequence may require 10^2-10^3 years [*Simon and Castro*, 2003].

Incised channels can form in any climatic or tectonic environment, but are particularly common in arid and semiarid regions [*Bull*, 1997; *Elliott et al.*, 1999]. Historical [*Webb et al.*, 1991]. Sedimentary records indicate that repeated episodes of alternating incision and aggradation characterize channels of varying size in dry environments. Incision and aggradation of these channels can be triggered by external drivers such as changes in flood magnitude and frequency [*Love*, 1979; *Webb and Baker*, 1987], other aspects of climate [*Leopold*, 1951, 1976], or land use practices including grazing and groundwater withdrawal [*Cooke and Reeves*, 1976; *Gonzalez*, 2001], as reviewed by *Graf* [1983] and *Phippen and Wohl* [2003]. Incision and aggradation can also be intrinsic to channels in dry environments, as first recognized by *Schumm and Hadley* [1957], who propose that limited transport capacity in dryland channels facilitates sediment accumulation and progressive increase in local bed gradient until a threshold is crossed and the channel begins to incise [*Patton and*

Schumm, 1975]. The localized nature of bed steepening typically promotes longitudinally discontinuous channel geometry such that one segment of a channel can be actively incising while another up- or downstream is aggrading [*Patton and Schumm*, 1975].

Incised channels are also likely to undergo multiple adjustments in response to a single external perturbation. The classic expression of *complex response* was conceptualized by *Schumm and Parker* [1973] as repeated alternations between incision and aggradation following a single drop in base level. As a knickpoint initiated at the channel mouth by base level fall translates upstream, increasing sediment supply to downstream reaches can cause them to aggrade while the channel upstream is still incising. Once knickpoint migration and incision stop, the reduction in sediment load can trigger another phase of incision in the lower channel that then migrates upstream. The result is that any segment of a channel can repeatedly aggrade and incise through time and different segments of the channel are likely to be out-of-phase from one another with respect to incision and aggradation (Figure 4.15). Although *Schumm* [1973; *Schumm and Parker*, 1973] originally developed the conceptual model of complex response to explain observations in relatively small ephemeral channels, the model has subsequently been applied to channels of diverse drainage area, climatic regime, and substrate composition [*Shimazu and Oguchi*, 1996; *Cohen and Brierley*, 2000; *Marston et al.*, 2005; *Harvey*, 2007].

Although large, expensive, highly engineered structures including check dams, block ramps, low-head sills, and artificial armor layers are commonly used to stabilize bed incision by creating local base levels, this approach typically has limited success [*Jaeggi and Zarn*, 1999]. Mitigation measures that address the causes or processes of incision, such as altering sediment and water yield to the channel, or restoring alluvial water tables and riparian vegetation and reconnecting side channels [*Bravard et al.*, 1999; *Shields et al.*, 1999] are more likely to be successful and sustainable. In some settings, simple, traditional brush structures that are inexpensive and quick to build can increase the incidence of overbank flow and thus limit channel incision and reconnect channels to floodplains and alluvial fans [*Norton et al.*, 2002].

In summary, incised alluvial channels can form in lower gradient reaches of mountain channel networks in response to changes in base level or water or sediment supply to the channel. Incised channel segments are likely to undergo multiple alterations of bed and bank erosion and aggradation as part of a complex response to external perturbation or intrinsic thresholds. These changes are described by channel evolution models.

4.5. Braided Channels

Braided channels have multiple flow paths of low sinuosity separated by islands or bars (Figure 4.16). This channel morphology characteristically occurs along valley reaches of more erodible banks, steeper gradient, higher bedload supply, and highly variable discharge. These channels have large width/depth ratios and abundant bedload and may have large fluctuations in water and sediment discharge. They also tend to be laterally unstable, with substantial channel movement at timescales of a few hours to a single flow season. Braided channels are thus the exception to the rule that mountain channels are supply limited and stable. *Leopold and Wolman* [1957] propose a slope-discharge threshold for differentiating channel patterns, although changes in other parameters such as bank strength and sediment transport capacity [*Simpson and Smith*, 2001] may dominate changes in channel pattern. Within the past few years, consensus has shifted to the idea that braided rivers are the default channel morphology in the absence of riparian vegetation, cohesive sediments, lateral valley confinement, or something else that stabilizes the stream banks and limits active channel width [*Paola*, 2001]. As *Lane* [2006] phrases it in his comprehensive review, "braiding appears to be the fundamental instability that arises when water moves over non-cohesive material in the absence of lateral confinement" (p. 107).

Braiding may be initiated by division of flow around a constructional bar formed within a single channel, by dissection of a transverse bar or point bar, or by avulsion and creation of a new braid channel [*Germanoski*, 1989], but the pool-bar unit is a fundamental building block that controls water and sediment routing and patterns of erosion and deposition [*Ashmore*, 1991, 1993; *Ashworth et al.*, 1992; *Ashworth*, 1996; *Nicholas and Sambrook Smith*, 1999]. Complex feedbacks between flow and the channel boundaries create regular fluctuations in the discharge distribution at the bifurcation node related to bar migration, and the influence of bars on bifurcation processes increases with bar amplitude [*Bertoldi et al.*, 2009]. Results from a one-dimensional model suggest that stable bifurcations are characterized by a strongly unbalanced partition of water and sediment discharges in the two branches [*Miori et al.*, 2006]. The control exerted by the cross-sectional distribution of flow velocity on flow distribution, sediment transport, and channel geometry appears to increase with increasing channel scale and width/depth ratio [*Szupiany et al.*, 2009]. Bedload transport in braided rivers is highly variable [*Griffiths*, 1979], as it is in physical models of braiding [*Warburton and Davies*, 1994], in association with processes such as migration of bedforms [*Young and Davies*, 1991]. Simple numerical models suggest that the apparently stochastic behavior of braided rivers reflects a simple non-linear relationship between stream power and bedload transport [*Murray and Paola*, 1994].

Avulsion frequency scales with the time required for bed sedimentation to produce a deposit equal to one channel depth, as reflected in a dimensionless mobility number based on the relative rates of bank erosion and channel sedimentation [*Jerolmack and Mohrig*, 2007]. Similarly, physical models demonstrate a strong, positive relationship between sediment supply and frequency of channel avulsion [*Ashworth et al.*, 2007]. *Aggett and Wilson* [2009] describe coupling a floodplain digital terrain model with a 1d hydraulic model in a GIS format to assess avulsion hazards. The dynamics of braid bar formation and channel change are discussed in more detail in *Lewin* [1976], *Davies* [1987], *Ferguson and Ashworth* [1992], *Laronne and Duncan* [1992], *Burge* [2006], *Bertoldi and Tubino* [2007], and *Tubino and Bertoldi* [2008].

Miall [1977] distinguishes three types of bars in braided rivers: longitudinal, composed of crudely bedded gravel sheets; transverse to linguoid, of sand or gravel deposited by downstream avalanche-face progradation; and point or side bars formed by bedform coalescence and chute and swale development in areas of low energy. He also distinguishes four main classes of facies assemblages and vertical sequences in braided rivers that are named for type localities. *Huggenberger and Regli* [2006] update this work by distinguishing seven different types of sedimentary textures and four types of depositional elements defined by the geometrical form of erosional and/or lithological boundaries and the character of the fill. *Rubin et al.* [2006] explore the links between depositional heterogeneity and river geometry and dynamics.

Braided channels tend to store sediment for shorter periods of time and to have more rapid turnover times for floodplains than meandering or straight channel segments in the same region [*Beechie et al.*, 2006]. Braided rivers undergoing aggradation can of course store more sediment for longer times than when they are in degrading phases [*Phillips et al.*, 2007]. Aerial photographs and ages of floodplain vegetation can be used to estimate sediment residence times over decades or longer [*Beechie et al.*, 2006], as can direct dating of sediments [*Phillips et al.*, 2007]. *Milan et al.* [2007] use a high-resolution 3D laser scanner to assess volumes of erosion and deposition over shorter time periods in a proglacial braided river. Based on repeat topographic surveys and automatic digital photography of a reach of the Tagliamento River in Italy, *Bertoldi et al.* [2010] distinguish flow pulses that occur more than once a year and alter the morphology of a few active channel branches and flood pulses with a return period larger than two years that induce a complete reworking of the network configuration. *Paola* [2001] hypothesizes that the tendency toward braiding is influenced by a river's ability to turn over its bed within the characteristic time for riparian vegetation establishment and growth to a mature, scour-resistant state. Subsequent field and laboratory tests confirm this hypothesis and indicate that this dimensionless time-scale parameter can predict whether a channel will braid [*Hicks et al.*, 2008].

Braided channel planforms can be quantified using a braiding index, although several indices have been proposed and there is no consistent standard of measurement. *Braiding intensity* is typically based on bar dimensions and frequency, the number of channels in the network, and the total channel length in a given river length [*Egozi and Ashmore*, 2008]. The most commonly used indices of braiding intensity are channel count (the mean number of links intersected by cross sections of the river) and total sinuosity (total length of channels per unit length of river) [*Egozi and Ashmore*, 2008]. *Germanoski and Schumm* [1993] observe that braiding intensity decreases in the upstream portion of an experimental channel when sediment supply is eliminated, but increases in the downstream portion. *Chew and Ashmore* [2001] attribute observed downstream increases in braiding intensity along a proglacial braided stream to decreasing grain size, and *Luchi et al.* [2007] note changes in braiding intensity associated with sharp longitudinal variations in grain size and gradient. Physical and numerical models of braiding indicate aggradation associated with channel multiplication and bar deposition, and degradation associated with channel pattern simplification [*Hoey and Sutherland*, 1991; *Thomas et al.*, 2007].

Physical models of braided rivers self-organize into a critical state with dynamic scaling such that small and large parts of the channel evolve statistically identically after self-adjustment of the profile and the braiding pattern [*Sapozhnikov and Foufoula-Georgiou*, 1999]. Other studies argue that braided rivers exhibit self-similar scaling, a type of fractal behavior in which symmetry does not depend on direction [*Walsh and Hicks*, 2002], or self-affine scaling, a fractal behavior in which the x- and y-dimensions scale by different amounts [*Sapozhnikov and Foufoula-Georgiou*, 1996]. *Hundey and Ashmore* [2009] demonstrate that confluence-bifurcation units scale at 4-5 times the channel width, analogous to the relationship between channel width and pool spacing in pool-riffle channels.

Braided channels can be found in proglacial [*Fahnestock*, 1963] to arid [*Laronne and Shlomi*, 2007] to tropical mountain environments. Braiding at the reach scale is commonly associated with a large point source of sediment from a hillslope mass movement [*Knighton*, 1976; *Hoffman and Gabet*, 2007] or from a glacier. Braided channel reaches are relatively common in mountain environments because of the relative abundance of point sources of coarse sediment from adjacent hillslopes and the ability of frequent bedload movement and associated channel instability to limit vegetation establishment [*Tal and Paola*, 2007]. These channel reaches are most likely to be viewed as hazardous or "degraded" and targeted for management or restoration [*Korpak*, 2007] (see chapter 6).

Proglacial braided rivers systems are known as *sandurs* where the channel system expands freely (Figure 4.17) and *valley sandurs* where the development of the channel network is confined by valley walls [*Krigstrom*, 1962]. Because the carrying capacity and competence of glacial ice and subglacial tunnels are commonly greater than the broad, shallow channels of the sandur, some of the coarse clasts deposited on the upper sandur are immobile [*Bogen*, 1995]. The steep upper portion of the sandur is composed of the coarse-grained fractions of the sediment supplied. Channel bifurcation is common in this proximal zone, as is lateral channel shifting, and vegetation is absent. The number of channels reaches a maximum in the intermediate zone where grain sizes in the channel bed and bars are smaller and some vegetation is present. Down-valley, in the distal zone, more abundant riparian vegetation stabilizes the channel banks and a layer of overbank sediments is deposited on the floodplain [*Church*, 1972]. Because velocity in the secondary channels in particular may vary substantially with discharge, secondary channels can act as sediment traps during low discharges and as sediment sources as discharge rises [*Bogen*, 1995].

Study of the dynamics of braided rivers has advanced significantly during the first decade of the 21st century through the development of new techniques for measuring flow and bed morphology [*Sambrook Smith et al.*, 2006] and through numerical modeling of braided river process and form [*Lane*, 2006; *Guin et al.*, 2010; *Ramanathan et al.*, 2010]. New techniques include flow measurement with acoustic Doppler current profiling, direct measurement of bed morphology with multibeam echo sounding, remote sensing of braided river morphology using synoptic digital photo-grammetric and airborne laser survey methods, and characterization of deposits using ground penetrating radar [*Sambrook Smith et al.*, 2006].

Murray and Paola [1994] develop the first numerical representation of spatially distributed, time-dependent evolution of river braiding, which *Lane* [2006] highlights as the most important contribution to braided river science during the preceding decade. The model is based on: mass conservation of water and sediment; a flow routing approximation; a downstream sediment transport law which can have a non-linear dependence on routed flow; and a lateral sediment transport law [*Lane*, 2006].

The Murray and Paola model produces what resembles a braided river without much of the hydraulic complexity considered to be necessary in other models. More recently, investigators have used two- and three-dimensional flow computational fluid dynamics modeling [*Nicholas and Sambrook Smith*, 1999] and cellular automata models [*Coulthard et al.*, 1996; *Murray and Paola*, 1997; *Thomas and Nicholas*, 2002; *Nicholas et al.*, 2006; *Van De Wiel et al.*, 2007], sometimes coupled with bedload transport models [*McArdell and Faeh*, 2001], to understand patterns and processes in braided rivers. These models suggest that the presence or absence of significant local redeposition distinguishes braiding from other channel types [*Murray and Paola*, 1997]. Expanding on *Paola*'s [1996] work, *Nicholas* [2000] develops a model that is able to accurately predict annual and longer-term fluctuations in bedload yield of the Waimakariri River in New Zealand. The model also suggests that increased intensity of braiding may promote higher rates of bedload transport.

In their very useful review of cellular automata models, *Coulthard et al.* [2007] note that the fundamental problems of (i) integrating sediment transport with fluid flow and (ii) temporal and spatial scaling continue to hamper the applicability of most numerical models. Although current reduced-complexity models include high-resolution cellular approaches that resolve process-form feedbacks at small time and space scales and models that use section-averaged representations of channel geometry and process [*Nicholas and Quine*, 2007a], there is a distinct gap between these detailed models at small spatial and temporal scales and models of catchment evolution over longer time spans. The lack of models addressing time scales of 1-100 years and spatial scales of 1-100 km^2 is particularly problematic for river management [*Coulthard et al.*, 2007].

River management is more likely to be aimed at braided channels than at other types of channels because of the potential for rapid and substantial lateral and vertical changes in braided channels. *Piégay et al.* [2006] review strategies undertaken to limit braiding, which include channelization, bank protection, aggregate mining, regulating sediment supply from tributaries, and afforesting the catchment, as well as strategies that can enhance braiding, which include limiting riparian vegetation within the channel, increasing coarse sediment supply, promoting bank erosion, and enhancing temporal flow variability. Increasing recognition of the environmental benefits provided by braided rivers has led to attempts to restore braiding, which requires knowledge of the interactions between flow, sediment and channel morphology [*Marti and Bezzola*,

2006] (Figure 4.18). The flow pulse concept of *Tockner et al.* [2000b] (see chapter 5) applies particularly to braided rivers, for example, which can contain diverse aquatic and riparian habitats.

In summary, braided channels with multiple, low-sinuosity flow paths separated by islands or bars typically occur in portions of a mountain river network with more erodible banks, higher bedload supply, and/or highly variable discharge. These unstable channels may be the default channel morphology where vegetation or cohesive substrate do not stabilize the stream banks. Diverse types of bars form in braided channels, and the pool-bar unit is a fundamental building block that controls water and sediment routing and patterns of erosion and deposition. The frequency of channel avulsion demonstrates a strong, positive relationship with sediment supply. Braided channel planforms can be quantified using a braiding index, such as braiding intensity, which is typically based on bar dimensions and frequency, the number of channels in the network, and the total channel length in a given river length. The first numerical representation of river braiding is based on: mass conservation of water and sediment; a flow routing approximation; a downstream sediment transport law which can have a non-linear dependence on routed flow; and a lateral sediment transport law. Subsequent models use two- and three-dimensional flow computational fluid dynamics and cellular automata approaches to understand patterns and processes in braided rivers. Braided rivers are likely to be targeted for stabilization and management, although these channels provide many environmental benefits.

4.6. Anabranching Channels

Anabranching channels differ from braided channels in that, although they include multiple flow paths with regular bifurcations and junctions, the individual subchannels are separated by semi-permanent islands or ridges. Anabranching channels range from straight to highly sinuous, from fine-grained, low-energy systems (sometimes called anastomosing; *Nanson and Knighton*, 1996) to coarse-grained, high-energy systems, and from two-channeled to many-channeled [*Nanson and Huang*, 1999]. Most studies of anabranching channels come from non-mountainous regions [e.g., *Schumm et al.*, 1996; *Van Niekerk et al.*, 1999]. Anabranching alluvial, bedrock, and mixed channel segments can occur in mountainous environments, but they are not as common as other channel types. Stable accumulations of wood created by fluvial processes or by beavers can promote anabranching [*Abbe and Montgomery*, 2003], as can abundant bedload supplied by hillslope mass movements and glaciers [*Clague et al.*, 2003]. *Nanson and Huang* [1999] interpret the formation of ana-branches as a form of channel adjustment where the ability to increase slope is limited. Conversion from a wide, single channel to multiple channels reduces total width and increases average flow depth, hydraulic radius, and velocity, thus maintaining or enhancing water and sediment throughput [*Nanson and Huang*, 1999; *Jansen and Nanson*, 2004; *Huang and Nanson*, 2007].

4.7. Spatial Distribution of Morphologic Types and Network Heterogeneity

The simplified downstream progression of cascade to step-pool, then plane-bed, pool-riffle, and finally dune-ripple channel morphology illustrated in *Montgomery and Buffington* [1997] and Figure 4.5 is not meant to imply that this regular sequence always occurs in mountain rivers. Because reach-scale gradient and local tributary or hillslope-derived sediment exert a strong influence on channel morphology [*Montgomery and Buffington*, 1997; *Halwas and Church*, 2002; *Wohl and Merritt*, 2005], and gradient and lateral sediment supply along mountain rivers are so longitudinally variable, channel morphologic types can alternate abruptly and repeatedly downstream, even as other variables influencing channel morphology, such as drainage area and discharge, increase progressively. Consequently, it is not necessarily straightforward to understand and predict the spatial distribution of channel morphology and associated processes throughout a mountainous drainage basin.

Investigators use different variables and statistical techniques to distinguish spatial variations in channel-reach morphology and in associated geomorphic and ecological attributes. *Jensen et al.* [2001], using a broad array of physical and biological variables to categorize valley bottoms and streams, are able to correctly assign 80 percent of their subwatersheds in the U.S. Pacific Northwest based on 13 variables such as average watershed slope, drainage density and two-year maximum 24-hour storm intensity. *Molnar et al.* [2002] use nested models of hydrology, hydrodynamics, sedimentology, and ecology to evaluate local changes in river morphology and habitat. *Wohl and Merritt* [2005] correctly identify channel type for 76 percent of the streams in a dataset from the U.S., Panama, and New Zealand using only channel gradient, D_{84}, and channel width. Working in mountainous southeastern Australian catchments, *Thompson et al.* [2006, 2008] distinguish channel types using channel gradient, lithology, and contributing drainage area. *Brardinoni and Hassan* [2007], working in glaciated valley networks in British Columbia, Canada, find that multivariate discriminant analyses successfully distinguish channel morphologic types based on channel gradient, shear stress, and relative roughness, and that glacially-induced variations in channel longitudinal profiles, along with the degree of colluvial-alluvial coupling imposed by glacial valley morphology, dictate spatial organization of channel types. *Altunkaynak and Strom* [2009] use multilayer perceptron methods to develop predictive models for channel morphology with an accuracy of ~80% based on gradient, relative grain submergence, and channel width. *Rot et al.* [2000] find strong interrelations among valley constraint, riparian landforms, riparian plant community, and channel morphology. Channel morphology (bedrock, plane-bed, forced pool-riffle) correlates strongly with wood volume at the smallest spatial scale, for example, and valley constraint at the largest. *Splinter et al.* [2010] demonstrate consistent differences in stream morphology across ecoregions in eastern Oklahoma, USA.

Approaches using parameters such as gradient, drainage area, and lithology that can be derived from existing sources such as DEMs and geologic maps are of course easier to apply and calibrate across entire catchments or larger regions than are approaches that rely on field-measured parameters. Our ability to remotely image and categorize river characteristics, even down to the scale of channel units such as riffles and pools, continues to improve with technology and the availability, resolution, and accuracy of spatially explicit data [*Wright et al.*, 2000; *Mertes*, 2002; *Richardson et al.*, 2009], although the steep terrain, forest cover, and small streams of mountainous regions present special challenges for imaging the land surface from space.

Montgomery [1999] proposes that regional climate, geology, vegetation and topography influence the geomorphic processes distributed over a landscape at large scales. Spatial variability in geomorphic processes in turn governs temporal patterns of disturbances that influence ecosystem structure and dynamics. Classification of stream channels can guide identification of functionally similar portions of a channel network. Geomorphic processes can be divided into spatially identifiable areas known as process domains and *Montgomery*'s [1999] *process domain concept* implies that channel networks can be divided into discrete regions in which biological community structure and dynamics respond to distinctly different disturbance regimes (Figure 4.19). This approach has been successfully used to characterize the spatial distribution, morphology, stability, and disturbance regime of diverse mountain channel networks in the western United States [*Buffington et al.*, 2003; *Wohl et al.*, 2007].

One of the purposes in using a conceptual framework such as the process domain concept is to assess and monitor channel condition in the context of resource management, restoration, or potential response to anthropogenic disturbances that alter water and sediment yield to the channel or channel geometry. *Fitzpatrick* [2001] reviews different approaches to habitat and geomorphic assessment and stream classification. As noted by *Montgomery and MacDonald* [2002], the development of specific diagnostic criteria and monitoring protocols must be tailored to specific geographic areas because of the variability in controls on channel processes within river basins and between regions.

4.8. Summary

Mountain river channels are less likely than low-gradient alluvial channels to approximate the graded longitudinal profile and progressive downstream changes in channel morphology that are commonly regarded as the stable conditions toward which rivers trend. Mountain rivers may exhibit downstream trends of decreasing gradient and grain size and greater channel stability, but these trends are likely to be interrupted as a result of changes in lithology and structure, glacial or climatic history, or coarse sediment input. Valley morphology sets the larger-scale template for channel morphology and can reflect glacial history, changes in base level, and asymmetries reflecting different hillslope aspects and associated differences in weathering and erosion.

Because of the relative lack of consistent downstream trends, mountain rivers are most appropriately classified at the reach scale. Using the *Montgomery and Buffington* [1997] classification, step-pool channels, plane-bed channels, and pool-riffle channels are the most common channel morphologic types among mountain rivers. Step-pool sequences occur at high gradients where sediment supply is limited. Steps and pools influence boundary roughness and tend to evolve toward a condition of maximum flow resistance. Step-pool sequences form transport reaches that are relatively insensitive to changes in water and sediment discharge, in contrast to pool-riffle sequences, which form response reaches. Pool volume and grain size are particularly sensitive. Differential velocity, shear stress and sediment movement through pools and riffles are interpreted as reflecting a velocity reversal between pools and riffles at differing discharges. An increase in coarse sediment supply associated with a hillslope mass movement or a glacier may create a braided or anabranching channel reach that interrupts the general downstream trends in channel morphology, and a change in water or sediment supply or an increase in gradient can result in incision along alluvial channel segments. Braided channels may be the default channel type in the absence of something that stabilizes the stream banks and limits active channel width.

Each type of channel morphology has characteristic distributions of substrate and hydraulics as well as characteristic stability and these are conceptualized in process domains. These characteristics provide the physical template for aquatic and riparian biota along the river.

5. MOUNTAIN RIVER BIOTA

5.1. River Ecology

5.1.1. Only Connect

Luna Leopold once described rivers as "the gutters down which flow the ruins of continents." If rivers were only gutters draining the products of bedrock weathering and erosion, a great deal of contemporary river management and restoration would be unnecessary, because there would be no consideration of ecosystem services or biodiversity. But of course rivers are much more than conduits for water and sediment. The most comprehensive and appropriate view of any river catchment or segment of river is as an ecosystem. Even the most contaminated rivers support some life and most rivers support an abundance and diversity of living organisms that is much larger than the relatively small percentage of the Earth's surface that they occupy. The term *riverscape* is coming into use as a means of recognizing a portion of the landscape that has, or had, a watercourse as its focus [*Fausch et al.*, 2002; *Haslam*, 2008].

Headwater streams exemplify the biological wealth of rivers. These streams offer a refuge from temperature and flow extremes, competitors, predators, and introduced species and serve as a source of colonists to downstream portions of the stream network. Headwater streams also provide spawning sites and rearing areas, as well as a rich source of food, and create migration corridors through the landscape [*Gomi et al.*, 2002; *Meyer et al.*, 2007] (Figure 5.1). Limited studies document numbers such as 1004 invertebrate taxa in the 1-m-wide Breitenbach, a first-order stream in Germany [*Allan*, 1995] and at least 293 taxa in three first-order headwater streams draining 5-7.5 ha in North Carolina, USA [*Meyer et al.*, 2007]. Even naturally intermittent or temporary streams, which constitute 62% of total river length in the United States [*Nadeau and Rains*, 2007], yet which many people dismiss in terms of ecological function, can support rich and distinctive biological communities, in part because dry periods are predictable [*Stanley et al.*, 1997; *Meyer et al.*, 2007; *Larned et al.*, 2010]. Because riparian habitats harbor significantly different species pools than upland ecosystems, they increase regional species richness by >50%, on average [*Sabo et al.*, 2005].

Mountain Rivers Revisited
Water Resources Monograph 19
Copyright 2010 by the American Geophysical Union.
10.1029/2010WM001043

The wealth of plants and animals living in and along river corridors perform critical ecosystem services including regulation of nitrogen, carbon, and other nutrients and mediation of flow volume and water quality, as well as other processes that we are just beginning to understand. Increasing recognition of the ecosystem services provided by even the smallest streams makes land uses such as mountain-top removal [*Palmer et al.*, 2010; see chapter 6] impossible to justify.

Any river ecosystem includes six degrees of connection with the adjacent watershed and the greater world (Figure 5.2). Vertical exchanges occur between the river and the atmosphere as precipitation, eolian sediment, and contaminants enter the river, and larval aquatic insects hatch into terrestrial adults. Atmospheric inputs can originate in distant locations and result in contamination of otherwise pristine, remote mountain streams with airborne toxics such as mercury [*Biswas et al.*, 2007].

Vertical exchanges also occur between the shallow subsurface, or *hyporheic zone*, and the channel, and between the deeper, groundwater aquifers and the river. The ecological importance of the hyporheic zone results from the use of interstitial habitat by fluvial organisms such as invertebrates [*Stanford and Ward*, 1988, 1993] and from its influence on water chemistry. Water temperature, dissolved oxygen, carbon dioxide, pH, nitrate, and ammonium are all influenced by groundwater-stream water exchanges.

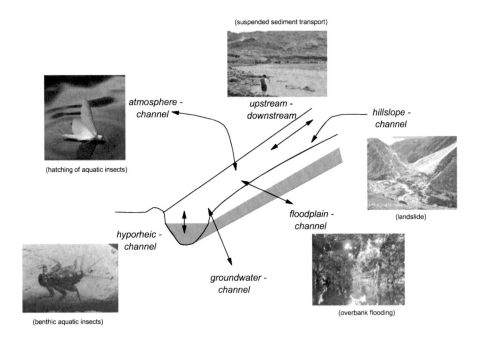

Figure 5.2 Six degrees of connection between a river ecosystem and the surrounding environment (light gray is water, dark gray is hyporheic zone). Inset photographs of aquatic insects courtesy of Jeremy Monroe of Freshwaters Illustrated.

Frequent hyporheic exchange enhances the exchange of heat between stream water and the hyporheic zone and moderates water temperatures in the stream [*Anderson et al.*, 2005]. Physicochemical gradients in the hyporheic zone result from the different properties of surface and groundwaters and from biogeochemical reactions associated with retention in areas of low hydraulic conductivity [*Findlay*, 1995; *Brunke and Gonser*, 1997]. Frequent hyporheic exchange in mountain streams likely increases the uptake of soluble nutrients in stream water [*Triska et al.*, 1989], as well as microbial processing of organic matter and nutrients [*Harvey and Wagner*, 2000; *Malard et al.*, 2002]. The hyporheic zone may act as a source or sink for dissolved organic matter, depending on the volume and direction of flow, dissolved organic carbon concentrations, and biotic activity [*Brunke and Gonser*, 1997]. Primary productivity in mountain streams may be nitrogen-limited [*Gregory*, 1980], so that the hyporheic transport of dissolved nitrogen and hyporheic biogeochemical processes may be especially important in these streams. The hyporheic zone along McRae Creek, a fourth-order mountain river in Oregon, USA, is a net source of nitrogen to the stream under all discharge conditions [*Wondzell and Swanson*, 1996b]. Studies along Sycamore Creek, Arizona, USA demonstrate that hyporheic respiration is consistently higher in downwelling zones than in upwelling zones [*Jones et al.*, 1995b], whereas algal standing crop in the channel is greater, and elevated concentrations of nitrate occur, in upwelling zones [*Valett et al.*, 1992].

Lateral exchanges between the river and floodplain occur as water, sediment, solutes, and organisms spread across the floodplain during high flows and then remain or return to the channel during lower flows [*Jenkins and Boulton*, 2003]. Although mountain rivers tend to be laterally confined and have limited floodplain development, wider valley segments can result from glaciation, beaver activity [*Westbrook et al.*, 2010], or reduced erosional resistance of the local bedrock [*Wohl*, 2008a], and these valley segments can have well-developed floodplains.

Lateral exchanges also take place between the river and uplands when water, sediment, wood, nutrients and contaminants enter the river from the uplands. In many regions these exchanges are episodic and driven by extreme precipitation associated with tropical storms, convective storms, or other meteorological patterns that can destabilize hillslopes and cause landslides or debris flows [*Korup et al.*, 2004; *Hilton et al.*, 2008a, 2008b].

Longitudinal exchanges occur as water, sediment [*Milliman and Syvitski*, 1992], nutrients [*McClain and Naiman*, 2008], contaminants and organisms move downstream into lakes and oceans, and as organisms move upstream [*Zhang et al.*, 2003b]. In mountain rivers, these exchanges can be mediated by longitudinal disconnectivity associated with geomorphic features such as knickpoints and gorges.

At smaller spatial scales, exchanges also occur across the four basic dimensions of vertical, lateral, longitudinal, and temporal within the river channel [*Ward*, 1989]. Vertical exchange within the water column can take the form of fine particulate matter stored in the interstices of the channel-bed gravel that is brought into suspension during a flood. Lateral exchange occurs when leaf litter stored along a channel-margin eddy is flushed into the main channel during a high discharge, or when a tributary enters the main channel.

Ecologists recognize basin- and landscape-scale hydrologic connectivity as a fundamental property of all rivers [*Pringle*, 2001, 2003; *Ward et al.*, 2002]. *Hydrologic connectivity* in rivers can be defined as the water-mediated fluxes of material, energy, and organisms within and among components of the ecosystem, including channel, floodplain and alluvial aquifer [*Kondolf et al.*, 2006]. Alteration of headwater streams through land uses such as flow regulation and channelization modifies hydrologic connectivity between uplands and channel networks and between headwater and downstream river segments, leading to lower secondary productivity of river systems, reduced viability of freshwater biota, and downstream eutrophication and coastal hypoxia [*Freeman et al.*, 2007]. Temporal and spatial dimensions of connectivity are crucial. Although connectivity is viewed as beneficial [*McGinness et al.*, 2002; *Jenkins and Boulton*, 2003; *Tetzlaff et al.*, 2007a], artificially enhanced connectivity can create problems, as when water diversions facilitate migration of exotic or invasive species. The most common human-induced alteration of river ecosystems, however, is a reduction in connectivity. Levees reduce lateral connectivity, dams reduce longitudinal connectivity, and siltation or channelization reduces vertical connectivity. Conceptualizing the form and magnitude of human-induced reductions in connectivity is critical to restoring connectivity and ecosystem function [*Kondolf et al.*, 2006].

In summary, river ecosystems are connected to the adjacent landscape and to geographically distant regions of the planet and its atmosphere at multiple levels. These diverse forms of connectivity are critical to maintaining the diverse, resilient river ecosystems, yet one of the primary human alterations of rivers is to change (typically reduce) connectivity in ways that reduce biodiversity and abundance.

5.1.2. The Physical and Chemical Environment of a River Ecosystem

The parameters of a river's physical and chemical environment most relevant to aquatic and riparian organisms include (i) hydrology (magnitude, duration, timing, and rate of change in discharge) [*Poff et al.*, 1997; *Rader and Belish*, 1999; *Doyle et al.*, 2005; *Rice et al.*, 2010], (ii) hydraulics (distribution and magnitude of velocity, turbulence, shear stress) [*Wellnitz et al.*, 2001], (iii) sediment supply and transport (bed stability, turbidity) [*Yarnell et al.*, 2006; *Harvey and Clifford*, 2010], (iv) substrate (grain size distribution, spatial heterogeneity of habitat patches and substrate types, stability) [*Johnson et al.*, 1998], (v) water chemistry (temperature, oxygen content, clarity, total dissolved solids, nutrient availability) [*Hopkins et al.*, 1989; *Clements et al.*, 2000], and (vi) characteristics of the adjacent riparian zone [*Stoddard and Hayes*, 2005]. These environmental factors partly govern the abundance and diversity of aquatic organisms [*Ricklefs*, 1987; *Swanson et al.*, 1988; *Allan*, 1995], along with biological factors such as competition, predation, and evolutionary history [*Lepori and Hjerdt*, 2006]. In some cases, the small-scale variations in these factors best explain the organisms present in a stream community; in other cases, a watershed-scale assessment provides more insight into biotic patterns [*Moyle and Randall*, 1998]. The spatial scale of correlations between organisms and environmental parameters depends on how specific organisms and environmental processes integrate across different scales [*Fausch et al.*, 2002; *Wiens*, 2002]. Physical organizational concepts, such as *hydraulic geometry*, can also be useful in conceptualizing spatial patterns among aquatic and riparian organisms [*Lamouroux*, 2008]. River restoration typically attempts to restore habitat diversity in streams, but is not likely to be effective unless it reproduces the specific aspects of stream structural heterogeneity that are most relevant to the target organisms [*Lepori et al.*, 2005].

Regardless of spatial scale, physical and chemical parameters vary through time and these variations define a *natural range of variability* that is commonly used as a target for restoration following human-induced disturbance. Because some of the parameters vary over decades to centuries, long-term records developed from proxy indicators (e.g., tree-ring records of wildfire and streamflow; *Hupp et al.*, 1988; *Woodhouse et al.*, 2006) are critical to accurately defining the natural range of variability in stream ecosystems [*Willis and Birks*, 2006]. Understanding how a river ecosystem responds to different scales of variation also involves quantifying disturbance thresholds. *Sousa* [1984] defines a *disturbance* as 'a discrete, punctuated killing, displacement, or damaging of one or more individuals (or colonies) that directly or indirectly creates an opportunity for new individuals (or colonies) to become established' (p. 356). An agent of disturbance can be physical or biological, and the regime of disturbance is typically described in terms of areal extent, magnitude, frequency, predictability, and turnover rate [*Sousa*, 1984].

When a river is in dynamic equilibrium, with conditions varying around long-term averages of flow, water quality, sediment transport, and channel morphology, biological communities evolve adaptations to the "normal" range of disturbance (Figure 5.3). Intensified disturbances associated with natural or human influences can stress ecosystem functions and biological communities to the point that a *disturbance threshold* is crossed and even slight changes in disturbance levels can cause major response in biological communities [*Groffman et al.*, 2006]. *Brenden et al.* [2008] review different approaches to quantifying disturbance thresholds in river ecosystems and *Nakamura et al.* [2008] review the role of various types of disturbance in creating habitats at different spatial scales.

Some of the physical and chemical parameters described above represent a one-way relationship in which organisms respond to physical and chemical parameters, but in many cases the relationship is reciprocal, with organisms both responding to and influencing the physical and chemical parameters. This is the focus of *biomorphodynamics*, which investigates the effects of organisms on physical processes and morphology, as well as how the biological processes depend on morphology and physical forcing [*Murray et al.*, 2008; *Reinhardt et al.*, 2010]. The community of aquatic macroinvertebrates present in a stream changes with substrate grain size [*Rice et al.*, 2001; *Cover et al.*, 2008], for example, yet macroinvertebrates such as net-spinning caddisfly larvae also consolidate sand and finer substrate particles and increase the coherent strength of streambed sediment [*Statzner et al.*, 1996; *Takao et al.*, 2006; *Nunokawa et al.*, 2008]. Bioturbation by invertebrates such as crayfish can reduce the percentage of sand and algal cover on streambeds [*Statzner and Peltret*, 2006; *Statzner and Sagnes*, 2008]. Algae and microorganisms can facilitate travertine deposition, which limits sediment mobility and enhances the formation of quasi-permanent step-pool sequences [*Marks et al.*, 2006]. Some species of fish alter streambed grain-size distribution, topography, and stability by creating spawning mounds or depressions [*Maurakis et al.*, 1992; *Sabaj et al.*, 2000] or by moving streambed sediments while feeding [*Statzner and Sagnes*, 2008]. Studies of salmonid habitat in western USA demonstrate that when the fish construct nests by excavating a hollow in the channel bed, finer sediments are selectively mobilized, leading to a coarser particle-size distribution than adjacent gravels not utilized for spawning [*Chapman*, 1988; *Kondolf et al.*, 1989, 1993]. Alteration of the streambed – in the form of bed coarsening, sorting, and increased form drag – by spawning salmon can reduce grain mobility and lessen the probability of bed scour and excavation of buried salmon embryos [*Montgomery et al.*, 1996b].

Seasonal or aperiodic exchanges of water, sediment, and nutrients between the channel and the adjacent floodplain [*Junk et al.*, 1989] and between the channel and the subsurface water of the hyporheic zone [*Stanford and Ward*, 1988; *Valett*, 1993] also strongly influence the physical and chemical conditions and nutrient availability of some channels and hence aquatic and riparian communities [*Boulton and Stanley*, 1995].

Finally, the influence of environmental variables on stream communities can be very persistent. Whole watershed land use during the 1950s is the best predictor of diversity among stream invertebrates and fish in the southern Appalachian Mountains of the USA, rather than contemporary land use [*Harding et al.*, 1998]. Over much longer time scales, genetic divergence among individuals of a fish species can provide insight into long-term drainage evolution and river capture [*Craw et al.*, 2007].

In summary, hydrology, hydraulics, sediment dynamics, substrate, water chemistry, and the adjacent riparian zone exert particularly strong influences on aquatic and riparian organisms. These organisms have adapted to a natural range of variability that includes disturbances such as floods and droughts. Changes in the disturbance regime of a river will likely result in changes in habitat and organisms. Organisms both respond to and influence physical and chemical parameters of a river.

5.2. Aquatic Communities

The energy resources of a river depend on both autochthonous and allochthonous production (Figure 5.4). *Autochthonous production* is primary production occurring within the river channel via photosynthesis by bryophytes (mosses and lichens), attached algae, and floating or rooted angiosperms [*Ward*, 1992]. Periphytic algae attached to the channel substrate and non-vascular plants such as mosses constitute most of the primary production in mountain rivers with streambanks that are not forested. These organisms are adapted to the specifics of substrate, nutrients, current, sunlight, and the presence of grazing organisms. Seasonal fluctuations in the relative importance of different types of runoff can cause corresponding fluctuations in the distribution and abundance of algae and other organisms [*Uehlinger et al.*, 1998; *Ward et al.*, 1998]. Spatial variations in runoff source can also create variations in water chemistry, hydrograph, and stream community in adjacent headwater catchments. *Füreder et al.* [1998] describe differences in aquatic invertebrate communities and trophic structure in glacier- and spring-fed streams in the Austrian Alps. They find that the glacier-fed stream has a lower number of invertebrate taxa and lower densities of insects, presumably because of stronger diurnal and season fluctuations in discharge, suspended sediment, and water chemistry.

A river ecosystem can be substantially altered when the source of autochthonous production changes, as illustrated by the invasive diatom *Didymosphenia geminata*, a single-celled algae commonly known as either Didymo or rock snot. A native of northern Eurasia and cold, low-nutrient waters, Didymo has been extending its range into warmer climates, other continents, and nutrient-rich waters during the past few years. Didymo cells can create large amounts of stalk material that form thick mats of cottony tissue over 20 cm thick on streambeds. Didymo can completely cover the stream substrate, smothering other plants and invertebrates and altering habitat and food for other organisms [*Mundie and Crabtree*, 1997].

Allochthonous production depends on detritus from outside the channel in the form of particulate and dissolved organic matter, much of which is derived from leaf litter and wood. Coarse particulate organic matter is broken down by fungi and by macroinvertebrate shredders. Wood and debris dams exert an important control on the retention of this particulate detritus [*Covich et al.*, 1991; *Bryant et al.*, 2007], as do other obstacles including bedrock outcrops and very large boulders [*Kiffney et al.*, 2000]. Seventy-five percent of the standing stock of coarse particulate organic matter (> 1 mm in diameter) in mountain rivers of New Hampshire, USA is stored by wood in first-order streams [*Bilby and Likens*, 1980]. Flow regime can also exert an important control on organic matter retention [*Quinn et al.*, 2007]; rapidly fluctuating hydrographs help to promote rapid downstream transport and limit processing of organic matter by aquatic organisms [*Merriam et al.*, 2002].

Inputs of fine and coarse allochthonous materials can fluctuate substantially through time and space along a mountain river. Temporal variations reflect short-term disturbances such as hillslope mass movements, wildfire, flooding and bank erosion, insect infestations, or tropical storms [*Vogt et al.*, 1996]. Spatial variations reflect differences in riparian structure and composition, which in turn reflect the history of forest disturbance by the processes that create temporal variations in allochthonous production.

The importance of allochthonous organic matter to aquatic communities is reflected in the presence of organisms. Leaf litter commonly supports a higher abundance and diversity of organisms than do many inorganic substrates. An interesting side note is that engineered wood structures consisting of an interlocking complex of small diameter poles may play a similar role. Comparison of macroinvertebrate abundance, distribution, and taxa diversity between natural and engineered wood and adjacent substrates reveals no statistically significant differences [*O'Neal and Sibley*, 1999].

The *primary consumers* of organic matter are biofilm assemblages, macroinvertebrates, and fish. *Biofilm* is an organic layer that coats solid surfaces. This layer includes attached algae, bacteria, fungi, protozoans and micrometazoans [*Lock et al.*, 1984]. Algal communities are controlled mostly by stream flow and light [*Gustina and Hoffmann*, 2000], although they can also be affected by physical disturbance (e.g., scouring) of the streambed or by contaminants such as heavy metals [*Leland and Carter*, 1985]. Dissolved organic matter, much of which reaches headwater streams via groundwater input [*Wallis et al.*, 1981], is particularly important in supporting bacteria and fungi within the channel.

Macroinvertebrates include aquatic insects, crustaceans, molluscs, leeches, amphipods, nematodes, and so forth. These organisms commonly constitute 98% of the stream biomass. Macroinvertebrates can be classified in terms of the physical niche they occupy within the channel, as benthic (closely associated with the channel bed); floating or swimming organisms residing in the open water; organisms associated with the air-water interface; or hyporheic organisms that inhabit the interstitial spaces between sediment grains below the channel-bed surface and laterally under the banks [*Ward*, 1992].

Benthic macroinvertebrates influence energy flow through stream food webs and nutrient cycling [*Covich et al.*, 1999]. These organisms accelerate detrital decomposition [*Wallace and Webster*, 1996] and release bound nutrients into solution through their feeding, excretion, and burrowing [*Cummins et al.*, 1995]. Benthic macroinvertebrates control the numbers, locations, and sizes of their prey [*Crowl and Covich*, 1990], supply food for both aquatic and terrestrial vertebrate consumers, and accelerate nutrient transfer to adjacent riparian zones [*Naiman and DéCamps*, 1997; *Wallace et al.*, 1997]. *Rabeni and Wallace* [1998] demonstrate the effect of flow regime on benthic macroinvertebrates by quantifying differences in community structure among perennial and intermittent stream segments within a drainage basin.

Although invertebrates use hyporheic habitat, its importance as a refuge during hydrological disturbance varies. Studying invertebrate response to floods of varying magnitude along a low-gradient (0.0015 m/m) sand-bed channel in Virginia, USA, *Palmer et al.* [1992] find that faunal migration into the hyporheic zone may be more important as a refuge along channels with more complex beds that are less susceptible to scour and fill during a flood.

Aquatic insects can also be classified via their feeding strategy as shredders, collector-gatherers, collector-filterers, grazers, and predators [*Allan*, 1995] (Figure 5.5). Shredders feed on non-woody coarse particulate organic matter such as leaves and other plant parts. Collectors feed on finer organic matter that they gather from surface deposits or filter from the water column. Grazers scrape diatoms from the biofilm assemblage. Predators eat other organisms in the stream.

Aquatic insects, which originated from terrestrial insects, appear to have initially invaded freshwater in headwater streams, which are rich in dissolved oxygen. More than 90% of the benthos in high-gradient temperate streams is aquatic insects. These insects also make up more than 80% of the benthos in high-gradient tropical streams. The insect fauna typically differs among large rivers, whereas the taxonomic composition of insects in mountain rivers is similar in being dominated by Ephemeroptera (mayflies), Plecoptera (stoneflies), and Trichoptera (caddisflies). Exceptions to this similarity occur in some glaciated regions such as the Alps, although it is not clear why [*Hynes*, 1970].

All of the types of aquatic insects present in mountain rivers have adapted to the swift current and turbulent flow characteristic of these rivers [*Hynes*, 1970]. Body morphology has adapted via flattening, streamlining, and reducing projections off the body. Holdfasts in the form of suckers, friction pads, hooks and claws, or sticky secretions help the organism to maintain position, as does ballast such as the pebbles in caddisfly cases. Behavioral adaptations include movement into crevices or protected sites of reduced flow velocity [*Rice et al.*, 2008], use of the current for downstream movement, and the creation of silk threads by Simuliidae larvae as "life lines" to regain original position if displaced or disturbed [*Wotton*, 1986; *Ward*, 1992]. Physiologically, these species lack adaptations to survive low oxygen conditions and warm water [*Ward*, 1992]. The number of generations per year and the timing of egg hatching and adult emergence into the atmosphere also reflect adaptations to the temperature and flow regime of a specific river.

Fish in a mountain river may be resident or may be migratory species that spawn in the headwaters after living in lower riverine reaches, lakes, or the ocean [*Ward*, 1992]. Fish may also be primarily bottom-dwellers or may spend most of their time swimming in open water. Fish may be further classified in terms of their primary food source as detrivores, herbivores, planktivores, insectivores, piscivores, or omnivores. Like aquatic insects, fish in mountain rivers have morphologies suited to life in fast currents (Figure 5.6). Fish are dorsoventrally flattened or have an arched profile with a flat ventrum, have more muscular fins, eyes located more dorsally, gill openings more lateral, and mouths more ventral, reduced swim bladders and decreased buoyancy, and mouth suckers or hydraulic discs near the mouth that may be used for attachment to rocks [*Ward*, 1992].

Drainage basin or stream morphology correlates with fish standing stock [*Lanka et al.*, 1987; *Stewart et al.*, 2005]. Consequently, maps of the distribution of habitat availability for fish and other aquatic species with specific habitat requirements in terms of water chemistry, substrate characteristics, or food supply can be used to predict the location of these species and to guide stream restoration and reintroduction of species [*Hogan and Church*, 1989; *Lunetta et al.*, 1997; *McDowell*, 2001; *Stoddard and Hayes*, 2005; *Wohl et al.*, 2007]. *Marcus et al.* [2003] describe the use of hyperspectral imagery for remote mapping of stream habitats. *McKenney* [2001] notes, however, that habitat diversity can change at much shorter timescales than other aspects of channel geometry in response to small to moderate floods.

In most situations, multiple factors interact to control habitat availability [*Sullivan et al.*, 1987; *Brown and Pasternack*, 2008]. Hydraulic roughness exerts a dominant control on distribution of spawning habitat for salmonids in the U.S. Pacific Northwest, for example, and hydraulic roughness in turn reflects flow depth, grain size, bedforms, and wood [*Buffington et al.*, 2004]. Habitat availability also does not necessarily correspond to presence of a particular species, because of regional-scale controls such as connectivity [*Hitt and Angermeier*, 2008] and biological interactions of competition and predation. Despite these multiple controls on fish abundance, some investigators demonstrate correlations between habitat suitability criteria and fish biomass [*Stewart et al.*, 2005]. Simple models of habitat suitability such PHABSIM (Physical Habitat Simulation) are used to predict the change in available habitat resulting from different land-use or restoration scenarios [*Gore*, 2001].

Fish tend to receive more attention than other aquatic organisms in terms of river restoration [*Schmutz et al.*, 2002], partly because fish are more likely to be listed as threatened or endangered than are aquatic insect species and partly because fish communities support recreational and commercial fisheries. The greater likelihood of listing reflects greater awareness of fish than of other organisms by most people, and the fact that fish integrate many of the negative alterations to stream ecosystems by accumulating contaminants from diverse food sources or by fragmenting into isolated subpopulations as longitudinal connectivity is limited by dams or diversions. This integrative role of fish makes it particularly alarming that, as of 2008, 39% of fish species in North America are imperiled. Of 700 listed fish taxa on the continent, 230 are vulnerable, 190 are threatened, and 280 are endangered, while 61 are already extinct [*Jelks et al.*, 2008]. Habitat degradation and introduced species are the primary threats to fish species at risk, many of which are restricted to small geographic ranges.

As discussed in the next chapter, mountain rivers in North America have been highly impacted by the activities of beaver (*Castor canadensis*). These large rodents build dams of wood and sediment which affect water and sediment flow along a river and across the valley bottom and greatly increase habitat diversity for other organisms [*Wright*, 1999; *Westbrook et al.*, 2010]. Beaver historically colonized North American rivers from the arctic tundra to the deserts of northern Mexico and the Atlantic to the Pacific coasts [*Naiman et al.*, 1988]. The European species of beaver (*Castor fiber*) was also widespread throughout Europe [*Gurnell*, 1998].

In summary, energy resources in a river come from autochthonous production occurring within the channel via photosynthesis and from allochthonous production via particulate and dissolved organic matter from outside the channel. The primary consumers of organic matter are biofilm assemblages of algae and bacteria, as well as macroinvertebrates and fish. Macroinvertebrates, which commonly constitute 98% of the stream biomass, include aquatic insects, crustaceans, molluscs, leeches, amphipods, and nematodes. These organisms are classified as benthic, floating/swimming, or hyporheic organisms. Aquatic insects can also be classified via their feeding strategy as shredders, collector-gatherers, collector-filterers, grazers, and predators. Because physical stream and valley characteristics can correlate with fish standing stocks, maps of habitat distribution can be used to predict fish distribution and to guide stream restoration and fish reintroduction.

5.3. Riparian Communities

5.3.1. Basic Characteristics

Adjacent to stream channels lie the zones of riparian (streamside) and floodplain vegetation. The actual species growing within these zones vary widely among rivers, but there is a general increase in tolerance for flooding, high water tables, and disturbance situated along a gradient from the floodplain to the channel. Species growing immediately adjacent to the channel tend to be herbaceous vegetation and shrubs, with small trees farther back from the channel and larger trees in the floodplain forest [*Hupp*, 1988], although many mountain channels are densely lined with woody species and have limited floodplains. In arid and semiarid regions, the floodplain vegetation may give way to upland species that are more tolerant of dry soils.

Riparian communities are arrayed in space and time along gradients in longitudinal, lateral and vertical dimensions [*Naiman et al.*, 2005]. At the most fundamental level, riparian vegetation is structured by antecedent geomorphic conditions, flood dynamics and animal activities, but the details of specific controls vary with climate and channel type [*Osterkamp and Hupp*, 1984; *Harris*, 1988; *Hupp*, 1988; *Hupp and Osterkamp*, 1996]. In humid regions, these patterns may be largely controlled by the frequency, duration, and intensity of floods [*Hack and Goodlett*, 1960; *Sigafoos*, 1961; *Walker et al.*, 1986; *Kalliola and Puhakka*, 1988]. *Harris* [1987], for example, finds that flood disturbance is the major environmental control on the spatial distribution of riparian communities along a sinuous alluvial channel in north-central California, USA. Vegetation patterns in arid and semiarid regions may be most closely related to patterns of water availability [*Auble et al.*, 1994; *Tabacchi et al.*, 1996], particularly minimum and maximum flows and soil moisture [*Shaw and Cooper*, 2008], although mechanical disturbances associated with flooding may also exert a strong control [*Baker and Walford*, 1995; *Parsons et al.*, 2005]. Species richness increases from the channel bed to less frequently disturbed sites [*Hupp and Rinaldi*, 2007]. *Camporeale and Ridolfi* [2006] and *Perona et al.* [2009] develop stochastic models of some of the interactions between hydrologic variability and patterns of riparian vegetation. *Muneepeerakul et al.* [2008] suggest that river networks display functional diversity signatures in riparian vegetation that reflect landscape structure and dispersal directionality.

Riparian vegetation is typically a mixed assemblage of obligate and facultative phreatophytes. *Obligate phreatophytes* send roots into or below the capillary fringe to use shallow groundwater, whereas *facultative species* can also survive where groundwater is not available [*Naiman et al.*, 2005]. Most riparian species are deeply rooted plants that obtain water directly from the stream or from groundwater. The reliability of groundwater encourages riparian trees to develop roots predominantly in the capillary fringe and saturated zone rather than throughout the soil profile, especially if precipitation during the growing season is unreliable [*Naiman et al.*, 2005].

Vegetation patterns along laterally mobile alluvial channels may be controlled by fluxes in sediment deposition and erosion [*Nilsson et al.*, 1993]. Channel (in)stability, as reflected in channel narrowing, meandering, and flood deposition, also promotes a variety of spatial and temporal patterns of vegetation establishment [*Scott et al.*, 1996]. During channel narrowing, vegetation becomes established on portions of the bed abandoned by the stream and establishment is associated with a period of low flow lasting one to several years. During channel meandering, vegetation may become established on point bars following moderate or higher peak flows. Following flood deposition, vegetation may become established on flood deposits high above the channel bed [*Scott et al.*, 1996].

Differences in riparian vegetation as a function of channel dynamics can also be illustrated by comparing different meandering rivers. Along the meandering Beatton River in northeastern British Columbia, Canada, rapid sedimentation on surfaces adjacent to the channel favors the establishment of a dense, non-reproducing balsam poplar (*Populus balsamifera*) stand of uniform age [*Nanson and Beach*, 1977]. As the active channel migrates further away and overbank sedimentation decreases (over a period of 50 years), white spruce (*Picea glauca*) colonize the soil beneath the poplar canopy and form a dense, relatively even-aged stand. As the surface reaches 100-150 years in age, the mature poplars die and the spruce develop into a mature, non-reproducing stand that persists until 350-400 years old. A second, less even-aged spruce stand develops on surfaces 500-550 years old, so that bands of differently-aged riparian species develop in regular patterns on the floodplain as a function of channel migration rate [*Nanson and Beach*, 1977]. Similarly, small floods along low-gradient meandering rivers in the southwestern United States facilitate frequent (< 5 yr) episodes of seedling establishment on point bars [*Hughes*, 1994]. On less sinuous rivers in this region, cottonwoods establish in large numbers at infrequent intervals in synchrony with large floods [*Stromberg et al.*, 1993]. Similar patterns appear along other rivers of varying sinuosity (Figure 5.7).

In mountainous regions, climate as a function of elevation may produce fairly regular longitudinal patterns of riparian vegetation from the headwaters to the piedmont region. Working on the western slope of the Colorado Rocky Mountains, USA, for example, *Baker* [1989] consistently finds wetlands at treeline or at the top of the watershed, upper subalpine carrs (shrub-dominated wetlands) or subalpine forests at elevations of 2620 to 3110 m, and montane forests below 2620 m. Modern patterns of plant distribution and fossil records suggest that riparian areas, which typically have cooler, wetter microclimates, may serve as refugia of Pleistocene and older flora. Plant distributions in the southern Rocky Mountains of the western USA indicate that the contemporary mountain flora once extended considerably downslope and eastward, and that woodland flora from the eastern United States extended westward, probably along the major watercourses. Today the mountain flora has retreated to the higher elevations, leaving relicts at scattered points on the Great Plains [*Weber*, 1965; *Axelrod and Raven*, 1985] where shaded canyons or north-facing slopes have microclimates reminiscent of colder, wetter regional climate. Along many mountain rivers, climate may thus influence longitudinal, basin-length patterns of riparian vegetation. *Acker et al.* [1999] find a downstream increase in the complexity of riparian forest stand from upland to low-order to mid-order streamside forests in the western Cascade Range of Oregon, USA. At the scale of the channel reach, vegetation can be primarily governed by the magnitude and frequency of disturbance by debris flows and floods [*White*, 1979; *Harris*, 1987; *Baker*, 1988b; *Malanson and Butler*, 1991; *Walford and Baker*, 1995], as filtered through channel and valley morphology, and as expressed in inundation, mechanical disturbances, substrate characteristics, and water availability.

Riparian vegetation can also be strongly influenced by biotic interactions. Changes in mammalian predator-prey dynamics in Yellowstone National Park, USA provide a fascinating example. Following extirpation of gray wolves (*Canis lupus*) from the region in the early 1900s, elk (*Cervis elaphus*) increasingly browsed riparian vegetation during winter. As elk numbers grew in the absence of predation, formerly extensive riparian willow (*Salix* spp.) communities began to disappear and streams widened, incised, or both [*Beschta and Ripple*, 2006]. Beavers also lost a primary food source, the number of beaver ponds decreased, and willow establishment on the fine sediments infilling abandoned beaver dams declined [*Wolf et al.*, 2007]. With the reintroduction of wolves in the mid-1990s, elk browsing in the riparian zone declined, cottonwood recruitment resumed for the first time in several decades, and willow communities began to spread [*Beschta and Ripple*, 2010].

Although the species present along rivers differ geographically, the underlying patterns of biophysical processes that create patterns of riparian vegetation are relatively predictable. Riparian plants have specific morphological, physiological and reproductive adaptations suiting them for life in high-energy and wet environments; some plants grow on instream wood, for example, establish upon mineral soils in the floodplain, or grow in saturated or flooded soils [*Naiman et al.*, 2005]. Riparian ecologists classify plants into guilds of similar life history strategies [*Naiman et al.*, 1998]: *invaders* such as willows (*Salix* spp.) produce large numbers of wind- and water-disseminated propagules that colonize alluvial substrates; *endurers* resprout after breakage or burial of either stem or roots from floods or after being partially eaten; *resisters* withstand flooding for weeks during growing season and withstand moderate fires or disease epidemics; and *avoiders* such as Sitka spruce (*Picea sitchensis*) lack adaptations to specific disturbance types [*Naiman et al.*, 2005]. Individual plants that germinate in an unfavorable habitat do not survive.

The *dynamic equilibrium model* of riparian vegetation conceptualizes the roles of disturbance and productivity in controlling the diversity of riparian vegetation [*Solbrig*, 1992]. Where potential productivity is low, and the rate of competitive exclusion is low, an increase in disturbance will decrease species richness. Where potential productivity is high, and the rate of competitive exclusion is high, an increase in disturbance will increase species richness [*DéCamps and Tabacchi*, 1994]. In between these extremes in productivity, an increase in disturbance will increase species richness to a maximum at some equilibrium level and then cause a decrease in species richness [*DéCamps and Tabacchi*, 1994].

The types of hydrologic and flow routing models discussed in chapter 3 can also be applied to modeling habitat suitability for riparian vegetation based on river and groundwater levels and channel morphology [*Büchele et al.*, 2006].

In summary, riparian communities are arrayed in space and time along gradients in longitudinal, lateral and vertical dimensions. At the most fundamental level, riparian vegetation is structured by antecedent geomorphic conditions, flood dynamics and animal activities, but the details of specific controls vary with climate and channel type. In dry regions, water availability is crucial, whereas in wetter regions the frequency, duration, and intensity of floods or fluxes in sediment deposition and erosion may exert a stronger influence. In mountains, climate as a function of elevation can strongly influence riparian communities. Although the species present along rivers differ geographically, the underlying patterns of biophysical processes that create patterns of riparian vegetation are relatively predictable.

5.3.2. Interactions Between Riparian Vegetation and Channel Processes

There are several levels of interaction among riparian vegetation and channel processes. As with the examples described for aquatic organisms, most of these interactions are not simple one-way relationships where riparian vegetation reacts to physical processes, or vice-versa. Among the most well documented interactions between riparian vegetation and channels are the progression from a freshly deposited or scoured, unvegetated surface, to a stable, vegetated surface. Vegetation succession occurs in four stages [*Naiman et al.*, 2005]. During *establishment* plants colonize sites; this is a developmental period before growing space is fully occupied. During the *stem exclusion stage* after all available growing space is occupied, species with a competitive advantage in size or growth expand into space utilized by other plants, eliminating them from the community. The new plants are mostly excluded from colonization by intra- and interspecific competition. A predictable vertical sorting of individuals (vertical stratification by height) in even-aged stands occurs during this stage, with some individuals growing faster than others. During the third stage, *understory initiation*, mortality in the overstory initiates understory development via the invasion of shade tolerant herbs, shrubs and trees. These can be the same species as those present during the stand-initiation phase, but now they grow slowly, creating a stand with multiple canopy layers. During the final, *mature stage*, individual trees die as a forest stand ages, opening up canopy space, which can be occupied by advanced regeneration in the understory. This is a period during which trees regenerate and grow without external disturbance.

Channel change and sedimentation may create habitat favorable to germination and growth of riparian vegetation [*Everitt*, 1980]. Seed dispersal of cottonwood species (*Populus* spp.), for example, typically coincides with declining river flows following spring snowmelt and storms, increasing the chance that seeds will be deposited on favorable sites along the channel [*Braatne et al.*, 1996]. The cottonwood seeds move by *hydrochory* [*Danvind and Nilsson*, 1997], or transport by water. *Anemochory* (transport by wind) and transport by animals can also disperse plant seeds. In the case of cottonwoods, the conditions necessary for seedling recruitment occur irregularly at intervals of 5-10 years or longer [*Baker*, 1990; *Stromberg et al.*, 1991, 1993; *Scott et al.*, 1996, 1997]. High flows during the dispersal phase may prevent exposure of recruitment sites until after the seeds have been dispersed or lost their viability [*Braatne et al.*, 1996]. High flows after the dispersal phase may bury or scour newly germinated seedlings. Along mountain rivers that are stable over periods of decades, channel reaches that are wider, shallower and more disturbed (such as tributary junctions, or sites of mass movement or logjams) may provide sites for establishment of seedlings, which then propagate clonally to more stable sites [*Roberts et al.*, 1998].

Deposition of riparian plant propagules along river margins during hydrochory depends on characteristics of the propagules [*Nilsson et al.*, 1991b], as well as the hydrograph and the channel configuration [*Gurnell et al.*, 2004]. More seed deposition occurs during the recessional limb of the hydrograph than during the ascending limb and stepped flows can enhance this effect [*Merritt and Wohl*, 2002]. Irregular channel margins that facilitate flow separation and seed retention in recirculation zones also promote seed deposition along channel margins [*Merritt and Wohl*, 2002].

Patterns of propagule dispersion and germination can be disrupted by dams and other forms of flow regulation. Along the braided Rekifune River in northern Japan, dams attenuate flooding, reduce flood frequency, and limit the development of gravel bars that serve as germination sites for pioneer species [*Nakamura and Shin*, 2001]. These changes cause a gradual shift from pioneer to late successional riparian species. Flow regulation can lower soil water availability in potential riparian recruitment sites [*Williams and Cooper*, 2005]. Dams can also disrupt hydrochory of plant species, particularly those which have propagules with poor floating capacity [*Andersson et al.*, 2000; *Jansson et al.*, 2000b]. Reduced hydrochory downstream from dams creates a seed-shadow that results in lower species richness of riparian vegetation below the dams [*Merritt and Wohl*, 2006] (Figure 5.8). The details of riparian vegetation response to flow regulation reflect the dispersal strategies of the species present, as well as the magnitude and structure of flow alteration [*Nilsson and Jansson*, 1995; *Nilsson et al.*, 1997]. Vegetative diversity is least affected along regulated but unim-pounded river reaches, for example, and more affected along rivers with storage reservoirs that created large fluctuations between seasonal high and low flows [*Jansson et al.*, 2000a].

The amount of time required for the effects of flow regulation to appear in riparian communities depends partly on the rates of recruitment, growth, and senescence of riparian species, and partly on the disturbance regime prior to damming. If the distribution of woody riparian species is "reset" every few decades by a large flood, then riparian communities along a regulated river may not differ from those along a nearby free-flowing river until the dam blocks or attenuates a large flood [*Katz et al.*,

2005]. Restoration of flood pulses may be necessary to maintain diverse riparian communities along regulated rivers [Bovee and Scott, 2002].

Once riparian vegetation becomes established, it typically alters bank resistance to erosion and in-channel and overbank hydraulics, as well as the trophic structure of the stream ecosystem. The roots of riparian vegetation increase soil cohesion [Coppin and Richards, 1990] by increasing the shear strength of the soil via root reinforcement [Gray and Barker, 2004]. Models that quantify these effects rely on parameterizations of root tensile strength and the diameter and spatial density of roots in the stream bank [Gray and Barker, 2004; Pollen et al., 2004; Pollen and Simon, 2005] (Figure 5.9). Riparian vegetation can also decrease bank stability when the weight of the vegetation increases the driving forces acting in the downslope direction [Pollen et al., 2004]. The above-ground portion of riparian vegetation intercepts precipitation and decreases infiltration that can decrease bank strength [Pollen et al., 2004], as well as removing water from the root zone [Dingman, 2001].

Above-ground portions of vegetation also increase hydraulic roughness on stream banks [Kean and Smith, 2004] and floodplains [Griffin and Smith, 2004; Smith, 2004], and may facilitate further sediment deposition and channel change [Graf, 1978; Nadler and Schumm, 1981; Tal et al., 2004]. Field and flume [Bennett, 2004; García et al., 2004] experiments quantify these effects of vegetation on hydraulics and on sediment deposition and these data inform numerical models of these processes [Kean and Smith, 2004; Wu and Wang, 2004; Wu et al., 2005]. The hydraulic effects of vegetation can be quite substantial; moderate to densely vegetated banks along the Rio Puerco, New Mexico, USA resulted in nearly a 40% reduction in perimeter-averaged boundary shear stress and a 20% reduction in boundary shear stress in the center of the channel [Griffin et al., 2005]. Laboratory experiments demonstrate that riparian vegetation can cause a braided channel to self-organize to, and maintain, a single-thread channel [Tal and Paola, 2010].

Several studies attempt to determine whether grass or forest vegetation create narrower and deeper channels [Sweeney, 1992; Stott, 1997; Trimble, 1997b, 2004; Hession et al., 2003; Sweeney et al., 2004; Hession et al., 2003; Allmendinger et al., 2005]. The consensus appears to be that channels with forested banks are wider because wood recruited to channels from the riparian zone tends to promote channel widening [Zimmerman et al., 1967] and because of how differences in grass cover on point bars influence rates of floodplain deposition [Allmendinger et al., 2005]. McBride et al. [2010] demonstrate increased channel width over 40 years in small headwater channels undergoing passive reforestation. Deforestation that causes channel narrowing can also result in fewer macroinvertebrates, less total ecosystem processing of organic matter, and less nitrogen uptake per unit channel length [Sweeney et al., 2004]. Brooks and Brierley [2002] describe channels that exist in mediated equilibrium when channel capacity, hydraulics, and rates of bedload transport, bank erosion, and in-channel deposition are substantially influenced by riparian vegetation and instream wood. Equilibrium conditions prevail as long as the mediating influences of vegetation are present, but substantial channel change occurs when vegetation is removed.

Bank erosion can also isolate the rootmass of a tree, creating a scalloped bank outline that alters near-bank hydraulics and stream habitat [*Rutherfurd and Grove*, 2004] (Figure 5.10). As the elevation of the vegetation surface increases through sediment deposition and flood inundation and scouring become less frequent, the vegetation influences soil development. Different plant communities develop over time [*Walker and Chapin*, 1986; *Kalliola and Puhakka*, 1988; *Malanson and Butler*, 1990; *Binkley et al.*, 1994].

Spatial changes in vegetation can alter downstream hydraulic geometry relationships [*Huang and Nanson*, 1997]. Temporal changes in vegetation type, extent and density along stream banks can drive substantial changes in channel planform and cross-sectional geometry, either alone or in combination with changes in flow regime. Reduced flood peaks and increased base flows, along with large increases in the extent of woody riparian vegetation, have caused many formerly braided channels of the western Great Plains in the USA to become either meandering or anastomosing during the past century [*Williams*, 1978b; *Nadler and Schumm*, 1981; *Johnson*, 1994]. Flume experiments suggest that changes in riparian vegetation can induce a straight channel to become sinuous [*Bennett et al.*, 2002]. Small rivers in arid and semiarid regions may alternate between sparsely vegetated, wide channels created in a period of hours during an infrequent flood, and well-vegetated, narrow channels developed over intervening decades [*Friedman and Lee*, 2002]. It is not yet clear whether changes in riparian vegetation affect channel morphology as strongly in coarse-grained mountain streams as in the finer-grained alluvium forming the channel boundaries in the examples cited above, although limited studies suggest that riparian vegetation is equally important in promoting bank stability in mountain rivers [*David et al.*, 2009] (Figure 5.11). Changes in riparian vegetation that alter wood dynamics in coarse-grained streams can certainly influence channel morphology and dynamics.

Interactions among flow energy, sediment supply, and riparian vegetation that control location and rate of bar or island formation [*Gurnell and Petts*, 2006; *Gilvear et al.*, 2008; *Corenblit et al.*, 2009] or stream width provide examples of riparian vegetation acting as an *ecosystem engineer*, which *Jones et al.* [1994] define as an organism that directly or indirectly modulates the availability of resources to other species by causing physical state changes in biotic or abiotic materials. Beavers are also an ecosystem engineer along rivers.

Riparian vegetation may also influence water quantity and quality in the river. In arid and semiarid regions, transpiration by phreatophytic vegetation can be a source of significant water transfer between the subsurface and the atmosphere [*Young et al.*, 1984]; measured evapotranspiration rates for riparian vegetation in the western United States range from 1 to 279 cm/yr [*Graf*, 1988]. Riparian plants can also increase evapotranspiration fluxes and affect catchment-scale water yields in humid climates [*Link et al.*, 2005]. Uptake of water from the surface and subsurface by riparian plants and transpiration of the water can affect local water table and stream discharge [*Dye et al.*, 2001; *Rowe and Pearce*, 1994], although the magnitude of water uptake and transpiration varies substantially with plant species, season and time of day.

Because riparian zones are hydraulically connected to both uplands and streams, and have an enhanced ability to take up and process nutrients, riparian zones form effective buffers against high levels of dissolved nutrients from uplands or streams [*Naiman et al.*, 2005]. The capacity of a riparian zone to retain dissolved and particulate nutrients such as N, P, Ca and Mg is controlled by hydrologic characteristics including water table depth, water residence time and degree of contact between soil and groundwater, and by biotic processes including plant uptake and denitrification [*Naiman et al.*, 2005].

Several studies document the influence of riparian vegetation on soil and water chemistry. A study of soil development on terraces along the Tanana River, Alaska, USA indicates that younger (< 5 yr), lower alluvial surfaces have soils high in calcium carbonate and sulfate [*Van Cleve et al.*, 1993]. Establishment of riparian vegetation between 5-10 years creates evapotranspiration that controls soil water potential. Newly-formed surface litter layers further help eliminate evaporation and concentration of salts. Soil properties change markedly as continued sedimentation and channel movement reduce inundation frequency and subsurface saturation in the alluvial surface, as vegetation continues to deposit organic matter, and as alder (*Alnus* spp.) fixes nitrogen in the soil [*Van Cleve et al.*, 1993].

An example of the effect of riparian vegetation on stream chemistry comes from a study at an upland site in Wales [*Fiebig et al.*, 1990]. Dissolved organic carbon (DOC) concentrations are consistently higher in riparian soil waters than in the stream, suggesting an average carbon input of 2500 g/C/m^2 per year to the stream, and an important influence on stream productivity. *Fiebig et al.* [1990] interpreted observed patterns to indicate that: (i) soil water DOC reaching the channel via saturated through-flow of the streambank and bed is mobilized within the hyporheic zone, prior to its discharge into the channel; (ii) during higher flows, DOC entering via the unsaturated zone of the streambank can directly influence the stream; and (iii) DOC in soil water further back from the channel can also directly influence the stream, reaching the channel by macropore flow.

Riparian vegetation directly influences aquatic habitat and trophic structure by providing shade, overhead cover, localized bed and bank scour around roots or trunks protruding into the stream, and organic matter in the form of leaves and twigs that fall into the channel. Indirect influences include providing habitat for insects on which aquatic organisms such as fish depend [*Naiman et al.*, 2005]. The structure and functional diversity of fish assemblages in streams of the southern Appalachian Mountains in the USA are lower in streams with less riparian forest cover during the recent past than in streams with a high degree of near-stream forest cover [*Burcher et al.*, 2008].

Because of the many connections between aquatic and riparian communities, any alteration has the potential to ripple through both communities in initially unexpected ways. *Baxter et al.* [2004] describe an example of such changes triggered by invasion of nonnative rainbow trout (*Oncorhynchus mykiss*) in a stream of northern Japan. The rainbow trout usurped terrestrial insects that fell into the stream from the riparian canopy, driving the native Dolly Varden charr (*Salvelinus malma*) to shift their foraging to macroinvertebrate grazers. Reduction in number of grazers led to greater algal biomass, but also reduced the biomass of adult aquatic insects emerging from the stream. Riparian-specialist spiders that preyed on these insects then declined significantly. *Spencer et al.* [1991] discuss how introduced possum shrimp (*Mysis relicta*) preyed on zooplankton populations in the Flathead River-Lake ecosystem in Montana, USA, contributing to the collapse of an important planktivorous fish population. Loss of the formerly abundant fish caused displacement of birds (bald eagles, gulls, mergansers, and others) and mammals (grizzly bears, coyotes, mink, river otter, and others) that had fed on the fish and fish eggs in upstream tributaries within Glacier National Park.

To paraphrase the 19th-century naturalist John Muir, you cannot touch anything in the universe without finding it connected to everything else. This idea is captured more technically in concepts of hydrologic connectivity and in the postulate of *ecohydrology* that a watershed is a super-organism or *ecological macrosystem* in which an action in one local ecosystem generates a reaction in another local ecosystem, through the hydrographical network, which is both a receiving and a transmitting system [*Gouder de Beauregard et al.*, 2002].

In summary, riparian vegetation interacts with channels in numerous ways. The progression from a freshly deposited or scoured, unvegetated surface to a stable, vegetated surface is particularly well documented. Once established, riparian vegetation alters bank resistance to erosion and in-channel and overbank hydraulics, as well as the trophic structure of the stream ecosystem. The roots of riparian vegetation increase soil cohesion by increasing the shear strength of the soil via root reinforcement. The above-ground portion of riparian vegetation intercepts precipitation and decreases infiltration that can decrease bank strength, as well as removing water from the root zone. Above-ground portions of vegetation also increase hydraulic roughness on stream banks and floodplains and may facilitate further sediment deposition and channel change. Interactions among flow energy, sediment supply, and riparian vegetation that control location and rate of bar or island formation or stream width provide examples of riparian vegetation acting as an ecosystem engineer. Riparian vegetation may also influence water quantity and quality in the river via transpiration and uptake and processing of nutrients. Riparian vegetation directly influences aquatic habitat and trophic structure by providing shade, overhead cover, localized bed and bank scour around roots or trunks protruding into the stream, and organic matter in the form of leaves and twigs that fall into the channel.

5.4. Conceptual Models

The *river continuum concept* is an important conceptual model of river ecosystems that emphasizes downstream gradients and is thus very applicable to mountain rivers. The river continuum concept [*Vannote et al.*, 1980] was developed from stable, unperturbed rivers in northern temperate, forested watersheds, but has been applied with at least partial success to rivers around the world [*Johnson et al.*, 1995]. Channels are differentiated on the basis of stream order [*Strahler*, 1957], with first- to third-order being small streams, fourth- to sixth-order being medium-sized rivers, and channels greater than sixth-order considered large rivers [*Vannote et al.*, 1980] (Figure 5.12). The relative importance of the three primary sources of energy for biological production (allochthonous, autochthonous, and transport of organic matter from upstream) varies along the river, with autochthonous sources and transport becoming increasingly important downstream. The proportions of benthic invertebrates in each of the four feeding groups also vary longitudinally as a result of changes in the physical channel characteristics and organic matter sources. Shredders and collectors dominate the small streams and grazers and collectors the medium-sized rivers, whereas collectors are most important in large rivers. Mid-sized rivers are predicted to have the highest biological diversity because they have the widest range of temperature and hydraulic conditions and energy sources (especially benthic algae), although the highest fish biological diversity occurs in the downstream reaches of large basins [*Vannote et al.*, 1980; *Stanford et al.*, 1996]. At least one study indicates that riparian species richness per site is higher along the main channel than along the tributaries, reaching a maximum at intermediate altitudes [*Nilsson et al.*, 1994]. This pattern is attributed to characteristics of mean annual discharge and variety of substrates.

Mountain rivers in tropical regions may differ from the temperate rivers described in the river continuum concept in that: first- and second-order tropical streams receive a continuous (rather than seasonal) supply of forest-produced leaf litter; daily discharge is unpredictable over annual time scales; extremely variable frequencies and magnitudes of rainfall-generated floods remove leaf detritus and scour the channel bed, thereby disrupting life history patterns and predator-prey dynamics; and freshwater vertebrates are low in abundance relative to decapods [*Covich and McDowell*, 1996]. Although regional patterns of community structure are apparent in temperate-zone streams, no such patterns are clearly evident in stream communities across latitude [*Covich*, 1988].

Related to the river continuum concept is the idea that resources are stored periodically in biological reservoirs such as organisms or detritus, rather than flowing continuously downstream [*Elwood et al.*, 1983]. *Nutrient cycling* describes the passage of an atom or element from a phase in which it is a dissolved available nutrient, through its incorporation into living tissue and passage through various links in the food chain, to its eventual release by excretion and decomposition into a form in which it is once again a dissolved available nutrient [*Allan*, 1995]. In terrestrial ecosystems this cycling may occur largely in place, but in rivers transport becomes an important component because the element may be transported downstream as a solute, then incorporated into the biota, and eventually returned to the water column as a solute. Hence, in rivers the phrase *nutrient spiraling* is used to describe the passage of nutrients through the river ecosystem [*Webster and Patten*, 1979]. Small streams tend to conserve or store resources because of the high organic matter retention behind debris dams, whereas large rivers tend to export relatively more organic matter [*Minshall et al.*, 1983] unless there is substantial storage on the floodplain.

Dams along the river [*Ward and Stanford*, 1983] and the local effects of a tributary junction [*Bruns et al.*, 1984; *Osborne and Wiley*, 1992] may disrupt the longitudinal continuity of physical and biological features predicted by the river continuum concept. Rivers in arid or semiarid regions that have sparse riparian vegetation [*Wiley et al.*, 1990; *Davies et al.*, 1994] or large rivers with extensive floodplains [*Sedell et al.*, 1989; *Welcomme et al.*, 1989; *Bayley and Li*, 1992] also may not be adequately described by the river continuum concept. In general, the concept seems to best describe large rivers confined to their channels, but not large floodplain rivers [*Johnson et al.*, 1995].

As an alternative to the continuous downstream gradations of the river continuum concept, *Ward and Stanford* [1983] propose the *serial discontinuity concept* (Figure 5.13), which recognizes the tendency of riverine response variables to reset or to shift in predictable patterns as a consequence of natural or human-induced regulation by on-channel impoundments. The relative impact of impoundments can be evaluated using the discontinuity distance, or the longitudinal shift of a given variable in terms of stream order or other units of distance, and intensity, which is the absolute change in a particular variable as a consequence of regulation [*Stanford et al.*, 1988]. A dam and reservoir that interrupt hydrochory, for example, can create a downstream seed shadow that alters the characteristics of riparian vegetation for some distance downstream from the dam [*Merritt and Wohl*, 2006]. Or a beaver dam and pond can host very different aquatic insect species than free-flowing channel segments up- and downstream [*Rolauffs et al.*, 2001].

Much attention has been given in recent years to large rivers because these rivers are perceived as being the most substantially altered by human activities [*Johnson et al.*, 1995]. One outcome of this attention has been the development of a model known as the *flood-pulse concept*, which includes some of the processes of lateral exchange not accounted for in the river continuum concept. The flood-pulse concept is based on the recognition that floods which inundate at least a portion of the floodplain are critical as a mechanism for providing clear, shallow water in overbank areas for autochthonous production; for fish nursery habitat and feeding; and for exchanging nutrients and organic matter between the channel and floodplain [*Junk et al.*, 1989; *Bayley*, 1991]. The floodplain is an aquatic-terrestrial transition zone that has a gradient of plant species adapted to seasonal gradients of inundation, nutrients, and light [*Junk et al.*, 1989]. As a flood spreads out from the channel across the floodplain, it creates a *moving littoral*, where the littoral consists of the inshore zone from the water's edge to a few meters depth [*Bayley*, 1995]. The moving littoral traverses a floodplain that may be much larger than the area of permanent moving or standing water associated with the channel. As a result, high primary production rates occur during the rising limb of the flood, with high decomposition rates during the flood peak, and nutrient runoff and concentration during the descending limb [*Bayley*, 1995]. As *Mertes* [1997, 2000] documents, some rivers may inundate dry floodplains with water from the main channel, other rivers may inundate fully saturated floodplains that have contributions from local water (groundwater, and waters from the hyporheic zone, from tributary channels and from precipitation). Waters from each source will have different sediment loads, chemistry and nutrients, and the *perirheic zone* that is created by the mixing of river and local water may form an ecotone [*Mertes*, 1997].

Under the flood-pulse concept [*Junk et al.*, 1989], biotic communities should exhibit a dynamic equilibrium with physical features of the flood pulse, such as timing, duration, and rate of rise and fall in water level [*Bayley*, 1995]. Rivers in both tropical and temperate regions have a higher multispecies fish yield per unit mean water area when a natural, predictable flood pulse occurs along the river because the flood pulse creates floodplain lakes that are connected to the main channel [*Bayley*, 1991]. And a regular flooding regime and associated floodplain sedimentation help maintain high biotic diversity in riparian forests [*Salo et al.*, 1986; *DéCamps et al.*, 1988]. The flood-pulse concept is most relevant for lowland rivers with broad floodplains, but may also describe lower gradient reaches of mountain rivers, such as broad glacial troughs, which have more extensive floodplains than confined reaches of mountain rivers (Figure 5.14).

Some of the habitats and processes associated with floodplains may also occur in the slower, shallower flows of secondary channels [*Amoros*, 1991] or in areas of flow separation along a single, confined channel. Working on a braided, gravel-bed river in Quebec, Canada, *Payne and Lapointe* [1997] find that a stream reach with secondary channels offers up to five times more potential habitat for juvenile fish than do single channel reaches. Backwater habitats associated with eddies along confined rivers also provide low-velocity, seasonally warm refuges for juvenile fish, and groundwater enriched in ammonium and dissolved organic carbon that is utilized by riparian plants [*Kaplinski et al.*, 1998]. *Tockner et al.* [2000b] build on observations such as these to extend the flood-pulse concept to *flow pulses*, which they define as fluctuations in surface waters below bankfull. In the braided, gravel-bed Tagliamento River of the Italian Alps, frequent and irregular flow pulses lead to rapid expansions and contractions of the wetted channel area, creating a dynamic and harsh environment for aquatic organisms [*Doering et al.*, 2007].

In addition to lateral and longitudinal variability, geomorphic and ecological characteristics of a river vary temporally in response to disturbance patterns such as floods and droughts [*White*, 1979; *Bender et al.*, 1984; *Forman and Godron*, 1986; *Resh et al.*, 1988; *Wissmar and Swanson*, 1990; *Johnson et al.*, 1998]. *Frissell et al.* [1986] differentiate recovery time and sensitivity to disturbance on the basis of spatial scale: microhabitats (e.g., gravel patches) are very sensitive to disturbance but recover quickly, whereas a watershed has a low sensitivity and long recovery time. A micro-habitat (10^{-1} m), for example, could be disturbed by scour during a small flood, but would likely recover within a few months or a year. A channel reach (10^1 m) might be disturbed by a debris flow, and require 10^1-10^2 years to recover. A watershed (10^3 m) disturbed by a climatic change might require 10^2-10^4 years to recover [*Naiman et al.*, 1992]. Biotic recovery is likely to occur more rapidly in rivers with a history of episodic disturbance because of selection for species within the community that can rapidly recolonize following displacement [*Poff and Ward*, 1990; *Milner*, 1994].

Larger rivers may also be likely to recover more rapidly from a natural disturbance such as a flood because of a greater upstream source of colonizers and a greater availability of different types of food resources [*Gore and Milner*, 1990], as well as the increasing likelihood that not all of the upstream drainage will serve as a contributing area and that erosional and depositional effects from the flood will be spatially discontinuous (see chapter 3). A human-induced disturbance (chemical spill, or dam), however, may affect a large river with stable communities more than a smaller river that has disturbance-tolerant species.

Fisher et al. [1998a] conceptualize the influence of disturbances on nutrient processing in rivers in the form of a *telescoping ecosystem model*. The processing length, or linear distance required to biogeochemically transform organic materials in transport, is specific to each subsystem (surface stream, hyporheic zone, riparian zone, etc.) within the river ecosystem. A disturbance causes the processing length to increase. In this model, the river ecosystem is composed of several connected subsystems which increase and decrease in processing length as a function of time since disturbance, analogous to the cylinders of a telescope.

The ecological role of disturbances along rivers can also be conceptualized in terms of *process domains* [*Montgomery*, 1999] or *process-disturbance corridors* [*Butler*, 2001], each of which has distinct geomorphic and ecological features as a function of the type, magnitude, and frequency of geomorphic processes such as floods or debris flows. Process domains also facilitate conceptualizing rivers as longitudinally and laterally discrete segments, rather than longitudinally continuous gradients. This approach is particularly useful for mountain rivers, which tend to be longitudinally segmented because of factors such as glacial history, tectonics, or lithologic variability. These larger-scale factors create reach-scale variability in valley and channel geometry, which in turn influences disturbance regime, habitat, and biological communities. *Pike* [2008], for example, use gradient analysis to demonstrate that geomorphic features structuring aquatic habitat along mountain rivers in Puerto Rico tend to be patchy at all scales, rather than varying systematically throughout the stream network. They also demonstrate strong correlations between species presence and abundance and geomorphic characteristics including pool size and substrate type.

One of the challenges to integrating our understanding of physical and biological processes in rivers has been determining which physical factors best describe, or correlate with, biotic communities [*Heede and Rinne*, 1990; *Milner et al.*, 1991; *Naiman et al.*, 1992; *Poff*, 1997]. Studying four streams in Europe, the United Kingdom, and the USA, *Statzner and Borchardt* [1994] find that longitudinal patterns of hydraulic shear stress explain far more than 50% of the variability in ecological responses by fish and invertebrates. Flow regime (magnitude, frequency, duration, timing, rate of change) has also been identified as a key variable that governs water quality, energy sources, physical habitat, and biotic interactions and, ultimately, the distribution and abundance of riverine species [*Resh et al.*, 1988; *Karr*, 1991; *Power et al.*, 1995; *Poff et al.*, 1997]. High and low flows that constitute disturbances may be particularly ecologically relevant because they serve as "bottlenecks" that present critical stresses and opportunities for riverine species [*Poff and Ward*, 1989].

Poff [1996] categorizes streams in the USA using ten ecologically-relevant hydrological characteristics (Table 5.1). Cluster analysis using these ten characteristics from 806 sites identifies ten distinctive stream types that *Poff* [1996] hypothesizes will share ecological features. An analogous study indicates that hydrological data for 34 sites in the north-central U.S. can clearly separate two ecologically defined groups of fish assemblages [*Poff and Allan*, 1995]. These studies reflect the importance of hydrological disturbances in structuring aquatic communities.

Frissell et al. [1986] identify spatially nested levels of resolution (Figure 5.15) at which to characterize form and process and *Cupp* [1989] adapts this concept to an actual classification system at the stream segment level for use in the state of Washington, USA. *Poff* [1997] uses a similar spatial hierarchy to describe multi-scale filters, or habitat selective forces, that govern distribution and abundance of species. For example, the watershed-scale filter *seasonality of flow* influences which fish life history strategies succeed, whereas the microhabitat-scale filter *hydraulic stress* influences which fish body morphologies will predominate.

Additional research is needed to understand how habitat features at different scales are functionally linked; which habitat filters most strongly constrain potential keystone species; and how species traits respond to a particular filter [*Poff*, 1997]. *Doyle* [2006] addresses some of these issues with a heuristic model that couples reach-scale channel morphology and hydraulics with trophic dynamics by assuming rules such as increasing predation with flow depth beyond a threshold depth. Most of the proposed approaches to linking organisms and habitat incorporate the assumptions that (i) physical structure can be related to biotic and physical functions in the channel, and (ii) characteristics of the riparian forest reflect in-channel dynamics [*Naiman et al.*, 1992]. Figure 5.16 illustrates two tropical streams in which differences in habitat features result in different biotic communities.

In summary, although the river continuum concept, which emphasizes downstream gradients, continues to exert a strong influence on how we conceptualize river ecosystems, several other models are useful in conceptualizing specific aspects of mountain rivers. Nutrient cycling describes the passage of a dissolved available nutrient, through its incorporation into living tissue and passage through various links in the food chain, to its eventual release by excretion and decomposition into a form in which it is once again a dissolved available nutrient. In rivers, transport becomes an important component because the element may be transported downstream as a solute, then incorporated into the biota, and eventually returned to the water column as a solute. Hence, nutrient spiraling best describes nutrient dynamics in rivers. The telescoping ecosystem model incorporates the influence of disturbances on nutrient processing in rivers. The serial discontinuity concept recognizes the tendency of riverine response variables to reset or to shift in predictable patterns as a consequence of natural or human-induced regulation by on-channel impoundments. The flood-pulse concept emphasizes processes of lateral exchange between the river and floodplain not accounted for in the river continuum concept. Flow pulses focuses on fluctuations in surface waters below bankfull. Process domains and process-disturbance corridors conceptualize the ecological role of disturbances along rivers in a manner especially suited to the longitudinal variations of mountain rivers.

5.5. Biological Stream Classifications

Closely linked to the correlation of physical and biological characteristics are biological classifications for rivers. As with the geomorphic classifications reviewed in chapter 4, biological river classifications are designed to facilitate comparison and generalization. A key aspect of any river classification scheme is whether it emphasizes a downstream zonation of river attributes such that fairly distinct changes may be identified, or a downstream continuum in which attributes change gradually and continuously.

Initial attempts at biological river classification were zonation schemes [*Hawkes*, 1975; *Naiman et al.*, 1992] based on: (i) physical factors such as substrate (rocky, gravel, sand/silt), gradient, temperature, or channel geomorphic units (e.g., *Pennak*, 1971; *Hawkins et al.*, 1993]; (ii) biotic assemblages such as the European scheme of trout-grayling-barbel-bream [*Huet*, 1954], or the biomass, numbers, or presence/absence of macroinvertebrates [*Dodds and Hisaw*, 1925; *Carpenter*, 1928]; or (iii) combined physical, chemical, and biological factors (e.g., *Ilies*, 1961; *Ilies and Botosaneanu*, 1963]. Hynes' [1975] paper "The stream and its valley" highlights the importance of controls beyond the channel on biotic characteristics and led river ecologists to give increased attention to geomorphic ideas of how flow energy dissipation affected downstream attributes. The river continuum concept grew out of this shift in perspective. Although this continuum-based classification is not very accurate at predicting differences between specific sites, it does accurately predict downstream trends and emphasizes that downstream patterns depend on upstream processes [*Allan*, 1995]. The importance of subsurface hydrological processes [*Frissell*, 1999], reach-based in-channel disturbance regime [*Poff and Wohl*, 1999], and hydrological response to watershed-scale disturbance [*Goodman et al.*, 1999] are among other criteria that have been proposed as a basis for fluvial ecosystem classification.

Managers attempting to evaluate or monitor fluvial aquatic ecosystems have used some form of instream habitat unit classification [e.g., *Frissell et al.*, 1986; *Hawkins et al.*, 1993]. This type of classification, however, may be inappropriate for quantifying aquatic habitat or channel morphology when monitoring the response of individual rivers to human activities because (i) the subjectivity of the measure introduces observer bias, (ii) geomorphological and ecological changes may not be manifested as changes in habitat-unit frequency or characteristics, and (iii) classification data are nominal and thus of limited use for statistical analysis [*Poole et al.*, 1997]. Systems also exist for evaluating the functionality and condition of river ecosystems or riparian zones using limited indicators [*Hylander et al.*, 2002] or multiple characteristics [*Platts et al.*, 1987; *Myers*, 1989; *Leonard et al.*, 1992; *Fryirs et al.*, 2008]. *Innis et al.* [2000] review and compare indicators and assessment methods for riparian zones.

5.6. Mountain River Ecosystems

In an excellent summary of mountain river ecology, *Ward* [1992] describes the ecologically-relevant characteristics of mountain rivers (Table 5.2). Added to these is the instability of some mountain rivers relative to lowland river systems. Mountain rivers may be subject to more frequent and diverse disturbances and usually have stronger landform controls [*Wissmar and Swanson*, 1990]. Natural disturbances include forest fire, slope failure, avalanches, floods, tree blowdown, glacial activity, and geomorphic changes in stream channels. These disturbances occur at different time-scales and have different effects spatially, so that each level of *Frissell et al.*'s [1986] hierarchical classification may be influenced by different disturbances. Floods, for example, have been shown to affect the distribution, successional stages, and size of riparian forest patches in rivers of the Cascade Mountains in the northwestern United States [*Wissmar and Swanson*, 1990; *Swanson et al.*, 1992] at the reach scale (Figure 5.15). Floods may destroy and subsequently exclude plants, create new areas for vegetation colonization, and form elevational gradients where plants show varying tolerances to flows and sediment movement [*Menges and Waller*, 1983]. Flood disturbances along mountain rivers are dominated by mechanical damage to stream and riparian habitats. Flood peaks may be brief and high, accompanied by hillslope debris flows and rapid movement of coarse sediment and wood along the channel [*Swanson et al.*, 1998]. Although the flood peak moves along the entire channel, disturbance patterns are very patchy because of local changes in channel and valley geometry and boundary resistance [*Lake*, 2000].

Studies along a montane stream in the Luquillo Experimental Forest of Puerto Rico following Hurricane Hugo in 1989 characterize the effects of disturbance at the watershed to reach scales. Hugo was a moderate intensity storm that produced up to 200 mm of rain and 10-30 year recurrence-interval floods in eastern Puerto Rico [*Scatena and Larsen*, 1991]. The storm produced blowdown of trees in the floodplain forest, greatly increasing the accumulation of wood on the forest floor, and consequently the long-term reservoir of organic mass and nutrients available to floodplain organisms [*Frangi and Lugo*, 1991]. One month after the hurricane, the densities of freshwater shrimp were reduced on average by 50% in the headwater reaches of the channel (apparently from washout), and increased by 80% in the middle reaches of the channel because of unusually abundant food resources in the form of decomposing leaves and algae [*Covich et al.*, 1999]. The combination of moderate flood discharges, wide and deep pools that served as flood refuges, rapid formation of debris dams that retained food resources, and a food web dominated by detrivores and herbivores, appears to have promoted recovery of the benthic community in this forested mountain river following disturbance [*Covich et al.*, 1991]. Five years after the hurricane, however, carbon and nutrient fluxes in the watershed had not recovered to pre-disturbance levels [*Vogt et al.*, 1996].

Landform controls on mountain rivers include narrow valley bottoms and smaller floodplains that reduce the development of riparian vegetation patches [*Pinay et al.*, 1990]. Variable sequences of constrained (e.g., landslides and canyons) and unconstrained channel reaches control erosional and depositional patterns along mountain rivers and hence substrate and disturbance for aquatic and riparian communities [*Harris*, 1988; *Mussetter et al.*, 1999]. Tributary junctions increase downstream habitat heterogeneity [*Benda et al.*, 2003; *Wallis et al.*, 2008]. Slope steepness and soil moisture govern fire occurrence and slope stability [*Wissmar and Swanson*, 1990]. The downstream variability in hydraulics and channel morphology imposed by these landform controls creates downstream variability in habitats and in nutrient availability, and consequently in biotic assemblages.

As described previously, organisms inhabiting mountain rivers have specific morphological, behavioral, and physiological adaptations to the high-velocity and coarse substrate of these rivers [*Ward*, 1992]. Most organisms have morphological adaptations to maintain position in fast water or to avoid the full force of the current, such as attachment to solid surfaces for benthic algae and insects. Aquatic and riparian biota may also have specific responses to the disturbances characteristic of mountain rivers. Species may be *resistant* to change during the disturbance, taking refuge in secondary channels or streambed interstices, for example, and they may be *resilient*, recovering quickly after the disturbance [*Swanson et al.*, 1998]. Post-disturbance biological responses depend on the distribution of disturbance patches and refuges, on species-habitat relations, on dispersal among patches, on biotic interactions such as competition, and on the availability of food resources [*Swanson et al.*, 1998]. As an example of the latter factor, algae at hyporheic upwelling zones recover from a flash-flood disturbance along a desert channel significantly faster than algae at downwelling zones [*Valett et al.*, 1994]. Hyporheic zones of desert streams may also provide a refuge for microbiota during flash floods [*Grimm et al.*, 1991].

A downstream progression along a river in the Colorado Front Range, USA illustrates the structure of mountain river ecosystems [*Ward*, 1992; K. Bestgen, Larval Fish Laboratory, Colorado State University, Fort Collins, pers. comm., 1992]. The biomass of attached flora is greatest in the headwaters (approx. 2500-2800 m elevation), where mosses and lichens are well developed, and at the base of the mountains (below approx. 1700 m elevation), where filamentous algae and angiosperms proliferate. Biomass levels are low at mid-elevation sites (2800-1700 m elevation), where a variety of epilithic microalgae are dominant. Zoobenthic fauna is dominated by insects, which constitute 79 to 99% of all zoobenthic organisms. Zoobenthic abundance is lowest in the headwaters, increases to the mid-elevation sites (approx. 2100 m), levels off, and then increases markedly beyond the mountains until the channel substrate becomes relatively mobile sand, at which point abundance again decreases. Zoobenthic diversity is low in the headwaters, increases steeply downstream, and levels off at the mid-elevation sites (approx. 2000 m).

The fish assemblage is controlled primarily by water temperature. Cold-water species of trout and sculpin dominate the higher elevations. Coolwater species of suckers and minnows appear at the lower mid-elevation sites and warmwater species dominate beyond the mountain front. No more than three and usually only one species exist in a relatively short stream reach (500 m) in the headwaters, whereas up to 15-20 species may be present along such a reach beyond the mountains. This increasing fish species diversity probably results from increasing stream size and habitat complexity, and from the relatively greater number of warmwater fish species.

Riparian vegetation also tends to become more abundant and diverse downstream because of increasing valley-bottom width and habitat complexity. Riparian communities at the higher elevations are dominated by dwarf birch and willow (*Salix* spp.), which give way to alder (*Alnus incana*), river birch (*Betula fontinalis*), larger willows, and narrowleaf cottonwood (*Populus angustifolia*) at intermediate elevations, and finally to box elder (*Acer negundo*) and cottonwood (*Populus* spp.) near the base of the mountains.

Disturbance regime also varies with elevation along the rivers of the Colorado Front Range. The highest elevation, steepest channels may be subject to infrequent but intensive disturbance from debris flows because of close coupling to hillslopes. Proceeding downstream, average channel gradient decreases and debris flows have less influence on channels because of at least slight buffering by wider valley bottoms. Snowmelt floods create seasonally regular, moderate disturbances, so that these second- to fourth-order channels are actually quite stable. Below approximately 2300 m in elevation, channels are subject to flashy, thunderstorm-driven floods that occur irregularly and substantially modify channel configuration. More frequent wildfires and associated debris flows also create disturbances along the channels and instream wood is more mobile than in the smallest channels. Wider channel segments will be more geomorphically buffered from hillslope and channel disturbances than more narrowly confined channel segments. Riparian and aquatic communities reflect these disturbance regimes, as well as elevation-related changes in climate.

5.7. Case Studies of Human Impacts to Mountain River Ecosystems

The impact of human actions on the aquatic and riparian biota of mountain rivers is of great concern in many mountainous areas [*Nakamura et al.*, 2008]. The various types of human impacts on mountain rivers will be discussed more thoroughly in the next chapter, but the two case studies outlined here provide examples of how multiple human impacts within a watershed can affect the biotic communities of mountain rivers.

5.7.1. The Columbia River Basin

Fish of the family Salmoniidae include salmon, trout, and whitefishes. Many of these commercially valuable species in the northwestern United States live a portion of their adult life in the Pacific Ocean, returning to freshwater to spawn. The migration of these fish upstream along the steep rivers of the Cascade Mountains is an amazing feat. Once the fish reach a suitable channel segment, they construct a redd (nest) in gravel substrates [*Everest et al.*, 1987]. Redds are typically located at the tails of pools where water movement through the gravel will be continuous [*Burner*, 1951]. The female selects and excavates the redd site, which is buried by upstream excavations after spawning. Low flows, dams, and weirs may prevent access to upstream spawning sites [*Baxter*, 1977; *Reiser and Bjornn*, 1979; *Goldsmith and Hildyard*, 1984; *Nakano et al.*, 1990; *Stanford and Hauer*, 1992], and high velocity during high flows may prevent the female from constructing a redd. The adult fish die after spawning, and the eggs incubate in the gravel for up to nine months. During this period the eggs are vulnerable to high flows that mobilize the bed gravel and destroy the eggs [*Montgomery et al.*, 1996b], to low flows that expose the bed and dessicate or freeze the eggs, and to silt- and clay-size particles that infiltrate the gravel matrix, reducing the flow of water and oxygen past the eggs and suffocating the embryos.

After hatching, the sac fry (young fish) live in the gravel. They remain vulnerable to reduced concentrations of dissolved oxygen, desiccation, or freezing during low flows [*McNeil*, 1966], and to crushing by bedload movement during high flows [*Reiser and Bjornn*, 1979]. When the young fish emerge from the gravel, they must continually swim against the current to avoid being swept downstream. Shallow stream margins, pool eddies, and other sites of low velocity and protective cover provide ideal sites for newly emerged fish [*Everest and Chapman*, 1972]. The fish then move to summer rearing sites that provide adequate food and living space [*Chapman*, 1966]. Individual fish will choose sites that minimize energy costs of defense against other fish and of maintaining position in a current, while maximizing energy gain and growth through feeding [*Fausch*, 1984]. Juvenile salmonids tend to face into the current, waiting for food items to drift within sight [*Wickham*, 1967; *Fausch*, 1984]. Cover, in the form of instream wood, overhanging vegetation near the water surface, boulders, undercut banks, or deep water, is important to provide protection from predators, competitors, and variations in streamflow [*Platts et al.*, 1983]. During the autumn, fish move to sites with more cover and refuge from high velocity flow for winter rearing, before moving out to sea [*Sullivan et al.*, 1987].

During the mid-1800s, 16 million salmon annually returned to the Columbia River system [*Dietrich*, 1995]. Eighty percent of the 2.5 million fish that return today are hatchery fish, and the numbers of both wild and hatchery fish continue to drop. This precipitous decline in salmonids is attributed to several factors [*Palmer*, 1994]. From the 1930s to the 1970s, 59 major dams were constructed in the 648,000 km^2 Columbia River drainage basin (Figure 5.17), primarily for hydroelectric power generation. Although an estimated $500 million has been spent on hatcheries, and on fish ladders and other structures to improve fish passage across the dams [*Dietrich*, 1995], the dams remain a substantial barrier to salmonid migration. The alteration of flow regime associated with dam operation has also impacted the salmonids, from flooding of spawning and rearing sites by reservoirs to exposure of redds by unnaturally low flows [*Ligon et al.*, 1995].

Overfishing by commercial operators is also implicated in salmonid decline. The commercial catch of chinook salmon peaked at 19.5 million kg in 1883, and then plunged from overfishing [*Dietrich*, 1995]. Native American fishermen noted that the huge salmon runs had disappeared from Priest Rapids on the middle Columbia River by 1905 because of unregulated commercial fishing; this was more than 30 years before the first major dam on the river was completed [*Dietrich*, 1995].

A third major impact on the salmonids of the Columbia River basin is the timber harvest in this region of humid climate, tectonic uplift, high relief, and naturally unstable slopes. Timber harvest exacerbates slope instability, leading to mass movements that greatly increase the sediment load of mountain rivers, causing siltation of spawning gravels and infilling of pools [*Slaney et al.*, 1977; *Sullivan et al.*, 1987; *Kasran*, 1988; *Platts et al.*, 1989; *Overton et al.*, 1993]. The removal of surrounding upland and riparian vegetation reduces the supply of wood to the channel and removes overhanging vegetation, leading to increased water temperatures, reduced nutrient input, and loss of cover [*Heede*, 1991b; *Fausch and Northcote*, 1992; *Richmond and Fausch*, 1995].

The cumulative impact of all of these human activities has been to substantially alter many of the stream channel reaches that provide habitat critical to salmonids. Channels that naturally had a diverse morphology and substrate, with pool-riffle sequences, sand- to boulder-size clasts, and wood, have become more uniform runs or riffles with fewer low velocity areas, less cover and nutrients, and a flow regime that may bear little resemblance to natural precipitation cycles. Such complete alteration of mountain river systems is unfortunately common in developed nations, and is proceeding rapidly in developing countries [*Wohl*, 2006]. Recommended strategies for facilitating salmonid recovery include modification of human-imposed disturbance regimes to create and maintain a range of habitat conditions in space and time within and among watersheds containing salmon (Figure 5.18), and establishment of watershed reserves that contain the best existing habitats and include the most ecologically intact watersheds [*Reeves et al.*, 1995].

5.7.2. The Danube River Basin

The Danube, which drains 817,000 km^2 of central Europe, formally begins at the confluence of the Brigach and Breg rivers in Germany's Black Forest (Figure 5.19). From its confluence the river drops approximately 680 meters vertically to its mouth in the Black Sea. The tributaries that enter lower along the river's course begin in higher, steeper headlands; the Inn River descends from 4,000 m in Switzerland and northern and eastern tributaries of the Danube drain heights of more than 2,400 m in the Carpathian Mountains. These tributaries produce much of the sediment carried by the Danube. Historically, the mainstem Danube and its larger tributaries alternated downstream between braided or anastomosing segments where the river flows across broad basins and narrow gorges where the river flows through mountain ranges.

The diversity of habitats associated with the swift, narrow portions of the Danube and the broad and diverse floodplains helps to support the richest fish fauna of any European river [*Avis et al.*, 2000; *Tockner et al.*, 2000a; *Bloesch*, 2002]. River lamprey (*Lampetra fluviatilis*), salmon (*Salmo salar*), sea trout (*Salmo trutta trutta*), barbel (*Barbus barbus*), dace (*Leuciscus leuciscus*), and chub (*Leuciscus cephalus*) prefer the swifter current and deeper flow of the main channel. Burbot (*Lota lota*) spend some stages of their life history in tributaries or backwater channels connected to the main flow. Smelt (*Osmerus eperlanus*) and flounder (*Platichthys flesus*) spend a portion of their lives in the slowly flowing brackish water of poorly connected secondary channels. Brown bullhead (*Ictalurus nebulosus*) and crucian carp (*Carassius carassius*) spend their entire lives in still waters that have rooted aquatic vegetation.

One hundred species of fish spend at least part of their lives in the waters of the Danube drainage basin and thirty of these species are commercially important [*Bloesch*, 2003]. Approximately 30,000 tons of fish are caught yearly by commercial and sport fishermen, and records of Danube fisheries date back to 335 BC, when Greek traders commercialized the fish of the lower Danube. The Danube is also rich in fish because of its orientation and glacial history. The predominantly east-west alignment of the drainage basin made it a corridor of migration and recolonization both before and after the ice ages as individual fish moved between the Ponto-Caspic and Central Asian fish communities to the east and the alpine regions to the west. The mainstem of the Danube was unglaciated, and served as a refuge during periods when the continental ice sheets advanced. As the ice sheets retreated, fish expanded from this refuge to the rest of Europe.

Fish and other aquatic and riparian communities of the Danube have been substantially altered by river engineering [*Bloesch and Sieber*, 2003]. The three primary forms of river engineering in the basin are sediment control in the mountainous headwater tributaries, dams for hydroelectric power and navigation in the gorge segments, and channelization in the braided segments. The intensity of engineering and other human alterations of the Danube vary with location in the drainage. The upstream Danube countries of western Europe have relatively clean water flowing through highly engineered channels. The downstream countries of eastern Europe have polluted water flowing through more natural channels.

The channel of the Hopfgartnergraben in the Austrian Alps exemplifies sediment control efforts (Figure 5.20). The channel drains 2 km^2 of mountainside above the village of Hopfgarten. Mountain peaks rise 2,000 m above the village of 4,600 people, and along the flanks of those peaks lie weathered rock and soil that in places reach thicknesses of almost 12 m. During intense rainstorms, infiltrating water builds up pressure that can trigger hillslope instability. The rains of 1965-66 destabilized 30,000 m^3 of sediment.

Hillslope and channel protective measures date back to the early 16th century in the Austrian Alps, but systematic government programs for hazard zoning and associated hazard reduction really began during the 1950s-1970s. As population densities and land use in the alpine regions increased during the latter 20th century, mainly in association with tourism, expenditures on engineering structures designed to decrease natural hazards rose into the billions of Euros.

A chronology of mass movements of sediment and engineering responses in the Hopfgartnergraben during the 20th century makes the hazards abundantly clear: debris flows in 1917 and 1921; construction of 23 check dams in the upper catchment in 1930-33; debris flow in 1945; more protective structures built the same year, as well as in 1956 and 1958; largest sediment disaster in 1965; further protection measures in 1966-1971, 1972-74, 1978, 1981, 1991, and 2000. Picking out just one year of protection measures, 1981 – neither the most nor least expensive – required an estimated 1.75 million Euro to protect this one small catchment in a region thick with similarly active mountain sides.

Simultaneous with sediment control, people of the alpine valleys have tried for centuries to restrain the downstream braided river segments within progressively narrower channels. The Drau River, an Austrian tributary of the Danube, illustrates some of the challenges of this form of river engineering [*Habersack and Piégay*, 2008; *Muhar et al.*, 2008]. Narrowing and bank stabilization along the Drau River date to the 1880s. The initial measures were enhanced following large floods in 1930 and 1965-66. These measures largely eliminated the broad tangle of channels that branched and rejoined across much of the valley bottom, storing and transmitting large quantities of gravel- and cobble-sized bed sediment. The control measures worked so well that today the lack of sediment creates problems as the Drau erodes its bed below bridge footings and protective structures along the channel banks. Scientists have restored about 10 of the 70 most impacted kilometers of the Drau since 1991 by widening the channel, allowing bars and islands to form, and allowing the river to occupy former side channels that had been blocked for decades. Results thus far indicate that river-bed erosion has stopped in the restored segments.

In addition to channel instability, channelization has resulted in substantial reduction in floodplain habitat. At the start of the 21st century, 95% of the original floodplain had been lost in the Upper Danube, 75% in the Middle Danube, 72% in the Lower Danube, and 30% in the delta [*Bloesch*, 2003].

River habitat has also been restored along the Drau and other rivers where connectivity has been re-established between the main channel and secondary channels [*Hohensinner et al.*, 2004]. Smaller fish need refuges such as small side arms and backwaters in order to survive and grow larger [*Guti*, 2002]. The slow water, shallow depth, warmer temperature, finer streamed sediment, and dense aquatic vegetation of these refuges are particularly important for the larval and young fish of many species, but it is just these refuges that are lost when complex river channels are simplified to straight, uniform canals. Bank stabilization and narrowing reduce the connectivity of river side-arms and narrow the main channel, eliminating islands and backwaters. Also lost is the ability of the river channel to provide small refuges that buffer the effect of rapid changes in water level from either natural floods or the flow regime associated with hydroelectric power generation [*Hirzinger et al.*, 2004].

As habitat diversity declined in connection with river engineering along the Danube, the distribution of several fish species was drastically reduced and some of these species are now close to extinction [*Aarts et al.*, 2004]. The Upper Danube is dominated by salmonids, the lower portions of the river by cyprinids such as non-native carp originally introduced to Europe from China many centuries ago. Five species of salmonid once existed in the Danube drainage, but now they are mostly either extinct like one type of lake trout (*Salmo schiefermuelleri*), critically endangered like another type of lake trout (*Salmo labrax*) and Danube salmon (*Hucho hucho*), or endangered like the Alpine charr (*Salvelinus umbla*). Austria is the center of the natural distribution of the Danube salmon, for example, yet the fish today occupies only ten percent of its former distribution along 4,500 km of the river basin. The decline of the salmon, which can live for more than fifteen years and attain a weight of 60 kg, is attributed primarily to channelization and hydropower development.

Most of the fish species historically present along the Danube in Austria are still present today, except for the sturgeons that used to migrate more than 2,500 km upstream from the Black Sea. This suggests that restoring habitat could substantially increase fish abundance. The return of fish to restored zones is limited, however, by the presence of hydroelectric power dams. These dams create physical obstacles to upstream-downstream migratory movements by fish and eliminate the flood peaks that maintain the secondary channels. Operation of the dams creates strange flow regimes with surges of discharge during periods of power generation, mostly three times per day, and extremely low water conditions in between. This is so stressful to river organisms that scientists have recorded ninety-five percent decreases in abundance of organisms downstream from power dams. From the German headwaters down to the massive Djerdap dams in Romania, dams built in confined, formerly swift-flowing portions of the river create ponded water that is then taken over by reed beds and other rooted aquatic plants more characteristic of broad, slower-flowing stretches of river [*Lenhardt et al.*, 2004]. Dams also alter spawning habitat and reduce the seasonal peak flows on which many fish rely for spawning cues.

Trends such as these have spurred the restoration of side arms and floodplains along the Danube [*Habersack and Nachtnebel*, 1995; *Tockner et al.*, 1999]. These restoration measures are at present very limited in spatial extent, at least in part because much of the historical floodplain and secondary channels along the Danube and its tributaries are now occupied by towns or agricultural lands. Changes in flow regime associated with dams and hydropower generation also limit the efficacy of the restoration measures. Some scientists propose that large hydropower facilities be replaced with smaller dams that generate only up to ten megawatts of electricity [*Gouder de Beauregard et al.*, 2002]. The argument in favor of these so-called "ecohydro" facilities is that, because they are smaller, they backflood smaller areas upstream and disrupt the natural flow regime of shorter downstream segments (Figure 5.21). Some ecologists, however, argue that there is no basis for the 10 Mw limit, and that numerous ecohydro facilities can be just as disruptive as the larger dams.

Some skeptics dismiss the current attempts in upstream Danube countries to reconnect side channels and floodplains to the main channel as "gardening" because these projects do not address the flow regulation that will ultimately limit habitat viability and colonization of the reconnected areas by plants, insects, and fish. The presence of more than five hundred larger dams and reservoirs with a capacity of > 5 million m^3 of water in the Danube drainage indicates that changes in dam operation must ultimately occur to restore natural function along the Danube. But existing restoration projects represent an important start that can help to demonstrate the positive benefits of river restoration and thus build public support for more widespread and substantial changes in river form and function.

5.8. Summary

Connectivity is a critical component of river ecosystems. Any segment of a mountain river has six degrees of connection with the rest of the world, and maintenance of naturally occurring levels of connectivity is critical to maintaining ecosystem health. Hydrology, hydraulics, substrate, water chemistry, and riparian characteristics, including the spatial and temporal variations in these parameters, all strongly influence river ecosystems. In some cases this is a one-way interaction in which physical and chemical characteristics govern the abundance and diversity of organisms, but in many cases there are feedbacks by which organisms also alter the physical and chemical parameters of the river and riparian zone.

The aquatic and riparian communities of mountain rivers have some distinctive characteristics relative to lowland rivers. Much of the primary production in mountain rivers is allochthonous, with biofilm assemblages and macroinvertebrate shredders and collectors processing the organic matter and then serving as prey for other organisms. Flow regime and the configuration of the channel boundaries exert important controls on the retention and processing of allochthonous organic matter, and the presence of small-scale obstacles to flow such as instream wood or large boulders tends to increase retention. Although fish typically get more attention, invertebrates constitute the vast majority of biomass in mountain rivers. Aquatic insects and fish have behavioral strategies and morphologies that facilitate survival in swift, turbulent flow.

Riparian communities are structured by gradients of water availability and physical disturbance (inundation, bank erosion, channel migration) along the channel, as well as by regional factors such as elevation and associated climate. Riparian vegetation alters bank resistance to erosion as well as overbank hydraulic roughness, and thus can strongly influence rates and processes of channel adjustment. Riparian vegetation can also influence water quantity and quality, and introduce wood to channels.

Aquatic and riparian species along mountain rivers tend to be those that can tolerate natural disturbances such as floods or debris flows. The disturbance regime of a mountain river is partly a function of the spatial scale being considered; microhabitats are subject to more frequent disturbances than are stream segments (Figure 5.15). Depending on the scale of the disturbances, river characteristics from hyporheic zone chemistry and nutrient availability [*Valett et al.*, 1990; *Fisher et al.*, 1998b] to reach-scale riparian vegetation [*Swanson et al.*, 1998] may be affected. A critical component of understanding and predicting the behavior of aquatic and riparian ecosystems is being able to elucidate the mechanisms, including disturbance, that generate patterns at different spatial and temporal scales [*Hildrew and Giller*, 1994; *Raffaelli et al.*, 1994]. The downstream gradients described in the river continuum concept characterize many mountain rivers well, although these gradients and the processes which produce them are increasingly being disrupted by human activities. Models of nutrient spiraling, serial discontinuity, flood pulses, and process domains also help to conceptualize process and pattern in mountain river ecosystems.

Increasing emphasis on reciprocal interactions between living organisms and physical and chemical processes in rivers and other landscape components has given rise to questions such as whether there is a topographic signature of life [*Dietrich and Perron*, 2006]. As *Dietrich et al.* [2003] note, all landscapes follow an equation for the conservation of mass:

$$\frac{\partial z}{\partial t} = U - E - \nabla q_s \tag{5.1}$$

where z is the elevation of the ground surface, t is time, U is uplift rate, E is the incision rate into bedrock, and q_s is the volume flux of stored sediment per unit width. *Dietrich and Perron* [2006] argue that biota influence all three terms on the right-hand side of equation 5.1. *Reinhardt et al.* [2010] summarize the results of a workshop focused on dynamic interactions between living organisms and landscapes, highlighting the state of the art, knowledge gaps, and ways forward for topics related to this theme.

6. MOUNTAIN RIVERS AND HUMANS

Humans inevitably modify any landscape that they occupy. As long as this modification is of low intensity or limited extent or duration, the processes operating along hillslopes or in river channels are unlikely to cross a substantial threshold as a result. Hunter-gatherer societies may start wildfires or cultivate small patches of land, but are less likely than sedentary cultures to alter hillslope water and sediment yield enough to cause a change in channel morphology. Even relatively non-intensive and short-term agriculture can transform landscapes and ecosystems for a very long time [*Briggs et al.*, 2006], however, and the development of sustained, intensive agriculture (crops and grazing) marked a turning point for many societies in terms of impact on the landscape. This development was followed by channel alluviation in regions as diverse as Mesolithic Britain [*Limbrey*, 1983], Neolithic Poland [*Starkel*, 1988], Neolithic China [*Mei-e and Xianmo*, 1994], the 18th century eastern United States [*Chorley et al.*, 1984; *Kearney and Stevenson*, 1991], and contemporary Greece [*Astaras*, 1984]. Although much of this alluviation occurred in the transport and depositional zones of the middle and lower portions of drainage basin [*Hooke and Redmond*, 1992], much of the sediment originated in the uplands and was transported along headwater channels.

At the start of the 21st century, about 10% of the world's population lives in mountainous regions [*Grötzbach and Stadel*, 1997]. While some mountainous regions, such as central Europe, experienced decreasing population densities and gradual recovery of native upland vegetation during the latter 20th century [*Marston et al.*, 2003; *Latocha and Migon*, 2006; *Klimek and Latocha*, 2007], many low-latitude mountainous regions now undergo increasing removal of native vegetation associated with timber harvest and mining. Other mountainous regions experience increasing alteration of geomorphic processes as large-scale movement of sediment associated with land uses such as open-pit mining becomes more extensive [*Hooke*, 1999]. *Hooke* [2000] argues that humans are now the premier geomorphic agent sculpting the landscape globally. A 'mass-action' metric representing the product of mass displaced, distance moved, and mean speed of displacement indicates that technological mass-action exceeds that of all land-based geomorphic systems except rivers [*Haff*, 2010]. *Haff* [2003] proposes that the combination of physical and social forces driving

Mountain Rivers Revisited
Water Resources Monograph 19
Copyright 2010 by the American Geophysical Union.
10.1029/2010WM001044

modern landscape change be known as the Anthropic Force and *Zalasiewicz et al.* [2008] suggest that geologists formally adopt the term Anthropocene to designate the time period since the start of the Industrial Revolution. *Wainwright and Millington* [2010] highlight the absence of human effects in geomorphic models and suggest that agent-based models provide a useful tool for overcoming this gap.

As increasing world population and demand for food continue to bring formerly marginal, upland regions into cultivation, human impacts have spread throughout the world's mountainous regions, exacerbating changes that have already resulted from other human activities. Reviewing the types and relative magnitude of human activities in mountainous regions around the world, *Wohl* [2006] finds that very few mountains have streams not at least moderately affected by land use. The least-affected mountainous regions are those at very high or very low latitudes, although relative scientific ignorance of conditions in low-latitude mountains means that streams in these regions might be more altered than is widely recognized [*Harden*, 2006].

The hydrologic and geomorphic responses of mountain river catchments to land use partly depend on *landscape sensitivity*. *Brunsden and Thornes* [1979] define the sensitivity of a landscape to change as "the likelihood that a given change in the controls of a system will produce a sensible, recognizable and persistent response." Changes in controlling factors must overcome five basic sources of resistance to change in order to produce a landscape response. These are strength resistance, morphological resistance, structural resistance, filter resistance, and system-state resistance [*Brunsden*, 1993]. *Strength resistance* depends on rock properties such as strength and erodibility, which produce a limited range of geometrical outcomes such as joint spacing, orientation, and continuity. *Morphological resistance* depends on the slope, relief, and elevation of a landscape, which vary across space and over time. *Structural resistance* depends on the closeness of sensitive elements to the processes initiating change (e.g., closeness to base level) and the ability of the landscape to transmit the impulse of change. *Filter resistance* is governed by the way in which kinetic energy is transmitted through or absorbed by the landscape. *System-state resistance* is largely a function of history and trend of the system at the time of disturbance. Because of the great variety in these sources of resistance among the world's mountainous regions, mountain-river responses to land-use changes have also varied greatly. *Fryirs et al.* [2009] describe a spatially fragmented pattern of river adjustment in the upper Hunter catchment of Australia, for example, with localized adjustments to human disturbance reflecting the distribution of bedrock channel segments that are resilient to change.

6.1. Types of Impact

Human impacts on rivers can be indirect if they involve changes in water and sediment yield to the channel as a result of changes in hillslope processes. Returning to the concepts of landscape connectivity discussed in the first chapter, the main role of many human impacts on rivers is to either intensify or weaken the connectivity between hillslope and river systems [*Latocha*, 2009]. Human impacts on rivers can also be direct in that the discharge of water or sediment, or channel morphology, is modified by human activities within the channel [*Knighton*, 1984]. The human activities that principally impact mountain rivers are discussed in the following sections. An increase in natural hazards is closely associated in many mountainous regions with changes in hillslope and channel processes resulting from human activities. An activity such as timber harvest, for example, not only increases hillslope water and sediment yields, but can sufficiently destabilize the slopes that debris-flow magnitude and frequency increase, leading to channel aggradation and an increase in out-of-bank flooding [*Wemple et al.*, 2001; *Marston et al.*, 2003].

It is also important to recognize that land use does not necessarily accelerate geomorphic processes occurring on hillslopes and in channels. The celebrated controversy over whether deforestation in the Himalaya exacerbated local hillslope failures and floods, and ultimately caused more flooding, sedimentation and channel change in the Ganges River basin beyond the mountain front, rested on assumptions about land-use effects that were thoroughly grounded in numerous studies elsewhere [*Ives*, 2006]. Himalayan research over a period of decades, however, demonstrates that in this region of high uplift, deforestation has a minor effect on hillslope processes compared to the influences exerted by lithology, structure and tectonics, and slope aspect [*Marston*, 2008]. Other land uses, including road construction, do accelerate slope failure even in the Himalaya [*Marston et al.*, 1998].

6.1.1. Indirect Impacts

1. *Timber harvest and road building*. The cutting of trees and the associated road building can greatly increase hillslope sediment yield over the short term (approximately 1-10 years) and water yield over longer periods (approximately 1-30 years) until vegetation grows back [*Bosch and Hewlett*, 1982; *Haigh et al.*, 1990; *Roberts et al.*, 1994; *Bierman et al.*, 1997; *Constantine et al.*, 2005; *Douglas et al.*, 1999a, 1999b]. Logging in the mountains of Peninsular Malaysia, for example, resulted in 70-90% increases in suspended sediment yield [*Kasran*, 1988] and a 470% increase in water yield [*Nik*, 1988] over a period of 4-6 years. The magnitude of these changes will depend on climate and on hillslope characteristics of substrate type, steepness, length and aspect, as well as on the methods used in timber harvest and the intensity and extent of forest removal and subsequent recovery [*MacDonald et al.*, 2000; *Grip et al.*, 2005; *Douglas*, 1999]. The elevational distribution of timber harvest relative to precipitation type and magnitude also influences the magnitude of the effects on water yield. Studies in the snowmelt-dominated Redfish Creek catchment of Canada indicate that logging in the lower catchment causes little or no change in peak flow because of the small low-elevation snowpack and the timing of snowmelt, whereas logging at higher elevations creates a much greater increase in peak flow [*Whitaker et al.*, 2002; *Schnorbus and Alila*, 2004]. Large-scale deforestation in the tropics can also decrease runoff because of the feedbacks between the vegetation and atmosphere that reduce

precipitation over the cleared area [*Costa*, 2005]. *Drigo* [2005] identifies 'hot spots' of accelerated and widespread deforestation in the three main humid tropical forest regions of South America, Africa, and southeast Asia. *Alila et al.* [2009] caution that the effect of deforestation on flood magnitude and frequency can only be evaluated when floods of similar frequency are paired, rather than pairing floods by meteorological input.

Mineral soils exposed during timber harvest become more susceptible to surface erosion [*Megahan and Kidd*, 1972; *Johnson and Beschta*, 1980]. Compaction reduces infiltration rates and may increase shallow landsliding, overland flow, sheetwash, rilling and channel initiation and gullying [*Montgomery*, 1994a; *Magilligan and Stamp*, 1997; *Reid et al.*, 2010; *Wolter et al.*, 2010]. Reduction in interception and evapotranspiration may cause elevated soil moisture levels that increase the weight of the soil and decrease slope stability [*Megahan and Bohn*, 1989], as well as increasing soil pipeflow and subsurface pore pressures [*Ziemer*, 1992; *Keppeler and Brown*, 1998]. As positive pore water pressures become more likely, landslides may increase [*Swanston and Swanson*, 1976]. Removal of vegetation causes a decline in the tensile strength of root material [*Gray and Megahan*, 1981], leading to lower soil shear strength. Bank erosion and channel headward expansion in cut areas, and windthrow along streams for which a fringe of riparian trees has been left, can also increase sediment yield to the channel [*Lewis*, 1998]. *Price and Leigh* [2006] find an inverse correlation between forest cover and percent of fine sediment on the bed in streams of the southern Blue Ridge Mountains, USA.

Roads constructed in association with timber harvest increase the slope angle at the cut and fill slope and redistribute the weight on hillslopes, thus increasing slope instability [*Sah and Mazari*, 1998; *Fransen et al.*, 2001; *Wemple et al.*, 2001; *Chappell et al.*, 2004, 2009]. Roads alter the downslope flow of water by reducing infiltration and increasing runoff and conveyance via culverts [*Swanston and Swanson*, 1976; *La Marche and Lettenmaier*, 2001] and alter drainage density and network patterns by initiating gullies [*Wemple et al.*, 1996; *Croke and Mockler*, 2001; *Tague and Band*, 2001]. Roads also contribute substantial fine sediment [*Reid and Dunne*, 1984; *Froehlich*, 1995; *MacDonald et al.*, 2001; *Motha et al.*, 2003; *Croke et al.*, 2005; *Sheridan and Noske*, 2007] long after vegetative cover has regrown. Modeling of hillslope hydrology by *Dutton et al.* [2005] supports the widespread hypothesis that a near-surface permeability contrast, caused by surface compaction associated with roads, can result in diverted subsurface flow paths that produce increased up-slope pore pressures and slope failure. *Jones et al.* [2000] summarize the effects of roads on hydrology, geomorphology, and disturbance patterns in stream networks.

Numerous studies quantify the effects of road-related erosion on sediment yield in diverse mountain environments. Studies in the humid-tropical mountains of Puerto Rico indicate that landslide frequency within 85 m of a road is five times the frequency found at greater distances from a road [*Larsen and Parks*, 1997]. In the nearby U.S. Virgin Islands, road erosion since 1950 has caused at least a fourfold increase in island-wide sediment yields [*MacDonald et al.*, 1997]. For the 44,000 km of roads constructed in the mountains of India during a period of 25 years, 40,000-80,000 m³ of debris were moved for each kilometer of road, creating an average 550 m³ of erosion material per kilometer of road [*Valdiya*, 1987]. In the Upper Konto watershed of Indonesia, roads occupy only 3% of the total area, but contribute 15% of the sediment [*Rijskijk and Bruijnzeel*, 1991]. Road construction in the mountainous drainage of North Halawa Valley, Hawaii, USA produced 90% of the fluvial fine-sediment load during the period of construction [*Hill et al.*, 1998]. Sediment yield from roads may be greatly reduced if roadside vegetation is retained and roadside ditches are not cleaned [*Luce and Black*, 1999]. *Madej et al.* [2006] develop two optimization models that can be applied to determining the most cost-effective treatment for reducing sediment yield from roads and stream crossings.

A related effect of forest roads is to restrict the passage of aquatic organisms at sites where roads cross streams and streamflow is routed through culverts. The extent of this problem is significant; national forest lands in the U.S. states of Washington and Oregon, for example, include more than 6,250 road-stream crossings on fish-bearing streams and approximately 90 percent of these crossings create at least partial barriers to anadromous fish passage [*Gubernick et al.*, 2008]. Culverts can create barriers to passage of fish and other aquatic organisms by confining flow and producing high velocities that restrict upstream movement and by creating vertical drops through bed erosion downstream from the culvert which limit upstream swimming, jumping, or crawling (Figure 6.1). Culverts can also enhance channel erosion downstream from the culvert and thus reduce overhead cover or change the bed grain-size distribution, so that organisms such as salamanders are more exposed to predation [*Gubernick et al.*, 2008]. *Gubernick et al.* [2008] provide guidelines for designing stable road crossings that minimize the effects on passage of aquatic organisms. Roads parallel to a channel also reduce lateral connectivity across the valley bottom, particularly in mountainous regions where transportation routes are most likely to follow river corridors along valley bottoms [*Blanton and Marcus*, 2009].

Several mechanisms may act to increase water yields following timber harvest. The removal of vegetation reduces evapotranspiration and interception [*Cline et al.*, 1977] and may increase snowmelt [*Berris and Harr*, 1987]. Additional water stored in the soil may contribute to higher water tables [*Hotta et al.*, 2010] and summer base-flow, and previously intermittent stream reaches and soil pipes may become perennial [*Keppeler*, 1998]. Burning during site preparation may form temporarily hydrophobic soil layers that decrease infiltration [*DeByle*, 1973]. Compaction along roads decreases infiltration and intercepts subsurface flow. Increased water yields may in turn enhance peak flows in the channel [*Ziemer*, 1981; *Harr*, 1986; *Cheng*, 1989].

Numerous studies describe the processes by which altered hillslope water and sediment yields cause change in mountain channel geometry. The combined increase in water and sediment yields following timber harvest commonly results in channel aggradation, bank erosion, loss of pool habitat and wood, and an increased potential for overbank flow and channel change during floods [*Ryan and Grant*, 1991; *Madsen*, 1995]. *Heede* [1991a] documents a 62% increase in the peak discharge of the South

Fork of Thomas Creek, Arizona, USA, a small mountain catchment, during an eight-year period after basal area of timber stand was reduced to 28%. This increase in peak discharge caused channel cross-sectional area to increase by 10% and removed nearly half of the previously existing log and clast steps within the channel. During a 15-year study of the 10 km^2 Carnation Creek watershed in western Canada, bank erosion, channel width, and channel mobility increased within clearcut reaches, but not along uncut channel segments [*Hartman and Scrivener*, 1990]. Increased flow magnitude and frequency following timber harvest in Idaho, USA increase the depth and frequency of streambed scour and likely increase the mortality of endangered salmonid embryos [*Tonina et al.*, 2008]. Forest clearing in the Ruahine Range of New Zealand caused the Tanaki River to increase in width from 5 m to 10 m in the early 1920s, to 54 m by 1942, and to 60 m by 1976 [*Mosley*, 1978]. Channels may subsequently grow narrower if forest cover is replaced by meadow vegetation [*Baillie and Davies*, 2002a]. Timber harvest in the Caspar Creek watershed of California, USA initially caused increased inputs of bed material, although this trend is expected to reverse as existing wood decays and wood inputs from depleted riparian sources decrease [*Lisle and Napolitano*, 1998]. Changes in channel morphology associated with the effects of timber harvest can persist for at least several decades [*Madej and Ozaki*, 2009] and in some cases more than a century after logging ceases [*Napolitano*, 1998] and recruitment of instream wood following clearcutting may not return to pre-cutting levels for more than 200 years [*Bragg et al.*, 2000]. *Cognard-Plancq et al.* [2001] provide an example of modeling the effects of changes in forest cover on streamflow.

In addition to altered channel morphology, the increase of hillslope water yield following timber harvest can cause increased flooding [*Marston et al.*, 2003]. Deforestation, combined with minor climatic variability, caused an increase in the number of damaging floods in Switzerland during the 19th century [*Vischer*, 1989].

Finally, timber harvest can affect stream chemistry. Forest-floor and soil organic matter decline, leading to a net loss of nitrogen from the soil and increased export of inorganic nitrogen in streams [*Dahlgren*, 1998], as demonstrated at the Hubbard Brook Experimental Forest in New Hampshire and at Coweeta, North Carolina, both in the USA [*Bormann and Likens*, 1979; *Nihlgard et al.*, 1994; *Waide et al.*, 1998]. Increased availability and loss of NO_3^- also increase the loss of cations from the ecosystem [*Likens et al.*, 1977]. Solute fluxes of Ca and K required five years to return to pre-logging levels [*Swank*, 1986; *Johnson et al.*, 1994]. The magnitude of these responses is largely determined by soil type, site climate, and harvesting intensity [*Reynolds et al.*, 1995b]. Studies of logging in Malaysian rainforests show similar increases in solutes [*Johnson et al.*, 1994]. Removal of the forest canopy adjacent to streams increases solar heating of surface waters and thus increases maximum summer water temperatures [*Swank and Johnson*, 1994]. Changes in water and sediment yield and shading of the river corridor can increase nutrient loading and algal productivity of stream periphyton and alter substrate distributions that affect macroinvertebrates and fish [*Hauer et al.*, 2007]. *Connolly and Pearson* [2005] review the effects of timber harvest on the ecology of streams in the humid tropics.

Although selective removal of particular trees, rather than clearcutting entire swaths of forest, limits the magnitude of changes in water, sediment, and nutrient yields to streams following timber harvest, such changes do occur even under selective forestry [*Chappell et al.*, 1999, 2005b]. *Cassells and Bruijnzeel* [2005] and *Thang and Chappell* [2005] suggest guidelines for minimizing adverse effects of timber harvest on streams in the humid tropics; most of these suggestions, such as retention of streamside buffer strips, apply to any environment.

Regrowth of forest cover, whether natural or human-induced, can reverse some of these trends and promote increased infiltration and base flow and decreased runoff [*Hugen and Jiaquan*, 1993; *Buttle*, 1996] and sediment yield [*Liébault et al.*, 2002; *Vanacker et al.*, 2007], as well as channel narrowing and deepening [*Liébault and Piégay*, 2001, 2002; *Liébault et al.*, 2002; *Piégay et al.*, 2004], although *Mount et al.* [2005] document increased sediment yields following upland afforestation in Britain. Working in a 26 km^2 watershed in Puerto Rico, *Larsen and Román* [2001] caution that the massive amount of sediment stripped from hillslopes by landslides following deforestation will remain available for subsequent remobilization from colluvial and alluvial storage sites for many decades despite reforestation and other soil conservation measures.

Changes in the magnitude, frequency, and intensity of wildfires through fire suppression or the use of controlled burns typically accompany timber harvest and other forms of land use in forested regions. Wildfires may be a primary form of hillslope disturbance in arid and semiarid regions; several investigators have used stratigraphic records to demonstrate periods of high sediment transport and aggradation during floods and debris flows following fires, followed by long (10^2-10^3 yr) periods of quiescence and relative stability [*Wohl and Pearthree*, 1991; *Meyer et al.*, 1992, 2001; *Elliott and Parker*, 2001]. As with timber harvest, the initial response following wildfire is typically an increase in runoff and sediment yield [*Benavides-Solorio and MacDonald*, 2001; *Moody and Martin*, 2001; *Kunze and Stednick*, 2006; *Larsen and MacDonald*, 2007; *Reneau et al.*, 2007] as formation of a hydrophobic layer and deposition of ash inhibit infiltration [*Woods et al.*, 2007; *Gabet and Sternberg*, 2008], the extent and stability of first- and second-order channels change via the development of rill networks [*Moody and Kinner*, 2006], and debris flows increase in magnitude and frequency [*Pierce and Meyer*, 2008]. If the riparian zone is burned, enhanced bank erosion can result in major channel morphologic changes even in the absence of substantial changes in water and sediment yield [*Eaton et al.*, 2010a,c].

In summary, timber harvest is one of the most widespread and intensive human impacts on mountain rivers around the world. The cutting of trees, along with associated construction of roads and compaction of soil by heavy equipment, increase water and sediment yields to nearby streams over periods ranging from months to decades. These changes typically result in channel aggradation, bank erosion, loss of aquatic habitat diversity and stability, increased overbank flooding and channel change during floods, and increased nutrient export.

2. *Commercial ski resorts.* Although fewer studies focus on the impacts of commercial ski resorts on mountain rivers, these impacts can be severe and extensive. The combined influence of deforestation, road construction, snowmaking, and machine-grading of slopes can increase water and sediment yield to nearby streams [*Keller et al.*, 2004; *Wipf et al.*, 2005], which can then experience bank erosion, bed scour, channel incision, and changes in bed-material size. Streams affected by ski resorts in the White River National Forest of Colorado, USA have a higher percentage of fine sediment, smaller pool residual depth, and a higher percentage of unstable banks in cascade and step-pool streams than in otherwise comparable reference streams [*David et al.*, 2009]. Streams most affected by these changes tend to flow over lithologies that produce less fine (silt, clay) sediment and tend to have less dense riparian vegetation. Additional effects of commercial ski resorts can include degraded water quality and enhanced runoff from paved surfaces, analogous to other forms of urbanization. *Todd et al.* [2003] describe how snowmaking that draws from water bodies contaminated by mining can spread heavy metals into adjacent, uncontaminated streams when the snow melts.

3. *Grazing.* In upland regions of a drainage basin, heavy grazing compacts the soil, reduces infiltration, and increases runoff substantially enough that the runoff regime may be altered from variable source area to Hortonian overland flow [*Trimble and Mendel*, 1995; *Naeth and Chanasyk*, 1996; *De Jong et al.*, 2008; *Descroix et al.*, 2008]. One study in the southeastern US demonstrates up to 80% decreases in infiltration rate as a result of upland grazing [*Holtan and Kirkpatrick*, 1950]. *Hamilton and Bruijnzeel* [1997] note that, worldwide, overgrazing characterizes mountain livestock husbandry, leading to land degradation. Erosion and sediment yield increase by up to 100 times [*Walling and Webb*, 1983; *Bari et al.*, 1993] as vegetation cover is reduced [*Hofmann and Ries*, 1991], fertility and organic matter content decrease [*Trimble and Mendel*, 1995], and soil aggregate stability is reduced by trampling. An aerial study of soil erosion in Nepal found severe erosion associated with overgrazed land immediately above cultivated zones [*Pereira*, 1989]. The effects of increased water and sediment yield on channel morphology, stability, and flooding are as described for timber harvest and road construction. In extreme cases, sediment yields may decline because all erodible material has been removed from hillslopes [*Gale and Haworth*, 2005].

Grazing within the riparian zone decreases channel bank resistance by reducing vegetation [*Platts*, 1981; *Marlow and Pogacnik*, 1985; *Myers and Swanson*, 1996b] and exposing the substrate, and by trampling that directly erodes banks [*Kauffman and Krueger*, 1984; *Trimble and Mendel*, 1995]. Grazing animals may create ramps along steep or wooded channel banks and these bank irregularities enhance turbulence and bank erosion during high flows [*Trimble*, 1994]. Trails along the floodplain may also channel flow and promote erosion during overbank flows [*Cooke and Reeves*, 1976]. Enhanced bank and overbank erosion can lead to deposition of fine sediments on the channel bed, reducing pool volume and spawning habitat [*Myers and Swanson*, 1996a]. Channels with banks of low cohesivity are most susceptible to these effects

[*Myers and Swanson*, 1992] (Figure 6.2). All of these effects are exacerbated by continuous grazing, but short-duration grazing can reduce adverse impacts to stream corridors [*Magner et al.*, 2008].

Both riverine and upland areas are commonly grazed simultaneously, so that channel erosion is enhanced by increased water yield and bank weakening [*Trimble and Mendel*, 1995]. Channel erosion is most pronounced during flood flows. Associated flood hazards to aquatic communities include siltation along the channel bed [*Myers and Swanson*, 1991] and loss of bank cover by overhanging and riparian vegetation, as well as reduced water quality from fecal material of grazing animals or from water temperature and dissolved oxygen changed because of loss of riparian shading.

Where riparian-zone grazing is limited by exclosures, channel recovery may still be inhibited by a lag time for vegetation re-establishment and sediment deposition or by continued grazing in upland regions [*Kondolf*, 1993]. Removal of grazing animals from riparian zones, or use of grazing exclosures, can reduce bankfull channel dimensions, increase pool area, promote more woody riparian vegetation, and enhance bank stability [*Marston et al.*, 1995; *Magilligan and McDowell*, 1997; *McDowell and Magilligan*, 1997].

In summary, both upland and riparian grazing tend to increase water and sediment yield to channels by reducing infiltration. As described for timber harvest, this can result in channel aggradation, bank erosion, loss of aquatic habitat diversity and stability, and changes in water chemistry.

4. *Crops*. The planting of crops alters the infiltration capacity of the soil and thus the water and sediment yields from hillslopes [*Dunne*, 1979; *Boardman*, 1995, *Kao and Milliman*, 2008], analogous to the changes accompanying timber harvest and grazing. *Ziegler et al.* [2001] identify two mechanisms that increase the potential for Hortonian overland flow in cropped areas, for example; the first is associated with the transition period of 12-18 months after crop fields are abandoned and until sufficient non-domesticated vegetation regrows to intercept rainfall and pond surface water, and the second is associated with compacted footpath or road surfaces. The combined effects of increases in runoff-generating areas, runoff-concentrating features, and connectivity caused extensive gully erosion in agricultural lands of mountainous northern Thailand [*Turkelboom et al.*, 2008]. *Wang et al.* [2008] document faster rates of gully-head erosion in arid agricultural lands of China.

The magnitude of alterations in water and sediment yield will vary with the spatial extent and type of crops [*Boardman*, 1993; *Boardman et al.*, 1996], as well as the soil and topographic characteristics of the site [*Verstraeten and Poesen*, 1999; *Lespez*, 2003]. *Klimek* [1987] correlates enhanced sediment yields, increased flood peaks, and a change from meandering to braided channel pattern following the rapid development of "potato plantations" in the Polish Carpathian Mountains during the latter half of the 19th century. *Rawat and Rawat* [1994] find that agricultural land in the central Himalaya of India has an erosion rate of 0.18 mm/yr, as compared to rates of 0.02-0.04 mm/yr under natural forests. During the month of heaviest rainfall, suspended sediment yields are 7 to 26 times higher in the most disturbed agricultural land than in the forest. The timing of crop cycles may also be important. For example, the pre-monsoon rains in Nepal's Middle Mountains occur when the fields are not yet adequately vegetated (Figure 6.3). As a result, soil loss during this period can be greater than during the heavier monsoon rains [*Carver and Nakarmi*, 1995]. Land drainage associated with crops can either increase or decrease runoff, depending on factors such as soil characteristics, position within the drainage basin, and slope properties [*Newson and Robinson*, 1983].

In general, the most common effect of converting natural vegetation to crops appears to be increases in sediment yield and associated changes in channel pattern and stability [*Astaras*, 1984; *Clark and Wilcock*, 2000; *Kukulak*, 2003; *Ambers et al.*, 2006; *Starkel et al.*, 2006], as well as enhanced sedimentation on alluvial fans [*Jennings et al.*, 2003; *Gomez-Villar et al.*, 2006]. These changes can be documented in sedimentary records extending back into prehistory [*Mei-e and Xianmo*, 1994] and studies of these effects date back decades [e.g., *Gottschalk*, 1945]. Soil erosion under conventional agriculture exceeds rates of soil production and geological erosion rates by from several times to several orders of magnitude, and these patterns cannot be sustained in the long term [*Montgomery*, 2007]. Soil nutrients such as N, P, and K are also typically lost from agricultural lands [*Kothyari et al.*, 2004], causing increases in nutrient flux to rivers [*Wang et al.*, 2004].

Much of the increase in sediment yield resulting from mountain agriculture during the last 100, and particularly the last 50, years has been associated with the migration of people from the lowlands to the mountains in developing nations [*Wiesmann*, 1992; *Hamilton and Bruijnzeel*, 1997]. In contrast to long-term mountain residents who successfully practice soil and water conservation via techniques such as terracing and mulching [*Pereira*, 1989; *Liniger*, 1992], people from the lowlands typically apply inappropriate agricultural methods and then move on to clear new lands when the existing land becomes degraded [*Wiesmann*, 1992]. Terracing can reduce soil loss by 80% in the cultivated mountainous regions of the Philippines [*Daño and Siapno*, 1992] and *Froehlich et al.* [1993] demonstrate analogous effects from other regions. When terraced fields are abandoned and the terraces are no longer maintained, however, these sites can become "major hotspots of erosion" [*Hooke*, 2006] when walls or banks are breached by overflow or ponded water results in large-scale piping [*Inbar and Llerena*, 2000; *Hooke*, 2006].

In contrast to trends of increasing population density in some of the world's mountains, populations have decreased since the mid-20th century throughout the Alps and other mountains of western and central Europe [*Taillefumier and Piégay*, 2003; *Klimek and Latocha*, 2007]. As native vegetation cover gradually returns to these regions, hillslope stability increases and sediment yields to rivers decline, leading to river incision [*Keesstra et al.*, 2005; *Latocha and Migoń*, 2006]. *Coulthard et al.* [2000] describe numerical simulations of how changes in land use affect sediment yields and *Shrestha et al.* [2004] develop a procedure for assessing agricultural land degradation and assessing the magnitude of soil loss.

In summary, the planting of crops, like timber harvest, is a widespread land use that indirectly affects mountain rivers. Clearing native vegetation for crops alters infiltration capacity and thus water and sediment yields in ways that depend on the extent and type of crops and the agricultural methods employed. Adoption of intensive agriculture has resulted in increased sediment yields and channel aggradation and planform change in diverse regions at different times.

5. *Urbanization*. Although human population densities are generally not as high in mountainous regions as in adjacent lowland and piedmont areas, many mountain rivers do pass through large urbanized areas such as Kingston, Jamaica [*Barker and McGregor*, 1988], Quito, Ecuador (Figure 6.4), or Yogyakarta, Java. Increasing residential use of mountainous regions during the 21st century is also likely to increase the impacts of smaller urban centers on mountain rivers [*Hansen et al.*, 2002].

The transition from natural or agricultural vegetation to buildings and roads dramatically effects water and sediment yield from a drainage basin. During the initial phase of construction, vegetation is largely removed and the land surface is leveled or artificially contoured. This results in very high yields as the unconsolidated surface is exposed to rainbeat, sheet flow, and rilling [*Goldman et al.*, 1986; *Gellis*, 1991; *Ruslan*, 1995]. After the completion of construction, when ground surfaces are stabilized beneath roads, buildings, and lawns, sediment yield drops to a negligible value that is commonly lower than pre-urbanization values.

Over the longer term, urbanization results in a substantial increase in water yield as impervious surface area in the basin is greatly increased [*Schick et al.*, 1999]. Urbanization appears to induce greater hydrologic responses than similar proportions of agricultural land cover in watersheds [*Poff et al.*, 2006]. The increase in surface runoff may cause sheetflooding and damage to low-lying structures unless the runoff is quickly channelized and removed from the surface. The installation of storm sewers rapidly drains road surfaces, but also further decreases infiltration and time for water to move from hillslopes and into channels. This results in flood peaks of higher magnitude and shorter duration [*Konrad et al.*, 2005], which can trigger further channel instability [*Trimble*, 1997a] that results in wider, less sinous channels [*Pizzuto et al.*, 2000]. Channels generally enlarge to 2-3 times and as much as 15 times their size prior to urbanization [*Chin*, 2006]. The response of channels to urbanization depends in part on the percent of the catchment that becomes impervious, as well as the connectivity and conveyance of impervious surfaces and the characteristics of the receiving channel [*Bledsoe and Watson*, 2001]. Road crossings can also fragment urban channel networks [*Chin and Gregory*, 2001] and undersized urban structures such as culverts and bridges that cannot convey the large volumes of bedload occasionally mobilized in mountain streams can exacerbate hazards during floods and debris flows [*Anthony and Julian*, 1999]. Traction sand and gravel spread on roads in cold climates can locally overwhelm fluvial transport capacity, resulting in changes in grain-size distribution and bedforms [*Lorch*, 1998].

The classic study of the effects of urbanization on sediment yield and channel response was conducted on channels in the Piedmont region of Maryland, USA

[*Wolman*, 1967; *Wolman and Schick*, 1967]. Increases in sediment yield following timber harvest in the early 1800s and the initial phase of construction in the early 1960s were documented from the sedimentary record, with channel aggradation accompanying the increased sediment yields. Subsequent reductions in sediment yield were associated with channel erosion. Although the detailed history of the Piedmont streams has been updated as a result of recognizing the role of legacy sediments filling closely spaced, abandoned milldams [*Walter and Merritts*, 2008], the fundamental response of initially increased sediment yields followed by decreases in sediment and increases in water yield after urbanization has been supported by numerous subsequent studies from a range of environments [*Miller et al.*, 1971; *Lvovich and Chernogaeva*, 1977; *Park*, 1977a; *Morisawa and LaFlure*, 1979; *Nanson and Young*, 1981; *Harvey et al.*, 1983; *Balamurugan*, 1991; *Urbonas and Benik*, 1995; *Leopold et al.*, 2005; *Smith et al.*, 2005; *Chin*, 2006].

Urbanization can also involve more direct alterations of stream channels through channelization, bank protection, channel realignment, containing all flow within subterranean pipes, or moving channels to new locations [*Grable and Harden*, 2006]. *Gurnell et al.* [2007] review the changes to channels associated with urbanization. *Gregory and Chin* [2002] and *Chin and Gregory* [2005] describe a strategy for categorizing urban channel segments with respect to response to urbanization and hazard management.

Another important aspect of urbanization is the introduction of contaminants into stream channels via runoff and sediment. These contaminants include biodegradable organics that reduce dissolved oxygen in the stream. Contaminants may also take the form of nutrients that cause eutrophication or solids that contribute to turbidity and toxicity. Bacteria that pose a hazard to human health can be contaminants. Contaminants can also be metals and hydrocarbons toxic to a variety of organisms [*Ellis*, 1987]. The treatment or purification of urban runoff varies greatly among cities, from no treatment through primary treatment (an infiltration basin designed to remove primarily solids and bacterial contaminants) to tertiary treatment plants that remove solids, microbes, and hydrocarbons, and use flocculants such as alum to remove colloidal solids including metals. Contaminants reaching the stream channel may be transported in solution or as particulate matter in suspended or bedload. The contaminants may also be stored in bed, bank, and floodplain sediments. Benthic organisms, phytoplankton, fish and other aquatic organisms may accumulate contaminants from any of these reservoirs, with effects as described in the section below on in-channel mining. Airborne contaminants including nitrates can also enter mountainous watersheds from neighboring urbanized regions (see section 3.3.3). *Inyan and Williams* [2001] use maps of landscape types and water-quality parameters to assess sensitivity to acidification and nutrient enrichment and to recommend restrictions for urban development in sensitive high-elevation areas of the San Miguel River basin in Colorado, USA.

In summary, the single greatest effect of urbanization is to increase runoff and flood magnitude, which typically results in dramatic channel erosion. Additional effects include reduced longitudinal and lateral connectivity of river networks and the introduction of a wide array of contaminants.

6. *Upland mining.* Upland mining refers to mining that occurs within the watershed but outside of the channel and floodplain. Upland mining can take the form of lode or hard-rock mining for metals or surface or subsurface excavation of coal. Upland mining can increase sediment yield to channels if the surface is extensively disturbed, as in the case of rock quarrying near the Sungai Relau (river) in Malaysia [*Ismail and Rahaman*, 1994]. Nearly 30 m of weathered rock has to be removed to reach fresh rock in this humid-tropical environment and the disruption of slope vegetation and substrate produces suspended sediment concentrations 1200 times greater than those under natural forest cover. Tailings piles associated with the mining activities may also contribute sediment directly or via hillslope erosion to stream channels [*Vincent and Elliott*, 1999; *Harden*, 2006). Tailings discharge into the Ok Tedi River in Papua New Guinea increase annual suspended sediment load in the river anywhere from 1.1 to 38 times pre-mining levels at different locations along the river [*Markham and Day*, 1994]. Upland mining can also increase peak runoff within a catchment by changing runoff pathways [*McCormick et al.*, 2009].

At sites where mine tailings have been chemically treated to extract ore minerals, or at sites where naturally occurring but toxic metals are being mined, the sediment and runoff from the tailings piles may be a source of toxic contaminants to the river and to groundwater [*Meyer and Watt*, 1999; *Vincent and Elliott*, 1999; *Serban et al.*, 2004; *Segura et al.*, 2006; *Church et al.*, 2007; *Mendoza et al.*, 2008]. Drainage from mine adits and tailings can be very acidic and the bed of the receiving river can become coated with hydrated ferric oxide slimes [*Davies*, 1983; *Church et al.*, 2007]. Acid oxidizing conditions may mobilize other metals from the mining waste and these metals may be transported downstream in solution or complexed and adsorbed onto suspended sediment [*Merefield*, 1995] (see also chapter 3). Metals in solution may continue to be transported from the drainage basin decades or centuries after mining has ceased [*Grimshaw et al.*, 1976]. In the Carson River valley of Nevada, USA, for example, mercury used to process gold in the mid- to late-1800s is now associated with the fine-grained (< 63 µm) sediment fraction of the channel alluvium. The sediment and mercury are preferentially eroded, redeposited, and stored along channel reaches characterized by low gradients and wide valley floors, as occurred during a 1997 flood [*Miller et al.*, 1999]. *Macklin et al.* [2006] outline a geomorphically-based assessment and management scheme for rivers contaminated by metals. *Johnson* [2007] describes application of the groundwater model MODFLOW to understanding subsurface spread of mining contaminants in the Animas River basin of Colorado, USA.

Although upland mining can involve enormous open pits that completely disrupt geomorphic and ecological processes, the greatest impacts from upland mining are likely associated with mountain-top removal, a technique practiced in the eastern United States. The peer-reviewed literature contains little mention of the geomorphic effects of this mining [*Palmer et al.*, 2010], although it has been practiced since the 1970s in the Appalachian Mountains of the eastern United States. Mountain-top removal is concentrated in the states of West Virginia, Kentucky, Virginia, and Tennessee. Material overlying coal seams is removed to vertical thicknesses as great as 300 m, then dumped into adjacent headwater valleys, obliterating surface flow and valley topography. The U.S. Environmental Protection Agency estimates that as of 2010, some 5700 km^2 of terrain were disturbed by mountain-top removal. Although a federal judge has twice ruled that most valley fills are illegal under the Clean Water Act, his rulings have been overturned on technicalities and the practice continues as of 2009.

Surface mining for coal began in the United States in 1866, but large-scale surface mining did not become widespread until the 1940s [*Starnes and Gasper*, 1995]. Although the U.S. Congress imposed national standards regulating mining and reclamation in 1977, the effectiveness of reclamation varies widely [*Toy*, 1984; *Marston and Furin*, 2004]. Physical and biological characteristics of soils can require decades to recover following reclamation [*Mummey et al.*, 2002], which limits establishment of plant communities [*Ashby*, 2006]. Runoff and sediment yield can remain much greater from reclaimed areas than from adjacent undisturbed lands for decades following reclamation [*Lusby and Toy*, 1976], which can lead to gullying and unstable drainage networks [*Soulliere and Toy*, 1986; *Ritter and Gardner*, 1993].

Early reclamation designs used the geomorphic characteristics of adjacent, undisturbed lands, including vegetation cover, drainage density, hillslope gradient, and channel form and stability, as targets for restoration [*Toy and Hadley*, 1989]. Modified forms of the Universal Soil Loss Equation were used to estimate sediment yield for mined, reclaimed, and undisturbed lands [*Toy and Osterkamp*, 1995; *Toy et al.*, 1999]. More recent approaches employ topography designed to allow natural slope evolution within contained areas [*Martín-Duque et al.*, 2010] or use landform evolution models to assess different reclamation designs and the time necessary to achieve landscape stability following reclamation [*Evans*, 2000; *Evans et al.*, 2000].

In summary, upland mining typically results in increased sediment yields and introduction of contaminants to mountain rivers. Extensive topographic alteration, such as occurs during mountain-top removal, can obliterate large portions of mountainous headwater stream networks and completely disrupt longitudinal connectivity for water, sediment, nutrients, and organisms.

7. Climate change. The final indirect form of human impact on mountain rivers is less understood and predictable than those described previously. Both systematic sampling of atmospheric gases during the past forty years and air bubbles trapped in polar ice sheets indicate that atmospheric concentrations of CO_2 and other compounds (CH_4, N_2O, chlorofluorocarbons) have increased since the start of the Industrial Revolution. These so-called greenhouse gases permit a given volume of air to absorb more infrared radiation and contribute to global warming. Continued increases of CO_2 have been projected to cause a rise of 1-6 °C in average global temperature during the 21st century [*U.S. National Academy of Sciences*, 1983; *IPCC*, 2008].

As discussed in chapter 2, the climates of mountain regions are especially complex because of their temporal and spatial variability, and are poorly recorded by existing systematic records. The General Circulation Models (GCMs) used to simulate potential global warming scenarios presently do not have sufficient resolution to adequately address mountain climates. As a result, two approaches have been adopted for projecting climate changes in mountains [*Price and Barry*, 1997]. Statistical down-scaling is used to relate local climate variables to large-scale meteorological predictors [*Von Storch et al.*, 1993; *Gyalistras et al.*, 1997; *Andreasson et al.*, 2003]. Alternatively, a more detailed regional model may be nested within a GCM [*Giorgi et al.*, 1994; *Marinucci et al.*, 1995].

Relating predicted climate change to hydrology is also problematic for mountain regions. In addition to uncertainty of model simulations for regional to local scales, processes of precipitation, soil moisture, and runoff may be difficult to model because of orographic modifications of cloud microphysics and mesoscale weather systems, and interrelated changes in snowpack, snow-line elevation, and the timing of accumulation and ablation seasons [*Price and Barry*, 1997]. Local-scale models (10^1-10^2 km^2) suggest that even a small increase in air temperature may significantly influence the distribution, volume, snow water equivalent, and snowmelt timing of mountain snowpacks [*Keller and Goyette*, 2005; *López-Moreno et al.*, 2009; *Gillan et al.*, 2010], with associated effects on hydrologic and geomorphic processes and mountain biota. General predictions are that precipitation totals will increase in high latitudes and in mid-latitudes during winter. The variability of precipitation and possibly the frequency of extreme precipitation events will increase in the tropics. Snow cover duration will decrease and mid-latitude soil moisture may decrease in summer [*Kattenberg et al.*, 1996].

Climate changes already impact mountain rivers. Fluctuations in flood frequency across Switzerland reflect changes in large-scale atmospheric circulation [*Schmocker-Fackel and Naef*, 2010], and changes in seasonal streamflow patterns associated with earlier snowmelt and the reduction in snow accumulation are reported from the European Alps, Himalaya, and western North America, and are among the most well-documented and already observable impacts of climate change [*Kundzewicz et al.*, 2007]. *Rood et al.* [2005] document decreased streamflows in western North America during the 20th century. Up to 60% of the climate-related trends of decreasing river flow in the arid western United States between 1950 and 1999 are human-induced [*Barnett et al.*, 2008]. Differences in subsurface drainage rates and seasonal distribution of precipitation within the Cascade and Sierra mountain ranges of the USA exert a first-order control on between-watershed differences in sensitivity of streamflow to climate warming [*Tague et al.*, 2008a; *Tague*, 2009].

Climate changes affect all aspects of mountain hydrology and geomorphology [*Goudie*, 2006]. As vegetation and soil-forming processes respond to changes in temperature and precipitation, this causes changes in infiltration capacity, water yield, hydrologic flow paths, stream chemistry, and slope stability [*Onodera et al.*, 2007; *Wang et al.*, 2007; *Christensen et al.*, 2008]. Because temperature or precipitation thresholds influence the magnitude and frequency of mass movements such as debris flows [*Menounos*, 2000; *Jakob and Lambert*, 2009], climate change may substantially alter sediment yields to mountain rivers, as indicated by preliminary evidence from the Himalaya and Tibetan Plateau [*Lu et al.*, 2010]. Changes in air temperature, moisture, and thunderstorms can also alter the frequency and severity of wildfires [*Westerling et al.*, 2006], leading to changes in forest structure and wood recruitment and sediment yield to streams [*Hauer et al.*, 1997]. Sediment transport patterns may change in response to changes in sediment supply and flow characteristics. Changes in the magnitude, frequency, duration, temperature, and predictability of flow play an important role in regulating river ecological processes and patterns [*Poff*, 1992], and aquatic organisms will be affected by climate change. A study of Alaskan streams that assesses the impact of warming-induced increases in glacial runoff, for example, demonstrates that as flow regime, water temperature, and sediment discharge change, corresponding changes occur in channel substrate, bedforms, channel stability, leaf litter quantity and quality, and habitat complexity [*Oswood et al.*, 1992]. Changes in flood magnitude and frequency may trigger channel incision along arid-region rivers, removing the deep hyporheic sediments that support microbial communities, while changes in precipitation and runoff alter the availability of nitrogen, which is a limiting element in these rivers [*Grimm and Fisher*, 1992]. Finally, by altering water availability to human communities, climate change may impact water management and use [*Chang et al.*, 1992].

Warming-induced changes in the snowmelt runoff of the Rocky Mountains in western North America provide an example [*Sridhar and Nayak*, 2010]. Spring and summer snowmelt in mountain rivers of the region typically comprises 50-80% of the total flow [*Hauer et al.*, 1997]. The peak of snowmelt runoff shifted toward earlier in the water year by more than 20 days at some measurement sites during 1948-2000 [*Stewart et al.*, 2004]. Warmer air temperatures during springtime drive these changes and air temperatures are likely to continue to increase, leading to predictions of peak runoff 30-40 days earlier than average timing prior to 1948. Snowmelt-dominated basins at lower elevations in the mountain ranges might change to rain-dominated if cold season temperature increases are sufficiently large [*Stewart et al.*, 2004] and this will create challenges for aquatic and riparian organisms, as well as water managers. Sites with multidecadal climate and streamflow records already show a decrease in the proportion of snow to rain and in the snow water equivalent, as well as a decrease in the length of the snow season, and corresponding shifts in streamflow [*Chang and Jung*, 2010; *Nayak et al.*, 2010]. As storm tracks and the seasonal distribution of precipitation change, rivers in the central and southern Rockies may decline to half of their historical spring runoff volumes, whereas spring runoff may substantially increase in rivers of the northern Rockies [*Byrne et al.*, 1999]. *Tague and Grant* [2009] suggest that geologically-influenced groundwater dynamics will mediate streamflow response to warming. Figure 6.5 illustrates some of the implications in these changes for mountain river ecosystems.

Numerous studies assess the potential impacts of climate change on water resources in mountain regions from the scale of small drainage basins to entire mountain ranges [*Price and Barry*, 1997]. *McCabe and Hay* [1995], for example, assume an increase of 4 °C in mean annual temperature for the 775 km^2 East River basin of Colorado, USA. This results in a 4-5% increase in annual precipitation and larger, earlier runoff peaks for this snowmelt-dominated basin. *Rango and Martinec* [2000] model nearly a doubling of winter runoff and a decrease in summer runoff in humid regions of the northwestern U.S. and southwestern Canada. *Eckhardt and Ulbrich* [2003] model smaller proportions of precipitation occurring as snowfall, which results in increased flood hazards from winter rains and up to 50% decreases in groundwater recharge and streamflow in summer for mountains in central Europe. Analogous studies have been conducted in Australia [*Whetton et al.*, 1996], Europe [*Bultot et al.*, 1992], New Zealand [*Garr and Fitzharris*, 1994], the United States [*Nash and Gleick*, 1991], China [*Wang et al.*, 2001b], Greece [*Mimikou et al.*, 1991; *Kaleris et al.*, 2001], Canada [*Muzik*, 2002], India [*Roy et al.*, 2005], the German Alps (*Kunstmann and Stadler*, 2005], and many other mountainous regions. Regional-scale studies also focus on integrated effects across large drainage basins such as the Upper Danube [*Ludwig et al.*, 2003], which include mountainous regions. Predictions of hydrological changes can also be coupled with morphodynamic models to simulate channel response [*Verhaar et al.*, 2008].

Although the usual assumption is that climate-related changes in streamflow will stress aquatic and riparian ecosystems and human water-supply networks, in cases where human consumptive uses have reduced streamflow and climate change may increase flow, the effects of climate change are anticipated with optimism. *Deng* [2006], for example, discusses how increased discharge can be used for river restoration in the Tarim River of China.

Changes in glacial mass balance will be particularly important in altering the hydrology of many mountain streams. Alpine glaciers have retreated with increasing speed during the past several decades in most of the world's mountains as increasing summer melt overcomes simultaneous lesser increases in winter precipitation [*Ohmura*, 2006]. A current focus of research is to monitor [*Krishna*, 2005; *Lillquist and Walker*, 2006; *Koboltschnig et al.*, 2007a; *Villa et al.*, 2008], simulate [*Mark and Seltzer*, 2005; *Braithwaite et al.*, 2006; *Juen et al.*, 2007; *Xiao et al.*, 2007], and predict [*Schneeberger et al.*, 2003; *Singh and Bengtsson*, 2004, 2005; *Hagg and Braun*, 2005; *Hagg et al.*, 2006; *Rees and Collins*, 2006b; *Fujita et al.*, 2007; *Huss et al.*, 2008] the processes and variables controlling this retreat. As these enormous reserves of freshwater disappear, they influence timing, magnitude [*Han et al.*, 2007], water chemistry, and ecology [*Hodson et al.*, 2008] of rivers fed by meltwater. Once the glaciers are gone – a future that is very close or already here in some mountain drainages – summer peak flows and autumn base flows are likely to be much lower [*Hagg and Braun*, 2005; *Koboltschnig et al.*, 2007b]. Modeling suggests that many of the lower-altitude glaciers in the Andes will completely disappear within the next 10-20 years [*Ramírez rez et al.*, 2001; *Bradley et al.*, 2006], resulting in substantial ecological, social, and economic impacts [*Vergara et al.*, 2007]. Disappearance of Ecuador's Cotacachi glacier has already caused declines in biodiversity, agriculture, tourism [*Rhoades et al.*, 2006]. Other models suggest that summer streamflow in the 150 km^2 Bridge River basin of British Columbia, Canada declines strongly even under steady-climate scenarios, with much larger decreases under warming-climate scenarios [*Stahl et al.*, 2008]. These types of changes will likely be a source of social upheaval in regions such as India, which presently relies on snow- and glacier-melt from headwater tributaries of the Ganges, Indus, and Brahmaputra rivers, obtaining more than half of its water supply from these mountain catchments [*Hasnain*, 2002; *Singh et al.*, 2006]. The threat to water supply posed by glacial melting facilitates collaborative international mapping of glacial retreat in regions such as South Asia, where aerial photographs and discharge data have traditionally been treated as sensitive information not readily available [*Cyranoski*, 2008].

Changes in type and amount of precipitation will likely create non-linear glacial responses to warming climate. The presence of snowpack on a glacier, for example, reduces absorbed solar radiation and melt rate [*Oerlemans and Klok*, 2004] and its removal can increase daily discharge amplitude by >1000% [*Willis et al.*, 2002]. This suggests that upward retreat of rainfall-snow elevation limits can substantially increase rates of glacial melting. Also, mountain glaciers at low latitudes are much more influenced by fluctuations in precipitation associated with circulation patterns such as El Niño-La Niña than are glaciers at higher latitudes [*Leiva et al.*, 2007; *Vuille et al.*, 2008] and are likely to melt more rapidly than high-latitude glaciers [*Ren et al.*, 2007]. Glaciers with cold, dry climates and short melt seasons are less sensitive to air temperature than are glaciers in regions with warm, wet climates and longer melt seasons [*Braithwaite et al.*, 2002]. Non-linearities in glacial retreat may also arise from the effects of sediment cover on stagnating glaciers [*Mihalcea et al.*, 2008].

The melting of mountain glaciers under climatic warming also creates hazards well beyond the mountain environment. Thermal expansion and melting of mountain glaciers are projected to make a greater contribution to sea level rise than are the Greenland and Antarctic ice sheets [*Raper and Braithwaite*, 2006].

In addition to simulating future climate-related changes in mountain hydrology, numerous studies quantify such changes during the latter half of the 20th century, and *Barry* [2005] concludes that historical and recent changes in climate in many mountainous regions are at least comparable to, and locally can be greater than, those observed in adjacent lowlands. The European Greater Alpine region has experienced more extensive summer droughts as a result of warmer air temperatures [*van der Schrier et al.*, 2007]. The altitudinal lower boundary of permafrost in the Tien Shan Mountains of central Asia has shifted upward by 150-200 m during the 20th century [*Marchenko et al.*, 2007]. Permafrost has warmed during the past few decades throughout Alaska [*Osterkamp*, 2005]. Some catchments in the arid Qilian Mountains of China have experienced increased streamflow, whereas others show decreases as seasonality of precipitation changes and air temperatures increase [*Ding et al.*, 2000]. Changes in precipitation, temperature, snow accumulation and vegetation density have led to a marked reduction in water availability in the Spanish Pyrenees, causing water managers to alter dam operations [*Lopez-Moreno et al.*, 2008]. Significant warming and decreased precipitation during 1912-2002 have resulted in widespread glacier recession in northwestern Patagonia [*Masiokas et al.*, 2008]. Glaciers have receded steadily and alpine treeline has advanced to higher elevations in the Rocky Mountains of Montana, USA and southern Canada [*Selkowitz et al.*, 2002]. Ice cover in the high mountains of western North America has decreased by 25-50% and alpine permafrost is thinning and disappearing in many regions [*Clague*, 2008]. Some large, mountainous catchments such as the Mackenzie River basin of Canada seem to have sufficient buffering that observed changes in air temperature have not yet translated to statistically significant changes in volume or timing of streamflow [*Woo and Thorne*, 2003], but this appears to be the exception to observed climate-related changes in mountainous catchments.

In summary, numerous studies clearly indicate that climate change is already affecting mountainous drainage basins around the world, through changes in glacier dynamics, precipitation inputs, and temperature regimes. Changes in temperature and precipitation will in turn alter: water and sediment yield to channels; the magnitude, frequency, duration and timing of streamflow; sediment and nutrient transport; channel dynamics; and aquatic and riparian communities in ways that are highly complex to model and predict. Changes in glacier extent and melt rate and in snowpack and snowmelt are particularly important in many high-elevation and high-latitude mountain catchments.

6.1.2. Direct Impacts

8. *Beaver trapping.* Beavers are herbivorous rodents that build dams and canals along waterways in North America and Europe. Beaver density averages 2-3 colonies per kilometer along streams with suitable habitat of permanent and relatively constant water flow, valley widths of approximately 45 m, channel gradients less than 15%, and aspen or willow growing nearby [*Allen*, 1983; *DeByle*, 1985]. Beaver (*Castor canadensis*) were present historically throughout much of North America. The beaver population prior to the coming of Europeans is estimated at between 60 and 400 million animals, or a range of 6 to 40 animals per kilometer of stream [*Naiman et al.*, 1986]. The European beaver, *Castor fiber*, was historically found from Britain in the west across the whole Eurasian continent, and from the Mediterranean Sea in the south to the tundras in the north [*Hartman*, 1996].

Channels inhabited by beavers have a stepped appearance (Figure 6.6). Water ponds behind the dams of wood and sediment constructed by the beavers, creating reaches of low gradient that are punctuated by abrupt drops of 1-2 m over the dams. The ponds serve to decrease flow velocity, store sediment, reduce channel bed and bank erosion, promote more uniform streamflows during periods of high and low discharge, and diversify aquatic and riparian habitat [*Naiman et al.*, 1988; *Olson and Hubert*, 1994]. Because beaver ponds are important sites for processing of organic matter, channel networks containing beaver ponds have substantially different biogeochemical processes than those without beaver [*Naiman et al.*, 1986]. Beaver-modified portions of channel networks can accumulate $\sim 10^3$ times more nitrogen than other channel segments [*Naiman and Melillo*, 1984]. Beaver dams and ponds can also greatly increase the extent, depth, and duration of inundation associated with overbank floods, as well as greatly elevate the floodplain water table during periods of high and low flow [*Westbrook et al.*, 2006] (Figure 6.7). Because these effects extend downstream, as well as upstream and near the dam, beavers can help to create and maintain hydrologic regimes that support the formation and maintenance of valley-bottom wetlands [*Westbrook et al.*, 2006]. By maintaining relatively high valley-bottom water tables, beavers can create a self-sustaining environment that limits the encroachment of upland trees and plants adapted to more xeric conditions. Studying geomorphic changes upstream from beaver dams in Oregon, USA, *Pollock et al.* [2007] note that vertical aggradation rates are initially as rapid as 0.47 m/yr where a previously incised channel fills, then decrease by an order of magnitude over a period of a few years as sediment starts to accumulate across the width of the valley bottom. As sediment gradually fills the beaver pond, meadow grasses and riparian shrubs and trees take over the site, providing a broad, stable floodplain that continues to store sediment and slow the passage of floodwaters [*Butler and Malanson*, 1995] (Figure 6.8). Although an early, widely cited paper suggests that beaver-meadow complexes might account for an important proportion of the alluvium in mountain valleys of Colorado, USA [*Ives*, 1942], and subsequent work confirms this interpretation [*Westbrook et al.*, 2010], work in the Yellowstone region of the USA suggests that in some environments beaver-induced sedimentation may be relatively minor in volume compared to other depositional processes [*Persico and Meyer*, 2009].

Hunting and change of land use drove the European beaver close to extinction throughout its range by the 19th century [*Hartman*, 1996; *Gurnell*, 1998]. Similarly, beaver trapping for the fur trade nearly eradicated beavers in North America. The eastern portions of the continent were trapped out as early as the 1600s, the interior and western regions by the 1830s [*Sandoz*, 1964]. Beaver populations are presently 15-50% of those prior to trapping [*Naiman et al.*, 1986]. In the upper Mississippi River basin, 99% of the beaver ponds present in 1600 A.D. were gone by 1990. At a depth of 1 m, the original area ponded by beaver dams in this 21 million ha region could have stored more than three floods the size of the destructive summer 1993 flood on the upper Mississippi River, USA [*Hey and Philippi*, 1995].

The removal of beaver resulted in channel incision and bank erosion, increased channel gradient, and higher sediment discharge. Hydrographs became more peaked, and aquatic and riparian habitat became less diverse along the channels affected. As the extent and duration of overbank flooding decreased and alluvial water tables declined, xeric upland plants encroached on valley bottoms, altering the riparian community and the introduction of allochthonous organic matter to the stream. Although many of these changes occurred prior to systematic study of river systems, recent re-introductions of beaver indicate the magnitude of change that can occur along rivers, as exemplified by Currant Creek, Wyoming, USA. Following beaver reestablishment along this channel, daily sediment transport decreased from 30,000 to 3,600 kg [*Brayton*, 1984; *Parker*,

1986]. Re-introduction of beavers as a restoration strategy for degraded channel and riparian environments is now being explored [*Pollock et al.*, 2007].

Conversely, introduction of beavers during the 20th century to areas in which they are not native has disrupted riparian communities by altering tree mortality and hydrologic regimes. Beavers introduced to the Rio Claro of Isla Grande in Tierra del Fuego, Argentina in 1946, for example, have extensively colonized the island during the past few decades [*Lizarralde et al.*, 2004] and spread to adjacent parts of Tierra del Fuego [*Skewes et al.*, 2006].

9. *Tie drives, log floating, and removal of wood.* Wood has been removed from river channels to reduce flooding and to enhance river navigation and fish passage. Historic records indicate that rivers across a wide range of drainage basin sizes, channel gradients, and climatic regimes contained substantial wood. Some of the largest logjams were described for lowland rivers. *Veatch* [1906] wrote of more than 20 lakes formed by logjam impoundments along a 260-km reach of the Red River, Louisiana, USA. The entire structure was called the Great Raft and constantly migrated downstream as new wood was added to the upper reaches and wood about 200 years older decayed and dispersed at the downstream end. These logs were cleared from the Red River in 1873 and the lakes began draining. Eight hundred thousand snags were removed along a 16-km stretch of the lower Mississippi River from 1870-1920 [*Sedell et al.*, 1982] and similarly massive logjams occurred on rivers in the northwestern U.S. [*Sedell and Froggatt*, 1984], on the Ohio River, USA [*Scarpino*, 1985], and in Europe [*Piégay and Gurnell*, 1997]. Personal observation in old-growth forest of the Colorado Rocky Mountains and the Cascade Range of the northwestern U.S. indicates that logjams spanning the channel may extend several channel widths upstream and individual pieces of wood or smaller jams are ubiquitous (Figure 6.9).

The consequences of wood removal on rivers both large and small are similar to those described following the eradication of beaver. The channel initially becomes unstable, with more rapid passage of flood waves, enhanced bed and bank erosion, and higher sediment discharge [*Heede*, 1985]. When the channel does stabilize, the earlier hydraulic and substrate diversity are replaced by a more uniform run or riffle-type channel that provides much less habitat for aquatic and riparian organisms. The numerous functions of wood in mountain rivers are described in chapter 3.

The use of river channels for navigation or transport of goods may have effects similar to those described from removing logjams. The use of streams in the Rocky Mountains of Colorado and Wyoming, USA for the transport of railroad cross-ties provides an example. During the 1870s and 1880s, timber cut in the mountains was sent downstream to lumber mills via the stream channels. Many of the rivers did not carry a sufficient volume of water to float timber except during the annual spring snowmelt floods. Logs cut during autumn and winter were stored beside the rivers and in the channels. The timing for releasing the logs at high water had to be carefully planned; insufficient water would prevent the logs from reaching their destination, but an unexpected flood could scatter broken logs along the valley [*Wroten*, 1956]. In an attempt to control water flow, splash dams were built along many streams to store water that could then be released at appropriate times to ease the logs through difficult channel reaches (Figure 6.10). It was also common practice to facilitate the downstream movement of logs by altering irregularities along the rivers. Such alternations in the lower gradient or meandering rivers included blocking off sloughs, swamps, low meadows, and banks along wider sections with log cribbing [*Schmal and Wesche*, 1989]. In higher-gradient reaches, large boulders, logs, debris, and encroaching riparian vegetation were cut or blasted to facilitate passage. The logs still became jammed along the rivers, in some cases creating jams 9-12 m tall with water running over them in waterfalls [*Wroten*, 1956]. The base of the jam then had to be blasted or broken up with long pikes. At some point downstream, usually where a railroad crossed the river, a boom was built to catch the logs. The distance from cutting area to boom varied from a few kilometers on short streams to more than 80 km on the main rivers. During the winter of 1868-69, more than 200,000 logs were cut and floated down the Poudre River, Colorado, USA (drainage area 5230 km^2), with a similar rate of operations maintained the next year [*Wroten*, 1956].

Research on mountain streams in the Medicine Bow National Forest of southern Wyoming, USA compares analogous streams that did and did not have tie drives. The streams that had drives tend to be 1 to 3.6 times wider, with minimal bank development, riparian or bank cover for fish, and habitat diversity [*Schmal and Wesche*, 1989]. Tie-driven streams have more riffles, fewer well-developed pools, less instream wood, and fewer wood-related habitats for aquatic organisms [*Young et al.*, 1990, 1994]. The riparian zones of these streams are dominated by uniformly-aged lodgepole pine (*Pinus contorta*) dating to the time of the last drive, in contrast to the uneven-aged, old-growth vegetation of the stream without tie drives [*Young et al.*, 1990].

Other examples of tie drives or log floating and associated modification of channels and riparian areas come from New Zealand [*Mackay*, 1991], Sweden [*Törnlund and Östlund*, 2002], and Italy [*Comiti et al.*, 2006]. Construction of massive channelization features to facilitate log floating was particularly widespread along rivers of northern Sweden, where more than 30,000 km of rivers were affected by the early 1900s [*Nilsson et al.*, 2005a]. Subsequent restoration measures attempt to widen and increase the sinuosity of rivers, as well as increasing hydraulic roughness to facilitate retention of sediment, organic matter, and nutrients [*Nilsson et al.*, 2005a].

10. *Dams and flow regulation.* The construction of water-storage dams and the regulation of river flow by reservoirs and diversions have been undertaken for purposes of flood control, water supply, navigation, and hydroelectric power generation. Dams have been built since circa 2800 B.C. [*Smith*, 1971], but the construction of dams larger than 15 m tall has accelerated substantially since the 1950s [*Goldsmith and Hildyard*, 1984; *Petts*, 1984; *Graf*, 1999; *Nilsson et al.*, 2005b]. Dams in the European Alps, for example, provide an important power supply because of the abundance of water and the high energy associated with steep channel gradients. In the winter of 1972-73 such dams supplied 24% of Switzerland's power consumption [*Wilhelm*, 1994] and by 2006 the supply of hydroelectric power generation was close to the maximum possible, producing 31% of the country's consumption [*Tunç et al.*, 2006].

In some cases, such as the Grande Dixence hydropower scheme in southern Switzerland, which involves 35 glacier basins, meltwater intake structures divert virtually all of the meltwater from proglacial streams into tunnels to a large storage reservoir [*Gurnell et al.*, 1990] (Figure 6.11). As of 1970, 322 reservoirs were present in the Alps, with an effective volume of approximately 10.4 km³ [*Wilhelm*, 1994] (Figure 6.12). Austria alone has 30 major reservoirs and these generate 23% of the country's electric power [*Pircher*, 1990]. Large dams greater than 150 m tall were being completed at a rate of one every 1.65 years in Europe as of 1989 [*Petts*, 1989] and these dams are increasingly being sited in mountainous regions. Although the rate of dam construction has now diminished substantially in Europe because of environmental concerns and few remaining suitable sites, large dams continue to be built in other mountainous regions of the world.

The effect of a dam or flow regulation on a river will depend on the nature of the alterations in water and sediment discharge and on the characteristics of the channel [*Grant et al.*, 2003; *Salant et al.*, 2006]. Dam construction commonly reduces the mean and the coefficient of variation of annual peak flow; increases minimum flows; shifts the seasonal flow variability; and, in the case of hydroelectric dams, may greatly increase diurnal flow fluctuations [*Surian*, 1999; *Magilligan and Nislow*, 2001; *Bonacci and Roje-Bonacci*, 2003]. Average annual floods downstream from 29 dams in the central and western United States range from 3% to 90% of pre-dam values, with an average of 40% [*Williams and Wolman*, 1984]. Many dams operate under minimum-release requirements that increase the magnitude of low flows occurring within a given recurrence interval. Annual seven-day low flows on the Columbia River in southwestern Canada nearly doubled following construction of a reservoir [*Hirsch et al.*, 1990]. The high flows released during peak production times (approximately 6 hours per day, 5 days per week) from hydroelectric dams may be nearly as large as the mean annual flood that occurred in pre-dam conditions [*Hirsch et al.*, 1990]. During evening and weekend low production, flow release may drop precipitously. *Grant et al.* [2003] propose that geomorphic response to dams can be predicted based on the ratio of sediment supply below the dam to that above the dam and the fractional change in frequency of sediment-transporting flows (Figure 6.13). *Do Carmo* [2007] reviews dam effects and considerations for mitigating these effects.

One of the challenges along regulated rivers is quantifying changes in flow regime caused by regulation. Methods used to quantify changes include (i) the range of variability approach, sometimes called the index of hydrologic alteration, in which the interannual variability of 67 streamflow parameters that reflect magnitude, timing, frequency, duration and rate of change of discharge are compared pre- and post-regulation [*Richter et al.*, 1996, 1997, 1998], (ii) wavelet transform, a mathematical tool used to extract the dominant modes of variability from statistically nonstationary signals [*Zolezzi et al.*, 2009], and (iii) comparison of the seasonal probability density function of regulated and natural streamflows [*Botter et al.*, 2010].

The channel downstream from a dam commonly adjusts its morphology to the altered flow regime and sediment supply caused by the sediment-trapping properties of the dam. Along channels with high suspended sediment concentrations, the change in sediment transport may be substantial. Three major dams were finished between 1952 and 1955 on the Missouri River, USA. The post-dam annual suspended sediment load 8 km downstream from the lowest dam was less than 1% of the pre-dam load and the load was only 30% of pre-dam levels 1200 km downstream [*Williams and Wolman*, 1985]. Construction of the Mosul Dam on the Tigris River in Iraq reduced suspended sediment load 60 km downstream from the dam by more than 95%, causing channel incision and coarsening of the bed sediment [*Al-Taiee*, 1990]. The downstream impact of dams on sediment supply may be relatively minor or of short extent if other sources of sediment such as tributaries or hillslope processes are abundant below the dam [*Salant et al.*, 2007]. Channel adjustment may take the form of incision that occurs during the first few years after dam construction and may extend hundreds of kilometers downstream from the dam [*Galay*, 1983; *Weiss*, 1994; *Jiongxin*, 1996]. The reduction in peak flows below a dam may also reduce sediment transport capacity and facilitate channel-bed fining or channel aggradation or narrowing as a result of continuing tributary inputs [*Petts*, 1977; *Allen et al.*, 1989; *Wilcock et al.*, 1996b; *Webb et al.*, 1999; *Magirl et al.*, 2005; *Yanites et al.*, 2006; *Dubinski and Wohl*, 2007; *Curtis et al.*, 2010]. The reduction in flood scouring may enhance the establishment of riparian vegetation, so that bars, islands, and channel banks are stabilized and the channel narrows. Channel adjustment varies with distance downstream from the dam [*Lagasse*, 1981; *Andrews*, 1986], but a common theme of case studies focusing on channel adjustment to flow regulation is a loss of physical and ecological complexity [*Surian*, 1999; *Graf et al.*, 2002]. Where numerous dams trap sediment and alter flow regime on rivers throughout a mountainous region [*Radoane and Radoane*, 2005], the cumulative downstream effects can cause substantial incision on larger rivers with headwater tributaries in the mountains. *DeJong* [2007] presents a toolbox to analyze basin-wide resilience to dams along mountain rivers.

In some cases low dams that were historically present may be removed or buried in sediment, although the sediment stored upstream from the dam continues to affect channel form and process. Water-powered mills were extensively used throughout Europe and the United States until the 19th century and were then largely abandoned. The continuing effect of milldam sediments was only recently recognized [*Walter and Merritts*, 2008].

Dams in mountainous regions can be particularly susceptible to landslides or other mass movements from adjacent valley walls that rapidly introduce large volumes of sediment. This can generate a wave that either overtops the dam or causes the dam to fail, producing an outburst flood downstream. The flood generated by a wave that overtopped the Vajont dam in the Italian Dolomites in 1963 and killed about 2500 people is a commonly cited example of this scenario, although landslides have triggered massive outburst floods at the site of the Tehri Dam in the Garhwal Himalaya of India [*Valdiya*, 1997] and other sites. China's recently completed Three Gorges Dam is potentially susceptible to this type of hazard because of the numerous sites of hillslope instability within the upstream gorges [*Liu et al.*, 2004b].

Dams not only interfere with the movement of water and sediment, but also with the migration of fish, with water temperature regime and chemistry, and with the movement of nutrients, plant propagules, and the instream wood that enhances pool formation and the diversity of aquatic habitat [*Baxter*, 1977; *Brooker*, 1981; *Ligon et al.*, 1995; *Stanford et al.*, 1996]. Fish may also be severely impacted by the loss of floodplain spawning and nursery habitat when reduced peak flows limit overbank flooding [*Jubb*, 1972; *Chikova*, 1974; *Lake*, 1975; *Whitley and Campbell*, 1974], or by reductions in minimum flow that prevent spawning [*Bundi et al.*, 1990]. Loss of flow variability and alterations in sediment supply and transport, as well as reduced channel diversity and mobility, affect all components of aquatic and riparian ecosystems [*Power et al.*, 1996; *Nilsson and Berggren*, 2000; *Bunn and Arthington*, 2002; *Nilsson and Svedmark*, 2002]. *Braatne et al.* [2008] review different methods of assessing the impacts of flow regulation on riparian communities. Syntheses across numerous case studies and databases indicate that aquatic and riparian species are becoming increasingly homogeneous as a result of flow regulation [*Poff et al.*, 2007; *Moyle and Mount*, 2007].

The impact of flow regulation on aquatic biota is illustrated by the Colorado pikeminnow (*Ptychocheilus lucius*). The pikeminnow is an endangered species inhabiting the Colorado River basin in the United States. The pikeminnow spawns at a limited number of gravel bars during the recessional limb of the annual snowmelt hydrograph. Sediment deposition and bar formation along a channel reach studied by *Harvey et al.* [1993] occur at discharges greater than 280 m^3/s. Spawning habitat is formed by bar dissection and erosion of sand and finer sediments at flows of 10-140 m^3/s. The pikeminnow cannot successfully spawn if the historical May-June peak flow is prolonged too late into the season by dam releases because prolonged sand deposition restricts adult fish access to spawning gravel and may smother developing embryos [*Osmundson et al.*, 2002]. This has been a problem since the construction of a dam upstream from the study reach in 1962 and, along with other changes in channel complexity and dynamics related to flow regulation, contributes to the endangered status of the pikeminnow [*Pitlick and Van Steeter*, 1998; *Van Steeter and Pitlick*, 1998].

Another example of the impacts of dam-related flow regulation on aquatic ecosystems comes from the Flathead River basin of Montana, USA, which includes both unregulated channel reaches and three dams operated for flood control and hydroelectric power generation. Wetting and subsequent drying of the channel edges and floodplain (the varial zone) occur once a year during spring snowmelt floods along the unregulated channels. Along regulated channels, the variable zone is unpredictably flooded and dried so that aquatic and riparian biota have little chance of naturally colonizing new areas as the stage rises or emigrating when the stage falls. As a result, aquatic biodiversity is drastically reduced along the regulated reaches [*Stanford and Hauer*, 1992].

The changes in hydrology and channel morphology associated with dams may impact riparian vegetation in various manners [*Pearlstine et al.*, 1985; *Merritt and*

Wohl, 2006; *Andersen et al.*, 2007; see also chapter 5]. Altered channel morphology or disconnection of the channel and floodplain because of reduced peak flows can reduce or eliminate the freshly-scoured surfaces necessary for riparian seedling establishment (Figure 6.14). Seedlings may be killed by prolonged submersion because of increased base flows. Low flows occurring at the wrong time of year may not adequately transport riparian seeds downstream. Changing grain-size distribution along the banks and bars of the channel may inhibit seedling establishment. Changes in the water table beneath the floodplain as a result of changed flow regime may cause soil salinization that kills riparian vegetation in arid and semiarid regions [*Jolly*, 1996]. Numerous studies document lower species-richness and percentage cover of riparian vegetation along regulated rivers than along unregulated rivers [*Nilsson et al.*, 1991a; *Dynesius and Nilsson*, 1994]. *Burke et al.* [2009] describe the effects of dams on recruitment of riparian trees as third-order effects; changes in hydrology constitute first-order effects and changes in channel hydraulics and bed mobility constitute second-order effects.

As dams continue to be constructed at an accelerating pace along mountain rivers in developing countries, other countries attempt to mitigate the negative impacts of dams by operating the dams so as to mimic a more natural flow regime or by removing the dam. At least 121 dams were removed between 1930 and 1999 in the United States, with the majority of these removals occurring in the 1990s [*Draft Report*, 1999]. An estimated 750 dams were decommissioned by 2008, with between 20 and 50 dams joining the list each year during the first decade of the 21st century [*O'Connor et al.*, 2008]. Removal of the Grangeville Dam on the South Fork of the Clearwater River, Idaho, USA (*A* 27,000 km^2) provides an example. The 17-m tall, 134-m wide arched concrete structure was built in 1903 for hydroelectric power. After the dam's fish ladder failed in 1949, migration of salmon along the river ended. By 1963 the reservoir behind the dam was filled with silt and the dam was removed. Reservoir sediment was flushed downstream within a few weeks and 70 km of main channel habitat and more than 170 km of tributaries were opened to migratory fish, revitalizing the local fish stocks [*Draft Report*, 1999]. Numerous studies based on field observations and flume experiments now describe the geomorphic and ecological effects of dam removal [*Stanley et al.*, 2002; *Doyle et al.*, 2003; *Cantelli et al.*, 2004; *Lorang and Aggett*, 2005; *Auble et al.*, 2007; *Cheng and Granata*, 2007; *Major et al.*, 2008a, 2008b]. Synthesis papers review the geomorphic [*Doyle et al.*, 2002; *Pizzuto*, 2002] and ecological effects [*Gregory et al.*, 2002] of dam removal, and develop frameworks for assessing removal versus other options [*Hart et al.*, 2002; *Poff and Hart*, 2002; *Stanley and Doyle*, 2003].

Although it is typically not economically or politically feasible to remove large dams that are still in use, modification of dam operation can reduce the disruption of the hydrograph and sediment dynamics [*Graf*, 2001]. Experimental flood releases from Glen Canyon Dam on the Colorado River in the United States provide ideal conditions for testing conceptual and numerical models of river dynamics and response to altered dam operation [*Collier et al.*, 1997; *Wright et al.*, 2008] and modified dam operations are also being used on other regulated rivers to create targeted effects on river ecosystem dynamics [*Bednarek and Hart*, 2005; *Jorde et al.*, 2008]. Target flow regimes are sometimes referred to as *environmental flows* [*Richter et al.*, 2003; *Rathburn et al.*, 2009; *Richter*, 2010].

Flow regulation via diversions has many of the same effects on water and sediment discharge and on channel morphology as do dams and reservoirs. Diversion for irrigation or for flood control commonly reduces base flows and flood peaks [*Schleusener et al.*, 1962; *Richards and Wood*, 1977] (Figure 6.15) and thus impacts sediment transport, bed grain-size distribution, bedforms, channel morphology [*Bray and Kellerhals*, 1979; *Angelaki and Harbor*, 1995; *Ryan*, 1997; *Parker et al.*, 2003], and riparian and aquatic communities [*Johnson*, 1978; *Rader and Belish*, 1999]. The downstream extent of these impacts depends partly on tributary and groundwater inputs below the point of diversion [*McCarthy*, 2008]. Diversions can also result in flow augmentation if water is transferred between drainage basins. Flow augmentation that increases peak flow magnitude and duration can result in changed channel grain size, geometry, and planform [*Dominick and O'Neill*, 1998].

In summary, some form of flow regulation affects even the smallest mountain rivers in many regions of the world. The resulting alterations of flow regime and channel form and process vary widely between sites, but typically include reducing the mean and coefficient of variation of annual peak flow, increasing the minimum flow, and shifting the seasonal flow variability. Reduction in downstream sediment transport results in channel erosion, whereas substantial reduction in downstream flow in the presence of continuing sediment inputs below a dam can result in aggradation or narrowing. Altered water chemistry and nutrient transport, along with reduced longitudinal connectivity and altered channel morphology, commonly stress aquatic and riparian communities.

11. *Check dams and bank stabilization; river training.* The type of channelization undertaken in the mountains of Europe and Asia for centuries is commonly referred to as river training. The primary objectives are typically to limit downstream sediment movement in very steep reaches, to reduce overbank flooding in reaches of slightly lower gradient and broader valley bottom, and to restrict multi-thread, braided channels into single channels with narrower and deeper cross-sectional geometry as a means of improving navigation and flood conveyance and making floodplain lands available for farming or settlement. These efforts have achieved their original intent to some degree, but the unanticipated costs of continual maintenance and ecosystem degradation have been exorbitant in many cases. Consequently, much of the river restoration currently being done in mountain rivers is designed to mitigate some of the negative impacts of river training. The case study of the Danube River basin summarized in this chapter 5 provides an example of river training and subsequent restoration.

Check dams are among the structures most widely used to stabilize mountain streams and reduce flood hazards. These structures can be designed to: create a local base level that limits upstream migration of bed incision and associated bank destabilization; store sediment and limit deposition and overbank flooding or channel avulsion downstream; or enhance profile irregularity and associated dissipation of flow energy

[*Lenzi*, 2002; *Castillo et al.*, 2007; *Shieh et al.*, 2007]. *Closed check dams* are designed to trap all sediment until the structure fills, whereas *open check dams* can pass finer sediment downstream. Closed check dams, which are typically made of concrete, rock (grouted or loose), or gabions (wire cages filled with rock), are traditionally arranged in a cascade, analogous to a natural step-pool sequence (Figure 6.16). More recently, construction of closed check dams in the European Alps utilizes large boulders, with some concrete dams partially buried beneath the boulders, to enhance the ecological and esthetic value of the dams [*Lenzi*, 2002]. Although the selection of the location and design of check dams mostly focuses on hazard reduction, ecological and esthetic considerations are becoming increasingly important. Open check dams were traditionally designed based primarily on experience, with limited analysis of their efficiency after construction [*Catella et al.*, 2005] (Figure 6.17). *Armanini and Larcher* [2001] develop a theoretical approach to check-dam design based on the conservation of mass of water and sediment, as well as the energy balance under steady conditions. *Busnelli et al.* [2001] apply a 1d unsteady coupled numerical mobile-bed model that can handle rapidly varying flows and discontinuities such as hydraulic jumps to the design of check dams.

At the most fundamental geomorphic level, check dams create a local base level that promotes sediment deposition upstream from the structure and a vertical drop that induces scour immediately downstream. Transverse depositional features such as boulder berms are typically included at the downstream end of the scour hole and upstream of the next check dam. The installation of check dams can also produce changes in channel cross-sectional geometry, grain-size distribution, bedforms, bed slope, and riparian vegetation [*Bombino et al.*, 2009; *Martín-Vide and Andreatta*, 2009]. The details of these changes vary with each site as a result of the interactions among multiple control variables. These control variables include height and longitudinal spacing of successive check dams, water and sediment supply in the channel, and substrate resistance of the channel [*Lai et al.*, 1998]. Sediment retention behind check dams can facilitate channel narrowing, armoring, and/or incision downstream [*Boix-Fayos et al.*, 2007]. Decreased flow velocity upstream from the check dam can cause substrate fining and associated changes in stream habitat and trophic structure [*Shieh et al.*, 2007]. By creating unnaturally high barriers, check dams can also potentially limit longitudinal movement of stream organisms [*Jaeggi and Zarn*, 1999; *Shieh et al.*, 2007].

As bank stabilization restricts formerly multi-thread rivers to a single channel and check dams in the headwaters decrease sediment supply, 'trained' rivers tend to incise [*Wyżga*, 1991]. This commonly necessitates further engineering to maintain channel stability, as well as altering aquatic and riparian habitat [*Wyżga*, 2008]. As *Korpak* [2007] notes in her review of river training in the mountains of Europe, river training distorts the ability of a river to adjust to changes in water and sediment supply by dividing the river channel into artificial reaches and constraining the degrees of freedom within each reach. Because river managers tend to consider each channel segment in isolation, most training schemes on mountain rivers are ineffective in the long term [*Korpak*, 2007]. Trained rivers are particularly widespread in the mountains of Europe and Japan (Figure 6.18). *Talbot and Lapointe* [2002] provide an example of how numerical models can be used to predict and simulate channel responses to river training.

12. *In-channel mining.* In-channel mining for sand and gravel used in construction and for placer deposits of precious metals has occurred for millennia in some regions of the world. This may be particularly common in mountain rivers because of the coarser bed-material and the proximity to *in situ* ore deposits. Systematic studies of the effects of these activities on rivers were not undertaken until the early 20th century, starting with *Gilbert*'s [1917] work on mountain river channels impacted by placer gold mining in the Sierra Nevada, California, USA.

It is now widely recognized that mining disrupts water and sediment transport, causing various channel adjustments in response. Large, localized excavations in the channel may initiate a knickpoint that migrates upstream from the excavation [*Kondolf and Larson*, 1995; *Marston et al.*, 2003; *Wishart et al.*, 2008] and bed and bank erosion may be accelerated downstream from the excavation as sediment transport is disrupted [*Lagasse et al.*, 1980]. Removal of sediments from within the channel can also cause coarsening of bed sediment and loss of spawning gravels for fish [*Kondolf*, 1994, 1997]. A 50-year flood along Tujunga Wash, California, USA triggered knickpoint migration from a 300 m X 460 m X 15-23 m deep gravel pit [*Bull and Scott*, 1974]. The knickpoint caused the failure of three major highway bridges and lateral scour downstream from the pit destroyed seven homes and a long section of a major highway. The upper 100 km of the Arno River in Italy incised 1-3 m between 1945 and 1960 as a result of gravel mining and the construction of two dams [*Rinaldi and Simon*, 1998]. Bed incision can also result in lowering of alluvial water tables adjacent to the channel [*Kondolf*, 1994]. *Chang* [1987] uses a 1d model of water and sediment routing to evaluate stream channel changes induced by gravel mining. Mining can be so extensive that it becomes the primary determinant of local channel form and function [*Graf*, 2000].

Placer mining that disrupts a coarse, stable channel-bed surface may increase sediment mobility and preferential transport of fine sediments, leading to downstream aggradation and bank instability associated with braiding [*Gilvear et al.*, 1995; *Hilmes and Wohl*, 1995]. The banks of many mountain rivers are resistant to erosion because either the bank is composed of very coarse, tightly-packed sediment, or the bank is capped by a layer of silt and clay held in place by densely growing roots of riparian vegetation. If the erosional resistance of the channel bank is lowered through the disruption associated with mining, the channel commonly becomes laterally unstable, repeatedly shifting back and forth across the valley bottom in a braided pattern [*Jackson and Van Haveren*, 1984]. The rapid lateral movement of braided channels discourages thick accumulations of fine-grained overbank sediment, making it difficult for riparian vegetation to become established. The continual shifting also increases the amount of sediment introduced to the river, causing further increases in suspended load. In addition, the instability of the channel substrate is likely to affect aquatic macroinverte-brates by mimicking the effect of frequent large, erosive floods. The erosion and deposition of bed sediments during periods of high discharge (or channel change) can destroy sessile, stone-surface communities and cause the downstream displacement of bottom-dwelling animals, which need refuges such as backwaters in order to survive disruptive events. In general, streams which flood or are disrupted frequently may show a marked succession of stream flora and fauna with time, in contrast to those that are not continually disturbed [*Winterbottom and Townsend*, 1991].

The Middle Fork of the South Platte River in the Colorado Rocky Mountains, USA provides an example of channel instability associated with mining [*Hilmes and Wohl*, 1995] (Figure 6.19). Historical documents and photographs of the Middle Fork prior to mining, and analogous, unmined rivers nearby (such as the South Fork of the South Platte) suggest that, prior to mining, the Middle Fork was a relatively deep, narrow stream that meandered through broad meadows of willows and grasses. The channel bed was gravel to cobble size material and the banks had an upper layer of fine sediment held in place by dense vegetation. Today, reaches of the river that were mined are less sinuous, in many cases having changed from meandering to braided. Mined reaches have been more mobile, as observed in comparisons of aerial photographs from 1930 to 1990, and have coarser bed sediments than unmined reaches, although it has been 65 to 80 years since mining occurred in some of these reaches [*Hilmes*, 1993; *Hilmes and Wohl*, 1995].

The mined portions of the Middle Fork flow through a broad intermontane valley. In contrast, where mined channels are closely constrained by bedrock valley walls, as in many mountain canyons, the channel may respond to increased sediment primarily through changes in bed configuration and gradient, rather than planimetric form [*Graf*, 1979]. Other examples of mining-induced channel instability include the American River of California, USA, which aggraded anywhere from 1.5 to 9 m during the primary hydraulic mining period (1861-1884), causing channel avulsions [*James*, 1991, 1994] and overbank flooding and deposition [*Fischer and Harvey*, 1991]. The river later degraded as sediment input decreased [*Gilbert*, 1917; *James*, 1997]. Mercury contamination persists in association with this mining history [*Prokopovich*, 1984]. More than 5 m of aggradation, increases of up to 300% in channel width, and enhanced braiding resulted from placer tin mining along the Ringarooma River in Tasmania, Australia, and bridges had to be frequently replaced following floods [*Knighton*, 1989]. Heavy sedimentation of tailings and waste rock along the Ok Tedi River system in Papua New Guinea resulted in aggradation of the channel bed, flooding problems, and the loss of a substantial area of riparian forest [*Higgins et al.*, 1987; *Parker et al.*, 1997] (Figure 6.20). The long-term response of river systems in which mining dramatically increases sediment load can be influenced by management responses. *James et al.* [2009] describe contrasting responses of the Yuba and Feather Rivers in California as a result of the width between levees. Widely spaced levees along the Yuba River facilitate sediment storage and the persistence of a braided channel, whereas narrowly spaced levees along the Feather River facilitate downstream transport of mining sediment [*James et al.*, 2009].

The adverse impact of mining on aquatic organisms is illustrated by studies of sediment increases associated with placer mining along Alaskan streams. These studies indicate that mining-related increases in turbidity, total residue concentration, and settleable solids reduce gross primary productivity to undetectable levels, primarily by eliminating algal production [*Bjerklie and LaPerriere*, 1985; *Van Nieuwenhuyse and LaPerriere*, 1986]. Similarly, the suspended sediment decreases the density and biomass of invertebrates [*Wagener and LaPerriere*, 1985]. Increased suspended sediment also affects fish by impairing feeding activity, reducing growth rates, and causing downstream displacement, color changes, decreased resistance to toxins and, in extreme cases, death [*McLeay et al.*, 1987; *Reynolds et al.*, 1989].

Placer mining may also indirectly cause diminution or augmentation of river discharge as water is diverted to work the placer deposits. Diminution of flow results in decreased ability to transport sediment, which can lead to channel aggradation as sediment coming from the hillslopes accumulates in the channel. Diminution of flow can also stress aquatic organisms through decreased oxygen and nutrients and increased temperatures. In the extreme cases where flow is completely diverted so that the former channel bed can be thoroughly worked over by the miners, it seems appropriate to consider the river as ceasing to exist for a time.

In contrast, augmented channel flow can increase the river's sediment transport capability. This might be offset by increased sediment introduction, or it might cause erosion of the channel bed and banks. Prolonged high levels of flow during seasons other than the traditional high-flow period can also stress aquatic biota by affecting water temperature and clarity and the necessity to expend energy. Some fish, for example, do not efficiently process the lactic acid produced by activity. If they are forced to be active (e.g., to maintain a position in a strong current) for too long, they will die from exhaustion. Most aquatic organisms have adapted to the flow regime of the specific river they inhabit. Numerous studies indicate that indigenous organisms are stressed or extirpated in rivers where the flow regime abruptly changes as a result of flow regulation [*Stanford and Hauer*, 1992].

The final direct effect of mining comes from the introduction of toxic materials used in mining [*Graf et al.*, 1991; *Langedal*, 1997; *Miller et al.*, 1999; *Ciszewski*, 2001; *Hren et al.*, 2001; *Marcus et al.*, 2001; *Taylor and Kesterton*, 2002]. Toxins such as heavy metals may enter the channel in solution or adsorbed to sediment. Metals in solution may come from abandoned tailings piles, which have been found to contain substantial concentrations of heavy metals that may enter the ground and surface water by leaching of precipitation through the pile material [*Ralston and Morilla*, 1974].

Toxic materials have a deleterious effect on stream biota, interfering with the respiratory, growth, and reproductive functions of members of the entire food chain [*LaPerriere et al.*, 1985]. The toxic materials act as a time bomb, for they have an impact at many spatial and temporal levels. There is the initial introduction, followed by processes of bioaccumulation and biomagnification over a period of years. In *biomagnification*, some toxic materials are not expelled by organisms, but accumulate in fatty tissues [*Burton*, 1992]. Any predator thus ingests all of the toxins accumulated by each of its prey organisms, so that concentrations of toxins increase with trophic level, culminating in raptors or humans. Longer-lived organisms may also continually ingest more of the toxin without expelling it, leading to *bioaccumulation* [*Burton*, 1992]. In addition, the toxins may be adsorbed onto clay or silt particles, lie buried in a sedimentary deposit, and then be re-mobilized at some later date by channel-bed erosion or lateral channel shifting during a flood [*Lewin et al.*, 1977].

The most general effects of any pollutant are to reduce community diversity within a channel [*Mackenthun and Ingram*, 1966; *Haslam*, 1990] and to reduce biomass and cover provided by rooted aquatic plants [*Haslam*, 1990]. Pollutants may be excess organic matter (such as that introduced by human or domestic animal wastes), excess sediment, heavy metals, or synthetic chemicals. Degradable organic matter reduces dissolved oxygen and light penetration, while increasing nitrite, nitrate, ammonia, and turbidity [*Haslam*, 1990; *Ward and Kondratieff*, 1992]. Excess sediment can smother aquatic plants, abrade or clog respiratory surfaces and collect on the feeding parts of macroinvertebrates [*Haslam*, 1990], and smother fish eggs and larvae.

The toxicity of any substance varies with the species under consideration, as well as with the organic matter, oxygen, temperature, and combination of pollutants present in the river [*Haslam*, 1990]. Acute toxicity kills organisms outright [*Buhl and Hamilton*, 1990], whereas chronic toxicity is caused by long-continued exposure to sublethal levels of a toxin. The most toxic heavy metals are mercury, copper, cadmium, and zinc [*Haslam*, 1990]. Creatures from unicellular to higher organisms absorb heavy metals from their food or directly from the water. Some organisms can excrete the toxins and are generally more tolerant of polluted environments, but other organisms absorb them permanently [*Bryan*, 1976]. Among the organisms that most efficiently concentrate metals are molluscs, followed by many types of aquatic plants [*Haslam*, 1990]. Molluscs tend to store metals in their tissues and digestive glands; crustaceans in their exoskeleton and hepatopancreas; fish in liver and muscle; and mammals in bone, kidneys, and liver [*Bryan*, 1976]. The metals can be enzyme inhibitors [*Bryan*, 1976] and they may damage the gill surfaces of fish and interfere with respiration [*Hughes*, 1976]. Metals can also affect the biomass, density, and diversity of riparian vegetation [*Stoughton and Marcus*, 2000].

Heavy metal pollution (Mn, Fe, Cu, Zn, Pb, Mo, Cd) associated with historic mining along the Arkansas River of the Colorado Rocky Mountains, USA provides an example of toxic contamination. The U.S. Environmental Protection Agency designated California Gulch, a mining area within the Arkansas River basin, as a Superfund site. Concentrations of metals in the Arkansas River are highly dependent on flow and tend to increase during high spring runoff. This suggests that these metals are abundant in the fine sediments that are carried as suspended load during high flows [*Roline and Boehmke*, 1981]. Much of the historic mining activity within the basin occurred along tributary streams. Today, the diversity and total abundance of aquatic macroinvertebrates are lower downstream from the junction of the contaminated tributary streams and higher below the junction of clean tributaries. The bottom-dwelling invertebrate communities below the contaminated tributaries are shifted toward more metal-tolerant species [*Rees*, 1994], whereas the fish population is reduced or absent [*Roline and Boehmke*, 1981]. The invertebrates at polluted sites also have higher concentrations of zinc, cadmium, and copper than those upstream, as do the gut and gill tissue of trout [*Rees*, 1994]. Gills and gut are the primary route for uptake of heavy metals by fish and the brown trout at the study sites rarely reached the age of four years or greater. The condition of trout in all age classes is generally poor [*Rees*, 1994]. Brown trout along the length of the channel downstream from polluted tributaries show bioaccumulation of copper and zinc in their livers and chronically high concentrations of these metals are present in the water [*Roline and Boehmke*, 1981].

In summary, in-channel mining of construction aggregate and placer metals typically alter channel-bed stability, resulting in downstream increases in sediment supply and turbidity, altered channel planform, and reduced aquatic habitat. These physical effects can persist for many decades, as can the toxic effects of contaminants introduced in association with mining.

13. *Exotic riparian and aquatic species.* Physical scientists are perhaps more likely to recognize the effects of exotic woody riparian species than other forms of plant and animal life that are introduced to rivers because of the extensive and well-documented effects of woody riparian plants on river form and process (see section 5.3.1). Around the world, however, a wide variety of exotic aquatic and riparian species have been actively introduced to mountain rivers through human intervention or have colonized mountain river systems on their own. As with terrestrial communities, islands can be particularly vulnerable to invasive species; only 26 of the 46 freshwater fish species in New Zealand are native [*Giller and Malmqvist*, 1998]. Salmonid fish have been actively introduced to mountain rivers beyond their natural distribution in many regions where people are interested in sport fishing. Beaver (*Castor canadensis*) introduced to South America have altered native *Nothofagus* riparian forests and stream processes [*Lizarralde et al.*, 2004]. Physical disturbances such as floods and debris flows can facilitate the spread of exotic shrub and herb species by creating new germination sites [*Watterson and Jones*, 2006].

Tamarisk (*Tamarix* spp.) and Russian olive (*Elaeagnus angustifolia*) are among the exotic woody riparian species of particular interest in the western United States because of the speed with which they colonize new regions and their high stand densities and ability to out-compete native species. *Tamarix ramosissima* is now the second most dominant woody riparian species in the western United States, although its sensitivity to minimum temperature will limit its spread into northern portions of the region [*Friedman et al.*, 2005b]. Russian olive is the fifth most dominant woody species in the region and is not limited by minimum temperatures [*Friedman et al.*, 2005b]. Each of these species can alter bank resistance to erosion and overbank roughness and sedimentation by growing more densely along stream channels than native species [*Graf*, 1978; *Allred and Schmidt*, 1999].

6.2. Contemporary Status of Mountain Rivers

In-channel alterations such as channelization, the construction of levees, the introduction of exotic vegetation, and channel stabilization affect portions of some mountain rivers, typically those in which streams affect local urban areas or transportation corridors [*Talbot and Lapointe*, 2002], but these activities commonly focus on channels in the middle and lower portions of drainage basins. Where channelization has been carried out for debris flow or flood control, as in the Austrian Alps or the Polish Carpathian Mountains, channel incision has also occurred [*Habersack and Nachtnebel*, 1994], leading to flashier flood peak discharges and increased flood hazards [*Wyzga*, 1996]. Worldwide, it is difficult to estimate which, if any, single type of activity has had the greatest impact on mountain rivers. In-channel mining has dramatically affected the Sierra Nevada Mountains of the western United States, for example, and beaver trapping has substantially altered many rivers in the Rocky Mountains. Overall, deforestation has probably had the single greatest effect on mountain rivers [*Wohl*, 2006], although the rapid and continuing construction of dams in the world's mountainous is also creating significant impacts. Rather than focusing on the human activity with the single greatest extent and impact to mountain rivers, it is important to recognize the *cumulative effects* – the combined effects of multiple land activities over space and/or time – of humans on mountain rivers [*Schnackenberg and MacDonald*, 1998]. Cumulative effects can be additive or synergistic (the combined effect is greater than the sum of the individual effects), although additive effects are more common [*MacDonald*, 2000].

As reviewed by *James and Marcus* [2006] and *Gregory* [2006], investigation of human affects on rivers has steadily accelerated during the past few decades, although recognition of human alterations dates back more than a century. Research focused on human alterations is increasingly driven by institutional needs related to resource management, by environmental regulations, by the desire to protect and restore river ecosystems, and by the need to develop sustainable human societies [*James and Marcus*, 2006].

Physical and biological scientists are still coming to grips with the idea that almost no natural rivers remain around the world, if we define natural as not even indirect human influence. With increasing recognition of the extent and pace of global warming, one can reasonably argue that no mountain rivers are completely unaffected by humans. Even the more direct alterations of rivers by human activities are commonly not recognized if they occurred earlier in history (although the effects of these activities may be continuing) or if they are so widespread that no unaffected, reference river systems are available for comparison [*Wohl*, 2001; *Wohl and Merritts*, 2007]. Human alterations must be borne in mind, however, if we are to understand river process and form and to effectively manage and restore river systems.

6.3. Hazards

The occurrence of hazards along mountain rivers is inextricably tied to human impacts on these systems. In some cases, increased hazards result from attempts to mitigate naturally occurring hazards, as when a flash flood in the Arás drainage basin of Spain destroyed more than 30 check dams full of sediment, flushing a huge amount of debris downstream to an alluvial fan inhabited by people [*Benito et al.*, 1998; *Gutiérrez et al.*, 1998; *Batalla et al.*, 1999]. The combined effect of a high-energy environment and human land use is that a disproportionate number of the world's natural disasters occur in mountain lands [*Hewitt*, 1997]. Of mountain disasters, 60% are associated with earthquakes and floods (including debris flows), with a high percentage occurring in southern and southwestern Asia and in south-central America [*OFDA*, 1988].

Hazards associated with mountain rivers occur principally in the form of floods and debris flows. Lahars may also be significant hazards in volcanically active areas [*Fairchild*, 1987] and other forms of slope instability (e.g., landslides, rockfalls, avalanches, sturzstroms) may or may not be associated with river channels. Worldwide, these mass movements kill hundreds to thousands of people and cause extensive property damage each year [*Ericksen et al.*, 1970; *Williams and Guy*, 1971; *Cooley et al.*, 1977; *Alexander*, 1983; *Zicheng and Jing*, 1987; *Benitez*, 1989; *Willi*, 1991]. The floods and debris flows may occur during fairly predictable, seasonally-driven periods or they may be unpredictable, aperiodic events. Much of the research devoted to hazards along mountain rivers focuses on (i) recognizing hazardous locations, (ii) understanding the controls on the occurrence of hazardous processes, (iii) predicting magnitude and frequency of hazardous processes, and (iv) mitigating the unwanted effects of these processes.

A key first step in mitigating hazards involves recognizing the potential for hazardous processes at a given location. Many people appear to regard the landscape as a static entity. They do not recognize that substantial landscape change, in the form of forest fire or flood, is likely to occur during their tenure at a specific location. Consequently, these people take no precautions to protect themselves or their property from natural hazards. One of the first responsibilities of natural scientists and engineers is therefore to increase public and governmental awareness of potential hazards.

6.3.1. Debris Flows

Hazards from debris flows, landslides, and floods are increasing in many mountainous regions [*Modaressi*, 2006] as a result of: land uses that enhance the conditions which initiate these processes; increasing population density and building in areas such as floodplains and alluvial fans that are likely to be affected by mass movements and floods [*Sepulveda et al.*, 2006]; changes in climate and forest management that increase wildfires [*Cannon and Gartner*, 2005]; and climate warming that enhances retreat of glaciers [*Chiarle et al.*, 2007]. The literature devoted to understanding and mitigating these hazards has increased enormously since the first edition of this book in 2000, with more contributions to international journals from investigators in Asia and Europe. *Jakob* [2005] provides a thorough review of hazard analysis for debris flows. *Jakob and Hungr* [2005] include numerous case studies and hazard assessment and mitigation measures employed in each case.

Various criteria may be used to identify the past occurrence of debris flows and debris flows are conservative in the sense that they tend to recur in the same locations.

Historical records may include both the location, approximate magnitude, and date of occurrence of past debris flows [*Eisbacher and Clague*, 1984; *Clark*, 1987b; *Skermer and VanDine*, 2005; *Marchi and Cavalli*, 2007]. Debris flow stratigraphy along channels and on debris fans, when constrained by various chronologic techniques, may be used to estimate debris-flow frequency in a drainage basin [*Kochel*, 1987; *Osterkamp and Hupp*, 1987; *Shlemon et al.*, 1987; *Wohl and Pearthree*, 1991]. The presence of debris flow scars and levees [*Eisbacher and Clague*, 1984] and specific landscape characteristics, such as alluvial fan gradient, surface morphology, or drainage basin ruggedness number [*Jackson et al.*, 1987; Whipple, 1993], may indicate areas prone to mass movements. Topographic maps, surficial geologic maps, aerial photographs, multi-temporal DEMs [*Zhang et al.*, 2006), satellite data [*Pack*, 2005; *Abdallah et al.*, 2007], and field evaluations may be used to locate past debris flows or potentially unstable terrain [*Chatwin et al.*, 1991; *Wieczorek et al.*, 1997] based on characteristic slope morphology, detection of movement, statistical techniques such as logistic regression analysis [*Tunusluoglu et al.*, 2008], or analytical methods such as Artificial Neural Networks [*Chang*, 2007]. The potential for debris flows and other forms of hillslope instability can also be monitored using remote sensing data on rainfall characteristics [*Chen et al.*, 2007]. High-frequency seismic noise analysis can be used to monitor debris-flow generation and the evacuation of sediment [*Burtin et al.*, 2009]. The physical characteristics that govern debris-flow occurrence (e.g., climate and vegetation, lithology, aspect, slope angle) can be compared with past debris-flow magnitude and frequency using Geographic Information Systems (GIS) technology [*Gupta and Joshi*, 1990; *Mejia-Navarro and Wohl*, 1994; *Mejia-Navarro et al.*, 1994; *Rowbotham and Dudycha*, 1998; *Wohl and Oguchi*, 2004; *Kneisel et al.*, 2007] or digital elevation models [*Dietrich and Montgomery*, 1992; *Mark*, 1992; *Glaze and Baloga*, 2003] to develop a hazards map for a region [*Akojima*, 1996; *Diodato*, 2004; *Lin et al.*, 2006a; *Magirl et al.*, 2010]. Models that predict the location, occurrence, and characteristics (e.g., volume, velocity) of debris flows can be statistical or physically based [*Carrara et al.*, 2008; *Toyos et al.*, 2008; *Taboada and Estrada*, 2009]. An important caveat to the use of historical data in predicting future debris flow magnitude, frequency, recurrence interval, or triggering factors is that characteristics of debris flows and other mass movements in mountain drainages are likely to change or are already changing in association with global climate changes and changing air temperatures, precipitation, vegetation, and patterns of frozen ground [*Menounos*, 2000; *Jomelli et al.*, 2004; *Geertsema et al.*, 2006; *Liu et al.*, 2008].

In addition to the use of past debris-flow indicators, regional trends may be used to delineate hazard areas at the basin scale. A study of Japanese mountain rivers, for example, demonstrates that rivers flowing through areas of high relief have longitudinal profiles that can be described by power or linear functions. Sediment is transported by debris flows and floods all the way to the lower portions of these drainages. In contrast, rivers flowing through areas of lower relief have longitudinal profiles that can be approximated by exponential functions, and these basins are likely to have substantial sediment deposition in their middle reaches [*Ohmori and Shimazu*, 1994]. As another example, recently deglaciated valleys may be particularly prone to mass movements because glacial erosion has created u-shaped valleys with steep flanks and glacial deposition has left moraine sediments perched along the valley walls [*Kuhle et al.*, 1998] or temporarily impounding meltwater [*Shroder et al.*, 1998].

Once the past, or probable future, occurrence of debris flows is recognized, it becomes important to define the thresholds governing that occurrence. These thresholds may be governed by precipitation intensity and duration [*Campbell*, 1975; *Caine*, 1980; *Kobashi and Suzuki*, 1987; *Cannon*, 1988; *Wilson et al.*, 1992; *Larsen and Simon*, 1993; *Luino*, 2005; *Cannon et al.*, 2008; *Coe et al.*, 2008] or by snowmelt [*Wieczorek et al.*, 1989]. Thresholds may be dependent on the volume of colluvium accumulated in hollows [*Johnson and Sitar*, 1990; *Dengler and Montgomery*, 1989; *Montgomery et al.*, 1991], on the combination of precipitation and sediment accumulation [*Church and Miles*, 1987; *Neary and Swift*, 1987; *Wieczorek*, 1987; *Thornes and Alcántara-Ayala*, 1998], on the rate of regolith development and subsurface water accumulation [*Roesli and Schindler*, 1990; *Chatwin et al.*, 1991], or on the slope morphometry associated with landslides [*Di Crescenzo and Santo*, 2005]. Forest fires may trigger debris flows in some regions [*Parrett*, 1987; *Wells*, 1987; *Garcia-Ruiz et al.*, 1988; *Wohl and Pearthree*, 1991; *Cannon et al.*, 2001; *Larsen et al.*, 2006] and *Cannon and Gartner* [2005] review hazards associated with fire-triggered debris flows. In addition to defining thresholds for debris-flow initiation, empirically-based models can also be used to predict debris-flow volume [*Gartner et al.*, 2008] or magnitude-frequency relationships [*Hungr et al.*, 2008]. As *Glade* [2005] notes, many hazard assessment models assume unlimited sediment supply, but models can be more realistically constrained by including available sediment stored within debris-flow contributing areas. Physical models are also used to study the conditions under which debris flows initiate [*Cook et al.*, 2008] and the processes occurring during debris flows [*Brufau et al.*, 2000; *Papa et al.*, 2004]. Instrumented catchments in which debris flows recur at relatively short intervals can also provide data used to calibrate models of debris flow initiation and processes [*Marchi et al.*, 2002; *Huerlimann et al.*, 2003].

Predicting the magnitude and frequency of hazardous debris-flow processes depends in part on the ability of models to simulate debris-flow velocity, hydrograph, flow depth, runout length, and depositional areas. One- and two-dimensional mathematical models [*Gallino and Pierson*, 1985; *Chen*, 1987; *Mizuyama et al.*, 1987; *Takahashi et al.*, 1987; *Rickenmann*, 1990; *Takahashi*, 1991a; *O'Brien*, 1993; *Rickenmann and Koch*, 1997; *Bathurst et al.*, 2003; *Lenzi et al.*, 2003a; *Malet et al.*, 2005; *Sosio et al.*, 2007; *Nakatani et al.*, 2008), including non-uniform and unsteady flow models [*Hungr*, 2000], cellular automata models [*Iovine et al.*, 2005; *D'Ambrosio et al.*, 2006], flume experiments [*Melosh*, 1987; *O'Brien and Julien*, 1988; *Takahashi*, 1991a; *Costa*, 1992; *Iverson and LaHusen*, 1992] and empirical relations between peak discharge and flow volume or sediment concentration as well as other debris-flow metrics [*Parrett*, 1987; *Webb et al.*, 1988; *Gavrilovic and Matovic*, 1991; *Mizuyama et al.*, 1992; *Miller and Burnett*, 2008], are each used to study the mechanical processes of debris flows. *Takahashi* [2007] reviews various approaches to modeling debris flows.

Methods of debris-flow hazard mitigation include (i) passive measures of hazard mapping and zoning, building codes, and warning inhabitants once a debris flow starts [*Itakura et al.*, 2000; *Bessason et al.*, 2007] and (ii) active measures that focus either on preventing a debris flow from occurring or on containing the debris flow and keeping it separate from areas of human occupation [*Eisbacher and Clague*, 1984; *Hungr et al.*, 1987; *Wieczorek*, 1993; *Prochaska et al.*, 2008]. Warning and evacuation systems may rely on indirect methods such that a warning is issued when rainfall enters a critical magnitude-intensity domain defined by the rainfall which has triggered debris flows in the past. Direct warning systems use either wires installed across debris-flow channels that are cut when a debris flow occurs or vibration sensors that are activated by the movement of a debris flow. These direct methods allow only a few minutes to evacuate potentially hazardous areas [*Mizuyama*, 1993]. *LaHusen* [2005] reviews instrumentation methods for detecting hazardous conditions that may lead to debris flows, as well as methods for characterizing debris flows while they occur.

Prediction of debris-flow runout is important for determining affected areas and flow intensity parameters [*D'Agostino et al.*, 2010; *Scheidl and Rickenmann*, 2010], as reviewed by *Rickenmann* [2005]. Runout properties reflect rate of slope decrease, degree of flow confinement in the runout zone, material composition and water content of the flow, momentum loss in bends [*Benda and Cundy*, 1990; *Scheidl and Rickenmann*, 2009], amount of entrained wood [*Lancaster et al.*, 2003], and the presence of obstacles in the runout zone [*Corominas*, 1996; *Guthrie et al.*, 2010]. *Rickenmann* [2005] reviews methods of predicting debris-flow runout.

Active debris-flow mitigation measures may focus on debris-flow source, transport, or depositional zones. Hillslope hollows that accumulate debris may be artificially drained to prevent the build-up of pore-water pressure that can trigger slope failure [*Reneau and Dietrich*, 1987; *Montgomery et al.*, 1991], or debris can be artificially removed on a controlled basis [*Baldwin et al.*, 1987; *Chatwin et al.*, 1991]. Reforestation measures can be used to stabilize cut or burned slopes, and timber harvest and road construction can be controlled [*Croft*, 1967; *Eisbacher*, 1982; *Hungr et al.*, 1987; *Swanson et al.*, 1987; *Chatwin et al.*, 1991]. Hillslope reinforcement with rock riprap, retaining walls, hanging metal nets, fences, rock bolts, or concrete lattices is also used [*Baldwin et al.*, 1987; *Chatwin et al.*, 1991; *De Wolfe et al.*, 2008] (Figure 6.21). The channel bed and side slopes can be stabilized with channel linings or check dams in source areas [*Liu*, 1992; *De Wolfe et al.*, 2008]. In the transportation zone of the debris-flow channel, the passage of the debris flow can be trained by chutes, artificial channels (Figure 6.22), deflecting walls, or dikes, and transportation routes across the channel can use bypass tunnels or bridges sited well above the likely maximum debris-flow height [*Hungr et al.*, 1987; *Willi*, 1991]. Free-standing baffles of timber, steel, or concrete can reduce debris-flow velocity and promote deposition at sites accessible to maintenance equipment [*Baldwin et al.*, 1987]. Deposition basins, retention barriers, and tunnels or sheds over transportation routes can be used in the debris-flow deposition zone [*Hungr et al.*, 1987] (Figure 6.23). *Prochaska et al.* [2008] note that published guidelines for detention basins and deflection berms are rare, resulting in little standardization between sites, and they recommend various design improvements. *Huebl and Fiebiger* [2005] review measures for mitigating debris-flow hazards.

In many societies, the traditional belief has been that gods or other supernatural forces govern the occurrence of floods and debris flows [*Eisbacher*, 1982; *Bjonness*, 1986]. However, active structural measures such as check dams have been extensively used for centuries in the European Alps [*Jaeggi and Zarn*, 1999] and in Japan. A Japanese imperial decree of 806 A.D. prohibited cutting of trees along riverbanks in an effort to control channel erosion, simultaneous with the construction of the first sabo dams, or sediment-retention structures [*Japanese Ministry of Construction*, 1993]. George Perkins Marsh's 1864 book *Man and Nature* highlights the importance of watershed protection and forest preservation using examples from the State of New York, but the U.S. government did not take national steps to preserve watershed characteristics until the establishment of national forest reserves in the 1890s and the 1935 establishment of the U.S. Soil Conservation Service. Governments in the European Alps were more proactive. The French government began afforestation programs along Alpine rivers in the 1890s [*Clark*, 1987a]. The Austro-Hungarian Empire established in 1884 a state service for debris-flow control [*Armanini et al.*, 1991]. Austria publishes a journal devoted entirely to the control of mountain torrents and snow avalanches (*Wildbach und Lawinenverbau*), and several national geological surveys (e.g., France, the Czech Republic, Canada, Italy, USA, Japan) have devoted considerable effort to mapping potentially unstable terrain and debris sources [*Eisbacher*, 1982]. Journals such as *Mountain Research and Development*, *Arctic, Antarctic, and Alpine Research*, the *Bulletin of the Association of Engineering Geologists* (USA), *Zeitschrift für Geomorphologie*, and various national geological surveys regularly publish research on mountain-channel hazards. The journals *Mountain Research and Development* and *Geomorphology* each devoted a special issue to mountain hazard geomorphology in 1994 and 2005, respectively. Countries with mountainous regions are increasingly developing research programs focused on debris-flow hazards [*Wang and Han*, 1993].

The Japanese have been especially active in developing methods of mitigating the hazards from debris flows. Following extensive deforestation during World War II, Japan's Erosion-Control Engineering Society was established in 1947. Society membership in the early 21st century is over 3000. The activities of the Disaster Prevention Research Institute at Kyoto University, which has numerous field stations throughout the Japanese islands, illustrate Japan's approach to debris-flow hazard mitigation. At the Institute's Hodaka Sedimentation Observatory in central Honshu, for example, an elaborate network of rain gages, sediment samplers, discharge gages, television cameras, and radar velocity meters are connected in a telemeterized observation system that continuously records channel conditions and the passage of floods and debris flows in the 7.2-km^2 Ashiaraidani Creek drainage basin (Figure 6.24). These types of process studies help to identify thresholds of slope instability and magnitude and frequency of mass movements. Analogous instrumented watersheds exist in several catchments in the mountains of Europe.

Kyoto University's Disaster Prevention Research Institute also actively investigates the effectiveness of various types of sabo dams and debris-flow training structures. The combination in Japan of a wet climate, active tectonic uplift, volcanism, and high relief on the one hand, and a high population density and healthy economy on the other hand, has led to Japanese mountain rivers being among the most intensively engineered in the world (Figure 6.25). The Japanese typically design integrated debris-flow control structures in mountainous drainage basins, with structures placed in appropriate locations to control, capture, disperse, channelize, and deflect debris flows (Figure 6.26). The volcanic mountain Sakurajima serves as an example. Sakurajima is an 80-km^2 stratovolcano in southern Kyushu that has had 4800 minor eruptions since 1965 [*Ohsumi Works Office*, 1995]. An average of 70 debris flows occur each year on Sakurajima and the Japanese Ministry of Construction has implemented extensive monitoring and control facilities. Sabo, or check, dams were first constructed on the mountain by the local government in 1943 (Figure 6.27). These are designed to temporarily trap sediment which can then be gradually transported by subsequent water flows with less hazard [*Armanini et al.*, 1991; *Mizuyama*, 1993]. The open-type sabo dams have the advantage over closed sabo dams (gravity dams of massive concrete) of not being filled by sediment from normal water discharges or by a single debris flow.

In 1965 the implementation at Sakurajima of wire-triggered on-site automatic recording of debris flows using video cameras began and this was expanded to 11 sites by 1971 [*Ohsumi Works Office*, 1995]. As of 1995, the system included 21 rain gages, 23 gages for volcanic ash, 16 cameras, 8 debris-flow detector lines, 8 discharge gages and current meters, and more than 200 sabo dams or other control structures [*Ohsumi Works Office*, 1995] (Figure 6.28). The intensity of this monitoring program is representative of similar efforts in many mountainous regions of Japan and Europe.

Although either debris flows or floods may occur throughout mountainous drainage basins, the prevalent geologic hazard is likely to change from debris flows to floods in the lower parts of a drainage basin. *Ohmori and Shimazu* [1994] find that this transition commonly occurs at channel gradients between 0.08 and 0.001 in Japan. Misidentifying debris flows as water floods may lead to over-estimation of discharge and to inappropriate design of mitigation structures [*Costa and Jarrett*, 1981].

6.3.2. Landslides

Landslides differ from debris flows in the mechanics of downslope movement of material. In the context of hazards along mountain rivers, landslides also differ from debris flows in that the slides do not follow the river course, but rather affect the river by creating a local base level and ponding [*Panek et al.*, 2007] and by altering sediment supply.

As with debris flows, the potential for hillslope instability in the form of landslides can be assessed using empirical relations developed from past landslides [*Galli et al.*,

2008] and manipulated in a GIS format or via statistical models [*Gritzner et al.*, 2001; *McKean and Roering*, 2004; *Wohl and Oguchi*, 2004; *Moreiras*, 2005; *Zêzere et al.*, 2008; *Kawabata and Bandibas*, 2009; *Saito et al.*, 2009; *Wu and Chen*, 2009; *Das et al.*, 2010; *Rossi et al.*, 2010]. This facilitates regional studies rather than site-based case studies [*Alexander*, 2008]. Numerical models of slope instability typically couple hydrological inputs and slope morphologic and geomechanical parameters [e.g., *Chemenda et al.*, 2009; *S. Cohen et al.*, 2009; *Lee and Ho*, 2009; *Santini et al.*, 2009; *Stark and Guzzetti*, 2009]. *Bathurst et al.* [2006] describe the SHETRAN basin-scale model of landslide erosion and sediment yield and *Simoni et al.* [2008] describe GEOtop-FS, a coupled, distributed, hydrological-geotechnical model which simulates the probability of occurrence of shallow landslides and debris flows. *Salciarini et al.* [2006] model rainfall infiltration and slope stability to predict landslide initiation and *Wu* [2003] models combined loading from precipitation, earthquakes, and land use to predict slope stability. Artificial neural networks can be used to forecast landslide susceptibility based on lithology, contributing area, slope morphometry and land cover [*Ermini et al.*, 2005; *Melchiorre et al.*, 2008]. *Iverson* [2003] reviews models for landslide runout.

Slope, lithology, elevation, and some index of moisture are typically associated with thresholds for landslide occurrence [*Gritzner et al.*, 2001; *Conoscenti et al.*, 2008; *Ruff and Czurda*, 2008; *Gao and Maro*, 2010], although structural and lithological controls that promote greater sediment production on hillslopes can also strongly influence regional patterns of landsliding [*Roering et al.*, 2005]. Because landslides can be progressive [*Glastonbury and Fell*, 2008; *Wang et al.*, 2010], with periodic reactivation associated with controls such as gradual loss of rock mass strength from shear stress release effects following deglaciation [*Guglielmi and Cappa*, 2010], hazard assessment also involves monitoring the activity of existing landslides using remotely sensed imagery [*Hervás et al.*, 2003; *Catani et al.*, 2005; *A. L. Booth et al.*, 2009; *A. M. Booth et al.*, 2009; *Kasai et al.*, 2009] and field-based benchmarks and instrumentation [*Corsini et al.*, 2005; *Petley et al.*, 2005]. Although many landslide models assume that landslides are uncorrelated in time, *Witt et al.* [2010] demonstrate temporal clustering of landslides for the Emilia-Romagna region of Italy. *Larsen* [2008] and *Wieczorek and Leahy* [2008] review landslide hazard mitigation strategies and *Liu et al.* [2006] and *Ni et al.* [2006] describe rapid mass movement hazard zonation for China. *Canuti et al.* [2007] review the integration of remote sensing techniques into landslide hazard mitigation. One aspect of landslide hazards is that the triggering processes are less understood by the general public than processes associated with other hazards such as flash floods [*Wagner*, 2007]. Another aspect is that because landslides, like debris flows and floods, are strongly influenced by precipitation and subsurface moisture, changing climates are likely to change landslide frequency [*Jakob and Lambert*, 2009].

In addition to the hazards associated directly with landslides, landslide-generated dams can overtop or fail catastrophically, causing debris flows and floods [*Marui and Yoshimatsu*, 2007]. *Haritashya et al.* [2006] describe a "perfect storm" scenario from the Bhagirathi River in India when intense rainfall triggered landslides that dammed the river, quickly creating a new lake of storm runoff and glacial meltwater, which then failed catastrophically. *Davies* [2002] uses numerical simulations of landslide-dambreak floods to assess flood hazards in New Zealand and *Korup* [2005a] describes GIS-based modeling of virtual landslide dams as a means of assessing the hazards associated with these features.

6.3.3. Floods

Many of the issues already discussed for debris flows also apply to floods along mountain rivers. Both active and passive hazard mitigation procedures [*Lin et al.*, 2005] may be employed to reduce flood hazards. Flood hazards commonly take the form of changes in channel configuration and changes in sediment transport. Both types of changes may occur very rapidly during a single flood or they may occur over a period of years to decades as progressive, small alterations resulting from a shift in the magnitude, frequency, and sediment transport of floods. The rate and magnitude of channel change during a flood will be a function of flood hydrology and hydraulics, particularly as these compare to more frequent, sustained flows. Channel change will also be governed by channel boundary resistance as this determines the flow energy necessary to modify channel boundaries and the form of boundary change (e.g., cavitation of bedrock or slumping of unconsolidated banks). Sediment supply, as it influences the erosive forces exerted on the channel boundaries during a flood, will influence channel change. Finally, channel change will relate to the sequence of flows preceding a flood as these govern how much adjustment is required for the channel to convey a flood flow [*Wohl*, 2000a]. Flood hazards may result from sediment deposition, erosion, lateral channel movement and planform change, or overbank flow of water.

Transport of wash load, suspended load, and bedload each generally increases during a flood. For example, studies of a small high-relief catchment in Hong Kong indicate that 85% of the bedload transported during a five-year period moved during two storms [*Peart et al.*, 1994]. Deposition resulting from this type of increased sediment transport may take the form of aggradation or in-channel deposition that raises the level of the channel bed [*Lisle*, 1982; *Nolan and Marron*, 1985; *Vuichard and Zimmermann*, 1987]. Deposition may also occur as the growth of bars or islands within a channel [*Scott and Gravlee*, 1968] or as lateral boulder berms beside the low-flow channel [*Scott and Gravlee*, 1968; *Carling and Glaister*, 1987; *Cenderelli and Cluer*, 1998; *Cenderelli and Wohl*, 1998]. Flood deposition may occur via the formation of an alluvial fan where a steep, confined channel joins a broader valley [*Williams and Guy*, 1973; *Jarrett and Costa*, 1986; *Schick and Lekach*, 1987]. Deposition may also occur via lateral accretion of the floodplain through the construction of point bars as a meander bend migrates, and vertical accretion of the floodplain as sediment settles from suspension in overbank flows. The in-channel forms of deposition are particularly prevalent along mountain rivers.

Erosion is more likely to occur along particularly steep or confined portions of a channel [*Shroba et al.*, 1979]. The alluvial veneer may be partially or wholly stripped from the channel [*Vuichard and Zimmermann*, 1987; *Church*, 1988] and the bedrock substrate may be quarried [*Tinkler*, 1993], eroded via cavitation [*Eckley and Hinchcliff*, 1986], or abraded. Bed scour at the base of obstacles such as bridge piers is particularly hazardous [*Williams and Guy*, 1973; *Landers et al.*, 1994]. During a 1982 flood in the upper Segre Valley of the southern Pyrenees, for example, more than half of the bridges were destroyed because of scouring more than 4 m deep at bridge-pier foundations and damming by debris [*Corominas and Alonso*, 1990]. Channel banks may be eroded by the collapse of blocks of sediment [*Warburton*, 1994; *Russell et al.*, 1995] or by the continuous removal of individual clasts or small aggregates. Areas of overbank flow may also be eroded via longitudinal grooves [*Smith and Zawada*, 1988; *Miller and Parkinson*, 1993], widespread stripping of alluvium [*Inbar*, 1987], or solution in carbonate terrains [*Hyatt and Jacobs*, 1996]. Within-channel deposition- and erosion-related flood hazards are likely to be most severe along mountain rivers, as these rivers are less likely to overflow their banks during floods than are lowland rivers.

Lateral channel movement may occur as predictable but rapid change, such as meander migration, or may occur as less predictable meander cutoff or avulsion along any type of channel [*Scott and Gravlee*, 1968]. Mountain rivers are unlikely to be freely meandering for more than short segments and the valley confinement common along mountain rivers will likely limit avulsion. Planform change commonly takes the form of (i) a single channel becoming braided, which may be particularly widespread downstream from glaciers [*Fahnestock*, 1963; *Desloges and Church*, 1992; *Warburton*, 1994] or in other regions of high sediment supply and low bank resistance or (ii) meander cutoff and channel straightening [*Corominas and Alonso*, 1990].

Rainfall, snowmelt, and rain-on-snow may cause flooding in mountainous regions. As reviewed in chapter 3, the magnitude of precipitation-related flooding in relation to normal flows varies as a function of regional climate [*Pitlick*, 1994b] and elevation [*Jarrett*, 1989]. The largest floods ever measured (in terms of discharge versus drainage area) in temperate latitudes have occurred in semiarid to arid drainages [*Costa*, 1985]. Although most models of meteorological causes and recurrence intervals of hazardous floods focus on rainfall floods, models of snowmelt flood risk have also been developed [*Graybeal and Leathers*, 2006].

In addition to precipitation-related causes of flooding, damburst floods are particularly important along mountain rivers because of the more widespread occurrence of natural dams. Natural dams may be glacial ice or moraine dams, or landslide/debris flow dams. (Catastrophic drainage of caldera lakes [*Waythomas et al.*, 1996] may produce similar results.) Ice- or moraine-dammed lakes are widespread in Iceland, the Alps, Scandinavia, the Himalaya, and the coastal ranges of northwestern North America. The failure of these dams may be a periodic phenomenon related to the increase in pressure behind the dam, as in ice-jam jökulhlaups [*Young*, 1980; *Driedger and Fountain*, 1989]. The dam failure may also be a one-time occurrence resulting from gradual erosion of the dam sediments, or overtopping by a wave generated from an icefall or landslide into the lake, or seismic activity [*Blown and Church*, 1985; *Vuichard and Zimmermann*, 1987; *Clague and Evans*, 2000]. *Dong et al.* [2009b] use discriminant analysis to determine that peak flow and dam dimensions exert the greatest influence on the stability of landslide dams. As with other types of hazards along mountain rivers, climate warming is causing an increase in moraine-dam failures and outburst floods by destabilizing moraine dams with interstitial or ice cores, as well as increasing rate of glacial melting [*Clague and Evans*, 2000; *Chalise et al.*, 2006]. *Huss et al.* [2007] find that glacier-dammed lake outburst floods are occurring earlier in the melt season at the Gornersee in Switzerland.

Landslide-dammed lakes may occur in any mountainous region with sufficiently large mass movements [e.g., *Russell*, 1972; *Ouchi and Mizuyama*, 1989; *Reneau and Dethier*, 1996; *Becker et al.*, 2007]. *Costa and Schuster* [1988] subdivide landslide dams into six types based on their distance of movement from the source area, morphology, and complexity (i.e., single or multiple dams from the same source). Of 73 documented landslide-dam failures, 27% occurred less than one day after formation and approximately 50% failed within ten days. Overtopping was by far the most common cause of failure [*Costa and Schuster*, 1988].

The peak discharge produced by an outburst flood commonly exceeds by at least an order of magnitude the "normal" snowmelt- or rain-induced flood peak for a particular drainage basin. These floods are also particularly hazardous in that dam failure may be unpredictable and produces a very steep rising limb [*Young*, 1980; *Ives*, 1986]. The mode of dam failure significantly controls rate of discharge rise and peak discharge magnitude; ice-marginal drainage or mechanical failure of part of the ice dam, for example, produces a significantly higher peak discharge than tunnel-drainage at the base of the ice [*Walders and Costa*, 1996]. If a dammed lake is located, flood hazard can be reduced by monitoring lake level and dam condition [*Braun and Fiener*, 1995], artificially draining the lake, or stabilizing the dam [*Eisbacher*, 1982]. Other flood hazard mitigation strategies are similar to those described for debris flows; passive measures of mapping, zoning, and warning, and active measures of prevention and containment for flood waters [*Huggel et al.*, 2003] and for wood that might be entrained during a flood [*Mao et al.*, 2008b]. Examples of hazardous outburst floods in mountainous terrain caused by the failure of artificial dams include the Johnstown, Pennsylvania, USA flood in 1874 (2200 deaths), the Vajont, Italy dam failure in 1963 (2600 deaths) (Figure 6.29), and the Sempor, Indonesia dam failure in 1967 (> 2000 deaths) [*Hewitt*, 1997]. *Spreafico* [2006] reviews hazard mitigation for flash floods resulting from either meteorological events or dam failures.

In many settings, floods are indirectly enhanced by wildfire, urbanization and other changes in land cover that increase overland flow from rainfall [*Migon et al.*, 2002; *Conedera et al.*, 2003; *Hicks et al.*, 2005]. Flood damages and hazards can also be directly enhanced by structures that reduce channel conveyance [*Arnaud-Fassetta et al.*, 2005] or interfere with the flood's ability to entrain sediment from the channel boundaries. Longitudinally discontinuous bank protection, for example, typically exacerbates bank erosion immediately downstream of the protected reach.

6.4. River Management

People have been deliberately manipulating river process and form for thousands of years. The earliest efforts were typically taken to limit hazards associated with processes such as flooding or to enhance water supply. These continue to be the focus of much river management, but a significant component of contemporary management is also directed toward mitigating the negative effects of past management actions. Careful integration of geomorphic and ecological knowledge is critical to the success of most river management actions [*Kondolf et al.*, 2003c; *Brierley and Fryirs*, 2008], as exemplified by attempts to restore and rehabilitate rivers.

6.4.1. Restoration and Rehabilitation

The implementation of active hazard mitigation measures can substantially alter the form and function of mountain rivers. Channelized and dammed streams lose riparian vegetation, allochthonous nutrient input, and wood. Water temperature increases as a result of lack of shade and increased surface area. Pool-riffle sequences and other bedforms may be destroyed, reducing substrate and habitat diversity [*Stanford et al.*, 1996], and providing fish no protection from high velocities during high flows [*Barclay*, 1978; *Nunnally*, 1978; *Beschta and Platts*, 1986]. Fish passage and the flow of nutrients along the channel may be blocked [*Nakano et al.*, 1990]. Hyporheic exchange and associated effects on water temperature and chemistry may be reduced or lost [*Hester and Gooseff*, 2010].

As humans increasingly recognize the value and crucial importance of naturally functioning river ecosystems, concern has grown over our alteration of these ecosystems. More efforts are made to restore or rehabilitate river channels using bioengineering (e.g., riparian vegetation to stabilize channel banks) [*Coppin and Richards*, 1990; *Nagle*, 2007; *Tague et al.*, 2008b]; environmentally sensitive structures (e.g., fish passage on dams, or log grade-control structures) [*U.S. Dept. of Transportation*, 1979]; intentional breaches of structures such as levees [*Florsheim and Mount*, 2002] or bank stabilization that cut off secondary channels [*Helfield et al.*, 2007]; gravel augmentation or slope creation [*Elkins et al.*, 2007]; controlled floods (e.g., the 1996 Grand Canyon of the Colorado River flood; *Collier et al.*, 1997, or experimental floods on the River Spöl in Swiss National Park, *Mürle et al.*, 2003; *Ortlepp and Mürle*, 2003); or a return to more natural hydrographs that include seasonally fluctuating flows [*Kattelman*, 1996; *Stanford et al.*, 1996; *Poff et al.*, 1997; *Rathburn et al.*, 2009]. Ultimately, these activities are designed to restore healthy riparian systems, which may be defined as those performing their ecological functions of protecting river banks against erosion, filtering diffuse pollution, and maintaining high levels of aquatic and terrestrial biodiversity [*DéCamps and DéCamps*, 1999].

Restoration is defined as a return to a close approximation of the river condition prior to disturbance; *rehabilitation* refers to improvements of a visual nature, or "putting the channel back into good condition" [*National Academy*, 1992]. Too often, attempts to restore or rehabilitate a river channel are site-specific measures that do not give sufficient consideration to upstream-downstream or channel-hillslope interactions – to the connectivity of the river ecosystem – or to the impacts of the rehabilitation measures on all aspects of the river ecosystem [*Montgomery et al.*, 1995b; *Cullum et al.*, 2008]. Successful river management must maintain or restore the vitally important lateral, vertical, and longitudinal connectivity of water, sediment, nutrients, and organisms in river ecosystems [*Bunn and Arthington*, 2002; *Brierley and Fryirs*, 2005; *Kondolf et al.*, 2006; *Bracken and Croke*, 2007; *Fryirs et al.*, 2007; *Hillman et al.*, 2008; *Mika et al.*, 2008], but must also consider artificial connectivity (e.g., diversions across drainage divides that introduce exotic species or contaminants) and pollutant source areas [*Qiu*, 2009].

Syntheses of existing restoration projects in the USA indicate that many restoration projects lack a set of measurable objectives, as well as systematic post-project monitoring to assess effectiveness of the project [*Bernhardt et al.*, 2005, 2007; *Palmer et al.*, 2005; *Follstad Shah et al.*, 2007]. *Boulton et al.* [2008] and *Ryder et al.* [2008] suggest strategies for setting restoration goals and evaluating success.

It seems appropriate to approach all attempts at channel manipulation within the framework used by physicians; first, do no harm (K. Prestegaard, pers. comm., 1998). It is also appropriate to use a series of questions to guide restoration efforts: What are you restoring to? How much can be inferred regarding a reference state? [*Sear and Arnell*, 2006] How have the controlling factors of water and sediment yield to the channel changed? What are the goals of restoration – channel stability, flood conveyance, habitat enhancement? How will the proposed restoration measures likely affect other aspects of river function beyond those explicit in the restoration design? How will the restoration design likely perform over time? No river channel exists isolated in time or space. The channel has a history of interactions between process and form that is at least partially recorded in local stratigraphy and historical records such as aerial photographs, and this record may provide critical insights into how channel form results from process and may respond to future changes in process [*Kondolf and Larson*, 1995; *Brierley and Fryirs*, 2005; *Brierley et al.*, 2008]. If the channel is incised, for example, the bankfull discharge may be meaningless and it becomes important to understand the recurrence interval and shear stress exerted by floods of varying magnitude in order to predict channel stability.

Ultimately, river management is not really restoration or rehabilitation, but another form of channel design. Contemporary phrases such as 'river repair' or 'integrative river science' reflect the desire to take a more holistic approach that includes geomorphic and ecological, as well as engineering, considerations, in the hope of creating self-sustaining river ecosystems [*Montgomery*, 2001a; *Brierley and Fryirs*, 2008; *Cullum et al.*, 2008]. The River Styles framework of *Brierley and Fryirs* [2005] provides a good example of this type of approach in that it emphasizes linkages between river form and process and their capacity to adjust in any given setting and recommends applying procedures at the catchment scale. The framework also suggests that appraisals of condition and recovery potential build on evolutionary trajectories of each channel reach within the catchment.

Palmer et al. [1997] and *McDonald et al.* [2004] distinguish between three basic approaches to river restoration. The 'field of dreams' approach relies on traditional engineering techniques to engineer river form to the desired condition in the hope that this will create the processes necessary to maintain that form. In the 'system function' approach, the initial conditions required to achieve a restoration project's goals are identified and ameliorated. The 'keystone' method is based on the idea of identifying and incorporating crucial components of form and function, with the recognition that there must be an open-ended view of what the project achieves.

One of the great challenges of river restoration is that, unlike other projects undertaken by engineers, the dynamic nature of river form and process [*Ward et al.*, 2001] create a large and unavoidable component of uncertainty [*Nilsson et al.*, 2003]. As *Hillman and Brierley* [2008] and *Sear et al.* [2008] emphasize, expressing uncertainty as a statement of plausible outcome and/or significance is an informative statement of knowledge, rather than an admission of ignorance, but most restoration projects do not yet effectively incorporate uncertainty. *Wilcock et al.* [2003] review some of the advantages and difficulties of applying geomorphic models to prediction and management, including river restoration.

River restoration, rehabilitation, and environmental assessment have received increasing attention since the 1970s [*Orsborn and Anderson*, 1986; *Brookes*, 1988, 1990; *National Academy*, 1992; *Brierley and Fryirs*, 2008]. Although much of this attention has been directed toward the lowland rivers that have been most altered by human activities, mountain rivers have also been the focus of studies [*Jaeggi and Zarn*, 1999; *Loheide and Gorelick*, 2007]. The headwater reaches of many river basins control not only downstream water and sediment movement, but also the biological integrity of the whole river ecosystem, as in the case of salmonid fish in western North America [*Salo and Cundy*, 1987]. At the same time, the growing worldwide pressure on land and its renewable resources is forcing people into geomorphically unstable mountainous regions and thus increasing human exposure to natural hazards and the consequent need to mitigate those hazards [*Eisbacher*, 1982; *Davies*, 1991]. The 6100 km^2 Province of Trent, Italy, for example, has more than 10,000 check dams on 4,000 km of mountain river channels, and an annual budget of more than $15 million for debris-flow control [*Armanini et al.*, 1991]. Attempts to rehabilitate these intensely engineered channels may be complicated by remaining structures that can be damaged by the associated increase in sediment movement [*Bechteler et al.*, 1994] or by a lack of attention to potential upstream and downstream influences on the rehabilitation measures [*Newson and Leeks*, 1987].

In many cases, it is not enough to simply re-create a historical channel planform. This planform may no longer be stable or functional because of changes in flow, sediment supply, bank stability, or local water table [*Kondolf et al.*, 2002; *Clarke et al.*, 2003; *Long et al.*, 2003; *Brooks and Brierley*, 2004; *James*, 2006; *Merz et al.*, 2006; *Loheide and Gorelick*, 2007]. It is more important, but also more difficult, to restore at least some function to the river rather than focusing restoration solely on form [*Pitlick and Wilcock*, 2001; *Smith and Prestegaard*, 2005; *Wohl et al.*, 2005; *Brierley and Fryirs*, 2008]. *Lepori et al.* [2005] find no difference in fish and invertebrate diversity between channelized and restored stream reaches in northern Sweden, for example, even though placement of wood and boulders in the restored streams had increased structural heterogeneity. These results imply that restored form had not translated into restored function. Function must also include ecological processes. A constructed stream in the Canadian Arctic had lower species richness and density of macroinvertebrates than unimpacted streams, for example, because the damaged riparian zone of the constructed stream did not supply authochthonous organic matter that was critical to the trophic structure of the aquatic ecosystem [*Jones et al.*, 2008]. Connectivity is a critical component of restored function; *Gowan and Fausch* [1996] demonstrate that fish habitat enhancement projects increase trout abundance on streams in the U.S. Rocky Mountains as long as fish dispersal remains unimpeded within the drainage. Building on definitions of *biological integrity* that incorporate maintenance of biodiversity and productivity of aquatic and riparian ecosystems [*SER*, 2002], *Graf* [2001] proposes assessing the *physical integrity* of rivers, which he defines as 'a set of active fluvial processes and landforms wherein the channel, flood plains, sediments, and overall spatial configuration maintain a dynamic equilibrium, with adjustments not exceeding limits of change defined by societal values' (p. 6). The phrase regarding limits of change defined by societal values acknowledges that it is not possible to completely restore process and form in many regulated rivers, but *Wheaton et al.* [2004a, 2004b] provide an example of restoring limited but important process and form in rivers supporting salmonid fish. *Brierley and Fryirs* [2008] also provide a useful overview of these concepts and the related ideas of *ecological resilience* (the capacity of an ecosystem to sustain all critical measures of functionality, such that the self-sustaining capacity of the system is regained or maintained) and *river health* (a societal perspective that frames biophysical concerns for river condition in relation to human values for river goods and associated ecosystem services).

Typically, restoration projects will be designed based on the characteristics of otherwise analogous but undisturbed stream reaches, known as *reference reaches*. *Trainor and Church* [2003] propose the use of a Euclidean distance measure of dissimilarity for selecting channel pairs that are sufficiently similar to act as reference reaches. The characteristics of the river prior to disturbance can also be used for reference. This approach is controversial, however, for several reasons [*Hughes et al.*, 2005; *Newson and Large*, 2006; *McAllister*, 2008]. Appropriate references may not be available. Catchment parameters may have changed since the time chosen for historical reference. Climate change has been continuous throughout the Holocene and is currently accelerating. Exotic species are present in many areas. Reference sites provide relatively short-term representations of current conditions in a few locations and may not reveal the natural range of variability. Finally, definitions of reference conditions may be too static to support sustainable strategies for restoration and management. Despite these potential complications, the use of relatively unaltered reference reaches to provide at least basic guidelines or targets for restoration remains extremely popular and widespread. The rationale behind this approach is that reference streams provide some basis for understanding (i) how regional climate, geology, topography, and biota interact to create spatial and temporal variability in stream ecosystems, (ii) the hydrologic and geomorphic conditions to which native riparian and aquatic biota have adapted, and (iii) specific examples of values for stream parameters such as channel width/depth ratio or bedform dimensions.

Mountain rivers in the European Alps illustrate some of the challenges and contemporary responses for river restoration in steep channels. Form and process along Alpine rivers have been altered by centuries of changes in land cover for agriculture, river alteration for navigation, flood control, and gravel mining, and flow regulation for hydroelectric power (Figure 6.30). Channelization, bank stabilization and levees along braided reaches and impoundment behind dams have caused incision and narrowing, as well as limiting lateral and longitudinal connectivity along the rivers [*Muhar et al.*, 2008]. The resulting channel instability, flood hazards, and decreases in biodiversity have brought river restoration to the forefront in the Alps. The emphasis on restoration is reinforced by the European Water Directive Frame-work and its objective of attaining good ecological status in rivers by 2015 [*Newson and Large*, 2006]. Restoration of Alpine rivers relies on identifying the critical processes and parameters for improving geomorphic and ecological conditions under restricted boundary conditions. The entire historical width of the river corridor typically cannot be returned to a natural condition, for example, although localized channel widening and levee setbacks can be used to at least partially reconnect main and secondary channels and floodplains and to restore sediment movement within the hillslope-floodplain-channel system. *Habersack and Piégay* [2008] provide an over-view of human impacts and river restoration in the Alps.

Attempts to manage mountain rivers must balance historical land-use activities and environmental impacts [*Wohl et al.*, 2008]. As public pressure is brought to bear on the government agencies that manage large tracts of public land in the United States, the agencies have developed standardized procedures for assessing and monitoring channel condition on these lands [e.g., *Ziemer*, 1998; *Fitzpatrick*, 2001]. The U.S. Bureau of Land Management (BLM), for example, manages 110 million ha of public lands, 9% of which is riparian-wetland areas. The BLM developed a field-based checklist that is used to rank river channels as being in (i) proper functioning condition, (ii) functional but at risk, or (iii) non-functional, with respect to ecological status based on characteristics of hydrology, vegetation, and erosion/deposition [*USDI*, 1993]. This approach is also used by the U.S. Forest Service, which relies on *Rosgen*'s [1994] stream classification (see chapter 4) for inventorying rivers and for predicting likely channel response to land use activities. One of the great challenges for these agencies is to develop a field-based classification system that can be applied by technicians with minimal training, but that is based on geomorphically meaningful categories [*Miller and Ritter*, 1996; *Wohl et al.*, 2007].

Once a portion of a river is perceived as being nonfunctional or at risk, passive or active river restoration may be implemented. *Passive restoration* may involve removing structures within the channel or altering hillslope conditions within the drainage basin so as to control water and sediment yield to the channel. *Active restoration* involves introducing structures into the channel or artificially re-shaping the channel to a desired configuration. Active restoration may implement traditional engineering materials such as concrete, or may implement "soft" or "bio" engineering methods in which natural materials such as instream wood are used [*Newbury and Gaboury*, 1993; *Abbe et al.*, 1998; *Bolton et al.*, 1998; *Drury et al.*, 1998; *Pess et al.*, 1998; *Lester and Boulton*, 2008] or beavers are re-introduced [*Beechie et al.*, 2008]. *Bennett et al.* [2008] use flume experiments and numerical simulations, for example, to evaluate how meanders can be introduced along a channel by using managed vegetation plantings as an alternative to the widespread practice of creating a sinuous planform using heavy machinery and then anchoring the channel banks with riprap. *Evette et al.* [2009] review the use of bioengineering techniques along rivers in western Europe.

Although restoration structures built with natural materials may be more esthetically pleasing than concrete, they may not be any more effective than traditional, 'hard' engineering approaches. Structures designed to enhance fish habitat date back to the 1890s in the United States [*Thompson and Stull*, 2002]. Despite a lack of any evidence that these structures actually increased habitat and fish populations, their use continued with little modification into the 1970s [*Thompson*, 2006b] and many contemporary deflector and bank-cover structures still closely reflect these original designs. Where detailed examinations have been conducted, as on structures emplaced during the 1930s and 1950s along the Blackledge River in Connecticut, USA, they indicate that the structures have actually decreased habitat by interfering with natural channel processes that create and maintain habitat [*Thompson*, 2002b; Figure 6.31]. As noted by *Wilcock* [1997], active re-shaping of channels is unlikely to succeed if the new channel configuration is based on water and sediment yield estimates pre-watershed disturbance or on measurements from a nearby, undisturbed channel. The artificially restored channel must be allowed to equilibrate with contemporary water and sediment yield.

Many channel restoration projects have not had the desired effect or been able to withstand even normal flood flows [*Kondolf et al.*, 2001; *Thompson*, 2003; *Kondolf*, 2006]. The small mountain watershed of Tenmile Creek, west of Denver, Colorado, USA, provides an example [*Babcock*, 1986]. During the 1880s two narrow-gauge railroads changed the creek's location. In the course of the 20th century a gas pipeline, several power distribution systems, and a buried telephone cable, all with associated roads and construction, further disturbed the creek. Upstream mining introduced toxic drainage and mill tailings into the creek and during the 1970s five kilometers of channel were relocated for Interstate Highway 70. Following this relocation, half a million dollars were spent on channel rehabilitation, which included log and rock check dams and deflectors, and habitat rocks (large boulders). The rehabilitated channel remained stable for approximately 1.5 years, until a large flood removed most of the check dams, and returned the channel to its pre-rehabilitation configuration [*Babcock*, 1986]. It is worthwhile to remember the difficulties of re-creating naturally functioning river ecosystems as we continue to alter mountain drainage basins and river channels.

Examples of successful restoration in steep mountain streams are appearing in the literature. Successful restoration of step-pool streams, as summarized in *Chin et al.* [2009a,b], suggests that the principle of maximum flow resistance [*Abraham et al.*, 1995] provides a useful guiding framework and that restoration designs that allow for self-organization through natural adjustments following initial engineering installation are more likely to remain stable under changing conditions in the drainage. Some of the geomorphic and ecological effects of step-pool sequences can also be restored by using morphologically based artificial steps constructed of large boulders rather than traditional concrete check dams [*Comiti et al.*, 2009].

Different types of models can be used to predict the likely effects of restoration on parameters such as available habitat [*Gore*, 2001; *Hauer and Lorang*, 2004; *Singer and Dunne*, 2006]. Just as human effects on mountain rivers are most effectively considered in the context of cumulative effects, so multiple restoration projects within large watersheds can be prioritized and coordinated using an integrated approach that considers cumulative benefits of the projects [*Kondolf et al.*, 2008].

A new aspect of river restoration in the United States involves the concept of stream mitigation banking, which gives developers the option to offset construction impacts by purchasing 'credits' generated by for-profit companies that restore damaged streams [*Lave et al.*, 2008]. The commodity is quantified as a stream mitigation unit, which can be complex to define because it requires establishing the quantity and quality of the commodity to be traded. In some states this approach is a major private-sector source of funding for stream restoration [*Lave et al.*, 2008].

6.4.2. Instream and Channel Maintenance Flows and the Natural Flow Regime

A large component of river restoration involves mitigating the impacts of flow regulation on physical processes and river ecosystems. A variety of hydrologic and hydraulic models are used to simulate the effects of differing flow regime on aquatic and riparian habitat [*Carter et al.*, 1998]. *Giers et al.* [1998], for example, present a methodology for assessing the ecological and hydrological effects of dam construction on headwater channels.

Specification of minimum flow levels, commonly known as *instream flow*, has typically been undertaken to protect water quality, river recreation, riparian vegetation, or fish. Expanding this concept to include a range of flow levels that mimic features of the natural hydrograph enhances the potential for maintaining diverse river processes and forms [*Rathburn et al.*, 2009]. River scientists increasingly focus on the need to maintain the entire trophic structure of a stream in order to protect fish populations, for example. *Buffagni* [2001] describe the BENHFOR (BENthic Habitat FOR optimum flow reckoning) procedure, which is designed to identify optimum and minimum flow conditions for macroinvertebrate communities.

Channel maintenance flows typically focus on the characteristics of flow neces-sary to maintain specific physical characteristics of a channel, such as sediment transport or flood conveyance. Channel maintenance flows are thus analogous to bankfull, effective, or dominant discharges. Channel maintenance flows may specify particular magnitude and frequency of flow to achieve a very limited objective or may incorporate a broader range of flows. *Doyle et al.* [2007] recommend constructing a cumulative sediment discharge curve to estimate the effective discharge for a particular channel reach.

One of the challenges to restoring instream flow or a more natural flow regime, particularly in arid and semiarid regions, is water supply. Water banks, in which water rights are temporarily purchased from willing sellers on an as-needed basis, provide one mechanism for providing sufficient flows for environmental needs [*Burke et al.*, 2004]. Assuming that water can be made available, there remains the challenge of specifying flow magnitude, frequency, duration, and timing – specifying an interannual hydrograph – that achieves restoration or preservation targets, as first conceptualized by *Poff et al.* [1997]. *Rathburn et al.* [2009] provide an example of identifying targets related to bedload transport, pool-riffle maintenance, and overbank flooding/bank erosion for riparian vegetation along the North Fork Poudre River, Colorado, USA and then designing an appropriate flow regime to achieve these targets.

Although ecologists tend to focus on the critical importance of natural variability in water discharge, geomorphologists increasingly emphasize the equally vital role of sediment input and output in maintaining channel complexity and habitat heterogene-ity [*Pitlick and Wilcock*, 2001]. Sediment flux involves both sufficient, but not excessive, sediment supply, and the flow necessary to transport supplied sediment. Water and sediment supply can be separate or closely interrelated issues, depending partly on sediment sources within a catchment.

6.5. Summary

People have directly and indirectly altered mountain rivers for centuries as they have used the land, water, and other resources of mountainous regions. In many cases, patterns of resource use have exacerbated hazards associated with debris flows, floods, and other processes in mountain drainage basins. Increasing hazards, along with resource scarcity, foster attempts to restore and rehabilitate mountain rivers. Although successful restoration projects exist, too often restoration is compromised by the lack of a clearly articulated model of how the river system functions, how restoration will manipulate that function to achieve desired goals, and of how historical or contempo-rary conditions outside of the immediate restoration site might influence the success of restoration activities.

The world's mountains are very diverse. They provide more than half of the world's fresh water, much of its timber, mineral, and grazing resources, and a spiritual refuge for humans [*Ives et al.*, 1997]. Many of the world's industrialized nations recognized more than a century ago that to protect the natural processes and organisms of mountain regions was also to protect the lives and livelihoods of humans living in the surrounding lowlands. One can easily argue, however, that the industrialized nations have not yet sufficiently protected their mountain drainage basins, even as the developing nations rush to exploit the natural resources in their own mountainous regions. And once people have come to depend on exploitation of a resource, protection of that resource from alteration becomes much more difficult. Norway, Switzerland and Canada each generate a substantial portion of their electricity from dams on mountain rivers [*Bandyopadhyay et al.*, 1997]. Colorado and many other western states in the USA obtain the majority of their water via dams and diversions on rivers in the Rocky Mountains (Figure 6.32). Although the citizens and governments of these regions are now well aware of the ecological and geomorphic disruptions along dammed rivers, only limited attempts are being made to alter flow regime from dams and relatively few dams are being removed in mountains of the western United States. As climate change and continuing population growth further stress existing water supplies, new dams and diversions are proposed for mountainous headwater streams.

At the most basic level, human communities in each mountain watershed and each nation will have to decide what balance they find acceptable between resource protection and resource use. In the worst-case scenario, such decisions will not be consciously made, but will instead simply result from accidents of development, the increasing pressure of human population, and the relative strengths of contending factions. In the best-case scenario, contemporary scientific understanding of mountain rivers and watersheds will be systematically used to achieve the desired balance. In either case, the professional responsibility of river scientists extends to both understanding how mountain rivers operate and disseminating that understanding to the wider public so as to facilitate informed decisions. Let us hope and work for the best-case scenario.

7. FIELD DATA, EXPERIMENTS, COMPUTERS, BRAINS

The diversity of conditions present among mountain rivers makes it difficult to create a single, grand synthesis. Differences in lithology, tectonic regime, and climate create characteristics that vary systematically by region, as well as unique conditions for individual drainage basins. Whatever coherent picture we may be able to develop is still emerging as a result of continuing advances in historical and process studies. Recent developments in geochronology, numerical modeling, and remote sensing have allowed our understanding of rates and styles of Quaternary tectonism to expand rapidly, including our understanding of the feedbacks among tectonism, erosion, topography, and climate. Simultaneously, field studies, lab experiments, and numerical modeling are enhancing our insight into the mechanics of sediment production on hillslopes and the movement of water and sediment down slope and through drainage networks.

Among the generalizations we can make about mountain rivers is that there is likely to be a strong structural influence on drainage development. Tectonic regime likely influences the arrangement of channels in the drainage network, sediment supply to the channel, the rate of channel incision, and the longitudinal profile of the channel. Bedrock is more likely to be present along mountain than along lowland rivers and mountain rivers have relatively high rates of bedrock channel incision. These rates are spatially variable because of different erosional-process domains. Riparian and aquatic communities correlate strongly with process domains, although the communities are also influenced by biological history and processes.

Mountain rivers are likely to have straight or convex longitudinal profiles with knickpoints. Longitudinal changes in lithology, climate, tectonic regime, or Quaternary history tend to create longitudinal changes in channel form and process, so that mountain rivers are most appropriately characterized at the reach scale. For a reach with a given gradient, discharge regime, and sediment supply, channel morphology and process domain are relatively predictable.

Mountain rivers typically have non-logarithmic velocity distributions and highly turbulent flow which is more commonly close to critical conditions than flow in lowland rivers. Mountain channels tend to be coarse-grained, with high boundary resistance to erosion and high flow resistance coefficients. The system is commonly

Mountain Rivers Revisited
Water Resources Monograph 19
Copyright 2010 by the American Geophysical Union.
10.1029/2010WM001045

supply-limited, with low suspended sediment loads and episodic bedload discharge, although high sediment yields may occur following a disturbance such as a volcanic eruption or forest fire. Mountain rivers exhibit high sediment connectivity: Tectonic, climatic, and lithologic changes are more likely to influence water and sediment yields in mountain rivers in the absence of filtering mechanisms associated with thick, stable regolith and broad valley bottoms and floodplains that buffer water and sediment changes. Drainage area-discharge relations are not linear because of downstream variations in runoff-generating mechanisms, which tend to be strongly seasonal and spatially variable. Instream wood plays a critical role in form and process along forested mountain rivers. Hyporheic zones are of limited extent relative to those associated with lower gradient rivers, but nonetheless exert an important influence on water chemistry and aquatic ecology in mountain rivers.

Some mountain rivers are strongly influenced by the history of Quaternary climate change or Pleistocene glaciation. Pleistocene or contemporary glaciation may control discharge, river chemistry, sediment supply, and valley morphology. The history of high-magnitude floods, such as landslide-dam or glacier-lake outburst floods, may also dominate channel and valley morphology. Debris flows and floods tend to exert a stronger influence on mountain river morphology and process than they do on lowland rivers because these extreme events that can overcome the high boundary resistance of mountain channels are more frequent in mountain than lowland catchments.

What do we *not* know about mountain river form and process that we would like to know? In the first edition I listed four basic categories here; hydraulics, sediment transport, bedforms and channel morphology, and river incision and longitudinal profile development. We have made impressive progress in each of these areas since 2000 and I no longer find it necessary to list bedforms and channel morphology as a basic category. Field and flume studies of step-pool and pool-riffle sequences allow us predict with reasonable accuracy the conditions under which these bedforms will occur and to quantify the processes by which they are created and maintained. The other three categories remain in the list, to which I now add river restoration.

1. *Hydraulics.* Recent syntheses such as that by *Ferguson* [2007] highlight the progress made in measuring hydraulic variables including the distribution of velocity and shear stress, sources of flow resistance, forms of turbulence, and interactions between turbulent flow and the channel boundaries through field and flume measurements during the past two decades. The picture emerging from these measurements suggests that hydraulic processes differ significantly by channel form and that we need to distinguish by channel form when developing predictive equations. Although all steep channels obey the same physics, the interactions among gradient, discharge, and sediment supply result in distinct forms that behave differently with respect to sources and magnitude of flow resistance and mechanics of sediment movement. The distinction among cascade, step-pool, plane-bed, and pool-riffle channel types [*Montgomery and Buffington*, 1997] appears at present to provide an effective framework for distinguishing hydraulic regimes among steep channels because these categories reflect differences in clast size and mobility relative to flow depth

that both influence and reflect hydraulics. We still need more detailed field data sets on the distribution of hydraulic forces under differing flow magnitudes across the range of channel forms present in mountain rivers. Although we have made good progress in developing field datasets of bedload movement (see below), we have no analogous collection of spatially and temporally detailed hydraulic measurements.

2. *Sediment transport.* Three aspects of sediment transport along mountain rivers remain particularly difficult to measure and predict: particle entrainment under highly spatially and temporally variable conditions; the impact of limitations in sediment supply on entrainment and transport; and the stability or persistence of the coarse bed-surface layer and of bedforms. Until recently, the many equations proposed for sediment transport along gravel-bed channels were based on only a very few field datasets, such as the dataset from Oak Creek, Oregon, USA [*Milhous*, 1973]. The development of highly instrumented watersheds that include sites for measuring water and sediment discharge (e.g., Rio Cordon, Italy) and the dissemination of results from these sites in the peer-reviewed literature have increased the basic data available, but mountainous watersheds remain under-represented in sediment transport records and many sets of direct measurements are too short in duration to include the large-magnitude fluctuations in water and sediment yield that we know characterize mountainous regions. Instrumented watersheds are expensive to maintain, but they yield extremely valuable data that can provide the basis for new analyses for years to come.

3. *River incision and longitudinal profile development.* Field studies, flume experiments, and numerical modeling have increased our understanding of the spatial distribution and relative importance of different incisional processes in mountain drainage networks. We now have quantitative expressions for incision by debris flows, knickpoint retreat, and bedload abrasion, as well as equations for predicting where each of these processes acts. We still need field tests of these equations in environments different than those for which they were developed. We have constrained the conditions under which the stream-power law accurately describes bedrock erosion and the conditions under which it does not. We have substantially increased the number of site-specific estimates of incision rates over Holocene and Quaternary time spans. Several competing numerical models describe how bedrock channels erode through time. Current investigations focus on the specific mechanics of bedrock erosion and the balance between incision and channel widening in relation to base level stability, substrate resistance, and sediment supply. We also need more field-based studies of the spatial differentiation among bedrock channel incisional processes across a range of field settings.

4. *River restoration.* We have made great strides in broad quantitative understanding of mountain rivers. We can constrain rates of incision over geologic time scales, for example, or predict the spatial distribution of channel morphologic types across a drainage network. It remains challenging to apply our understanding within a contemporary, management-oriented context, however, because of uncertainty.

Engineers have noted for decades that rivers have more dependent than independent variables, making it very difficult to make precise, deterministic predictions of change. Concepts such as the *probabilistic river* [*Graf*, 2001] reflect this uncertainty. Uncertainty is inherent and acceptable in scientific research, but it can become problematic in applied settings where, for example, there is limited physical space for river adjustment and uncertainty in our ability to predict river adjustment creates difficulties. Consequently, we still need better ways of expressing uncertainty when communicating with other stakeholders in river management and we need to become more effective in communicating to non-scientists our understanding of spatial and temporal variability in river form and process. Given the scientific community's understanding of rivers in general, including mountain rivers, the limited success of river restoration projects to date is disappointing. Unsuccessful river restoration projects are not only a waste of time, money, and human energy that can do more harm than good; they also discredit the idea of river restoration and the geomorphic community's understanding of rivers.

A common theme to the four topics outlined above is a lack of field data; limited spatially and temporally detailed measurements of hydraulics, limited datasets of bed-load from diverse field settings, limited characterization of the process domains and mechanics of bedrock erosion in different field settings, and limited case studies of successfully communicating uncertainty and implementing river restoration projects that achieve the desired goals. Tremendous progress has been made in field-based studies of mountain rivers during the past decade and many of these results have been incorporated into quantitative expressions of process [*Dietrich et al.*, 2003]. The field remains our ultimate source of information and, when combined with experimental studies, numerical modeling, and our own intelligence, has yielded powerful insights into mountain rivers. Our understanding of steep rivers remains much more limited, however, than our knowledge of low-gradient, alluvial rivers, and this creates a critical challenge as growing human populations continue to expand into mountain environments. Hydrological data on water levels, discharge, sediment and water quality, for example, are indispensable for water engineering works such as dams and for water-quality protection, water resources management, zoning, insurance, standards, legislation, and flood forecasting, yet these data are lacking for mountain regions because of difficult access, high spatial variability, sparse settlement with limited services, and a harsh physical environment [*Kundzewicz*, 1997]. It seems appropriate to have some understanding of how mountain rivers operate as integrated physical-biological systems [*Piégay and Schumm*, 2003] before we completely disrupt the river ecosystem. We might thereby avoid some of the costly mistakes that we have made along lowland rivers and better equip ourselves for the increasingly intensive river management of the 21st century.

GLOSSARY

Alluvial fan	a depositional feature created by fluvial process, debris flows, rockfall, and landslides either in the piedmont environment at the base of a mountain range or within wider valley segments of mountain rivers
Base level	the lowest point to which a river can flow; local base levels can be stable points that are intermediate along the river's course (e.g., lakes, waterfalls that are stationary or eroding very slowly), whereas the base level for each river is its downstream-most point, whether this be another, larger river, a basin with internal drainage, or the ocean
Bedload	sediment moving in contact with the streambed via rolling, sliding or saltating
Bed material load	sediment composed of particle sizes that are found in abundance on the streambed; this sediment can move as bedload or suspended load
Catchment	the area defined by drainage divides and occupied by the channels that drain to any particular reference point; also know as drainage basin or watershed
Channel head	the upstream boundary of concentrated water flow and sediment transport between definable banks
Channel reach	a length of channel, typically tens to hundreds of meters long, with consistent morphology and gradient
Cirque	ampitheater-shaped depressions in valley headwalls formed by freeze-thaw weathering and glacial erosion
Complex response	a reaction to an external perturbation, such as base level fall, that involves multiple oscillations in the variables responding to the perturbation
Creep	slow downslope movement of material in response to gravity, typically facilitated by wetting and drying or freezing and thawing
D_{50}	the grain size in a cumulative frequency distribution for which 50% of the distribution is smaller in size; D_x refers to the grain size for x% of the distribution is smaller in size
Disturbance	in ecology, a discrete, punctuated killing, displacement or damaging of one or more individuals or colonies that directly or indirectly creates an opportunity for new individuals or colonies to become established; in geomorphology, the term is typically used to refer to events beyond the mean condition, such as floods or debris flows
Drainage basin	the area defined by drainage divides and occupied by the channels that drain to any particular reference point; also know as catchment or watershed

Fluvial system	the sediment-source area, transportation network, and depositional sites of a river network; synonymous with catchment, drainage basin, or watershed
Geomorphic transport laws	a mathematical statement derived from a physical principle or mechanism, which expresses the mass flux or erosion caused by one or more processes
Geotherm	a curve that illustrates the temperature change with depth within the Earth
Gorge	a narrow, steep-walled bedrock canyon
Gradient (stream)	the downstream slope of a streambed
Hanging valley	a tributary valley that enters the main valley part way up the main valley wall as a result of lesser incision
Headcut	a sudden change in elevation or knickpoint at the upstream end of an incised channel, typically a headcut migrates upstream
Human impact	alterations to river channel process and form (direct), or to the water, sediment, nutrients, and other materials entering rivers (indirect), as a result of human activities
Hydraulic jump	an abrupt rise in the water surface when flow transitions from critical or supercritical to subcritical conditions; the jump typically takes the form of a standing wave
Isostatic rebound	an increase in crustal elevation as mass is removed through processes such as glacial melting or erosion of rock and sediment
Knickpoint	a relatively short, steep or nearly vertical section of a channel; a steeper section that is not vertical is sometimes known as a knickzone
Longitudinal profile	a plot of streambed elevation versus distance downstream from a reference point, typically the channel head or stream head
Orogen	an extensive belt of rocks deformed by mountain building; a large-scale mountain range such as the Himalaya
Overland flow	water flowing on the ground surface in direct response to precipitation, also known as runoff
Primary consumers	in ecology, organisms that eat the autotrophs (autotrophs are organisms that produce complex organic compounds from simple inorganic molecules using energy from light or inorganic chemical reactions)
Process domains	areas in which distinct geomorphic processes create a disturbance pattern that influences ecosystem structure and dynamics; discrete process domains are spatially differentiated based on differences in geomorphic processes
Resistance	hydraulic resistance results from water flowing over a rough boundary (see roughness)
Roughness	hydraulic roughness is a measure of the surface irregularity of the streambed and banks
Runoff	water flowing on the ground surface in direct response to precipitation, also known as overland flow

Sediment yield	the volume of sediment transported past a point, typically expressed in volume per unit area per unit time
Shear stress	the force exerted on a surface such as a streambed by flowing water, commonly estimated as the product of the specific weight of water, the mean depth of flow, and the local streambed gradient
Sorting	a measure of the uniformity of grain size distribution of sediment; well sorted sediments have a narrow range of grain sizes and poorly sorted sediment have a wide range of grain sizes
Stationarity	the assumption that any hydrologic variable (e.g., annual streamflow, annual maximum flood) has a time-invariant probability density function with statistical properties such as mean and standard deviation that do not vary with time
Stream head	the upstream boundary of perennial flow within a channel network
Stream order	numerical classification of stream segments, most commonly defined such that channel segments without tributaries are first order, and the junction of two channel segments of equal order creates a channel segment of higher order (i.e., two first order channels join to form a second order channel)
Stream power	a measure of the rate of stream energy expenditure, typically defined as the product of discharge, streambed gradient, and the specific weight of water (total stream power) or the product of shear stress and velocity (stream power per unit area)
Suspended load	sediment transported within the water column, with minimal contact with the streambed
Tectonism	the processes of folding, faulting or other deformation of the lithosphere that typically result in vertical movement of the lithosphere
Terrace	former floodplain surface that is now isolated from the active channel and is infrequently or never inundated by overbank floods – Fill: terrace formed in unconsolidated materials – Strath: terrace formed on bedrock
Thalweg	a line connecting the deepest parts of a channel
Threshold	the limit of equilibrium or stable river form and process; when a threshold is exceeded, major change of form and/or process is likely to occur
Turbulence	in turbulent flow, water does not move in parallel layers; instead, velocity fluctuates continuously in all directions within the fluid, which can include components of flow directed laterally and vertically, as well as longitudinally (or downstream)
Velocity profile	a plot of velocity magnitude and direction at different depths within the water column
Watershed	the area defined by drainage divides and occupied by the channels that drain to any particular reference point; also know as catchment or drainage basin

REFERENCES

Aalto, R., and W. Dietrich (2005), Sediment accumulation determined with ^{210}Pb geochronology for Strickland River flood plains, Papua New Guinea, in *Sediment Budgets I*, edited by D. E. Walling and A. J. Horowitz, *IAHS Publ., 291*, 303–309.

Aarts, B. G. W., F. W. B. van den Brink, and P. H. Nienhuis (2004), Habitat loss as the main cause of the slow recovery of fish faunas of regulated large rivers in Europe: The transversal floodplain gradient, *River Res. Appl., 20*, 3–23.

Abbe, T. B., and D. R. Montgomery (1996), Large woody debris jams, channel hydraulics and habitat formation in large rivers, *Reg. Rivers Res. Manage., 12*, 201–221.

Abbe, T. B., and D. R. Montgomery (2003), Patterns and processes of wood debris accumulation in the Queets River basin, Washington, *Geomorphology, 51*, 81–107.

Abbe, T. B., D. R. Montgomery, G. R. Pess, T. Drury, and C. Petroff (1998), Emulating organized chaos: Engineered log jams as a tool for rehabilitating fluvial environments, *Eos Trans. AGU, 79*, F345.

Abbott, J. T. (1976), Geologic map of the Big Narrows quadrangle, Larimer County, Colorado, map, scale 1:24,000, U.S. Geol. Surv., Denver, Colo.

Abdallah, C., J. Chorowicz, R. Boukheir, and D. Dhont (2007), Comparative use of processed satellite images in remote sensing of mass movements; Lebanon as a case study, *Int. J. Remote Sens., 28*, 4409–4427.

Aberle, J., and V. Nikora (2006), Statistical properties of armored gravel bed surfaces, *Water Resour. Res., 42*, W11414, doi:10.1029/2005WR004674.

Aberle, J., V. Nikora, M. Henning, B. Ettmer, and B. Hentschel (2010), Statistical characterization of bed roughness due to bed forms: A field study in the Elbe River at Aken, Germany, *Water Resour. Res., 46*, W03521, doi:10.1029/2008WR007406.

Aberle, J., and G. M. Smart (2003), The influence of roughness structure on flow resistance on steep slopes, *J. Hydraul. Res., 41*, 259–269.

Abernethy, B., and I. D. Rutherfurd (1998), Where along a river's length will vegetation most effectively stabilise stream banks?, *Geomorphology, 23*, 55–75.

Abernethy, B., and I. D. Rutherfurd (2001), The distribution and strength of riparian tree roots in relation to riverbank reinforcement, *Hydrol. Processes, 15*, 63–79.

Abrahams, A. D., and P. Gao (2006), A bed-load transport model for rough turbulent open-channel flows on plane beds, *Earth Surf. Processes Landforms, 31*, 910–928.

Abrahams, A. D., G. Li, and J. F. Atkinson (1995), Step-pool streams: Adjustment to maximum flow resistance, *Water Resour. Res., 31*, 2593–2602.

Abrahams, A. D., and A. J. Parsons (1991), Relation between sediment yield and gradient on debris-covered hillslopes, Walnut Gulch, Arizona, *Geol. Soc. Am. Bull., 103*, 1109–1113.

Acker, S. A., S. L. Johnson, F. J. Swanson, and G. S. Kennedy (1999), Interactions of riparian vegetation and geomorphology from the stand to watershed scale in the western Cascade Range of Oregon, paper presented at Annual Meeting, Ecol. Soc. of Am., Spokane, Wash.

Ackroyd, P., and R. J. Blakely (1984), En masse debris transport in a mountain stream, *Earth Surf. Processes Landforms, 9*, 307–320.

Adams, C. C. (1901), Base-leveling and its faunal significance, *Am. Nat., 5*, 839–852.

Adams, J. (1979), Sediment loads of North Island rivers, New Zealand - a reconnaissance, *J. Hydrol. N. Z., 18*, 36–48.

Adams, K. D., W. W. Locke, and R. Rossi (1992), Obsidian-hydration dating of fluvially reworked sediments in the West Yellowstone region, Montana, *Quat. Res., 38*, 180–195.

Adams, R. K., and J. A. Spotila (2005), The form and function of headwater streams based on field and modeling investigations in the southern Appalachian Mountains, *Earth Surf. Processes Landforms*, *30*, 1521–1546.

Adenlof, K. A., and E. E. Wohl (1994), Controls on bedload movement in a subalpine stream of the Colorado Rocky Mountains, USA, *Arct. Alp. Res.*, *26*, 77–85.

Agata, Y. (1994), Change in runoff characteristics of a mountain river caused by a gigantic failure and debris flow, paper presented at International Symposium on Forest Hydrology, Univ. of Tokyo, Tokyo.

Aggett, G. R., and J. P. Wilson (2009), Creating and coupling a high-resolution DTM with a 1-D hydraulic model in a GIS for scenario-based assessment of avulsion hazard in a gravel-bed river, *Geomorphology*, *113*, 21–34.

Agirre, U., M. Goni, J. J. Lopez, and F. N. Gimena (2005), Application of a unit hydrograph based on subwatershed division and comparison with Nash's instantaneous unit hydrograph, *Catena*, *64*, 321–332.

Aizen, V., E. Aizen, and J. Melack (1995), Characteristics of runoff formation at the Kirgizskiy Alatoo, Tien Shan, in *Biogeochemistry of Seasonally Snow-covered Catchments*, edited by K. A. Tonnessen, M. W. Williams, and M. Tranter, *IAHS Publ.*, *228*, 413–430.

Ajward, M. H., and I. Muzik (2000), A spatially varied unit hydrograph model, *J. Environ. Hydrol.*, *8*, 1–8.

Akiyama, M., and T. Muto (2006), The origin of stream terraces in the alluvial lower reaches of rivers; an experimental examination of the Complex Response model, *J. Geol. Soc. Jpn.*, *112*, 315–330.

Akojima, I. (1996), Designs of medium scale hazard maps of mountain slopes in Japan, *GeoJournal*, *38*, 365–372.

Aksoy, H., M. L. Kavvas, and J. Yoon (2003), Physically-based mathematical formulation for hillslope-scale prediction of erosion in ungauged basins, in *Erosion Prediction in Ungauged Basins: Integrating Methods and Techniques*, edited by D. de Boer et al., *IAHS Publ.*, *279*, 101–108.

Alexander, D. (1983), Gods handy-worke in wonders - landslide dynamics and natural hazard implications of a sixteenth century disaster, *Prof. Geogr.*, *35*, 314–323.

Alexander, D. E. (2008), A brief survey of GIS in mass-movement studies, with reflections on theory and methods, *Geomorphology*, *94*, 261–267.

Alexandrov, Y., H. Cohen, J. B. Laronne, and I. Reid (2009), Suspended sediment load, bed load, and dissolved load yields from a semiarid drainage basin: A 15-year study, *Water Resour. Res.*, *45*, W08408, doi:10.1029/2008WR007314.

Al-Farraj, A., and A. M. Harvey (2005), Morphometry and depositional style of late Pleistocene alluvial fans; Wadi Al-Bih, northern UAE and Oman, in *Alluvial fans; Geomorphology, Sedimentology, Dynamics*, edited by A. M. Harvey, A. E. Mather, and M. Stokes, *Geol. Soc. Spec. Publ.*, *251*, 85–94.

Alho, P., A. Kukko, H. Hyyppä, H. Kaartinen, J. Hyyppä, and A. Jaakkola (2009), Application of boat-based laser scanning for river survey, *Earth Surf. Processes Landforms*, *34*, 1831–1838.

Ali, G. A., and A. G. Roy (2009), Revisiting hydrologic sampling strategies for an accurate assessment of hydrologic connectivity in humid temperate systems, *Geogr. Compass*, *3*, 350–374.

Ali, K. F., and D. De Boer (2003), Construction of sediment budgets in large-scale drainage basins: The case of the upper Indus River, in *Erosion Prediction in Ungauged Basins: Integrating Methods and Techniques*, edited by D. de Boer et al., *IAHS Publ.*, *279*, 206–215.

Alila, Y., P. K. Kuraś, M. Schnorbus, and R. Hudson (2009), Forests and floods: A new paradigm sheds light on age-old controversies, *Water Resour. Res.*, *45*, W08416, doi:10.1029/2008WR007207.

Allan, J. D. (1995), *Stream Ecology: Structure and Function of Running Waters*, 388 pp., Chapman and Hall, London.

Allen, A. W. (1983), Habitat suitability index models: Beaver, report, 20 pp, U.S. Fish and Wildlife Serv., Ft. Collins, Colo.

Allen, P. M., R. Hobbs, and N. D. Maier (1989), Downstream impacts of a dam on a bedrock fluvial system, Brazos River, central Texas, *Bull. Assoc. Eng. Geol.*, *26*, 165–189.

Alley, R. B., D. E. Lawson, G. J. Larson, E. B. Evenson, and G. S. Baker (2003), Stabilizing feedbacks in glacier-bed erosion, *Nature, 424,* 758–760.

Allmendinger, N. E., J. E. Pizzuto, N. Potter, T. E. Johnson, and W. C. Hession (2005), The influence of riparian vegetation on stream width, eastern Pennsylvania, USA, *Geol. Soc. Am. Bull., 117,* 229–243.

Allred, T. M., and J. C. Schmidt (1999), Channel narrowing by vertical accretion along the Green River near Green River, Utah, *Geol. Soc. Am. Bull., 111,* 1757–1772.

Almedeij, J., and P. Diplas (2005), Bed load sediment transport in ephemeral and perennial gravel bed streams, *Eos Trans. AGU, 86,* 429–434.

Almond, P., J. Roering, and T. C. Hales (2007), Using soil residence time to delineate spatial and temporal patterns of transient landscape response, *J. Geophys. Res., 112,* F03S17, doi:10.1029/2006JF000568.

Alonso, C. V. (2004), Transport mechanics of stream-borne logs, in *Riparian Vegetation and Fluvial Geomorphology, Water Sci. Appl.,* vol. 8, edited by S. J. Bennett and A. Simon, pp. 59–69, AGU, Washington, D. C.

Alonso, C. V., S. J. Bennett, and O. R. Stein (2002), Predicting head cut erosion and migration in concentrated flows typical of upland areas, *Water Resour. Res., 38*(12), 1303, doi:10.1029/2001WR001173.

Al-Rawas, G., and C. Valeo (2009), Characteristics of rainstorm temporal distributions in arid mountainous and coastal regions, *J. Hydrol., 376,* 318–326.

Al-Taiee, T. M. (1990), The influence of a dam on the downstream degradation of a riverbed: Case study of the Tigris River, in *Hydrology in Mountainous Regions II. Artificial Reservoirs, Water and Slopes,* edited by R. O. Sinniger and M. Monbaron, *IAHS Publ., 194,* 3–10.

Altunkaynak, A., and K. B. Strom (2009), A predictive model for reach morphology classification in mountain streams using multilayer perceptron methods, *Water Resour. Res., 45,* W12501, doi:10.1029/2009WR008055.

Alvera, B., and J. M. Garcia-Ruiz (2000), Variability of sediment yield from a high mountain catchment, central Spanish Pyrenees, *Arct. Antarct. Alp. Res., 32,* 478–484.

Ambers, R. K. R., D. L. Druckenbrod, and C. P. Ambers (2006), Geomorphic response to historical agriculture at Monument Hill in the Blue Ridge foothills of central Virginia, *Catena, 65,* 49–60.

Amerson, B. E., D. R. Montgomery, and G. Meyer (2008), Relative size of fluvial and glaciated valleys in central Idaho, *Geomorphology, 93,* 537–547.

Amin, I. E., and A. M. Jacobs (2007), Accounting for sediment sources and sinks in the linear regression analysis of the suspended sediment load of streams: The Rio Puerco, New Mexico, as an example, *Environ. Geosci., 14,* 1–14.

Amoros, C. (1991), Changes in side-arm connectivity and implications for river system management, *Rivers, 2,* 105–112.

Amos, C. B., and D. W. Burbank (2007), Channel width response to differential uplift, *J. Geophys. Res., 112,* F02010, doi:10.1029/2006JF000672.

Anders, A. M., S. G. Mitchell, and J. H. Tomkin (2010), Cirques, peaks, and precipitation patterns in the Swiss Alps: Connections among climate, glacial erosion, and topography, *Geology, 38,* 239–242.

Anders, A. M., G. H. Roe, B. Hallet, D. R. Montgomery, N. J. Finnegan, and J. Putkonen (2006), Spatial patterns of precipitation and topography in the Himalaya, in *Tectonics, Climate, and Landscape Evolution,* edited by S. D. Willett et al., *Spec. Pap. Geol. Soc. Am., 398,* 39–53.

Anders, N. S., A. C. Seijmonsbergen, and W. Bouten (2009), Modelling channel incision and alpine hillslope development using laser altimetry data, *Geomorphology, 113,* 35–46.

Andersen, D. C., D. J. Cooper, and K. Northcott (2007), Dams, floodplain land use, and riparian forest conservation in the semiarid Upper Colorado River basin, USA, *Environ. Manage., 40,* 453–475.

Anderson, D. E. (2005), Holocene fluvial geomorphology of the Amargosa River through Amargosa Canyon, California, *Earth Sci. Rev., 73,* 291–307.

Anderson, J. K., S. M. Wondzell, M. N. Gooseff, and R. Haggerty (2005), Patterns in stream longitudinal profiles and implications for hyporheic exchange flow at the H.J. Andrews Experimental Forest, Oregon, USA, *Hydrol. Processes*, *19*, 2931–2949.

Anderson, R. S. (1994), Evolution of the Santa Cruz Mountains, California, through tectonic growth and decay, *J. Geophys. Res.*, *99*(B10), 20,161–20,179.

Anderson, R. S., and S. P. Anderson (2010), *Geomorphology: The Mechanics and Chemistry of Landscapes*, 654 pp., Cambridge Univ. Press, Cambridge, U. K.

Anderson, R. S., J. L. Repka, and G. S. Dick (1996), Explicit treatment of inheritance in dating depositional surfaces using in situ ^{10}Be and ^{26}Al, *Geology*, *24*, 47–51.

Anderson, R. S., P. Molnar, and M. A. Kessler (2006a), Features of glacial valley profiles simply explained, *J. Geophys. Res.*, *111*, F01004, doi:10.1029/2005JF000344.

Anderson, R. S., C. A. Riihimaki, E. B. Safran, and K. R. MacGregor (2006b), Facing reality: Late Cenozoic evolution of smooth peaks, glacially ornamented valleys, and deep river gorges of Colorado's Front Range, in *Tectonics, Climate, and Landscape Evolution*, edited by S. D. Willett et al., *Spec. Pap. Geol. Soc. Am.*, *398*, 397–418.

Anderson, S. P., and W. E. Dietrich (2001), Chemical weathering and runoff chemistry in a steep headwater catchment, *Hydrol. Processes*, *15*, 1791–1815.

Anderson, S. P., W. E. Dietrich, D. R. Montgomery, R. Torres, M. E. Conrad, and K. Loague (1997), Subsurface flow paths in a steep, unchanneled catchment, *Water Resour. Res.*, *33*, 2637–2653.

Anderson, S. P., S. A. Longacre, and E. R. Kraal (2003), Patterns of water chemistry and discharge in the glacier-fed Kennicott River, Alaska; evidence for subglacial water storage cycles, *Chem. Geol.*, *202*, 297–312.

Andersson, E., C. Nilsson, and M. E. Johansson (2000), Effects of river fragmentation on plant dispersal and riparian flora, *Reg. Rivers Res. Manage.*, *16*, 83–89.

Anderton, S. P., S. M. White, and B. Alvera (2002), Micro-scale spatial variability and the timing of snow melt runoff in a high mountain catchment, *J. Hydrol.*, *268*, 158–176.

Anderton, S. P., S. M. White, and B. Alvera (2004), Evaluation of spatial variability in snow water equivalent for a high mountain catchment, *Hydrol. Processes*, *18*, 435–453.

Andreadis, K. M., P. Storck, and D. P. Lettenmaier (2009), Modeling snow accumulation and ablation processes in forested environments, *Water Resour. Res.*, *45*, W05429, doi:10.1029/2008WR007042.

Andreasson, J., S. Bergstrom, B. Carlsson, and L. P. Graham (2003), The effect of downscaling techniques on assessing water resources impacts from climate change scenarios, in *Water Resources Systems; Water Availability and Global Change*, edited by S. Franks et al., *IAHS Publ.*, *280*, 160–164.

Andreoli, A., F. Comiti, and M. A. Lenzi (2007), Characteristics, distribution and geomorphic role of large woody debris in a mountain stream of the Chilean Andes, *Earth Surf. Processes Landforms*, *32*, 1675–1692.

Andrews, E. D. (1979), *Scour and fill in a stream channel, East Fork River, western Wyoming, U.S. Geol. Surv. Prof. Pap.*, vol. 1117, 49 pp.

Andrews, E. D. (1983), Entrainment of gravel from naturally sorted riverbed material, *Geol. Soc. Am. Bull.*, *94*, 1225–1231.

Andrews, E. D. (1984), Bed-material entrainment and hydraulic geometry of gravel-bed rivers in Colorado, *Geol. Soc. Am. Bull.*, *95*, 371–378.

Andrews, E. D. (1986), Downstream effects of Flaming Gorge Reservoir on the Green River, Colorado and Utah, *Geol. Soc. Am. Bull.*, *97*, 1012–1023.

Andrews, E. D. (2000), Bed material transport in the Virgin River, Utah, *Water Resour. Res.*, *36*, 585–596.

Andrews, E. D., and D. C. Erman (1986), Persistence in the size distribution of surficial bed material during an extreme snowmelt flood, *Water Resour. Res.*, *22*, 191–197.

Andrews, E. D., and J. M. Nelson (1989), Topographic response of a bar in the Green River, Utah to variation in discharge, in *River Meandering, Water Resour. Monogr.*, vol. 12, edited by S. Ikeda and G. Parker, pp. 463–484, AGU, Washington, D. C.

Andrews, E. D., and J. D. Smith (1992), A theoretical model for calculating marginal bedload transport rates of gravel, in *Dynamics of Gravel-Bed Rivers*, edited by P. Billi et al., pp. 41–52, John Wiley, Chichester, U. K.

Angelaki, V., and J. M. Harbor (1995), Impacts of flow diversion for small hydroelectric power plants on sediment transport, northwest Washington, *Phys. Geogr.*, *16*, 432–443.

Annandale, G. W. (2002), Quantification of the relative ability of rock to resist scour, in *Rock Scour due to Falling High-Velocity Jets*, edited by A. J. Schleiss and E. Bollaert, pp. 201–212, Swets and Zeitlinger, Lisse, Netherlands.

Anthony, D. M., and D. E. Granger (2007), A new chronology for the age of Appalachian erosional surfaces determined by cosmogenic nuclides in cave sediments, *Earth Surf. Processes Landforms*, *32*, 874–887.

Anthony, E. J., and M. Julian (1999), Source-to-sink sediment transfers, environmental engineering and hazard mitigation in the steep Var River catchment, French Riviera, southeastern France, *Geomorphology*, *31*, 337–354.

Applegarth, M. T. (2004), Assessing the influence of mountain slope morphology on pediment form, south-central Arizona, *Phys. Geogr.*, *25*, 225–236.

Arbellay, E., M. Stoffel, and M. Bollschweiler (2010), Dendrogeomorphic reconstruction of past debris-flow activity using injured broad-leaved trees, *Earth Surf. Processes Landforms*, *35*, 399–406.

Arboleya, M.-L., J. Babault, L. A. Owen, A. Teixell, and R. C. Finkel (2008), Timing and nature of Quaternary fluvial incision in the Ouarzazate foreland basin, Morocco, *J. Geol. Soc. London*, *165*, 1059–1073.

Arcement, G. J., Jr., and V. R. Schneider (1989), *Guide for selecting Manning's roughness coefficients for natural channels and flood plains*, U.S. Geol. Surv. Water Supply Pap., vol. 2339, 38 pp.

Armanini, A. (1992), Variation of bed and sediment load mean diameters due to erosion and deposition processes, in *Dynamics of Gravel-Bed Rivers*, edited by P. Billi et al., pp. 351–359, John Wiley, Chichester, U. K.

Armanini, A., F. Dellagiacoma, and L. Ferrari (1991), From the check dam to the development of functional check dams, in *Fluvial Hydraulics of Mountain Regions*, edited by A. Armanini and G. Di Silvio, pp. 331–344, Springer, Braunschweig, Germany.

Armanini, A., and G. Di Silvio (Eds.) (1991), *Fluvial Hydraulics of Mountain Regions*, 468 pp., Springer, Berlin.

Armanini, A., and C. Gregoretti (2005), Incipient sediment motion at high slopes in uniform flow condition, *Water Resour. Res.*, *41*, W12431, doi:10.1029/2005WR004001.

Armanini, A., and M. Larcher (2001), Rational criterion for designing opening of slit-check dam, *J. Hydraul. Eng.*, *127*, 94–104.

Armbruster, M., M. Abiy, and K.-H. Feger (2003), The biogeochemistry of two forested catchments in the Black Forest and the Eastern Ore Mountains (Germany), *Biogeochemistry*, *65*, 341–368.

Arnaud-Fassetta, G., E. Cossart, and M. Fort (2005), Hydro-geomorphic hazards and impact of man-made structures during the catastrophic flood of June 2000 in the Upper Guil catchment (Queyras, Southern French Alps), *Geomorphology*, *66*, 41–67.

Arp, C. D., J. C. Schmidt, M. A. Baker, and A. K. Myers (2007), Stream geomorphology in a mountain lake district: Hydraulic geometry, sediment sources and sinks, and downstream lake effects, *Earth Surf. Processes Landforms*, *32*, 525–543.

Arnett, R. R. (1971), Slope form and geomorphological process: An Australian example, in *Slopes: Form and Process*, edited by D. Brunsden, Inst. Br. Geogr. Spec. Publ., *3*, 81–92.

Arrigoni, A. S., G. C. Poole, L. A. K. Mertes, S. J. O'Daniel, W. W. Woessner, and S. A. Thomas (2008), Buffered, lagged, or cooled? Disentangling hyporheic influences on temperature cycles in stream channels, *Water Resour. Res.*, *44*, W09418, doi:10.1029/2007WR006480.

Arsenault, A. M., and A. J. Meigs (2005), Contribution of deep-seated bedrock landslides to erosion of a glaciated basin in southern Alaska, *Earth Surf. Processes Landforms*, *30*, 1111–1125.

Arzani, N. (2005), The fluvial megafan of Abarkoh Basin (central Iran): An example of flash-flood sedimentation in arid lands, in *Alluvial Fans: Geomorphology, Sedimentology, Dynamics*, edited by A. M. Harvey, A. E. Mather, and M. Stokes, *Geol. Soc. Spec. Publ.*, *251*, 41–59.

Asano, Y., T. Uchida, Y. Mimasu, and N. Ohte (2009), Spatial patterns of stream solute concentrations in a steep mountainous catchment with a homogeneous landscape, *Water Resour. Res.*, *45*, W10432, doi:10.1029/2008WR007466.

ASCE Task Committee on Sediment Transport and Aquatic Habitats, Sedimentation Committee (1992), Sediment and aquatic habitat in river systems, *J. Hydraul. Eng.*, *118*, 669–687.

Ashby, W. C. (2006), Sustainable stripmine reclamation, *Int. J. Mining Reclam. Environ.*, *20*, 87–95.

Ashida, K., S. Egashira, and N. Ando (1984), Generation and geometric features of step-pool bed forms, *Bull. Disaster Prev. Res. Inst. Kyoto Univ.*, *27*(B-2), 341–353.

Ashida, K., T. Takahashi, and T. Sawada (1976), Sediment yield and transport on a mountain small watershed, *Bull. Disaster Prev. Res. Inst. Kyoto Univ.*, *26*, 119–144.

Ashiq, M., J. C. Doering, and T. Hosoda (2006), Bed-load transport model based on fractional size distribution, *Can. J. Civ. Eng.*, *33*, 69–80.

Ashmore, P. E. (1982), Lab modelling of gravel braided stream morphology, *Earth Surf. Processes Landforms*, *7*, 201–225.

Ashmore, P. E. (1988), Bed load transport in braided gravel-bed stream models, *Earth Surf. Processes Landforms*, *13*, 677–695.

Ashmore, P. E. (1991), How do gravel-bed rivers braid?, *Can. J. Earth Sci.*, *28*, 326–341.

Ashmore, P. E. (1993), Anabranch confluence kinetics and sedimentation processes in gravel-braided streams, in *Braided Rivers*, edited by J. L. Best and C. S. Bristow, *Geol. Soc. Spec. Publ.*, *75*, 129–146.

Ashworth, P. J. (1996), Mid-channel bar growth and its relationship to local flow strength and direction, *Earth Surf. Processes Landforms*, *21*, 103–123.

Ashworth, P. J., J. L. Best, and M. A. Jones (2007), The relationship between channel avulsion, flow occupancy and aggradation in braided rivers: Insights from an experimental model, *Sedimentology*, *54*, 497–513.

Ashworth, P. J., and R. I. Ferguson (1989), Size-selective entrainment of bedload in gravel bed streams, *Water Resour. Res.*, *25*, 627–634.

Ashworth, P. J., R. I. Ferguson, and D. M. Powell (1992), Bedload transport and sorting in braided channels, in *Dynamics of Gravel-Bed Rivers*, edited by P. Billi et al., pp. 497–513, John Wiley, Chichester, U. K.

Astaras, T. (1984), Drainage basins as process-response systems: An example from central Macedonia, North Greece, *Earth Surf. Processes Landforms*, *9*, 333–341.

Attal, M., and J. Lavé (2006), Changes of bedload characteristics along the Marsyandi River (central Nepal): Implications for understanding hillslope sediment supply, sediment load evolution along fluvial networks, and denudation in active orogenic belts, in *Tectonics, Climate, and Landscape Evolution*, edited by S. D. Willett et al., *Spec. Pap. Geol. Soc. Am.*, *398*, 143–171.

Attal, M., and J. Lavé (2009), Pebble abrasion during fluvial transport: Experimental results and implications for the evolution of the sediment load along rivers, *J. Geophys. Res.*, *114*, F04023, doi:10.1029/2009JF001328.

Attal, M., G. E. Tucker, A. C. Whittaker, P. A. Cowie, and G. P. Roberts (2008), Modeling fluvial incision and transient landscape evolution: Influence of dynamic channel adjustment, *J. Geophys. Res.*, *113*, F03013, doi:10.1029/2007JF000893.

Aubert, D., A. Probst, P. Stille, and D. Viville (2002), Evidence of hydrological control of Sr behavior in stream water (Strengbach catchment, Vosges Mountains, France), *Appl. Geochem.*, *17*, 285–300.

Auble, G. T., J. M. Friedman, and M. L. Scott (1994), Relating riparian vegetation to present and future streamflows, *Ecol. Appl.*, *4*, 544–554.

Auble, G. T., P. B. Shafroth, M. L. Scott, and J. E. Roelle (2007), Early vegetation development on an exposed reservoir: Implications for dam removal, *Environ. Manage.*, *39*, 806–818.

Aulenbach, B. T., R. P. Hooper, and O. P. Bricker (1996), Trends in the chemistry of precipitation and surface water in a national network of small watersheds, *Hydrol. Processes*, *10*, 151–181.

Axelrod, D. I., and P. H. Raven (1985), Origins of the Cordilleran flora, *J. Biogeogr.*, *12*, 21–47.

Axelson, J. N., D. J. Sauchyn, and J. Barichivich (2009), New reconstructions of streamflow variability in the South Saskatchewan River basin from a network of tree ring chronologies, Alberta, Canada, *Water Resour. Res.*, *45*, W09422, doi:10.1029/2008WR007639.

Axtmann, E. V., and R. F. Stallard (1995), Chemical weathering in the South Cascade glacier basin, comparison of subglacial and extra-glacial weathering, in *Biogeochemistry of Seasonally Snow-covered Catchments*, edited by K. A. Tonnessen, M. W. Williams, and M. Tranter, *IAHS Publ.*, *228*, 431–439.

Avis, C., J. Van Wetten, J. Seffer, G. Tinchev, and P. Weller (2000), Danube River basin: Wetlands and floodplains, in *The Root Causes of Biodiversity Loss*, edited by A. Wood, P. Stedman-Edwards, and J. Mang, pp. 183–212, Earthscan Publications, London.

Avouac, J. P., and E. B. Burov (1996), Erosion as a driving mechanism of intracontinental mountain growth, *J. Geophys. Res.*, *101*(B8), 747–769.

Babcock, W. H. (1986), Tenmile Creek: A study of stream relocation, *Water Resour. Bull.*, *22*, 405–415.

Babiński, Z. (2005), The relationship between suspended and bed load transport in river channels, in *Sediment Budgets I*, edited by D. E. Walling and A. J. Horowitz, *IAHS Publ.*, *291*, 182–188.

Bacchi, B., and V. Villi (2005), Runoff and floods in the Alps: An overview, in *Climate and Hydrology in Mountain Areas*, edited by C. de Jong, D. Collins, and R. Ranzi, pp. 217–220, John Wiley, Chichester, U. K.

Baetzing, W., and P. Messerli (1992), The Alps: An ecosystem in transformation, *The State of the World's Mountains: A Global Report*, edited by P. B. Stone, pp. 46–91, Zen Books, London.

Bagnold, R. A. (1956), The flow of cohesionless grains in fluids, *Philos. Trans. R. Soc. London, Ser. A*, *249*, 235–297.

Bagnold, R. A. (1966), *An approach to the sediment transport problem from general physics*, U.S. Geol. Surv. Prof. Pap., vol. 422, 37 pp.

Bagnold, R. A. (1977), Bed load transport by natural rivers, *Water Resour. Res.*, *13*, 303–312.

Bagnold, R. A. (1980), An empirical correlation of bedload transport rates in flumes and natural rivers, *Proc. R. Soc. London, Ser. A*, *372*, 453–473.

Baillie, B. R., and T. R. Davies (2002a), Effects of land use on the channel morphology of streams in the Moutere Gravels, Nelson, New Zealand, *J. Hydrol. N. Z.*, *41*, 19–45.

Baillie, B. R., and T. R. Davies (2002b), Influence of large woody debris on channel morphology in native forest and pine plantation streams in the Nelson region, New Zealand, *N. Z. J. Mar. Freshwater Res.*, *36*, 763–774.

Bailly, J.-S., Y. Le Coarer, P. Languille, C.-J. Stigermark, and T. Allouis (2010), Geostatistical estimations of bathymetric LiDAR errors on rivers, *Earth Surf. Processes Landforms*, *35*, 1199–1210.

Baker, J. P., and C. L. Schofield (1982), Aluminum toxicity to fish in acid waters, *Water Air Soil Pollut.*, *18*, 289–309.

Baker, V. R. (1973), Erosional forms and processes for the catastrophic Pleistocene Missoula floods in eastern Washington, in *Fluvial Geomorphology*, edited by M. Morisawa, pp. 123–148, State Univ. of New York, Binghamton.

Baker, V. R. (1974), Paleohydraulic interpretation of Quaternary alluvium near Golden, Colorado, *Quat. Res.*, *4*, 94–112.

Baker, V. R. (1977), Stream-channel response to floods with examples from central Texas, *Geol. Soc. Am. Bull.*, *88*, 1057–1071.

Baker, V. R. (1978), Large-scale erosional and depositional features of the scabland floods, in *The Channeled Scabland*, edited by V. R. Baker and D. Nummedal, pp. 81–115, NASA, Washington, D. C.

Baker, V. R. (1983), Paleoflood hydrologic techniques for the extension of streamflow records, in *Improving estimates from flood studies, Transportation Research Record 922*, pp. 18–23, Transp. Res. Board, Natl. Res. Counc., Washington, D. C.

Baker, V. R. (1987), Paleoflood hydrology and extraordinary flood events, *J. Hydrol.*, *96*, 79–99.

Baker, V. R. (1988a), Flood erosion, in *Flood Geomorphology*, edited by V. R. Baker, R. C. Kochel, and P. C. Patton, pp. 81–95, John Wiley, New York.

Baker, V. R., G. Benito, and A. N. Rudoy (1993), Paleohydrology of late Pleistocene superflooding, Altay Mountains, Siberia, *Science*, *259*, 348–350.

Baker, V. R., and R. C. Bunker (1985), Cataclysmic late Pleistocene flooding from glacial Lake Missoula, a review, *Quat. Sci. Rev.*, *4*, 1–41.

Baker, V. R., L. L. Ely, J. E. O'Connor, and J. B. Partridge (1987), Paleoflood hydrology and design applications, in *Regional Flood Frequency Analysis*, edited by V. P. Singh, pp. 339–353, D. Reidel, Boston, Mass.

Baker, V. R., and R. C. Kochel (1988), Flood sedimentation in bedrock fluvial systems, in *Flood Geomorphology*, edited by V. R. Baker, R. C. Kochel, and P. C. Patton, pp. 123–137, John Wiley, New York.

Baker, V. R., and G. Pickup (1987), Flood geomorphology of the Katherine Gorge, Northern Territory, Australia, *Geol. Soc. Am. Bull.*, *98*, 635–646.

Baker, V. R., and D. F. Ritter (1975), Competence of rivers to transport coarse bedload material, *Geol. Soc. Am. Bull.*, *86*, 975–978.

Baker, W. L. (1988b), Size-class structure of contiguous riparian woodlands along a Rocky Mountain river, *Phys. Geogr.*, *9*, 1–14.

Baker, W. L. (1989), Classification of the riparian vegetation of the montane and subalpine zones in western Colorado, *Great Basin Nat.*, *49*, 214–228.

Baker, W. L. (1990), Climatic and hydrologic effects on the regeneration of Populus angustifolia along the Animas River, Colorado, *J. Biogeogr.*, *17*, 59–73.

Baker, W. L., and G. M. Walford (1995), Multiple stable states and models of riparian vegetation succession on the Animas River, Colorado, *Ann. Assoc. Am. Geogr.*, *85*, 320–338.

Balamurugan, G. (1991), Some characteristics of sediment transport in the Sungai Kelang Basin, Malaysia, *J. Inst. Engineers, Malaysia*, *48*, 31–52.

Baldwin, J. E., H. F. Donley, and T. R. Howard (1987), On debris flow/avalanche mitigation and control, San Francisco Bay area, California, in *Debris Flows/Avalanches: Process, Recognition, and Mitigation*, edited by J. E. Costa and G. F. Wieczorek, pp. 223–236, Geol. Soc. of Am., Boulder, Colo.

Bales, R. C., N. P. Molotch, T. H. Painter, M. D. Dettinger, R. Rice, and J. Dozier (2006), Mountain hydrology of the western United States, *Water Resour. Res.*, *42*, W08432, doi:10.1029/2005WR004387.

Bali, R., D. D. Awasthi, and N. K. Tiwari (2004), Record of neotectonic activity in the Gangotri Glacier valley; some significant observations, *Geol. Surv. India Spec. Publ. Ser.*, *80*, 79–86.

Balin, D., C. Joerin, and A. Musy (2006), Hydrological conceptual models of two Haute-Mentue subcatchments through environmental tracing and TDR soil moisture measurements, in *Predictions in Ungauged Basins: Promise and Progress*, edited by M. Sivapalan et al., *IAHS Publ.*, *303*, 7–14.

Ball, J., and S. Trudgill (1995), Overview of solute modelling, in *Solute Modelling in Catchment Systems*, edited by S. T. Trudgill, pp. 3–56, John Wiley, Chichester, U. K.

Ballantyne, C. K. (2002), Debris flow activity in the Scottish Highlands: Temporal trends and wider implications for dating, *Stud. Geomorphol. Carpatho-Balcanica*, *36*, 7–27.

Banasik, K., M. Barszcz, and J. Brański (2005), Major components of a sediment budget for four river catchments in Poland, in *Sediment Budgets 2*, edited by A. J. Horowitz and D. E. Walling, *IAHS Publ.*, *292*, 32–36.

Banasik, K., and D. Bley (1994), An attempt at modelling suspended sediment concentration after storm events in an Alpine torrent, in *Dynamics and Geomorphology of Mountain Rivers*, edited by P. Ergenzinger and K.-H. Schmidt, pp. 161–170, Springer, Berlin.

Band, L. E., P. Patterson, N. Ramakrishna, and S. W. Running (1993), Forest ecosystem processes at the watershed scale: Incorporating hillslope hydrology, *Agric. For. Meteorol.*, *63*, 93–126.

Band, L. E., C. L. Tague, P. Groffman, and K. Belt (2001), Forest ecosystem processes at the watershed scale: Hydrological and ecological controls of nitrogen export, *Hydrol. Processes*, *15*, 2013–2028.

Bandyopadhyay, J., J. C. Rodda, R. Kattelmann, Z. W. Kundzewicz, and D. Kraemer (1997), Highland waters - A resource of global significance, in *Mountains of the World: A Global Priority*, edited by B. Messerli and J. D. Ives, pp. 131–155, The Parthenon, London.

Bänzinger, R., and H. Burch (1990), Acoustic sensors (hydrophones) as indicators for bed load transport in a mountain torrent, in *Hydrology in Mountainous Regions I. Hydrological Measurements: The Water Cycle*, edited by H. Lang and A. Musy, *IAHS Publ.*, *193*, 207–214.

Barclay, J. S. (1978), The effects of channelization on riparian vegetation and wildlife in south central Oklahoma, *Strategies for Protection and Management of Floodplain Wetlands and Other Riparian Ecosystems, USDA Forest Service General Tech. Rep. WO-12*, pp. 129–138, USDA Forest Service, Washington, D. C.

Bardou, E., and M. Jaboyedoff (2008), Debris flows as a factor of hillslope evolution controlled by a continuous or a pulse process?, *Geol. Soc. Spec. Publ.*, *296*, 63–78.

Bari, F., M. K. Wood, and L. Murray (1993), Livestock grazing impacts on infiltration rates in a temperate range of Pakistan, *J. Range Manage.*, *46*, 367–372.

Barker, D., and D. F. M. McGregor (1988), Land degradation in the Yallahs basin, Jamaica: Historical notes and contemporary observations, *Geography*, *73*(2), 116–124.

Barnard, P. L., L. A. Owen, and R. C. Finkel (2006), Quaternary fans and terraces in the Khumbu Himal south of Mount Everest; their characteristics, age and formation, *J. Geol. Soc. London*, *163*, 383–399.

Barnard, P. L., L. A. Owen, M. C. Sharma, and R. C. Finkel (2001), Natural and human-induced landsliding in the Garhwal Himalaya of northern India, *Geomorphology*, *40*, 21–35.

Barnard, P. L., L. A. Owen, M. C. Sharma, and R. C. Finkel (2004), Late Quaternary (Holocene) landscape evolution of a monsoon-influenced high Himalayan valley, Gori Ganga, Nanda Devi, NE Garhwal, *Geomorphology*, *61*, 91–110.

Barnes, C., and M. Bonell (2005), How to choose an appropriate catchment model, in *Forests, Water and People in the Humid Tropics*, edited by M. Bonell and L. A. Bruijnzeel, pp. 717–741, Cambridge Univ. Press, Cambridge, U. K.

Barnes, H. H., Jr. (1967), *Roughness characteristics of natural channels*, U.S. Geol. Surv. Water Supply Pap., vol. 1849, 213 pp.

Barnett, T. P., et al. (2008), Human-induced changes in the hydrology of the western United States, *Science*, *319*, 1080–1083.

Baron, J. (Ed.) (1992), *Biogeochemistry of a Subalpine Ecosystem: Loch Vale Watershed*, 8, pp. 142–186, Springer, New York.

Baron, J. S., E. J. Allstott, and B. K. Newkirk (1995), Analysis of long term sulfate and nitrate budgets in a Rocky Mountain basin, in *Biogeochemistry of Seasonally Snow-covered Catchments*, edited by K. A. Tonnessen, M. W. Williams, and M. Tranter, *IAHS Publ.*, *228*, 255–261.

Baron, J. S., D. S. Ojima, E. A. Holland, and W. J. Parton (1994), Analysis of nitrogen saturation potential in Rocky Mountain tundra and forest: Implications for aquatic systems, *Biogeochemistry*, *27*, 61–82.

Baron, J. S., H. M. Rueth, A. N. Wolfe, K. R. Nydick, E. J. Allstott, J. T. Minear, and B. Moraska (2000), Ecosystem responses to nitrogen deposition in the Colorado Front Range, *Ecosystems*, *3*, 352–368.

Barontini, S., A. Clerici, R. Ranzi, and B. Bacchi (2005), Saturated hydraulic conductivity and water retention relationships for alpine mountain soils, in *Climate and Hydrology in Mountain Areas*, edited by C. de Jong, D. Collins, and R. Ranzi, pp. 101–121, John Wiley, Chichester, U. K.

Barros, A. P., S. Chiao, T. J. Lang, D. Burbank, and J. Putkonen (2006), From weather to climate – Seasonal and interannual variability of storms and implications for erosion processes in the Himalaya, in *Tectonics, Climate, and Landscape Evolution*, edited by S. D. Willett et al., *Spec. Pap. Geol. Soc. Am.*, *398*, 17–38.

Barros, A. P., and D. P. Lettenmaier (1994), Dynamic modelling of orographically induced precipitation, *Rev. Geophys.*, *32*, 265–284.

Barry, J. J., J. M. Buffington, P. Goodwin, J. G. King, and W. W. Emmett (2008), Performance of bed-load transport equations relative to geomorphic significance: Predicting effective discharge and its transport rate, *J. Hydraul. Eng.*, *134*, 601–615.

Barry, J. J., J. M. Buffington, and J. G. King (2004), A general power equation for predicting bed load transport rates in gravel bed rivers, *Water Resour. Res.*, *40*, W10401, doi:10.1029/2004WR003190.

Barry, R. G. (2005), Alpine climate change and cryospheric responses: An introduction, in *Climate and Hydrology in Mountain Areas*, edited by C. de Jong, D. Collins, and R. Ranzi, pp. 1–4, John Wiley, Chichester, U. K.

Barry, R. G. (2008), *Mountain Weather and Climate*, 3rd ed., 512 pp., Cambridge Univ. Press, Cambridge, U. K.

Barry, R. G., and R. J. Chorley (1987), *Atmosphere, Weather, and Climate*, 5th ed., 460 pp., Methuen, New York.

Barsch, D. (Ed.) (1984), *High mountain research*, *Mt. Res. Dev.*, vol. 4, pp. 286–374.

Barsch, D., M. Gude, R. Mäusbacher, G. Schukraft, and A. Schulte (1994a), Sediment transport and discharge in a high arctic catchment (Liefdefjorden, NW Spitsbergen), in *Dynamics and Geomorphology of Mountain Rivers*, edited by P. Ergenzinger and K.-H. Schmidt, pp. 225–237, Springer, Berlin.

Barsch, D., H. Happoldt, R. Mäusbacher, L. Schrott, and G. Schukraft (1994b), Discharge and fluvial sediment transport in a semi-arid high mountain catchment, Agua Negra, San Juan, Argentina, in *Dynamics and Geomorphology of Mountain Rivers*, edited by P. Ergenzinger and K.-H. Schmidt, pp. 213–224, Springer, Berlin.

Barta, A. F., P. R. Wilcock, and C. C. C. Shea (1994), The transport of gravels in boulder-bed streams, in *Proceedings, ASCE Hydraulic Engineering '94 Conference, August 1–5, Buffalo, New York*, edited by G. V. Cotroneo and R. R. Rumer, pp. 780–784, Am. Soc. of Civ. Eng., New York.

Bartholomew, M. J., and H. H. Mills (1991), Old courses of the New River: Its late Cenozoic migration and bedrock control inferred from high-level stream gravels, southwestern Virginia, *Geol. Soc. Am. Bull.*, *103*, 73–81.

Bartnik, W., and A. Michalik (1994), Fluvial hydraulics of streams and mountain rivers with mobile bed, in *Proceedings, ASCE Hydraulic Engineering '94 Conference, August 1–5, Buffalo, New York*, edited by G. V. Cotroneo and R. R. Rumer, pp. 767–771, Am. Soc. of Civ. Eng., New York.

Batalla, R. J., C. De Jong, P. Ergenzinger, and M. Sala (1999), Field observations on hyperconcentrated flows in mountain torrents, *Earth Surf. Processes Landforms*, *24*, 247–253.

Batalla, R. J., and D. Vericat (2009), Hydrological and sediment transport dynamics of flushing flows: Implications for management in large Mediterranean rivers, *River Res. Appl.*, *25*, 297–314.

Bates, P. D. (2004), Remote sensing and flood inundation modeling, *Hydrol. Processes*, *18*, 2593–2597.

Bates, P. D., M. G. Anderson, J.-M. Hervouet, and J. C. Hawkes (1997), Investigating the behaviour of two-dimensional finite element models of compound channel flow, *Earth Surf. Processes Landforms*, *22*, 3–17.

Bates, P. D., and A. P. J. De Roo (2000), A simple raster-based model for flood inundation simulation, *J. Hydrol.*, *236*, 54–77.

Bates, P. D., and J.-M. Hervouet (1999), A new method for moving-boundary hydrodynamic problems in shallow water, *Proc. R. Soc. London, Ser. A*, *455*, 3107–3128.

Bates, P. D., M. Horritt, and J.-M. Hervouet (1998), Investigating two-dimensional, finite element predictions of floodplain inundation using fractal generated topography, *Hydrol. Processes*, *12*, 1257–1277.

Bates, P. D., M. S. Horritt, C. N. Smith, and D. Mason (1997), Integrating remote sensing observations of flood hydrology and hydraulic modelling, *Hydrol. Processes*, *11*, 1777–1795.

Bathurst, J. C. (1978), Flow resistance of large-scaleI roughness, *J. Hydraul. Div. Am. Soc. Civ. Eng.*, *104*, 1587–1603.

Bathurst, J. C. (1985), Flow resistance estimation in mountain rivers, *J. Hydraul. Eng.*, *111*, 625–641.

Bathurst, J. C. (1987a), Critical conditions for bed material movement in steep, boulder-bed streams, in *Erosion and Sedimentation in the Pacific Rim, Proceedings of the Corvallis Symposium, Aug. 1987*, edited by R. L. Beschta et al., *IAHS Publ., 165*, 309–318.

Bathurst, J. C. (1987b), Measuring and modelling bedload transport in channels with coarse bed materials, in *River Channels: Environment and Process*, edited by K. S. Richards, pp. 272–294, Basil Blackwell, Oxford.

Bathurst, J. C. (1988), Velocity profile in high-gradient, boulder-bed channels, *Proceedings of International Conference on Fluvial Hydraulics, Budapest, 30 May–9 June*, pp. 29–34, IAHR, Madrid.

Bathurst, J. C. (1990), Tests of three discharge gauging techniques in mountain rivers, in *Hydrology of Mountainous Areas*, edited by L. Molnar, *IAHS Publ., 190*, 93–100.

Bathurst, J. C. (1993), Flow resistance through the channel network, in *Channel Network Hydrology*, edited by K. Beven and M. J. Kirkby, pp. 69–98, John Wiley, Chichester, U. K.

Bathurst, J. C. (1994), At-a-site mountain river flow resistance variation, in *Proceedings, ASCE Hydraulic Engineering '94 Conference, August 1–5, Buffalo, New York*, edited by G. V. Cotroneo and R. R. Rumer, pp. 682–686, Am. Soc. of Civ. Eng., New York.

Bathurst, J. C. (2002), At-a-site variation and minimum flow resistance for mountain rivers, *J. Hydrol., 269*, 11–26.

Bathurst, J. C., and M. Ashiq (1998), Dambreak flood impact on mountain stream bedload transport after 13 years, *Earth Surf. Processes Landforms, 23*, 643–649.

Bathurst, J. C., G. B. Crosta, J. M. Garci-Ruiz, F. Guzzetti, M. A. Lenzi, and S. R. Aragüés (2003), DAMOCLES: Debris-fall assessment in mountain catchments for local end-users, in *Debris-flow Hazards Mitigation: Mechanics, Prediction, and Assessment*, edited by D. Rickenmann and C. Chen, pp. 1073–1083, Millpress, Rotterdam.

Bathurst, J. C., W. H. Graf, and H. H. Cao (1987), Bed load discharge equations for steep mountain streams, in *Sediment Transport in Gravel-Bed Rivers*, edited by C. R. Thorne, J. C. Bathurst, and R. D. Hey, pp. 453–477, John Wiley, Chichester, U. K.

Bathurst, J. C., L. Hubbard, G. J. L. Leeks, M. D. Newson, and C. R. Thorne (1990), Sediment yield in the aftermath of a dambreak flood in a mountain stream, in *Hydrology in Mountainous Regions II. Artificial Reservoirs, Water and Slopes*, edited by R. O. Sinniger and M. Monbaron, *IAHS Publ., 194*, 287–294.

Bathurst, J. C., A. Burton, B. G. Clarke, and F. Gallart (2006), Application of the SHETRAN basin-scale, landslide sediment yield model to the Llobregat Basin, Spanish Pyrenees, *Hydrol. Processes, 20*, 3119–3138.

Battin, T. J., L. A. Kaplan, S. Findlay, C. S. Hopkinson, E. Marti, A. I. Packman, J. D. Newbold, and F. Sabater (2008), Biophysical controls on organic carbon fluxes in fluvial networks, *Nat. Geosci., 1*, 95–100.

Battin, T. J., S. Luyssaert, L. A. Kaplan, A. K. Aufdenkampe, A. Richter, and L. J. Tranvik (2009), The boundless carbon cycle, *Nat. Geosci., 2*, 598–600.

Baumgart-Kotarba, M. (1986), Formation of coarse gravel bars and alluvial channels, braided Bialka River, Carpathians, Poland, in *International Geomorphology 1986, Part I*, edited by V. Gardiner, pp. 633–648, John Wiley, Chichester, U. K.

Baumgart-Kotarba, M., J.-P. Bravard, M. Chardon, V. Jomelli, S. Kedzia, A. Kotarba, P. Pech, and Z. Raczkowska (2003), High-mountain valley floors evolution during recession of alpine glaciers in the Massif des Ecrins, France, *Geogr. Pol., 76*, 65–87.

Baxter, C. V., K. D. Fausch, M. Murakami, and P. L. Chapman (2004), Fish invasion restructures stream and forest food webs by interrupting reciprocal prey subsidies, *Ecology, 85*, 2656–2663.

Baxter, R. M. (1977), Environmental effects of dams and impoundments, *Annu. Rev. Ecol. Syst., 8*, 255–283.

Bayard, D., and M. Stähli (2005), Effects of frozen soil on the groundwater recharge in alpine areas, in *Climate and Hydrology in Mountain Areas*, edited by C. de Jong, D. Collins, and R. Ranzi, pp. 73–83, John Wiley, Chichester, U. K.

Bayley, P. B. (1991), The flood-pulse advantage and the restoration of river-floodplain systems, *Reg. Rivers Res. Manage.*, *6*, 75–86.

Bayley, P. B. (1995), Understanding large river-floodplain ecosystems, *BioScience*, *45*, 153–158.

Bayley, P. B., and H. W. Li (1992), Riverine fishes, in *The Rivers Handbook*, edited by P. Calow and G. E. Petts, pp. 251–281, Blackwell Sci., Oxford, U. K.

Beaty, C. B. (1989), Great big boulders I have known, *Geology*, *17*, 349–352.

Beaty, C. B. (1990), Anatomy of a White Mountain debris-flow – the making of an alluvial fan, in *Alluvial Fans: A Field Approach*, edited by A. H. Rachocki and M. Church, pp. 69–89, John Wiley, New York.

Beaumont, C., and G. Quinlan (1994), A geodynamic framework for interpreting crustal-scale seismic reflectivity patterns in compressional orogens, *Int. J. Geophys.*, *116*, 754–783.

Becht, M., F. Haas, T. Heckmann, and V. Wichmann (2005), Investigating sediment cascades using field measurements and spatial modelling, in *Sediment Budgets I*, edited by D. E. Walling and A. J. Horowitz, *IAHS Publ.*, *291*, 206–213.

Bechteler, W., H.-J. Vollmers, and S. Wieprecht (1994), Model investigations into the influence of renaturalization on sediment transport, in *Dynamics and Geomorphology of Mountain Rivers*, edited by P. Ergenzinger and K.-H. Schmidt, pp. 37–52, Springer, Berlin.

Becker, A., and J. J. McDonnell (1998), Topographical and ecological controls of runoff generation and lateral flows in mountain catchments, in *Hydrology, Water Resources and Ecology in Headwaters*, edited by K. Kovar et al., *IAHS Publ.*, *248*, 199–206.

Becker, J. S., D. M. Johnston, D. Paton, G. T. Hancox, T. R. Davies, M. J. McSaveney, and V. R. Manville (2007), Response to landslide dam failure emergencies; issues resulting from the October 1999 Mount Adams landslide and dam-break flood in the Poerua River, Westland, New Zealand, *Nat. Hazards Rev.*, *8*, 35–42.

Bedford, D. R., and E. E. Small (2008), Spatial patterns of ecohydrologic properties on a hillslope-alluvial fan transect, central New Mexico, *Catena*, *73*, 34–48.

Bednarek, A. T., and D. D. Hart (2005), Modifying dam operations to restore rivers: Ecological responses to Tennessee River dam mitigation, *Ecol. Appl.*, *15*, 997–1008.

Beechie, T. J., M. Liermann, M. M. Pollock, S. Baker, and J. Davies (2006), Channel pattern and river-floodplain dynamics in forested mountain river systems, *Geomorphology*, *78*, 124–141.

Beechie, T. J., G. Pess, P. Kennard, R. E. Bilby, and S. Bolton (2000), Modeling recovery rates and pathways for woody debris recruitment in northwestern Washington streams, *N. Am. J. Fish. Manage.*, *20*, 436–452.

Beechie, T. J., M. M. Pollock, and S. Baker (2008), Channel incision, evolution and potential recovery in the Walla Walla and Tucannon River basins, northwestern USA, *Earth Surf. Processes Landforms*, *33*, 784–800.

Beechie, T. J., and T. H. Sibley (1997), Relationships between channel characteristics, woody debris, and fish habitat in northwestern Washington streams, *Trans. Am. Fish. Soc.*, *126*, 217–229.

Begin, Z. B., and M. Inbar (1984), Relationship between flows and sediment size in some gravel streams of the arid Negev, Israel, in *Sedimentology of Gravels and Conglomerates*, edited by E. H. Koster and R. J. Steel, *Can. Soc. Petrol. Geol., Mem.*, *10*, 69–75.

Benavides-Solorio, J., and L. H. MacDonald (2001), Post-fire runoff and erosion from simulated rainfall on small plots, Colorado Front Range, *Hydrol. Processes*, *15*, 2931–2952.

Bencala, K. E., and R. A. Walters (1983), Simulation of solute transport in a mountain pool-and-riffle stream: A transient storage model, *Water Resour. Res.*, *19*, 718–724.

Benda, L. E. (1990), The influence of debris flows on channels and valley floors in the Oregon Coast Range, USA, *Earth Surf. Processes Landforms*, *15*, 457–466.

Benda, L. E., and T. W. Cundy (1990), Predicting deposition of debris flows in mountain channels, *Can. Geotech. J.*, *27*, 409–417.

Benda, L. E., and T. Dunne (1987), Sediment routing by debris flow, in *Erosion and Sedimentation in the Pacific Rim*, edited by R. L. Beschta et al., *IAHS Publ.*, *165*, 213–223.

Benda, L. E., and T. Dunne (1997a), Stochastic forcing of sediment routing and storage in channel networks, *Water Resour. Res.*, *33*, 2865–2880.

Benda, L. E., and T. Dunne (1997b), Stochastic forcing of sediment supply to channel networks from landsliding and debris flow, *Water Resour. Res.*, *33*, 2849–2863.

Benda, L. E., and J. C. Sais (2003), A quantitative framework for evaluating the mass balance of in-stream organic debris, *For. Ecol. Manage.*, *172*, 1–16.

Benda, L. E., C. Veldhuisen, and J. Black (2003a), Debris flows as agents of morphological heterogeneity at low-order confluences, Olympic Mountains, Washington, *Geol. Soc. Am. Bull.*, *115*, 1110–1121.

Benda, L. E., D. Miller, J. Sais, D. Martin, R. Bilby, C. Veldhuisen, and T. Dunne (2003b), Wood recruitment processes and wood budgeting, in *The Ecology and Management of Wood in World Rivers*, edited by S. V. Gregory, K. L. Boyer, and A. M. Gurnell, pp. 49–73, Am. Fish. Soc., Bethesda, Md.

Benda, L. E., K. Andras, D. Miller, and P. Bigelow (2004), Confluence effects in rivers: Interactions of basin scale, network geometry, and disturbance regimes, *Water Resour. Res.*, *40*, W05402, doi:10.1029/2003WR002583.

Benda, L. E., M. A. Hassan, M. Church, and C. L. May (2005), Geomorphology of steepland headwaters: The transition from hillslopes to channels, *J. Am. Water Resour. Assoc.*, *41*, 835–851.

Ben-David, R., Y. Eyal, E. Zilberman, and D. Bowman (2002), Fluvial systems response to rift margin tectonics; Makhtesh Ramon area, southern Israel, *Geomorphology*, *45*, 147–163.

Bender, E. A., T. J. Case, and M. E. Gilpin (1984), Perturbation experiments in community ecology: Theory and practice, *Ecology*, *65*, 1–13.

Bendix, J., and C. M. Cowell (2010), Fire, floods and woody debris: Interactions between biotic and geomorphic processes, *Geomorphology*, *116*, 297–304.

Benischke, R., and T. Harum (1990), Determination of discharge rates in turbulent streams by salt tracer dilution applying a microcomputer system. Comparison with current meter measurements, in *Hydrology in Mountainous Regions I. Hydrological Measurements: The Water Cycle*, edited by H. Lang and A. Musy, *IAHS Publ.*, *193*, 215–221.

Benitez, S. (1989), Landslides: Extent and economic significance in Ecuador, in *Landslides: Extent and Economic Significance*, edited by E. E. Brabb and B. L. Harrod, pp. 123–126, A. A. Balkema, Rotterdam.

Benito, G., T. Grodek, and Y. Enzel (1998), The geomorphic and hydrologic impacts of the catastrophic failure of flood-control-dams during the 1996-Biescas flood (Central Pyrenees, Spain), *Z. Geomorphol.*, *42*, 417–437.

Benito, G., M. J. Machado, and A. Perez-Gonzalez (1996), Climate change and flood sensitivity in Spain, in *Global Continental Changes: The Context of Palaeohydrology*, edited by J. Branson, A. G. Brown, and K. J. Gregory, *Geol. Soc. Spec. Publ.*, *115*, 85–98.

Bennett, S. J. (2004), Effects of emergen riparian vegetation on spatially averaged and turbulent flow within an experimental channel, in *Riparian Vegetation and Fluvial Geomorphology*, edited by S. J. Bennett and A. Simon, pp. 29–41, AGU, Washington, D. C.

Bennett, S. J., and C. V. Alonso (2005), Kinematics of flow within headcut scour holes on hillslopes, *Water Resour. Res.*, *41*, W09418, doi:10.1029/2004WR003752.

Bennett, S. J., and J. Casali (2001), Effect of initial step height on headcut development in upland concentrated flows, *Water Resour. Res.*, *37*, 1475–1484.

Bennett, S. J., C. V. Alonso, S. N. Prasad, and M. J. M. Römkens (2000a), Experiments on headcut growth and migration in concentrated flows typical of upland areas, *Water Resour. Res.*, *36*, 1911–1922.

Bennett, S. J., J. Casalí, K. M. Robinson, and K. C. Kadavy (2000b), Characteristics of actively eroding ephemeral gullies in an experimental channel, *Trans. ASAE*, *43*, 641–649.

Bennett, S. J., T. Pirim, and B. D. Barkdoll (2002), Using simulated emergent vegetation to alter stream flow direction within a straight experimental channel, *Geomorphology*, *44*, 115–126.

Bennett, S. J., W. Wu, C. V. Alonso, and S. S. Y. Wang (2008), Modeling fluvial response to in-stream woody vegetation: Implications for stream corridor restoration, *Earth Surf. Processes Landforms*, *33*, 890–909.

Benson, M. A. (1950), Use of historical data in flood-frequency analysis, *Eos Trans. AGU*, *31*, 419–424.

Berg, N., A. Carlson, and D. Azuma (1998), Function and dynamics of woody debris in stream reaches in the central Sierra Nevada, California, *Can. J. Fish. Aquat. Sci.*, *55*, 1807–1820.

Berg, N. H. (1990), Water from the alpine: Niwot Ridge and the Green Valley Lakes, Colorado, in *Cold Regions Hydrology and Hydraulics*, edited by W. L. Ryan and R. D. Crissman, pp. 641–667, Am. Soc. of Civ. Eng., New York.

Berger, C., B. W. McArdell, B. Fritschi, and F. Schlunegger (2010), A novel method for measuring the timing of bed erosion during debris flows and floods, *Water Resour. Res.*, *46*, W02502, doi:10.1029/2009WR007993.

Berger, R. C., and R. L. Stockstill (1995), Finite-element model for high-velocity channels, *J. Hydraul. Eng.*, *121*, 710–716.

Bergeron, N. E. (1994), An analysis of flow velocity profiles, stream bed roughness, and resistance to flow in natural gravel bed streams, in *Proceedings, ASCE Hydraulic Engineering '94 Conference, August 1–5, Buffalo, New York*, edited by G. V. Cotroneo and R. R. Rumer, pp. 692–696, Am. Soc. of Civ. Eng., New York.

Berlin, M. M., and R. S. Anderson (2007), Modeling of knickpoint retreat on the Roan Plateau, western Colorado, *J. Geophys. Res.*, *112*, F03S06, doi:10.1029/2006JF000553.

Berlin, M. M., and R. S. Anderson (2009), Steepened channels upstream of knickpoints: Controls on relict landscape response, *J. Geophys. Res.*, *114*, F03018, doi:10.1029/2008JF001148.

Berner, E. K., and R. A. Berner (1987), *The Global Water Cycle: Geochemistry and Environment*, 397 pp., Prentice Hall, Englewood Cliffs, N. J.

Bernhardt, E. S., et al. (2005), Synthesizing US river restoration efforts, *Science*, *308*, 636–637.

Bernhardt, E. S., et al. (2007), Restoring rivers one reach at a time: Results from a survey of U.S. river restoration practitioners, *Restor. Ecol.*, *15*, 482–493.

Berris, S., and R. D. Harr (1987), Comparative snow accumulation and melt during rainfall in forested and clearcut plots in the western Cascades of Oregon, *Water Resour. Res.*, *23*, 135–142.

Berry, J. W. (1974), The climate of Colorado, in *Climates of the States*, vol. 2, pp. 595–613, Water Information Center, Port Washington, N. YX.

Berti, M., and A. Simoni (2007), Prediction of debris flow inundation areas using empirical mobility relationships, *Geomorphology*, *90*, 144–161.

Berti, M., R. Genevois, A. Simoni, and P. Rosella Tecca (1999), Field observations of a debris flow event in the Dolomites, *Geomorphology*, *29*, 265–274.

Bertoldi, W., and M. Tubino (2007), River bifurcations: Experimental observations on equilibrium configurations, *Water Resour. Res.*, *43*, W10437, doi:10.1029/2007WR005907.

Bertoldi, W., L. Zanoni, S. Miori, R. Repetto, and M. Tubino (2009), Interaction between migrating bars and bifurcations in gravel bed rivers, *Water Resour. Res.*, *45*, W06418, doi:10.1029/2008WR007086.

Bertoldi, W., L. Zanoni, and M. Tubino (2010), Assessment of morphological changes induced by flow and flood pulses in a gravel bed braided river: The Tagliamento River (Italy), *Geomorphology*, *114*, 348–360.

Beschta, R. L. (1983), The effects of large organic debris upon channel morphology: A flume study, in *Symposium on Erosion and Sedimentation*, edited by D. B. Simons, pp. 8-63–8-78, Simons, Li, and Associates, Ft. Collins, Colo.

Beschta, R. L. (1987), Conceptual models of sediment transport in streams, in *Sediment Transport in Gravel-Bed Rivers*, edited by C. R. Thorne, J. C. Bathurst, and R. D. Hey, pp. 387–419, John Wiley, Chichester, U. K.

Beschta, R. L., T. Blinn, G. E. Grant, G. G. Ice, and F. J. Swanson (1987), *Erosion and Sedimentation in the Pacific Rim, IAHS Publ.*, *165*.

Beschta, R. L., and W. L. Jackson (1979), The intrusion of fine sediment into a stable gravel bed, *J. Fish. Res. Board Can.*, *36*, 202–210.

Beschta, R. L., and W. S. Platts (1986), Morphological features of small streams: Significance and function, *Water Resour. Bull.*, *22*, 369–379.

Beschta, R. L., and W. J. Ripple (2006), River channel dynamics following extirpation of wolves in northwestern Yellowstone National Park, USA, *Earth Surf. Processes Landforms*, *31*, 1525–1539.

Beschta, R. L., and W. J. Ripple (2010), Recovering riparian plant communities with wolves in northern Yellowstone, USA, *Restor. Ecol.*, *18*, 380–389.

Bessason, B., G. Eiriksson, O. Thorarinsson, A. Thorarinsson, and S. Einarsson (2007), Automatic detection of avalanches and debris flows by seismic methods, *J. Glaciol.*, *53*, 461–472.

Best, J. L. (1992), On the entrainment of sediment and initiation of bed defects: Insights from recent developments within turbulent boundary layer research, *Sedimentology*, *39*, 797–811.

Best, J. L. (1993), On the interactions between turbulent flow structure, sediment transport and bedform development: Some considerations from recent experimental research, in *Turbulence: Perspectives on Flow and Sediment Transport*, edited by N. J. Clifford, J. R. French, and J. Hardisty, pp. 61–92, John Wiley, Chichester, U. K.

Bevan, J. (Ed.) (1983), *The Compleat Angler, 1653–1676/Izaak Walton*, 435 pp., Clarendon, Oxford, U. K.

Beven, K. J. (2001), *Rainfall-runoff Modeling: The Primer*, Wiley, Chichester, U. K.

Beven, K. J., and A. Binley (1992), The future of distributed models: Model calibration and uncertainty prediction, *Hydrol. Processes*, *6*, 279–298.

Beven, K. J., and N. J. Kirby (1979), A physically based variable contributing area model of basin hydrology, *Hydrol. Sci. Bull.*, *24*, 43–69.

Beven, K. J., and M. J. Kirby (1984), Testing a physically based flood forecasting model (TOPMODEL) for three UK catchments, *J. Hydrol.*, *69*, 119–143.

Beverage, J. P., and J. K. Culbertson (1964), Hyperconcentrations of suspended sediment, *J. Hydraul. Div. Am. Soc. Civ. Eng.*, *90*, 117–126.

Beylich, A. A., and D. Gintz (2004), Effects of high-magnitude/low-frequency fluvial events generated by intense snowmelt or heavy rainfall in arctic periglacial environments in northern Swedish Lapland and northern Siberia, *Geogr. Ann., Ser. A, Phys. Geogr.*, *86*, 11–29.

Beylich, A. A., O. Sandberg, U. Molau, and S. Wache (2006), Intensity and spatio-temporal variability of fluvial sediment transfers in an Arctic-oceanic periglacial environment in northernmost Swedish Lapland (Latnjavagge catchment), *Geomorphology*, *80*, 114–130.

Bezinge, A. (1987), Glacial meltwater streams, hydrology and sediment transport: The case of the Grande Dixence Hydroelectricity Scheme, in *International Geomorphology 1986, Part I*, edited by V. Gardiner, pp. 473–498, John Wiley, Chichester, U. K.

Bhowmik, N. G. (1984), Hydraulic geometry of floodplain, *J. Hydrol.*, *68*, 369–374.

Bhowmik, N. G., and M. Demissie (1982), Bed material sorting in pools and riffles, *J. Hydraul. Div. Am. Soc. Civ. Eng.*, *108*, 1227–1231.

Biedenharn, D. S. (1989), Knickpoint migration characteristics in the Loess Hills of northern Mississippi, USA, *US-China Sedimentation Symposium*, 8 pp.

Bierman, P. R., and K. K. Nichols (2004), Rock to sediment – slope to sea with [10]Be – rates of landscape change, *Annu. Rev. Earth Planet. Sci.*, *32*, 215–255.

Bierman, P. R., A. Lini, P. Zehfuss, A. Church, P. T. Davis, J. Southon, and L. Baldwin (1997), Postglacial ponds and alluvial fans: Recorders of Holocene landscape history, *GSA Today*, *7*(10), 1–8.

Bierman, P. R., J. M. Reuter, M. Pavich, A. C. Gellis, M. W. Caffee, and J. Larsen (2005), Using cosmogenic nuclides to contrast rates of erosion and sediment yield in a semi-arid, arroyo-dominated landscape, Rio Puerco Basin, New Mexico, *Earth Surf. Processes Landforms*, *30*, 935–953.

Biggs, B. J. F., M. J. Duncan, S. N. Francoeur, and W. D. Meyer (1997), Physical characterization of microform bed cluster refugia in 12 headwater streams, New Zealand, *N. Z. J. Mar. Freshwater Res.*, *31*, 413–422.

Bigi, A., L. E. Hasbargen, A. Montanari, and C. Paola (2006), Knickpoints and hillslope failures: Interactions in a steady-state experimental landscape, in *Tectonics, Climate, and Landscape Evolution*, edited by S. D. Willett et al., *Spec. Pap. Geol. Soc. Am.*, *398*, 295–307.

Bilby, R. E. (1981), Role of organic debris dams in regulating the export of dissolved and particulate matter from a forested watershed, *Ecology*, *62*, 1234–1243.

Bilby, R. E. (1984), Removal of wood debris may affect stream channel stability, J. For., 609–613.

Bilby, R. E. (1985), Influence of stream size on the function and characteristics of large organic debris. Proceedings, West Coast Meeting of National Council of Paper Industry for Air and Stream Improvement, Portland, Oreg.

Bilby, R. E. (2003), Decomposition and nutrient dynamics of wood in streams and rivers, in *The Ecology and Management of Wood in World Rivers*, edited by S. V. Gregory, K. L. Boyer, and A. M. Gurnell, pp. 135–147, Am. Fish. Soc., Bethesda, Md.

Bilby, R. E., and G. E. Likens (1980), Importance of organic debris dams in the structure and function of stream ecosystems, *Ecology*, *61*, 1107–1113.

Bilby, R. E., and J. W. Ward (1989), Changes in characteristics and function of woody debris with increasing size of streams in western Washington, *Trans. Am. Fish. Soc.*, *118*, 368–378.

Bilby, R. E., B. R. Fransen, and P. A. Bisson (1996), Incorporation of nitrogen and carbon from spawning coho salmon into the trophic system of small streams: Evidence from stable isotopes, *Can. J. Fish. Aquat. Sci.*, *53*, 164–173.

Bilby, R. E., J. T. Heffner, B. R. Fransen, J. W. Ward, and P. A. Bisson (1999), Effects of immersion in water on deterioration of wood from five species of trees used for habitat enhancement projects, *N. Am. J. Fish. Manage.*, *19*, 687–695.

Billi, P. (1988), A note on cluster bedform behaviour in a gravel-bed river, *Catena*, *15*, 473–481.

Billi, P. (1994), Streambed dynamics and grain-size characteristics of two gravel rivers of the Northern Apennines, Italy, in *Dynamics and Geomorphology of Mountain Rivers*, edited by P. Ergenzinger and K.-H. Schmidt, pp. 197–212, Springer, Berlin.

Billi, P., and P. Tacconi (1987), Bedload transport processes monitored at Virginio Creek measuring station, Italy, in *International Geomorphology 1986, Part I*, edited by V. Gardiner, pp. 549–559, John Wiley, Chichester, U. K.

Billi, P., R. D. Hey, and C. R. Thorne (1992), *Dynamics of Gravel-Bed Rivers*, 688 pp., John Wiley, Chichester, U. K.

Binder, P., and C. Schär (1996), *Mesoscale Alpine Programme: Design Proposal*, 2nd ed., 77 pp., MAP Programme Office, Zurich, Switzerland.

Binkley, D. R., F. Stottlemyer, F. Suarez, and J. Cortina (1994), Soil nitrogen availability in some arctic ecosystems in northwestern Alaska: Responses to temperature and moisture, *Ecoscience*, *1*, 64–70.

Binnie, S. A., W. M. Phillips, M. A. Summerfield, L. K. Fifield, and J. A. Spotila (2010), Tectonic and climatic controls od denudation rates in active orogens: The San Bernardino Mountains, California, *Geomorphology*, *118*, 249–261.

Bird, S., D. Hogan, and J. Schwab (2010), Photogrammetric monitoring of small streams under a riparian forest canopy, *Earth Surf. Processes Landforms*, *35*, 952–970.

Birkeland, P. W. (1968), Mean velocities and boulder transport during Tahoe-age floods of the Truckee River, California-Nevada, *Geol. Soc. Am. Bull.*, *79*, 137–142.

Birkeland, P. W. (1982), Subdivision of Holocene glacial deposits, Ben Oahu Range, New Zealand, using relative-dating methods, *Geol. Soc. Am. Bull.*, *93*, 433–449.

Birkeland, P. W., R. R. Shroba, S. F. Burns, A. B. Price, and P. J. Tonkin (2003), Integrating soils and geomorphology in mountains – an example from the Front Range of Colorado, *Geomorphology*, *55*, 329–344.

Biron, P. M., A. Richer, A. D. Kirkbride, A. G. Roy, and S. Han (2002), Spatial patterns of water surface topography at a river confluence, *Earth Surf. Processes Landforms*, *27*, 913–928.

Bischetti, G. B., C. Gandolfi, and M. J. Whelan (1998), The definition of stream channel head location using digital elevation data, in *Hydrology, Water Resources and Ecology in Headwaters*, edited by K. Kovar et al., *IAHS Publ.*, *248*, 545–552.

Bishop, M. P., J. D. Colby, J. C. Luvall, D. Quattrochi, and D. L. Rickman (2004), Remote-sensing science and technology for studying mountain environments, in *Geographic Information Science and Mountain Geomorphology*, edited by M. P. Bishop and J. F. Shroder, pp. 147–187, Praxis, Chichester, U. K.

Bishop, M. P., J. F. Shroder, R. Bonk, and J. Olsenholler (2002), Geomorphic change in high mountains: A western Himalayan perspective, *Global Planet. Change*, *32*, 311–329.

Bishop, M. P., J. F. Shroder, and J. D. Colby (2003), Remote sensing and geomorphometry for studying relief production in high mountains, *Geomorphology*, *55*, 345–361.

Bishop, P. (2007), Long-term landscape evolution: Linking tectonics and surface processes, *Earth Surf. Processes Landforms*, *32*, 329–365.

Bishop, P., and G. Goldrick (1992), Morphology, processes and evolution of two waterfalls near Cowra, New South Wales, *Aust. Geogr.*, *23*, 116–121.

Bishop, P., T. B. Hoey, J. D. Jansen, and I. L. Artza (2005), Knickpoint recession rate and catchment area: The case of uplifted rivers in eastern Scotland, *Earth Surf. Processes Landforms*, *30*, 767–778.

Bisson, P. A., J. L. Nielson, R. A. Palmason, and L. E. Grove (1982), A system of naming habitat types in small streams, with examples of habitat utilization by salmonids during low streamflow, in *Acquisition and Utilization of Aquatic Habitat Inventory Information*, edited by N. B. Armantrout, pp. 62–73, Am. Fish. Soc., Portland, Oreg.

Bisson, P. A., R. E. Bilby, M. D. Bryant, C. A. Dolloff, G. B. Grette, R. A. House, M. L. Murphy, K. V. Koski, and J. R. Sedell (1987), Large woody debris in forested streams in the Pacific Northwest: Past, present, and future, in *Streamside Management: Forestry and Fishery Interactions*, edited by E. O. Salo, and T. W. Cundy, pp. 143–190, Contribution No. 57, Univ. of Washington, Institute of Forest Resources, Seattle.

Biswas, A., J. D. Blum, B. Klaue, and G. J. Keeler (2007), Release of mercury from Rocky Mountain forest fires, *Global Biogeochem. Cycles*, *21*, GB1002, doi:10.1029/2006GB002696.

Bjerklie, D. M., and J. D. LaPerriere (1985), Gold-mining effects on stream hydrology and water quality, Circle Quadrangle, Alaska, *Water Resour. Bull.*, *21*, 235–243.

Bjerklie, D. M., S. L. Dingman, and C. H. Bolster (2005), Comparison of constitutive flow resistance equations based on the Manning and Chezy equations applied to natural rivers, *Water Resour. Res.*, *41*, W11502, doi:10.1029/2004WR003776.

Bjonness, I.-M. (1986), Mountain hazard perception and risk-avoiding strategies among the Sherpas of Khumbu Himal, Nepal, *Mt. Res. Dev.*, *6*, 277–292.

Björnsson, H. (1992), Jökulhlaups in Iceland: Prediction, characteristics and simulation, *Ann. Glaciol.*, *16*, 95–106.

Blackwelder, E. (1928), Mudflow as a geologic agent in semi-arid mountains, *Geol. Soc. Am. Bull.*, *39*, 465–484.

Blahut, J., C. J. van Westen, and S. Sterlacchini (2010), Analysis of landslide inventories for accurate prediction of debris-flow source areas, *Geomorphology*, *119*, 36–51.

Blainey, J. B., R. H. Webb, M. E. Moss, and V. R. Baker (2002), Bias and information content of paleoflood data in flood- frequency analysis, in *Ancient Floods, Modern Hazards: Principles and Applications of Paleoflood Hydrology*, edited by P. K. House et al., pp. 161–174, AGU, Washington, D. C.

Blair, T. C. (1987), Sedimentary processes, vertical stratification sequences, and geomorphology of the Roaring River alluvial fan, Rocky Mountain National Park, Colorado, *J. Sediment. Petrol.*, *57*, 1–18.

Blair, T. C. (1999), Alluvial fan and catchment initiation by rock avalanching, Owens Valley, California, *Geomorphology*, *28*, 201–221.

Blair, T. C. (2001), Outburst flood sedimentation on the proglacial Tuttle Canyon alluvial fan, Owens Valley, California, USA, *J. Sediment. Res.*, *71*, 657–679.

Blair, T. C., and J. G. McPherson (1998), Recent debris-flow processes and resultant form and facies of the Dolomite alluvial fan, Owens Valley, California, *J. Sediment. Res.*, *68*, 800–818.

Blake, W. H., I. G. Droppo, P. J. Wallbrink, S. H. Doerr, R. A. Shakesby, and G. S. Humphreys (2005), Impacts of wildfire on effective sediment particle size: Implications for post-fire sediment budgets, in *Sediment Budgets I*, edited by D. E. Walling and A. J. Horowitz, *IAHS Publ.*, *291*, 143–150.

Blanton, P., and W. A. Marcus (2009), Railroads, roads and lateral disconnection in the river landscapes of the continental United States, *Geomorphology*, *112*, 212–227.

Bledsoe, B. P., and C. C. Watson (2001), Effects of urbanization on channel instability, *J. Am. Water Resour. Assoc.*, *37*, 255–270.

Blench, T. (1951), Regime theory for self-formed sediment-bearing channels, *Proc. Am. Soc. Civ. Eng.*, *77*, 1–18.

Blijenberg, H. M. (2007), Application of physical modelling of debris flow triggering to field conditions; limitations posed by boundary conditions, *Eng. Geol.*, *91*, 25–33.

Blisniuk, P. M., and W. D. Sharp (2003), Rates of late Quaternary normal faulting in central Tibet from U-series dating of pedogenic carbonate in displaced fluvial gravel deposits, *Earth Planet. Sci. Lett.*, *215*, 169–186.

Blizard, C. R. (1994), Hydraulic variables and bedload transport in East St. Louis Creek, Rocky Mountains, Colorado, unpublished MS thesis, 176 pp., Colo. State Univ., Ft. Collins.

Blizard, C. R., and E. E. Wohl (1998), Relationships between hydraulic variables and bedload transport in a subalpine channel, Colorado Rocky Mountains, USA, *Geomorphology*, *22*, 359–371.

Blocken, B., J. Poesen, and J. Carmeliet (2006), Impact of wind on the spatial distribution of rain over micro-scale topography; numerical modelling and experimental verification, *Hydrol. Processes*, *20*, 345–369.

Bloesch, J. (2002), The unique ecological potential of the Danube and its tributaries, *Arch. Hydrobiol. Suppl.*, *141*, 175–188.

Bloesch, J. (2003), Flood plain conservation in the Danube River Basin, the link between hydrology and limnology, *Arch. Hydrobiol. Suppl.*, *147*, 347–362.

Bloesch, J., and U. Sieber (2003), The morphological destruction and subsequent restoration programmes of large rivers in Europe, *Arch. Hydrobiol. Suppl.*, *147*, 363–385.

Bloom, A. L. (1998), *Geomorphology: A Systematic Analysis of Late Cenozoic Landforms*, 482 pp., Prentice Hall, Englewood Cliffs, N. J.

Blown, I., and M. Church (1985), Catastrophic lake drainage within the Homathko River basin, British Columbia, *Can. Geotech. J.*, *22*, 551–563.

Blyth, K., and J. C. Rodda (1973), A stream length study, *Water Resour. Res.*, *9*, 1454–1461.

Boardman, J. (1993), The sensitivity of Downland arable land to erosion by water, in *Landscape Sensitivity*, edited by D. S. G. Thomas and R. J. Allison, pp. 211–228, John Wiley, Chichester, U. K.

Boardman, J. (1995), Damage to property by runoff from agricultural land, South Downs, southern England, 1976–93, *Geogr. J.*, *161*, 177–191.

Boardman, J., T. P. Burt, R. Evans, M. C. Slattery, and H. Shuttleworth (1996), Soil erosion and flooding as a result of a summer thunderstorm in Oxfordshire and Berkshire, May 1993, *Appl. Geogr.*, *16*, 21–34.

Bocchiola, D., M. C. Rulli, and R. Rosso (2006a), Flume experiments on wood entrainment in rivers, *Adv. Water Resour.*, *29*, 1182–1195.

Bocchiola, D., M. C. Rulli, and R. Rosso (2006b), Transport of large woody debris in the presence of obstacles, *Geomorphology*, *76*, 166–178.

Bocchiola, D., M. C. Rulli, and R. Rosso (2008), A flume experiment on the formation of wood jams in rivers, *Water Resour. Res.*, *44*, W02408, doi:10.1029/2006WR005846.

Bogen, J. (1980), The hysteresis effect of sediment transport systems, *Norsk Geogr. Tidsskrq.*, *34*, 45–54.

Bogen, J. (1995), Sediment transport and deposition in mountain rivers, in *Sediment and Water Quality in River Catchments*, edited by I. Foster, A. Gurnell, and B. Webb, pp. 437–451, John Wiley, Chichester, U. K.

Bogen, J., and T. E. Bønsnes (2003), Erosion prediction in ungauged glacierized basins, in *Erosion Prediction in Ungauged Basins: Integrating Methods and Techniques*, edited by D. de Boer et al., *IAHS Publ.*, *279*, 13–23.

Bogen, J., and K. Møen (2003), Bed load measurements with a new passive acoustic sensor, in *Erosion and Sediment Transport Measurement in Rivers: Technological and Methodological Advances*, edited by J. Bogen, T. Fergus, and D. E. Walling, *IAHS Publ.*, *283*, 181–192.

Boix-Fayos, C., G. G. Barbera, F. Lopez-Bermudez, and V. M. Castillo (2007), Effects of check dams, reforestation and land-use changes on river channel morphology: Case study of the Rogativa Catchment (Murcia, Spain), *Geomorphology*, *91*, 103–123.

Boll, J., T. J. M. Thewessen, E. L. Meijer, and S. B. Kroonenberg (1988), A simulation of the development of river terraces, *Z. Geomorphol.*, *32*, 31–45.

Bollschweiler, M., M. Stoffel, and D. M. Scheuwly (2008), Dynamics in debris-flow activity on a forested cone: A case study using different dendroecological approaches, *Catena*, *72*, 67–78.

Bolton, S., A. Watts, T. Sibley, and J. Dooley (1998), A pilot study examining the effectiveness of engineered large woody debris (Elwd) as an interim solution to lack of LWD in streams, *Eos Trans. AGU*, *79*, F346.

Bombino, G., A. M. Gurnell, V. Tamburino, D. A. Zema, and S. M. Zimbone (2009), Adjustments in channel form, sediment caliber and vegetation around check-dams in the headwater reaches of mountain torrents, Calabria, Italy, *Earth Surf. Processes Landforms*, *34*, 1011–1021.

Bonacci, O., and T. Roje-Bonacci (2003), The influence of hydroelectrical development on the flow regime of the karstic river Cetina, *Hydrol. Processes*, *17*, 1–15.

Bonell, M. (2005), Runoff generation in tropical forests, in *Forests, Water and People in the Humid Tropics*, edited by M. Bonell and L. A. Bruijnzeel, pp. 314–406, Cambridge Univ. Press, Cambridge, U. K.

Bonniwell, E. C., G. Matisoff, and P. J. Whiting (1999), Determining the times and distances of particle transit in a mountain stream using fallout radionuclides, *Geomorphology*, *27*, 75–92.

Booker, D. J., D. A. Sear, and A. J. Payne (2001), Modelling three-dimensional flow structures and patterns of boundary shear stress in a natural pool-riffle sequence, *Earth Surf. Processes Landforms*, *26*, 553–576.

Bookhagen, B., and D. W. Burbank (2010), Toward a complete Himalayan hydrological budget: Spatiotemporal distribution of snowmelt and rainfall and their impact on river discharge, *J. Geophys. Res.*, *115*, F03019, doi:10.1029/2009JF001426.

Bookhagen, B., D. Fleitmann, K. Nishiizumi, M. R. Strecker, and R. C. Thiede (2006), Holocene monsoonal dynamics and fluvial terrace formation in the northwest Himalaya, India, *Geology*, *34*, 601–604.

Boon, P. J. (1988), The impact of river regulation on invertebrate communities in the U.K, *Reg. Rivers Res. Manage.*, *2*, 389–409.

Boon, S., M. Sharp, and P. Nienow (2003), Impact of an extreme melt event on the runoff and hydrology of a high Arctic glacier, *Hydrol. Processes*, *17*, 1051–1072.

Booth, A. L., C. P. Chamberlain, W. S. F. Kidd, and P. K. Zeitler (2009), Constraints on the metamorphic evolution of the eastern Himalayan syntaxis from geochronologic and petrologic studies of Namche Barwa, *Geol. Soc. Am. Bull.*, *121*, 385–407.

Booth, A. M., J. J. Roering, and J. T. Perron (2009), Automated landslide mapping using spectral analysis and high-resolution topographic data: Puget Sound lowlands, Washington, and Portland Hills, Oregon, *Geomorphology*, *109*, 132–147.

Bordas, M. P., and J. H. Silvestrini (1992), Threshold of sediment deposition in medium stream power flow, in *Erosion, Debris Flows and Environment in Mountain Regions (Proceedings, Chengdu Symposium, July 1992)*, edited by D. E. Walling, T. R. Davies, and B. Hasholt, *IAHS Publ.*, *209*, 3–13.

Bormann, F. H., and G. E. Likens (1979), *Pattern and Process in a Forested Ecosystem*, 253 pp., Springer, New York.

Bosch, J. M., and J. D. Hewlett (1982), A review of catchment experiments to determine the effects of vegetation changes on water yield and evapotranspiration, *J. Hydrol.*, *55*, 3–23.

Bottacin-Busolin, A., S. J. Tait, A. Marion, A. Chegini, and M. Tregnaghi (2008), Probabilistic description of grain resistance from simultaneous flow field and grain motion measurements, *Water Resour. Res.*, *44*, W09419, doi:10.1029/2007WR006224.

Botter, G., S. Basso, A. Porporato, I. Rodriguez-Iturbe, and A. Rinaldo (2010), Natural streamflow regime alterations: Damming of the Piave river basin (Italy), *Water Resour. Res.*, *46*, W06522, doi:10.1029/2009WR008523.

Boulton, A., H. Piégay, and M. D. Sanders (2008), Turbulence and train wrecks: Using knowledge strategies to enhance the application of integrative river science to effective river management, in *River Futures: An Integrative Scientific Approach to River Repair*, edited by G. J. Brierley and K. A. Fryirs, pp. 28–39, Island Press, Washington, D. C.

Boulton, A. J., and E. H. Stanley (1995), Hyporheic processes during flooding and drying in a Sonoran Desert stream. II. Faunal dynamics, *Arch. Hydrobiol.*, *134*, 27–52.

Bovee, K. D., and M. L. Scott (2002), Implications of flood pulse restoration for Populus regeneration on the Upper Missouri River, *River Res. Appl.*, *18*, 287–298.

Bovis, M. J., and B. R. Dagg (1987), Mechanisms of debris supply to steep channels along Howe Sound, southwestern British Columbia, in *Erosion and Sedimentation in the Pacific Rim*, edited by R. L. Beschta et al., *IAHS Publ.*, *165*, 191–200.

Bowman, D. (1977), Stepped-bed morphology in arid gravelly channels, *Geol. Soc. Am. Bull.*, *88*, 291–298.

Bowman, D., T. Svoray, S. Devora, I. Shapira, and J. B. Laronne (2010), Extreme rates of channel incision and shape evolution in response to a continuous, rapid base-level fall, the Dead Sea, Israel, *Geomorphology*, *114*, 227–237.

Boyce, R. C. (1975), *Sediment routing with sediment delivery ratios, Agricultural Research Service ARS-S-*, *40*, pp. 61–65.

Boyer, E. W., G. M. Hornberger, K. E. Bencala, and D. M. McKnight (1995), Variation of dissolved organic carbon during snowmelt in soil and stream waters of two headwater catchments, Summit County, Colorado, in *Biogeochemistry of Seasonally Snow-covered Catchments*, edited by K. A. Tonnessen, M. W. Williams, and M. Tranter, *IAHS Publ.*, *228*, 303–312.

Boyer, E. W., G. M. Hornberger, K. E. Bencala, and D. M. McKnight (2000), Response characteristics of DOC flushing in an alpine catchment, *Hydrol. Processes*, *11*, 1635–1647.

Boyer, E. W., C. L. Goodale, N. A. Jaworski, and R. W. Howarth (2002), Anthropogenic nitrogen sources and relationships to riverine nitrogen export in the northeastern U.S.A, *Biogeochemistry*, *57/58*, 137–169.

Boyer, E. W., R. W. Howarth, J. N. Galloway, F. J. Dentener, P. A. Green, and C. J. Vörösmarty (2006), Riverine nitrogen export from the continents to the coasts, *Global Biogeochem. Cycles*, *30*, GB1S91, doi:10.1029/2005GB002537.

Braatne, J. H., S. B. Rood, and P. E. Heilman (1996), Life history, ecology, and conservation of riparian cottonwoods in North America, in *Biology of Populus and its Implications for Management and Conservation*, edited by R. F. Stettler et al., pp. 57–85, NRC Res. Press, Ottawa, Canada.

Braatne, J. H., S. B. Rood, L. A. Goater, and C. L. Blair (2008), Analyzing the impacts of dams on riparian ecosystems: A review of research strategies and their relevance to the Snake River through Hells Canyon, *Environ. Manage.*, *41*, 267–281.

Bracken, L. J., and J. Croke (2007), The concept of hydrological connectivity and its contribution to understanding runoff-dominated geomorphic systems, *Hydrol. Processes*, *21*, 1749–1763.

Bradley, J. B., and S. C. McCutcheon (1987), Influence of large suspended-sediment concentrations in rivers, in *Sediment Transport in Gravel-Bed Rivers*, edited by C. R. Thorne, J. C. Bathurst, and R. D. Hey, pp. 645–689, John Wiley, Chichester, U. K.

Bradley, J. B., and D. T. Williams (1993), Sediment budgets in gravel bed streams, in *Hydraulic Engineering '93*, edited by H. W. Shen, S. T. Su, and F. Wen, pp. 713–718, Am. Soc. of Civ. Eng., New York.

Bradley, R., M. Vuille, H. Diaz, and W. Vergara (2006), Threats to water supplies in the tropical Andes, *Science*, *312*, 1755–1756.

Bradley, W. C., and A. I. Mears (1980), Calculations of flow needed to transport coarse fraction of Boulder Creek alluvium at Boulder, Colorado: Summary, *Geol. Soc. Am. Bull.*, *91*, 135–138.

Bragg, D. C., J. L. Kershner, and D. W. Roberts (2000), Modeling large woody debris recruitment for small streams of the central Rocky Mountains, *General Tech. Rep. RMRS-GTR-55*, USDA Forest Service, Fort Collins, Colo.

Braithwaite, R. J., Y. Zhang, and S. C. B. Raper (2002), Temperature sensitivity of the mass balance of mountain glaciers and ice caps as a climatological characteristic, *Z. Gletscherkd. Glazialgeol.*, *38*, 35–61.

Braithwaite, R. J., S. C. B. Raper, and K. Chutko (2006), Accumulation at the equilibrium-line altitude of glaciers inferred from a degree-day model and tested against field observations, *Ann. Glaciol.*, *43*, 329–344.

Brandes, D., C. J. Duffy, and J. P. Cusumano (1998), Stability and damping in a dynamical model of hillslope hydrology, *Water Resour. Res.*, *34*, 3303–3313.

Brański, J., and K. Banasik (1996), Sediment yields and denudation rates in Poland, in *Erosion and Sediment Yield: Global and Regional Perspectives*, edited by D. E. Walling and B. W. Webb, *IAHS Publ.*, *236*, 133–138.

Brardinoni, F., and M. A. Hassan (2007), Glacially induced organization of channel-reach morphology in mountain streams, *J. Geophys. Res.*, *112*, F03013, doi:10.1029/2006JF000741.

Brasington, J., J. Langham, and B. Rumsby (2003), Methodological sensitivity of morphometric estimates of coarse fluvial sediment transport, *Geomorphology*, *53*, 299–316.

Brath, A., A. Montanari, and E. Toth (2004), Analysis of the effects of different scenarios of historical data availability on the calibration of a spatially-distributed hydrological model, *J. Hydrol.*, *291*, 232–253.

Braudrick, C. A., and G. E. Grant (2000), When do logs move in rivers?, *Water Resour. Res.*, *36*, 571–583.

Braudrick, C. A., and G. E. Grant (2001), Transport and deposition of large woody debris in streams: A flume experiment, *Geomorphology*, *41*, 263–283.

Braudrick, C. A., G. E. Grant, Y. Ishikawa, and H. Ikeda (1997), Dynamics of wood transport in streams: A flume experiment, *Earth Surf. Processes Landforms*, *22*, 669–683.

Braun, J., A. M. Heimsath, and J. Chappell (2001), Sediment transport mechanisms on soil-mantled hillslopes, *Geology*, *29*, 683–686.

Braun, L. N., C. Hottelet, and W. Grabs (1998), Measurement and simulation of runoff from Nepalese head watersheds, in *Hydrology, Water Resources and Ecology in Headwaters*, edited by K. Kovar et al., *IAHS Publ.*, *248*, 9–18.

Braun, M., and P. Fiener (1995), *Report on the GLOF Hazard Mapping Project in the Imja Khola/Dudh Kosi Valley, Nepal*, 30 pp., ICIMOD, Kathmandu.

Bravard, J.-P., G. M. Kondolf, and H. Piégay (1999), Environmental and societal effects of channel incision and remedial strategies, in *Incised River Channels: Processes, Forms, Engineering and Management*, edited by S. E. Darby and A. Simon, pp. 303–341, John Wiley, Chichester, U. K.

Bray, D. I. (1979), Estimating average velocity in gravel-bed rivers, *J. Hydraul. Div. Am. Soc. Civ. Eng.*, *105*, 1103–1122.

Bray, D. I., and M. Church (1980), Armored versus paved gravel beds, *J. Hydraul. Div. Am. Soc. Civ. Eng.*, *106*, 1937–1940.

Bray, D. I., and R. Kellerhalls (1979), Some Canadian examples of the response of rivers to man-made changes, in *Adjustments of the Fluvial System*, edited by D. D. Rhodes and G. P. Williams, pp. 351–372, George Allen and Unwin, London.

Brayshaw, A. C. (1984), Characteristics and origin of cluster bedforms in coarse-grained alluvial channels, in *Sedimentology of Gravels and Conglomerates*, edited by E. H. Koster and R. J. Steel, *Can. Soc. Petrol. Geol., Mem.*, *10*, 77–85.

Brayshaw, A. C. (1985), Bed microtopography and entrainment thresholds in gravel-bed rivers, *Geol. Soc. Am. Bull.*, *96*, 218–223.

Brayshaw, A. C., L. F. Frostick, and I. Reid (1983), The hydrodynamics of particle clusters and sediment entrainment on coarse alluvial channels, *Sedimentology*, *30*, 137–143.

Brayshaw, D., and M. A. Hassan (2009), Debris flow initiation and sediment recharge in gullies, *Geomorphology*, *109*, 122–131.

Brayton, S. D. (1984), The beaver and the stream, *J. Soil Water Conserv.*, *39*, 108–109.

Brazier, V., and C. K. Ballantyne (1989), Late Holocene debris cone evolution in Glen Feshie, western Cairngorm Mountains, Scotland, *Trans. R. Soc. Edinburgh Earth Sci.*, *80*, 17–24.

Brazier, V., G. Whittington, and C. K. Ballantyne (1988), Holocene debris cone evolution in Glen Etive, western Grampian Highlands, Scotland, *Earth Surf. Processes Landforms*, *13*, 525–531.

Brenden, T. O., L. Wang, and Z. Su (2008), Quantitative identification of disturbance thresholds in support of aquatic resource management, *Environ. Manage.*, *42*, 821–832.

Bridge, J. S. (2003), *Rivers and Floodplains: Forms, Processes, and Sedimentary Record*, 504 pp., Blackwell, Malden, Mass.

Bridges, E. M. (1990), *World Geomorphology*, 260 pp., Cambridge Univ. Press, Cambridge, U. K.

Brierley, G. J., and K. A. Fryirs (2005), *Geomorphology and river management: Applications of the River Styles Framework*, 398 pp., Blackwell, Oxford, U. K.

Brierley, G. J., and K. A. Fryirs (2008), Moves toward an era of river repair, in *River Futures: An Integrative Scientific Approach to River Repair*, edited by G. J. Brierley and K. A. Fryirs, pp. 3–15, Island Press, Washington, D. C.

Brierley, G. J., and E. J. Hickin (1985), The downstream gradation of particle sizes in the Squamish River, British Columbia, *Earth Surf. Processes Landforms*, *10*, 597–606.

Brierley, G. J., K. A. Fryirs, and V. Jain (2006), Landscape connectivity: The geographic basis of geomorphic applications, *Area*, *38*, 165–174.

Brierley, G. J., K. A. Fryirs, A. Boulton, and C. Cullum (2008), Working with change: The importance of evolutionary perspectives in framing the trajectory of river adjustment, in *River Futures: An Integrative Scientific Approach to River Repair*, edited by G. J. Brierley and K. A. Fryirs, pp. 65–84, Island Press, Washington, D. C.

Briggs, J. M., K. A. Spielmann, H. Schaafsma, K. W. Kintigh, M. Kruse, K. Morehouse, and K. Schollmeyer (2006), Why ecology needs archaeologists and archaeology needs ecologists, *Front. Ecol. Environ.*, *4*, 180–188.

Briggs, M. A., M. N. Gooseff, C. D. Arp, and M. A. Baker (2009), A method for estimating surface transient storage parameters for streams with concurrent hyporheic exchange, *Water Resour. Res.*, *45*, W00D27, doi:10.1029/2008WR006959.

Brocard, G. Y., and P. A. van der Beek (2006), Influence of incision rate, rock strength, and bedload supply on bedrock river gradients and valley-flat widths: Field-based evidence and calibrations from western Alpine rivers (southeast France), in *Tectonics, Climate, and Landscape Evolution*, edited by S. D. Willett et al., *Spec. Pap. Geol. Soc. Am.*, *398*, 101–126.

Brocard, G. Y., P. A. van der Beek, D. L. Bourles, L. L. Siame, and J.-L. Mugnier (2003), Long-term fluvial incision rates and postglacial river relaxation time in the French Western Alps from [10]Be dating of alluvial terraces with assessment of inheritance, soil development and wind ablation effects, *Earth Planet. Sci. Lett.*, *209*, 197–214.

Brocklehurst, S. H., and K. X. Whipple (2002), Glacial erosion and relief production in the eastern Sierra Nevada, California, *Geomorphology*, *42*, 1–24.

Bronstert, A., and E. J. Plate (1997), Modelling of runoff generation and soil moisture dynamics for hillslopes and micro-catchments, *J. Hydrol.*, *198*, 177–195.

Brooker, M. P. (1981), The impact of impoundments on the downstream fisheries and general ecology of rivers, in *Advances in Applied Biology*, vol. 6, edited by T. H. Coaker, pp. 91–152, Academic Press, London.

Brookes, A. (1988), *Channelized Rivers: Perspectives for Environmental Management*, 326 pp., John Wiley, Chichester, U. K.

Brookes, A. (1990), Restoration and enhancement of engineered river channels: Some European experiences, *Reg. Rivers Res. Manage.*, *5*, 45–56.

Brookfield, M. E. (1998), The evolution of the great river systems of southern Asia during the Cenozoic India-Asia collision: Rivers draining southwards, *Geomorphology*, *22*, 285–312.

Brooks, A. P., and G. J. Brierley (2002), Mediated equilibrium: The influence of riparian vegetation and wood on the long-term evolution and behaviour of a near-pristine river, *Earth Surf. Processes Landforms*, *27*, 343–367.

Brooks, A. P., and G. J. Brierley (2004), Framing realistic river rehabilitation targets in light of altered sediment supply and transport relationships: Lessons from East Gippsland, Australia, *Geomorphology*, *58*, 107–123.

Brooks, A. P., P. C. Gehrke, J. D. Jansen, and T. B. Abbe (2004), Experimental reintroduction of woody debris on the Williams River, NSW; Geomorphic and ecological responses, *River Res. Appl.*, *20*, 513–536.

Brooks, A. P., T. Howell, T. B. Abbe, and A. H. Arthington (2006), Confronting hysteresis: Wood based river rehabilitation in highly altered riverine landscapes of south-eastern Australia, *Geomorphology*, *79*, 395–422.

Brooks, S. M. (2003), Slopes and slope processes: Research over the past decade, *Prog. Phys. Geogr.*, *27*, 130–141.

Broscoe, A. J., and S. Thomson (1969), Observations on an alpine mudflow, Steel Creek, Yukon, *Can. J. Earth Sci.*, *6*, 219–229.

Brown, C. B. (1950), Sediment transportation, in *Engineering Hydraulics*, edited by H. Rouse, pp. 769–857, John Wiley, New York.

Brown, D. J. A. (1982), The effect of pH and calcium on fish and fisheries, *Water Air Soil Pollut.*, *18*, 343–351.

Brown, E. T., R. Bendick, D. L. Bourles, V. Gaur, P. Molnar, G. M. Raisbeck, and F. Yiou (2003), Early Holocene climate recorded in geomorphological features in western Tibet, *Palaeogeogr. Palaeoclimatol. Palaeoecol.*, *199*, 141–151.

Brown, G. H., and R. Fuge (1998), Trace element chemistry of glacial meltwaters in an Alpine headwater catchment, in *Hydrology, Water Resources and Ecology in Headwaters*, edited by K. Kovar et al., *IAHS Publ.*, *248*, 435–442.

Brown, L., R. Thorne, and M.-K. Woo (2008), Using satellite imagery to validate snow distribution simulated by a hydrological model in large northern basins, *Hydrol. Processes*, *22*, 2777–2787.

Brown, R. A., and G. B. Pasternack (2008), Engineered channel controls limiting spawning habitat rehabilitation success on regulated gravel-bed rivers, *Geomorphology*, *97*, 631–654.

Brozovic, N., D. W. Burbank, and A. J. Meigs (1997), Climatic limits on landscape development in the northwestern Himalaya, *Science*, *276*, 571–574.

Brufau, P., P. Garcia-Navarro, P. Ghilardi, L. Natale, and F. Savi (2000), 1D mathematical modelling of debris flow, *J. Hydraul. Res.*, *38*, 435–446.

Bruijnzeel, L. A. (2005), Tropical montane cloud forest: A unique hydrological case, in *Forests, Water and People in the Humid Tropics*, edited by M. Bonell and L. A. Bruijnzeel, pp. 462–483, Cambridge Univ. Press, Cambridge, U. K.

Brummer, C. J., T. B. Abbe, J. R. Sampson, and D. R. Montgomery (2006), Influence of vertical channel change associated with wood accumulations on delineating channel migration zones, Washington, USA, *Geomorphology*, *80*, 295–309.

Brummer, C. J., and D. R. Montgomery (2003), Downstream coarsening in headwater channels, *Water Resour. Res.*, *39*(10), 1294, doi:10.1029/2003WR001981.

Brunke, M., and T. Gonser (1997), The ecological significance of exchange processes between rivers and groundwater, *Freshwater Biol.*, *37*, 1–33.

Bruns, D. A., G. W. Minshall, C. E. Cushing, K. W. Cummins, J. T. Brock, and R. L. Vannote (1984), Tributaries as modifiers of the river-continuum concept: Analysis by polar ordination and regression models, *Arch. Hydrobiol.*, *99*, 208–220.

Brunsden, D. (1993), Barriers to geomorphological change, in *Landscape Sensitivity*, edited by D. S. G. Thomas and R. J. Allison, pp. 7–12, John Wiley, Chichester, U. K.

Brunsden, D., and D. K. C. Jones (1976), The evolution of landslide deposits in Dorset, *Philos. Trans. R. Soc. London, Ser. A*, *283*, 605–631.

Brunsden, D., and J. B. T. Homes (1979), Landscape sensitivity and change, *Trans. Inst. Br. Geogr.*, *4*, 463–484.

Bryan, G. W. (1976), Some aspects of heavy metal tolerance in aquatic organisms, in *Effects of Pollutants on Aquatic Organisms*, edited by A. P. M. Lockwood, pp. 7–34, Cambridge Univ. Press, Cambridge, U. K.

Bryan, K. (1940), Gully gravure, a method of slope retreat, *J. Geomorphol.*, *3*, 89–106.

Bryant, M. D., T. Gomi, and J. J. Piccolo (2007), Structures linking physical and biological processes in headwater streams of the Maybeso watershed, southeast Alaska, *For. Sci.*, *53*, 371–383.

Büchele, B., P. Burek, R. Baufeld, and I. Leyer (2006), Modelling flood plain vegetation based on long-term simulations of daily river-groundwater dynamics, in *Predictions in Ungauged Basins: Promise and Progress*, edited by M. Sivapalan et al., *IAHS Publ.*, *303*, 318–333.

Büdel, J. (1982), *Climatic Geomorphology*, 443 pp., Princeton Univ. Press, Princeton, N. J.

Buffagni, A. (2001), The use of benthic invertebrate production for the definition of ecologically acceptable flows in mountain rivers, in *Hydro-ecology: Linking Hydrology and Aquatic Ecology*, edited by M. C. Acreman, *IAHS Publ.*, *266*, 31–41.

Buffin-Bélanger, T., and A. G. Roy (1998), Effects of a pebble cluster on the turbulent structure of a depth-limited flow in a gravel-bed river, *Geomorphology*, *25*, 249–267.

Buffin-Bélanger, T., and A. G. Roy (2005), 1 min in the life of a river: Selecting the optimal record length for the measurement of turbulence in fluvial boundary layers, *Geomorphology*, *68*, 77–94.

Buffin-Bélanger, T., A. G. Roy, and A. D. Kirkbride (2000), On large-scale flow structures in a gravel-bed river, *Geomorphology*, *32*, 417–435.

Buffin-Bélanger, T., S. Rice, I. Reid, and J. Lancaster (2006), Spatial heterogeneity of near-bed hydraulics above a patch of river gravel, *Water Resour. Res.*, *42*, W04413, doi:10.1029/2005WR004070.

Buffington, J. M. (1999), The legend of A.F. Shields, *J. Hydraul. Eng.*, *125*, 376–387.

Buffington, J. M., T. E. Lisle, R. D. Woodsmith, and S. Hilton (2002), Controls on the size and occurrence of pools in coarse-grained forest rivers, *River Res. Appl.*, *18*, 507–531.

Buffington, J. M., and D. R. Montgomery (1999a), A procedure for classifying textural facies in gravel-bed rivers, *Water Resour. Res.*, *35*, 1903–1914.

Buffington, J. M., and D. R. Montgomery (1999b), Effects of hydraulic roughness on surface textures of gravel-bed rivers, *Water Resour. Res.*, *35*, 3507–3521.

Buffington, J. M., and D. R. Montgomery (1999c), Effects of sediment supply on surface textures of gravel-bed rivers, *Water Resour. Res.*, *35*, 3523–3530.

Buffington, J. M., D. R. Montgomery, and H. M. Greenberg (2004), Basin-scale availability of salmonid spawning gravel as influenced by channel type and hydraulic roughness in mountain catchments, *Can. J. Fish. Aquat. Sci.*, *61*, 2085–2096.

Buffington, J. M., and D. Tonina (2009a), Hyporheic exchange in mountain rivers I: Mechanics and environmental effects, *Geogr. Compass*, *3*, 1063–1086.

Buffington, J. M., and D. Tonina (2009b), Hyporheic exchange in mountain rivers II: Effects of channel morphology on mechanics, scales, and rates of exchange, *Geogr. Compass*, *3*, 1038–1062.

Buffington, J. M., R. D. Woodsmith, D. B. Booth, and D. R. Montgomery (2003), Fluvial processes in Puget Sound rivers and the Pacific Northwest, in *Restoration of Puget Sound Rivers*, edited by D. R. Montgomery et al., pp. 46–78, Univ. of Washington Press, Seattle.

Bugosh, N., and S. G. Custer (1989), The effect of a log-jam burst on bedload transport and channel characteristics in a headwaters stream, in *Proceedings of the Symposium on Headwaters Hydrology*, edited by W. W. Woessner and D. F. Potts, pp. 203–211, Am. Water Resour. Assoc., Middleburg, Va.

Buhl, K. J., and S. J. Hamilton (1990), Comparative toxicity of inorganic contaminants released by placer mining to early life stages of salmonids, *Ecotoxicol. Environ. Safety*, *20*, 325–342.

Bull, W. B. (1962), Relations of alluvial-fan size and slope to drainage-basin size and lithology in western Fresno County, California, *U.S. Geol. Surv. Prof. Pap.*, vol. 450-B, pp. 51–53.

Bull, W. B. (1964), *Geomorphology of segmented alluvial fans in western Fresno County, California*, U.S. *Geol. Surv. Prof. Pap.*, vol. 352-E, pp. 89–129.

Bull, W. B. (1977), The alluvial fan environment, *Prog. Phys. Geogr.*, *1*, 222–270.

Bull, W. B. (1979), Threshold of critical power in streams, *Geol. Soc. Am. Bull.*, *90*, 453–464.

Bull, W. B. (1988), Floods; degradation and aggradation, in *Flood Geomorphology*, edited by V. R. Baker, R. C. Kochel, and P. C. Patton, pp. 157–165, John Wiley, New York.

Bull, W. B. (1991), *Geomorphic Responses to Climatic Change*, 326 pp., Oxford Univ. Press, New York.

Bull, W. B. (1997), Discontinuous ephemeral streams, *Geomorphology*, *19*, 227–276.

Bull, W. B., and P. L. K. Knuepfer (1987), Adjustments by the Charwell River, New Zealand, to uplift and climatic changes, *Geomorphology*, *1*, 15–32.

Bull, W. B., and K. M. Scott (1974), Impact of mining gravel from urban stream beds in the southwestern US, *Geology*, *2*, 171–174.

Bultot, F., D. Gellens, M. Spreafico, and B. Schädler (1992), Repercussions of a CO_2 doubling on the water balance - a case study in Switzerland, *J. Hydrol.*, *137*, 199–208.

Bunch, M. A., R. Mackay, J. H. Tellam, and P. Turner (2004), A model for simulating the deposition of water-lain sediments in dryland environments, *Hydrol. Earth Syst. Sci.*, *8*, 122–134.

Bundi, U., E. Eichenberger, and A. Peter (1990), Water flow regime as the driving force for the formation of habitats and biological communities in Alpine rivers, in *Hydrology in Mountainous Regions II. Artificial Reservoirs, Water and Slopes*, edited by R. O. Sinniger and M. Monbaron, *IAHS Publ.*, *194*, 197–204.

Bunn, J. T., and D. R. Montgomery (2004), Patterns of wood and sediment storage along debris-flow impacted headwater channels in old-growth and industrial forests of the western Olympic Mountains, Washington, in *Riparian Vegetation and Fluvial Geomorphology*, edited by S. J. Bennett and A. Simon, pp. 99–112, AGU, Washington, D. C.

Bunn, S. E., and A. H. Arthington (2002), Basic principles and ecological consequences of altered flow regimes for aquatic biodiversity, *Environ. Manage.*, *30*, 492–507.

Bunte, K. (1990), Experiences and results from using a big-frame bed load sampler for coarse material bed load, in *Hydrology in Mountainous Regions I. Hydrological Measurements; the Water Cycle*, edited by H. Lang and A. Musy, *IAHS Publ.*, *193*, 223–230.

Bunte, K. (1992), Particle number grain-size composition of bedload in a mountain stream, in *Dynamics of Gravel-Bed Rivers*, edited by P. Billi et al., pp. 55–72, John Wiley, Chichester, U. K.

Bunte, K. (1996), Analyses of the temporal variation of coarse bedload transport and its grain size distribution: Squaw Creek, Montana, USA, *General Tech. Rep. RM-GTR-288*, 123 pp, USDA Forest Service.

Bunte, K., and S. R. Abt (2005), Effect of sampling time on measured gravel bed load transport rates in a coarse-bedded stream, *Water Resour. Res.*, *41*, W11405, doi:10.1029/2004WR003880.

Bunte, K., S. R. Abt, J. P. Potyondy, and S. E. Ryan (2004), Measurement of coarse gravel and cobble transport using portable bedload traps, *J. Hydraul. Eng.*, *130*, 879–893.

Bunte, K., S. R. Abt, J. P. Potyondy, and K. W. Swingle (2008), A comparison of coarse bedload transport measured with bedload traps and Helley-Smith samplers, *Geodin. Acta*, *21*, 53–66.

Bunte, K., K. W. Swingle, and S. R. Abt (2007), Guidelines for using bedload traps in coarse-bedded mountain streams: Construction, installation, operation, and sample processing, *USDA Forest Service General Tech. Rep. RMRS-GTR-191*, Rocky Mtn. Res. Sta., Fort Collins, Colo.

Burbank, D. W., J. Leland, E. Fielding, R. S. Anderson, N. Brozovic, M. R. Reid, and C. Duncan (1996), Bedrock incision, rock uplift and threshold hillslopes in the northwestern Himalayas, *Nature*, *379*, 505–510.

Burcher, C. L., M. E. McTammany, E. F. Benfield, and G. S. Helfman (2008), Fish assemblage responses to forest cover, *Environ. Manage.*, *41*, 336–346.

Burge, L. M. (2006), Stability, morphology and surface grain size patterns of channel bifurcation in gravel-cobble bedded anabranching rivers, *Earth Surf. Processes Landforms*, *31*, 1211–1226.

Burke, M., K. Jorde, and J. M. Buffington (2009), Application of a hierarchical framework for assessing environmental impacts of dam operation: Changes in streamflow, bed mobility and recruitment of riparian trees in a western North American river, *J. Environ. Manage.*, *90*(Suppl. 3), S224–S236.

Burke, B. C., A. M. Heimsath, and A. F. White (2007), Coupling chemical weathering with soil production across soil-mantled landscapes, *Earth Surf. Processes Landforms*, *32*, 853–873.

Burke, S. M., R. M. Adams, and W. W. Wallender (2004), Water banks and environmental water demands: Case of the Klamath Project, *Water Resour. Res.*, *40*, W09S02, doi:10.1029/2003WR002832.

Burner, C. J. (1951), Characteristics of spawning nests of Columbia River salmon, *U.S. Fish Wildlife Serv. Bull.*, *61*, 97–110.

Burnett, B. N., G. A. Meyer, and L. D. McFadden (2008), Aspect-related microclimatic influences on slope forms and processes, northeastern Arizona, *J. Geophys. Res.*, *113*, F03002, doi:10.1029/2007JF000789.

Burns, D. A. (2004), The effects of atmospheric nitrogen deposition in the Rocky Mountains of Colorado and southern Wyoming, USA – a critical review, *Environ. Pollut.*, *127*, 257–269.

Burns, D. A., J. J. McDonnell, R. P. Hooper, N. E. Peters, J. E. Freer, C. Kendall, and K. Beven (2001), Quantifying contributions to storm runoff through end-member mixing analysis and hydrologic measurements at the Panola Mountain Research Watershed (Georgia, USA), *Hydrol. Processes*, *15*, 1903–1924.

Burtin, A., L. Bollinger, R. Cattin, J. Vergne, and J. L. Nábělek (2009), Spatiotemporal sequence of Himalayan debris flow from analysis of high-frequency seismic noise, *J. Geophys. Res.*, *114*, F04009, doi:10.1029/2008JF001198.

Burton, A., and J. C. Bathurst (1998), Physically based modelling of shallow landslide sediment yield at a catchment scale, *Environ. Geol.*, *35*, 89–99.

Burton, A., T. J. Arkell, and J. C. Bathurst (1998), Field variability of landslide model parameters, *Environ. Geol.*, *35*, 100–114.

Buscombe, D., D. M. Rubin, and J. A. Warrick (2010), A universal approximation of grain size from images of noncohesive sediment, *J. Geophys. Res.*, *115*, F02015, doi:10.1029/2009JF001477.

Bush, A. B. G., M. L. Prentice, M. P. Bishop, and J. F. Shroder (2004), Modeling global and regional climate systems: Climate forcing and topography, in *Geographic Information Science and Mountain Geomorphology*, edited by M. P. Bishop and J. F. Shroder, pp. 403–423, Praxis, Chichester, U. K.

Burton, G. A., Jr. (Ed.) (1992), *Sediment Toxicity Assessment*, 480 pp., Lewis Publishers, Boca Raton, Fla.

Busnelli, M. M., G. S. Stelling, and M. Larcher (2001), Numerical morphological modeling of open-check dams, *J. Hydraul. Eng.*, *127*, 105–114.

Busskamp, R. (1994), The influence of channel steps on coarse bed load transport in mountain torrents: Case study using the radio tracer technique 'PETSY', in *Dynamics and Geomorphology of Mountain Rivers*, edited by P. Ergenzinger and K.-H. Schmidt, pp. 129–139, Springer, Berlin.

Butler, D. R. (2001), Geomorphic process-disturbance corridors: A variation on a principle of landscape ecology, *Prog. Phys. Geogr.*, *25*, 237–248.

Butler, D. R., and G. P. Malanson (1995), Sedimentation rates and patterns in beaver ponds in a mountain environment, *Geomorphology*, *13*, 255–269.

Buttle, J. M. (1996), Identifying hydrological responses to basin restoration: An example from southern Ontario, in *Watershed Restoration Management: Physical, Chemical, and Biological Considerations*, edited by J. J. McDonnell et al., pp. 5–13, Am. Water Resour. Assoc., Herndon, Va.

Buttle, J. M., and J. J. McDonnell (2005), Isotope tracers in catchment hydrology in the humid tropics, in *Forests, Water and People in the Humid Tropics*, edited by M. Bonell and L. A. Bruijnzeel, pp. 770–789, Cambridge Univ. Press, Cambridge, U. K.

Buzin, V. A. (2000), Floods caused by ice jams on rivers, *Water Resour.*, *27*, 476–481.

Byrd, T. C. (1997), Dynamical analysis of nonlogarithmic velocity profiles in steep, rough channels, unpublished MS thesis, 74 pp., Florida State Univ., Tallahassee, Fla.

Byrne, J. M., A. Berg, and I. Townshend (1999), Linking observed and general circulation model upper air circulation patterns to current and future snow runoff for the Rocky Mountains, *Water Resour. Res.*, *35*, 3793–3802.

Caamaño, D., P. Goodwin, J. M. Buffington, J. C. P. Liou, and S. Daley-Laursen (2009), Unifying criterion for the velocity reversal hypothesis in gravel-bed rivers, *J. Hydraul. Eng.*, *135*, 66–70.

Cadol, D., E. Wohl, J. R. Goode, and K. L. Jaeger (2009), Wood distribution in neotropical forested headwater streams of La Selva, Costa Rica, *Earth Surf. Processes Landforms*, *34*, 1198–1215.

Caine, N. (1974), The geomorphic processes of the alpine environment, in *Arctic and Alpine Environments*, edited by J. D. Ives and R. G. Barry, pp. 721–748, Methuen, London.

Caine, N. (1976), Summer rainstorms in an alpine environment and their influence on soil erosion, San Juan Mountains, Arizona, *Arct. Alp. Res.*, *8*, 183–196.

Caine, N. (1980), The rainfall intensity-duration control of shallow landslides and debris flows, *Geogr. Ann.*, *62A*, 23–27.

Caine, N., and P. K. Mool (1981), Channel geometry and flow estimates for two small mountain streams in the Middle Hills, Nepal, *Mt. Res. Dev.*, *1*, 231–243.

Calhoun, R. S., and C. H. Fletcher (1999), Measured and predicted sediment yield from a subtropical, heavy rainfall, steep-sided river basin: Hanalei, Kauai, Hawaiian Islands, *Geomorphology*, *30*, 213–226.

Calvo, L. E., F. L. Ogden, and J. M. H. Hendrickx (2005), Infiltration in the Upper Rio Chagres basin, Panama, in *The Rio Chagres, Panama: A Multidisciplinary Profile of a Tropical Watershed*, edited by R. S. Harmon, pp. 139–147, Springer, Dordrecht, The Netherlands.

Campbell, A. J., and R. C. Sidle (1985), Bedload transport in a pool-riffle sequence of a coastal Alaska stream, *Water Resour. Bull.*, *21*, 579–590.

Campbell, D., and M. Church (2003), Reconnaissance sediment budgets for the Lynn Valley, British Columbia; Holocene and contemporary time scales, *Can. J. Earth Sci.*, *40*, 701–713.

Campbell, D. H., D. W. Clow, G. P. Ingersoll, M. A. Mast, N. E. Spahr, and J. T. Turk (1995a), Nitrogen deposition and release in alpine watersheds, Loch Vale, Colorado, USA, in *Biogeochemistry of Seasonally Snow-covered Catchments*, edited by K. A. Tonnessen, M. W. Williams, and M. Tranter, *IAHS Publ.*, *228*, 243–253.

Campbell, D. H., D. W. Clow, G. P. Ingersoll, M. A. Mast, N. E. Spahr, and J. T. Turk (1995b), Processes controlling the chemistry of two snowmelt-dominated streams in the Rocky Mountains, *Water Resour. Res.*, *31*, 2811–2821.

Campbell, E. P. (2005), Physical-statistical models for predictions in ungauged basins, in *Predictions in Ungauged Basins: International Perspectives on the State of the Art and Pathways Forward*, edited by S. Franks et al., *IAHS Publ.*, *301*, 292–298.

Campbell, R. H. (1975), Soil slips, debris flows, and rainstorms in the Santa Monica Mountains and vicinity, southern California, *U.S. Geol. Surv. Prof. Pap.*, vol. 851, 51 pp.

Camporeale, C., and L. Ridolfi (2006), Riparian vegetation distribution induced by river flow variability: A stochastic approach, *Water Resour. Res.*, *42*, W10415, doi:10.1029/2006WR004933.

Candy, I., S. Black, and B. W. Sellwood (2004), Interpreting the response of a dryland river system to late Quaternary climate change, *Quat. Sci. Rev.*, *23*, 2513–2523.

Cannon, S. H. (1988), Regional rainfall-threshold conditions for abundant debris-flow activity, in *Landslides, floods, and marine effects of the storm of January 3–5, 1982, in the San Francisco bay region, California*, edited by S. D. Ellen and G. F. Wieczorek, *U.S. Geol. Surv. Prof. Pap.*, *1434*, 35–42.

Cannon, S. H. (2001), Debris-flow generation from recently burned watersheds, *Environ. Eng. Geosci.*, *7*, 321–341.

Cannon, S. H., and J. E. Gartner (2005), Wildfire-related debris flow from a hazards perspective, in *Debris-flow Hazards and Related Phenomena*, edited by M. Jakob and O. Hungr, pp. 363–385, Springer, Berlin.

Cannon, S. H., R. M. Kirkham, and M. Parise (2001), Wildfire-related debris-flow initiation processes, Storm King Mountain, Colorado, *Geomorphology*, *39*, 171–188.

Cannon, S. H., J. E. Gartner, R. C. Wilson, J. C. Bowers, and J. L. Laber (2008), Storm rainfall conditions for floods and debris flows from recently burned areas in southwestern Colorado and southern California, *Geomorphology*, *96*, 250–269.

Canovaro, F., and L. Solari (2007), Dissipative analogies between a schematic macro-roughness arrangement and step-pool morphology, *Earth Surf. Processes Landforms*, *32*, 1628–1640.

Canovaro, F., E. Paris, and L. Solari (2007), Effects of macro-scale bed roughness geometry on flow resistance, *Water Resour. Res.*, *43*, W10414, doi:10.1029/2006WR005727.

Cantelli, A., C. Paola, and G. Parker (2004), Experiments on upstream-migrating erosional narrowing and widening of an incised channel caused by dam removal, *Water Resour. Res.*, *40*, W03304, doi:10.1029/2003WR002940.

Canuti, P., N. Casagli, F. Catani, G. Falorni, and P. Farina (2007), Integration of remote sensing techniques in different stages of landslide reponse, in *Progress in Landslide Science*, edited by K. Sassa et al., pp. 251–260, Springer, Berlin.

Cao, Z., P. Carling, and R. Oakey (2003), Flow reversal over a natural pool-riffle sequence: A computational study, *Earth Surf. Processes Landforms*, *28*, 689–705.

Cao, Z., S. Egashira, and P. A. Carling (2003), Role of suspended-sediment particle size in modifying velocity profiles in open channel flows, *Water Resour. Res.*, *39*(2), 1029, doi:10.1029/2001WR000934.

Carbonneau, P. E. (2005), The threshold effect of image resolution on image-based automated grain size mapping in fluvial environments, *Earth Surf. Processes Landforms*, *30*, 1687–1693.

Carbonneau, P. E., and N. E. Bergeron (2000), The effect of bedload transport on mean and turbulent flow properties, *Geomorphology*, *35*, 267–278.

Carbonneau, P. E., N. Bergeron, and S. N. Lane (2005), Automated grain size measurements from airborne remote sensing for long profile measurements of fluvial grain sizes, *Water Resour. Res.*, *41*, W11426, doi:10.1029/2005WR003994.

Cardenas, M. B. (2010), Thermal skin effect of pipes in streambeds and its implications on groundwater flux estimation using diurnal temperature signals, *Water Resour. Res.*, *46*, W03536, doi:10.1029/2009WR008528.

Cardenas, M. S., and M. N. Gooseff (2008), Comparison of hyporheic exchange under covered and uncovered channels based on linked surface and groundwater flow simulations, *Water Resour. Res.*, *44*, W03418, doi:10.1029/2007WR006506.

Cardenas, M. B., J. L. Wilson, and V. A. Zlotnik (2004), Impact of heterogeneity, bed forms, and stream curvature on subchannel hyporheic exchange, *Water Resour. Res.*, *40*, W08307, doi:10.1029/2004WR003008.

Carey, A. E., S.-J. Kao, D. M. Hicks, C. A. Nezat, and W. B. Lyons (2006), The geochemistry of rivers in tectonically active areas of Taiwan and New Zealand, in *Tectonics, Climate, and Landscape Evolution*, edited by S. D. Willett et al., *Spec. Pap. Geol. Soc. Am.*, *398*, 339–351.

Carling, P. A. (1983), Threshold of coarse sediment transport in broad and narrow natural streams, *Earth Surf. Processes Landforms*, *8*, 1–18.

Carling, P. A. (1984a), Comparison of suspended sediment rating curves obtained using two sampling methods, in *Channel Processes - Water, Sediment, Catchment Controls*, edited by A. P. Schick, *Catena Suppl.*, *5*, 43–49.

Carling, P. A. (1984b), Deposition of fine and coarse sand in an open-work gravel, *Can. J. Fish. Aquat. Sci.*, *41*, 263–270.

Carling, P. A. (1989a), Bedload transport in two gravel-bedded streams, *Earth Surf. Processes Landforms*, *14*, 27–39.

Carling, P. A. (1989b), Hydrodynamic models of boulder berm deposition, *Geomorphology*, *2*, 319–340.

Carling, P. A. (1990), Particle over-passing on depth-limited gravel bars, *Sedimentology*, *37*, 345–355.

Carling, P. A. (1991), An appraisal of the velocity-reversal hypothesis for stable pool-riffle sequences in the River Severn, England, *Earth Surf. Processes Landforms*, *16*, 19–31.

Carling, P. A. (1994a), Palaeohydraulic reconstruction of floods in upland UK bedrock streams: Progress, problems and prospects, in *Proceedings, ASCE Hydraulic Engineering '94 Conference, August 1–5, Buffalo, New York*, edited by G. V. Cotroneo and R. R. Rumer, pp. 860–864, Am. Soc. of Civ. Eng., New York.

Carling, P. A. (1994b), Particle dynamics and bed level adjustments in a mountain stream, in *Proceedings, ASCE Hydraulic Engineering '94 Conference, August 1–5, Buffalo, New York*, edited by G. V. Cotroneo and R. R. Rumer, pp. 839–843, Am. Soc. of Civ. Eng., New York.

Carling, P. A. (1995), Flow-separation berms downstream of a hydraulic jump in a bedrock channel, *Geomorphology, 11*, 245–253.

Carling, P. A., Z. Cao, M. J. Holland, D. A. Ervine, and K. Babaeyan-Koopaei (2002), Turbulent flow across a natural compound channel, *Water Resour. Res., 38*(12), 1270, doi:10.1029/2001WR000902.

Carling, P. A., and M. S. Glaister (1987), Rapid deposition of sand and gravel mixtures downstream of a negative step: The role of matrix-infilling and particle-overpassing in the process of bar-front accretion, *J. Geol. Soc. London, 144*, 543–551.

Carling, P. A., and T. Grodek (1994), Indirect estimation of ungauged peak discharges in a bedrock channel with reference to design discharge selection, *Hydrol. Processes, 8*, 497–511.

Carling, P. A., and M. A. Hurley (1987), A time-varying stochastic model of the frequency and magnitude of bed load transport events in two small trout streams, in *Sediment Transport in Gravel-Bed Rivers*, edited by C. R. Thorne, J. C. Bathurst, and R. D. Hey, pp. 897–920, John Wiley, Chichester, U. K.

Carling, P. A., and H. G. Orr (2000), Morphology of riffle-pool sequences in the River Severn, England, *Earth Surf. Processes Landforms, 25*, 369–384.

Carling, P. A., and N. A. Reader (1982), Structure, composition and bulk properties of upland stream gravels, *Earth Surf. Processes Landforms, 7*, 349–365.

Carling, P. A., and K. Tinkler (1998), Conditions for the entrainment of cuboid boulders in bedrock streams: An historical review of literature with respect to recent investigations, in *Rivers Over Rock: Fluvial Processes in Bedrock Channels, Geophys. Monogr. Ser.*, vol. 107, edited by K. J. Tinkler and E. E. Wohl, pp. 19–34, AGU, Washington, D. C.

Carling, P. A., and N. Wood (1994), Simulation of flow over pool-riffle topography: A consideration of the velocity reversal hypothesis, *Earth Surf. Processes Landforms, 19*, 319–332.

Carling, P. A., A. Kelsey, and M. S. Glaister (1992), Effect of bed roughness, particle shape and orientation on initial motion criteria, in *Dynamics of Gravel-Bed Rivers*, edited by P. Billi et al., pp. 23–39, John Wiley, Chichester, U. K.

Carling, P. A., H. G. Orr, and M. S. Glaister (1994), Preliminary observations and significance of dead zone flow structures for solute and fine particle dynamics, in *Mixing and Transport in the Environment*, edited by K. J. Beven, P. C. Chatwin, and J. H. Millbank, pp. 139–157, John Wiley.

Carling, P. A., M. Hoffmann, A. S. Blatter, and A. Dittrich (2002), Drag of emergent and submerged rectangular obstacles in turbulent flow above bedrock surface, in *Rock Scour due to Falling High-velocity Jets*, edited by A. J. Schleiss and E. Bollaert, pp. 83–94, Swets and Zeitlinger, Lisse.

Carling, P. A., W. Tych, and K. Richardson (2005), The hydraulic scaling of step-pool systems, in *River, Coastal and Estuarine Morphodynamics*, vol. 1, edited by G. Parker and M. H. Garcia, pp. 55–63, Balkema, Taylor, and Francis, New York.

Carling, P. A., L. Whitcombe, I. A. Benson, B. G. Hankin, and A. M. Radecki-Pawlik (2006), A new method to determine interstitial flow patterns in flume studies of sub-aqueous gravel bedforms such as fish nests, *River Res. Appl., 22*, 691–701.

Carlson, J. Y., C. W. Andrus, and H. A. Froehlich (1990), Woody debris, channel features, and macro-invertebrates of streams with logged and undisturbed riparian timber in northeastern Oregon, U.S.A, *Can. J. Fish. Aquat. Sci., 47*, 1103–1111.

Carpenter, K. E. (1928), *Life in Inland Waters*, 267 pp., Macmillan Company, New York.

Carrara, A., G. Crosta, and P. Frattini (2008), Comparing models of debris-flow susceptibility in the alpine environment, *Geomorphology, 94*, 353–378.

Carrivick, J. L., A. J. Russell, and F. S. Tweed (2004), Geomorphological evidence for jökulhlaups from Kverkfjöll volcano, Iceland, *Geomorphology*, *63*, 81–102.

Carson, E. C., J. C. Knox, and D. M. Mickelson (2007), Response of bankfull flood magnitudes to Holocene climate change, Uinta Mountains, northeastern Utah, *Geol. Soc. Am. Bull.*, *119*, 1066–1078.

Carson, M. A., and G. A. Griffiths (1987), Bedload transport in gravel channels, *J. Hydrol. N. Z.*, *26*, 1–151.

Carson, M. A., and M. J. Kirkby (1972), *Hillslope Form and Process*, 475 pp., Cambridge Univ. Press, London.

Carter, C. L., and R. S. Anderson (2006), Fluvial erosion of physically modeled abrasion-dominated slot canyons, *Geomorphology*, *81*, 89–113.

Carter, G., M. Duncan, and B. Biggs (1998), Numerical hydrodynamic modelling of mountain streams for assessing instream habitat, in *Hydrology, Water Resources and Ecology in Headwaters*, edited by K. Kovar et al., *IAHS Publ.*, *248*, 217–223.

Carver, M., and G. Nakarmi (1995), The effect of surface conditions on soil erosion and stream suspended sediments, in *Challenges in Mountain Resource Management in Nepal*, edited by H. Schreier, P. B. Shah, and S. Brown, pp. 155–162, ICIMOD, Kathmandu.

Casas, M. A., S. N. Lane, R. J. Hardy, G. Benito, and P. J. Whiting (2010), Reconstruction of subgrid-scale topographic variability and its effect upon the spatial structure of three-dimensional river flow, *Water Resour. Res.*, *46*, W03519, doi:10.1029/2009WR007756.

Cashman, E., and K. Potter (2006), Modelling the impacts of climate variability on sediment transport, in *Sediment Dynamics and the Hydromorphology of Fluvial Systems*, edited by J. S. Rowan, R. W. Duck, and A. Werritty, *IAHS Publ.*, *306*, 611–619.

Cassells, D. S., and L. A. Bruijnzeel (2005), Guidelines for controlling vegetation, soil and water impacts of timber harvesting in the humid tropics, in *Forests, Water and People in the Humid Tropics*, edited by M. Bonell and L. A. Bruijnzeel, pp. 840–851, Cambridge Univ. Press, Cambridge, U. K.

Castillo, V. M., W. M. Mosch, C. C. Garcia, G. G. Barbera, J. A. N. Cano, and F. Lopez-Bermudez (2007), Effectiveness and geomorphological impacts of check dams for soil erosion control in a semiarid Mediterranean catchment; El Carcavo (Murcia, Spain), *Catena*, *70*, 416–427.

Castro, J. M., and P. L. Jackson (2001), Bankfull discharge recurrence intervals and regional hydraulic geometry relationships: Patterns in the Pacific Northwest, USA, *J. Am. Water Resour. Assoc.*, *37*, 1249–1262.

Catani, F., P. Farina, S. Moretti, and G. Nico (2003), Spaceborne radar interferometry; a promising tool for hydrological analysis in mountain alluvial fan environments, in *Erosion Prediction in Ungauged Basins: Integrating Methods and Techniques*, edited by D. H. de Boer et al., *IAHS Publ.*, *279*, 241–248.

Catani, F., P. Farina, S. Moretti, G. Nico, and T. Strozzi (2005), On the application of SAR interferometry to geomorphological studies: Estimation of landform attributes and mass movements, *Geomorphology*, *66*, 119–131.

Catella, M., E. Paris, and L. Solari (2005), Case study; efficiency of slit-check dams in the mountain region of Versilia Basin, *J. Hydraul. Eng.*, *131*, 145–152.

Cencetti, C., P. Tacconi, M. Del Prete, and M. Rinaldi (1994), Variability of gravel movement on the Virginio gravel-bed stream (central Italy) during some floods, in *Variability in Stream Erosion and Sediment Transport*, edited by L. J. Olive, R. J. Loughran, and J. A. Kesby, *IAHS Publ.*, *224*, 3–11.

Cenderelli, D. A. (1998), Glacial-lake outburst floods in the Mount Everest region of Nepal: Flow processes, flow hydraulics, and geomorphic effects, unpublished Ph.D. dissertation, 247 pp., Colo. State Univ., Ft. Collins.

Cenderelli, D. A. (2000), Floods from natural and artificial dam failures, in *Inland Flood Hazards: Human, Riparian, and Aquatic Communities*, edited by E. E. Wohl, pp. 73–103, Cambridge Univ. Press, Cambridge, U. K.

Cenderelli, D. A., and B. L. Cluer (1998), Depositional processes and sediment supply in resistant-boundary channels: Examples from two case studies, in *Rivers Over Rock: Fluvial Processes in Bedrock Channels*,

Geophys. Monogr. Ser., vol. 107, edited by K. J. Tinkler and E. E. Wohl, pp. 105–131, AGU, Washington, D. C.

Cenderelli, D. A., and J. S. Kite (1994), Erosion and deposition by debris flows in mountainous channels on North Fork Mountain, eastern West Virginia, in *Proceedings, ASCE Hydraulic Engineering '94 Conference, August 1–5, Buffalo, New York*, edited by G. V. Cotroneo and R. R. Rumer, pp. 772–776, Am. Soc. of Civ. Eng., New York.

Cenderelli, D. A., and J. S. Kite (1998), Geomorphic effects of large debris flows on channel morphology at North Fork Mountain, eastern West Virginia, USA, *Earth Surf. Processes Landforms*, *23*, 1–19.

Cenderelli, D. A., and E. E. Wohl (1998), Sedimentology and clast orientation of deposits produced by glacial-lake outburst floods in the Mount Everest region, Nepal, in *Geomorphological Hazards in High Mountain Areas*, edited by J. Kalvoda and C. L. Rosenfeld, pp. 1–26, Kluwer Acad., The Netherlands.

Cenderelli, D. A., and E. E. Wohl (2001), Peak discharge estimates of glacial-lake outburst floods and "normal" climatic floods in the Mount Everest region, Nepal, *Geomorphology*, *40*, 57–90.

Cenderelli, D. A., and E. E. Wohl (2003), Flow hydraulics and geomorphic effects of glacial-lake outburst floods in the Mount Everest region, Nepal, *Earth Surf. Processes Landforms*, *28*, 385–407.

Chacho, E. F., Jr., W. W. Emmett, and R. L. Burrows (1994), Monitoring grain movement using radio transmitters, in *Proceedings, ASCE Hydraulic Engineering '94 Conference, August 1–5, Buffalo, New York*, edited by G. V. Cotroneo and R. R. Rumer, pp. 785–789, Am. Soc. of Civ. Eng., New York.

Chakrabarti, U., and R. Abhinaba (2007), Sedimentary processes and facies of upper Pleistocene alluvial fans in the Purna Valley basin of central India, *J. Geol. Soc. India*, *69*, 916–924.

Chakraborty, T., and P. Ghosh (2010), The geomorphology and sedimentology of the Tista megafan, Darjeeling Himalaya: Implications for megafan building processes, *Geomorphology*, *115*, 252–266.

Chalise, S. R., M. L. Shrestha, O. R. Bajracharya, and A. B. Shrestha (2006), Climate change impacts on glacial lakes and glacierized basins in Nepal and implications for water resources, in *Climate Variability and Change – Hydrological Impacts*, edited by S. Demuth et al., *IAHS Publ.*, *308*, 460–465.

Chandler, J., P. Ashmore, C. Paola, M. Gooch, and F. Varkaris (2002), Monitoring river-channel change using terrestrial oblique digital imagery and automated digital photogrammetry, *Ann. Assoc. Am. Geogr.*, *92*, 631–644.

Chang, H., and I.-W. Jung (2010), Spatial and temporal changes in runoff caused by climate change in a complex large river basin in Oregon, *J. Hydrol.*, *388*, 186–207.

Chang, H. H. (1980), Geometry of gravel streams, *J. Hydraul. Div. Am. Soc. Civ. Eng.*, *105*, 1443–1456.

Chang, H. H. (1987), Modelling fluvial processes in streams with gravel mining, in *Sediment Transport in Gravel-Bed Rivers*, edited by C. R. Thorne, J. C. Bathurst, and R. D. Hey, pp. 977–988, John Wiley, Chichester, U. K.

Chang, H. H. (1988), *Fluvial Processes in River Engineering*, 432 pp., Wiley Interscience, New York.

Chang, L. H., C. T. Hunsaker, and J. D. Draves (1992), Recent research on effects of climate change on water resources, *Water Resour. Bull.*, *28*, 273–286.

Chang, T.-C. (2007), Risk degree of debris flow applying neural networks, *Nat. Hazards*, *42*, 209–224.

Chanson, H. (1994), Hydraulics of nappe flow regime above stepped chutes and spillways, *Aust. Civ. Eng. Trans.*, *CE 36*, 69–76.

Chanson, H. (1995), *Hydraulic Design of Stepped Cascades, Channels, Weirs and Spillways*, 292 pp., Pergamon, Tarrytown, N. J.

Chanson, H. (1996), Comment on 'Step-pool streams: Adjustment to maximum flow resistance' by A.D. Abrahams, G. Li and J.F. Atkinson, *Water Resour. Res.*, *32*, 3401–3402.

Chanson, H. (2004), Experimental study of flash flood surges down a rough sloping channel, *Water Resour. Res.*, *40*, W03301, doi:10.1029/2003WR002662.

Chaplot, V., and C. Walter (2003), Subsurface topography to enhance the prediction of the spatial distribution of soil wetness, *Hydrol. Processes*, *17*, 2567–2580.

Chapman, D. W. (1966), Food and space as regulators of salmonid populations in streams, *Am. Nat., 100*, 345–357.

Chapman, D. W. (1988), Critical review of variables used to define effects of fines in redds of large salmonids, *Trans. Am. Fish. Soc., 117*, 1–21.

Chaponniere, A., G. Boulet, A. Chehbouni, and M. Aresmouk (2008), Understanding hydrological processes with scarce data in a mountain environment, *Hydrol. Processes, 22*, 1908–1921.

Chappell, N. A., S. W. Franks, and J. Larenus (1998), Multi-scale permeability estimation for a tropical catchment, *Hydrol. Processes, 12*, 1507–1523.

Chappell, N. A., P. McKenna, K. Bidin, I. Douglas, and R. P. D. Walsh (1999), Parsimonious modelling of water and suspended sediment flux from nested catchments affected by selective tropical forestry, *Philos. Trans. R. Soc. London, Ser. B, 354*, 1831–1846.

Chappell, N. A., I. Douglas, J. M. Hanapi, and W. Tych (2004), Sources of suspended sediment within a tropical catchment recovering from selective logging, *Hydrol. Processes, 18*, 685–701.

Chappell, N. A., K. Bidin, M. D. Sherlock, and J. W. Lancaster (2005a), Parsimonious spatial representation of tropical soils within dynamic rainfall-runoff models, in *Forests, Water and People in the Humid Tropics*, edited by M. Bonell and L. A. Bruijnzeel, pp. 756–769, Cambridge Univ. Press, Cambridge, U. K.

Chappell, N. A., W. Tych, Z. Yusop, N. A. Rahim, and B. Kasran (2005b), Spatially significant effects of selective tropical forestry on water, nutrient and sediment flows: A modelling-supported review, in *Forests, Water and People in the Humid Tropics*, edited by M. Bonell and L. A. Bruijnzeel, pp. 513–532, Cambridge Univ. Press, Cambridge, U. K.

Charlton, R. A. (1999), Initial stages in the development of a coupled hillslope hydrology-floodplain inundation model, *Phys. Chem. Earth, Part B, 24*, 37–41.

Chartrand, S. M., and P. J. Whiting (2000), Alluvial architecture in headwater streams with special emphasis on step-pool topography, *Earth Surf. Processes Landforms, 25*, 583–600.

Chase, K. J. (1994), Thresholds for gravel and cobble motion, in *Proceedings, ASCE Hydraulic Engineering '94 Conference, August 1–5, Buffalo, New York*, edited by G. V. Cotroneo and R. R. Rumer, pp. 790–794, Am. Soc. of Civ. Eng., New York.

Chatanantavet, P., and G. Parker (2009), Physically based modeling of bedrock incision by abrasion, plucking, and macroabrasion, *J. Geophys. Res., 114*, F04018, doi:10.1029/2008JF001044.

Chatanantavet, P., E. Lajeunesse, G. Parker, L. Malverti, and P. Meunier (2010), Physically based model of downstream fining in bedrock streams with lateral input, *Water Resour. Res., 46*, W02518, doi:10.1029/2008WR007208.

Chatwin, S. C., D. E. Howes, J. W. Schwab, and D. N. Swanston (1991), *A guide for management of landslide-prone terrain in the Pacific Northwest, British Columbia Ministry of Forests, Land Management Handbook, 18*, 212 pp.

Che, Z.-X., M. Jin, X.-L. Zhangm, Y. Niu, and X.-L. Dong (2008), Effect of vegetation type on the ablation of snow cover in the Qilian Mountains, *J. Glaciol. Geocryol., 30*, 392–397.

Chemenda, A. I., T. Bois, S. Bouissou, and E. Tric (2009), Numerical modelling of the gravity-induced destabilization of a slope: The example of the La Clapière landslide, southern France, *Geomorphology, 109*, 86–93.

Chen, C.-L. (1987), Comprehensive review of debris flow modeling concepts in Japan, in *Debris Flows/avalanches: Process, Recognition, and Mitigation*, edited by J. E. Costa and G. F. Wieczorek, *Rev. Eng. Geol., 7*, 13–29.

Chen, C. W., S. A. Gherini, J. D. Dean, R. J. M. Hudson, and R. A. Goldstein (1984), Development and calibration of the Integrated Lake-Watershed Acidification study model, in *Modeling of Total Acid Precipitation Impacts*, edited by J. Schoor, pp. 175–203, Ann Arbor Science, Ann Arbor, Mich.

Chen, C.-Y., L.-Y. Lin, F.-C. Yu, C.-S. Lee, C.-C. Tseng, A.-H. Wang, and K.-W. Cheung (2007), Improving debris flow monitoring in Taiwan by using high-resolution rainfall products from QPESUMS, *Nat. Hazards, 40*, 447–461.

Chen, J., and A. Ohmura (1990), On the influence of Alpine glaciers on runoff, in *Hydrology in Mountainous Regions I. Hydrological Measurements, the Water Cycle*, edited by H. Lang and A. Musy, *IAHS Publ.*, *193*, 117–125.

Chen, L., and M. C. Stone (2008), Influence of bed material size heterogeneity on bedload transport uncertainty, *Water Resour. Res.*, *44*, W01405, doi:10.1029/2006WR005483.

Cheng, F., and T. Granata (2007), Sediment transport and channel adjustments associated with dam removal: Field observations, *Water Resour. Res.*, *43*, W03444, doi:10.1029/2005WR004271.

Cheng, J. D. (1989), Streamflow changes after clearcut logging of a pine-beetle infested watershed in southern British Columbia, Canada, *Water Resour. Res.*, *25*, 449–456.

Cheng, J. D., Y. C. Huang, H. L. Wu, J. L. Yeh, and C. H. Chang (2005), Hydrometeorological and land use attributes of debris flows and debris floods during typhoon Toraji, July 29–20, 2001 in central Taiwan, *J. Hydrol.*, *306*, 161–173.

Cherkauer, D. S. (1973), Minimization of power expenditure in a riffle-pool alluvial channel, *Water Resour. Res.*, *9*, 1613–1628.

Cherry, J., and R. L. Beschta (1989), Coarse woody debris and channel morphology: A flume study, *Water Resour. Bull.*, *25*, 1031–1036.

Chew, L. C., and P. E. Ashmore (2001), Channel adjustment and a test of rational regime theory in a proglacial braided stream, *Geomorphology*, *37*, 43–63.

Chiarle, M., S. Iannotti, G. Mortara, and P. Deline (2007), Recent debris flow occurrences associated with glaciers in the Alps, *Global Planet. Change*, *56*, 123–136.

Chikita, K. A., R. Kemnitz, and R. Kumai (2002), Characteristics of sediment discharge in the subarctic Yukon River, Alaska, *Catena*, *48*, 235–253.

Chikita, K. A., T. Wada, I. Kudo, D. Kido, Y.-I. Narita, and Y. Kim (2007), Modelling discharge, water chemistry and sediment load from a subarctic river basin; The Tanana River, Alaska, in *Water Quality and Sediment Behaviour of the Future: Predictions for the 21st Century*, edited by B. W. Webb and D. de Boer, *IAHS Publ.*, *314*, 45–56.

Chikova, V. M. (1974), Species and age composition of fishes in the lower reach (downstream) of the V.I. Lenin Hydroelectric Station, in *Biological and Hydrological Factors of Local Movements of Fish in Reservoirs*, edited by B. S. Kuzin, pp. 185–192, Amerind, New Delhi, India.

Chin, A. (1989), Step pools in stream channels, *Prog. Phys. Geogr.*, *13*, 391–407.

Chin, A. (1998), On the stability of step-pool mountain streams, *J. Geol.*, *106*, 59–69.

Chin, A. (1999a), On the origin of step-pool sequences in mountain streams, *Geophys. Res. Lett.*, *26*, 231–234.

Chin, A. (1999b), The morphologic structure of step-pools in mountain streams, *Geomorphology*, *27*, 191–204.

Chin, A. (2002), The periodic nature of step-pool mountain streams, *Am. J. Sci.*, *302*, 144–167.

Chin, A. (2006), Urban transformation of river landscapes in a global context, *Geomorphology*, *79*, 460–487.

Chin, A., and K. J. Gregory (2001), Urbanization and adjustment of ephemeral stream channels, *Ann. Assoc. Am. Geogr.*, *91*, 595–608.

Chin, A., and K. J. Gregory (2005), Managing urban river channel adjustments, *Geomorphology*, *69*, 28–45.

Chin, A., and J. D. Phillips (2007), The self-organization of step-pools in mountain streams, *Geomorphology*, *83*, 346–358.

Chin, A., and E. Wohl (2005), Toward a theory for step pools in stream channels, *Prog. Phys. Geogr.*, *29*, 275–296.

Chin, A., et al. (2008), Perceptions of wood in rivers and challenges for stream restoration in the United States, *Environ. Manage.*, *41*, 893–903.

Chin, A., et al. (2009a), Linking theory and practice for restoration of step-pool streams, *Environ. Manage.*, *43*, 645–661.

Chin, A., A. H. Purcell, J. W. Y. Quan, and V. H. Resh (2009b), Assessing geomorphological and ecological responses in restored step-pool systems, in *Management and Restoration of Fluvial Systems with Broad Historical Changes and Human Impacts*, edited by L. A. James, S. L. Rathburn, and G. R. Whittecar, pp. 199–214, Geol. Soc. of Am., Boulder, Colo.

Chiverell, R. C., A. M. Harvey, and G. C. Foster (2007), Hill slope gullying in the Solway Firth, Morecambe Bay region, Great Britain; responses to human impact and/or climatic deterioration?, *Geomorphology, 84*, 317–343.

Chorley, R. J. (1962), Geomorphology and the general systems theory, *U.S. Geol. Surv. Prof. Pap., 500-B*.

Chorley, R. J., and B. A. Kennedy (1971), *Physical Geography, A Systems Approach*, 370 pp., Prentice-Hall, London.

Chorley, R. J., S. A. Schumm, and D. E. Sugden (1984), *Geomorphology*, 605 pp., Methuen, London.

Chow, V. T. (1959), *Open-channel Hydraulics*, 679 pp., McGraw-Hill, New York.

Christensen, L., C. L. Tague, and J. S. Baron (2008), Spatial patterns of simulated transpiration response to climate variability in a snow-dominated mountain ecosystem, *Hydrol. Processes, 22*, 3576–3588.

Christophersen, N., and R. F. Wright (1981), Sulfate budget and a model for sulfate concentrations in stream-water at Birkenes, a small forested catchment in southernmost Norway, *Water Resour. Res., 17*, 377–389.

Church, M. (1972), *Baffin Island Sandurs. A study of arctic fluvial processes*, Bull. Geol. Surv. Can., *216*, 208 pp.

Church, M. (1978), Palaeohydrological reconstructions from a Holocene valley fill, in *Fluvial Sedimentology*, edited by A. D. Miall, *Memoir, 5*, 743–772.

Church, M. (1988), Floods in cold climates, in *Flood Geomorphology*, edited by V. R. Baker, R. C. Kochel, and P. C. Patton, pp. 205–229, John Wiley, New York.

Church, M. (1992), Channel morphology and typology, in *The Rivers Handbook*, vol. 1, edited by P. Calow and G. E. Petts, pp. 126–143, Blackwell Sci., London.

Church, M. (2002), Geomorphic thresholds in riverine landscapes, *Freshwater Biol., 47*, 541–557.

Church, M. (2008), Multiple scales in rivers, in *Gravel-Bed Rivers VI: From Process Understanding to River Restoration*, edited by H. Habersack, H. Piégay, and M. Rinaldi, pp. 3–32, Elsevier, Amsterdam.

Church, M., and M. A. Hassan (1992), Size and distance of travel of unconstrained clasts on a streambed, *Water Resour. Res., 28*, 299–303.

Church, M., and M. A. Hassan (2002), Mobility of bed material in Harris Creek, *Water Resour. Res., 38*(11), 1237, doi:10.1029/2001WR000753.

Church, M., and M. A. Hassan (2005), Estimating the transport of bed material at low rate in gravel armoured channels, in *Geomorphological Processes and Human Impacts in River Basins*, edited by R. J. Batalla and C. Garcia, *IAHS Publ., 299*, 141–153.

Church, M., and M. J. Miles (1987), Meteorological antecedents to debris flow in southwestern British Columbia; some case studies, in *Debris Flows/avalanches: Process, Recognition, and Mitigation*, edited by J. E. Costa and G. F. Wieczorek, *Rev. Eng. Geol., 7*, 63–79.

Church, M., and A. Zimmermann (2007), Form and stability of step-pool channels: Research progress, *Water Resour. Res., 43*, W03415, doi:10.1029/2006WR005037.

Church, M., J. F. Wolcott, and W. K. Fletcher (1991), A test of equal mobility in fluvial sediment transport: Behavior of the sand fraction, *Water Resour. Res., 27*, 2941–2951.

Church, M., M. A. Hassan, and J. F. Wolcott (1998), Stabilizing self-organized structures in gravel-bed stream channels: Field and experimental observations, *Water Resour. Res., 34*, 3169–3179.

Church, M. A., D. G. McLean, and J. F. Wolcott (1987), River bed gravels: Sampling and analysis, in *Sediment Transport in Gravel-Bed Rivers*, edited by C. R. Thorne, J. C. Bathurst, and R. D. Hey, pp. 43–88, John Wiley, Chichester, U. K.

Church, S. E., J. R. Owen, P. von Guerard, P. L. Verplanck, B. A. Kimball, and D. B. Yager (2007), The effects of acidic mine drainage from historical mines in the Animas River watershed, San Juan County, Colorado; what is being done and what can be done to improve water quality?, *Rev. Eng. Geol., 17*, 47–83.

Ciszewski, D. (2001), Flood-related changes in heavy metal concentrations within sediments of the Biała Przemsza River, *Geomorphology*, *40*, 205–218.

Clague, J. J. (2008), Effects of recent climate change on high mountains of western North America, *Permafrost: Proceedings of the 9th International Conference on Permafrost*, pp. 269–274, Inst. of Northern Eng., Univ. of Alaska, Fairbanks.

Clague, J. J., and S. G. Evans (1997), The 1994 jökulhlaup at Farrow Creek, British Columbia, Canada, *Geomorphology*, *19*, 77–87.

Clague, J. J., and S. G. Evans (2000), A review of catastrophic drainage of moraine-dammed lakes in British Columbia, *Quat. Sci. Rev.*, *19*, 1763–1783.

Clague, J. J., R. J. W. Turner, and A. V. Reyes (2003), Record of recent river channel instability, Cheakamus Valley, British Columbia, *Geomorphology*, *53*, 317–332.

Clapp, E. M., P. R. Bierman, A. P. Schick, J. Lekach, Y. Enzel, and M. Caffee (2000), Sediment yield exceeds sediment production in arid region drainage basins, *Geology*, *28*, 995–998.

Clark, C. (1987a), Deforestation and floods, *Environ. Conserv.*, *14*, 67–69.

Clark, G. M. (1987b), Debris slide and debris flow historical events in the Appalachians south of the glacial border, in *Debris Flows/avalanches: Process, Recognition, and Mitigation*, edited by J. E. Costa and G. F. Wieczorek, *Rev. Eng. Geol.*, *7*, 125–138.

Clark, J. J., and P. R. Wilcock (2000), Effects of land-use change on channel morphology in northeastern Puerto Rico, *Geol. Soc. Am. Bull.*, *112*, 1763–1777.

Clark, M. K., L. M. Schoenbohm, L. H. Royden, K. X. Whipple, B. C. Burchfiel, X. Zhang, W. Tang, E. Wang, and L. Chen (2004), Surface uplift, tectonics, and erosion of eastern Tibet from large-scale drainage patterns, *Tectonics*, *23*, TC1006, doi:10.1029/2002TC001402.

Clarke, G. K. C. (1982), Glacier outburst floods from 'Hazard Lake', Yukon Territory, and the problem of flood magnitude prediction, *J. Glaciol.*, *28*, 3–21.

Clarke, L., T. A. Quine, and A. Nicholas (2010), An experimental investigation of autogenic behavior during alluvial fan evolution, *Geomorphology*, *115*, 278–285.

Clarke, S. J., L. Bruce-Burgess, and G. Wharton (2003), Linking form and function: Towards an eco-hydromorphic approach to sustainable river restoration, *Aquat. Conserv. Mar. Freshwater Ecosyst.*, *13*, 439–450.

Clayton, J. A. (2010), Local sorting, bend curvature, and particle mobility in meandering gravel bed rivers, *Water Resour. Res.*, *46*, W02601, doi:10.1029/2008WR007669.

Clayton, J. A., and J. C. Knox (2008), Catastrophic flooding from Glacial Lake Wisconsin, *Geomorphology*, *93*, 384–397.

Clayton, J. A., and J. Pitlick (2007), Spatial and temporal variations in bed load transport intensity in a gravel bed river bend, *Water Resour. Res.*, *43*, W02426, doi:10.1029/2006WR005253.

Clayton, J. A., and J. Pitlick (2008), Persistence of the surface texture of a gravel-bed river during a large flood, *Earth Surf. Processes Landforms*, *33*, 661–673.

Clemence, K. T. (1988), Influence of stratigraphy and structure on knickpoint erosion, *Bull. Assoc. Eng. Geol.*, *25*, 11–15.

Clements, W. H., D. M. Carlisle, J. M. Lazorchak, and P. C. Johnson (2000), Heavy metals structure benthic communities in Colorado mountain streams, *Ecol. Appl.*, *10*, 626–638.

Clevis, Q., P. de Boer, and M. Wachter (2003), Numerical modelling of drainage basin evolution and three-dimensional alluvial fan stratigraphy, *Sediment. Geol.*, *163*, 85–110.

Clifford, N. J. (1993a), Differential bed sedimentology and the maintenance of riffle-pool sequences, *Catena*, *20*, 447–468.

Clifford, N. J. (1993b), Formation of riffle-pool sequences: Field evidence for an autogenetic process, *Sediment. Geol.*, *85*, 39–51.

Clifford, N. J. (1996), Morphology and stage-dependent flow structure in a gravel-bed river, in *Coherent Flow Structures in Open Channels*, edited by P. J. Ashworth et al., pp. 545–566, John Wiley, Chichester, U. K.

Clifford, N. J., and J. R. French (1993a), Monitoring and analysis of turbulence in geophysical boundaries: Some analytical and conceptual issues, in *Turbulence: Perspectives on Flow and Sediment Transport*, edited by N. J. Clifford, J. R. French, and J. Hardisty, pp. 93–120, John Wiley, Chichester, U. K.

Clifford, N. J., and J. R. French (1993b), Monitoring and modelling turbulent flow: Historical and contemporary perspectives, in *Turbulence: Perspectives on Flow and Sediment Transport*, edited by N. J. Clifford, J. R. French, and J. Hardisty, pp. 1–34, John Wiley, Chichester, U. K.

Clifford, N. J., K. S. Richards, and A. Robert (1992a), The influence of microform bed roughness elements on flow and sediment transport in gravel bed rivers: Comment on a paper by Marwan A. Hassan and Ian Reid, *Earth Surf. Processes Landforms*, *17*, 529–534.

Clifford, N. J., A. Robert, and K. S. Richards (1992b), Estimation of flow resistance in gravel-bedded rivers: A physical explanation of the multiplier of roughness length, *Earth Surf. Processes Landforms*, *17*, 111–126.

Clifford, N. J., J. Hardisty, J. R. French, and S. Hart (1993), Downstream variation in bed material characteristics in the braided River Swale: A turbulence-controlled form-process feedback mechanism, in *Braided Rivers: Form, Process and Economic Applications*, edited by J. L. Best and C. S. Bristow, *Geol. Soc. Spec. Publ.*, *75*, 89–104.

Cline, R. G., H. Haupt, and G. Campbell (1977), Potential water yield response following clearcut harvesting on north and south slopes in northern Idaho, *USDA Forest Service Research Paper INT-191*, 16 pp, United States Intermountain Forest and Range Experiment Station, Ogden, Utah.

Clow, D. W., and M. A. Mast (1995), Composition of precipitation, bulk deposition, and runoff at a granitic bedrock catchment in the Loch Vale watershed, Colorado, USA, in *Biogeochemistry of Seasonally Snow-covered Catchments*, edited by K. T. Tonnessen, M. W. Williams, and M. Tranter, *IAHS Publ.*, *228*, 235–242.

Clow, D. W., and J. K. Sueker (2000), Relations between basin characteristics and stream water chemistry in alpine/subalpine basins in Rocky Mountain National Park, Colorado, *Water Resour. Res.*, *36*, 49–61.

Clow, D. W., L. Schrott, R. Webb, D. H. Campbell, A. Torizzo, and M. M. Dornblaser (2003), Ground water occurrence and contributions to streamflow in an alpine catchment, Colorado Front Range, *Ground Water*, *41*, 937–950.

Cluer, B. L. (1995), Cyclic fluvial processes and bias in environmental monitoring, Colorado River in Grand Canyon, *J. Geol.*, *103*, 411–421.

Code, J. A., and S. Sirhindi (1986), Engineering implications of impoundment of the Indus River by an earthquake-induced landslide, in *Landslide Dams: Processes, Risk, and Mitigation*, edited by R. L. Schuster, pp. 97–110, Am. Soc. of Civ. Eng., New York.

Coe, J. A., D. A. Kinner, and J. W. Godt (2008), Initiation conditions for debris flows generated by runoff at Chalk Cliffs, central Colorado, *Geomorphology*, *96*, 270–297.

Cognard-Plancq, A.-L., V. Marc, J.-F. Didon-Lescot, and M. Normand (2001), The role of forest cover on streamflow down sub-Mediterranean mountain watersheds: A modelling approach, *J. Hydrol.*, *254*, 229–243.

Cohen, D., P. Lehmann, and D. Or (2009), Fiber bundle model for multiscale modeling of hydromechanical triggering of shallow landslides, *Water Resour. Res.*, *45*, W10436, doi:10.1029/2009WR007889.

Cohen, S., G. Willgoose, and G. Hancock (2009), The mARM spatially distributed soil evolution model: A computationally efficient modeling framework and analysis of hillslope soil surface organization, *J. Geophys. Res.*, *114*, F03001, doi:10.1029/2008JF001214.

Cohen, T. J., and G. J. Brierley (2000), Channel instability in a forested catchment: A case study from Jones Creek, East Gippsland, Australia, *Geomorphology*, *32*, 109–128.

Collier, M. P., R. H. Webb, and E. D. Andrews (1997), Experimental flooding in Grand Canyon, *Sci. Am.*, *276*, 66–73.

Collins, B. D., D. R. Montgomery, and A. D. Haas (2002), Historical changes in the distribution and functions of large wood in Puget Lowland rivers, *Can. J. Fish. Aquat. Sci.*, *59*, 66–76.

Collins, D. B. G., and R. L. Bras (2010), Climatic and ecological controls of equilibrium drainage density, relief, and channel concavity in dry lands, *Water Resour. Res.*, *46*, W04508, doi:10.1029/2009WR008615.

Collins, D. N. (1990), Seasonal and annual variations of suspended sediment transport in meltwaters draining from an Alpine glacier, in *Hydrology in Mountainous Regions I. Hydrological Measurements, the Water Cycle*, edited by H. Lang and A. Musy, *IAHS Publ.*, *193*, 439–446.

Collins, D. N. (1995a), Daily patterns of discharge, solute content and solute flux in meltwaters draining from two alpine glaciers, in *Biogeochemistry of Seasonally Snow-covered Catchments*, edited by K. A. Tonnessen, M. W. Williams, and M. Tranter, *IAHS Publ.*, *228*, 371–378.

Collins, D. N. (1995b), Diurnal variations of flow-through velocity and transit time of meltwaters traversing moulin-conduit systems in an alpine glacier, in *Biogeochemistry of Seasonally Snow-covered Catchments*, edited by K. A. Tonnessen, M. W. Williams, and M. Tranter, *IAHS Publ.*, *228*, 363–369.

Collins, D. N. (1996), Sediment transport derived from glacierized basins in the Karakoram mountains, in *Erosion and Sediment Yield: Global and Regional Perspectives*, edited by D. E. Walling and B. W. Webb, *IAHS Publ.*, *236*, 85–96.

Collins, D. N. (1998), Rainfall-induced high-magnitude runoff events in highly-glacierized Alpine basins, in *Hydrology, Water Resources and Ecology in Headwaters*, edited by K. Kovar et al., *IAHS Publ.*, *248*, 69–78.

Collins, D. N. (2006a), Climatic variation and runoff in mountain basins with differing proportions of glacier cover, *Nord. Hydrol.*, *37*, 315–326.

Collins, D. N. (2006b), Variability of runoff from Alpine basins, in *Climate Variability and Change – Hydrological Impacts*, edited by S. Demuth et al., *IAHS Publ.*, *308*, 466–472.

Collins, D. N. (2008), Climatic warming, glacier recession and runoff from alpine basins after the Little Ice Age maximum, *Ann. Glaciol.*, *48*, 119–124.

Collins, D. N., and D. P. Taylor (1990), Variability of runoff from partially-glacierised Alpine basins, in *Hydrology in Mountainous Regions I. Hydrological Measurements, the Water Cycle*, edited by H. Lang and A. Musy, *IAHS Publ.*, *193*, 365–372.

Collins, P. E., D. J. Rust, and M. S. Bayraktutan (2008), Geomorphological evidence for a changing tectonic regime, Pasinler Basin, Turkey, *J. Geol. Soc. London*, *165*, 849–857.

Colman, S. M., and K. L. Pierce (1981), Weathering rinds on andesitic and basaltic stones as a Quaternary age indicator, western United States, *U.S. Geol. Surv. Prof. Pap.*, *1210*, 56 pp.

Colombo, F. (2005), Quaternary telescopic-like alluvial fans, Andean Ranges, Argentina, in *Alluvial Fans: Geomorphology, Sedimentology, Dynamics*, edited by A. M. Harvey, A. E. Mather, and M. Stokes, *Geol. Soc. Spec. Publ.*, *251*, 69–84.

Coltorti, M., S. Ravani, G. Cornamusini, A. Ielpi, and F. Verrazani (2009), A sagging along the eastern Chianti Mts., Italy, *Geomorphology*, *112*, 15–26.

Colwell, R. (1998), Balancing the biocomplexity of the planet's living systems: A twenty-first century task for science, *BioScience*, *48*, 786–787.

Comiti, F., A. Andreoli, and M. A. Lenzi (2005), Morphological effects of local scouring in step-pool streams, *Earth Surf. Processes Landforms*, *30*, 1567–1581.

Comiti, F., A. Andreoli, M. A. Lenzi, and L. Mao (2006), Spatial density and characteristics of woody debris in five mountain rivers of the Dolomites (Italian Alps), *Geomorphology*, *78*, 44–63.

Comiti, F., L. Mao, A. Wilcox, E. E. Wohl, and M. A. Lenzi (2007), Field-derived relationships for flow velocity and resistance in high-gradient streams, *J. Hydrol.*, *340*, 48–62.

Comiti, F., A. Andreoli, L. Mao, and M. A. Lenzi (2008a), Wood storage in three mountain streams of the southern Andes and its hydro-morphological effects, *Earth Surf. Processes Landforms*, *33*, 244–262.

Comiti, F., L. Mao, E. Preciso, L. Picco, L. Marchi, and M. Borga (2008b), Large wood and flash floods: Evidence from the 2007 event in the Davč;a basin (Slovenia), in *Monitoring, Simulation, Prevention and Retention of Dense and Debris Flow II*, edited by D. de Wrachien, C. A. Brebbia, and M. A. Lenzi, *WIT Trans. Eng. Sci.*, *60*, 173–182.

Comiti, F., D. Cadol, and E. Wohl (2009a), Flow regimes, bed morphology, and flow resistance in self-formed step-pool channels, *Water Resour. Res.*, *45*, W04424, doi:10.1029/2008WR007259.

Comiti, F., L. Mao, M. A. Lenzi, and M. Siligardi (2009b), Artificial steps to stabilize mountain rivers: A post-project ecological assessment, *River Res. Appl.*, *25*, 639–659.

Conedera, M., L. Peter, P. Marxer, F. Forster, D. Rickenmann, and L. Re (2003), Consequences of forest fires on the hydrogeological response of mountain catchments: A case study of the Riale Buffaga, Ticino, Switzerland, *Earth Surf. Processes Landforms*, *28*, 117–129.

Connolly, N. M., and R. G. Pearson (2005), Impacts of forest conversion on the ecology of streams in the humid tropics, in *Forests, Water and People in the Humid Tropics*, edited by M. Bonell and L. A. Bruijnzeel, pp. 811–835, Cambridge Univ. Press, Cambridge, U. K.

Conoscenti, C., C. Di Maggio, and E. Rotigliano (2008), GIS analysis to assess landslide susceptibility in a fluvial basin of NW Sicily (Italy), *Geomorphology*, *94*, 325–339.

Constantine, C. R., J. F. Mount, and J. L. Florsheim (2003), The effects of longitudinal differences in gravel mobility on the downstream fining pattern in the Cosumnes River, California, *J. Geol.*, *111*, 233–241.

Constantine, J. A., G. B. Pasternack, and M. L. Johnson (2005), Logging effects on sediment flux observed in a pollen-based record of overbank deposition in a northern California catchment, *Earth Surf. Processes Landforms*, *30*, 813–821.

Constantz, J. (2008), Heat as a tracer to determine streambed water exchanges, *Water Resour. Res.*, *44*, W00D10, doi:10.1029/2008WR006996.

Conway, S. J., A. Decaulne, M. R. Balme, J. B. Murray, and M. C. Towner (2010), A new approach to estimating hazard posed by debris flows in the Westfjords of Iceland, *Geomorphology*, *114*, 556–572.

Cook, D. I., P. M. Santi, J. D. Higgins, and R. D. Short (2008), Field-scale measurement of groundwater profiles in a drained slope, *Environ. Eng. Geosci.*, *14*, 167–182.

Cook, K. L., K. X. Whipple, A. M. Heimsath, and T. C. Hanks (2009), Rapid incision of the Colorado River in Glen Canyon – insights from channel profiles, local incision rates, and modeling of lithologic controls, *Earth Surf. Processes Landforms*, *34*, 994–1010.

Cooke, R. U., and R. W. Reeves (1976), *Arroyos and Environmental Change in the American South-West*, 213 pp., Clarendon Press, Oxford.

Cooley, M. E., B. N. Aldridge, and R. C. Euler (1977), Effects of the catastrophic flood of December 1966, North Rim area, eastern Grand Canyon, Arizona, *U.S. Geol. Surv. Prof. Pap.*, *980*, 43 pp.

Coon, W. F. (1994), Roughness coefficients for high-gradient channels in New York State, in *Proceedings, ASCE Hydraulic Engineering '94 Conference, August 1–5, Buffalo, New York*, edited by G. V. Cotroneo and R. R. Rumer, pp. 722–726, Am. Soc. of Civ. Eng., New York.

Cooper, J. R., and S. J. Tait (2010), Examining the physical components of boundary shear stress for water-worked gravel deposits, *Earth Surf. Processes Landforms*, *35*, 1240–1246.

Cooper, J. R., S. J. Tait, and K. V. Horoshenkov (2006), Determining hydraulic resistance in gravel-bed rivers from the dynamics of their water surfaces, *Earth Surf. Processes Landforms*, *31*, 1839–1848.

Coppin, N. J., and I. G. Richards (1990), *Use of Vegetation in Civ. Eng.*, 292 pp., Butterworths, London.

Corbel, J. (1959), Vitesse de l érosion, *Z. Geomorphol.*, *3*, 1–28.

Corenblit, D., J. Steiger, A. M. Gurnell, E. Tabacchi, and L. Roques (2009), Control of sediment dynamics by vegetation as a key function driving biogeomorphic succession within fluvial corridors, *Earth Surf. Processes Landforms*, *34*, 1790–1810.

Corn, P. S., and R. B. Bury (1989), Logging in western Oregon: Response of headwater habitats and stream amphibians, in *For. Ecol. Manage.*, pp. 39–57, Elsevier, Amsterdam.

Cornwell, K., D. Norsby, and R. Marston (2003), Drainage, sediment transport, and denudation rates on the Nanga Parbat Himalaya, Pakistan, *Geomorphology*, *55*, 25–43.

Corominas, J. (1996), The angle of reach as a mobility index for small and large landslides, *Can. Geotech. J.*, *33*, 260–271.

Corominas, J., and E. E. Alonso (1990), Geomorphological effects of extreme floods (November 1982) in the southern Pyrenees, in *Hydrology in Mountainous Regions II. Artificial Reservoirs, Water and Slopes*, edited by R. O. Sinniger and M. Monbaron, *IAHS Publ., 194*, 295–302.

Corripio, J. G., and R. S. Purves (2005), Surface energy balance of high altitude glaciers in the central Andes: The effect of snow penitentes, in *Climate and Hydrology in Mountain Areas*, edited by C. de Jong, D. Collins, and R. Ranzi, pp. 15–27, John Wiley, Chichester, U. K.

Corsini, A., A. Pasuto, M. Soldati, and A. Zannoni (2005), Field monitoring of the Corvara landslide (Dolomites, Italy) and its relevance for hazard assessment, *Geomorphology, 66*, 149–165.

Cortecci, G., and A. Longinelli (1970), Isotopic composition of sulfate in rain water, Pisa, Italy, *Earth Planet. Sci. Lett., 8*, 36–40.

Cosby, B. J., G. M. Hornberger, J. N. Galloway, and R. F. Wright (1985), Modeling the effects of acid deposition: Assessment of a lumped parameter model of soil water and streamwater chemistry, *Water Resour. Res., 21*, 51–63.

Costa, J. E. (1978), Holocene stratigraphy in flood frequency analysis, *Water Resour. Res., 14*, 626–632.

Costa, J. E. (1983), Paleohydraulic reconstruction of flash-flood peaks from boulder deposits in the Colorado Front Range, *Geol. Soc. Am. Bull., 94*, 986–1004.

Costa, J. E. (1984), Physical geomorphology of debris flows, in *Developments and Applications of Geomorphology*, edited by J. E. Costa and P. J. Fleisher, pp. 268–317, Springer, Berlin.

Costa, J. E. (1985), Interpretation of the largest rainfall-runoff floods measured by indirect methods on small drainage basins in the conterminous United States, paper presented at China-U.S. Bilateral Symposium on the Analysis of Extraordinary Flood Events, Nanjing, China.

Costa, J. E. (1987), Hydraulics and basin morphometry of the largest flash floods in the conterminous United States, *J. Hydrol., 93*, 313–338.

Costa, J. E. (1992), Characteristics of a debris fan formed at the U.S. Geological Survey debris-flow flume, H.J. Andrews Experimental Forest, Blue River, Oregon, *Eos Trans. AGU, 73*, 227.

Costa, J. E., and R. D. Jarrett (1981), Debris flows in small mountain stream channels of Colorado and their hydrologic implications, *Bull. Assoc. Eng. Geol., 18*, 309–322.

Costa, J. E., and J. E. O'Connor (1995), Geomorphically effective floods, in *Natural and Anthropogenic Influences in Fluvial Geomorphology*, edited by J. E. Costa et al., pp. 45–56, AGU, Washington, D. C.

Costa, J. E., and R. L. Schuster (1988), The formation and failure of natural dams, *Geol. Soc. Am. Bull., 100*, 1054–1068.

Costa, M. H. (2005), Large-scale hydrological impacts of tropical forest conversion, in *Forests, Water and People in the Humid Tropics*, edited by M. Bonell and L. A. Bruijnzeel, pp. 590–597, Cambridge Univ. Press, Cambridge, U. K.

Cotroneo, G. V., and R. R. Rumer (Eds.) (1994), *Proceedings, ASCE Hydraulic Engineering '94 Conference, August 1'5, Buffalo, New York*, pp. 634–880 and 1257–1300, Am. Soc. of Civ. Eng., New York.

Cotton, J. A., G. Wharton, J. A. B. Bass, C. M. Heppell, and R. S. Wotton (2006), The effects of seasonal changes to in-stream vegetation cover on patterns of flow and accumulation of sediment, *Geomorphology, 77*, 320–334.

Coulthard, T. J., D. M. Hicks, and M. J. Van De Wiel (2007), Cellular modelling of river catchments and reaches: Advantages, limitations and prospects, *Geomorphology, 90*, 192–207.

Coulthard, T. J., M. J. Kirkby, and M. G. Macklin (1996), A cellular automaton fluvial and slope model of landscape evolution, in *Proceedings of the First International Conference on Geocomputation*, edited by R. J. Abrahart, pp. 168–185, Univ. of Leeds, Leeds, U. K.

Coulthard, T. J., M. K. Kirkby, and M. G. Macklin (2000), Modelling geomorphic response to environmental change in an upland catchment, *Hydrol. Processes, 14*, 2031–2045.

Coulthard, T. J., M. G. Macklin, and M. J. Kirkby (2002), A cellular model of Holocene upland river basin and alluvial fan evolution, *Earth Surf. Processes Landforms, 27*, 269–288.

Coulthard, T. J., J. Lewin, and M. G. Macklin (2008), Non-stationarity of basin scale sediment delivery in response to climate change, in *Gravel-Bed Rivers VI: From Process Understanding to River Restoration*, edited by H. Habersack, H. Piégay, and M. Rinaldi, pp. 315–335, Elsevier, Amsterdam.

Cover, M. R., C. L. May, W. E. Dietrich, and V. H. Resh (2008), Quantitative linkages among sediment supply, streambed fine sediment, and benthic macroinvertebrates in northern California streams, *J. North Am. Benthol. Soc.*, *27*, 135–149.

Covich, A. P. (1988), Geographical and historical comparisons of neotropical streams: Biotic diversity and detrital processing in highly variable habitats, *J. North Am. Benthol. Soc.*, *7*, 361–386.

Covich, A. P., and T. A. Crowl (1990), Effects of hurricane storm flow on transport of woody debris in a rain forest stream (Luquillo Experimental Forest, Puerto Rico), in *Tropical Hydrology and Caribbean Water Resources*, edited by J. Hari Krishna et al., pp. 197–205, Am. Water Resour. Assoc., Bethesda, Md.

Covich, A. P., and W. H. McDowell (1996), The stream community, in *The Food Web of a Tropical Rain Forest*, edited by D. P. Reagan and R. B. Waide, pp. 433–459, The Univ. of Chicago Press, Chicago, Ill.

Covich, A. P., T. A. Crowl, S. L. Johnson, D. Varza, and D. L. Certain (1991), Post-Hurricane Hugo increases in atyid shrimp abundances in a Puerto Rican montane stream, *Biotropica*, *23*, 448–454.

Covich, A. P., M. A. Palmer, and T. A. Crowl (1999), The role of benthic invertebrate species in freshwater ecosystems, *BioScience*, *49*, 119–127.

Cowan, W. L. (1956), Estimating hydraulic roughness coefficients, *Agric. Eng.*, *37*, 473–475.

Cowie, P. A., M. Attal, G. E. Tucker, A. C. Whittaker, M. Naylor, A. Ganas, and G. P. Roberts (2006), Investigating the surface process response to fault interaction and linkage using a numerical modelling approach, *Basin Res.*, *18*, 231–266.

Craig, R. (1998), Changing climate and mountain streams – what do we need to know?, in *Hydrology, Water Resources and Ecology in Headwaters*, edited by K. Kovar et al., *IAHS Publ.*, *248*, 19–25.

Crandell, D. R. (1971), *Postglacial lahars from Mount Rainier Volcano, Washington*, U.S. Geol. Surv. Prof. Pap., *677*, 75 pp.

Craw, D., and J. Waters (2007), Geological and biological evidence for regional drainage reversal during lateral tectonic transport, Marlborough, New Zealand, *J. Geol. Soc.*, *164*, 785–793.

Craw, D., C. Burridge, L. Anderson, and J. M. Waters (2007), Late Quaternary river drainage and fish evolution, Southland, New Zealand, *Geomorphology*, *84*, 98–110.

Crawford, N. H., and R. K. Lindsley (1962), The synthesis of continuous streamflow hydrographs on a digital computer, *Tech. Rep. 12*, Dep. of Civ. Eng., Stanford Univ., Stanford, Calif.

Creed, I. F., and F. D. Beall (2009), Distributed topographic indicators for predicting nitrogen export from headwater catchments, *Water Resour. Res.*, *45*, W10407, doi:10.1029/2008WR007285.

Croft, A. R. (1967), *Rainstorm Debris Floods: A Problem in Public Welfare*, 36 pp., Univ. of Arizona, Agricultural Experiment Station, Tucson.

Croke, J., and S. Mockler (2001), Gully initiation and road-to-stream linkage in a forested catchment, southeastern Australia, *Earth Surf. Processes Landforms*, *26*, 205–217.

Croke, J., S. Mockler, P. Fogarty, and I. Takken (2005), Sediment concentration changes in runoff pathways from a forest road network and the resultant spatial pattern of catchment connectivity, *Geomorphology*, *68*, 257–268.

Cronin, S. J., V. E. Neall, J. A. Lecointre, and A. S. Palmer (1999), Dynamic interactions between lahars and stream flow: A case study from Ruapehu volcano, New Zealand, *Geol. Soc. Am. Bull.*, *111*, 28–38.

Crosby, B. T., and K. X. Whipple (2006), Knickpoint initiation and distribution within fluvial networks: 236 waterfalls in the Waipaoa River, North Island, New Zealand, *Geomorphology*, *82*, 16–38.

Crosby, B. T., K. X. Whipple, N. M. Gasparini, and C. W. Wobus (2007), Formation of fluvial hanging valleys: Theory and simulation, *J. Geophys. Res.*, *112*, F03S10, doi:10.1029/2006JF000566.

Crosta, G. B., and P. Frattini (2004), Controls on modern alluvial fan processes in the Central Alps, northern Italy, *Earth Surf. Processes Landforms*, *29*, 267–293.

Crosta, G. B., and P. Frattini (2008), Rainfall-induced landslides and debris flows, *Hydrol. Processes, 22,* 473–477.

Crowder, D., and P. Diplas (1994), Some benefits of grid by number sampling, in *Proceedings, ASCE Hydraulic Engineering '94 Conference, August 1–5, Buffalo, New York*, edited by G. V. Cotroneo and R. R. Rumer, pp. 795–799, Am. Soc. of Civ. Eng., New York.

Crowder, D. W., and H. V. Knapp (2005), Effective discharge recurrence intervals of Illinois streams, *Geomorphology, 64,* 167–184.

Crowl, T. A., and A. P. Covich (1990), Predator-induced life history shifts in a freshwater snail, *Science, 247,* 949–951.

Crowley, P. D., P. W. Reiners, J. M. Reuter, and G. D. Kaye (2002), Laramide exhumation of the Bighorn Mountains, Wyoming: An apatite (U/Th)/He thermochronology study, *Geology, 30,* 27–30.

Cubito, A., V. Ferrara, and G. Pappalardo (2005), Landslide hazard in the Nebrodi Mountains (northeastern Sicily), *Geomorphology, 66,* 359–372.

Cudden, J. R., and T. B. Hoey (2003), The causes of bedload pulses in a gravel channel: The implications of bedload grain-size distributions, *Earth Surf. Processes Landforms, 28,* 1411–1428.

Cudennec, C., Y. Fouad, I. Sumarjo Gatot, and J. Duchesne (2004), A geomorphological explanation of the unit hydrograph concept, *Hydrol. Processes, 18,* 603–621.

Cudney, J. J. (1995), The influence of variable width on flow structure in a steep mountain stream, unpublished M.S. thesis, 103 pp., Florida State Univ., Tallahassee, Fla.

Cui, Y., G. Parker, T. E. Lisle, J. Gott, M. E. Hansler-Ball, J. E. Pizzuto, N. E. Allmendinger, and J. M. Reed (2003a), Sediment pulses in mountain rivers: 1. Experiments, *Water Resour. Res., 39*(9), 1239, doi:10.1029/2002WR001803.

Cui, Y., G. Parker, J. Pizzuto, and T. E. Lisle (2003b), Sediment pulses in mountain rivers: 2. Comparison between experiments and numerical predictions, *Water Resour. Res., 39*(9), 1240, doi:10.1029/2002WR001805.

Cuker, B. E., P. T. Gama, and J. M. Burkholder (1990), Type of suspended clay influences lake productivity and phytoplankton community response to phosphorus loading, *Limnol. Oceanogr., 35,* 830–839.

Cullmann, J., G. H. Schmitz, and W. Görner (2006), PAI-OFF: A new strategy for online flood forecasting in mountainous catchments, in *Predictions in Ungauged Basins: Promise and Progress*, edited by M. Sivapalan et al., *IAHS Publ., 303,* 57–68.

Cullum, C., G. J. Brierley, and M. Thoms (2008), The spatial organization of river systems, in *River Futures: An Integrative Scientific Approach to River Repair*, edited by G. J. Brierley and K. A. Fryirs, pp. 43–64, Island Press, Washington, D. C.

Cummins, K. W. (1974), Structure and function of stream ecosystems, *BioScience, 24,* 631–641.

Cummins, K. W., C. E. Cushing, and G. W. Minshall (1995), Introduction: An overview of stream ecosystems, in *River and Stream Ecosystems*, edited by C. E. Cushing, K. W. Cummins, and G. W. Minshall, pp. 1–8, Elsevier, Amsterdam.

Cunha, P. P., A. A. Martins, S. Daveau, and P. F. Friend (2005), Tectonic control of the Tejo river fluvial incision during the late Cenozoic, in Ródäo – central Portugal (Atlantic Iberian border), *Geomorphology, 64,* 271–298.

Cunningham, D., A. H. Dijkstra, J. Howard, A. Quarles, and G. Badarch (2003), Active intraplate strike-slip faulting and transpression uplift in the Mongolian Altai, in *Intraplate Strike-slip Deformation Belts*, edited by F. Storti, R. E. Holdsworth, and F. Salivin, *Geol. Soc. Spec. Publ., 210,* 65–87.

Cupp, C. E. (1989), *Stream Corridor Classification for Forested Lands of Washington*, 24 pp., Washington Forest Protection Association, Olympia, Wash.

Curran, J. C., and P. R. Wilcock (2005), Characteristics dimensions of the step-pool bed configuration: An experimental study, *Water Resour. Res., 41,* W02030, doi:10.1029/2004WR003568.

Curran, J. H. (1999), Hydraulics of large woody debris in step-pool channels, Cascade Range, Washington, unpublished M.S. thesis, 197 pp., Colo. State Univ., Ft. Collins.

Curran, J. H., and E. E. Wohl (2003), Large woody debris and flow resistance in step-pool channels, Cascade Range, Washington, *Geomorphology*, *51*, 141–147.

Curry, R. R. (1966), Observation of alpine mudflows in the Tenmile Range, central Colorado, *Geol. Soc. Am. Bull.*, *77*, 771–776.

Curtis, J. A., L. E. Flint, C. N. Alpers, and S. M. Yarnell (2005), Conceptual model of sediment processes in the upper Yuba River watershed, Sierra Nevada, CA, *Geomorphology*, *68*, 149–166.

Curtis, K. E., C. E. Renshaw, F. J. Magilligan, and W. B. Dade (2010), Temporal and spatial scales of geomorphic adjustments to reduced competency following flow regulation in bedload-dominated systems, *Geomorphology*, *118*, 105–117.

Cyranoski, D. (2008), Glacial melt thaws South Asian rivalry, *Nature*, *452*(7189), 793.

Czarnomski, N. M., D. M. Dreher, K. U. Snyder, J. A. Jones, and F. J. Swanson (2008), Dynamics of wood in stream networks of the western Cascades Range, Oregon, *Can. J. For. Res.*, *38*, 2236–2248.

Dade, W. B., and M. E. Verdeyen (2007), Tectonic and climatic controls of alluvial-fan size and source-catchment relief, *J. Geol. Soc.*, *164*, 353–358.

Dadson, S. J., et al. (2003), Links between erosion, runoff variability and seismicity in the Taiwan orogen, *Nature*, *426*, 648–651.

D'Agostino, V., M. Cesca, and L. Marchi (2010), Field and laboratory investigations of runout distances of debris flows in the Dolomites (Eastern Italian Alps), *Geomorphology*, *115*, 294–304.

Dahlgren, R. A. (1998), Effects of forest harvest on stream-water quality and nitrogen cycling in the Caspar Creek watershed, in *Proceedings of the Conference on Coastal Watersheds: The Caspar Creek Story*, edited by R. R. Ziemer, pp. 45–53, Pacific Southwest Res. Sta., USDA, Albany, Calif.

Dahlström, N., K. Jönsson, and C. Nilsson (2005), Long-term dynamics of large woody debris in a managed boreal forest stream, *For. Ecol. Manage.*, *210*, 363–373.

Dai, F. C., C. F. Lee, J. H. Deng, and L. G. Tham (2005), The 1786 earthquake-triggered landslide dam and subsequent dam-break flood on the Dadu River, southwestern China, *Geomorphology*, *65*, 205–221.

Dal Cin, R. (1968), Pebble clusters: Their origin and utilization in the study of palaeocurrents, *Sediment. Geol.*, *2*, 233–241.

Dalla Fontana, G., and L. Marchi (1998), GIS indicators for sediment sources study in Alpine basins, in *Hydrology, Water Resources and Ecology in Headwaters*, edited by K. Kovar et al., *IAHS Publ.*, *248*, 553–560.

Dalla Fontana, G., and L. Marchi (2003), Slope-area relationships and sediment dynamics in two alpine streams, *Hydrol. Processes*, *17*, 73–87.

Dalrymple, T., and M. A. Benson (1967), Measurement of peak discharge by the slope-area method, *Techniques of Water-Resources Investigations of the U.S. Geological Survey*, Book 3, chap. A2, pp. 1–12.

D'Ambrosio, D., W. Spataro, and G. Iovine (2006), Parallel genetic algorithms for optimising cellular automata models of natural complex phenomena; an application to debris flows, *Comput. Geosci.*, *32*, 861–875.

Daniels, M. D., and B. L. Rhoads (2004), Effect of large woody debris configuration on three-dimensional flow structure in two low-energy meander bends at varying stages, *Water Resour. Res.*, *40*, W11302, doi:10.1029/2004WR003181.

Daño, A. M., and F. E. Siapno (1992), The effectiveness of soil conservation structures in steep cultivated mountain regions of the Philippines, in *Erosion, Debris Flows and Environment in Mountain Regions*, edited by D. E. Walling, T. R. Davies, and B. Hasholt, *IAHS Publ.*, *209*, 399–405.

Danvind, M., and C. Nilsson (1997), Seed floating ability and distribution of alpine plants along a Swedish river, *J. Veg. Sci.*, *8*, 271–276.

Darby, S. E., and C. R. Thorne (1996), Development and testing of riverbank-stability analysis, *J. Hydraul. Eng.*, *122*, 443–454.

Darby, S. E., and M. J. Van De Wiel (2003), Models in fluvial geomorphology, in *Tools in Fluvial Geomorphology*, edited by G. M. Kondolf and H. Piégay, pp. 503–537, John Wiley, Chichester, U. K.

Darby, S. E., D. Gessler, and C. R. Thorne (2000), Computer program for stability analysis of steep, cohesive riverbanks, *Earth Surf. Processes Landforms*, *25*, 175–190.

Darwin, C. (1859), *The Origin of Species by Means of Natural Selection*, Murray, London.

Das, I., S. Sahoo, C. van Westen, A. Stein, and R. Hack (2010), Landslide susceptibility assessment using logistic regression and its comparison with a rock mass classification system, along a road section in the northern Himalayas (India), *Geomorphology*, *114*, 627–637.

Das, T., G. Moretti, A. Bárdossy, and A. Montanari (2006), Assessing the predictive ability of the spatially distributed conceptual AFFDEF model for a mesoscale catchment, in *Predictions in Ungauged Basins: Promise and Progress*, edited by M. Sivapalan et al., *IAHS Publ.*, *303*, 351–359.

Daubert, O., J.-M. Hervouet, and A. Jami (1989), Description of some numerical tools for solving incompressible turbulent and free surface flows, *Int. J. Numer. Methods Eng.*, *27*, 3–20.

David, G. C. L., B. P. Bledsoe, D. M. Merritt, and E. Wohl (2009), The impacts of ski slope development on stream channel morphology in the White River National Forest, Colorado, USA, *Geomorphology*, *103*, 375–388.

David, G. C. L., E. Wohl, S. E. Yochum, and B. P. Bledsoe (2010), Controls on spatial variations in flow resistance along steep mountain streams, *Water Resour. Res.*, *46*, W03513, doi:10.1029/2009WR008134.

Davies, B. E. (1983), Heavy metal contamination from base metal mining and smelting: Implications for man and his environment, in *Applied Environmental Geochemistry*, edited by I. Thornton, pp. 425–462, Academic Press, London.

Davies, B. R., M. C. Thoms, K. F. Walker, J. H. O'Keefe, and J. A. Gore (1994), Dryland rivers: Their ecology, conservation, and management, in *The Rivers Handbook*, edited by P. Calow and G. E. Petts, pp. 484–511, Blackwell Sci., Oxford, U. K.

Davies, T. R. H. (1980), Bedform spacing and flow resistance, *J. Hydraul. Div. Am. Soc. Civ. Eng.*, *106*, 423–433.

Davies, T. R. H. (1987), Problems of bedload transport in braided gravel-bed rivers, in *Sediment Transport in Gravel-Bed Rivers*, edited by C. R. Thorne, J. C. Bathurst, and R. D. Hey, pp. 793–828, John Wiley, Chichester, U. K.

Davies, T. R. H. (1991), Research of fluvial processes in mountains: A change of emphasis, in *Fluvial Hydraulics of Mountain Regions*, edited by A. Armanini and G. DiSilvio, pp. 251–266, Springer, Berlin.

Davies, T. R. H. (2002), Landslide-dambreak floods at Franz Josef Glacier township, Westland, New Zealand: A risk assessment, *J. Hydrol. N. Z.*, *41*, 1–17.

Davies, T. R. H., and O. Korup (2007), Persistent alluvial fan head trenching resulting from large, infrequent sediment inputs, *Earth Surf. Processes Landforms*, *32*, 725–742.

Davies, T. R. H., and M. J. McSaveney (2001), Anthropogenic fanhead aggradation, Waiho River, Westland, New Zealand, in *Gravel-Bed Rivers V*, edited by M. P. Mosley, pp. 531–553, N. Z. Hydrol. Soc., Wellington, New Zealand.

Davies, T. R. H., and A. J. Sutherland (1983), Extremal hypotheses for river behaviour, *Water Resour. Res.*, *19*, 141–148.

Davies, T. R. H., M. J. McSaveney, and P. J. Clarkson (2003), Anthropic aggradation of the Waiho River, New Zealand; microscale modelling, *Earth Surf. Processes Landforms*, *28*, 209–218.

Davis, W. M. (1898), The grading of mountain slopes, *Science*, *7*, 81.

Davis, W. M. (1899), The geographical cycle, *J. Geogr.*, *14*, 481–504.

Davis, W. M. (1902), Base-level, grade, and peneplain, in *Geographical Essays*, vol. 18, pp. 381–412, Ginn, Boston, Mass.

Davy, P., and D. Lague (2009), Fluvial erosion/transport equation of landscape evolution models revisited, *J. Geophys. Res.*, *114*, F03007, doi:10.1029/2008JF001146.

Dawdy, D. R., and W. C. Wang (1993), Prediction of gravel transport using Parker's algorithm, in *Hydraulic Engineering '93*, edited by H. W. Shen, S. T. Su, and F. Wen, pp. 1523–1528, Am. Soc. of Civ. Eng., New York.

Day, D. G. (1978), Drainage density changes during rainfall, *Earth Surf. Processes*, *3*, 319–326.

Day, T. J. (1972), The channel geometry of mountain streams, in *Mountain Geomorphology: Geomorphological Processes in the Canadian Cordillera*, edited by H. O. Slaymaker and H. J. McPherson, pp. 141–149, Tantalus Res. Ltd., Vancouver, Canada.

Day, T. J. (1977), Field procedures and evaluation of a slug dilution gauging method in mountain streams, *J. Hydrol. N. Z.*, *16*, 113–133.

Deangelis, M. L., and E. F. Wood (1998), A detailed model to simulate heat and moisture transport in a frozen soil, in *Hydrology, Water Resources and Ecology in Headwaters*, edited by K. Kovar et al., *IAHS Publ.*, *248*, 141–148.

De Blasio, F. V., and M.-B. S×ter (2009), Small-scale experimental simulation of talus evolution, *Earth Surf. Processes Landforms*, *34*, 1685–1692.

De Boer, D. H. (1997), Changing contributions of suspended sediment sources in small basins resulting from European settlement on the Canadian Prairies, *Earth Surf. Processes Landforms*, *22*, 623–639.

De Boer, D. H. (2005), Predicting sediment rating curves with a cellular landscape model, in *Sediment Budgets 2*, edited by A. J. Horowitz and D. E. Walling, *IAHS Publ.*, *292*, 75–84.

DeByle, N. V. (1973), Broadcast burning of logging residues and the water repellency of soils, *Northwest Sci.*, *47*, 77–87.

DeByle, N. V. (1985), Wildlife, in *Aspen: Ecology and Management in the Western US*, edited by N. V. DeByle and R. P. Winokur, pp. 135–152, U.S. Fish and Wildlife Service, Rocky Mountain Forest and Range Experiment Station, Ft. Collins, Colo.

DéCamps, H. (1984), Towards a landscape ecology of river valleys, in *Trends in Ecological Research for the 1980s*, edited by J. H. Cooley and F. G. Golley, pp. 163–178, Plenum, New York.

DéCamps, H., and O. DéCamps (1999), The relevance of the concept of health in managing riparian landscapes, *Ecological Society of America, 1999 Annual Meeting Abstracts*, p. 13.

DéCamps, H., M. Fortuné, F. Gazelle, and G. Pautou (1988), Historical influence of man on the riparian dynamics of a fluvial landscape, *Landscape Ecol.*, *1*, 163–173.

DéCamps, H., and E. Tabacchi (1994), Species richness in vegetation along river margins, in *Aquatic Ecology: Scale, Pattern and Process*, edited by P. S. Giller, A. G. Hildrew, and D. G. Raffaelli, pp. 1–20, Blackwell Sci., Oxford.

De Chant, L. J., P. P. Pease, and V. P. Tchakerian (1999), Modelling alluvial fan morphology, *Earth Surf. Processes Landforms*, *24*, 641–652.

Dedkov, A. P., and V. I. Moszherin (1992), Erosion and sediment yield in mountain regions of the world, in *Erosion, Debris Flows and Environment in Mountain Regions*, edited by D. E. Walling, T. R. Davies, and B. Hasholt, *IAHS Publ.*, *209*, 29–36.

Degenhardt, J. J. (2009), Development of tongue-shaped and multilobate rock glaciers in alpine environments – interpretations from ground penetrating radar surveys, *Geomorphology*, *109*, 94–107.

De Jong, C. (1991), A reappraisal of the significance of obstacle clasts in cluster bedform dispersal, *Earth Surf. Processes Landforms*, *16*, 737–744.

De Jong, C. (1994), The significance of extreme events in the development of mountain river beds, in *Variability in Stream Erosion and Sediment Transport*, edited by L. J. Olive, R. J. Loughran, and J. A. Kesby, *IAHS Publ.*, *224*, 13–24.

De Jong, C. (2007), River resilience and dams in mountain areas, *14 Deutsches Tasperrensymposium; 7th ICOLD European Club dam symposium*, pp. 75–80, Technische Universität München, Munich, Germany.

De Jong, C., S. Cappy, M. Finckh, and D. Funk (2008), A transdisciplinary analysis of water problems in the mountainous karst areas of Morocco, *Eng. Geol.*, *99*, 228–238.

De Jong, C., P. Ergenzinger, M. Borufka, A. Köcher, and M. Dresen (2005a), Geomorphological zoning: An improvement to coupling alpine hydrology and meteorology?, in *Climate and Hydrology in Mountain Areas*, edited by C. de Jong, D. Collins, and R. Ranzi, pp. 247–260, John Wiley, Chichester, U. K.

De Jong, C., M. Mundelius, and K. Migała (2005b), Comparison of evaportranspiration and condensation measurements between the Giant Mountains and the Alps, in *Climate and Hydrology in Mountain Areas*, edited by C. de Jong, D. Collins, and R. Ranzi, pp. 161–183, John Wiley.

Demers, J. D., C. T. Driscoll, and J. B. Shanley (2010), Mercury mobilization and episodic stream acidification during snowmelt: Role of hydrologic flow paths, source areas, and supply of dissolved organic carbon, *Water Resour. Res.*, 46, W01511, doi:10.1029/2008WR007021.

Demir, T., and R. P. D. Walsh (2005), Shape and size characteristics of bedload transported during winter storm events in the Cwm Treweryn Stream Brecon Beacons, South Wales, *Turkish J. Earth Sci.*, 14, 105–121.

Demoulin, A., B. Bovy, G. Rixhon, and Y. Cornet (2007), An automated method to extract fluvial terraces from digital elevation models: The Vesdre valley, a case study in eastern Belgium, *Geomorphology*, 91, 51–64.

Deng, M. (2006), Changes in climate and runoff in the Tarim River basin and ecosystem restoration in the lower reaches of the Tarim River, *J. Glaciol. Cryol. (China)*, 28, 694–702.

Dengler, L., A. K. Lehre, and C. J. Wilson (1987), Bedrock geometry of unchannelized valleys, in *Erosion and Sedimentation in the Pacific Rim*, edited by R. L. Beschta et al., *IAHS Publ.*, 165, 81–90.

Dengler, L., and D. R. Montgomery (1989), Estimating thickness of colluvial fill in unchanneled valleys from surface topography, *Bull. Assoc. Eng. Geol.*, 26, 333–342.

Denning, A. S., J. Baron, M. A. Mast, and M. Arthur (1991), Hydrologic pathways and chemical composition of runoff during snowmelt in Loch Vale watershed, Rocky Mountain National Park, Colorado, USA, *Water Air Soil Pollut.*, 59, 107–123.

Densmore, A. L., P. A. Allen, and G. Simpson (2007), Development and response of a coupled catchment fan system under changing tectonic regime and climatic forcing, *J. Geophys. Res.*, 112, F01002, doi:10.1029/2006JF000474.

Deroanne, C., and F. Petit (1999), Longitudinal evaluation of the bed load size and of its mobilisation in a gravel bed river, in *Floods and Landslides*, edited by R. Casale and C. Margottini, pp. 335–342, Springer, Berlin.

De Scally, F. A., and I. F. Owens (2004), Morphometric controls and geomorphic responses on fans in the Southern Alps, New Zealand, *Earth Surf. Processes Landforms*, 29, 311–322.

De Scally, F. A., and I. F. Owens (2005), Depositional processes and particle characteristics on fans in the Southern Alps, New Zealand, *Geomorphology*, 69, 46–56.

De Scally, F. A., I. F. Owens, and J. Louis (2010), Controls on fan depositional processes in the schist ranges of the Southern Alps, New Zealand, and implications for debris-flow hazard assessment, *Geomorphology*, 122, 99–116.

De Scally, F. E. S., O. Slaymaker, and I. Owens (2001), Morphometric controls and basin response in the Cascade Mountains, *Geogr. Ann., Ser. A, Phys. Geogr.*, 83, 117–130.

Descroix, L., E. Gautier, A. L. Besnier, O. Amogu, D. Viramontes, and J. L. Gonzalez-Barrios (2005), Sediment budget as evidence of land-use changes in mountainous areas: Two stages of evolution, in *Sediment Budgets 2*, edited by A. J. Horowitz and D. E. Walling, *IAHS Publ.*, 292, 262–270.

Descroix, L., J. L. Gonzalez Barrios, D. Viramontes, J. Poulenard, E. Anaya, M. Esteves, and J. Estrada (2008), Gully and sheet erosion on subtropical mountain slopes; their respective roles and the scale effect, *Catena*, 72, 325–339.

De Serres, B., A. G. Roy, P. M. Biron, and J. L. Best (1999), Three-dimensional structure of flow at a confluence of river channels with discordant beds, *Geomorphology*, 26, 313–335.

Desloges, J. R., and M. Church (1992), Geomorphic implications of glacier outburst flooding: Noieck River valley, British Columbia, *Can. J. Earth Sci.*, 29, 551–564.

Dessert, C., B. Dupré, J. Gaillardet, L. M. François, and C. J. Allègre (2003), Basalt weathering laws and the impact of basalt weathering on the carbon cycle, *Chem. Geol.*, 202, 257–273.

Detert, M., M. Klar, T. Wenka, and G. H. Jirka (2008), Pressure- and velocity-measurements above and within a porous gravel bed at the threshold of stability, in *Gravel-Bed Rivers VI: From Process Understanding to River Restoration*, edited by H. Habersack, H. Piégay, and M. Rinaldi, pp. 85–107, Elsevier, Amsterdam.

Dethier, D. P. (2001), Pleistocene incision rates in the western United States calibrated using Lava Creek B Tephra, *Geology, 29*, 783–786.

De Villiers, G. D. T. (1990), Rainfall variations in mountainous regions, in *Hydrology in Mountainous Regions I. Hydrological Measurements: The Water Cycle*, edited by H. Lang and A. Musy, *IAHS Publ., 193*, 33–41.

De Vries, M. (1971), On accuracy of bed-material sampling, *J. Hydraul. Res., 8*, 523–533.

DeVries, P., S. J. Burges, J. Daigneau, and D. Stearns (2001), Measurement of the temporal progression of scour in a pool-riffle sequence in a gravel-bed stream using an electronic scour monitor, *Water Resour. Res., 37*, 2805–2816.

De Vries, P. P. E. (2002), Bedload layer thickness and disturbance depth in gravel bed streams, *J. Hydraul. Eng., 128*, 983–991.

De Wolfe, V. G., P. M. Santi, J. Ey, and J. E. Gartner (2008), Effective mitigation of debris flows at Lemon Dam, La Plata County, Colorado, *Geomorphology, 96*, 366–377.

Dhakal, A. S., T. Amada, M. Aniya, and R. R. Sharma (2002), Detection of areas associated with flood and erosion caused by a heavy rainfall using multitemporal Landsat TM data, *Photogramm. Eng. Remote Sens., 68*, 233–239.

Di Crescenzo, G., and A. Santo (2005), Debris slides-rapid earth flows in the carbonate massifs of the Campania region (southern Italy): Morphological and morphometric data for evaluating triggering susceptibility, *Geomorphology, 66*, 255–276.

Diepenbroek, M., and C. De Jong (1994), Quantification of textural particle characteristics by image analysis of sediment surfaces - examples from active and paleosurfaces in steep, coarse grained mountain environments, in *Dynamics and Geomorphology of Mountain Rivers*, edited by P. Ergenzinger and K.-H. Schmidt, pp. 301–314, Springer, Berlin.

Dietrich, W. (1995), *Northwest Passage: The Great Columbia River*, 448 pp., Univ. of Washington Press, Seattle.

Dietrich, W. E., and R. Dorn (1984), Significance of thick deposits of colluvium on hillslopes: A case study involving the use of pollen analysis in the coastal mountains of northern California, *J. Geol., 92*, 147–158.

Dietrich, W. E., and T. Dunne (1978), Sediment budget for a small catchment in mountainous terrain, *Z. Geomorphol., 29*, 191–206.

Dietrich, W. E., and T. Dunne (1993), The channel head, in *Channel Network Hydrology*, edited by K. Beven and M. J. Kirkby, pp. 175–219, John Wiley, Chichester, U. K.

Dietrich, W. E., and M. R. Montgomery (1992), A digital terrain model for predicting debris flow source areas, *Eos Trans. AGU, 73*, 227.

Dietrich, W. E., and J. T. Perron (2006), The search for a topographic signature of life, *Nature, 439*, 411–418.

Dietrich, W. E., and P. J. Whiting (1989), Boundary shear stress and sediment transport in river meanders of sand and gravel, in *River Meandering*, edited by S. Ikeda and G. Parker, pp. 1–50, AGU, Washington, D. C.

Dietrich, W. E., D. G. Bellugi, L. S. Sklar, J. D. Stock, A. M. Heimsath, and J. J. Roering (2003), Geomorphic transport laws for predicting landscape form and dynamics, in *Prediction in Geomorphology*, edited by P. R. Wilcock and R. M. Iverson, pp. 103–132, AGU, Washington, D. C.

Dietrich, W. E., J. W. Kirchner, H. Ikeda, and F. Iseya (1989), Sediment supply and the development of the coarse surface layer in gravel-bedded rivers, *Nature, 340*, 215–217.

Dietrich, W. E., C. J. Wilson, D. R. Montgomery, and J. McKean (1993), Analysis of erosion thresholds, channel networks, and landscape morphology using a digital terrain model, *J. Geol., 101*, 259–278.

Dietrich, W. E., C. J. Wilson, D. R. Montgomery, J. McKean, and R. Bauer (1992), Erosion thresholds and land surface morphology, *Geology, 20*, 675–679.

Dill, L. M., R. C. Ydenberg, and A. H. G. Fraser (1981), Food abundance and territory size in juvenile coho salmon (Oncorhynchus kisutch), *Can. J. Zool., 59*, 1801–1809.

Dinehart, R. L. (1999), Correlative velocity fluctuations over a gravel bed river, *Water Resour. Res., 35*, 569–582.

Ding, Y., B. Ye, and S. Liu (2000), Impact of climate change on the alpine streamflow during the past 40 a in the middle part of the Qilian Mountains, northwest China, *J. Glaciol. Geocryol., 22*, 193–199.

Dingman, S. L. (1991), *Fluvial Hydrology*, 383 pp., W. H. Freeman, New York.

Dingman, S. L. (2001), *Physical Hydrology*, 2nd ed., 646 pp., Prentice Hall, Englewood Cliffs, N. J.

Dingwall, P. R. (1972), Erosion by overland flow on an alpine debris slope, in *Mountain Geomorphology: Geomorphological Processes in the Canadian Cordillera*, edited by H. O. Slaymaker and H. J. McPherson, pp. 113–120, Tantalus Res. Ltd., Vancouver, Canada.

Diodato, N. (2004), Local models for rainstorm-induced hazard analysis on Mediterranean river-torrential geomorphological systems, *Nat. Hazards Earth Syst. Sci., 4*, 389–397.

Diplas, P., and J. B. Fripp (1992), Properties of various sediment sampling procedures, *J. Hydraul. Eng., 118*, 955–970.

Diplas, P., and G. Parker (1992), Deposition and removal of fines in gravel-bed streams, in *Dynamics of Gravel-Bed Rivers*, edited by P. Billi et al., pp. 313–329, John Wiley, Chichester, U. K.

Diplas, P., and H. Shaheen (2008), Bed load transport and streambed structure in gravel streams, in *Gravel-Bed Rivers VI: From Process Understanding to River Restoration*, edited by H. Habersack, H. Piégay, and M. Rinaldi, pp. 291–312, Elsevier, Amsterdam.

Diplas, P., and A. J. Sutherland (1988), Sampling techniques of gravel sized sediments, *J. Hydraul. Eng., 114*, 484–501.

Diplas, P., C. L. Dancey, A. O. Celik, M. Valyrakis, K. Greer, and T. Akar (2008), The role of impulse on the initiation of particle movement under turbulent flow conditions, *Science, 322*, 717–720.

DiSilvio, G. (1994), Floods and sediment dynamics in mountain rivers, in *Coping with Floods*, edited by G. Rossi, N. Harmancioglu, and V. Yevjevich, pp. 375–392, Kluwer, The Netherlands.

Di Silvio, G., and S. Brunelli (1991), Experimental investigations on bed-load and suspended transport in mountain streams, in *Fluvial Hydraulics of Mountain Regions*, edited by A. Armanini and G. DiSilvio, pp. 443–457, Springer, Berlin.

Dittrich, A., F. Nestmann, and P. Ergenzinger (1996), Ratio of lift and shear forces over rough surfaces, in *Coherent Flow Structures in Open Channels*, edited by P. J. Ashworth et al., pp. 125–146, John Wiley, Chichester, U. K.

Dixon, J. L., A. M. Heimsath, J. Kaste, and R. Amundson (2009), Climate-driven processes of hillslope weathering, *Geology, 37*, 975–978.

Djorovic, M. (1992), Ten-years of sediment discharge measurement in the Jasenica research drainage basin, Yugoslavia, in *Erosion, Debris Flows and Environment in Mountain Regions*, edited by D. E. Walling, T. R. Davies, and B. Hasholt, *IAHS Publ., 209*, 37–40.

Do Carmo, J. S. A. (2007), The environmental impact and risks associated with changes in fluvial morphodynamic processes, in *Water in Celtic Countries: Quantity, Quality and Climate Variability*, edited by J. P. Lobo Ferreira and J. M. P. Viera, *IAHS Publ., 310*, 307–319.

Docker, B. B., and T. C. T. Hubble (2008), Quantifying root-reinforcement of river bank soils by four Australian tree species, *Geomorphology, 100*, 401–418.

Dodds, G. S., and F. L. Hisaw (1925), Ecological studies of aquatic insects. IV. Altitudinal range and zonation of mayflies, stoneflies and caddisflies in the Colorado Rockies, *Ecology, 6*, 380–390.

D'Odorico, P., and R. Rigon (2003), Hillslope and channel contributions to the hydrologic response, *Water Resour. Res., 39*(5), 1113, doi:10.1029/2002WR001708.

Dodov, B., and E. Foufoula-Georgiou (2005), Fluvial processes and streamflow variability: Interplay in the scale-frequency continuum and implications for scaling, *Water Resour. Res.*, *41*, W05005, doi:10.1029/2004WR003408.

Doering, M., U. Uehlinger, A. Rotach, D. R. Schlaepfer, and K. Tockner (2007), Ecosystem expansion and contraction dynamics along a large Alpine alluvial corridor (Tagliamento River, northeast Italy), *Earth Surf. Processes Landforms*, *32*, 1693–1704.

Dogan, E., S. Tripathi, D. A. Lyn, and R. S. Govindaraju (2009), From flumes to rivers: Can sediment transport in natural alluvial channels be predicted from observations at the laboratory scale?, *Water Resour. Res.*, *45*, W08433, doi:10.1029/2008WR007637.

Dolan, R., A. Howard, and D. Trimble (1978), Structural control of the rapids and pools of the Colorado River in the Grand Canyon, *Science*, *202*, 629–631.

Doll, B. A., G. L. Grabow, K. R. Hall, J. Halley, W. A. Harman, G. D. Jennings, and D. E. Wise (2003), *Stream Restoration: A Natural Channel Design Handbook*, 128 pp., North Carolina Stream Restoration Institute, North Carolina State Univ., Raleigh, N. C.

Dolling, R. K. (1968), Occurrence of pools and riffles: An element in the quasi-equilibrium state of river channels, *Ont. Geogr.*, *2*, 3–11.

Dominick, D. S., and M. P. O'Neill (1998), Effect of flow augmentation on stream channel morphology and riparian vegetation; upper Arkansas River basin, Colorado, *Wetlands*, *18*, 591–607.

Dong, J.-J., C.-T. Lee, Y.-H. Tung, C.-N. Liu, K.-P. Lin, and J.-F. Lee (2009a), The role of the sediment budget in understanding debris flow susceptibility, *Earth Surf. Processes Landforms*, *34*, 1612–1624.

Dong, J.-J., Y.-H. Tung, C.-C. Chen, J.-J. Liao, and Y.-W. Pan (2009b), Discriminant analysis of the geomorphic characteristics and stability of landslide dams, *Geomorphology*, *110*, 162–171.

Donnell, B. D., J. L. Finnie, J. V. Letter, Jr., W. H. McAnally, Jr., L. C. Roig, and W. A. Thomas (1997), *Users Guide to RMA2 WES Version 4.3*, U.S. Army Corps of Engineers, Waterways Experiment Station Hydraulics Laboratory, Vicksburg, Miss.

Dooge, J. C. I. (1959), A general theory of the unit hydrograph, *J. Geophys. Res.*, *64*, 241–256.

Dornes, P. F., J. W. Pomeroy, A. Pietroniro, and D. L. Verseghy (2008), Effects of spatial aggregation of initial conditions and forcing data on modeling snowmelt using a land surface scheme, *J. Hydrometeorol.*, *9*, 789–803.

Doten, C. O., L. C. Bowling, J. S. Lanini, E. P. Maurer, and D. P. Lettenmaier (2006), A spatially distributed model for the dynamic prediction of sediment erosion and transport in mountainous forested watersheds, *Water Resour. Res.*, *42*, W04417, doi:10.1029/2004WR003829.

Douglas, I. (1967), Man, vegetation and the sediment yield of rivers, *Nature*, *215*, 925–928.

Douglas, I. (1996), The impact of land-use changes, especially logging, shifting cultivation, mining and urbanization on sediment yields in humid tropical Southeast Asia: A review with special reference to Borneo, in *Erosion and Sediment Yield: Global and Regional Perspectives*, edited by D. E. Walling and B. W. Webb, *IAHS Publ.*, *236*, 463–471.

Douglas, I. (1999), Hydrological investigations of forest disturbance and land cover impacts in south-east Asia: A review, *Philos. Trans. R. Soc. London, Ser. B*, *354*, 1725–1738.

Douglas, I. (2003), Predicting road erosion rates in selectively logged tropical rain forests, in *Erosion Prediction in Ungauged Basins: Integrating Methods and Techniques*, edited by D. de Boer et al., *IAHS Publ.*, *279*, 199–205.

Douglas, I., and J. L. Guyot (2005), Erosion and sediment yield in the humid tropics, in *Forests, Water and People in the Humid Tropics*, edited by M. Bonell and L. A. Bruijnzeel, pp. 407–421, Cambridge Univ. Press, Cambridge, U. K.

Douglas, I., K. Bidin, G. Balamurugan, N. A. Chappell, R. P. D. Walsh, T. Greer, and W. Sinun (1999a), The role of extreme events in the impacts of selective tropical forestry on erosion during harvesting and recovery phases at Danum Valley, Sabah, *Philos. Trans. R. Soc. London, Ser. B*, *354*, 1749–1761.

Douglas, I., T. Spencer, T. Greer, K. Bidin, W. Sinun, and W. W. Meng (1999b), The impact of selective commercial logging on stream hydrology, chemistry and sediment loads in the Ulu Segama rain forest, Sabah, Malaysia, *Philos. Trans. R. Soc. London, Ser. B*, *335*, 397–406.

Douglass, J., and M. Schmeeckle (2007), Analogue modeling of transverse drainage mechanisms, *Geomorphology*, *874*, 22–43.

Downing, J., P. J. Farley, K. Bunte, K. Swingle, S. E. Ryan, and M. Dixon (2003), Acoustic gravel-transport sensor: Description and field tests in the Little Granite Creek, Wyoming, USA, in *Erosion and Sediment Transport Measurement in Rivers: Technological and Methodological Advances*, edited by J. Bogen, T. Fergus, and D. E. Walling, *IAHS Publ.*, *283*, 193–200.

Downs, P. W., and G. Priestnall (2003), Modelling catchment processes, in *Tools in Fluvial Geomorphology*, edited by G. M. Kondolf and H. Piégay, pp. 205–230, John Wiley, Chichester, U. K.

Doyle, M. W. (2006), A heuristic model for potential geomorphic influences on trophic interactions in streams, *Geomorpohlogy*, *77*, 235–248.

Doyle, M. W., E. H. Stanley, and J. M. Harbor (2002), Geomorphic analogies for assessing probable channel response to dam removal, *J. Am. Water Resour. Assoc.*, *38*, 1567–1579.

Doyle, M. W., E. H. Stanley, and J. M. Harbor (2003), Channel adjustments following two dam removals in Wisconsin, *Water Resour. Res.*, *39*(1), 1011, doi:10.1029/2002WR001714.

Doyle, M. W., E. H. Stanley, D. L. Strayer, R. B. Jacobson, and J. C. Schmidt (2005), Effective discharge analysis of ecological processes in streams, *Water Resour. Res.*, *41*, W11411, doi:10.1029/2005WR004222.

Doyle, M. W., D. Shields, K. F. Boyd, P. B. Skidmore, and D. Dominick (2007), Channel-forming discharge selection in river restoration design, *J. Hydraul. Eng.*, *133*, 831–837.

Draft Report (1999), *Dam removal Success Stories, American Rivers and Trout Unlimited*, unpublished report. (Available at http://www.amrivers.org/success-intro.html)

Drake, T. G., R. L. Shreve, W. E. Dietrich, P. J. Whiting, and L. B. Leopold (1988), Bedload transport of fine gravel observed by motion-picture photography, *J. Fluid Mech.*, *192*, 193–217.

Drever, J. I. (1982), *The Geochemistry of Natural Waters*, 388 pp., Prentice Hall, Englewood Cliffs, N. J.

Drever, J. I. (1988), *The Geochemistry of Natural Waters*, 2nd ed., 437 pp., Prentice Hall, Englewood Cliffs, N. J.

Drever, J. I., and D. R. Hurcomb (1986), Neutralization of atmospheric acidity by chemical weathering in an alpine drainage basin in the North Cascade Mountains, *Geology*, *14*, 221–224.

Driedger, C. L., and A. G. Fountain (1989), Glacier outburst floods at Mount Rainier, Washington State, U.S. A, *Ann. Glaciol.*, *13*, 51–55.

Drigo, R. (2005), Trends and patterns of tropical land use change, in *Forests, Water and People in the Humid Tropics*, edited by M. Bonell and L. A. Bruijnzeel, pp. 9–39, Cambridge Univ. Press, Cambridge, U. K.

Droppo, I. G. (2003), A new definition of suspended sediment: Implications for the measurement and prediction of sediment transport, in *Erosion and Sediment Transport Measurement in Rivers: Technological and Methodological Advances*, edited by J. Bogen, T. Fergus, and D. E. Walling, *IAHS Publ.*, *283*, 3–12.

Drury, T. A., T. B. Abbe, G. R. Pess, C. Petroff, and D. R. Montgomery (1998), Experimental application of engineered log jams: North Fork Stillaguamish River, Washington, *Eos Trans. AGU*, *79*, F346.

Dubinski, I. M., and E. Wohl (2007), Assessment of coarse sediment mobility in the Black Canyon of the Gunnison River, Colorado, *Environ. Manage.*, *40*, 147–160.

Duckson, D. W., Jr., and L. J. Duckson (1995), Morphology of bedrock step pool systems, *Water Resour. Bull.*, *31*, 43–51.

Duckson, D. W., and L. J. Duckson (2001), Channel bed steps and pool shapes along Soda Creek, Three Sisters Wilderness, Oregon, *Geomorphology*, *38*, 267–279.

Dugdale, S. J., P. E. Carbonneau, and D. Campbell (2010), Aerial photosieving of exposed gravel bars for the rapid calibration of airborne grain size maps, *Earth Surf. Processes Landforms*, *35*, 627–639.

Dunkerley, D. L. (1990), The development of armour in the Tambo River, Victoria, Australia, *Earth Surf. Processes Landforms*, *15*, 405–412.

Dunne, T. (1979), Sediment yield and land use in tropical catchments, *J. Hydrol.*, *42*, 281–300.

Dunne, T. (1980), Formation and controls of channel networks, *Prog. Phys. Geogr.*, *4*, 211–239.

Dunne, T. (2001), Problems in measuring and modeling the influence of forest management on hydrologic and geomorphic processes, in *Land Use and Watersheds: Human Influence on Hydrology and Geomorphology in Urban and Forest Areas*, edited by M. S. Wigmosta et al., pp. 77–83, AGU, Washington, D. C.

Dunne, T., and R. G. Black (1970a), An experimental investigation of runoff production in permeable soils, *Water Resour. Res.*, *6*, 478–490.

Dunne, T., and R. G. Black (1970b), Partial area contributions to storm runoff in a small New England watershed, *Water Resour. Res.*, *6*, 1296–1311.

Dunne, T., and L. B. Leopold (1978), *Water in Environmental Planning*, W. H. Freeman, New York.

Dunne, T., K. X. Whipple, and B. F. Aubry (1995), Microtopography of hillslopes and initiation of channels by Horton overland flow, in *Natural and Anthropogenic Influences in Fluvial Geomorphology*, Geophys. Monogr. Ser., vol. 89, edited by J. E. Costa et al., pp. 27–44, AGU, Washington, D. C.

Dunne, T., D. V. Malmon, and S. M. Mudd (2010), A rain splash transport equation assimilating field and laboratory measurements, *J. Geophys. Res.*, *115*, F01001, doi:10.1029/2009JF001302.

Dutton, A. L., K. Loague, and B. C. Wemple (2005), Simulated effect of a forest road on near-surface hydrologic response and slope stability, *Earth Surf. Processes Landforms*, *30*, 325–338.

Duvall, A., E. Kirby, and D. Burbank (2004), Tectonic and lithologic controls on bedrock channel profiles and processes in coastal California, *J. Geophys. Res.*, *109*, F03002, doi:10.1029/2003JF000086.

Dühnforth, M., R. S. Anderson, D. Ward, and G. M. Stock (2010), Bedrock fracture control of glacial erosion processes and rates, *Geology*, *38*, 423–426.

Dühnforth, M., A. L. Densmore, S. Ivy-Ochs, P. A. Allen, and P. W. Kubik (2007), Timing and patterns of debris flow deposition on Shepherd and Symmes creek fans, Owens Valley, California, deduced from cosmogenic ^{10}Be, *J. Geophys. Res.*, *112*, F03S15, doi:10.1029/2006JF000562.

Dühnforth, M., A. L. Densmore, S. Ivy-Ochs, and P. A. Allen (2008), Controls on sediment evacuation from glacially modified and unmodified catchments in the eastern Sierra Nevada, California, *Earth Surf. Processes Landforms*, *33*, 1602–1613.

Dupré, B., C. Dessert, P. Oliva, Y. Goddéris, J. Viers, L. François, R. Millot, and J. Gaillardet (2003), Rivers, chemical weathering and Earth's climate, *C. R. Geosci.*, *335*, 1141–1160.

Dye, P., G. Moses, P. Vilakazi, R. Ndlela, and M. Royappen (2001), Comparative water use of wattle thickets and indigenous plant communities at riparian sites in the Western Cape and KwaZulu-Natal, *Water S.A.*, *27*, 529–538.

Dykes, A. P. (2002), Weathering-limited rainfall-triggered shallow mass movements in undisturbed steepland tropical rainforest, *Geomorphology*, *46*, 73–93.

Dykes, A. P., and J. B. Thornes (2000), Hillslope hydrology in tropical rainforest steeplands in Brunei, *Hydrol. Processes*, *14*, 215–235.

Dynesius, M., and C. Nilsson (1994), Fragmentation and flow regulation of river systems in the northern third of the world, *Science*, *266*, 753–762.

Earle, C. J. (1993), Asynchronous droughts in California streamflow as reconstructed from tree rings, *Quat. Res.*, *39*, 290–299.

Easterbrook, D. J. (1993), *Surface Processes and Landforms*, 520 pp., Macmillan, New York.

Eaton, B. C. (2006), Bank stability analysis for regime models of vegetated gravel bed rivers, *Earth Surf. Processes Landforms*, *31*, 1438–1444.

Eaton, B. C., and M. F. Lapointe (2001), Effects of large floods on sediment transport and reach morphology in the cobble-bed Sainte Marguerite River, *Geomorphology*, *40*, 291–309.

Eaton, B. C., and R. G. Millar (2004), Optimal alluvial channel width under a bank stability constraint, *Geomorphology*, *62*, 35–45.

Eaton, B. C., M. Church, and R. G. Millar (2004), Rational regime model of alluvial channel morphology and response, *Earth Surf. Processes Landforms*, *29*, 511–529.

Eaton, B. C., C. A. E. Andrews, T. R. Giles, and J. C. Phillips (2010a), Wildfire, morphologic change and bed material transport at Fishtrap Creek, British Columbia, *Geomorphology*, *118*, 409–424.

Eaton, B. C., R. G. Millar, and S. Davidson (2010b), Channel patterns: Braided, anabranching, and single-thread, *Geomorphology*, *120*, 353–364.

Eaton, B. C., R. D. Moore, and T. R. Giles (2010c), Forest fire, bank strength and channel instability: The 'unusual' response of Fishtrap Creek, British Columbia, *Earth Surf. Processes Landforms*, *35*, 1167–1183.

Eaton, L. S., B. A. Morgan, C. R. Kochel, and A. D. Howard (2003a), Quaternary deposits and landscape evolution of the central Blue Ridge of Virginia, *Geomorphology*, *56*, 139–154.

Eaton, L. S., B. A. Morgan, R. C. Kochel, and A. D. Howard (2003b), Role of debris flows in long-term landscape denudation in the central Appalachians of Virginia, *Geology*, *31*, 339–342.

Ebisemiju, F. S. (1987), The effects of environmental heterogeneity on the interdependence of drainage basin morphometric properties and its implications for applied studies, *Singapore J. Trop. Geogr.*, *8*, 114–128.

Eckhardt, K., and U. Ulbrich (2003), Potential impacts of climate change on groundwater recharge and streamflow in a central European low mountain range, *J. Hydrol.*, *284*, 244–252.

Eckley, M. S., and D. L. Hinchliff (1986), Glen Canyon Dam's quick fix, *Civ. Eng.*, *56*, 46–48.

Eddins, W. H., and T. J. Zembrzuski, Jr. (1994), Factors affecting accuracy of slope-area discharge determination of the Sept. 1992 flood in Raven Fork, western North Carolina, in *Proceedings, ASCE Hydraulic Engineering '94 Conference, August 1–5, Buffalo, New York*, edited by G. V. Cotroneo and R. R. Rumer, pp. 645–649, Am. Soc. of Civ. Eng., New York.

Eden, D. N., A. S. Palmer, S. J. Cronin, M. Marden, and K. R. Berryman (2001), Dating the culmination of river aggradation at the end of the last glaciation using distal tephra compositions, eastern North Island, New Zealand, *Geomorphology*, *38*, 133–151.

Egashira, S., and K. Ashida (1991), Flow resistance and sediment transportation in streams with step-pool bed morphology, in *Fluvial Hydraulics of Mountain Regions*, edited by A. Armanini and G. DiSilvio, pp. 45–58, Springer, Berlin.

Egli, M., D. Brandová, R. Böhlert, F. Favilli, and P. W. Kubik (2010a), [10]Be inventories in Alpine soils and their potential for dating land surfaces, *Geomorphology*, *119*, 62–73.

Egli, M., G. Sartori, A. Mirabella, and D. Giaccai (2010b), The effects of exposure and climate on the weathering of late Pleistocene and Holocene Alpine soils, *Geomorphology*, *114*, 466–482.

Egozi, R., and P. Ashmore (2008), Defining and measuring braiding intensity, *Earth Surf. Processes Landforms*, *33*, 2121–2138.

Ehlen, J., and E. Wohl (2002), Joints and landform evolution in bedrock canyons, *Trans. Jpn. Geomorphol. Union*, *23*, 237–255.

Einstein, H. A. (1937), *Der Geschliebetrieb also Wahrscheinlichkeitsproblem [Bedload transport as a probability problem]*, Mitteilungen der Versuchsantallt für Wasserbau an der Eidgenössischen Technischen Hochschule in Zürich.

Einstein, H. A. (1942), Formulas for the transportation of bedload, *Trans., ASCE*, *107*, 561–577.

Einstein, H. A. (1950), *The bedload function for sediment transportation in open channel flows*, *USDA Technical Bulletin*, *1026*, 70 pp.

Eisbacher, G. H. (1982), Mountain torrents and debris flows, *Episodes*, *4*, 12–17.

Eisbacher, G. H., and J. J. Clague (1984), Destructive mass movements in high mountains: Hazard and management, *Geol. Surv. of Can., Paper*, *84-16*, 75 pp.

Eisler, R. (1989), Atrazine hazards to fish, wildlife, and invertebrates: A synoptic review, *U. S. Fish Wildl. Serv. Biol. Rep.*, vol. 85, U.S. Fish and Wildl. Serv., Washington, D. C.

Eisler, R. (1993), Zinc hazards to fish, wildlife, and invertebrates: A synoptic review, *U. S. Fish Wildl. Serv. Biol. Rep.*, vol. 10, U.S. Fish and Wildl. Serv., Washington, D. C.

Ekes, C., and P. Friele (2003), Sedimentary architecture and post-glacial evolution of Cheekye Fan, southwestern British Columbia, Canada, in *Ground Penetrating Radar in Sediments*, edited by C. S. Bristow and H. M. Jol, *Geol. Soc. Spec. Publ.*, *211*, 87–98.

Elder, K. (1995), Modeling the spatial distribution of seasonal snow accumulation on Teton Glacier, Wyoming, USA, in *Biogeochemistry of Seasonally Snow-covered Catchments*, edited by K. A. Tonnessen, M. W. Williams, and M. Tranter, *IAHS Publ.*, *228*, 445–454.

Elder, K., R. Kattelmann, and R. Ferguson (1990), Refinements in dilution gauging for mountain streams, in *Hydrology in Mountainous Regions I. Hydrological Measurements: The Water Cycle*, edited by H. Lang and A. Musy, *IAHS Publ.*, *193*, 247–254.

Elevatorski, E. A. (1959), *Hydraulic Energy Dissipaters*, 214 pp., McGraw-Hill, New York.

Elfström, A. (1987), Large boulder deposits and catastrophic floods. A case study of the Båldakatj area, Swedish Lapland, *Geogr. Ann.*, *69A*, 101–121.

Elkins, E. M., G. B. Pasternack, and J. E. Merz (2007), Use of slope creation for rehabilitating incised, regulated, gravel bed rivers, *Water Resour. Res.*, *43*, W05432, doi:10.1029/2006WR005159.

Elliott, A. H., and N. H. Brooks (1997), Transfer of nonsorbing solutes to a streambed with bed forms: Theory, *Water Resour. Res.*, *33*, 123–136.

Elliott, J. G., and L. A. Hammack (1999), Geomorphic and sedimentologic characteristics of alluvial reaches in the Black Canyon of the Gunnison National Monument, Colorado, *Water Resour. Invest. Rep. 99-4082*, U.S. Geol. Surv., Denver, Colo.

Elliott, J. G., and R. S. Parker (1992), Potential climate-change effects on bed-material entrainment, the Gunnison Gorge, Colorado, in *Managing Water Resources During Global Change*, pp. 751–759, Am. Water Resources Assoc.

Elliott, J. G., and R. S. Parker (2001), Developing a post-fire flood chronology and recurrence probability from alluvial stratigraphy in the Buffalo Creek watershed, Colorado, USA, *Hydrol. Processes*, *15*, 3039–3051.

Elliott, J. G., A. C. Gellis, and S. B. Aby (1999), Evolotion of arroyos: Incised channels of the southwestern United States, in *Incised River Channels: Processes, Forms, Engineering and Management*, edited by S. E. Darby and A. Simon, pp. 153–185, John Wiley, Chichester, U. K.

Ellis, E. R., and M. Church (2005), Hydraulic geometry of secondary channels of lower Fraser River, British Columbia, from acoustic Doppler profiling, *Water Resour. Res.*, *41*, W08421, doi:10.1029/2004WR003777.

Ellis, J. B. (1987), Sediment-water quality interactions in urban rivers, in *International Geomorphology 1986, Part I*, edited by V. Gardiner, pp. 287–301, John Wiley, Chichester, U. K.

Elsenbeer, H. (2001), Hydrological flowpaths in tropical rainforest soilscapes—a review, *Hydrol. Processes*, *15*, 1751–1759.

Elsenbeer, H., and A. Lack (1996), Hydrometric and hydrochemical evidence for fast flowpaths at La Cuenca, western Amazonia, *J. Hydrol.*, *180*, 237–250.

Ely, L. L. (1997), Response of extreme floods in the southwestern United States to climatic variations in the late Holocene, *Geomorphology*, *19*, 175–201.

Ely, L. L., Y. Enzel, V. R. Baker, and D. R. Cayan (1993), A 5000-year record of extreme floods and climate change in the southwestern United States, *Science*, *262*, 410–412.

Ely, L. L., Y. Enzel, V. R. Baker, V. S. Kale, and S. Mishra (1996), Changes in the magnitude and frequency of late Holocene monsoon floods on the Narmada River, central India, *Geol. Soc. Am. Bull.*, *108*, 1134–1148.

Ely, L. L., R. H. Webb, and Y. Enzel (1992), Accuracy of post-bomb ^{137}Cs and ^{14}C in dating fluvial deposits, *Quat. Res.*, *38*, 196–204.

Elwood, J. W., J. D. Newbold, R. V. O'Neill, and W. Van Winkle (1983), Resource spiraling: An operational paradigm for analyzing lotic ecosystems, in *Dynamics of Lotic Ecosystems*, edited by T. D. Fontaine and S. M. Bartell, pp. 3–27, Ann Arbor Science, Ann Arbor, Mich.

Emmett, W. W. (1978), Overland flow, in *Hillslope Hydrology*, edited by M. J. Kirkby, pp. 145–176, Wiley, Chichester, U. K.

Emmett, W. W. (1980), A field calibration of the sediment-trapping characteristics of the Helley-Smith bedload sampler, *U.S. Geol. Surv. Prof. Pap.*, *1139*, 44 pp.

Emmett, W. W., R. L. Burrows, and E. F. Chacho, Jr. (1996), Coarse-particle transport in a gravel-bed river, *Int. J. Sediment Res.*, *11*, 8–21.

Emmett, W. W., and M. G. Wolman (2001), Effective discharge and gravel-bed rivers, *Earth Surf. Processes Landforms*, *26*, 1369–1380.

England, J. F., R. D. Jarrett, and J. D. Salas (2003a), Data-based comparisons of moments estimators using historical and paleoflood data, *J. Hydrol.*, *278*, 172–196.

England, J. F., J. D. Salas, and R. D. Jarrett (2003b), Comparisons of two moments-based estimators that utilize historical and paleoflood data for the log Pearson type III distribution, *Water Resour. Res.*, *39*(9), 1243, doi:10.1029/2002WR001791.

Engman, E. T. (1997), Soil moisture, the hydrologic interface between surface and ground waters, in *Remote Sensing and Geographic Information Systems for Design and Operation of Water Resource Systems*, edited by M. F. Baumgartner, G. A. Schultz, and A. I. Johnson, *IAHS Publ.*, *242*, 129–140.

Enzel, Y., L. L. Ely, P. K. House, V. R. Baker, and R. H. Webb (1993), Paleoflood evidence for a natural upper bound to flood magnitudes in the Colorado River basin, *Water Resour. Res.*, *29*, 2287–2297.

Ergenzinger, P. (1992), Riverbed adjustments in a step-pool system: Lainbach, Upper Bavaria, in *Dynamics of Gravel-Bed Rivers*, edited by P. Billi et al., pp. 415–430, John Wiley, Chichester, U. K.

Ergenzinger, P. (1994), The susceptibility of valley slopes and river beds to erosion and accretion under the impact of climatic change - Alpine examples, in *Variability in Stream Erosion and Sediment Transport*, edited by L. J. Olive, R. J. Loughran, and J. A. Kesby, *IAHS Publ.*, *224*, 43–53.

Ergenzinger, P., and C. De Jong (1994), Monitoring and modeling the transport of coarse single particles in mountain rivers, in *Proceedings, ASCE Hydraulic Engineering '94 Conference, August 1–5, Buffalo, New York*, edited by G. V. Cotroneo and R. R. Rumer, pp. 634, Am. Soc. of Civ. Eng., New York.

Ergenzinger, P., and C. de Jong (2003), Perspectives on bed load measurement, in *Erosion and Sediment Transport in Mountain Rivers: Technological and Methodological Advances*, edited by J. Bogen, T. Fergus, and D. E. Walling, *IAHS-AISH Publ.*, *283*, 113–125.

Ergenzinger, P., and K.-H. Schmidt (1990), Stochastic elements of bed load transport in a step-pool mountain river, in *Hydrology in Mountainous Regions II. Artificial Reservoirs, Water and Slopes*, edited by R. O. Sinniger and M. Monbaron, *IAHS Publ.*, *194*, 39–46.

Ergenzinger, P., and K.-H. Schmidt (Eds.) (1994), *Dynamics and Geomorphology of Mountain Rivers*, 326 pp., Springer, Berlin.

Ergenzinger, P., C. De Jong, and G. Christaller (1994a), Interrelationships between bedload transfer and river-bed adjustment in mountain rivers: An example from Squaw Creek, Montana, in *Process Models and Theoretical Geomorphology*, edited by M. J. Kirkby, pp. 141–158, John Wiley, Chichester, U. K.

Ergenzinger, P., C. De Jong, J. Laronne, and I. Reid (1994b), Short term temporal variations in bedload transport rates: Squaw Creek, Montana, USA and Nahal Yatir and Nahal Estemoa, Israel, in *Dynamics and Geomorphology of Mountain Rivers*, edited by P. Ergenzinger and K.-H. Schmidt, pp. 251–264, Springer, New York.

Ericksen, G. E., G. Pflaker, and J. F. Concha (1970), Preliminary report on the geologic events associated with the May 31, 1970 Peru earthquake, *U.S. Geol. Surv. Circ.*, *639*, 25 pp.

Erickson, T. A., M. W. Williams, and A. Winstral (2005), Persistence of topographic controls on the spatial distribution of snow in rugged mountain terrain, Colorado, United States, *Water Resour. Res.*, *41*, W04014, doi:10.1029/2003WR002973.

Ermini, L., F. Catani, and N. Casagli (2005), Artificial neural networks applied to landslide susceptibility assessment, *Geomorphology*, *66*, 327–343.

Ersi, K., A. Ohmura, and H. Lang (1995), Simulation of runoff processes of a continental mountain glacier in the Tian Shan, China, in *Biogeochemistry of Seasonally Snow-covered Catchments*, edited by K. A. Tonnessen, M. W. Williams, and M. Tranter, *IAHS Publ.*, *228*, 455–465.

Erskine, W. D., and M. J. Saynor (1996), Effects of catastrophic floods on sediment yields in southeastern Australia, in *Erosion and Sediment Yield: Global and Regional Perspectives*, edited by D. E. Walling and B. W. Webb, *IAHS Publ.*, *236*, 381–388.

Etheredge, D., D. S. Gutzler, and F. J. Pazzaglia (2004), Geomorphic response to seasonal variations in rainfall in the Southwest United States, *Geol. Soc. Am. Bull.*, *116*, 606–618.

Ettema, R. (1984), Sampling armor-layer sediments, *J. Hydraul. Eng.*, *110*, 992–996.

Evans, K. G. (2000), Methods for assessing mine site rehabilitation design for erosion impact, *Aust. J. Soil Res.*, *38*, 231–247.

Evans, S. G. (1986), The maximum discharge of outburst floods caused by the breaching of man-made and natural dams, *Can. Geotech. J.*, *23*, 385–387.

Evans, S. G., and J. J. Clague (1994), Recent climatic change and catastrophic geomorphic processes in mountain environments, *Geomorphology*, *10*, 107–128.

Evans, S. G., M. J. Saynor, G. R. Willgoose, and S. J. Riley (2000), Post-mining landform evolution modelling: 1. Derivation of sediment transport model and rainfall-runoff model parameters, *Earth Surf. Processes Landforms*, *25*, 743–763.

Everest, F. H., and D. W. Chapman (1972), Habitat selection and spatial interaction by juvenile chinook salmon and steelhead trout in two Idaho streams, *J. Fish. Res. Board Can.*, *29*, 91–100.

Everest, F. H., R. L. Beschta, J. C. Scrivener, K. V. Koski, J. R. Sedell, and C. J. Cederholm (1987), Fine sediment and salmonid production: A paradox, in *Streamside Management: Forestry and Fishery Implications*, edited by E. O. Salo and T. W. Cundy, pp. 98–142, Univ. of Washington, Institute of Forest Resources, Seattle.

Everitt, B. L. (1968), Use of the cottonwood in an investigation of the recent history of a flood plain, *Am. J. Sci.*, *266*, 417–439.

Everitt, B. L. (1980), Ecology of saltcedar - a plea for research, *Environ. Geol.*, *3*, 77–84.

Evette, A., S. Labonne, F. Rey, F. Liebault, O. Jancke, and J. Girel (2009), History of bioengineering techniques for erosion control in rivers in western Europe, *Environ. Manage.*, *43*, 972–984.

Ewing, R. (1996), Postfire suspended sediment from Yellowstone National Park, Wyoming, *J. Am. Water Resour. Assoc.*, *32*, 605–627.

Fabel, D., J. Harbor, and C. Steele (1998), Valley-scale glacial erosion rates from [10]Be and [26]Al inheritance in bedrock, *Eos Trans. AGU*, *79*, F336.

Faeh, A., S. Scherrer, and F. Naef (1997), A combined field and numerical approach to investigate flow processes in natural macroporous soils under extreme precipitation, *Hydrol. Earth Syst. Sci.*, *1*, 787–800.

Fahnestock, R. K. (1963), Morphology and hydrology of a glacial stream - White River, Mt. Rainier, Washington, U.S, *Geol. Surv. Prof. Pap.*, *422A*, 61.

Fairchild, L. H. (1987), The importance of lahar initiation processes, in *Debris Flows/Avalanches: Process, Recognition, and Mitigation*, edited by J. E. Costa and G. F. Wieczorek, *Rev. Eng. Geol.*, *7*, 51–61.

Fanelli, R. M., and L. K. Lautz (2008), Patterns of water, heat, and solute flux through streambeds around small dams, *Ground Water*, *46*, 671–687.

Fanok, S. F., and E. E. Wohl (1997), Assessing the accuracy of paleohydrologic indicators, Harpers Ferry, West Virginia, *J. Am. Water Resour. Assoc.*, *33*, 1091–1102.

Farnworth, E. G., et al. (1979), Impacts of sediment and nutrients on biota in surface waters of the United States, *Rep. EPA-60013-79-105USEPA*, Athens, Ga.

Farnsworth, K. L., and J. D. Milliman (2003), Effects of climatic and anthropogenic change on small mountainous rivers; the Salinas River example, *Global Planet. Change*, *39*, 53–64.

Fassnacht, S. R., K. A. Dressler, and R. C. Bales (2003), Snow water equivalent interpolation for the Colorado River Basin from snow telemetry (SNOTEL) data, *Water Resour. Res.*, *39*(8), 1208, doi:10.1029/2002WR001512.

Fausch, K. D. (1984), Profitable stream positions for salmonids: Relating specific growth rate to net energy gain, *Can. J. Zool.*, *62*, 441–451.

Fausch, K. D., and T. G. Northcote (1992), Large woody debris and salmonid habitat in a small coastal British Columbia stream, *Can. J. Fish. Aquat. Sci.*, *49*, 682–693.

Fausch, K. D., C. E. Torgersen, C. V. Baxter, and H. W. Li (2002), Landscapes to riverscapes: Bridging the gap between research and conservation of stream fishes, *BioScience*, *52*, 483–498.

Faustini, J. M., and J. A. Jones (2003), Influence of large woody debris on channel morphology and dynamics in steep, boulder-rich mountain streams, western Cascades, Oregon, *Geomorphology*, *51*, 187–205.

Fay, H. (2002), Formation of kettle holes following a glacial outburst flood (jökulhlaup), Skeiäarársandur, southern Iceland, in *The Extremes of the Extremes: Extraordinary Floods*, edited by A. Snorrason, H. P. Finnsdóttir, and M. E. Moss, *IAHS Publ.*, *271*, 205–210.

Fengjing, L., M. W. Williams, Y. Daqing, and J. Melack (1995), Snow and water chemistry of a headwater alpine basin, Urumqi River, Tian Shan, China, in *Biogeochemistry of Seasonally Snow-covered Catchments*, edited by K. A. Tonnessen, M. W. Williams, and M. Tranter, *IAHS Publ.*, *228*, 207–219.

Fenn, C. R. (1987), Sediment transfer processes in alpine glacier basins, in *Glacio-fluvial Sediment Transfer*, edited by A. M. Gurnell and M. J. Clark, pp. 59–85, John Wiley, Chichester, U. K.

Fenn, C. R., and A. M. Gurnell (1987), Proglacial channel processes, in *Glacio-fluvial Sediment Transfer*, edited by A. M. Gurnell and M. J. Clark, pp. 423–472, John Wiley, Chichester, U. K.

Fenn, M. E., et al. (2003), Ecological effects of nitrogen deposition in the western United States, *Bioscience*, *53*, 404–420.

Fenton, C. R., R. H. Webb, T. E. Cerling, R. J. Poreda, and B. P. Nash (2002), Cosmogenic ^{3}He ages and geochemical discrimination of lava-dam outburst-flood deposits in western Grand Canyon, Arizona, in *Ancient Floods, Modern Hazards: Principles and Applications of Paleoflood Hydrology*, edited by P. K. House et al., pp. 191–215, AGU, Washington, D. C.

Ferguson, R. I. (1986), Hydraulics and hydraulic geometry, *Prog. Phys. Geogr.*, *10*, 10–31.

Ferguson, R. I. (1994), Critical discharge for entrainment of poorly sorted gravel, *Earth Surf. Processes Landforms*, *19*, 179–186.

Ferguson, R. I. (2003), The missing dimension: Effects of lateral variation on 1-D calculations of fluvial bedload transport, *Geomorphology*, *56*, 1–14.

Ferguson, R. I. (2005), Estimating critical stream power for bedload transport calculations in gravel-bed rivers, *Geomorphology*, *70*, 33–41.

Ferguson, R. (2007), Flow resistance equations for gravel- and boulder-bed streams, *Water Resour. Res.*, *43*, W05427, doi:10.1029/2006WR005422.

Ferguson, R. I. (2008), Gravel-bed rivers at the reach scale, in *Gravel-Bed Rivers VI: From Process Understanding to River Restoration*, edited by H. Habersack, H. Piégay, and M. Rinaldi, pp. 33–60, Elsevier, Amsterdam.

Ferguson, R. I., and P. J. Ashworth (1991), Slope-induced changes in channel character along a gravel-bed stream: The Allt Dubhaig, Scotland, *Earth Surf. Processes Landforms*, *16*, 65–82.

Ferguson, R. I., and P. J. Ashworth (1992), Spatial patterns of bedload transport and channel change in braided and near-braided rivers, in *Dynamics of Gravel-Bed Rivers*, edited by P. Billi et al., pp. 477–496, John Wiley, Chichester, U. K.

Ferguson, R., and M. Church (2009), A critical perspective on 1-D modeling of river processes: Gravel load and aggradation in lower Fraser River, *Water Resour. Res.*, *45*, W11424, doi:10.1029/2009WR007740.

412   References

Ferguson, R. I., and T. B. Hoey (2002), Long-term slowdown of river tracer pebbles: Generic models and implications for interpreting short-term tracer studies, *Water Resour. Res.*, *38*(8), 1142, doi:10.1029/2001WR000637.

Ferguson, R. I., and C. Paola (1997), Bias and precision of percentiles of bulk grain size distributions, *Earth Surf. Processes Landforms*, *22*, 1061–1077.

Ferguson, R. I., and S. J. Wathen (1998), Tracer-pebble movement along a concave river profile: Virtual velocity in relation to grain size and shear stress, *Water Resour. Res.*, *34*, 2031–2038.

Ferguson, R. I., A. D. Kirkbride, and A. G. Roy (1996), Markov analysis of velocity fluctuations in gravel-bed rivers, in *Coherent Flow Structures*, edited by P. J. Ashworth et al., pp. 165–183, John Wiley, Chichester, U. K.

Ferguson, R. I., K. L. Prestegaard, and P. J. Ashworth (1989), Influence of sand on hydraulics and gravel transport in a braided gravel river, *Water Resour. Res.*, *25*, 635–643.

Ferguson, R. I., D. J. Bloomer, T. B. Hoey, and A. Werritty (2002), Mobility of river tracer pebbles over different timescales, *Water Resour. Res.*, *38*(5), 1045, doi:10.1029/2001WR000254.

Fernandez, R., J. Best, and F. López (2006), Mean flow, turbulence structure, and bed form superposition across the ripple-dune transition, *Water Resour. Res.*, *42*, W05406, doi:10.1029/2005WR004330.

Fernandez Luque, R., and R. Van Beek (1976), Erosion and transport of bedload sediment, *J. Hydraul. Res.*, *14*, 127–144.

Ferrier, K. L., J. W. Kirchner, and R. C. Finkel (2005), Erosion rates over millennial and decadal timescales at Caspar Creek and Redwood Creek, northern California Coast Ranges, *Earth Surf. Processes Landforms*, *30*, 1025–1038.

Ferro, V. (2003), ADV measurements of velocity distributions in a gravel-bed flume, *Earth Surf. Processes Landforms*, *28*, 707–722.

Ferro, V., and R. Pecoraro (2000), Incomplete self-similarity and flow velocity in gravel bed channels, *Water Resour. Res.*, *36*, 2761–2769.

Fiebig, D. M., M. A. Lock, and C. Neal (1990), Soil water in the riparian zone as a source of carbon for a headwater stream, *J. Hydrol.*, *116*, 217–237.

Field, J. (2001), Channel avulsion on alluvial fans in southern Arizona, *Geomorphology*, *37*, 93–104.

Fierz, C., P. Riber, E. E. Adams, A. R. Curran, P. M. B. Foehn, M. Lehning, and C. Pleuss (2003), Evaluation of snow-surface energy balance models in alpine terrain, *J. Hydrol.*, *282*, 76–94.

Findlay, S. (1995), Importance of surface-subsurface exchange in stream ecosystems: The hyporheic zone, *Limnol. Oceanogr.*, *40*, 159–164.

Finley, J. B., J. I. Drever, and J. T. Turk (1995), Sulfur isotope dynamics in a high-elevation catchment, West Glacier Lake, Wyoming, *Water Air Soil Pollut.*, *79*, 227–241.

Finnegan, N. J., G. Roe, D. R. Montgomery, and B. Hallet (2005), Controls on the channel width of rivers: Implications for modeling fluvial incision of bedrock, *Geology*, *33*, 229–232.

Finnegan, N. J., L. S. Sklar, and T. K. Fuller (2007), Interplay of sediment supply, river incision, and channel morphology revealed by the transient evolution of an experimental bedrock channel, *J. Geophys. Res.*, *112*, F03S11, doi:10.1029/2006JF000569.

Fiori, A., D. Russo, and M. Di Lazzaro (2009), Stochastic analysis of transport in hillslopes: Travel time distribution and source zone dispersion, *Water Resour. Res.*, *45*, W08435, doi:10.1029/2008WR007668.

Fischer, H. (1993), Geomorphology in Austria, in *The Evolution of Geomorphology*, edited by H. J. Walker and W. E. Grabau, pp. 45–49, John Wiley, New York.

Fischer, K. J., and M. D. Harvey (1991), Geomorphic response of lower Feather River to 19th century hydraulic mining operations, in *Inspiration: Come to the Headwaters, Proceedings, 15th Annual Conference of Association of State Floodplain Managers, Denver, Colo., June 10–14, 1991*, vol. SP24, pp. 128–132, Natl. Hazards Cent., Boulder, Colo.

Fisher, G. B., F. J. Magilligan, J. M. Kaste, and K. H. Nislow (2010), Constraining the timescales of sediment sequestration associated with large woody debris using cosmogenic ^{7}Be, *J. Geophys. Res.*, *115*, F01013, doi:10.1029/2009JF001352.

Fisher, S. G., N. B. Grimm, E. Martí, R. M. Holmes, and J. B. Jones, Jr. (1998a), Material spiraling in stream corridors: A telescoping ecosystem model, *Ecosystems*, *1*, 19–34.

Fisher, S. G., N. B. Grimm, E. Martí, and R. Gomez (1998b), Hierarchy, spatial configuration, and nutrient cycling in a desert stream, *Aust. J. Ecol.*, *23*, 41–52.

Fitzpatrick, F. A. (2001), A comparison of multi-disciplinary methods for measuring physical conditions of streams, in *Geomorphic Processes and Riverine Habitat*, edited by J. M. Dorava et al., pp. 7–18, AGU, Washington, D. C.

Flerchinger, G. N., and K. R. Cooley (2000), A ten-year water balance of a mountainous semi-arid watershed, *J. Hydrol.*, *237*, 86–99.

Flint, A. L., L. E. Flint, and M. D. Dettinger (2008), Modeling soil moisture processes and recharge under a melting snowpack, *Vadose Zone J.*, *7*, 350–357.

Flint, J. J. (1973), Development of headward growth of channel networks, *Geol. Soc. Am. Bull.*, *84*, 1087–1094.

Flores, A. N., B. P. Bledsoe, C. O. Cuhayican, and E. E. Wohl (2006), Channel-reach morphology dependence on energy, scale, and hydroclimatic processes with implications for prediction using geospatial data, *Water Resour. Res.*, *42*, W06412, doi:10.1029/2005WR004226.

Florsheim, J. L., E. A. Keller, and D. W. Best (1991), Fluvial sediment transport in response to moderate storm flows following chaparral wildfire, Ventura County, southern California, *Geol. Soc. Am. Bull.*, *103*, 504–511.

Florsheim, J. L., and J. F. Mount (2002), Restoration of floodplain topography by sand-splay complex formation in response to intentional levee breaches, Lower Cosumnes River, California, *Geomorphology*, *44*, 67–94.

Flowers, G. E. (2008), Subglacial modulation of the hydrograph from glacierized basins, *Hydrol. Processes*, *22*, 3903–3918.

Foley, M. G. (1980a), Quaternary diversion and incision, Dearborn River, Montana, *Geol. Soc. Am. Bull., Part 1*, *91*, 2152–2188.

Foley, M. G. (1980b), Bedrock incision by streams, *Geol. Soc. Am. Bull., Part 2*, *91*, 2189–2213.

Follstad Shah, J. J., C. N. Dahm, S. P. Gloss, and E. S. Bernhardt (2007), River and riparian restoration in the Southwest: Results of the National River Restoration Science Synthesis project, *Restor. Ecol.*, *15*, 550–562.

Fonstad, M. A. (2003), Spatial variation in the power of mountain streams in the Sangre de Cristo Mountains, New Mexico, *Geomorphology*, *55*, 75–96.

Fonstad, M. A., and W. A. Marcus (2003), Self-organized criticality in riverbank systems, *Ann. Assoc. Am. Geogr.*, *93*, 281–296.

Fonstad, M. A., and W. A. Marcus (2005), Remote sensing of stream depths with hydraulically assisted bathymetry (HAB) models, *Geomorphology*, *72*, 320–339.

Fonstad, M. A., and W. A. Marcus (2010), High resolution, basin extent observations and implications for understanding river form and process, *Earth Surf. Processes Landforms*, *35*, 680–698.

Font, M., D. Amorese, and J.-L. Lagarde (2010), DEM and GIS analysis of the stream gradient index to evaluate effects of tectonics: The Normany intraplate area (NW France), *Geomorphology*, *119*, 172–180.

Fontaine, T. A., T. S. Cruickshank, J. G. Arnold, and R. H. Hotchkiss (2002), Development of a snowfall-snowmelt routine for mountainous terrain for the Soil Water Assessment Tool (SWAT), *J. Hydrol.*, *262*, 209–223.

Ford, D. C., and P. W. Williams (1989), *Karst Geomorphology and Hydrology*, pp. 601, Unwin Hyman, London.

Forman, R. T. T., and M. Godron (1986), *Landscape Ecology*, pp. 640, John Wiley, New York.

Formento-Trigilio, M. L., D. W. Burbank, A. Nicol, J. Shulmeister, and U. Rieser (2003), River response to an active fold-and-thrust belt in a convergent margin setting, North Island, New Zealand, *Geomorphology*, *49*, 125–152.

Foster, D., S. H. Brocklehurst, and R. L. Gawthorpe (2008), Small valley glaciers and the effectiveness of the glacial buzzsaw in the northern Basin and Range, USA, *Geomorphology*, *102*, 624–639.

Foster, D., S. H. Brocklehurst, and R. L. Gawthorpe (2010), Glacial-topographic interactions in the Teton Range, Wyoming, *J. Geophys. Res.*, *115*, F01007, doi:10.1029/2008JF001135.

Foster, I. D. L., H. Dalgleish, J. A. Dearing, and E. D. Jones (1994), Quantifying soil erosion and sediment transport in drainage basins; some observations on the use of ^{137}Cs, in *Variability in Stream Erosion and Sediment Transport*, edited by L. J. Olive, R. J. Loughran, and J. A. Kesby, *IAHS Publ.*, *224*, 55–64.

Foufoula-Georgiou, E., V. Ganti, and W. E. Dietrich (2010), A nonlocal theory of sediment transport on hillslopes, *J. Geophys. Res.*, *115*, F00A16, doi:10.1029/2009JF001280.

Fournier, F. (1960), Debit solide des cours d'eau. Essai d' estimation de la perte en terre subie par l'ensemble du globe terrestre, *IAHS Publ.*, *53*, 19–22.

Fox, A. M., I. C. Willis, and N. S. Arnold (2008), Modification and testing of a one-dimensional energy and mass balance model for supraglacial snowpacks, *Hydrol. Processes*, *22*, 3194–3209.

Fox, G. A., G. V. Wilson, A. Simon, E. Langendoen, O. Akay, and J. W. Fuchs (2007), Measuring streambank erosion due to ground water seepage: Correlation to bank pore water pressure, precipitation and stream stage, *Earth Surf. Processes Landforms*, *32*, 1558–1573.

Fox, J. F., and A. N. Papanicolaou (2008), Application of the spatial distribution of nitrogen stable isotopes for sediment tracing at the watershed scale, *J. Hydrol.*, *358*, 46–55.

Fox, M., and S. Bolton (2007), A regional and geomorphic reference for quantities and volumes of instream wood in unmanaged forest basins of Washington state, *N. Am. J. Fish. Manage.*, *27*, 342–359.

Francis, R. A., G. E. Petts, and A. M. Gurnell (2008), Wood as a driver of past landscape change along river corridors, *Earth Surf. Processes Landforms*, *33*, 1622–1626.

Frangi, J. L., and A. E. Lugo (1991), Hurricane damage to a flood plain forest in the Luquillo Mountains of Puerto Rico, *Biotropica*, *23*, 324–335.

Frankl, K. L., and F. J. Pazzaglia (2006), Mountain fronts, base-level fall, and landscape evolution; insights from the Southern Rocky Mountains, in *Tectonics, Climate, and Landscape Evolution*, edited by S. D. Willett et al., *Spec. Pap. Geol. Soc. Am.*, *398*, 419–434.

Frankel, K. L., and J. F. Dolan (2007), Characterizing arid region alluvial fan surface roughness with airborne laser swath mapping digital topographic data, *J. Geophys. Res.*, *112*, F02025, doi:10.1029/2006JF000644.

Frankel, K. L., F. J. Pazzaglia, and J. D. Vaughn (2007), Knickpoint evolution in a vertically bedded substrate, upstream-dipping terraces, and Atlantic slope bedrock channels, *Geol. Soc. Am. Bull.*, *119*, 476–486.

Franks, S., M. Sivapalan, K. Takeuchi, and Y. Tachikawa (2005), *Predictions in Ungauged Basins: International Perspectives on the State of the Art and Pathways Forward*, IAHS Publ., *301*.

Fransen, P. J. B., C. J. Phillips, and B. D. Fahey (2001), Forest road erosion in New Zealand: Overview, *Earth Surf. Processes Landforms*, *26*, 165–174.

Fredlund, D. G., N. R. Morgenstern, and R. A. Widger (1978), The shear strength of unsaturated soils, *Can. Geotech. J.*, *15*, 313–321.

Freeman, M. C., C. M. Pringle, and C. R. Jackson (2007), Hydrologic connectivity and the contribution of stream headwaters to ecological integrity at regional scales, *J. Am. Water Resour. Assoc.*, *43*, 5–14.

Frei, S., J. H. Fleckenstein, S. J. Kollet, and R. M. Maxwell (2009), Patterns and dynamics of river-aquifer exchange with variably-saturated flow using a fully-coupled model, *J. Hydrol.*, *375*, 383–393.

Fremier, A. K., J. I. Seo, and F. Nakamura (2010), Watershed controls on the export of large wood from stream corridors, *Geomorphology*, *117*, 33–43.

French, R. H., J. J. Miller, and S. Curtis (2001), Estimating the depth of deposition (erosion) at slope transitions on alluvial fans, *J. Hydraul. Eng.*, *127*, 780–782.

Friedman, J. M., and V. J. Lee (2002), Extreme floods, channel change, and riparian forests along ephemeral streams, *Ecol. Monogr.*, *72*, 409–425.

Friedman, J. M., K. R. Vincent, and P. B. Shafroth (2005a), Dating floodplain sediments using tree-ring response to burial, *Earth Surf. Processes Landforms*, *30*, 1077–1091.

Friedman, J. M., G. T. Auble, P. B. Shafroth, M. L. Scott, M. F. Merigliano, M. D. Freehling, and E. R. Griffin (2005b), Dominance of non-native riparian trees in western USA, *Biol. Invasions*, *7*, 747–751.

Friend, D. A., F. M. Phillips, S. W. Campbell, T. Liu, and P. Sharma (2000), Evolution of desert colluvial boulder slopes, *Geomorphology*, *36*, 19–45.

Fripp, J. B., and P. Diplas (1993), Surface sampling in gravel streams, *J. Hydraul. Eng.*, *119*, 473–490.

Frissell, C. A. (1999), Groundwater processes and stream classification in the montane West, *1999 Annual Meeting Abstracts*, pp. 17, Ecological Society of America.

Frissell, C. A., W. J. Liss, C. E. Warren, and M. D. Hurley (1986), A hierarchical framework for stream classification: Viewing streams in a wateshed context, *Environ. Manage.*, *10*, 199–214.

Froehlich, W. (1995), Sediment dynamics in the Polish Flysch Carpathians, in *Sediment and Water Quality in River Catchments*, edited by I. Foster, A. Gurnell, and B. Webb, pp. 453–461, John Wiley, Chichester, U. K.

Froehlich, W. (2003), Monitoring bed load transport using acoustic and magnetic devices, in *Erosion and Sediment Transport Measurement in Rivers: Technological and Methodological Advances*, edited by J. Bogen, T. Fergus, and D. E. Walling, *IAHS Publ.*, *283*, 201–210.

Froehlich, W., and L. Starkel (1987), Normal and extreme monsoon rains - their role in the shaping of the Darjeeling Himalaya, *Stud. Geomorphol. Carpatho-Balcanica*, *21*, 129–158.

Froehlich, W., and D. E. Walling (1992), The use of fallout radionuclides in investigations of erosion and sediment delivery in the Polish Flysh Carpathians, in *Erosion, Debris Flows and Environment in Mountain Regions*, edited by D. E. Walling, T. R. Davies, and B. Hasholt, *IAHS Publ.*, *209*, 61–76.

Froehlich, W., E. Gil, I. Kasza, and L. Starkel (1990), Thresholds in the transformation of slopes and river channels in the Darjeeling Himalaya, India, *Mt. Res. Dev.*, *10*, 301–312.

Froehlich, W., D. L. Higgitt, and D. E. Walling (1993), The use of caesium-137 to investigate soil erosion and sediment delivery from cultivated slopes in the Polish Carpathians, in *Farm Land Erosion: In Temperate Plains Environment and Hills*, edited by S. Wicherek, pp. 271–283, Elsevier, Amsterdam.

Froehlich, W., and D. E. Walling (2005), Using environmental radionuclides to elucidate sediment sources within a small drainage basin in the Polish Flysch Carpathians, in *Sediment Budgets I*, edited by D. E. Walling and A. J. Horowitz, *IAHS Publ.*, *291*, 102–112.

Froehlich, W., and D. E. Walling (2006), The use of ^{137}Cs and ^{210}Pb$_x$ to investigate sediment sources and overbank sedimentation rates in the Teesta River basin, Sikkim Himalaya, India, in *Sediment Dynamics and Hydromorphology of Fluvial Systems*, edited by J. S. Rowan, R. W. Duck, and A. Werritty, *IAHS Publ.*, *306*, 380–388.

Froese, D. G., D. G. Smith, and D. T. Clement (2005), Characterizing large river history with shallow geophysics; middle Yukon River, Yukon Territory and Alaska, *Geomorphology*, *67*, 391–406.

Frostick, L. E., B. Murphy, and R. Middleton (2006), Unravelling flood history using matrices in fluvial gravel deposits, in *Sediment Dynamics and the Hydromorphology of Fluvial Systems*, edited by J. S. Rowan, R. W. Duck, and A. Werritty, *IAHS Publ.*, *306*, 425–433.

Fryirs, K. A., and G. J. Brierley (2001), Variability in sediment delivery and storage along river courses in Bega catchment, NSW, Australia: Implications for geomorphic river recovery, *Geomorphology*, *38*, 237–265.

Fryirs, K. A., and G. J. Brierley (2010), Antecedent controls on river character and behaviour ni partly confined valley settings: Upper Hunter catchment, NSW, Australia, *Geomorphology*, *117*, 106–120.

Fryirs, K. A., G. J. Brierley, N. J. Preston, and J. Spencer (2007), Catchment scale (dis)connectivity in sediment flux in the upper Hunter catchment, New South Wales, Australia, *Geomorphology*, *84*, 297–316.

Fryirs, K. A., A. Arthington, and J. Grove (2008), Principles of river condition assessment, in *River Futures: An Integrative Scientific Approach to River Repair*, edited by G. J. Brierley and K. A. Fryirs, pp. 100–124, Island Press, Washington, D. C.

Fryirs, K. A., A. Spink, and G. Brierley (2009), Post-European settlement response gradients of river sensitivity and recovery across the upper Hunter catchment, Australia, *Earth Surf. Processes Landforms*, *34*, 897–918.

Fu, B., Y. Awata, J. Du, and W. He (2005), Late Quaternary systematic stream offsets caused by repeated large seismic events along the Kunlun Fault, northern Tibet, *Geomorphology*, *71*, 278–292.

Fuchs, M., and A. Lang (2009), Luminescence dating of hillslopes – A review, *Geomorphology*, *109*, 17–26.

Fujimoto, M., N. Ohte, and M. Tani (2008), Effects of hill slope topography on hydrological responses in a weathered granite mountain, Japan: Comparison of the runoff response between the valley head and the side slope, *Hydrol. Processes*, *22*, 2581–2594.

Fujita, K., T. Ohta, and Y. Ageta (2007), Characteristics and climatic sensitivities of runoff from a cold-type glacier on the Tibetan Plateau, *Hydrol. Processes*, *21*, 2882–2891.

Fujita, M., M. Michine, and K. Ashida (1991), Simulation of reservoir sedimentation in mountain regions, in *Fluvial Hydraulics of Mountain Regions*, edited by A. Armanini and G. DiSilvio, pp. 209–222, Springer, Berlin.

Fuller, I. C. (2007), Geomorphic work during a '150-year' storm; contrasting behaviors of river channels in a New Zealand catchment, *Ann. Assoc. Am. Geogr.*, *97*, 665–676.

Fuller, T. K., L. A. Perg, J. K. Willenbring, and K. Lepper (2009), Field evidence for climate-driven changes in sediment supply leading to strath terrace formation, *Geology*, *37*, 467–470.

Furbish, D. J. (1993), Flow structure in a bouldery mountain stream with complex bed topography, *Water Resour. Res.*, *29*, 2249–2263.

Furbish, D. J. (2003), Using the dynamically coupled behavior of land-surface geometry and soil thickness in developing and testing hillslope evolution models, in *Prediction in Geomorphology*, edited by P. R. Wilcock and R. M. Iverson, pp. 169–181, AGU, Washington, D. C.

Furbish, D. J., E. M. Childs, P. K. Haff, and M. W. Schmeeckle (2009a), Rain splash of soil grains as a stochastic advection-dispersion process, with implications for desert plant-soil interactions and land-surface evolution, *J. Geophys. Res.*, *114*, F00A03, doi:10.1029/2009JF001265.

Furbish, D. J., P. K. Haff, W. E. Dietrich, and A. M. Heimsath (2009b), Statistical description of slope-dependent soil transport and the diffusion–like coefficient, *J. Geophys. Res.*, *114*, F00A05, doi:10.1029/2009JF001267.

Füreder, L., C. Schütz, R. Burger, and M. Wallinger (1998), High Alpine streams as models for ecological gradients, in *Hydrology, Water Resources and Ecology in Headwaters*, edited by K. Kovar et al., *IAHS Publ.*, *248*, 387–394.

Fushimi, H., K. Ikegami, and K. Higuchi (1985), Nepal case study: Catastrophic floods, in *Techniques for Prediction of Runoff from Glacierized Areas*, edited by G. J. Young, *IAHS Publ.*, *149*, 125–130.

Gabet, E. J. (2003), Post-fire thin debris flows; sediment transport and numerical modelling, *Earth Surf. Processes Landforms*, *28*, 1341–1348.

Gabet, E. J., and A. Bookter (2008), A morphometric analysis of gullies scoured by post-fire progressively bulked debris flows in southwest Montana, USA, *Geomorphology*, *96*, 298–209.

Gabet, E. J., and P. Sternberg (2008), The effects of vegetative ash on infiltration capacity, sediment transport, and the generation of progressively bulked debris flows, *Geomorphology*, *101*, 666–673.

Gabris, G., and B. Nagy (2005), Climate and tectonically controlled river style changes on the Sajo-Hernad alluvial fan (Hungary), in *Alluvial Fans: Geomorphology, Sedimentology, Dynamics*, edited by A. Harvey, A. E. Mather, and M. Stokes, *Geol. Soc. Spec. Publ.*, *251*, 61–67.

Gaeuman, D. A., J. C. Schmidt, and P. R. Wilcock (2003), Evaluation of in-channel gravel storage with morphology-based gravel budgets developed from planimetric data, *J. Geophys. Res.*, *108*(F1), 6001, doi:10.1029/2002JF000002.

Gaeuman, D. A., E. D. Andrews, A. Krause, and W. Smith (2009), Predicting fractional bed load transport rates: Application of the Wilcock-Crowe equations to a regulated gravel bed river, *Water Resour. Res.*, *45*, W06409, doi:10.1029/2008WR007320.

Galay, V. J. (1983), Causes of river bed degradation, *Water Resour. Res.*, *19*, 1057–1090.

Gale, S. J., and R. J. Haworth (2005), Catchment-wide soil loss from pre-agricultural times to the present: Transport- and supply-limited, *Geomorphology*, *68*, 314–333.

Gallant, J. C., and T. I. Dowling (2003), A multiresolution index of valley bottom flatness for mapping depositional areas, *Water Resour. Res.*, *39*(12), 1347, doi:10.1029/2002WR001426.

Gallaway, J. M., Y. E. Martin, and E. A. Johnson (2009), Sediment transport due to tree root throw: Integrating tree population dynamics, wildfire and geomorphic response, *Earth Surf. Processes Landforms*, *34*, 1255–1269.

Galli, M., F. Ardizzone, M. Cardinali, F. Guzzetti, and P. Reichenbach (2008), Comparing landslide inventory maps, *Geomorphology*, *94*, 268–289.

Gallino, G. L., and T. C. Pierson (1985), Polallie Creek debris flow and subsequent dam-break flood of 1980, East Fork Hood River basin, Oregon, *U. S. Geol. Surv. Water Supply Pap.*, *2273*, 22 pp.

Galloway, J. N., et al. (2004), Nitrogen cycles: Past, present, and future, *Biogeochemistry*, *70*, 153–226.

Galy, A., and C. France-Lanord (2001), Higher erosion rates in the Himalaya; geochemical constraints on riverine fluxes, *Geology*, *29*, 23–26.

Galy, A., O. Beyssac, C. France-Lanord, and T. Eglinton (2008a), Recycling of graphite during Himalayan erosion: A geological stabilization of carbon in the crust, *Science*, *322*, 943–945.

Galy, A., C. France-Lanord, and B. Lartiges (2008b), Loading and fate of particulate organic carbon from the Himalaya to the Ganga-Brahmaputra delta, *Geochim. Cosmochim. Acta*, *72*, 1767–1787.

Gangodagamage, C., E. Barnes, and E. Foufoula-Georgiou (2007), Scaling in river corridor widths depicts organization in valley morphology, *Geomorphology*, *91*, 198–215.

Gao, J., and J. Maro (2010), Topographic controls on evolution of shallow landslides in pastoral Wairarapa, New Zealand, 1979-2003, *Geomorphology*, *114*, 373–381.

Gao, P., and A. D. Abrahams (2004), Bedload transport resistance in rough open-channel flows, *Earth Surf. Processes Landforms*, *29*, 423–435.

García, A. F. (2006), Thresholds of strath genesis deduced from landscape response to stream piracy by Pancho Rico Creek in the Coast Ranges of central California, *Am. J. Sci.*, *306*, 655–681.

García, A. F., and S. A. Mahan (2009), Sediment storage and transport in Pancho Rico Valley during and after the Pleistocene-Holocene transition, Coast Ranges of central California (Monterey County), *Earth Surf. Processes Landforms*, *34*, 1136–1150.

García, A. F., Z. Zhu, T. L. Ku, O. A. Chadwick, and J. Chacón Montero (2004), An incision wave in the geologic record, Alpujarran Corridor, southern Spain (Almería, *Geomorphology*, *60*, 37–72.

Garcia, C., J. B. Laronne, and M. Sala (2000), Continuous monitoring of bedload flux in a mountain gravel-bed river, *Geomorphology*, *34*, 23–31.

García, M. H., F. López, C. Dunn, and C. V. Alonso (2004), Flow, turbulence, and resistance in a flume with simulated vegetation, in *Riparian Vegetation and Fluvial Geomorphology*, edited by S. J. Bennett and A. Simon, pp. 11–27, AGU, Washington, D. C.

Garcia-Ruiz, J. M., J. Arnaez-Vadillo, L. O. Izquierdo, and A. Gomez-Villar (1988), Debris flows subsequent to a forest fire in the Najerilla River valley (Iberian system, Spain, *Pirineos*, *131*, 3–23.

Garde, R. J., A. Sahay, and S. Bhatnagar (2006), A simple mathematical model to predict the particle size distribution of the armour layer, *J. Hydraul. Res.*, *44*, 815–821.

Gardner, J. S. (1982), Alpine mass-wasting in contemporary time: Some examples from the Canadian Rocky Mountains, in *Space and Time in Geomorphology*, edited by C. E. Thorn, pp. 171–192, George Allen and Unwin, London.

Gardner, J. S., and N. K. Jones (1993), Sediment transport and yield at the Raikot Glacier, Nanga Parbat, Punjab Himalaya, in *Himalaya to the Sea: Geology, geomorphology, and the Quaternary*, edited by J. F. Shroder, pp. 184–197, Routledge, London.

Gardner, K. K., and B. L. McGlynn (2009), Seasonality in spatial variability and influence of land use/land cover and watershed characteristics on stream water nitrate concentrations in a developing watershed in the Rocky Mountain West, *Water Resour. Res.*, 45, W08411, doi:10.1029/2008WR007029.

Gardner, T. W. (1983), Experimental study of knickpoint and longitudinal profile evolution in cohesive, homogeneous material, *Geol. Soc. Am. Bull.*, 94, 664–672.

Garen, D. C., and D. Marks (2005), Spatially distributed energy balance snowmelt modelling in a mountainous river basin; estimation of meteorological inputs and verification of model results, *J. Hydrol.*, 315, 126–153.

Garfì, G., D. E. Bruno, D. Calcaterra, and M. Parise (2007), Fan morphodynamics and slope instability in the Mucone River basin (Sila Massif, southern Italy): Significance of weathering and role of land use changes, *Catena*, 69, 181–196.

Garner, H. F. (1974), *The Origin of Landscapes: A Synthesis of Geomorphology*, 734 pp., Oxford Univ. Press, New York.

Garr, C. E., and B. B. Fitzharris (1994), Sensitivity of mountain runoff and hydro-electricity to changing climate, in *Mountain Environments in Changing Climates*, edited by M. Beniston, pp. 366–381, Routledge, London.

Garrels, R. M., and F. T. Mackenzie (1971), *Evolution of Sedimentary Rocks*, 397 pp., W. W. Norton, New York.

Gartner, J. E., S. H. Cannon, P. M. Santi, and V. G. De Wolfe (2008), Empirical models to predict the volumes of debris flows generated by recently burned basins in the western U. S, *Geomorphology*, 96, 339–354.

Garvin, C. D., T. C. Hanks, R. C. Finkel, and A. M. Heimsath (2005), Episodic incision of the Colorado River in Glen Canyon, Utah, *Earth Surf. Processes Landforms*, 30, 973–984.

Gasparini, N. M., G. E. Tucker, and R. L. Bras (2004), Network-scale dynamics of grain-size sorting: Implications for downstream fining, stream-profile concavity, and drainage basin morphology, *Earth Surf. Processes Landforms*, 29, 401–421.

Gasparini, N. M., R. L. Bras, and K. X. Whipple (2006), Numerical modeling of non-steady-state river profile evolution using a sediment-flux-dependent incision model, in *Tectonics, Climate, and Landscape Evolution*, edited by S. D. Willett et al., *Spec. Pap. Geol. Soc. Am.*, 398, 127–141.

Gatto, L. W. (2000), Soil freeze-thaw-induced changes to a simulated rill: Potential impacts on soil erosion, *Geomorphology*, 32, 147–160.

Gavrilovic, Z., and Z. Matovic (1991), Review of disastrous torrent flood on the Vlasina River on June 26, 1988, including analysis of flood and the obtained results, in *Fluvial Hydraulis of Mountain Regions*, edited by A. Armanini and G. DiSilvio, pp. 235–250, Springer, Berlin.

Gazis, C. A., J. D. Blum, A. D. Jacobson, and C. P. Chamberlain (1998), Controls on the Strontium isotope geochemistry of the Indus River in northern Pakistan, *Eos Trans. AGU*, 79, F337.

Geertsema, M., J. J. Clague, J. W. Schwab, and S. G. Evans (2006), An overview of recent large catastrophic landslides in northern British Columbia, Canada, *Eng. Geol.*, 83, 120–143.

Gees, A. (1990), Flow measurement under difficult measuring conditions: Field experience with the salt dilution method, in *Hydrology in Mountainous Regions I. Hydrological Measurements; The Water Cycle*, edited by H. Lang and A. Musy, *IAHS Publ.*, 193, 255–262.

Gelabert, B., J. J. Fornós, and L. Gómez-Pujol (2003), Geomorphological characteristics and slope processes associated with different basins: Mallorca (western Mediterranean, *Geomorphology*, 52, 253–267.

Gellis, A. C. (1991), Construction effects on sediment for two basins, Puerto Rico, *U. S. Geological Survey Proceedings Paper, Fifth Federal Interagency Sedimentation Conference*, Las Vegas, Nevada, pp. 4-72–4-78.

Gellis, A. C. (1993), The effects of Hurricane Hugo on suspended-sediment loads, Lago Loiza basin, Puerto Rico, *Earth Surf. Processes Landforms*, *18*, 505–517.

Gellis, A. C. (2001), Factors controlling storm-generated suspended-sediment concentrations and loads in a humid-tropical basin, Quebrada Blanca, Puerto Rico, in *Proceedings of the Seventh Federal Interagency Sedimentation Conference*, Reno, Nevada, pp. 91–98.

Gellis, A. C., M. J. Pavich, P. R. Bierman, E. M. Clapp, A. Ellwein, and S. Aby (2004), Modern sediment yield compared to geologic rates of sediment production in a semi-arid basin, New Mexico: Assessing the human impact, *Earth Surf. Processes Landforms*, *29*, 1359–1372.

Gellis, A. C., M. J. Pavich, and A. Ellwein (2001), Erosion and sediment yields in two subbasins of contrasting land use, Rio Puerco, New Mexico, in *Proceedings of the Seventh Federal Interagency Sedimentation Conference*, Reno, Nevada, pp. 83–90.

Gellis, A. C., R. M. T. Webb, S. C. McIntyre, and W. J. Wolfe (2006), Land-use effects on erosion, sediment yields, and reservoir sedimentation: A case study in the Lago Loíza basin, Puerto Rico, *Phys. Geogr.*, *27*, 39–69.

Georgiev, B. V. (1990), Reliability of bed load measurements in mountain rivers, in *Hydrology in Mountainous Regions I: Hydrological Measurements: The Water Cycle*, edited by H. Lang and A. Musy, *IAHS Publ.*, *193*, 263–270.

Germanoski, D. (1989), The effects of sediment load and gradient on braided river morphology, unpublished Ph.D. dissertation, 407 pp., Colo. State Univ., Ft. Collins.

Germanoski, D., and M. D. Harvey (1993), Asynchronous terrace development in degrading braided channels, *Phys. Geogr.*, *14*, 16–38.

Germanoski, D., and S. A. Schumm (1993), Changes in braided morphology resulting from aggradation and degradation, *J. Geol.*, *101*, 451–466.

Gerrard, J. (1990), *Mountain Environments: An Examination of the Physical Geography of Mountains*, 317 pp., The MIT Press, Cambridge, Mass.

Gessler, J. (1965), *Der Geschiebetriebbeginn bei Mischungen Untersucht an Natürlichen Abpflästerungssterungserscheinungen in Kanälen*, *Mitteilungen der Versuchsanstalt für Wasserbau und Erdbau, Zurich*, vol. 69.

Geza, M., E. P. Poeter, and J. E. McCray (2009), Quantifying predictive uncertainty for a mountain-watershed model, *J. Hydrol.*, *376*, 170–181.

Ghisalberti, M., and H. M. Nepf (2004), The limited growth of vegetative shear layers, *Water Resour. Res.*, *40*, W07502, doi:10.1029/2003WR002776.

Ghose, B., S. Pandey, S. Singh, and G. Lal (1957), Quantitative geomorphology of the drainage basins in the central Luni Basin in western Rajasthan, *Z. Geomorphol.*, *1*, 146–160.

Gibbs, R. J. (1970), Mechanisms controlling world water chemistry, *Science*, *170*, 1088–1090.

Giers, A., E. Freistühler, and G. A. Schultz (1998), Methodology for assessment of ecohydrological effects of dam construction in a headwater region, in *Hydrology, Water Resources and Ecology in Headwaters*, edited by K. Kovar et al., *IAHS Publ.*, *248*, 509–514.

Gilbert, G. K. (1877), Report on the Geology of the Henry Mountains, 160 pp, Government Printing Office, Washington, D. C.

Gilbert, G. K. (1896), Niagara Falls and their history, *Nat. Geogr. Monogr.*, *1*, 203–236.

Gilbert, G. K. (1909), The convexity of hilltops, *J. Geol.*, *17*, 344–350.

Gilbert, G. K. (1914), Transportation of debris by running water, *U.S. Geol. Surv. Prof. Pap.*, *86*, 221 pp.

Gilbert, G. K. (1917), Hydraulic-mining debris in the Sierra Nevada, *U.S. Geol. Surv. Prof. Pap.*, *105*.

Giles, P. T. (2010), Investigating the use of alluvial fan volume to represent fan size in morphometric studies, *Geomorphology*, *121*, 317–328.

Gillan, B. J., J. T. Harper, and J. N. Moore (2010), Timing of present and future snowmelt from high elevations in northwest Montana, *Water Resour. Res.*, *46*, W01507, doi:10.1029/2009WR007861.

Giller, P. S., and B. Malmqvist (1998), *The Biology of Streams and Rivers*, Oxford Univ. Press, Oxford, U. K.

Gilman, K., and M. D. Newson (1980), *Soil Pipes and Pipeflow - A Hydrological Study in Upland Wales*, Geobooks, Norwich.

Gilvear, D., R. Francis, N. Willby, and A. Gurnell (2008), Gravel bars: A key habitat of gravel-bed rivers for vegetation, in *Gravel-Bed Rivers VI: From Process Understanding to River Restoration*, edited by H. Habersack, H. Piégay, and M. Rinaldi, pp. 677–700, Elsevier, Amsterdam.

Gilvear, D. J., T. M. Waters, and A. M. Milner (1995), Image analysis of aerial photography to quantify changes in channel morphology and instream habitat following placer mining in interior Alaska, *Freshwater Biol.*, *34*, 389–398.

Gintz, D., M. A. Hassan, and K.-H. Schmidt (1996), Frequency and magnitude of bedload transport in a mountain river, *Earth Surf. Processes Landforms*, *21*, 433–445.

Giorgi, F., C. Shields-Brodeur, and G. T. Bates (1994), Regional climate change scenarios over the United States produced with a nested regional climate model: Spatial and seasonal characteristics, *J. Clim.*, *7*, 375–399.

Givone, C., and X. Meignien (1990), Influence of topography on spatial distribution of rain, in *Hydrology of Mountainous Areas*, edited by L. Molnar, *IAHS Publ.*, *190*, 57–65.

Glade, T. (2005), Linking debris-flow hazard assessments with geomorphology, *Geomorphology*, *66*, 189–213.

Glancy, P. A., and R. P. Williams (1994), Problems with indirect determinations of peak streamflows in steep, desert stream channels, *Proceedings, ASCE Hydraulic Engineering '94 Conference, August 1–5, Buffalo, New York*, edited by G. V. Cotroneo and R. R. Rumer, pp. 635–639, Am. Soc. of Civ. Eng., New York.

Glastonbury, J., and R. Fell (2008), Geotechnical characteristics of large slow, very slow, and extremely slow landslides, *Can. Geotech. J.*, *45*, 984–1005.

Glaze, L. S., and S. M. Baloga (2003), DEM flow path prediction algorithm for geologic mass movements, *Environ. Eng. Geosci.*, *9*, 225–240.

Gleeson, T., and A. H. Manning (2008), Regional groundwater flow in mountainous terrain: Three-dimensional simulations of topographic and hydrogeologic controls, *Water Resour. Res.*, *44*, W10403, doi:10.1029/2008WR006848.

Glock, W. S. (1931), The development of drainage systems: A synoptic view, *Geogr. Rev.*, *21*, 475–482.

Glysson, G. D. (1993), U. S. Geological Survey bedload sampling policy, in *Hydraulic Engineering '93*, edited by H. W. Shen, S. T. Su, and F. Wen, pp. 701–706, Am. Soc. of Civ. Eng., New York.

Gob, F., N. Jacob, J.-P. Bravard, and F. Petit (2005), Determining the competence of mountainous Mediterranean streams using lichenometric techniques, in *Geomorphological Processes and Human Impacts in River Basins*, edited by R. J. Batalla and C. Garcia, *IAHS Publ.*, *299*, 161–170.

Gob, F., N. Jacob, J.-P. Bravard, and F. Petit (2008), The value of lichenometry and historical archives in assessing the incision of submediterranean rivers from the Little Ice Age in the Ardèche and upper Loire (France), *Geomorphology*, *94*, 170–183.

Godsey, S., H. Elsenbeer, and R. Stallard (2004), Overland flow generation in two lithologically distinct rainforest catchments, *J. Hydrol.*, *295*, 276–290.

Goetzmann, W. H. (1986), *New lands, New Men: America and the Second Great Age of Discovery*, 528 pp., Penguin, New York.

Golden, L. A., and G. S. Springer (2006), Channel geometry, median grain size, and stream power in small mountain streams, *Geomorphology*, *78*, 64–76.

Goldman, S., K. Jackson, and T. Bursktynsky (1986), *Erosion and Sediment Control Handbook*, 454 pp., McGraw Hill, New York.

Goldrick, G., and P. Bishop (1995), Differentiating the roles of lithology and uplift in the steepening of bedrock river long profiles: An example from southeastern Australia, *J. Geol.*, *103*, 227–231.

Goldrick, G., and P. Bishop (2007), Regional analysis of bedrock stream long profiles: Evaluation of Hack's SL form, and formulation and assessment of an alternative (the DS form), *Earth Surf. Processes Landforms*, *32*, 649–671.

Goldsmith, E., and N. Hildyard (1984), *The Social and Environmental Effects of Large Dams*, 404 pp., Sierra Club, San Francisco, Calif.

Gomez, B. (1983), Temporal variations in bedload trasnport rates: The effect of progressive bed armouring, *Earth Surf. Processes Landforms*, *8*, 41–54.

Gomez, B. (1987), Bedload, in *Glacio-fluvial Sediment Transfer*, edited by A. M. Gurnell and M. J. Clark, pp. 355–376, John Wiley, Chichester, U. K.

Gomez, B. (1993), Roughness of stable, armored gravel beds, *Water Resour. Res.*, *29*, 3631–3642.

Gomez, B., and M. Church (1989), An assessment of bed load sediment transport formulae for gravel bed rivers, *Water Resour. Res.*, *25*, 1161–1186.

Gomez, B., and B. M. Troutman (1997), Evaluation of process errors in bed load sampling using a dune model, *Water Resour. Res.*, *33*, 2387–2398.

Gomez, B., R. L. Naff, and D. W. Hubbell (1989), Temporal variations in bedload transport rates associated with the migration of bedforms, *Earth Surf. Processes Landforms*, *14*, 135–156.

Gomez, B., B. J. Rosser, D. H. Peacock, D. M. Hicks, and J. A. Palmer (2001), Downstream fining in a rapidly aggrading gravel bed river, *Water Resour. Res.*, *37*, 1813–1823.

Gomez, B., S. E. Coleman, V. W. K. Sy, D. H. Peacock, and M. Kent (2007), Channel change, bankfull and effective discharges on a vertically accreting, meandering, gravel-bed river, *Earth Surf. Processes Landforms*, *32*, 770–785.

Gomez-Peralta, D., S. F. Oberbauer, M. E. McClain, and T. E. Philippi (2008), Rainfall and cloud-water interception in tropical montane forests in the eastern Andes of central Peru, *For. Ecol. Manage.*, *255*, 1315–1325.

Gómez-Villar, A., and J. M. García-Ruiz (2000), Surface sediment characteristics and present dynamics in alluvial fans of the central Spanish Pyrenees, *Geomorphology*, *34*, 127–144.

Gómez-Villar, A., J. Álvarez-Martinez, and J. M. García-Ruiz (2006), Factors influencing the presence or absence of tributary-junction fans in the Iberian Range, Spain, *Geomorphology*, *81*, 252–264.

Gomi, T., and R. C. Sidle (2003), Bed load transport in managed steep-gradient headwater streams of southeastern Alaska, *Water Resour. Res.*, *39*(12), 1336, doi:10.1029/2003WR002440.

Gomi, T., R. C. Sidle, and J. S. Richardson (2002), Understanding processes and downstream linkages in headwater systems, *BioScience*, *52*, 905–916.

Gomi, T., R. C. Sidle, R. D. Woodsmith, and M. D. Bryant (2003), Characteristics of channel steps and reach morphology in headwater streams, southeast Alaska, *Geomorphology*, *51*, 225–242.

Gomi, T., R. C. Sidle, and D. N. Swanston (2004), Hydrogeomorphic linkages of sediment transport in headwater streams, Maybeso Experimental Forest, southeast Alaska, *Hydrol. Processes*, *18*, 667–683.

Gomi, T., R. C. Sidle, S. Noguchi, J. N. Negishi, A. R. Nik, and S. Sasaki (2006), Sediment and wood accumulations in humid tropical headwater streams: Effects of logging and riparian buffers, *For. Ecol. Manage.*, *224*, 166–175.

Gonzalez, M. A. (2001), Recent formation of arroyos in the Little Missouri Badlands of southwestern North Dakota, *Geomorphology*, *38*, 63–84.

Goode, J. R. (2009), Substrate controlled interactions between hydraulics, sediment transport, and erosional forms in bedrock rivers, Ph.D. dissertation, 179 pp., Colo. State Univ., Fort Collins.

Goode, J. R., and E. Wohl (2007), Relationships between land-use and forced-pool characteristics in the Colorado Front Range, *Geomorphology*, *83*, 249–265.

Goode, J. R., and E. Wohl (2010a), Coarse sediment transport in a bedrock channel with complex bed topography, *Water Resour. Res.*, *46*, W11524, doi:10.1029/2009WR008135.

Goode, J. R., and E. Wohl (2010b), Substrate controls on the longitudinal profile of bedrock channels: Implications for reach-scale roughness, *J. Geophys. Res.*, *115*, F03018, doi:10.1029/2008JF001188.

Goodman, I. A., M. E. Jensen, and D. P. Lettenmaier (1999), Watershed classification to assess vulnerability to disturbance, *Ecological Society of America, 1999 Annual Meeting Abstracts*, p. 18.

Gooseff, M. N. (2010), Defining hyporheic zones – advancing our conceptual and operational definitions of where stream water and groundwater meet, *Geogr. Compass*, *4*(8), 945–955.

Gooseff, M. N., S. M. Wondzell, R. Haggerty, and J. Anderson (2003), Comparing transient storage modeling and residence time distribution (RTD) analysis in geomorphically varied reaches in the Lookout Creek basin, Oregon, USA, *Adv. Water Resour.*, *26*, 925–937.

Gooseff, M. N., J. LaNier, R. Haggerty, and K. Kokkeler (2005), Determining in-channel (dead zone) transient storage by comparing solute transport in a bedrock channel-alluvial channel sequence, Oregon, *Water Resour. Res.*, *41*, W06014, doi:10.1029/2004WR003513.

Gooseff, M. N., J. K. Anderson, S. M. Wondzell, J. LaNier, and R. Haggerty (2006), A modeling study of hyporheic exchange pattern and sequence, size, and spacing of stream bedforms in mountain stream networks, Oregon, USA, *Hydrol. Processes*, *20*, 2443–2457.

Gordon, N. (1995), *Summary of technical testimony in the Colorado Water Division 1 trial*, USDA Forest Service General Tech. Rep. RM-GTR-270, 140 pp.

Gordon, W. S., J. S. Famiglietti, N. L. Fowler, T. G. F. Kittel, and K. A. Hibbard (2004), Validation of simulated runoff from six terrestrial ecosystem models: Results from VEMAP, *Ecol. Appl.*, *14*, 527–545.

Gore, J. A. (2001), Models of habitat use and availability to evaluate anthropogenic changes in channel geometry, in *Geomorphic Processes and Riverine Habitat*, edited by J. M. Dorava et al., pp. 27–36, AGU, Washington, D. C.

Gore, J. A., and A. M. Milner (1990), Island biogeographic theory: Can it be used to predict lotic recovery rates?, *Environ. Manage.*, *14*, 1491–1501.

Gottesfeld, A. S., and L. M. J. Gottesfeld (1990), Floodplain dynamics of a wandering river, dendro-chronology of the Morice River, British Columbia, Canada, *Geomorphology*, *3*, 159–179.

Gottesfeld, A. S., and J. Tunnicliffe (2003), Bed load measurements with a passive magnetic induction device, in *Erosion and Sediment Transport Measurement in Rivers: Technological and Methodological Advances*, edited by J. Bogen, T. Fergus, and D. E. Walling, *IAHS Publ.*, *283*, 211–221.

Gottfried, G. J., D. G. Neary, and P. F. Ffolliott (2002), Snowpack-runoff Relationships for Mid-elevation Snowpacks on the Workman Creek Watersheds of central Arizona, *Rocky Mountain Research Station Res. Pap. RMRS-RP-33*, USDA Forest Service, Fort Collins, Colo.

Gottschalk, L. C. (1945), Effects of soil erosion on navigation in Upper Chesapeake Bay, *Geogr. Rev.*, *35*, 219–238.

Götzinger, J. R., J. Barthel, Jagelke, and A. Bárdossy (2008), The role of groundwater recharge and baseflow in integrated models, in *Groundwater-surface Water Interaction: Process Understanding, Conceptuali-zation and Modelling*, edited by C. Abesser, T. Wagener, and G. Nuetzmann, *IAHS Publ.*, *321*, 103–109.

Gouder de Beauregard, A.-C., G. Torres, and F. Malaisse (2002), Ecohydrology: A new paradigm for bioengineers?, *Biotechnol. Agron. Soc. Environ.*, *6*, 17–27.

Goudie, A. S. (2006), Global warming and fluvial geomorphology, *Geomorphology*, *79*, 384–394.

Goudie, A. S. (2009), Dust storms: Recent developments, *J. Environ. Manage.*, *90*, 89–94.

Gough, L. P. (1993), *Understanding our fragile environment: Lessons from geochemical studies*, U.S. Geol. Surv. Circ., *1105*, 34 pp.

Gowan, C., and K. D. Fausch (1996), Long-term demographic responses of trout populations to habitat manipulation in six Colorado streams, *Ecol. Appl.*, *6*, 931–946.

Grable, J. L., and C. P. Harden (2006), Geomorphic response of an Appalachian Valley and Ridge stream to urbanization, *Earth Surf. Processes Landforms*, *31*, 1707–1720.

Graf, J. B. (1995), Measured and predicted velocity and longitudinal dispersion at steady and unsteady flow, Colorado River, Glen Canyon Dam to Lake Mead, *Water Resour. Bull.*, *31*, 265–281.

Graf, W. (1991), Flow resistance over a gravel bed: Its consequence on initial sediment movement, in *Fluvial Hydraulics of Mountain Regions*, edited by A. Armanini and G. DiSilvio, pp. 17–32, Springer, Berlin.

Graf, W. L. (1978), Fluvial adjustments to the spread of tamarisk in the Colorado Plateau region, *Geol. Soc. Am. Bull.*, *89*, 1491–1501.

Graf, W. L. (1979), Mining and channel response, *Ann. Assoc. Am. Geogr.*, *69*, 262–275.

Graf, W. L. (1983), The arroyo problem – palaeohydrology and palaeohydraulics in the short term, in *Background to Palaeohydrology*, edited by K. J. Gregory, pp. 279–302, John Wiley, Chichester, U. K.

Graf, W. L. (1988), *Fluvial Processes in Dryland Rivers*, 346 pp., Springer, Berlin.

Graf, W. L. (1996), Transport and deposition of plutonium-contaminated sediments by fluvial processes, Los Alamos Canyon, New Mexico, *Geol. Soc. Am. Bull.*, *108*, 1342–1355.

Graf, W. L. (1999), Dam nation: A geographic census of American dams and their large-scale hydrologic impacts, *Water Resour. Res.*, *35*, 1305–1311.

Graf, W. L. (2000), Locational probability for a dammed, urbanizing stream: Salt River, Arizona, USA, *Environ. Manage.*, *25*, 321–335.

Graf, W. L. (2001), Damage control: Restoring the physical integrity of America's rivers, *Ann. Assoc. Am. Geogr.*, *91*, 1–27.

Graf, W. L., S. L. Clark, M. T. Kammerer, T. Lehman, K. Randall, and R. Schroeder (1991), Geomorphology of heavy metals in the sediments of Queen Creek, Arizona, USA, *Catena*, *18*, 567–582.

Graf, W. L., J. Stromberg, and B. Valentine (2002), Rivers, dams, and willow flycatchers: A summary of their science and policy connections, *Geomorphology*, *47*, 169–188.

Graham, D. J., S. P. Rice, and I. Reid (2005), A transferable method for the automated grain sizing of river gravels, *Water Resour. Res.*, *41*, W07020, doi:10.1029/2004WR003868.

Graham, D. J., A.-J. Rollet, H. Piégay, and S. P. Rice (2010), Maximizing the accuracy of image-based surface sediment sampling techniques, *Water Resour. Res.*, *46*, W02508, doi:10.1029/2008WR006940.

Graham, W. F., and R. A. Duce (1979), Atmospheric pathways of the phosphorus cycle, *Geochim. Cosmochim. Acta*, *43*, 1195–1208.

Grams, P. E., and J. C. Schmidt (1999), Geomorphology of the Green River in the eastern Uinta Mountains, Dinosaur National Monument, Colorado and Utah, in *Varieties of Fluvial Form*, edited by A. J. Miller and A. Gupta, pp. 81–111, John Wiley, Chichester, U. K.

Grams, P. E., J. C. Schmidt, and D. J. Topping (2007), The rate and pattern of bed incision and bank adjustment on the Colorado River in Glen Canyon downstream from Glen Canyon Dam, 1956–2000, *Geol. Soc. Am. Bull.*, *119*, 556–575.

Gran, K. B., and D. R. Montgomery (2005), Spatial and temporal patterns in fluvial recovery following volcanic eruptions: Channel response to basin-wide sediment loading at Mount Pinatubo, Philippines, *Geol. Soc. Am. Bull.*, *117*, 195–211.

Gran, K. B., D. R. Montgomery, and D. G. Sutherland (2006), Channel bed evolution and sediment transport under declining sand inputs, *Water Resour. Res.*, *42*, W10407, doi:10.1029/2005WR004306.

Granger, D. E., D. Fabel, and A. N. Palmer (2001), Pliocene-Pleistocene incision of the Green River, Kentucky, Determined from radioactive decay of cosmogenic [26]Al and [10]Be in Mammoth Cave sediments, *Geol. Soc. Am. Bull.*, *113*, 825–836.

Grant, G. E. (1994), Hydraulics and sediment transport dynamics controlling step-pool formation in high gradient streams: A flume experiment, in *Dynamics and Geomorphology of Mountain Rivers*, edited by P. Ergenzinger and K.-H. Schmidt, pp. 241–250, Springer, Berlin.

Grant, G. (1997), Critical flow constrains flow hydraulics in mobile-bed streams: A new hypothesis, *Water Resour. Res.*, *33*, 349–358.

Grant, G. E., and T. Mizuyama (1991), Origin of step-pool sequences in high gradient streams: A flume experiment, in *Japan-US Workshop on Snow Avalanche, Landslide, Debris Flow Prediction and Control*, edited by M. Tominaga, pp. 523–532, Japan Sci. and Technol. Agency, Natl. Res. Inst. for Earth Sci. and Disaster Prev., Tsukuba, Japan.

Grant, G. E., and F. J. Swanson (1995), Morphology and processes of valley floors in mountain streams, western Cascades, Oregon, in *Natural and Anthropogenic Influences in Fluvial Geomorphology*, *Geophys. Monogr. Ser.*, vol. 89, edited by J. E. Costa et al., pp. 83–101, AGU, Washington, D. C.

Grant, G. E., F. J. Swanson, and M. G. Wolman (1990), Pattern and origin of stepped-bed morphology in high-gradient streams, western Cascades, Oregon, *Geol. Soc. Am. Bull.*, *102*, 340–352.

Grant, G. E., J. C. Schmidt, and S. L. Lewis (2003), A geological framework for interpreting downstream effects of dams on rivers, in *A Peculiar River: Geology, Geomorphology, and Hydrology of the Deschutes River, Oregon*, edited by J. E. O'Connor and G. E. Grant, pp. 203–219, AGU, Washington, D. C.

Grant, L., M. Seyfried, and J. McNamara (2004), Spatial variation and temporal stability of soil water in a snow-dominated, mountain catchment, *Hydrol. Processes*, *18*, 3493–3511.

Grass, A. J. (1971), Structural features of turbulent flow over smooth and rough boundaries, *J. Fluid Mech.*, *50*, 233–255.

Gray, D. H., and D. Barker (2004), Root-soil mechanics and interactions, in *Riparian Vegetation and Fluvial Geomorphology*, edited by S. J. Bennett and A. Simon, pp. 113–123, AGU, Washington, D. C.

Gray, D. H., and W. F. Megahan (1981), *Forest vegetation removal and slope stability in the Idaho batholith*, USDA Forest Service Res. Pap., vol. INT-271, 23 pp.

Graybeal, D. Y., and D. J. Leathers (2006), Snowmelt-related risk in Appalachia: First estimates from a historical snow climatology, *J. Appl. Meteorol. Climatol.*, *45*, 178–193.

Green, J. C. (2003), The precision of sampling grain-size percentiles using the Wolman method, *Earth Surf. Processes Landforms*, *28*, 979–991.

Green, P. A., C. J. Vörösmarty, M. Meybeck, J. N. Galloway, B. J. Peterson, and E. W. Boyer (2004), Pre-industrial and contemporary fluxes of nitrogen through rivers: A global assessment based on typology, *Biogeochemistry*, *68*, 71–105.

Greenbaum, N., and N. Bergman (2006), Formation and evacuation of a large gravel-bar deposited during a major flood in a Mediterranean ephemeral stream, Nahal Me'arot, NW Israel, *Geomorphology*, *77*, 169–186.

Greenbaum, N., Y. Enzel, and A. P. Schick (2001), Magnitude and frequency of paleofloods and historical floods in the Arava basin, Negev Desert, Israel, *Isr. J. Earth Sci.*, *50*, 159–186.

Gregory, K. J. (1976), Lichens and the determination of river channel capacity, *Earth Surf. Processes*, *1*, 273–285.

Gregory, K. J. (1987), The hydrogeomorphology of alpine proglacial areas, in *Glacio-fluvial Sediment Transfers*, edited by A. M. Gurnell and M. J. Clark, pp. 87–107, John Wiley, Chichester, U. K.

Gregory, K. J. (2006), The human role in changing river channels, *Geomorphology*, *79*, 172–191.

Gregory, K. J., and A. Chin (2002), Urban stream channel hazards, *Area*, *34*, 312–321.

Gregory, K. J., and V. Gardiner (1975), Drainage density and climate, *Z. Geomorphol.*, *19*, 287–298.

Gregory, S. (1991), Spatial and temporal patterns of woody debris retention and transport (abstract), *Bull. North Am. Benthol. Soc.*, *8*, 75.

Gregory, S., H. Li, and J. Li (2002), The conceptual basis for ecological responses to dam removal, *BioScience*, *52*, 713–723.

Gregory, S. V. (1980), Primary Production in Pacific Northwest Streams, unpublished Ph.D. dissertation, Oreg. State Univ., Corvallis.

Griffin, E. R., and J. D. Smith (2004), Floodplain stabilization by woody riparian vegetation during an extreme flood, in *Riparian Vegetation and Fluvial Geomorphology*, edited by S. J. Bennett and A. Simon, pp. 221–236, AGU, Washington, D. C.

Griffin, E. R., J. W. Kean, K. R. Vincent, J. D. Smith, and J. M. Friedman (2005), Modeling effects of bank friction and woody bank vegetation on channel flow and boundary shear stress in the Rio Puerco, New Mexico, *J. Geophys. Res.*, *110*, F04023, doi:10.1029/2005JF000322.

Griffiths, G. A. (1979), Recent sedimentation history of the Waimakariri River, New Zealand, *J. Hydrol. N. Z.*, *18*, 6–28.

Griffiths, G. A. (1980), Stochastic estimation of bed load yield in pool-and-riffle mountain streams, *Water Resour. Res.*, *16*, 931–937.

Griffiths, G. A. (1981), Flow resistance in coarse gravel bed rivers, *J. Hydraul. Div. Am. Soc. Civ. Eng., 107,* 899–918.

Griffiths, G. A. (1987), Form resistance in gravel channels with mobile beds, *J. Hydraul. Eng., 115,* 340–355.

Griffiths, G. A. (1989), Form resistance in gravel channels with mobile beds, *J. Hydraul. Eng., 115,* 340–355.

Griffiths, G. A., and M. A. Carson (2000), Channel width for maximum bedload transport capacity in gravel-bed rivers, South Island, New Zealand, *J. Hydrol. N. Z., 39,* 107–126.

Griffiths, P. G., R. Hereford, and R. H. Webb (2006), Sediment yield and runoff frequency of small drainage basins in the Mojave Desert, USA, *Geomorphology, 74,* 232–244.

Grimm, M. M. (1993), Paleoflood history and geomorphology of Bear Creek Basin, Colorado, unpublished M.S. thesis, 126 pp., Colo. State Univ., Ft. Collins.

Grimm, M. M., E. E. Wohl, and R. D. Jarrett (1995), Coarse-sediment distribution as evidence of an elevation limit for flash flooding, Bear Creek, Colorado, *Geomorphology, 14,* 199–210.

Grimm, N. B., and S. G. Fisher (1992), Responses of arid-land streams to changing climate, in *Global Climate Change and Freshwater Ecosystems,* edited by P. Firth and S. G. Fisher, pp. 211–233, Springer, New York.

Grimm, N. B., H. M. Valett, E. H. Stanley, and S. G. Fisher (1991), Contribution of the hyporheic zone to stability of an arid-land stream, *Verh. Internat. Verein. Limnol., 24,* 1595–1599.

Grimshaw, D. L., J. Lewin, and R. Fuge (1976), Seasonal and short-term variations in the concentration and supply of dissolved zinc to polluted aquatic environments, *Environ. Pollut., 11,* 1–7.

Grip, H., J.-M. Fritsch, and L. A. Bruijnzeel (2005), Soil and water impacts during forest conversion and stabilization to new land use, in *Forests, Water and People in the Humid Tropics,* edited by M. Bonell and L. A. Bruijnzeel, pp. 561–589, Cambridge Univ. Press, Cambridge, U. K

Griswold, J. P., and R. M. Iverson (2008), *Mobility statistics and automated hazard mapping for debris flows and rock avalanches,* U.S. Geol. Surv. Sci. Invest. Rep. 2007-5276, Reston, Va

Gritzner, M. L., W. A. Marcus, R. Aspinall, and S. G. Custer (2001), Assessing landslide potential using GIS, soil wetness modeling and topographic attributes, Payette River, Idaho, *Geomorphology, 37,* 149–165.

Grodek, T., M. Inbar, and A. P. Schick (1994), Step pool geometry and flow characteristics in low-sediment storage channel beds, in *Proceedings, ASCE Hydraulic Engineering '94 Conference, August 1–5, Buffalo, New York,* edited by G. V. Cotroneo and R. R. Rumer, *2,* 819–823.

Grodek, T., J. Lekach, and A. P. Schick (2000), Urbanizing alluvial fans as flood-conveying and flood-reducing systems; lessons from the October 1997 Eilat flood, in *The Hydrology-Geomorphology Interface: Rainfall, Floods, Sedimentation, Land Use,* edited by M. A. Hassan, O. Slaymaker, and S. M. Berkowicz, *IAHS Publ., 261,* 229–250.

Groffman, P. M., et al. (2006), Ecological thresholds: The key to successful environmental management or an important concept with no practical application?, *Ecosystems, 9,* 1–13.

Grötzbach, E., and C. Stadel (1997), Mountain peoples and cultures, in *Mountains of the World: A Global Priority,* edited by B. Messerli and J. D. Ives, pp. 17–38, The Parthenon, London.

Gu, C., G. M. Hornberger, J. S. Herman, and A. L. Mills (2008a), Effect of freshets on the flux of groundwater nitrate through streambed sediments, *Water Resour. Res., 44,* W05415, doi:10.1029/2007WR006488.

Gu, C., G. M. Hornberger, J. S. Herman, and A. L. Mills (2008b), Influence of stream-groundwater interactions in the streambed sediments on NO_3^- flux to a low-reflief coastal stream, *Water Resour. Res., 44,* W11432, doi:10.1029/2007WR006739.

Guadagno, F. M., R. Forte, P. Revellino, F. Fiorillo, and M. Focareta (2005), Some aspects of the initation of debris avalanches in the Campania region: The role of morphological slope discontinuities and the development of failure, *Geomorphology, 66,* 237–254.

Gubernick, R. A., D. A. Cenderelli, K. K. Bates, D. K. Johansen, and S. D. Jackson (2008), *Stream Simulation: An Ecological Approach to Providing Passage for Aquatic Organisms at Road-stream Crossings*, USDA Forest Service, San Dimas, Calif.

Gugliemi, Y., and F. Cappa (2010), Regional-scale relief evolution and large landslides: Insights from geomechanical analyses in the Tinée Valley (southern French Alps), *Geomorphology*, *117*, 121–129.

Guglielmin, M., B. Aldighieri, and B. Testa (2003), PERMACLIM: A model for the distribution of mountain permafrost, based on climatic observations, *Geomorphology*, *51*, 245–257.

Guin, A., R. Ramanathan, R. W. Ritzi, D. F. Dominic, I. A. Lunt, T. D. Scheibe, and V. L. Freedman (2010), Simulating the heterogeneity in braided channel belt deposits: 2. Examples of results and comparison to natural deposits, *Water Resour. Res.*, *46*, W04516, doi:10.1029/2009WR008112.

Gupta, A. (1988), Large floods as geomorphic events in the humid tropics, in *Flood Geomorphology*, edited by V. R. Baker, R. C. Kochel, and P. C. Patton, pp. 301–315, John Wiley, New York.

Gupta, A. (1995), Magnitude, frequency, and special factors affecting channel form and processes in the seasonal tropics, in *Natural and Anthropogenic Influences in Fluvial Geomorphology*, edited by J. E. Costa et al., *Geophys. Monogr. Res.*, *89*, 125–136.

Gupta, R. P., and B. C. Joshi (1990), Landslide hazard zoning using the GIS approach - a case study from Ramganga Catchment, Himalayas, *Eng. Geol.*, *28*, 119–132.

Gurnell, A. M. (1987a), Fluvial sediment yield from alpine, glacierized catchments, in *Glacio-fluvial Sediment Transfer*, edited by A. M. Gurnell and M. J. Clark, pp. 415–420, John Wiley, Chichester, U. K.

Gurnell, A. M. (1987b), Suspended sediment, in *Glacio-fluvial Sediment Transfer*, edited by A. M. Gurnell and M. J. Clark, pp. 305–354, John Wiley, Chichester, U. K.

Gurnell, A. M. (1995), Sediment yield from alpine glacier basins, in *Sediment and Water Quality in River Catchments*, edited by I. Foster, A. Gurnell, and B. Webb, pp. 407–435, John Wiley, Chichester, U. K.

Gurnell, A. M. (1998), The hydrogeomorphological effects of beaver dam-building activity, *Prog. Phys. Geogr.*, *22*, 167–189.

Gurnell, A. M. (2003), Wood storage and mobility, in *The Ecology and Management of Wood in World Rivers*, edited by S. V. Gregory, K. L. Boyer, and A. M. Gurnell, pp. 75–91, Am. Fish. Soc., Bethesda, Md.

Gurnell, A. M., and G. Petts (2006), Trees as riparian engineers: The Tagliamento River, Italy, *Earth Surf. Processes Landforms*, *31*, 1558–1574.

Gurnell, A. M., and J. Warburton (1990), The significance of suspended sediment pulses for estimating suspended sediment load and identifying suspended sediment sources in Alpine glacier basins, in *Hydrology in Mountainous Regions I. Hydrological Measurements: The Water Cycle*, edited by H. Lang and A. Musy, *IAHS Publ.*, *193*, 463–470.

Gurnell, A. M., M. J. Clark, and C. T. Hill (1990), The geomorphological impact of modified river discharge and sediment transport regimes downstream of hydropower scheme meltwater intake structures, in *Hydrology in Mountainous Regions II. Artificial Reservoirs, Water and Slopes*, edited by R. O. Sinniger and M. Monbaron, *IAHS Publ.*, *194*, 165–170.

Gurnell, A., D. Hannah, and D. Lawler (1996), Suspended sediment yield from glacier basins, in *Erosion and Sediment Yield: Global and Regional Perspectives*, edited by D. E. Walling and B. W. Webb, *IAHS Publ.*, *236*, 97–104.

Gurnell, A. M., H. Piégay, F. J. Swanson, and S. V. Gregory (2002), Large wood and fluvial processes, *Freshwater Biol.*, *47*, 601–619.

Gurnell, A. M., J.-L. Peiry, and G. E. Petts (2003), Using historical data in fluvial geomorphology, in *Tools in Fluvial Geomorphology*, edited by G. M. Kondolf and H. Piégay, pp. 77–101, John Wiley, Chichester, U. K.

Gurnell, A. M., J. M. Goodson, P. G. Angold, I. P. Morrissey, G. E. Petts, and J. Steiger (2004), Vegetation propagule dynamics and fluvial geomorphology, in *Riparian Vegetation and Fluvial Geomorphology*, edited by S. J. Bennett and A. Simon, pp. 209–219, AGU, Washington, D. C.

Gurnell, A. M., M. Lee, and C. Souch (2007), Urban rivers: Hydrology, geomorphology, ecology and opportunities for change, *Geogr. Compass*, *1*, 1–20.

Gustina, G. W., and J. P. Hoffmann (2000), Periphyton dynamics in a subalpine mountain stream during winter, *Arct. Antarct. Alp. Res.*, *32*, 127–134.

Guthrie, R. H., and S. G. Evans (2004), Analysis of landslide frequencies and characteristics in a natural system, coastal British Columbia, *Earth Surf. Processes Landforms*, *29*, 1321–1339.

Guthrie, R. H., and S. G. Evans (2007), Work, persistence, and formative events; the geomorphic impact of landslides, *Geomorphology*, *88*, 266–275.

Guthrie, R. H., A. Hockin, L. Colquhoun, T. Nagy, S. G. Evans, and C. Ayles (2010), An examination of controls on debris flow mobility: Evidence from coastal British Columbia, *Geomorphology*, *114*, 601–613.

Guti, G. (2002), Significance of side-tributaries and floodplains for Danubian fish populations, *Arch. Hydrobiol. Suppl.*, *141*, 151–163.

Gutiérrez, F., M. Gutiérrez, and C. Sancho (1998), Geomorphological and sedimentological analysis of a catastrophic flash flood in the Arás drainage basin (Central Pyrenees, Spain, *Geomorphology*, *22*, 265–283.

Guyot, J. L., J. Bourges, and J. Cortez (1994), Sediment transport in the Rio Grande, an Andean river of the Bolivian Amazon drainage basin, in *Variability in Stream Erosion and Sediment Transport*, edited by L. J. Olive, R. J. Loughran, and J. A. Kesby, *IAHS Publ.*, *224*, 223–231.

Gyalistras, D., C. Schaer, H. C. Davies, and H. Wanner (1997), Future Alpine climate, in *A View From the Alps: Regional Perspectives on Climate Change*, edited by P. Cebon et al., MIT Press, Boston, Mass.

Haas, F., T. Heckmann, V. Wichmann, and M. Becht (2004), Change of fluvial sediment transport rates after a high magnitude debris flow event in a drainage basin in the northern Limestone Alps, Germany, in *Sediment Transfer Through the Fluvial System*, edited by V. N. Golosov, V. R. Belyaev, and D. E. Walling, *IAHS-AISH Publ.*, *288*, 37–43.

Habersack, H. M. (2003), Use of radio-tracking techniques in bed load transport investigations, in *Erosion and Sediment Transport Measurement in Rivers: Technological and Methodological Advances*, edited by J. Bogen, T. Fergus, and D. E. Walling, *IAHS Publ.*, *283*, 172–180.

Habersack, H. M., and J. B. Laronne (2001), Bed load texture in an alpine gravel bed river, *Water Resour. Res.*, *37*, 3359–3370.

Habersack, H. M., and H. P. Nachtnebel (1994), Analysis of sediment transport developments in relation to human impacts, in *Variability in Stream Erosion and Sediment Transport*, edited by L. J. Olive, R. J. Loughran, and J. A. Kesby, *IAHS Publ.*, *224*, 385–393.

Habersack, H.-M., and H.-P. Nachtnebel (1995), Short-term effects of local river restoration on morphology, flow field, substrate and biota, *Reg. Rivers Res. Manage.*, *10*, 291–301.

Habersack, H. M., and H. Piégay (2008), River restoration in the Alps and their surroundings: Past experience and future challenges, in *Gravel-Bed Rivers VI: From Process Understanding to River Restoration*, edited by H. Habersack, H. Piégay, and M. Rinaldi, pp. 703–737, Elsevier, Amsterdam.

Habersack, H. M., H. P. Nachtnebel, and J. B. Laronne (2001), The continuous measurement of bedload discharge in a large alpine gravel bed river, *J. Hydraul. Res.*, *39*, 125–133.

Hack, J. T. (1957), Studies of longitudinal stream profiles in Virginia and Maryland, *U.S. Geol. Surv. Prof. Pap.*, *294-B*, pp. 45–97.

Hack, J. T. (1960), Interpretation of erosional topography in humid temperate regions, *Am. J. Sci.*, *258A*, 80–97.

Hack, J. T., and J. C. Goodlett (1960), Geomorphology and forest ecology of a mountain region in the central Appalachians, *U.S. Geol. Surv. Prof. Pap.*, *347*, 66 pp.

Haeberli, W. (1983), Frequency and characteristics of glacier floods in the Swiss Alps, *Ann. Glaciol.*, *4*, 85–90.

Haeberli, W., C. Benz, U. Gruber, M. Hoelzle, A. Kääb, and J. Schaper (2004), GIS applications for snow and ice in high-mountain areas: Examples from the Swiss Alps, in *Geographic Information Science and Mountain Geomorphology*, edited by M. P. Bishop and J. F. Shroder, pp. 381–402, Praxis, Chichester, U. K.

Haff, P. K. (2003), Neogeomorphology, prediction, and the Anthropic Force, in *Prediction in Geomorphology*, edited by P. R. Wilcock and R. M. Iverson, pp. 15–26, AGU, Washington, D. C.

Haff, P. K. (2010), Hillslopes, rivers, plows, and trucks: Mass transport on Earth's surface by natural and technological processes, *Earth Surf. Processes Landforms*, 35, 1157–1166.

Haga, H., T. Kumagai, K. Otsuki, and S. Ogawa (2002), Transport and retention of coarse woody debris in mountain streams: An in situ field experiment of log transport and a field survey of coarse woody debris distribution, *Water Resour. Res.*, 38(8), 1126, doi:10.1029/2001WR001123.

Hagg, W., and L. Braun (2005), The influence of glacier retreat on water yield from high mountain areas: Comparison of Alps and Central Asia, in *Climate and Hydrology in Mountain Areas*, pp. 263–275, John Wiley, Chichester, U. K

Hagg, W., L. N. Braun, M. Weber, and M. Becht (2006), Runoff modelling in glacierized central Asian catchments for present-day and future climate, *Nord. Hydrol.*, 37, 93–105.

Haggerty, R., and P. Reeves (2002), *STAMMT-L Version 1.0 User's Manual*, 76 pp., Sandia National Laboratory, Albuquerque, N. M.

Haigh, M. J., J. S. Rawat, and H. S. Bisht (1990), Hydrological impact of deforestation in the central Himalaya, in *Hydrology of Mountainous Areas*, edited by L. Molnar, *IAHS Publ.*, 190, 419–433.

Hairsine, P. B., L. Beuselinck, and G. C. Sander (2002), Sediment transport through an area of net deposition, *Water Resour. Res.*, 38(6), 1086, doi:10.1029/2001WR000265.

Hajdu, Z., and G. Füleky (2007), Distribution of nitrate pollution in the Niraj (Nyárád) River basin, *Carpathian J. Earth Environ. Sci.*, 2, 57–72.

Hales, T. C., and J. J. Roering (2007), Climatic controls on frost cracking and implications for the evolution of bedrock landscapes, *J. Geophys. Res.*, 112, F02033, doi:10.1029/2006JF000616.

Hales, T. C., and J. J. Roering (2009), A frost "buzzsaw" mechanism for erosion of the eastern Southern Alps, New Zealand, *Geomorphology*, 107, 241–253.

Hales, T. C., C. R. Ford, T. Hwang, J. M. Vose, and L. E. Band (2009), Topographic and ecologic controls on root reinforcement, *J. Geophys. Res.*, 114, F03013, doi:10.1029/2008JF001168.

Hall, R. O., and J. L. Tank (2003), Ecosystem metabolism controls nitrogen uptake in streams in Grand Teton National Park, Wyoming, *Limnol. Oceanogr.*, 48, 1120–1128.

Hallet, B., L. Hunter, and J. Bogen (1996), Rates of erosion and sediment evacuation by glaciers: A review of field data and their implications, *Global Planet. Change*, 12, 213–235.

Halwas, K. L., and M. Church (2002), Channel units in small, high gradient streams on Vancouver Island, British Columbia, *Geomorphology*, 43, 243–256.

Hamilton, L. S., and L. A. Bruijnzeel (1997), Mountain watersheds - integrating water, soils, gravity, vegetation, and people, in *Mountains of the World: A Global Priority*, edited by B. Messerli and J. D. Ives, pp. 337–370, The Parthenon, London.

Hammack, L., and E. Wohl (1996), Debris-fan formation and rapid modification at Warm Springs Rapid, Yampa River, Colorado, *J. Geol.*, 104, 729–740.

Hammann, K., and A. Dittrich (1994), Measurement systems to determine the velocity field in and close to the roughness sublayer, in *Dynamics and Geomorphology of Mountain Rivers*, edited by P. Ergenzinger and K.-H. Schmidt, pp. 265–288, Springer, Berlin.

Han, T., Y. Ding, C. Xie, B. Ye, Y. Shen, and K. Jiao (2007), Analysis on the facts of runoff increase in the Urumqi River basin, China, in *Glacier Mass Balance Changes and Meltwater Discharge*, edited by P. Ginot and J.-E. Sicart, *IAHS Publ.*, 318, 86–94.

Hancock, G. S., and R. S. Anderson (2002), Numerical modeling of fluvial strath-terrace formation in response to oscillating climate, *Geol. Soc. Am. Bull.*, 114, 1131–1142.

Hancock, G. S., R. S. Anderson, and K. X. Whipple (1998), Beyond power: Bedrock river incision process and form, in *Rivers Over Rock: Fluvial Processes in Bedrock Channels, Geophys. Monogr. Ser.*, vol. 107, edited by K. J. Tinkler and E. E. Wohl, pp. 35–60, AGU, Washington, D. C.

Hancock, G. S., R. S. Anderson, O. A. Chadwick, and R. C. Finkel (1999), Dating fluvial terraces with [10]Be and [26]Al profiles: Application to the Wind River, Wyoming, *Geomorphology*, 27, 41–60.

Hancock, G. R., J. B. C. Lowry, T. J. Coulthard, K. G. Evans, and D. R. Moliere (2010), A catchment scale evaluation of the SIBERIA and CAESAR landscape evolution models, *Earth Surf. Processes Landforms*, 35, 863–875.

Hannah, D. M., S. R. Kansakar, A. J. Gerrard, and G. Rees (2005), Flow regimes of Himalayan rivers of Nepal: Nature and spatial patterns, *J. Hydrol.*, 308, 18–32.

Hanrahan, T. P. (2007), Bedform morphology of salmon spawning areas in a large gravel-bed river, *Geomorphology*, 86, 529–536.

Hansen, A. J., R. Rasker, B. Maxwell, J. J. Rotella, J. D. Johnson, A. W. Parmenter, U. Langner, W. B. Cohen, R. L. Lawrence, and M. P. V. Kraska (2002), Ecological causes and consequences of demographic change in the New West, *BioScience*, 52, 151–162.

Hanson, P. R., J. A. Mason, and R. J. Goble (2006), Fluvial terrace formation along Wyoming's Laramie Range as a response to increased late Pleistocene flood magnitudes, *Geomorphology*, 76, 12–25.

Harbor, D. J. (1997), Landscape evolution at the margin of the Basin and Range, *Geology*, 25, 1111–1114.

Harbor, D. J., A. Bacastow, A. Heath, and J. Rogers (2005), Capturing variable knickpoint retreat in the central Appalachians, USA, *Geogr. Fis. Dinam. Quat.*, 28, 23–36.

Harbor, J., and J. Warburton (1993), Relative rates of glacial and nonglacial erosion in alpine environments, *Arct. Alp. Res.*, 25, 1–7.

Harden, C. P. (2006), Human impacts on headwater fluvial systems in the northern and central Andes, *Geomorphology*, 79, 249–263.

Harden, C. P., and P. D. Scruggs (2003), Infiltration on mountain slopes: A comparison of three environments, *Geomorphology*, 55, 5–24.

Harden, D. R. (1990), Controlling factors in the distribution and development of incised meanders in the central Colorado Plateau, *Geol. Soc. Am. Bull.*, 102, 233–242.

Harden, T., M. G. Macklin, and V. R. Baker (2010), Holocene flood histories in south-western USA, *Earth Surf. Processes Landforms*, 35, 707–716.

Harding, J. S., E. F. Benfield, P. V. Bolstad, G. S. Helfman, and E. B. D. Jones (1998), Stream biodiversity: The ghost of land use past, *Proc. Natl. Acad. Sci. U. S. A.*, 95, 14,843–14,847.

Hardy, R. J., J. L. Best, S. N. Lane, and P. E. Carbonneau (2010), Coherent flow structures in a depth-limited flow over a gravel surface: The influence of surface roughness, *J. Geophys. Res.*, 115, F03006, doi:10.1029/2009JF001416.

Hardy, R. J., S. N. Lane, R. I. Ferguson, and D. R. Parsons (2007), Emergence of coherent flow structures over a gravel surface: A numerical experiment, *Water Resour. Res.*, 43, W03422, doi:10.1029/2006WR004936.

Haritashya, U. K., A. Kumar, and P. Singh (2010), Particle size characteristics of suspended sediment transported in meltwater from the Gangotri Glacier, central Himalaya – an indicator of subglacial sediment evacuation, *Geomorphology*, 122, 140–152.

Haritashya, U. K., P. Singh, N. Kumar, and Y. Singh (2006), Hydrological importance of an unusual hazard in a mountainous basin; flood and landslide, *Hydrol. Processes*, 20, 3147–3154.

Harkins, N. W., D. J. Anastasio, and F. J. Pazzaglia (2005), Tectonic geomorphology of the Red Rock Fault, insights into segmentation and landscape evolution of a developing range front normal fault, *J. Struct. Geol.*, 27, 1925–1939.

Harkins, N., and E. Kirby (2008), Fluvial terrace riser degradation and determination of slip rates on strike-slip faults: An example from the Kunlun fault, China, *Geophys. Res. Lett.*, 35, L05406, doi:10.1029/2007GL033073.

Harman, W. A., G. D. Jennings, J. M. Patterson, D. R. Clinton, L. O. Slate, A. G. Jessup, J. R. Everhart, and R. E. Smith (1999), Bankfull hydraulic geometry relationships for North Carolina streams, in *Wildlife Hydrology*, edited by D. S. Olsen and J. P. Potyondy, *American Water Resources Association Tech. Publ. Ser.*, *99-3*, 401–408.

Harmon, M. E., J. F. Franklin, and F. J. Swanson (1986), Ecology of coarse woody debris in temperate ecosystems, *Adv. Ecol. Res.*, *15*, 133–302.

Harr, R. D. (1986), Effects of clearcutting on rain on snow runoff in western Oregon: A new look at old studies, *Water Resour. Res.*, *22*, 1095–1100.

Harris, R. R. (1987), Occurrence of vegetation on geomorphic surfaces in the active floodplain of a California alluvial stream, *Am. Midl. Nat.*, *118*, 393–405.

Harris, R. R. (1988), Associations between stream-valley geomorphology and riparian vegetation as a basis for landscape analysis in the eastern Sierra Nevada, California, USA, *Environ. Manage.*, *12*, 219–228.

Harrison, L. R., and E. A. Keller (2007), Modeling forced pool-riffle hydraulics in a boulder-bed stream, southern California, *Geomorphology*, *83*, 232–248.

Hart, D. D., T. E. Johnson, K. L. Bushaw-Newton, R. J. Horwitz, A. T. Bednarek, D. F. Charles, D. A. Kreeger, and D. J. Velinsky (2002), Dam removal: Challenges and opportunities for ecological research and river restoration, *BioScience*, *52*, 669–681.

Hart, D. R. (1995), Parameter estimation and stochastic interpretation of the transient storage model for solute transport in streams, *Water Resour. Res.*, *31*, 323–328.

Hart, E. A. (2002), Effects of woody debris on channel morphology and sediment storage in headwater streams in the Great Smoky Mountains, Tennessee-North Carolina, *Phys. Geogr.*, *23*, 492–510.

Hart, E. A., and S. G. Schurger (2005), Sediment storage and yield in an urbanized karst watershed, *Geomorphology*, *70*, 85–96.

Hartley, A. J., A. E. Mather, E. Jolley, and P. Turner (2005), Climatic controls on alluvial-fan activity, coastal Cordillera, northern Chile, in *Alluvial Fans: Geomorphology, Sedimentology, Dynamics*, edited by A. Harvey, A. E. Mather, and M. Stokes, *Geol. Soc. Spec. Publ.*, *251*, 95–116.

Hartman, G. (1996), Habitat selection by European beaver (Castor fiber) colonizing a boreal landscape, *J. Zool., London*, *240*, 317–325.

Hartman, G. F., and J. C. Scrivener (1990), *Impacts of Forestry Practices on a Coastal Stream Ecosystem, Carnation Creek, British Columbia*, Can. Bull. Fish. Aquat. Sci., *223*, 148 pp.

Hartshorn, K., N. Hovius, W. B. Dade, and R. L. Slingerland (2002), Climate-driven bedrock incision in an active mountain belt, *Science*, *297*, 2036–2038.

Hartvich, F., and P. Mentlík (2010), Slope development reconstruction at two sites in the Bohemian Forest Mountains, *Earth Surf. Processes Landforms*, *35*, 373–389.

Harvey, A. M. (1987), Sediment supply to upland streams; influence on channel adjustment, in *Sediment Transport in Gravel-Bed Rivers*, edited by C. R. Thorne, J. C. Bathurst, and R. D. Hey, pp. 121–150, John Wiley, Chichester, U. K.

Harvey, A. M. (1992), Controls on sedimentary style on alluvial fans, in *Dynamics of Gravel-Bed Rivers*, edited by P. Billi et al., pp. 519–535, John Wiley, Chichester, U. K.

Harvey, A. M. (2001), Coupling between hillslopes and channels in upland fluvial systems: Implications for landscape sensitivity, illustrated from the Howgill Fells, northwest England, *Catena*, *42*, 225–250.

Harvey, A. M. (2002a), Effective timescales of coupling within fluvial systems, *Geomorphology*, *44*, 175–201.

Harvey, A. M. (2002b), The role of base-level change in the dissection of alluvial fans: Case studies from southeast Spain and Nevada, *Geomorphology*, *45*, 67–87.

Harvey, A. M. (2005), Differential effects of base-level, tectonic setting and climatic change on Quaternary alluvial fans in the northern Great Basin, Nevada, USA, in *Alluvial Fans: Geomorphology, Sedimentology, Dynamics*, edited by A. M. Harvey, A. E. Mather, and M. Stokes, *Geol. Soc. Spec. Publ.*, *251*, 117–131.

Harvey, A. M. (2007), Differential recovery from the effects of a 100 year storm: Significance of long term hillslope channel coupling; Howgill Fells, northwest England, *Geomorphology*, *84*, 192–208.

Harvey, A. M., D. H. Hitchcock, and D. J. Hughes (1979), Event frequency and morphological adjustment of fluvial systems in upland Britain, in *Adjustments of the Fluvial System*, edited by D. D. Rhodes and G. P. Williams, pp. 139–167, George Allen and Unwin, London.

Harvey, G. L., and N. J. Clifford (2010), Experimental field assessment of suspended sediment pathways for characterizing hydraulic habitat, *Earth Surf. Processes Landforms*, *35*, 600–610.

Harvey, J. W., and K. E. Bencala (1993), The effect of streambed topography on surface-subsurface water exchange in mountain catchments, *Water Resour. Res.*, *29*, 89–98.

Harvey, J. W., and C. C. Fuller (1998), Effect of enhanced manganese oxidation in the hyporheic zone on basin-scale geochemical mass balance, *Water Resour. Res.*, *34*, 623–636.

Harvey, J. W., and B. J. Wagner (2000), Quantifying hydrologic interactions between streams and their subsurface hyporheic zones, in *Streams and Ground Waters*, edited by J. A. Jones and P. J. Mulholland, pp. 3–44, Academic Press, San Diego, Calif.

Harvey, M. D. (1980), Steepland channel response to episodic erosion, unpublished Ph.D. dissertation, 266 pp., Colo. State Univ., Ft. Collins.

Harvey, M. D., C. C. Watson, and S. A. Schumm (1983), Channelized streams: An analog for the effects of urbanization, *1983 International Symposium on Urban Hydrology, Hydraulics and Sediment Control*, pp. 401–409, Univ. of Ky., Lexington.

Harvey, M. D., J. Pitlick, and J. Laird (1987), Temporal and spatial variability of sediment storage and erosion in Ash Creek, Arizona, in *Erosion and Sedimentation in the Pacific Rim*, edited by R. L. Beschta et al., *IAHS Publ.*, *165*, 281–282.

Harvey, M. D., R. A. Mussetter, and E. J. Wick (1993), A physical process-biological response model for spawning habitat formation for the endangered Colorado squawfish, *Rivers*, *4*, 114–131.

Hasbargen, L. E., and C. Paola (2000), Landscape instability in an experimental drainage basin, *Geology*, *28*, 1067–1070.

Haschenburger, J. K. (1999), A probability model of scour and fill depths in gravel-bed channels, *Water Resour. Res.*, *35*, 2857–2869.

Haschenburger, J. K. (2006), Observations of event-based streambed deformation in a gravel bed channel, *Water Resour. Res.*, *42*, W11412, doi:10.1029/2006WR004985.

Haschenburger, J. K., and M. Church (1998), Bed material transport estimated from the virtual velocity of sediment, *Earth Surf. Processes Landforms*, *23*, 791–808.

Haschenburger, J. K., and S. P. Rice (2004), Changes in woody debris and bed material texture in a gravel-bed channel, *Geomorphology*, *60*, 241–267.

Haschenburger, J. K., and P. R. Wilcock (2003), Partial transport in a natural gravel bed channel, *Water Resour. Res.*, *39*(1), 1020, doi:10.1029/2002WR001532.

Haschenburger, J. K., S. P. Rice, and E. Voyde (2007), Evaluation of bulk sediment sampling criteria for gravel-bed rivers, *J. Sediment. Res.*, *77*, 415–423.

Hasegawa, K., and S. Kanbayashi (1996), Formation mechanism of step-pool systems in steep rivers and guide lines for the design of construction, *Ann. J. Hydraul. Eng. (Japan)*, *40*, 893–900.

Hasegawa, K., and T. Mizugaki (1994), Analysis of blocking phenomena in bifurcated channels found in mountainous rivers, in *Proceedings, ASCE Hydraulic Engineering '94 Conference, August 1–5, Buffalo, New York*, edited by G. V. Cotroneo and R. R. Rumer, pp. 829–833, Am. Soc. of Civ. Eng., New York.

Haslam, S. M. (1990), *River Pollution: An Ecological Perspective*, 218 pp., Belhaven Press, London.

Haslam, S. M. (2008), *The Riverscape and the River*, 420 pp., Cambridge Univ. Press, Cambridge, U. K.

Hasnain, S. I. (2002), Himalayan glaciers meltdown; impact on South Asian rivers, in *FRIEND 2002: Regional Hydrology; Bridging the Gap between Research and Practice*, edited by H. A. J. van Lanen et al., *IAHS Publ.*, *274*, 417–423.

Hassan, M. A. (1993), Bed material and bedload movement in two ephemeral streams, *Spec. Publ. Int. Assoc. Sedimentol.*, *17*, 37–49.

Hassan, M. A. (2005), Characteristics of gravel bars in ephemeral streams, *J. Sediment. Res.*, *75*, 29–42.

Hassan, M. A., and M. Church (1992), The movement of individual grains on the streambed, in *Dynamics of Gravel-Bed Rivers*, edited by P. Billi et al., pp. 159–175, John Wiley, Chichester, U. K.

Hassan, M. A., and M. Church (1994), Vertical mixing of coarse particles in gravel bed rivers: A kinematic model, *Water Resour. Res.*, *30*, 1173–1185.

Hassan, M. A., and M. Church (2001), Sensitivity of bed load transport in Harris Creek: Seasonal and spatial variation over a cobble-gravel bar, *Water Resour. Res.*, *37*, 813–825.

Hassan, M. A., and P. Ergenzinger (2003), Use of tracers in fluvial geomorphology, in *Tools in Fluvial Geomorphology*, edited by G. M. Kondolf and H. Piégay, pp. 397–423, John Wiley, Chichester, U. K.

Hassan, M. A., and I. Reid (1990), The influence of microform bed roughness elements on flow and sediment transport in gravel bed rivers, *Earth Surf. Processes Landforms*, *15*, 739–750.

Hassan, M. A., and R. D. Woodsmith (2004), Bed load transport in an obstruction-formed pool in a forest, gravelbed stream, *Geomorphology*, *58*, 203–221.

Hassan, M. A., A. P. Schick, and J. B. Laronne (1984), The recovery of flood-dispersed coarse sediment particles, in *Channel Processes - Water, Sediment, Catchment Controls*, edited by A. P. Schick, *Catena Suppl.*, *5*, 153–162.

Hassan, M. A., M. Church, and A. P. Schick (1991), Distance of movement of coarse particles in gravel bed streams, *Water Resour. Res.*, *27*, 503–511.

Hassan, M. A., A. P. Schick, and P. A. Shaw (1999), The transport of gravel in an ephemeral sandbed river, *Earth Surf. Processes Landforms*, *24*, 623–640.

Hassan, M. A., D. L. Hogan, S. A. Bird, C. L. May, T. Gomi, and D. Campbell (2005), Spatial and temporal dynamics of wood in headwater streams of the Pacific Northwest, *J. Am. Water Resour. Assoc.*, *41*, 899–919.

Hassan, M. A., R. Egozi, and G. Parker (2006), Experiments on the effect of hydrograph characteristics on vertical grain sorting in gravel bed rivers, *Water Resour. Res.*, *42*, W09408, doi:10.1029/2005WR004707.

Hassan, M. A., B. J. Smith, D. L. Hogan, D. S. Luzi, A. E. Zimmermann, and B. C. Eaton (2008), Sediment storage and transport in coarse bed streams: Scale considerations, in *Gravel-Bed Rivers VI: From Process Understanding to River Restoration*, edited by H. Habersack, H. Piégay, and M. Rinaldi, pp. 473–497, Elsevier, Amsterdam.

Hassan, M. A., M. Church, J. Rempel, and R. J. Enkin (2009), Promise, performance and current limitations of a magnetic Bedload Movement Detector, *Earth Surf. Processes Landforms*, *34*, 1022–1032.

Hatfield, R. G., and B. A. Maher (2009), Fingerprinting upland sediment sources: Particle size-specific magnetic linkages between soils, lake sediments and suspended sediments, *Earth Surf. Processes Landforms*, *34*, 1359–1373.

Hattanji, T., and Y. Matsushi (2006), Effect of runoff processes on channel initiation; comparison of four forested mountains in Japan, *Trans. Jpn. Geomorphol. Union*, *27*, 319–336.

Hattanji, T., and Y. Onda (2004), Coupling of runoff processes and sediment transport in mountainous watersheds underlain by different sedimentary rocks, *Hydrol. Processes*, *18*, 623–636.

Hattanji, T., Y. Onda, and Y. Matsukura (2006), Thresholds for bed load transport and channel initiation in a chert area in Ashio Mountains, Japan: An empirical approach from hydrogeomorphic observations, *J. Geophys. Res.*, *111*, F02022, doi:10.1029/2004JF000206.

Hauck, C. D., V. Mühll, and M. Hoelzle (2005), Permafrost monitoring in high mountain areas using a coupled geophysical and meteorological approach, in *Climate and Hydrology in Mountain Areas*, edited by C. de Jong, D. Collins, and R. Ranzi, pp. 59–71, John Wiley, Chichester, U. K.

Hauer, F. R., and M. S. Lorang (2004), River regulation, decline of ecological resources, and potential for restoration in a semi-arid lands river in the western USA, *Aquat Sci.*, *66*, 388–401.

Hauer, F. R., J. S. Baron, D. H. Campbell, K. D. Fausch, S. W. Hostetler, G. H. Leavesley, P. R. Leavitt, D. M. McKnight, and J. A. Stanford (1997), Assessment of climate change and freshwater ecosystems of the Rocky Mountains, USA and Canada, *Hydrol. Processes*, *11*, 903–924.

Hauer, F. R., J. A. Stanford, and M. S. Lorang (2007), Pattern and process in northern Rocky Mountain headwaters: Ecological linkages in the headwaters of the Crown of the Continent, *J. Am. Water Resour. Assoc.*, *43*, 104–117.

Haupt, H. F. (1967), Infiltration, overland flow and soil movement on frozen and snow covered plots, *Water Resour. Res.*, *3*, 145–161.

Haviv, I., Y. Enzel, K. X. Whipple, E. Zilberman, J. Stone, A. Matmon, and L. K. Fifield (2006), Amplified erosion above waterfalls and oversteepened bedrock reaches, *J. Geophys. Res.*, *111*, F04004, doi:10.1029/2006JF000461.

Hawkes, H. A. (1975), River zonation and classification, in *River Ecology*, edited by B. A. Whitton, pp. 312–374, Univ. of California Press, Berkeley.

Hawkins, C. P., et al. (1993), A hierarchical approach to classifying stream habitat features, *Fisheries*, *18*, 3–12.

Hay, L. E., and M. P. Clark (2003), Use of statistically and dynamically downscaled atmospheric model output for hydrologic simulations in three mountainous basins in the western United States, *J. Hydrol.*, *282*, 56–75.

Hay, L. E., R. Viger, and G. McCabe (1998), Precipitation interpolation in mountainous regions using multiple linear regression, in *Hydrology, Water Resources and Ecology in Headwaters*, edited by K. Kovar et al., *IAHS Publ.*, *248*, 33–38.

Hayakawa, Y. S., and Y. Matsukura (2003), Recession rates of waterfalls in Boso Peninsula, Japan, and a predictive equation, *Earth Surf. Processes Landforms*, *28*, 675–684.

Hayakawa, Y. S., and Y. Matsukura (2009), Factors influencing the recession rate of Niagara Falls since the 19th century, *Geomorphology*, *110*, 212–216.

Hayakawa, Y. S., and T. Oguchi (2006), DEM-based identification of fluvial knickzones and its application to Japanese mountain rivers, *Geomorphology*, *78*, 90–106.

Hayakawa, Y. S., and T. Oguchi (2009), GIS analysis of fluvial knickzone distribution in Japanese mountain watersheds, *Geomorphology*, *111*, 27–37.

Hayden, B. P. (1988), Flood climates, in *Flood Geomorphology*, edited by V. R. Baker, R. C. Kochel, and P. C. Patton, pp. 13–26, John Wiley, New York.

Hayward, J. A. (1979), Mountain stream sediments, in *Physical Hydrology: New Zealand Experience*, edited by D. L. Murray and P. Ackroyd, pp. 193–212, N. Z. Hydrol. Soc. Inc., North Wellington.

Hayward, J. A. (1980), *Hydrology and Stream Sediments from Torlesse Stream Catchment*, *Spec. Publ.*, vol. 17, Tussock Grasslands and Mtn. Lands Inst., Lincoln Coll., New Zealand.

Hayward, J. A., and A. J. Sutherland (1974), The Torlesse stream vortex-tube sediment trap, *J. Hydrol. N. Z.*, *13*, 41–53.

He, Q., and P. Owens (1995), Determination of suspended sediment provenance using Caesium-137, unsupported Lead-210 and Radium-226: A numerical mixing model approach, in *Sediment and Water Quality in River Catchments*, edited by I. Foster, A. Gurnell, and B. Webb, pp. 207–227, John Wiley, Chichester, U. K.

Hebeler, F., and R. S. Purves (2008), The influence of resolution and topographic uncertainty on melt modelling using hypsometric sub-grid parameterization, *Hydrol. Processes*, *22*, 3965–3979.

Heede, B. H. (1972), Influences of a forest on the hydraulic geometry of two mountain streams, *Water Resour. Bull.*, *8*, 523–530.

Heede, B. H. (1981), Dynamics of selected mountain streams in the western United States of America, *Z. Geomorphol.*, *25*, 17–32.

Heede, B. H. (1985), Channel adjustments to the removal of log steps: An experiment in a mountain streams, *Environ. Manage.*, *9*, 427–432.

Heede, B. H. (1991a), Increased flows after timber harvest accelerate stream disequilibrium. Erosion Control: A Global Perspective, Conference 22, Int. Erosion Control Assoc.Orlando, Fla.

Heede, B. H. (1991b), Response of a stream in disequilibrium to timber harvest, *Environ. Manage.*, *15*, 251–255.

Heede, B. H., and J. N. Rinne (1990), Hydrodynamic and fluvial morphologic processes: Implications for fisheries management and research, *N. Am. J. Fish. Manage.*, *10*, 249–268.

Heimsath, A. M., W. E. Dietrich, K. Nishiizumi, and R. C. Finkel (1997), The soil production function and landscape equilibrium, *Nature*, *388*, 358–361.

Heimsath, A. M., W. E. Dietrich, K. Nishiizumi, and R. C. Finkel (1999), Cosmogenic nuclides, topography, and the spatial variation of soil depth, *Geomorphology*, *27*, 151–172.

Heimsath, A. M., J. Chappell, W. E. Dietrich, K. Nishiizumi, and R. C. Finkel (2000), Soil production on a retreating escarpment in southeastern Australia, *Geology*, *28*, 787–790.

Heimsath, A. M., W. E. Dietrich, K. Nishiizumi, and R. C. Finkel (2001), Stochastic processes of soil production and transport: Erosion rates, topographic variation and cosmogenic nuclides in the Oregon Coast Range, *Earth Surf. Processes Landforms*, *26*, 531–552.

Heimsath, A. M., J. Chappell, N. A. Spooner, and D. G. Questiaux (2002), Creeping soil, *Geology*, *30*, 111–114.

Heimsath, A. M., D. J. Furbish, and W. E. Dietrich (2005), The illusion of diffusion: Field evidence for depth-dependent sediment transport, *Geology*, *33*, 949–952.

Helfield, J. M., S. J. Capon, C. Nilsson, R. Jansson, and D. Palm (2007), Restoration of rivers used for timber floating: Effects on riparian plant diversity, *Ecol. Appl.*, *17*, 840–851.

Helley, E. J., and V. C. LaMarche, Jr. (1973), Historic flood information for northern California streams from geological and botanical evidence, *U. S. Geol. Surv. Prof. Pap.*, *485-E*, pp. E1–E16.

Hem, J. D., A. Demayo, and R. A. Smith (1990), Hydrogeochemistry of rivers and lakes, in *Surface Water Hydrology*, edited by M. G. Wolman and H. C. Riggs, pp. 189–231, Geol. Soc. of Am., Boulder, Colo.

Hendrick, R. R., L. L. Ely, and A. N. Papanicolaou (2010), The role of hydrologic processes and geomorphology on the morphology and evolution of sediment clusters in gravel-bed rivers, *Geomorphology*, *114*, 483–496.

Hendrickx, J. M. H., D. Vega, J. B. J. Harrison, L. E. C. Gobbetti, P. Rojas, and T. W. Miller (2005), Hydrology of hillslope soils in the Upper Rio Chagres watershed, Panama, in *The Rio Chagres, Panama: A Multidisciplinary Profile of A Tropical Watershed*, edited by R. S. Harmon, pp. 113–138, Springer, Dordrecht, The Netherlands.

Hengl, T., and H. I. Reuter (Eds.) (2009), *Geomorphometry: Concepts, Software, Applications*, 772 pp., Elsevier, Amsterdam.

Henkle, J. (2010), Channel initiation in the semiarid Colorado Front Range, unpublished M.S. thesis, Colo. State Univ., Fort Collins, Colo.

Hennrich, K., and M. J. Crozier (2004), A hillslope hydrology approach for catchment-scale slope stability analysis, *Earth Surf. Processes Landforms*, *29*, 599–610.

Henriksen, A. (1980), Acidification of fresh waters – a large scale titration, in *Ecological Impact of Acid Precipitation*, edited by D. Drablos and A. Tollan, pp. 68–74, SNSF Project, Oslo-Ås.

Herget, J. (2004), Reconstruction of ice-dammed lake outburst floods in the Altai Mountains, Siberia: A review, *J. Geol. Soc. India*, *64*, 561–574.

Heritage, G. L., and D. Hetherington (2005), The use of high-resolution field laser scanning for mapping surface topography in fluvial systems, in *Sediment Budgets I*, edited by D. E. Walling and A. J. Horowitz, *IAHS Publ.*, *291*, 269–277.

Heritage, G. L., and D. J. Milan (2009), Terrestrial Laser Scanning of grain roughness in a gravel-bed river, *Geomorphology*, *113*, 4–11.

Herron, N., and C. Wilson (2001), A water balance approach to assessing the hydrologic buffering potential of an alluvial fan, *Water Resour. Res.*, *37*, 341–351.

Hervás, J., J. I. Barredo, P. L. Rosin, A. Pasuto, F. Mantovani, and S. Silvano (2003), Monitoring landslides from optical remotely sensed imagery: The case history of Tessina landslide, Italy, *Geomorphology, 54,* 63–75.

Hervouet, J.-M. (2000), TELEMAC modelling system: An overview, *Hydrol. Processes, 14,* 2209–2210.

Hession, W., J. E. Pizzuto, T. E. Johnson, and R. J. Horwitz (2003), Influence of bank vegetation on channel morphology in rural and urban watersheds, *Geology, 31,* 147–150.

Hester, E. T., and M. W. Doyle (2008), In-stream geomorphic structures as drivers of hyporheic exchange, *Water Resour. Res., 44,* W03417, doi:10.1029/2006WR005810.

Hester, E. T., and M. N. Gooseff (2010), Moving beyond the banks: Hyporheic restoration is fundamental to restoring ecological services and functions of streams, *Environ. Sci. Technol., 44,* 1521–1525.

Hetherington, D., G. Heritage, and D. Milan (2005), Daily fine sediment dynamics on an active Alpine glacier outwash plain, in *Sediment Budgets I,* edited by D. E. Walling and A. J. Horowitz, *IAHS Publ., 291,* 278–284.

Hetzel, R., S. Niedermann, S. Ivy-Ochs, P. W. Kubik, M. Tao, and B. Gao (2002), [21]Ne versus [10]Be and [26]Al exposure ages of fluvial terraces: The influence of crustal Ne in quartz, *Earth Planet. Sci. Lett., 201,* 575–591.

Heuer, K., K. A. Tonnessen, and G. P. Ingersoll (2000), Comparison of precipitation chemistry in the Central Rocky Mountains, Colorado, USA, *Atmos. Environ., 34,* 1713–1722.

Hewitt, K. (1982), Natural dams and outburst floods of the Karakoram Himalaya, in *Hydrological Aspects of Alpine and High Mountain Areas,* edited by J. W. Glen, *IAHS Publ., 138,* 259–269.

Hewitt, K. (1997), Risk and disasters in mountain lands, in *Mountains of the World: A Global Priority,* edited by B. Messerli and J. D. Ives, pp. 371–408, The Parthenon, London.

Hewitt, K. (1998), Catastrophic landslides and their effects on the Upper Indus streams, Karakoram Himalaya, northern Pakistan, *Geomorphology, 26,* 47–80.

Hey, R. D. (1979), Flow resistance in gravel-bed rivers, *J. Hydraul. Div. Am. Soc. Civ. Eng., 105,* 365–279.

Hey, R. D. (1987), River dynamics, flow regime and sediment transport, in *Sediment Transport in Gravel-Bed Rivers,* edited by C. R. Thorne, J. C. Bathurst, and R. D. Hey, pp. 17–40, John Wiley, Chichester, U. K.

Hey, R. D. (1988), Bar form resistance in gravel-bed rivers, *J. Hydraul. Eng., 114,* 1498–1508.

Hey, D. L., and N. S. Philippi (1995), Flood reduction through wetland restoration: The upper Mississippi River basin as a case history, *Restor. Ecol., 3,* 4–17.

Hey, R. D., and C. R. Thorne (1983), Accuracy of surface samples from gravel bed material, *J. Hydraul. Eng., 109,* 842–851.

Hey, R. D., C. R. Thorne, and J. C. Bathurst (1982), *Gravel-Bed Rivers,* John Wiley, Chichester, U. K.

Hicks, D. M., and B. Gomez (2003), Sediment transport, in *Tools in Fluvial Geomorphology,* edited by G. M. Kondolf and H. Piégay, pp. 425–461, John Wiley, Chichester, U. K.

Hicks, D. M., and P. D. Mason (1991), *Roughness Characteristics of New Zealand Rivers,* 336 pp., DSIR Water Resour. Surv., Wellington, New Zealand.

Hicks, D. M., J. Hill, and U. Shankar (1996), Variation of suspended sediment yields around New Zealand: The relative impotance of rainfall and geology, in *Erosion and Sediment Yield: Global and Regional Perspectives,* edited by D. E. Walling and B. W. Webb, *IAHS Publ., 236,* 149–156.

Hicks, D. M., M. J. Duncan, S. N. Lane, M. Tal, and R. Westaway (2008), Contemporary morphological change in braided gravel-bed rivers: New developments from field and laboratory studics, with particular reference to the influence of riparian vegetation, in *Gravel-Bed Rivers VI: From Process Understanding to River Restoration,* edited by H. Habersack, H. Piégay, and M. Rinaldi, pp. 557–586, Elsevier, Amsterdam.

Hicks, N. S., J. A. Smith, A. J. Miller, and P. A. Nelson (2005), Catastrophic flooding from an orographic thunderstorm in the central Appalachians, *Water Resour. Res., 41,* W12428, doi:10.1029/2005WR004129.

Higashino, M., J. J. Clark, and H. G. Stefan (2009), Pore water flow due to near-bed turbulence and associated solute transfer in a stream or lake sediment bed, *Water Resour. Res.*, *45*, W12414, doi:10.1029/2008WR007374.

Higashino, M., J. Clark, and H. Stefan (2010), Reply to comment by F. Boano et al. on "Pore water flow due to near-bed turbulence and associated solute transfer in a stream or lake sediment bed", *Water Resour. Res.*, *46*, W10802, doi:10.1029/2010WR009468.

Higgins, R. J., G. Pickup, and P. S. Cloke (1987), Estimating the transport and deposition of mining waste at Ok Tedi, in *Sediment Transport in Gravel-Bed Rivers*, edited by C. R. Thorne, J. C. Bathurst, and R. D. Hey, pp. 949–976, John Wiley, Chichester, U. K.

Hildebrandt, A., M. A. Al Aufi, M. Amerjeed, M. Shammas, and E. A. B. Eltahir (2007), Ecohydrology of a seasonal cloud forest in Dhofar: 1. Field experiment, *Water Resour. Res.*, *43*, W10411, doi:10.1029/2006WR005261.

Hildrew, A. G., and P. S. Giller (1994), Patchiness, species interactions and disturbance in the stream benthos, in *Aquatic Ecology: Scale, Pattern and Process*, edited by P. S. Giller, A. G. Hildrew, and D. G. Raffaelli, pp. 21–62, Blackwell Sci., Oxford, U. K.

Hill, B. R., E. H. Decarlo, C. C. Fuller, and M. F. Wong (1998), Using sediment 'fingerprints' to assess sediment-budget errors, North Halawa Valley, Oahu, Hawaii, 1991–92, *Earth Surf. Processes Landforms*, *23*, 493–508.

Hill, K. M., L. DellAngelo, and M. M. Meerschaert (2010), Heavy-tailed travel distance in gravel bed transport: An exploratory enquiry, *J. Geophys. Res.*, *115*, F00A14, doi:10.1029/2009JF001276.

Hillman, M., and G. J. Brierley (2008), Restoring uncertainty: Translating science into management practice, in *River Futures: An Integrative Scientific Approach to River repair*, edited by G. J. Brierley and K. A. Fryirs, pp. 257–272, Island Press, Washington, D. C.

Hillman, M., G. J. Brierley, and K. A. Fryirs (2008), Social and biophysical connectivity of river systems, in *River Futures: An Integrative Scientific Approach to River Repair*, edited by G. J. Brierley and K. A. Fryirs, pp. 125–145, Island Press, Washington, D. C.

Hilmes, M. M. (1993), Changes in channel morphology associated with placer mining along the Middle Fork of the South Platte River, Fairplay, Colorado, unpublished M.S. thesis, 261 pp., Colo. State Univ., Ft. Collins.

Hilmes, M. M., and E. E. Wohl (1995), Changes in channel morphology associated with placer mining, *Phys. Geogr.*, *16*, 223–242.

Hilton, R. G., A. Galy, and N. Hovius (2008a), Riverine particulate organic carbon from an active mountain belt: Importance of landslides, *Global Biogeochem. Cycles*, *22*, GB1017, doi:10.1029/2006GB002905.

Hilton, R. G., A. Galy, N. Hovius, M.-C. Chen, M.-J. Horng, and H. Chen (2008b), Tropical-cyclone-driven erosion of the terrestrial biosphere from mountains, *Nat. Geosci.*, *1*, 759–762.

Hilton, S., and T. E. Lisle (1993), *Measuring the fraction of pool volume filled with fine sediment*, USDA Forest Service Research Note PSW-RN-414, 11 pp.

Hirsch, R. M., J. F. Walker, J. C. Day, and R. Kallio (1990), The influence of man on hydrologic systems, in *Surface Water Hydrology*, edited by M. G. Wolman and H. C. Riggs, pp. 329–359, Geol. Soc. of Am., Boulder, Colo.

Hirschboeck, K. K. (1987), Hydroclimatically-defined mixed distributions in partial duration flood series, in *Hydrologic Frequency Modeling*, edited by V. P. Singh, pp. 195–205, D. Reidel, Boston, Mass.

Hirschboeck, K. K. (1988), Flood hydroclimatology, in *Flood Geomorphology*, edited by V. R. Baker, R. C. Kochel, and P. C. Patton, pp. 27–49, John Wiley, New York.

Hirzinger, V., H. Keckeis, H. L. Nemschkal, and F. Schiemer (2004), The importance of onshore areas for adult fish distribution along a free-flowing section of the Danube, Austria, *River Res. Appl.*, *20*, 137–149.

Hitt, N. P., and P. L. Angermeier (2008), River-stream connectivity affects fish bioassessment performance, *Environ. Manage.*, *42*, 132–150.

Hjulström, F. (1935), Studies of the morphological activity of rivers as illustrated by the River Fyris, *Bull. Geol. Inst. Univ. Uppsala*, *25*, 221–527.

Hobley, D. E. J., H. D. Sinclair, and P. A. Cowie (2010), Processes, rates, and time scales of fluvial response in an ancient postglacial landscape of the northwest Indian Himalaya, *Geol. Soc. Am. Bull.*, *122*, 1569–1584.

Hock, R. (2003), Temperature index melt modelling in mountain areas, *J. Hydrol.*, *282*, 104–115.

Hodel, H., T. P. Kersten, and I. Storchenegger (1998), A new approach for the estimation of extreme roughness in torrents by hydraulics and photogrammetry, in *Hydrology, Water Resources and Ecology in Headwaters*, edited by K. Kovar et al., *IAHS Publ.*, *248*, 225–230.

Hodge, R., J. Brasington, and K. Richards (2009), In situ characterization of grain-scale fluvial morphology using Terrestrial Laser Scanning, *Earth Surf. Processes Landforms*, *34*, 954–968.

Hodges, K. V., C. Wobus, K. Ruhl, T. Schildgen, and K. Whipple (2004), Quaternary deformation, river steepening, and heavy precipitation at the front of the Higher Himalayan ranges, *Earth Planet. Sci. Lett.*, *220*, 379–389.

Hodgkins, R. (1996), Seasonal trend in suspended-sediment transport from an Arctic glacier, and implications for drainage system structure, *Ann. Glaciol.*, *22*, 147–151.

Hodgkins, R. (1997), Glacier hydrology in Svalbard, Norwegian High Arctic, *Quat. Sci. Rev.*, *16*, 957–973.

Hodgkins, R., M. Tranter, and J. A. Dowdeswell (1997), Solute provenance, transport and denudation in a High Arctic glacierized catchment, *Hydrol. Processes*, *11*, 1813–1832.

Hodgkins, R., M. Tranter, and J. A. Dowdeswell (1998), The hydrochemistry of runoff from a 'cold-based' glacier in the High Arctic (Scott Turnerbreen, Svalbard), *Hydrol. Processes*, *12*, 87–103.

Hodgkins, R., R. Cooper, J. Wadham, and M. Tranter (2009), The hydrology of the proglacial zone of a high-Arctic glacier (Finsterwalderbreen, Svalbard): Atmospheric and surface water fluxes, *J. Hydrol.*, *378*, 150–160.

Hodson, A., A. M. Anesio, M. Tranter, A. Fountain, M. Osborn, J. Priscu, J. Laybourn-Parry, and B. Sattler (2008), Glacial ecosystems, *Ecol. Monogr.*, *78*, 41–67.

Hoehn, E. (1998), Solute exchange between river water and groundwater in headwater environments, in *Hydrology, Water Resources and Ecology in Headwaters*, edited by K. Kovar et al., *IAHS Publ.*, *248*, 165–171.

Hoey, T. (1992), Temporal variations in bedload transport rates and sediment storage in gravel-bed rivers, *Prog. Phys. Geogr.*, *16*, 319–338.

Hoey, T. B., and R. I. Ferguson (1997), Controls of strength and rate of downstream fining above a river base level, *Water Resour. Res.*, *33*, 2601–2608.

Hoey, T. B., and A. J. Sutherland (1991), Channel morphology and bedload pulses in braided rivers: A laboratory study, *Earth Surf. Processes Landforms*, *16*, 447–462.

Hoffman, D. F., and E. J. Gabet (2007), Effects of sediment pulses on channel morphology in a gravel-bed river, *Geol. Soc. Am. Bull.*, *119*, 116–125.

Hoffman, P. F., and J. P. Grotzinger (1993), Orographic precipitation, erosional unloading, and tectonic style, *Geology*, *21*, 195–198.

Hofmann, L., and R. E. Ries (1991), Relationship of soil and plant characteristics to erosion and runoff on pasture and range, *J. Soil Water Conserv.*, *46*, 143–147.

Hogan, D. L. (1986), Channel morphology of unlogged, logged and debris torrented streams in the Queen Charlotte Islands, *Land Management Rep. 49*, 94 pp, British Columbia Min. of For. and Lands.

Hogan, D. L., and M. Church (1989), Hydraulic geometry in small, coastal streams: Progress toward quantification of salmon habitat, *Can. J. Fish. Aquat. Sci.*, *46*, 844–852.

Hohberger, K., and G. Einsele (1979), Die Bedeutung des Lösungsabtrags verschiedener Gesteine für die Landschaftsentwicklung in Mitteleuropa, *Z. Geomorphol.*, *23*, 361–382.

Hohensinner, S., H. Habersack, M. Jungwirth, and G. Zauner (2004), Reconstruction of the characteristics of a natural alluvial river-floodplain system and hydromorphological changes following human
modifications: The Danube River (1812–1991), *River Res. Appl.*, *20*, 25–41.

Holland, H. D. (1978), *The Chemistry of the Atmosphere and Oceans*, 351 pp., Wiley, New York.

Holland, W. N., and G. Pickup (1976), Flume study of knickpoint development in stratified sediment, *Geol.
Soc. Am. Bull.*, *87*, 76–82.

Hollaus, M., W. Wagner, and K. Kraus (2005), Airborne laser scanning and usefulness for hydrological
models, *Adv. Geosci.*, *5*, 57–63.

Holliday, V., D. Higgitt, J. Warburton, and S. White (2003), Reconstructing upland sediment budgets in
ungauged catchments from reservoir sedimentation and rainfall records calibrated using short-term
streamflow monitoring, in *Erosion Prediction in Ungauged Basins: Integrating Methods and Techniques*,
edited by D. de Boer et al., *IAHS Publ.*, *279*, 59–67.

Holmes, R. M., S. G. Fisher, and N. B. Grimm (1994), Parafluvial nitrogen dynamics in a desert stream
ecosystem, *J. North Am. Benthol. Soc.*, *13*, 468–478.

Holtan, H. N., and M. H. Kirkpatrick (1950), Rainfall infiltration and hydraulics of flow in runoff compu-
tation, *Eos Trans. AGU*, *31*, 771–779.

Holzmann, H., H. P. Nachtnebel, and N. Sereinig (1998), Small-scale modelling of runoff components in an
Alpine environment, in *Hydrology, Water Resources and Ecology in Headwaters*, edited by K. Kovar
et al., *IAHS Publ.*, *248*, 231–238.

Hooke, J. M. (2003), Coarse sediment connectivity in river channel systems: A conceptual framework and
methodology, *Geomorphology*, *56*, 79–94.

Hooke, J. M. (2006), Human impacts on fluvial systems in the Mediterranean region, *Geomorphology*, *79*,
311–335.

Hooke, J. M., and J. M. Mant (2000), Geomorphological impacts of a flood event on ephemeral channels in
SE Spain, *Geomorphology*, *34*, 163–180.

Hooke, J. M., and C. E. Redmond (1989), Use of cartographic sources for analysing river channel change
with examples from Britain, in *Historical Change of Large Alluvial Rivers: Western Europe*, edited by
G. E. Petts, pp. 79–93, John Wiley, Chichester, U. K.

Hooke, J. M., and C. E. Redmond (1992), Causes and nature of river planform changes, in *Dynamics of
Gravel-Bed Rivers*, edited by P. Billi et al., pp. 557–571, John Wiley, Chichester, U. K.

Hooke, R. L. (1999), Spatial distribution of human geomorphic activity in the United States: Comparison
with rivers, *Earth Surf. Processes Landforms*, *24*, 687–692.

Hooke, R. L. (2000), On the history of humans as geomorphic agents, *Geology*, *28*, 843–846.

Hooke, R. L., and R. I. Dorn (1992), Segmentation of alluvial fans in Death Valley, California: New
insights from surface exposure dating and laboratory modelling, *Earth Surf. Processes Landforms*, *17*,
557–574.

Hooper, R. P. (2003), Diagnostic tools for mixing models of stream water chemistry, *Water Resour. Res.*,
39(3), 1055, doi:10.1029/2002WR001528.

Hopkins, P. S., K. W. Kratz, and S. D. Cooper (1989), Effects of an experimental acid pulse on invertebrates
in a high altitude Sierra Nevada stream, *Hydrobiologia*, *171*, 45–58.

Hooke, R. L. B. (1967), Processes on arid-region alluvial fans, *J. Geol.*, *75*, 438–460.

Hooke, R. L. B. (2000), Toward a uniform theory of clastic sediment yield in fluvial systems, *Geol. Soc. Am.
Bull.*, *112*, 1778–1786.

Hooke, R. L. B., and W. L. Rohrer (1979), Geometry of alluvial fans: Effect of discharge and sediment size,
Earth Surf. Processes, *4*, 147–166.

Hooper, R. P., B. T. Aulenbach, D. A. Burns, J. McDonnell, J. Freer, C. Kendall, and K. Beven
(1998), Riparian control of stream-water chemistry: Implications for hydrochemical basin models, in
Hydrology, Water Resources and Ecology in Headwaters, edited by K. Kovar et al., *IAHS Publ.*, *248*,
451–458.

Hopp, L., and J. J. McDonnell (2009), Connectivity at the hillslope scale: Identifying interactions between storm size, bedrock permeability, slope angle and soil depth, *J. Hydrol.*, *376*, 378–391.

Horecky, J., E. Stuchlik, P. Chvojka, D. W. Hardekopf, M. Mihaljevic, and J. Spacek (2006), Macroinvertebrate community and chemistry of the most atmospherically acidified streams in the Czech Republic, *Water Air Soil Pollut.*, *173*, 261–272.

Hornberger, G. M., P. F. Germann, and K. J. Beven (1991), Throughflow and solute transport in an isolated sloping block in a forested catchment, *J. Hydrol.*, *124*, 81–99.

Hornung, J., D. Pflanz, A. Hechler, A. Beer, M. Hinderer, M. Maisch, and U. Bieg (2010), 3-D architecture, depositional patterns and climate triggered sediment fluxes of an alpine alluvial fan (Samedan, Switzerland), *Geomorphology*, *114*, 202–214.

Horton, B. K. (1999), Erosional control on the geometry and kinematics of thrust belt development in the central Andes, *Tectonics*, *18*, 1292–1304.

Horton, R. E. (1932), Drainage basin characteristics, *Eos Trans. AGU*, *13*, 350–361.

Horton, R. E. (1945), Erosional development of streams and their drainage basins; hydrophysical approach to quantitative morphology, *Geol. Soc. Am. Bull.*, *56*, 275–370.

Hotta, N., N. Tanaka, S. Sawano, K. Kuraji, K. Shiraki, and M. Suzuki (2010), Changes in groundwater level dynamics after low-impact forest harvesting in steep, small watersheds, *J. Hydrol.*, *385*, 120–131.

Houghton, B. F., J. H. Latter, and W. R. Hackett (1987), Volcanic hazard assessment for Ruapehu composite volcano, Taupo volcanic zone, New Zealand, *Bull. Volcanol.*, *49*, 737–751.

Houjou, K., Y. Shimizu, and C. Ishii (1990), Calculation of boundary shear stress in open channel flow, *J. Hydrosci. Hydraul. Eng.*, *8*, 21–37.

House, P. K., and V. R. Baker (2001), Paleohydrology of flash floods in small desert watersheds in western Arizona, *Water Resour. Res.*, *37*, 1825–1839.

Hovius, N., C. P. Stark, M. A. Tutton, and L. D. Abbott (1998), Landslide-driven drainage network evolution in a pre-steady-state mountain belt: Finisterre Mountains, Papua New Guinea, *Geology*, *26*, 1071–1074.

Howard, A. D. (1967), Drainage analysis in geologic interpretation: A summation, *Am. Assoc. Pet. Geol. Bull.*, *51*, 2246–2259.

Howard, A. D. (1971), Simulation of stream networks by headward growth and branching, *Geogr. Anal.*, *3*, 29–50.

Howard, A. D. (1980), Thresholds in river regimes, in *Thresholds in Geomorphology*, edited by D. R. Coates and J. D. Vitek, pp. 227–258, George Allen and Unwin, London.

Howard, A. D. (1982), Equilibrium and time scales in geomorphology: Application to sand-bed alluvial streams, *Earth Surf. Processes Landforms*, *7*, 303–325.

Howard, A. D. (1987), Modelling fluvial systems: Rock-, gravel- and sand-bed channels, in *River Channels: Environment and Process*, edited by K. Richards, pp. 69–94, Blackwell, New York.

Howard, A. D. (1994), A detachment-limited model of drainage basin evolution, *Water Resour. Res.*, *30*, 2261–2285.

Howard, A. D. (1998), Long profile development of bedrock channels: Interaction of weathering, mass wasting, bed erosion, and sediment transport, in *Rivers Over Rock: Fluvial Processes in Bedrock Channels, Geophys. Monogr. Ser.*, vol. 107, edited by K. J. Tinkler and E. E. Wohl, pp. 297–319, AGU, Washington, D. C.

Howard, A. D., W. E. Dietrich, and M. A. Seidl (1994), Modelling fluvial erosion on regional to continental scales, *J. Geophys. Res.*, *99*(B7), 13,971–13,986.

Howarth, R. W., E. W. Boyer, W. J. Pabich, and J. N. Galloway (2002), Nitrogen use in the United States from 1961–2000 and potential future trends, *Ambio*, *31*, 88–96.

Howat, I. M., and S. Tulaczyk (2005), Climate sensitivity of spring snowpack in the Sierra Nevada, *J. Geophys. Res.*, *110*, F04021, doi:10.1029/2005JF000356.

Hoyle, J., G. Brierley, A. Brooks, and K. Fryirs (2008), Sediment organization along the upper Hunter River, Australia: A multivariate statistical approach, in *Gravel-Bed Rivers VI: From Process*

Understanding to River Restoration, edited by H. Habersack, H. Piégay, and M. Rinaldi, pp. 409–441, Elsevier, Amsterdam.

Hrachowitz, M., C. Soulsby, D. Tetzlaff, J. J. C. Dawson, and I. A. Malcolm (2009), Regionalization of transit time estimates in montane catchments by integrating landscape controls, *Water Resour. Res.*, *45*, W05421, doi:10.1029/2008WR007496.

Hren, M. T., C. P. Chamberlain, and F. J. Magilligan (2001), A combined flood surface and geochemical analysis of metal fluxes in a historically mined region: A case study from the New World Mining District, Montana, *Environ. Geol.*, *40*, 1334–1346.

Hsieh, M.-L., and P. L. K. Knuepfer (2001), Middle-late Holocene river terraces in the Erhjen River basin, southwestern Taiwan – implications of river response to climate change and active tectonic uplift, *Geomorphology*, *38*, 337–372.

Hsu, H.-L., B. J. Yanites, C.-C. Chen, and Y.-G. Chen (2010), Bedrock detection using 2D electrical resistivity imaging along the Peikang River, central Taiwan, *Geomorphology*, *114*, 406–414.

Hsu, Y.-S., and Y.-H. Hsu (2009), Impact of earthquake-induced dammed lakes on channel evolution and bed mobility: Case study of the Tsaoling landslide dammed lake, *J. Hydrol.*, *374*, 43–55.

Huang, H. Q., and G. C. Nanson (1997), Vegetation and channel variation; a case study of four small streams in southeastern Australia, *Geomorphology*, *18*, 237–249.

Huang, H. Q., and G. C. Nanson (2007), Why some alluvial rivers develop an anabranching pattern, *Water Resour. Res.*, *43*, W07441, doi:10.1029/2006WR005223.

Hubbell, D. W. (1987), Bed load sampling and analysis, in *Sediment Transport in Gravel-Bed Rivers*, edited by C. R. Thorne, J. C. Bathurst, and R. D. Hey, pp. 89–118, John Wiley, Chichester, U. K.

Hudson, J. A., R. C. Johnson, and J. R. Blackie (1990), Choice and calibration of streamflow structures for two mountain experimental basins, in *Hydrology in Mountainous Regions I. Hydrological Measurements: The Water Cycle*, edited by H. Lang and A. Musy, *IAHS Publ.*, *193*, 275–282.

Hudson, P. F. (2003), The influence of the El Niño Southern Oscillation on suspended sediment load variability in a seasonally humid tropical setting: Pánuco Basin, Mexico, *Geogr. Ann.*, *85A*, 263–275.

Huebl, J., and G. Fiebiger (2005), Debris-flow mitigation measures, in *Debris-flow Hazards and Related Phenomena*, edited by M. Jakob and O. Hungr, pp. 445–487, Springer, Berlin.

Huerlimann, M., D. Rickenmann, and C. Graf (2003), Field and monitoring data of debris-flow events in the Swiss Alps, *Can. Geotech. J.*, *40*, 161–175.

Huet, M. (1949), Appréciation de la valeur piscicole des eaux douces, *Trav. Stat. Rech. Faux Forêts, Groenendaal*, *D10*, 1–55.

Huet, M. (1954), Biologie, profils en long et en travers des eaux courantes, *Bull. Fr. Piscicult.*, *175*, 41–53.

Hugen, Z., and W. Jiaquan (1993), The effect of forests on flood control – some comments on the flood disaster in the Huaihe River basin of Anhui Province in 1991, *J. Environ. Hydrol.*, *1*, 38–43.

Huggel, C., A. Kaeaeb, W. Haeberli, and B. Krummenacher (2003), Regional-scale GIS-models for assessment of hazards from glacier lake outbursts: Evaluation and application in the Swiss Alps, *Nat. Hazards Earth Syst. Sci.*, *3*, 647–662.

Huggenberger, P., and C. Regli (2006), A sedimentological model to characterize braided river deposits for hydrogeological applications, in *Braided Rivers: Process, Deposits, Ecology and Management*, edited by G. H. Sambrook Smith et al., *Spec. Publ. Int. Assoc. Sedimentol.*, *36*, 51–74.

Hughes, F. M. R. (1994), Environmental change, disturbance, and regeneration in semi-arid floodplain forests, in *Environmental Change in Drylands: Biogeographical and Geomorphological Perspectives*, edited by A. C. Millington and K. Pye, pp. 321–345, John Wiley, New York.

Hughes, F. M. R., A. Colston, and J. O. Mountford (2005), Restoring riparian ecosystems: The challenge of accommodating variability and designing restoration trajectories, *Ecol. Soc.*, *10*(1), 12.

Hughes, G. M. (1976), Polluted fish respiratory physiology, in *Effects of Pollutants on Aquatic Organisms*, edited by A. P. M. Lockwood, pp. 163–183, Cambridge Univ. Press, Cambridge, U. K.

Hultberg, H., H. Apsimon, R. M. Church, P. Grenffelt, M. J. Mitchell, F. Moldan, and H. B. Ross (1994), Sulphur, in *Biogeochemistry of Small Catchments: A Tool for Environmental Research*, edited by B. Moldan and J. Cerny, pp. 229–254, John Wiley, Chichester, U. K.

Humphrey, N. F., and S. K. Konrad (2000), River incision or diversion in response to bedrock uplift, *Geology*, *28*, 43–46.

Hundey, E. J., and P. E. Ashmore (2009), Length scale of braided river morphology, *Water Resour. Res.*, *45*, W08409, doi:10.1029/2008WR007521.

Hungr, O. (2000), Analysis of debris flow surges using the theory of uniformly progressive flow, *Earth Surf. Processes Landforms*, *25*, 483–495.

Hung, O. (2005), Classification and terminology, in *Debris-Flow Hazards and Related Phenomena*, edited by M. Jakob and O. Hungr, pp. 9–23, Springer, Berlin.

Hungr, O., S. McDougall, and M. Bovis (2005), Entrainment of material by debris flows, in *Debris-Flow Hazards and Related Phenomena*, edited by M. Jakob and O. Hungr, pp. 135–158, Springer, Berlin.

Hungr, O., S. McDougall, M. Wise, and M. Cullen (2008), Magnitude-frequency relationships for debris flows and debris avalanches in relation to slope relief, *Geomorphology*, *96*, 355–365.

Hungr, O., G. C. Morgan, D. F. Van Dine, and D. R. Lister (1987), Debris flow defenses in British Columbia, in *Debris Flows/Avalanches: Process, Recognition, and Mitigation*, edited by J. E. Costa and G. F. Wieczorek, *Rev. Eng. Geol.*, *7*, 201–236.

Hunt, A. G., and A. N. Papanicolaou (2003), Tests of predicted downstream transport of clasts in turbulent flow, *Adv. Water Resour.*, *26*, 1205–1211.

Hunziker, R. P., and M. N. R. Jaeggi (2002), Grain sorting processes, *J. Hydraul. Eng.*, *128*, 1060–1068.

Hupp, C. R. (1988), Plant ecological aspects of flood geomorphology and paleoflood history, in *Flood Geomorphology*, edited by V. R. Baker, R. C. Kochel, and P. C. Patton, pp. 335–356, John Wiley, New York.

Hupp, C. R. (1990), Vegetation patterns in relation to basin hydrogeomorphology, in *Vegetation and Erosion*, edited by J. B. Thornes, pp. 217–237, John Wiley, Chichester, U. K.

Hupp, C. R., and G. Bornette (2003), Vegetation as a tool in the interpretation of fluvial geomorphic processes and landforms in humid temperate areas, in *Tools in Fluvial Geomorphology*, edited by G. M. Kondolf and H. Piégay, pp. 269–288, John Wiley, Chichester, U. K.

Hupp, C. R., and W. R. Osterkamp (1996), Riparian vegetation and fluvial geomorphic processes, *Geomorphology*, *14*, 277–295.

Hupp, C. R., and M. Rinaldi (2007), Riparian vegetation patterns in relation to fluvial landforms and channel evolution along selected rivers of Tuscany (central Italy), *Ann. Assoc. Am. Geogr.*, *97*, 12–30.

Hupp, C. R., and A. Simon (1991), Bank accretion and the development of vegetated depositional surfaces along modified alluvial channels, *Geomorphology*, *4*, 111–124.

Huss, M., A. Bauder, M. Werder, M. Funk, and R. Hock (2007), Glacier-dammed lake outburst events of Gornersee, Switzerland, *J. Glaciol.*, *53*, 189–200.

Huss, M., D. Farinotti, A. Bauder, and M. Funk (2008), Modelling runoff from highly glacierized alpine drainage basins in a changing climate, *Hydrol. Processes*, *22*, 3888–3902.

Hutton, J. (1795), *The Theory of the Earth*, vols. 1 and 2, William Creech, Edinburgh.

Hyatt, J. A., and P. M. Jacobs (1996), Distribution and morphology of sinkholes triggered by flooding following Tropical Storm Alberto at Albany, Georgia, USA, *Geomorphology*, *17*, 305–316.

Hyatt, T. L., and R. J. Naiman (2001), The residence time of large woody debris in the Queets River, Washington, USA, *Ecol. Appl.*, *11*, 191–202.

Hydrologic Engineering Center (1977), *HEC-6, Scour and Deposition in Rivers and Reservoirs, User's Manual*, U.S. Army Corps of Eng., Davis, Calif.

Hydrologic Engineering Center (1997), *HEC-RAS River Analysis System v. 2.0 User's Manual*, U.S. Army Corps of Eng., Davis, Calif.

Hygelund, B., and M. Manga (2003), Field measurements of drag coefficients for model large woody debris, *Geomorphology*, *51*, 175–185.

Hylander, K., B. G. Jonsson, and C. Nilsson (2002), Evaluating buffer strips along boreal streams using bryophytes as indicators, *Ecol. Appl.*, *12*, 797–806.

Hynes, N. (1970), *The Ecology of Running Waters*, 555 pp., Univ. of Toronto Press, Toronto, Ont., Canada.

Hynes, N. (1975), The stream and its valley, *Int. Assoc. Theor. Appl. Limnol.*, *19*, 1–15.

Ibbitt, R. P. (1997), Evaluation of optimal channel network and river basin heterogeneity concepts using measured flow and channel properties, *J. Hydrol.*, *96*, 119–138.

Ichim, I., and M. Radoane (1987), On the high erosion rate in the Vrancea region, Romania, in *International Geomorphology 1986, Part I*, edited by V. Gardiner, pp. 783–790, John Wiley, Chichester, U. K.

Iida, T., and K. Okunishi (1983), Development of hillslopes due to landslides, *Z. Geomorphol.*, *46*, 67–77.

Ijjasz-Vasquez, E. J., and R. L. Bras (1995), Scaling regimes of local slope versus contributing area in digital elevation models, *Geomorphology*, *12*, 299–311.

Ikeda, H. (1984), Flume experiments on the causes of superior mobility of sediment mixtures, *Annual Rep. 1053–56*, Inst. of Geosci., Univ. of Tsukuba, Japan.

Ikeda, H., and F. Iseya (1986), Thresholds in the mobility of sediment mixtures, in *International Geomorphology 1986, Part I*, edited by V. Gardiner, pp. 561–570, John Wiley, Chichester, U. K.

Ikeda, H., and F. Iseya (1988), *Experimental Study of Heterogeneous Sediment Transport, Environmental Research Center Paper*, vol. 12, 50 pp., Univ. of Tsukuba, Japan.

Ikeda, H., F. Iseya, and Y. Kodama (1993), Sedimentation on an alluvial cone in the Upper Oi River, central Japan, *Bull. Tsukuba Univ. For.*, *9*, 149–173.

Ilies, J. (1961), Versuch einer allegemein biozonitischen Gliederung der Fliessgewasser, *Verhandlungen der Internationalen Vereinigung für Theoretische und Angewandte Limnologie*, *13*, 834–844.

Ilies, J., and L. Botosaneanu (1963), Problèmes et méthodes de la classification et de la zonation écologique des eaux courantes considerées surtout du point de vue faunistique, *Mitteilungen der Internationalen Vereinigung für Theoretische und Angewandte Limnologie*, *12*, 1–57.

Imaizumi, F., and R. C. Sidle (2007), Linkage of sediment supply and transport processes in Miyagawa Dam catchment, Japan, *J. Geophys. Res.*, *112*, F03012, doi:10.1029/2006JF000495.

Imaizumi, F., R. C. Sidle, and R. Kamei (2008), Effects of forest harvesting on the occurrence of landslides and debris flows in steep terrain of central Japan, *Earth Surf. Processes Landforms*, *33*, 827–840.

Inbar, M. (1987), Effects of a high magnitude flood in a Mediterranean climate: A case study in the Jordan River basin, in *Catastrophic Flooding*, edited by L. Mayer and D. Nash, pp. 333–353, Allen and Unwin, Boston, Mass.

Inbar, M., and C. A. Llerena (2000), Erosion processes in high mountain agricultural terraces in Peru, *Mt. Res. Dev.*, *20*, 72–79.

Inbar, M., and A. P. Schick (1979), Bedload transport associated with high stream power, Jordan River, Israel, *Proc. Natl. Acad. Sci. U. S. A.*, *76*, 2515–2517.

Inbar, M., M. Tamir, and L. Wittenberg (1998), Runoff and erosion processes after a forest fire in Mount Carmel, a Mediterranean area, *Geomorphology*, *24*, 17–33.

Inman, D. L., and S. A. Jenkins (1999), Climate change and the episodicity of sediment flux of small California rivers, *J. Geol.*, *107*, 251–270.

Innes, J. L. (1983), Debris flows, *Prog. Phys. Geogr.*, *7*, 469–501.

Innis, S. A., R. J. Naiman, and S. R. Elliott (2000), Indicators and assessment methods for measuring the ecological integrity of semi-aquatic terrestrial environments, *Hydrobiologia*, *422/423*, 111–131.

Inyan, B. J., and M. W. Williams (2001), Protection of headwater catchments from future degradation; San Miguel River basin, Colorado, *Mt. Res. Dev.*, *21*, 54–60.

Iovine, G., D. D'Ambrosio, and S. Di Gregorio (2005), Applying genetic algorithms for calibrating a hexagonal cellular automata model for the simulation of debris flows characterised by strong inertial effects, *Geomorphology*, *66*, 287–303.

IPCC (Intergovernmental Panel on Climate Change) (2008), Climate Change 2007, *IPCC Fourth Assessment Report*, Cambridge Univ. Press, Cambridge, U. K.

Iroumé, A., A. Andreoli, F. Comiti, H. Ulloa, and A. Huber (2010), Large wood abundance, distribution and mobilization in a third order coastal mountain range river system, southern Chile, *For. Ecol. Manage.*, *260*, 480–490.

Iseya, F., and H. Ikeda (1987), Pulsations in bedload transport rates induced by a longitudinal sediment sorting: A flume study using sand and gravel mixtures, *Geogr. Ann.*, *69A*, 15–27.

Iseya, F., H. Ikeda, H. Maita, and Y. Kodama (1990), Fluvial deposits in a torrential gravel-bed stream by extreme sediment supply: Sedimentary structure and depositional mechanism. Third International Workshop on Gravel-Bed Rivers, Firenze, Italy.

Ismail, W. R., and Z. A. Rahaman (1994), The impact of quarrying activity on suspended sediment concentration and sediment load of Sungai Relau, Pulau Pinang, Malaysia, *Malays. J. Trop. Geogr.*, *25*, 45–57.

Iso, N., K. Yamakawa, H. Yonezawa, and T. Matsubara (1980), Accumulation rates of alluvial cones, constructed by debris-flow deposits, in the drainage basins of the Takahara River, Gifu prefecture, central Japan, *Geogr. Rev. Jpn.*, *53*, 699–720.

Istanbulluoglu, E., D. G. Tarboton, R. T. Pack, and C. Luce (2002), A probabilistic approach for channel initiation, *Water Resour. Res.*, *38*(12), 1325, doi:10.1029/2001WR000782.

Istanbulluoglu, E., R. L. Bras, H. Flores-Cervantes, and G. E. Tucker (2005), Implications of bank failures and fluvial erosion for gully development: Field observations and modeling, *J. Geophys. Res.*, *110*, F01014, doi:10.1029/2004JF000145.

Itakura, Y., N. Fujii, and T. Sawada (2000), Basic characteristics of ground vibration sensors for the detection of debris flow, *Phys. Chem. Earth, Part B*, *25*, 717–720.

Iverson, R. J. (2003), How should mathematical models of geomorphic processes be judged?, in *Prediction in Geomorphology*, edited by P. R. Wilcock and R. M. Iverson, pp. 83–94, AGU, Washington, D. C.

Iverson, R. J. (2005), Debris-flow mechanics, in *Debris-Flow Hazards and Related Phenomena*, edited by M. Jakob and O. Hungr, pp. 105–134, Springer, Berlin.

Iverson, R. M., and R. G. Lahusen (1992), Momentum transport in debris flows: Large-scale experiments, *Eos Trans. AGU*, *73*, 227.

Iverson, R. M., and M. E. Reid (1992), Gravity-driven groundwater flow and slope failure potential. 1. Elastic effective-stress model. And 2. Effects of slope morphology, material properties, and hydraulic heterogeneity, *Water Resour. Res.*, *28*, 925–950.

Ives, J. D. (1980), Introduction: A description of the Front Range, in *Geoecology of the Colorado Front Range: A Study of Alpine and Subalpine Environments*, edited by J. D. Ives, pp. 1–7, Westview Press, Boulder, Colo.

Ives, J. D. (1985), Mountain environments, *Prog. Phys. Geogr.*, *9*, 425–433.

Ives, J. D. (1986), *Glacial lake outburst floods and risk engineering in the Himalaya*, International Centre for Integrated Mountain Development (ICIMOD) Occasional Pap., vol. 5, 42 pp, Kathmandu, Nepal.

Ives, J. D. (2006), *Himalayan Perceptions: Environmental Change and the Well-Being of Mountain Peoples*, 2nd ed., Himalayan Association for the Advancement of Science, Lalitpur, Nepal.

Ives, J. D., and R. G. Barry (Eds.) (1974), *Arctic and Alpine Environments*, 999 pp., Methuen, London.

Ives, J. D., and B. Messerli (1989), *The Himalayan Dilemma: Reconciling Development and Conservation*, 295 pp., Routledge, London.

Ives, J. D., B. Messerli, and E. Spiess (1997), Mountains of the world - a global priority, in *Mountains of the World: A Global Priority*, edited by B. Messerli and J. D. Ives, pp. 1–15, The Parthenon, London.

Ives, R. L. (1942), The beaver-meadow complex, *J. Geomorphol.*, *5*, 191–203.

Jackman, A. P., R. A. Walters, and V. C. Kennedy (1984), Transport and concentration controls for chloride, strontium, potassium, and lead in Uvas Creek, a small cobble-bed stream in Santa Clara County, California, USA. 2. Mathematical modeling, *J. Hydrol.*, *75*, 111–141.

Jackson, L. E., Jr., R. A. Kostaschuk, and G. M. MacDonald (1987), Identification of debris flow hazard on alluvial fans in the Canadian Rocky Mountains, in *Debris Flows/Avalanches: Process, Recognition, and Mitigation*, edited by J. E. Costa and G. F. Wieczorek, *Rev. Eng. Geol.*, *7*, 115–124.

Jackson, W. L., and R. L. Beschta (1982), A model of two-phase bedload transport in an Oregon Coast Range stream, *Earth Surf. Processes Landforms*, *7*, 517–527.

Jackson, W. L., and B. P. Van Haveren (1984), Design for a stable channel in coarse alluvium for riparian zone restoration, *Water Resour. Bull.*, *20*, 695–703.

Jacobson, R. B., and K. B. Gran (1999), Gravel sediment routing from widespread, low-intensity landscape disturbance, Current River basin, Missouri, *Earth Surf. Processes Landforms*, *24*, 897–917.

Jacobson, R. B., J. P. McGeehin, E. D. Cron, C. E. Carr, J. M. Harper, and A. D. Howard (1993), Landslides triggered by the storm of November 3–5, 1985, Wills Mountain anticline, West Virginia and Virginia, *U. S. Geol. Surv. Bull. 1981, Part C*, C1–C33.

Jacobson, R. B., J. E. O'Connor, and T. Oguchi (2003), Surficial geologic tools in fluvial geomorphology, in *Tools in Fluvial Geomorphology*, edited by G. M. Kondolf and H. Piégay, pp. 25–57, John Wiley, Chichester, U. K.

Jaeger, K. L., D. R. Montgomery, and S. M. Bolton (2007), Channel and perennial flow initiation in headwater streams: Management implications of variability in source-area size, *Environ. Manage.*, *40*, 775–786.

Jaeggi, M. N. R. (1987), Interaction of bed load transport with bars, in *Sediment Transport in Gravel-Bed Rivers*, edited by C. R. Thorne, J. C. Bathurst, and R. D. Hey, pp. 829–841, John Wiley, Chichester, U. K.

Jaeggi, M. N. R., and B. Zarn (1999), Stream channel restoration and erosion control for incised channels in alpine environments, in *Incised River Channels: Processes, Forms, Engineering and Management*, edited by S. E. Darby and A. Simon, pp. 343–369, Wiley, Chichester, U. K.

Jaffé, R., D. McKnight, N. Maie, R. Cory, W. H. McDowell, and J. L. Campbell (2008), Spatial and temporal variations in DOM composition in ecosystems: The importance of long-term monitoring of optical properties, *J. Geophys. Res.*, *113*, G04032, doi:10.1029/2008JG000683.

Jain, R. K., and U. C. Kothyari (2009), Cohesion influences on erosion and bed load transport, *Water Resour. Res.*, *45*, W06410, doi:10.1029/2008WR007044.

Jain, V., N. Preston, K. Fryirs, and G. Brierley (2006), Comparative assessment of three approaches for deriving stream power plots along long profiles in the upper Hunter River catchment, New South Wales, Australia, *Geomorphology*, *74*, 297–317.

Jakab, S. (2007), Chrono-toposequences of soils on the river terraces in Transylvania (Romania), *Catena*, *71*, 406–410.

Jakob, M. (2005), Debris-flow hazard analysis, in *Debris-Flow Hazards and Related Phenomena*, edited by M. Jakob and O. Hungr, pp. 411–443, Springer, Berlin.

Jakob, M., and P. Friele (2010), Frequency and magnitude of debris flows on Cheekye River, British Columbia, *Geomorphology*, *114*, 382–395.

Jakob, M., and O. Hungr (Eds.) (2005), *Debris-Flow Hazards and Related Phenomena*, 739 pp., Springer, Berlin.

Jakob, M., and P. Jordan (2001), Design flood estimates in mountain streams; the need for a geomorphic approach, *Can. J. Civ. Eng.*, *28*, 425–439.

Jakob, M., and S. Lambert (2009), Climate change effects on landslides along the southwest coast of British Columbia, *Geomorphology*, *107*, 275–284.

James, L. A. (1991), Incision and morphologic evolution of an alluvial channel recovering from hydraulic mining sediment, *Geol. Soc. Am. Bull.*, *103*, 723–736.

James, L. A. (1993), Sustained reworking of hydraulic mining sediment in California: G. K. Gilbert's sediment wave model reconsidered, *Z. Geomorphol.*, *88*, 49–66.

James, L. A. (1994), Channel changes wrought by gold mining: Northern Sierra Nevada, California, in *Effects of Human-Induced Changes on Hydrologic Systems*, edited by R. Marston and V. R. Hasfurther, pp. 629–638, Am. Water Resour. Assoc., Bethesda, Md.

James, L. A. (1997), Channel incision on the lower American River, California, from streamflow gage records, *Water Resour. Res.*, *33*, 485–490.

James, L. A. (2006), Bed waves at the basin scale; implications for river management and restoration, *Earth Surf. Processes Landforms*, *31*, 1692–1706.

James, L. A., and W. A. Marcus (2006), The human role in changing fluvial systems: Retrospect, inventory and prospect, *Geomorphology*, *79*, 152–171.

James, L. A., M. B. Singer, S. Ghoshal, and M. Megison (2009), Historical channel changes in the lower Yuba and Feather Rivers, California: Long-term effects of contrasting river-management strategies, in *Management and Restoration of Fluvial Systems with Broad Historical Changes and Human Impacts*, edited by L. A. James, S. L. Rathburn, and G. R. Whittecar, pp. 57–81, Geol. Soc. of Am., Boulder, Colo.

Janda, R. J., K. M. Scott, K. M. Nolan, and H. A. Martinson (1981), Lahar movement, effects, and deposits, in *The 1980 Eruptions of Mount St. Helens, Washington*, edited by P. W. Lipman and D. R. Mullineaux, *U. S. Geol. Surv. Prof. Pap.*, *1250*, 461–478.

Jansen, J. D. (2006), Flood magnitude-frequency and lithologic control on bedrock river incision in post-orogenic terrain, *Geomorphology*, *82*, 39–57.

Jansen, J. D., and G. J. Brierley (2004), Pool-fills: A window to palaeoflood history and response in bedrock-confined rivers, *Sedimentology*, *51*, 901–925.

Jansen, J. D., and G. C. Nanson (2004), Anabranching and maximum flow efficiency in Magela Creek, northern Australia, *Water Resour. Res.*, *40*, W04503, doi:10.1029/2003WR002408.

Jansson, R., C. Nilsson, M. Dynesius, and E. Andersson (2000a), Effects of river regulation on river-margin vegetation: A comparison of eight boreal rivers, *Ecol. Appl.*, *10*, 203–224.

Jansson, R., C. Nilsson, and B. Renöfält (2000b), Fragmentation of riparian floras in rivers with multiple dams, *Ecology*, *81*, 899–903.

Japanese Ministry of Construction (1993), *Sabo*, 36 pp., Min. of Construct., Kobe, Japan.

Jarrett, R. D. (1984), Hydraulics of high-gradient streams, *J. Hydraul. Eng*, *110*, 1519–1539.

Jarrett, R. D. (1985), *Determination of roughness coefficients for streams in Colorado*, U.S. Geol. Surv. Water Resour. Invest. Rep. 85-4004, 54 pp.

Jarrett, R. D. (1987), Errors in slope-area computations of peak discharges in mountain streams, *J. Hydrol.*, *96*, 53–67.

Jarrett, R. D. (1989), Hydrology and paleohydrology used to improve the understanding of flood hydrometeorology in Colorado, in *Design of Hydraulic Structures 89: Proceedings of the Second International Symposium on Design of Hydraulic Structures, Fort Collins, CO, June 26–29, 1989*, edited by M. L. Albertson and R. A. Kia, pp. 9–16, A. A. Balkema, Rotterdam Netherlands.

Jarrett, R. D. (1990a), Hydrologic and hydraulic research in mountain rivers, *Water Resour. Bull.*, *26*, 419–429.

Jarrett, R. D. (1990b), Paleohydrologic techniques used to define the spatial occurrence of floods, *Geomorphology*, *3*, 181–195.

Jarrett, R. D. (1991), Wading measurements of vertical velocity profiles, *Geomorphology*, *4*, 243–247.

Jarrett, R. D. (1992), Hydraulics of mountain rivers, in *Channel Flow Resistance: Centennial of Manning's Formula*, edited by B. C. Yen, pp. 287–298, Water Resour. Publ., Littleton, Colo.

Jarrett, R. D. (1993), Flood elevation limits in the Rocky Mountains, in *Engineering Hydrology*, edited by C. Y. Kuo, pp. 180–185, ASCE Hydraulics Division.

Jarrett, R. D. (1994), Historic-flood evaluation and research needs in mountainous areas, in *Proceedings, ASCE Hydraulic Engineering '94 Conference, August 1–5, Buffalo, New York*, edited by G. V. Cotroneo and R. R. Rumer, pp. 875–879, Am. Soc. of Civ. Eng., New York.

Jarrett, R. D., and J. E. Costa (1986), Hydrology, geomorphology, and dam-break modeling of the July 15, 1982, Lawn Lake Dam and Cascade Dam failures, Larimer County, Colorado, *U.S. Geol. Surv. Prof. Pap.*, *1369*.

Jarrett, R. D., and J. E. Costa (1988), *Evaluation of the flood hydrology in the Colorado Front Range using precipitation, streamflow, and paleoflood data*, U.S. Geol. Surv. Water Resour. Invest. Rep. 87-4117, 37 pp.

Jarrett, R. D., and J. F. England (2002), Reliablity of paleostage indicators for paleoflood studies, in *Ancient Floods, Modern Hazards: Principles and Applications of Paleoflood Hydrology*, edited by P. K. House et al., pp. 91–109, AGU, Washington, D. C.

Jarrett, R. D., J. P. Capesius, D. Jarrett, and J. F. England, Jr. (1996), 1995: Where the past (paleoflood hydrology) meets the present, understanding maximum flooding, *Geological Society of America Annual Meeting, Abstracts with Programs*, p. A-110.

Jasper, K., J. Gurtz, and H. Lang (2002), Advanced flood forecasting in alpine watersheds by coupling meteorological observations and forecasts with a distributed hydrological model, *J. Hydrol., 267*, 40–52.

Jefferson, A., G. E. Grant, S. L. Lewis, and S. T. Lancaster (2010), Coevolution of hydrology and topography on a basalt landscape in the Oregon Cascade Range, USA, *Earth Surf. Processes Landforms, 35*, 803–816.

Jefferson, A., G. Grant, and T. Rose (2006), Influence of volcanic history on groundwater patterns on the west slope of the Oregon High Cascades, *Water Resour. Res., 42*, W12411, doi:10.1029/2005WR004812.

Jeffries, R., S. E. Darby, and D. A. Sear (2003), The influence of vegetation and organic debris on flood-plain sediment dynamics: Case study of a low-order stream in the New Forest, England, *Geomorphology, 51*, 61–80.

Jelks, H. L., et al. (2008), Conservation status of imperiled North American freshwater and diadromous fishes, *Fisheries, 33*, 372–407.

Jeník, J. (1997), The diversity of mountain life, in *Mountains of the World: A Global Priority*, edited by B. Messerli and J. D. Ives, pp. 199–231, The Parthenon, London.

Jenkins, A., N. E. Peters, and A. Rodhe (1994), Hydrology, in *Biogeochemistry of Small Catchments: A Tool for Environmental Research*, edited by B. Moldan and J. Cerny, pp. 31–54, John Wiley, Chichester, U. K.

Jenkins, K. M., and A. J. Boulton (2003), Connectivity in a dryland river: Short-term aquatic microinvertebrate recruitment following floodplain inundation, *Ecology, 84*, 2708–2723.

Jennings, K. L., P. R. Bierman, and J. Southon (2003), Timing and style of deposition on humid-temperate fans, Vermont, United States, *Geol. Soc. Am. Bull., 115*, 182–199.

Jensen, M. E., I. A. Goodman, P. S. Bourgeron, N. L. Poff, and C. K. Brewer (2001), Effectiveness of biophysical criteria in the hierarchical classification of drainage basins, *J. Am. Water Resour. Assoc., 37*, 1155–1167.

Jerolmack, D. J., and D. Mohrig (2007), Conditions for branching in depositional rivers, *Geology, 35*, 463–466.

Jimenez Sanchez, M. (2002), Slope deposits in the upper Nalon River basin (NW Spain): An approach to a quantitative comparison, *Geomorphology, 43*, 165–178.

Jin, H.-J., Q.-H. Yu, S.-L. Wang, and L.-Z. Lu (2008), Changes in permafrost environments along the Qinghai-Tibet engineering corridor induced by anthropogenic activities and climate warming, *Cold Reg. Sci. Technol., 53*, 317–333.

Jiongxin, X. (1996), Underlying gravel layers in a large sand bed river and their influence on downstream-dam channel adjustment, *Geomorphology, 17*, 351–359.

Johnejack, K. R., and W. F. Megahan (1991), Sediment transport in headwater channels in Idaho, in *Proceedings of the Fifth Federal Interagency Sedimentation Conference*, edited by S.-S. Fan and Y.-H. Kuo, pp. 4-155–4-161, Fed. Energy Regul. Comm., Washington, D. C.

Johnson, A. M., and J. R. Rodine (1984), Debris flow, in *Slope Instability*, edited by D. Brunsden and D. B. Prior, pp. 257–361, John Wiley, New York.

Johnson, B. L., W. B. Richardson, and T. J. Naimo (1995), Past, present, and future concepts in large river ecology, *BioScience, 45*, 134–141.

Johnson, C. E., M. I. Litaor, M. F. Billett, and O. P. Bricker (1994), Chemical weathering in small catchments: Climatic and anthropogenic influences, in *Biogeochemistry of Small Catchments: A Tool for Environmental Research*, edited by B. Moldan and J. Cerny, pp. 323–341, John Wiley, Chichester, U. K.

Johnson, J. P. L., and K. X. Whipple (2007), Feedbacks between erosion and sediment transport in experimental bedrock channels, *Earth Surf. Processes Landforms, 32*, 1048–1062.

Johnson, J. P. L., and K. X. Whipple (2010), Evaluating the controls of shear stress, sediment supply, alluvial cover, and channel morphology on experimental bedrock incision rate, *J. Geophys. Res.*, *115*, F02018, doi:10.1029/2009JF001335.

Johnson, J. P. L., K. X. Whipple, L. S. Sklar, and T. C. Hanks (2009), Transport slopes, sediment cover, and bedrock channel incision in the Henry Mountains, Utah, *J. Geophys. Res.*, *114*, F02014, doi:10.1029/2007JF000862.

Johnson, J. P. L., K. X. Whipple, and L. S. Sklar (2010), Contrasting bedrock incision rates from snowmelt and flash floods in the Henry Mountains, Utah, USA, *Geol. Soc. Am. Bull.*, *122*, 1600–1615.

Johnson, K. A., and N. Sitar (1990), Hydrologic conditions leading to debris-flow initiation, *Can. Geotech. J.*, *27*, 789–801.

Johnson, M. G., and R. L. Beschta (1980), Logging, infiltration and surface erodibility in western Oregon, *J. For.*, *78*, 334–337.

Johnson, R. H. (2007), Ground water flow modelling with sensitivity analyses to guide field data collection in a mountain watershed, *Ground Water Monit. Rem.*, *27*, 75–83.

Johnson, R. M., and J. Warburton (2002a), Annual sediment budget of a UK mountain torrent, *Geogr. Ann., Ser. A, Phys. Geogr.*, *84*, 73–88.

Johnson, R. M., and J. Warburton (2002b), Flooding and geomorphic impacts in a mountain torrent; Raise Beck, central Lake District, England, *Earth Surf. Processes Landforms*, *27*, 945–969.

Johnson, R. M., and J. Warburton (2006), Variability in sediment supply, transfer and deposition in an upland torrent system; Iron Crag, northern England, *Earth Surf. Processes Landforms*, *31*, 844–861.

Johnson, R. M., J. Warburton, and A. J. Mills (2008), Hillslope-channel sediment transfer in a slope failure event; Wet Swine Gill, Lake District, northern England, *Earth Surf. Processes Landforms*, *33*, 394–413.

Johnson, R. R. (1978), The lower Colorado River: A western system, in *Strategies for Protection and Management of Floodplain Wetlands and Other Riparian Ecosystems*, USDA Forest Service General Technical Report WO-12, pp. 41–55.

Johnson, S. L., A. P. Covich, T. A. Crowl, A. Estrada-Pinto, J. Bithorn, and W. A. Wurstbaugh (1998), Do seasonality and disturbance influence reproduction in freshwater atyid shrimp in headwater streams, Puerto Rico?, *Verh. Int. Verein. Limnol.*, *26*, 2076–2081.

Johnson, W. C. (1994), Woodland expansion in the Platte River, Nebraska: Patterns and causes, *Ecol. Monogr.*, *64*, 45–84.

Johnston, C. E., E. D. Andrews, and J. Pitlick (1998), In situ determination of particle friction angles of fluvial gravels, *Water Resour. Res.*, *34*, 2017–2030.

Jolly, I. D. (1996), The effects of river management on the hydrology and hydroecology of arid and semi-arid floodplains, in *Floodplain Processes*, edited by M. G. Anderson, D. E. Walling, and P. D. Bates, pp. 577–609, John Wiley, Chichester, U. K.

Jomard, H., T. Lebourg, Y. Guglielmi, and E. Tric (2010), Electrical imaging of sliding geometry and fluids associated with a deep seated landslide (La Clapière, France), *Earth Surf. Processes Landforms*, *35*, 588–599.

Jomelli, V., V. P. Pech, C. Chochillon, and D. Brunstein (2004), Geomorphic variations of debris flows and recent climatic change in the French Alps, *Clim. Change*, *64*, 77–102.

Jonas, T., C. Marty, and J. Magnusson (2009), Estimating the snow water equivalent from snow depth measurements in the Swiss Alps, *J. Hydrol.*, *378*, 161–167.

Jones, A. (1971), Soil piping and stream channel initiation, *Water Resour. Res.*, *7*, 602–610.

Jones, A. F., P. A. Brewer, E. Johnstone, and M. G. Macklin (2007), High-resolution interpretative geomorphological mapping of river valley environments using airborne LiDAR data, *Earth Surf. Processes Landforms*, *32*, 1574–1592.

Jones, A. P. (2000), Late Quaternary sediment sources, storages and transfers within mountain basins using clast lithological analysis; Pineta Basin, central Pyrenees, Spain, *Geomorphology*, *34*, 145–161.

Jones, C. G., J. H. Lawton, and M. Shackack (1994), Organisms as ecosystem engineers, *Oikos*, *69*, 373–386.

Jones, J. A., F. J. Swanson, B. C. Wemple, and K. U. Snyder (2000), Effects of roads on hydrology, geomorphology, and disturbance patches in stream networks, *Conserv. Biol.*, *14*, 76–86.

Jones, J. A. A. (1981), *The Nature of Soil Piping: A Review of Research*, Geobooks, Norwich.

Jones, J. B., Jr., S. G. Fisher, and N. B. Grimm (1995a), Nitrification in the hyporheic zone of a desert stream ecosystem, *J. North Am. Benthol. Soc.*, *14*, 249–258.

Jones, J. B., Jr., S. G. Fisher, and N. B. Grimm (1995b), Vertical hydrologic exchange and ecosystem metabolism in a Sonoran desert stream, *Ecology*, *76*, 942–952.

Jones, P. D., K. R. Briffa, and J. R. Pilcher (1984), Riverflow reconstruction from tree rings in southern Britain, *J. Climatol.*, *4*, 461–472.

Jones, N. E., G. J. Scrimgeour, and W. M. Tonn (2008), Assessing the effectiveness of a constructed Arctic stream using multiple biological attributes, *Environ. Manage.*, *42*, 1064–1076.

Jones, T. A., and L. D. Daniels (2008), Dynamics of large woody debris in small streams disturbed by the 2001 Dogrib fire in the Alberta foothills, *For. Ecol. Manage.*, *256*, 1751–1759.

Jorde, K., M. Burke, N. Scheidt, C. Welcker, S. King, and C. Borden (2008), Reservoir operations, physical processes, and ecosystem losses, in *Gravel-Bed Rivers VI: From Process Understanding to River Restoration*, edited by H. Habersack, H. Piégay, and M. Rinaldi, pp. 607–636, Elsevier, Amsterdam.

Jowett, I. G. (1997), Instream flow methods: A comparison of approaches, *Reg. Rivers Res. Manage.*, *13*, 115–127.

Jubb, R. A. (1972), The J. G. Strydom Dam: Pongolo River: Northern Zululand. The importance of floodplains below it, *Piscator*, *86*, 104–109.

Juen, I., G. Kaser, and C. Georges (2007), Modelling observed and future runoff from a glacierized tropical catchment (Cordillera Blanca, Peru), *Global Planet. Change*, *59*, 37–48.

Julien, P. Y. (1995), *Erosion and Sedimentation*, 280 pp., Cambridge Univ. Press, New York.

Jull, A. J. T., and M. Geertsema (2006), Over 16,000 years of fire frequency determined from AMS radiocarbon dating of soil charcoal in an alluvial fan at Bear Flat, northeastern British Columbia, *Radiocarbon*, *48*, 435–350.

Jungers, M. C., P. R. Bierman, A. Matmon, K. Nichols, J. Larsen, and R. Finkel (2009), Tracing hillslope sediment production and transport with in situ and meteoric ^{10}Be, *J. Geophys. Res.*, *114*, F04020, doi:10.1029/2008JF001086.

Júnior, G. W. (2005), Bed load transport described by a one-dimensional gamma functions model, in *Sediment Budgets I*, edited by D. E. Walling and A. J. Horowitz, *291*, 196–205.

Junk, W. J., P. B. Bayley, and R. E. Sparks (1989), The flood pulse concept in river-floodplain systems, *Can. Spec. Publ. Fish. Aquat Sci.*, *106*, 110–127.

Kaczka, R. J. (2003), The coarse woody debris dams in mountain streams of central Europe, structure and distribution, *Stud. Geomorphol. Carpatho-Balcanica*, *XXXVII*, 111–127.

Kaiser, K., G. Guggenberger, and L. Haumaier (2004), Changes in dissolved lignin-derived phenols, neutral sugars, uronic acids, and amino sugars with depth in forested Haplic Arenosols and Rendzic Leptosols, *Biogeochemistry*, *70*, 135–151.

Kaizuka, S., and T. Suzuki (1993), Geomorphology in Japan, in *The Evolution of Geomorphology*, edited by H. J. Walker and W. E. Grabau, pp. 255–271, Wiley, New York.

Kale, V. S. (2005), The sinuous bedrock channel of the Tapi River, central India; its form and processes, *Geomorphology*, *70*, 296–310.

Kale, V. S., and P. S. Hire (2007), Temporal variations in the specific stream power and total energy expenditure of a monsoonal river: The Tapi River, India, *Geomorphology*, *92*, 134–146.

Kaleris, V., D. Papanastasopoulos, and G. Lagas (2001), Case study on impact of atmospheric circulation changes on river basin hydrology: Uncertainty aspects, *J. Hydrol.*, *245*, 137–152.

Kalliola, R., and M. Puhakka (1988), River dynamics and vegetation mosaicism: A case study of the River Kamajohka, northernmost Finland, *J. Biogeogr.*, *15*, 703–719.

Kamp, U., K. Haserodt, and J. F. Shroder (2004), Quaternary landscape evolution in the eastern Hindu Kush, Pakistan, *Geomorphology*, *57*, 1–27.

Kao, S. J., and J. D. Milliman (2008), Water and sediment discharge from small mountainous rivers, Taiwan: The roles of lithology, episodic events, and human activities, *J. Geol.*, *116*, 431–448.

Kaplinski, M., J. Bennett, J. Cain, J. E. Hazel, M. Manone, R. Parnell, and L. E. Stevens (1998), Fluvial habitats developed on Colorado River sandbars in Grand Canyon, *Eos Trans. AGU*, *79*(45), F344.

Karr, J. R. (1991), Biological integrity: A long-neglected aspect of water resource management, *Ecol. Appl.*, *1*, 66–84.

Karr, J. R., and I. J. Schlosser (1977), Impact of nearstream vegetation and stream morphology on water quality and stream biota, *Rep. EPA-600/3-77-097*, USEPA, Athens, Ga.

Karssenberg, D., and J. S. Bridge (2008), A three-dimensional numerical model of sediment transport erosion and deposition within a network of channel belts, floodplain and hill slope; extrinsic and intrinsic controls on floodplain dynamics and alluvial architecture, *Sedimentology*, *55*, 1717–1745.

Kasahara, T., and S. M. Wondzell (2003), Geomorphic controls on hyporheic exchange flow in mountain streams, *Water Resour. Res.*, *39*(1), 1005, doi:10.1029/2002WR001386.

Kasai, M. (2006), Channel processes following land use changes in a degrading steep headwater stream in North Island, New Zealand, *Geomorphology*, *81*, 421–439.

Kasai, M., T. Marutani, L. M. Reid, and N. A. Trustrum (2001), Estimation of temporally averaged sediment delivery ratio using aggradational terraces in headwater catchments of the Waipaoa River, North Island, New Zealand, *Earth Surf. Processes Landforms*, *26*, 1–16.

Kasai, M., T. Marutani, and G. Brierley (2004), Channel bed adjustments following major aggradation in a steep headwater setting: Findings from Oyabu Creek, Kyushu, Japan, *Geomorphology*, *62*, 199–215.

Kasai, M., M. Ikeda, T. Asahina, and K. Fujisawa (2009), LiDAR-derived DEM evaluation of deep-seated landslides in a steep and rocky region of Japan, *Geomorphology*, *113*, 57–69.

Kasran, B. (1988), Effect of logging on sediment yield in a hill Dipterocarp forest in Peninsular Malaysia, *J. Trop. For. Sci.*, *1*, 56–66.

Kaste, J. M., A. M. Heimsath, and B. C. Bostick (2007), Short-term soil mixing quantified with fallout radionuclides, *Geology*, *35*, 243–246.

Kataoka, K. S., U. Urabe, V. Manville, and A. Kajiyama (2008), Breakout flood from an ignimbrite-dammed valley after the 5 ka Numazawako eruption, northeast Japan, *Geol. Soc. Am. Bull.*, *120*, 1233–1247.

Kattelmann, R. (1990), Floods in the high Sierra Nevada, California, USA, in *Hydrology in Mountainous Regions II. Artificial Reservoirs, Water and Slopes*, edited by R. O. Sinniger and M. Monbaron, *IAHS Publ.*, *194*, 39–46.

Kattelmann, R. (1996), A review of watershed degradation and rehabilitation throughout the Sierra Nevada, in *Watershed Restoration Management: Physical, Chemical, and Biological Considerations*, edited by J. J. McDonell et al., pp. 199–207, Am. Water Resources Association, Herndon, Va.

Kattenberg, A., et al. (1996), Climate models - projections of future climate, in *Climate Change 1995 - The Science of Climate Change*, edited by J. T. Houghton et al., pp. 285–357, Cambridge Univ. Press, Cambridge, U. K.

Katul, G., P. Wiberg, J. Albertson, and G. Hornberger (2002), A mixing layer theory for flow resistance in shallow streams, *Water Resour. Res.*, *38*(11), 1250, doi:10.1029/2001WR000817.

Katz, G. L., J. M. Friedman, and S. W. Beatty (2005), Delayed effects of flood control on a flood-dependent riparian forest, *Ecol. Appl.*, *15*, 1019–1035.

Kauffman, J. B., and W. C. Krueger (1984), Livestock impacts on riparian ecosystems and streamside management implications . . . a review, *J. Range Manage.*, *37*, 430–438.

Kaufman, P. R., J. M. Faustini, D. P. Larsen, and M. A. Shirazi (2008), A roughness-corrected index of relative bed stability for regional stream surveys, *Geomorphology*, *99*, 150–170.

Kawabata, D., and J. Bandibas (2009), Landslide susceptibility mapping using geological data, a DEM from ASTER images and an Artificial Neural Network (ANN), *Geomorphology, 113*, 97–109.

Kayastha, R. B., Y. Ageta, and K. Fujita (2005), Use of positive degree-day methods for calculating snow and ice melting and discharge in glacierized basins in the Langtang Valley, central Nepal, in *Climate and Hydrology in Mountain Areas*, edited by C. de Jong, D. Collins, and R. Ranzi, pp. 7–14, John Wiley, Chichester, U. K.

Kean, J. W., and J. D. Smith (2004), Flow and boundary shear stress in channels with woody bank vegetation, in *Riparian Vegetation and Fluvial Geomorphology*, edited by S. J. Bennett and A. Simon, pp. 237–252, AGU, Washington, D. C.

Kean, J. W., and J. D. Smith (2006a), Form drag in rivers due to small-scale natural topographic features: 1. Regular sequences, *J. Geophys. Res., 111*, F04009, doi:10.1029/2006JF000467.

Kean, J. W., and J. D. Smith (2006b), Form drag in rivers due to small-scale natural topographic features: 2. Irregular sequences, *J. Geophys. Res., 111*, F04010, doi:10.1029/2006JF000490.

Kearney, M. S., and J. C. Stevenson (1991), Island land loss and marsh vertical accretion rate evidence for historical sea-level changes in Chesapeake Bay, *J. Coastal Res., 7*, 403–415.

Keefer, D. K., M. E. Moseley, and S. D. deFrance (2003), A 38,000-year record of floods and debris flows in the Ilo region of southern Peru and its relation to El Nino events and great earthquakes, *Palaeogeogr. Palaeoclimatol. Palaeoecol., 194*, 41–77.

Keen-Zebert, A., and J. C. Curran (2009), Regional and local controls on the spatial distribution of bedrock reaches in the Upper Guadalupe River, Texas, *Geomorphology, 112*, 295–305.

Keesstra, S. D., L. A. Bruijnzeel, and J. van Huissteden (2009), Meso-scale catchment sediment budgets: Combining field surveys and modeling in the Dragonja catchment, southwest Slovenia, *Earth Surf. Processes Landforms, 34*, 1547–1561.

Keesstra, S. D., J. Van Huissteden, J. Vandenberghe, O. Van Dam, J. de Gier, and I. D. Pleizier (2005), Evolution of the morphology of the River Dragonja (SW Slovenia) due to land use changes, *Geomorphology, 69*, 191–207.

Keim, R. F., A. E. Skaugset, and D. S. Bateman (2000), Dynamics of coarse woody debris placed in three Oregon streams, *For. Sci., 46*, 13–22.

Keller, E. A. (1970), Bed-load movement experiments: Dry Creek, California, *J. Sediment. Petrol., 40*, 1339–1344.

Keller, E. A. (1971), Areal sorting of bedload material, the hypothesis of velocity reversal, *Geol. Soc. Am. Bull., 82*, 279–280.

Keller, E. A., and J. L. Florsheim (1993), Velocity-reversal hypothesis: A model approach, *Earth Surf. Processes Landforms, 18*, 733–740.

Keller, E. A., and W. N. Melhorn (1978), Rhythmic spacing and origin of pools and riffles, *Geol. Soc. Am. Bull., 89*, 723–730.

Keller, E. A., and F. J. Swanson (1979), Effects of large organic material on channel form and fluvial processes, *Earth Surf. Processes, 4*, 361–380.

Keller, E. A., and T. Tally (1979), Effects of large organic debris on channel form and fluvial processes in the coastal redwood environment, in *Adjustments of the Fluvial System*, edited by D. D. Rhodes and G. P. Williams, pp. 169–197, Kendall/Hunt, Dubuque, Iowa.

Keller, E. A., D. B. Seaver, D. L. Laduzinsky, D. L. Johnson, and T. L. Ku (2000), Tectonic geomorphology of active folding over buried reverse faults; San Emigdio Mountain front, southern San Joaquin Valley, California, *Geol. Soc. Am. Bull., 112*, 86–97.

Keller, F., and S. Goyette (2005), Snowmelt under different temperature increase scenarios in the Swiss Alps?, in *Climate and Hydrology in Mountain Areas*, pp. 277–289, John Wiley, Chichester, U. K.

Keller, K., J. D. Blum, and G. W. Kling (2007), Geochemistry of soils and streams on surfaces of varying ages in Arctic Alaska, *Arct. Antarct. Alp. Res., 39*, 84–98.

Keller, T., C. Pielmeier, C. Rixen, F. Gadient, D. Gustafsson, and M. Stahli (2004), Impact of artificial snow and ski-slope grooming on snowpack properties and soil thermal regime in a sub-alpine ski area, *Ann. Glaciol.*, *38*, 314–318.

Kellerhals, R., and D. I. Bray (1971), Sampling procedures for coarse fluvial sediments, *J. Hydraul. Div. Am. Soc. Civ. Eng.*, *97*, 1165–1180.

Kellerhals, R., and M. Church (1990), Hazard management on fans, with examples from British Columbia, in *Alluvial Fans: A Field Approach*, edited by A. H. Rachocki and M. Church, pp. 335–354, John Wiley, New York.

Kelsey, H. M., R. Lamberson, and M. A. Madej (1986), Modelling the transport of stored sediment in a gravel bed river, northwestern California, in *Drainage Basin Sediment Delivery*, edited by R. F. Hadley, *IAHS Publ.*, *15*, 367–391.

Kendall, C., J. J. McDonnell, and W. Gu (2001), A look inside 'black box' hydrograph separation models: A study at the Hydrohill catchment, *Hydrol. Processes*, *15*, 1877–1902.

Kendall, C., M. G. Sklash, and T. D. Bullen (1995), Isotope tracers of water and solute sources in catchments, in *Solute Modelling in Catchment Systems*, edited by S. T. Trudgill, pp. 261–303, John Wiley, Chichester, U. K.

Kennedy, J. F. (1963), The mechanics of dunes and antidunes in erodible-bed channels, *J. Fluid Mech.*, *16*, 521–544.

Keppeler, E. T. (1998), The summer flow and water yield response to timber harvest, in *Proceedings of the Conference on Coastal Watersheds: The Caspar Creek Story*, edited by R. R. Ziemer, pp. 35–43, Pacific Southwest Res. Sta., USDA, Albany, Calif.

Keppeler, E. T., and D. Brown (1998), Subsurface drainage processes and management impacts, in *Proceedings of the Conference on Coastal Watersheds: The Caspar Creek Story*, edited by R. R. Ziemer, pp. 25–34, Pacific Southwest Res. Sta., USDA, Albany, Calif.

Kershaw, J. A., J. J. Clague, and S. G. Evans (2005), Geomorphic and sedimentological signature of a two-phase outburst flood from moraine-dammed Queen Bess Lake, British Columbia, Canada, *Earth Surf. Processes Landforms*, *30*, 1–25.

Ketcheson, G. L., and W. F. Megahan (1991), Sediment tracing in step-pool granitic streams in Idaho, in *Proceedings of the Fifth Federal Interagency Sedimentation Conference*, edited by S.-S. Fan and Y.-H. Kuo, pp. 4-147–4-154, Las Vegas, Nev.

Kettner, A. J., B. Gomez, and J. P. M. Syvitski (2007), Modeling suspended sediment discharge from the Waipaoa River system, New Zealand: The last 3000 years, *Water Resour. Res.*, *43*, W07411, doi:10.1029/2006WR005570.

Khosrowshahi, F. B. (1990), Sediment transport in mountain streams, in *Hydrology in Mountainous Regions II - Artificial Reservoirs, Water and Slopes*, edited by R. O. Sinniger and M. Monbaron, *IAHS Publ.*, *194*, 59–66.

Kieffer, S. W. (1985), The 1983 hydraulic jump in Crystal Rapid: Implications for river-running and geomorphic evolution in the Grand Canyon, *J. Geol.*, *93*, 385–406.

Kieffer, S. W. (1987), *The rapids and waves of the Colorado River, Grand Canyon, Arizona*, U.S. Geol. Surv. Open File Rep. 87-096, 69 pp.

Kieffer, S. W. (1989), Geologic nozzles, *Rev. Geophys.*, *27*, 3–38.

Kieffer, S. W., J. B. Graf, and J. C. Schmidt (1989), Hydraulics and sediment transport of the Colorado River, in *Geology of the Grand Canyon, Northern Arizona*, edited by D. P. Elston, G. H. Billingsley, and R. A. Young, pp. 48–66, AGU, Washington, D. C.

Kiffney, P. M., J. S. Richardson, and M. C. Feller (2000), Fluvial and epilithic organic matter dynamics in headwater streams of southwestern British Columbia, Canada, *Arch. Hydrobiol.*, *148*, 109–129.

Kim, G., and A. P. Barros (2001), Quantitative flood forecasting using multisensor data and neural networks, *J. Hydrol.*, *246*, 45–62.

Kim, H. J., R. C. Sidle, R. D. Moore, and R. Hudson (2004), Throughflow variability during snowmelt in a forested mountain catchment, coasta British Columbia, Canada, *Hydrol. Processes*, *18*, 1219–1236.

Kim, J. C., and J. H. Kim (2007), Analysis of drainage structure based on the geometric characteristics of drainage density and source-basin, *J. Korea Water Resour. Assoc.*, *40*, 373–382.

Kim, S., and H. Lee (2004), A digital elevation analysis: A spatially distributed flow apportioning algorithm, *Hydrol. Processes*, *18*, 1777–1794.

King, T. V. V. (Ed.) (1995), *Environmental considerations of active and abandoned mine lands: Lessons from Summitville, Colorado*, U.S. Geol. Surv. Bull., *2220*, 38 pp.

Kirby, E., C. Johnson, K. Furlong, and A. Heimsath (2007), Transient channel incision along Bolinas Ridge, California: Evidence for differential rock uplift adjacent to the San Andreas fault, *J. Geophys. Res.*, *112*, F03S07, doi:10.1029/2006JF000559.

Kirchner, J. W. (1993), Statistical inevitability of Horton's laws and the apparent randomness of stream channel networks, *Geology*, *21*, 591–594.

Kirchner, J. W. (2003), A double paradox in catchment hydrology and geochemistry, *Hydrol. Processes*, *17*, 871–874.

Kirchner, J. W. (2006), Getting the right answers for the right reasons: Linking measurements, analysis, and models to advance the science of hydrology, *Water Resour. Res.*, *42*, W03S04, doi:10.1029/2005WR004362.

Kirchner, J. W., R. C. Finkel, C. S. Riebe, D. E. Granger, J. L. Clayton, J. G. King, and W. F. Megahan (2001), Mountain erosion over 10 yr, 10 k.y., and 10 m.y. time scales, *Geology*, *29*, 591–594.

Kirkbride, A. (1993), Observations of the influence of bed roughness on turbulence structure in depth limited flows over gravel beds, in *Turbulence: Perspectives on Flow and Sediment Transport*, edited by N. J. Clifford, J. R. French, and J. Hardisty, pp. 185–196, John Wiley, Chichester, U. K.

Kirkby, M. J. (1967), Measurement and theory of soil creep, *J. Geol.*, *75*, 359–378.

Kirkby, M. J. (1971), Hillslope process-response models based on the continuity equation, *Inst. Brit. Geogr. Spec. Publ.*, *3*, 15–30.

Kirkby, M. J. (1980), The stream head as a significant geomorphic threshold, in *Thresholds in Geomorphology*, edited by D. R. Coates and J. D. Vitek, pp. 53–73, George Allen and Unwin, London.

Kirkby, M. J. (1988), Hillslope runoff processes and models, *J. Hydrol.*, *100*, 315–339.

Kirkby, M. J. (2003), A consistent framework for modelling geomorphic process and landform evolution, in *Prediction in Geomorphology*, edited by P. R. Wilcock and R. M. Iverson, pp. 95–102, AGU, Washington, D. C.

Kirnbauer, R., and P. Haas (1998), Observations on runoff generation mechanisms in small Alpine catchments, in *Hydrology, Water Resources and Ecology in Headwaters*, edited by K. Kovar et al., *IAHS Publ.*, *248*, 239–247.

Kite, J. S., and R. C. Linton (1993), Depositional aspects of the November 1985 flood on Cheat River and Black Fork, West Virginia, in *Geomorphic Studies of the Storm and Flood of Nov. 3–5, 1985, in the Upper Potomac and Cheat River Basins in West Virginia and Virginia*, edited by R. B. Jacobson, U.S. Geol. Surv. Bull., *1981*, D1–D24.

Kite, J. S., T. W. Gebhardt, and G. S. Springer (2002), Slackwater deposits as paleostage indicators in canyon reaches of the central Appalachians: Reevaluation after the 1996 Cheat River flood, in *Ancient Floods, Modern Hazards: Principles and Applications of Paleoflood Hydrology*, edited by P. K. House et al., pp. 257–266, AGU, Washington, D. C.

Klaghofer, E., and W. Summer (1990), Estimation of soil erosion from a lower Alpine catchment, in *Hydrology in Mountainous Regions II. Artificial Reservoirs, Water and Slopes*, edited by R. O. Sinniger and M. Monbaron, *IAHS Publ.*, *194*, 67–74.

Klein, J. M., and H. E. Taylor (1980), Mount St. Helens, Washington, volcanic eruption: Part 2, Chemical variations in surface waters affected by volcanic activity, *Eos Trans. AGU*, *61*, 956.

Klein, R., R. Sonnevil, and D. Short (1987), Effects of woody debris removal on sediment storage in a northwestern California stream, in *Erosion and Sedimentation in the Pacific Rim*, edited by R. L. Beschta et al., *IAHS Publ.*, *165*, 403–404.

Kleinhans, M. G., A. W. E. Wilbers, A. de Swaaf, and J. H. van den Berg (2002), Sediment supply-limited bedforms in sand-gravel bed rivers, *J. Sediment. Res.*, *72*, 629–640.

Klemes, V. (1990), The modelling of mountain hydrology: The ultimate challenge, in *Hydrology of Mountainous Areas*, edited by L. Molnar, *IAHS Publ.*, *190*, 29–43.

Klimek, K. (1987), Man's impact on fluvial processes in the Polish Western Carpathians, *Geogr. Ann.*, *69A*, 221–225.

Klimek, K., and A. Latocha (2007), Response of small mid-mountain rivers to human impact with particular reference to the last 200 years; eastern Sudetes, central Europe, *Geomorphology*, *92*, 147–165.

Klingeman, P. C., and R. T. Milhous (1970), Oak Creek vortex bedload sampler. AGU, 17th Annual Pacific Northwest Regional Meeting, Tacoma, Washington.

Klinger, Y., J. P. Avouac, D. Bourles, and N. Tisnerat (2003), Alluvial deposition and lake-level fluctuations forced by late Quaternary climate change; the Dead Sea case example, *Sediment. Geol.*, *162*, 119–139.

Klok, E. J., K. Jasper, K. P. Roelofsma, J. Gurtz, and A. Badoux (2001), Distributed hydrological modelling of a heavily glaciated Alpine river basin, *Hydrol. Sci. J.*, *46*, 553–570.

Klose, C. (2006), Climate and geomorphology in the uppermost geomorphic belts of the Central Mountain Range, Taiwan, *Quat. Int.*, *147*, 89–102.

Kneisel, C., C. Rothenbuehler, F. Keller, and W. Haeberli (2007), Hazard assessment of potential periglacial debris flows based on GIS-based spatial modelling and geophysical field surveys: A case study in the Swiss Alps, *Permafrost Periglacial Processes*, *18*, 259–268.

Knighton, A. D. (1976), Stream adjustment in a small Rocky Mountain basin, *Arct. Alp. Res.*, *8*, 197–212.

Knighton, A. D. (1989), River adjustment to changes in sediment load: The effects of tin mining on the Ringarooma River, Tasmania, 1875–1984, *Earth Surf. Processes Landforms*, *14*, 333–359.

Knighton, A. D. (1999), Downstream variation in stream power, *Geomorphology*, *29*, 293–306.

Knighton, D. (1984), *Fluvial Forms and Processes*, 218 pp., Edward Arnold, London.

Knighton, D. (1998), *Fluvial Forms and Processes: A New Perspective*, 383 pp., Edward Arnold, London.

Knighton, D. (1999), The gravel-sand transition in a disturbed catchment, *Geomorphology*, *27*, 325–341.

Knuepfer, P. L. K. (1988), Estimating ages of late Quaternary stream terraces from analysis of weathering rinds and soils, *Geol. Soc. Am. Bull.*, *100*, 1224–1236.

Knust, A. E., and J. J. Warwick (2009), Using a fluctuating tracer to estimate hyporheic exchange in restored and unrestored reaches of the Truckee River, Nevada, USA, *Hydrol. Processes*, *23*, 1119–1130.

Kobashi, S., and M. Suzuki (1987), The critical rainfall (danger index) for disasters caused by debris flows and slope failures, in *Erosion and Sedimentation in the Pacific Rim*, edited by R. L. Beschta et al., *IAHS Publ.*, *165*, 201–211.

Kober, F., F. Schlunegger, G. Zeilinger, and H. Schneider (2006), Surface uplift and climate change: The geomorphic evolution of the Western Escarpment of the Andes of northern Chile between the Miocene and present, in *Tectonics, Climate, and Landscape Evolution*, edited by S. D. Willett et al., *Spec. Pap. Geol. Soc. Am.*, *398*, 75–86.

Koboltschnig, G. R., W. Schoener, and H. Holzmann (2007a), Extensive hydrological monitoring of a small, highly glacierized watershed in the Hohe Tauern region, Austrian Alps, in *Glacier Mass Balance Changes and Meltwater Discharge*, edited by P. Ginot and J.-E. Sicart, *IAHS Publ.*, *318*, 95–104.

Koboltschnig, G. R., W. Schoener, M. Zappa, and H. Holzmann (2007b), Contribution of glacier melt to stream runoff; if the climatically extreme summer of 2003 had happened in 1979, *Ann. Glaciol.*, *46*, 303–308.

Koboltschnig, G. R., W. Schoener, M. Zappa, C. Kroisleitner, and H. Holzmann (2008), Runoff modelling of the glacierized alpine upper Salzach Basin (Austria); multi-criteria result validation, *Hydrol. Processes*, *22*, 3950–3964.

Kobor, J. S., and J. J. Roering (2004), Systematic variation of bedrock channel gradients in the central Oregon Coast Range: Implications for rock uplift and shallow landsliding, *Geomorphology*, *62*, 239–256.

Kochel, R. C. (1987), Holocene debris flows in central Virginia, in *Debris Flows/Avalanches: Process, Recognition, and Mitigation*, edited by J. E. Costa and G. F. Wieczorek, *Rev. Eng. Geol.*, *7*, 139–155.

Kochel, R. C. (1988), Geomorphic impact of large floods: Review and new perspectives on magnitude and frequency, in *Flood Geomorphology*, edited by V. R. Baker, R. C. Kochel, and P. C. Patton, pp. 169–187, John Wiley, New York.

Kochel, R. C., R. A. Johnson, and S. Valastro (1982), Repeated episodes of Holocene debris avalanching in central Virginia, *Geol. Soc. Am. Abstr. Programs*, *14*, 31.

Kochel, R. C., D. F. Ritter, and J. Miller (1987), Role of tree dams in the construction of pseudo-terraces and variable geomorphic response to floods in Little River Valley, Virginia, *Geology*, *15*, 718–721.

Kochel, R. C., J. R. Miller, and D. F. Ritter (1997), Geomorphic response to minor cyclic climate changes, San Diego County, California, *Geomorphology*, *19*, 277–302.

Köhler, S. J., I. Buffam, J. Seibert, K. H. Bishop, and H. Laudon (2009), Dynamics of stream water TOC concentrations in a boreal headwater catchment: Controlling factors and implications for climate scenarios, *J. Hydrol.*, *373*, 44–56.

Komar, P. D. (1987a), Selective grain entrainment by a current from a bed of mixed sizes: A reanalysis, *J. Sediment. Petrol.*, *57*, 203–211.

Komar, P. D. (1987b), Selective grain entrainment and the empirical evaluation of flow competence, *Sedimentology*, *34*, 1165–1176.

Komar, P. D. (1989), Flow-competence evaluation of the hydraulic parameters of floods: An assessment of the technique, in *Floods: Hydrological, Sedimentological, and Geomorphological Implications*, edited by K. Beven and P. A. Carling, pp. 107–134, John Wiley, Chichester, U. K.

Komar, P. D., and P. A. Carling (1991), Grain sorting in gravel-bed streams and the choice of particle sizes for flow-competence evaluations, *Sedimentology*, *38*, 489–502.

Komar, P. D., and Z. Li (1986), Pivoting analyses of the selective entrainment of sediments by shape and size with application to grain threshold, *Sedimentology*, *33*, 425–436.

Komar, P. D., and S.-M. Shih (1992), Equal mobility versus changing bedload grain sizes in gravel-bed streams, in *Dynamics of Gravel-Bed Rivers*, edited by P. Billi et al., pp. 73–106, John Wiley, Chichester, U. K.

Kondolf, G. M. (1993), Lag in stream channel adjustment to livestock exclosure, White Mountains, California, *Restor. Ecol.*, *1*, 226–230.

Kondolf, G. M. (1994), Geomorphic and environmental effects of instream gravel mining, *Landscape Urban Plann.*, *28*, 225–243.

Kondolf, G. M. (1997), Hungry water: Effects of dams and gravel mining on river channels, *Environ. Manage.*, *21*, 533–551.

Kondolf, G. M. (2006), River restoration and meanders, *Ecol. Soc.*, *11*(2), 42.

Kondolf, G. M., and M. Larson (1995), Historical channel analysis and its application to riparian and aquatic habitat restoration, *Aquat. Conserv. Mar. Freshwater Ecosyst.*, *5*, 109–126.

Kondolf, G. M., and P. R. Wilcock (1996), The flushing flow problem: Defining and evaluating objectives, *Water Resour. Res.*, *32*, 2589–2599.

Kondolf, G. M., S. S. Cook, H. R. Maddux, and W. R. Persons (1989), Spawning gravels of rainbow trout in Glen and Grand Canyons, Arizona, *J. Ariz. Nev. Acad. Sci.*, *23*, 19–28.

Kondolf, G. M., M. J. Sale, and M. G. Wolman (1993), Modification of fluvial gravel size by spawning salmonids, *Water Resour. Res.*, *29*, 2265–2274.

Kondolf, G. M., M. W. Smeltzer, and S. F. Railsback (2001), Design and performance of a channel reconstruction project in a coastal California gravel-bed stream, *Environ. Manage.*, *28*, 761–776.

Kondolf, G. M., H. Piégay, and N. Landon (2002), Channel response to increased and decreased bedload supply from land use change: Contrasts between two catchments, *Geomorphology*, *45*, 35–51.

Kondolf, G. M., T. E. Lisle, and M. G. Wolman (2003a), Bed sediment measurement, in *Tools in Fluvial Geomorphology*, edited by G. M. Kondolf and H. Piégay, pp. 347–395, John Wiley, Chichester, U. K.

Kondolf, G. M., D. R. Montgomery, H. Piégay, and L. Schmitt (2003b), Geomorphic classification of rivers and streams, in *Tools in Fluvial Geomorphology*, edited by G. M. Kondolf and H. Piégay, pp. 171–204, John Wiley, Chichester, U. K.

Kondolf, G. M., H. Piégay, and D. Sear (2003c), Integrating geomorphological tools in ecological and management studies, in *Tools in Fluvial Geomorphology*, edited by G. M. Kondolf and H. Piégay, pp. 633–660, John Wiley, Chichester, U. K.

Kondolf, G. M., et al. (2006), Process-based ecological river restoration: Visualizing three-dimensional connectivity and dynamic vectors to recover lost linkages, *Ecol. Soc.*, *11*(2), 5.

Kondolf, G. M., et al. (2008), Projecting cumulative benefits of multiple river restoration projects: An example from the Sacramento-San Joaquin River system in California, *Environ. Manage.*, *42*, 933–934.

Konrad, C. P., D. B. Booth, S. J. Burges, and D. R. Montgomery (2002), Partial entrainment of gravel bars during floods, *Water Resour. Res.*, *38*(7), 1104, doi:10.1029/2001WR000828.

Konrad, C. P., D. B. Booth, and S. J. Burges (2005), Effects of urban development in the Puget Lowland, Washington, on interannual streamflow patterns: Consequences for channel form and streambed disturbance, *Water Resour. Res.*, *41*, W07009, doi:10.1029/2005WR004097.

Konrad, S. K. (1998), Possible outburst floods from debris-covered glaciers in the Sierra Nevada, California, *Geogr. Ann.*, *80A*, 183–192.

Konz, M., L. N. Braun, S. Uhlenbrook, S. Demuth, and A. Shrestha (2006), Regionalization of a distributed catchment model for highly glacierized Nepalese headwater catchments, in *Climate Variability and Change – Hydrological Impacts*, edited by S. Demuth et al., *IAHS Publ.*, *308*, 454–459.

Konz, M., and J. Seibert (2010), On the value of glacier mass balances for hydrological model calibration, *J. Hydrol.*, *385*, 238–246.

Koons, P. O. (2009), On the implications of low spatial correlation of tectonic and climate variables in the western European Alps, *Geology*, *37*, 863–864.

Koons, P. O., P. K. Zeitler, C. P. Chamberlain, D. Craw, and A. S. Meltzer (2002), Mechanical links between erosion and metamorphism in Nanga Parbat, Pakistan Himalaya, *Am. J. Sci.*, *302*, 749–773.

Koren, V. (2006), Parameterization of frozen ground effects: Sensitivity to soil properties, in *Predictions in Ungauged Basins: Promise and Progress*, edited by M. Sivapalan et al., *IAHS Publ.*, *303*, 125–133.

Korpak, J. (2007), The influence of river training on mountain channel changes (Polish Carpathian Mountains), *Geomorphology*, *92*, 166–181.

Korup, O. (2004), Landslide-induced river channel avulsions in mountain catchments of southwest New Zealand, *Geomorphology*, *63*, 57–80.

Korup, O. (2005a), Geomorphic hazard assessment of landslide dams in South Westland, New Zealand: Fundamental problems and approaches, *Geomorphology*, *66*, 167–188.

Korup, O. (2005b), Geomorphic imprint of landslides on alpine river systems, southwest New Zealand, *Earth Surf. Processes Landforms*, *30*, 783–800.

Korup, O., and D. R. Montgomery (2008), Tibetan plateau river incision inhibited by glacial stabilization of the Tsangpo gorge, *Nature*, *455*, 786–790.

Korup, O., M. J. McSaveney, and T. R. H. Davies (2004), Sediment generation and delivery from large historic landslides in the Southern Alps, New Zealand, *Geomorphology*, *61*, 189–207.

Korup, O., A. L. Densmore, and F. Schlunegger (2010), The role of landslides in mountain range evolution, *Geomorphology*, *120*, 77–90.

Korup, O., and F. Schlunegger (2007), Bedrock landsliding, river incision, and transience of geomorphic hillslope-channel coupling: Evidence from inner gorges in the Swiss Alps, *J. Geophys. Res.*, *112*, F03027, doi:10.1029/2006JF000710.

Korup, O., J. Schmidt, and M. J. McSaveney (2005), Regional relief characteristics and denudation pattern of the western Southern Alps, New Zealand, *Geomorphology*, *71*, 402–423.

Korup, O., A. L. Strom, and J. T. Weidinger (2006), Fluvial response to large rock-slope failures; examples from the Himalayas, the Tien Shan, and the Southern Alps in New Zealand, *Geomorphology*, *78*, 3–21.

Kostaschuk, R., J. Best, P. Villard, J. Peakall, and M. Franklin (2005), Measuring flow velocity and sediment transport with an acoustic Doppler current profiler, *Geomorphology*, *68*, 25–37.

Koster, E. H. (1978), Transverse ribs: Their characteristics, origin, and paleohydraulic singificance, in *Fluvial Sedimentology*, edited by A. D. Miall, pp. 161–186, Can. Soc. Petrol. Geologists.

Kostka, Z., and L. Holko (1994), Problems of the water balance determination in mountainous catchments, in *FRIEND: Flow Regimes from International Experimental and Network Data*, edited by P. Senna et al., *IAHS Publ.*, *221*, 433–438.

Kotarba, A. (2005), Geomorphic processes and vegetation pattern changes case study in the Zelene Pleso Valley, High Tatra, Slovakia, *Stud. Geomorphol. Carpatho-Balcanica*, *39*, 39–47.

Kothyari, B. P., P. K. Verma, B. K. Joshi, and U. C. Kothyari (2004), Rainfall-runoff-soil and nutrient loss relationships for plot size areas of Bhetagad watershed in central Himalaya, India, *J. Hydrol.*, *293*, 137–150.

Kraft, C. E., and D. R. Warren (2003), Development of spatial pattern in large woody debris and debris dams in streams, *Geomorphology*, *51*, 127–139.

Kraft, C. E., R. L. Schneider, and D. R. Warren (2002), Ice storm impacts on woody debris and debris dam formation in northeastern U. S. streams, *Can. J. Fish. Aquat. Sci.*, *59*, 1677–1684.

Krainer, K., and W. Mostler (2002), Hydrology of active rock glaciers: Examples from the Austrian Alps, *Arct. Antarct. Alp. Res.*, *34*, 142–149.

Krigstrom, A. (1962), Geomorphological studies of sandur plains and their braided rivers in Iceland, *Geogr. Ann.*, *44*, 328–346.

Krishna, A. P. (2005), Snow and glacier cover assessment in the high mountains of Sikkim Himalaya, *Hydrol. Processes*, *19*, 2375–2383.

Krzyszkowski, D., and R. Stachura (1998), Neotectonically controlled fluvial features, Walbrzych Upland, Middle Sudeten Mountains, southwestern Poland, *Geomorphology*, *22*, 73–91.

Kuhle, M. (2007), The past valley glacier network in the Himalayas and the Tibetan ice sheet during the last glacial period and its glacial-isostatic, eustatic and climatic consequences, *Tectonophysics*, *445*, 116–144.

Kuhn, M., and N. Batlogg (1998), Glacier runoff in Alpine headwaters in a changing climate, in *Hydrology, Water Resources and Ecology in Headwaters*, edited by K. Kovar et al., *IAHS Publ.*, *248*, 79–88.

Kuhle, M., S. Meiners, and L. Iturrizaga (1998), Glacier-induced hazards as a consequence of glacigenic mountain landscapes, in particular glacier- and moraine-dammed lake outbursts and Holocene debris production, in *Geomorphological Hazards in High Mountain Areas*, edited by J. Kalvoda and C. L. Rosenfeld, pp. 63–96, Kluwer Acad., The Netherlands.

Kuhn, N. J., and A. Yair (2004), Spatial distribution of surface conditions and runoff generation in small arid watersheds, Zin Valley Badlands, Israel, *Geomorphology*, *57*, 183–200.

Kuhn, T. S. (1962), *The Structure of Scientific Revolutions*, 210 pp., Univ. of Chicago Press, Chicago, Ill.

Kuhnle, R. A. (1992a), Bed load transport during rising and falling stages on two small streams, *Earth Surf. Processes Landforms*, *17*, 191–197.

Kuhnle, R. A. (1992b), Fractional transport rates of bedload on Goodwin Creek, in *Dynamics of Gravel-Bed Rivers*, edited by P. Billi, R. D. Hey, and C. R. Thorne, pp. 141–155, John Wiley, Chichester, U. K.

Kuhnle, R. A., and J. B. Southard (1988), Bedload transport fluctuations in a gravel bed laboratory channel, *Water Resour. Res.*, *24*, 247–260.

Kukulak, J. (2003), Impact of mediaeval agriculture on the alluvium in the San River headwaters (Polish Eastern Carpathians), *Catena*, *51*, 255–266.

Kumar, A., and D. Kumar (2005), Development of geomorphologic instantaneous unit hydrograph for prediction of direct runoff from a hilly watershed, *J. Appl. Hydrol.*, *18*, 98–106.

Kumar, R., A. K. Lohani, S. Kumar, C. Chatterjee, and R. K. Nema (2001), GIS based morphometric analysis of Ajay River basin up to Sarath gauging site of south Bihar, *J. Appl. Hydrol.*, *14*, 45–54.

Kundzewicz, Z. W. (1997), Water resources for sustainable development, *Hydrol. Sci. J.*, *42*, 467–480.

Kundezewicz, Z. W., L. J. Mata, N. W. Arnell, P. Döll, P. Kabat, B. Jiménez, K. A. Miller, T. Oki, Z. Sen, and I. A. Shiklomanov (2007), Freshwater resources and their management. Climate change 2007: Impacts, adaptation, and vulnerability, in *Contribution of Working Group II to the Fourth Assessment Report of the Intergovernmental Panel on Climate Change*, edited by M. L. Parry et al., pp. 173–210, Cambridge Univ. Press, Cambridge, U. K.

Kunstmann, H., and C. Stadler (2005), High resolution distributed atmospheric-hydrological modelling for Alpine catchments, *J. Hydrol.*, *314*, 105–124.

Kunze, M. D., and J. D. Stednick (2006), Streamflow and suspended sediment yield following the 2000 Bobcat fire, Colorado, *Hydrol. Processes*, *20*, 1661–1681.

Kurashige, Y. (1994), Mechanisms of suspended sediment supply to headwater rivers, *Trans., Jpn. Geomorphol. Union*, *15A*, 109–129.

Kurashige, Y. (1996), Process-based model of grain lifting from river bed to estimate suspended-sediment concentration in a small headwater basin, *Earth Surf. Processes Landforms*, *21*, 1163–1173.

Kurashige, Y. (1999), Monitoring of thickness of river-bed sediment in the Pankenai River, Hokkaido, Japan, *Trans. Jpn. Geomorphol. Union*, *20*, 21–33.

Kurashige, Y., H. Kibayashi, and G. Nakajima (2003), Chronology of alluvial sediment using the date of production of buried refuse: A case study in an ungauged river in central Japan, in *Erosion Prediction in Ungauged Basins: Integrating Methods and Techniques*, edited by D. de Boer et al., *IAHS Publ.*, *279*, 43–50.

Lacey, G. (1930), Stable channels in alluvium, *Inst. Civ. Eng. Proc.*, *229*, 259–384.

Lacey, R. W. J., P. Legendre, and A. G. Roy (2007), Spatial-scale partitioning of in situ turbulent flow data over a pebble cluster in a gravel-bed river, *Water Resour. Res.*, *43*, W03416, doi:10.1029/2006WR005044.

Lacoste, A., L. Loncke, F. Chanier, J. Bailleul, B. C. Vendeville, and G. Mahieux (2009), Morphology and structure of a landslide complex in an active margin setting: The Waitawhiti complex, North Island, New Zealand, *Geomorphology*, *109*, 184–196.

Laenen, A., and J. C. Risley (1997), *Precipitation-runoff and streamflow-routing models for the Willamette River Basin, Oregon*, U.S. Geol. Surv. Water Resour. Invest. Rep. 95-4284, 197 pp.

Lagasse, P. F. (1981), Geomorphic response of the Rio Grande to dam construction, *New Mexico Geol. Soc., Spec. Publ.*, vol. 10, pp. 27–46.

Lagasse, P. F., B. R. Winkley, and D. B. Simons (1980), Impact of gravel mining on river system stability, *J. Waterw. Port Coastal Ocean Div. Am. Soc. Civ. Eng.*, *106*, 389–404.

Lague, D. (2010), Reduction of long-term bedrock incision efficiency by short-term alluvial cover intermittency, *J. Geophys. Res.*, *115*, F02011, doi:10.1029/2008JF001210.

Lague, D., A. Crave, and P. Davy (2003), Laboratory experiments simulating the geomorphic response to tectonic uplift, *J. Geophys. Res.*, *108*(B1), 2008, doi:10.1029/2002JB001785.

Lague, D., N. Hovius, and P. Davy (2005), Discharge, discharge variability, and the bedrock channel profile, *J. Geophys. Res.*, *110*, F04006, doi:10.1029/2004JF000259.

LaHusen, R. (2005), Debris-flow instrumentation, in *Debris-Flow Hazards and Related Phenomena*, edited by M. Jakob and O. Hungr, pp. 291–304, Springer, Berlin.

Lai, F. S., I. Akkharath, and K. S. Low (2005), Impact of ground-based timber harvesting on suspended sediment yield in the Sungai Weng Experimental Watersheds, Kedah, Peninsular Malaysia, in *Sediment Budgets 2*, edited by A. J. Horowitz and D. E. Walling, *IAHS Publ.*, *292*, 253–261.

Lai, J.-S., T.-C. Tsay, and G.-F. Lin (1998), A study on equilibrium slope upstream of check dams, in *River Sedimentation; Theory and Applications*, edited by A. W. Jayawardena, J. H. Lee, and Z. Y. Wang, pp. 561–564, A. A. Balkema, Rotterdam, The Netherlands.

Laity, J. E., and M. C. Malin (1985), Sapping processes and the development of theater-headed valley networks on the Colorado Plateau, *Geol. Soc. Am. Bull.*, *96*, 203–217.

Lake, J. S. (1975), Fish of the Murray River, in *The Book of the Murray*, edited by G. C. Lawrence and G. K. Smith, pp. 213–223, Rigby, Adelaide, Australia.

Lake, P. S. (2000), Disturbance, patchiness, and diversity in streams, *J. North Am. Benthol. Soc.*, *19*, 573–592.

Lakshmi, V. (2005), Standing back looking forward: Role of satellite remote sensing in the prediction of ungauged basins, in *Predictions in Ungauged Basins: International Perspectives on the State of the Art and Pathways Forward*, edited by S. Franks et al., *IAHS Publ.*, *301*, 118–124.

La Marche, J. L., and D. P. Lettenmaier (2001), Effects of forest roads on flood flows in the Deschutes River, Washington, *Earth Surf. Processes Landforms*, *26*, 115–134.

LaMarche, V. C., Jr. (1968), Rates of slope degradation as determined from botanical evidence, White Mountains, California, *U.S. Geol. Surv. Prof. Pap.*, *352-I*.

Lamarre, H., and A. G. Roy (2005), Reach scale variability of turbulent flow characteristics in a gravel-bed river, *Geomorphology*, *68*, 95–113.

Lamarre, H., and A. G. Roy (2008), The role of morphology on the displacement of particles in a step-pool river system, *Geomorphology*, *99*, 270–279.

Lamb, H. R., M. Tranter, G. H. Brown, B. P. Hubbard, M. J. Sharp, S. Gordon, C. C. Smart, I. C. Willis, and M. K. Nielsen (1995), The composition of subglacial meltwaters sampled from boreholes at the Haut Glacier d'Arolla, Switzerland, in *Biogeochemistry of Seasonally Snow-Covered Catchments*, edited by K. A. Tonnessen, M. W. Williams, and M. Tranter, *IAHS Publ.*, *228*, 395–403.

Lamb, M. P., and W. E. Dietrich (2009), The persistence of waterfalls in fractured rock, *Geol. Soc. Am. Bull.*, *121*, 1123–1134.

Lamb, M. P., A. D. Howard, J. Johnson, K. X. Whipple, W. E. Dietrich, and J. T. Perron (2006), Can springs cut canyons into rock?, *J. Geophys. Res.*, *111*, E07002, doi:10.1029/2005JE002663.

Lamb, M. P., A. D. Howard, W. E. Dietrich, and J. T. Perron (2007), Formation of ampitheater-headed valleys by waterfall erosion after large-scale slumping on Hawai'i, *Geol. Soc. Am. Bull.*, *119*, 805–822.

Lamb, M. P., W. E. Dietrich, S. M. Aciego, D. J. DePaolo, and M. Manga (2008a), Formation of Box Canyon, Idaho, by megaflood: Implications for seepage erosion on Earth and Mars, *Science*, *320*, 1067–1070.

Lamb, M. P., W. E. Dietrich, and L. S. Sklar (2008b), A model for fluvial bedrock incision by impacting suspended and bed load sediment, *J. Geophys. Res.*, *113*, F03025, doi:10.1029/2007JF000915.

Lamb, M. P., W. E. Dietrich, and J. G. Venditti (2008c), Is the critical Shields stress for incipient motion dependent on channel-bed slope?, *J. Geophys. Res.*, *113*, F02008, doi:10.1029/2007JF000831.

Lamouroux, N. (2008), Hydraulic geometry of stream reaches and ecological implications, in *Gravel-Bed Rivers VI: From Process Understanding to River Restoration*, edited by H. Habersack, H. Piégay, and M. Rinaldi, pp. 661–676, Elsevier, Amsterdam.

Lana-Renault, N., D. Regüés, J. Latron, E. Nadal, P. Serrano, and C. Martí-Bono (2006), A volumetric approach to estimate bed load transport in a mountain stream (Central Spanish Pyrenees), in *Sediment Dynamics and the Hydromorphology of Fluvial Systems*, edited by J. S. Rowan, R. W. Duck, and A. Werritty, *IAHS Publ.*, *306*, 89–95.

Lana-Renault, N., D. Regüés, C. Marti-Bono, S. Begueria, J. Latron, E. Nadal, P. Serrano, and J. M. Garcia-Ruiz (2007), Temporal variability in the relationships between precipitation, discharge and suspended sediment concentration in a small Mediterranean mountain catchment, *Nord. Hydrol.*, *38*, 139–150.

Lancaster, S. T., and N. E. Casebeer (2007), Sediment storage and evacuation in headwater valleys at the transition between debris-flow and fluvial processes, *Geology*, *35*, 1027–1030.

Lancaster, S. T., and G. E. Grant (2006), Debris dams and the relief of headwater streams, *Geomorphology*, *82*, 84–97.

Lancaster, S. T., S. K. Hayes, and G. E. Grant (2001), Modeling sediment and wood storage and dynamics in small mountainous watersheds, in *Geomorphic Processes and Riverine Habitat*, edited by J. M. Dorava et al., pp. 85–102, AGU, Washington, D. C.

Lancaster, S. T., S. K. Hayes, and G. E. Grant (2003), Effects of wood on debris flow runout in small mountain watersheds, *Water Resour. Res.*, *39*(6), 1168, doi:10.1029/2001WR001227.

Lance, J. C., S. C. McIntyre, J. W. Naney, and S. S. Rousseva (1986), Measuring sediment movement at low erosion rates using Cesium-137, *Soil Sci. Soc. Am. J.*, *50*, 1303–1309.

Landers, M. N., J. S. Jones, and R. E. Trent (1994), Brief summary of national bridge scour data base, in *Proceedings, ASCE Hydraulic Engineering '94 Conference, August 1–5, Buffalo, New York*, edited by G. V. Cotroneo and R. R. Rumer, pp. 41–45, Am. Soc. of Civ. Eng., New York.

Lane, S. N. (1998), Hydraulic modelling in hydrology and geomorphology: A review of high resolution approaches, *Hydrol. Processes*, *12*, 1131–1150.

Lane, S. N. (2006), Approaching the system-scale understanding of braided river behaviour, in *Braided Rivers: Process, Deposits, Ecology and Management*, edited by G. H. Sambrook Smith et al., *Spec. Publ. Int. Assoc. Sedimentol.*, *36*, 107–135.

Lane, S. N., K. F. Bradbrook, K. S. Richards, P. A. Biron, and A. G. Roy (1999), The application of computational fluid dynamics to natural river channels: Three-dimensional versus two-dimensional approaches, *Geomorphology*, *29*, 1–20.

Lane, S. N., J. H. Chandler, and K. S. Richards (1994), Developments in monitoring and modelling small scale river bed topography, *Earth Surf. Processes Landforms*, *19*, 349–368.

Lane, S. N., V. Tayefi, S. C. Reid, D. Yu, and R. J. Hardy (2007), Interactions between sediment delivery, channel change, climate change and flood risk in a temperature upland environment, *Earth Surf. Processes Landforms*, *32*, 429–446.

Lane, S. N., S. M. Reaney, and A. L. Heathwaite (2009), Representation of landscape hydrological connectivity using a topographically driven surface flow index, *Water Resour. Res.*, *45*, W08423, doi:10.1029/2008WR007336.

Lane, S. N., P. E. Widdison, R. E. Thomas, P. J. Ashworth, J. L. Best, I. A. Lunt, G. H. Sambrook-Smith, and C. J. Simpson (2010), Quantification of braided river channel change using archival digital image analysis, *Earth Surf. Processes Landforms*, *35*, 971–985.

Lang, H., and D. Grebner (1998), On large-scale topographic control of the spatial distribution of extreme precipitation and floods in high mountain regions, in *Hydrology, Water Resources and Ecology in Headwaters*, edited by K. Kovar et al., *IAHS Publ.*, *248*, 47–50.

Lang, H., and A. Musy (Eds.) (1990), *Hydrology in Mountainous Regions I. Hydrological Measurements: The Water Cycle*, IAHS Publ., *193*, 810 pp.

Langbein, W. B., and L. B. Leopold (1964), Quasi-equilibrium states in channel morphology, *Am. J. Sci.*, *262*, 782–794.

Langbein, W. B., and L. B. Leopold (1966), River meanders - theory of minimum variance, *U.S. Geol. Surv. Prof. Pap.*, *422H*, 15 pp.

Langbein, W. B., and L. B. Leopold (1968), River channel bars and dunes - theory of kinematic waves, *U.S. Geol. Surv. Prof. Pap.*, *422L*.

Langbein, W. B., and S. A. Schumm (1958), Yield of sediment in relation to mean annual precipitation, *Eos Trans. AGU*, *39*, 1076–1084.

Lange, J., N. Greenbaum, S. Husary, M. Ghanem, C. Leibundgut, and A. P. Schick (2003), Runoff generation from successive simulated rainfalls on a rocky, semiarid, Mediterranean hillslope, *Hydrol. Processes*, *17*, 279–296.

Langedal, M. (1997), The influence of a large anthropogenic sediment source on the fluvial geomorphology of the Knabeåna-Kvina rivers, Norway, *Geomorphology*, *19*, 117–132.

Langlois, J. L., D. W. Johnson, and G. R. Mehuys (2005), Suspended sediment dynamics associated with snowmelt runoff in a small mountain stream of Lake Tahoe (Nevada), *Hydrol. Processes*, *19*, 3569–3580.

Lanka, R. P., W. A. Hubert, and T. A. Wesche (1987), Relations of geomorphology to stream habitat and trout standing stock in small Rocky Mountain streams, *Trans. Am. Fish. Soc.*, *116*, 21–28.

LaPerriere, J. D., S. M. Wagener, and D. M. Bjerklie (1985), Gold-mining effects on heavy metals in streams, Circle Quadrangle, Alaska, *Water Resour. Bull.*, *21*, 245–252.

Lapointe, M. F., Y. Secretan, S. N. Driscoll, N. Bergeron, and M. Leclerc (1998), Response of the Ha!Ha! River to the flood of July 1996 in the Saguenay Region of Quebec: Large-scale avulsion in a glaciated valley, *Water Resour. Res.*, *34*, 2382–2392.

Larned, S. T., T. Datry, D. B. Arscott, and K. Tockner (2010), Emerging concepts in temporary-river ecology, *Freshwater Biol.*, *55*, 717–738.

Laronne, J. B., and M. A. Carson (1976), Interrelationships between bed morphology and bed-material transport for a small, gravel-bed channel, *Sedimentology*, *23*, 67–85.

Laronne, J. B., and M. J. Duncan (1992), Bedload transport paths and gravel bar formation, in *Dynamics of Gravel-Bed Rivers*, edited by P. Billi et al., pp. 177–202, John Wiley, Chichester, U. K.

Laronne, J. B., and Y. Shlomi (2007), Depositional character and preservation potential of coarse-grained sediments deposited by flood events in hyper-arid braided channels in the Rift Valley, Arava, Israel, *Sediment. Geol.*, *195*, 21–37.

Laronne, J. B., I. Reid, Y. Yitschak, and L. E. Frostick (1992), Recording bedload discharge in a semiarid channel, Nahal Yatir, Israel, in *International Symposium on Erosion and Sediment Transport Monitoring Programmes in River Basins*, edited by J. Bogen, D. E. Walling, and T. Day, pp. 210, IAHS Press, Oslo.

Laronne, J. B., Y. Alexandrov, N. Bergman, H. Cohen, C. Garcia, H. Habersack, D. M. Powell, and I. Reid (2003), The continuous monitoring of bed load flux in various fluvial environments, in *Erosion and Sediment Transport Measurement in Rivers: Technological and Methodological Advances*, edited by J. Bogen, T. Fergus, and D. E. Walling, *IAHS Publ.*, *283*, 134–145.

Larsen, E. E., W. C. Bradley, and M. Ozima (1975), Development of the Colorado River system in northwestern Colorado during the late Cenozoic, in Canyonlands Country, in *A Guidebook of the Four Corners Geological Society, 8th Field Conference*, edited by J. E. Fassett, pp. 97–102, Canyonlands, Utah.

Larsen, I. J., and L. H. MacDonald (2007), Predicting postfire sediment yields at the hillslope scale: Testing RUSLE and Disturbed WEPP, *Water Resour. Res.*, *43*, W11412, doi:10.1029/2006WR005560.

Larsen, I. J., J. C. Schmidt, and J. A. Martin (2004), Debris-fan reworking during low-magnitude floods in the Green River canyons of the eastern Uinta Mountains, Colorado and Utah, *Geology*, *32*, 309–312.

Larsen, I. J., J. L. Pederson, and J. C. Schmidt (2006), Geologic versus wildfire controls on hillslope processes and debris flow initiation in the Green River canyons of Dinosaur National Monument, *Geomorphology*, *81*, 114–127.

Larsen, M. C. (2008), Rainfall-triggered landslides, anthropogenic hazards, and mitigation strategies, *Adv. Geosci.*, *14*, 147–153.

Larsen, M. C., and J. E. Parks (1997), How wide is a road? The association of roads and mass-wasting in a forested montane environment, *Earth Surf. Processes Landforms*, *22*, 835–848.

Larsen, M. C., and A. S. Román (2001), Mass wasting and sediment storage in a small montane watershed: An extreme case of anthropogenic disturbance in the humid tropics, in *Geomorphic Processes and Riverine Habitat*, edited by J. M. Dorava et al., pp. 119–138, AGU, Washington, D. C.

Larsen, M. C., and A. Simon (1993), A rainfall intensity-duration threshold for landslides in a humid-tropical environment, Puerto Rico, *Geogr. Ann.*, *75A*, 13–23.

Larsen, M. C., and A. J. Torres-Sanchez (1998), The frequency and distribution of recent landslides in three montane tropical regions of Puerto Rico, *Geomorphology*, *24*, 309–331.

Larsen, M. C., G. Wieczorek, L. S. Eaton, B. A. Morgan, and H. Torres-Sierra (2002), Natural hazards on alluvial fans: The Venezuela debris flow and flash flood disaster, *U.S. Geol. Surv. Fact Sheet, FS 0103-01*.

Larue, J.-P. (2008), Effects of tectonics and lithology on long profiles of 16 rivers of the southern Central Massif border between the Aude and the Orb (France), *Geomorphology*, *93*, 343–367.

Latocha, A. (2009), Land-use changes and longer-term human-environment interactions in a mountain region (Sudetes Mountains, Poland), *Geomorphology*, *108*, 48–57.

Latocha, A., and P. Migoń (2006), Geomorphology of medium-high mountains under changing human impact, from managed slopes to nature restoration: A study from the Sudetes, SW Poland, *Earth Surf. Processes Landforms*, *31*, 1657–1673.

Lauer, W. (Ed.) (1984), *Natural Environments and Man in Tropical Mountain Ecosystems*, Franz Steiner Verlag, Wiesbaden.

Lautz, L. K. (2010), Impacts of nonideal field conditions on vertical water velocity estimates from streambed temperature time series, *Water Resour. Res.*, *46*, W01509, doi:10.1029/2009WR007917.

Lautz, L. K., and R. M. Fanelli (2008), Seasonal biogeochemical hotspots in the streambed around restoration structures, *Biogeochemistry*, *91*, 85–104.

Lautz, L. K., D. I. Siegel, and R. L. Bauer (2006), Impact of debris dams on hyporheic interaction along a semi-arid stream, *Hydrol. Processes*, *20*, 183–196.

Lavé, J., and J. P. Avouac (2001), Fluvial incision and tectonic uplift across the Himalayas of central Nepal, *J. Geophys. Res.*, *105*(B11), 26,561–26,591.

Lave, R., M. M. Robertson, and M. W. Doyle (2008), Why you should pay attention to stream mitigation banking, *Ecol. Restor.*, *26*, 287–289.

Lawler, D. M. (1991), A new technique for the automatic monitoring of erosion and deposition rates, *Water Resour. Res.*, *27*, 2125–2128.

Lawler, D. M. (1992), Process dominance in bank erosion systems, in *Lowland Floodplain Rivers: Geomorphological Perspectives*, edited by P. A. Carling and G. E. Petts, pp. 117–143, Wiley, Chichester, U. K.

Lawler, D. M. (1993a), The measurement of river bank erosion and lateral channel change: A review, *Earth Surf. Processes Landforms*, *18*, 777–821.

Lawler, D. M. (1993b), Needle ice processes and sediment mobilization on river banks: The River Ilston, West Glamorgan, UK, *J. Hydrol.*, *150*, 81–114.

Lawler, D. M. (2005), The importance of high-resolution monitoring in erosion and deposition dynamics studies: Examples from estuarine and fluvial systems, *Geomorphology*, *64*, 1–23.

Lawler, D. M. (2008), Advances in the continuous monitoring of erosion and deposition dynamics: Developments and applications of the new PEEP-3T system, *Geomorphology*, *93*, 17–39.

Lawless, M., and A. Robert (2001a), Scales of boundary resistance in coarse-grained channels: Turbulent velocity profiles and implications, *Geomorphology*, *39*, 221–238.

Lawless, M., and A. Robert (2001b), Three-dimensional flow structure around small-scale bedforms in a simulated gravel-bed environment, *Earth Surf. Processes Landforms*, *26*, 507–522.

Lawrence, D. E. (1991), Woody debris budgets in 'burn' and 'reference' headwater streams: Long-term predictions for debris accumulation processes and geomorphic heterogeneity (abstract), *Bull. North Am. Benthol. Soc.*, *8*, 75.

Lawrence, G. B., G. M. Lovett, and Y. H. Baevsky (2000), Atmospheric deposition and watershed nitrogen export along an elevational gradient in the Catskill Mountains, New York, *Biogeochemistry*, *50*, 21–43.

Leavesley, G., and L. Hay (1998), The use of coupled atmospheric and hydrological models for water-resources management in headwater basins, in *Hydrology, Water Resources and Ecology in Headwaters*, edited by K. Kovar et al., *IAHS Publ.*, *248*, 259–265.

Lebedeva, M. I., R. C. Fletcher, and S. L. Brantley (2010), A mathematical model for steady-state regolith production at constant erosion rate, *Earth Surf. Processes Landforms*, *35*, 508–524.

Lecomte, K. L., M. G. García, S. M. Fórmica, and P. J. Depetris (2009), Influence of geomorphological variables on mountainous stream water chemistry (Sierras Pampeanas, Córdoba, Argentina), *Geomorphology*, *110*, 195–202.

Lee, A. J., and R. I. Ferguson (2002), Velocity and flow resistance in step-pool streams, *Geomorphology*, *46*, 59–71.

Lee, K. T., and C.-H. Chang (2005), Incorporating subsurface flow mechanism into geomorphology-based IUH modeling, *J. Hydrol.*, *311*, 91–105.

Lee, K. T., and J.-Y. Ho (2009), Prediction of landslide occurrence based on slope-instability analysis and hydrological model simulation, *J. Hydrol.*, *375*, 489–497.

Leeder, M. R., and P. H. Bridges (1975), Flow separation in meander bends, *Nature*, *253*, 338–339.

Leeder, M. R., and G. H. Mack (2001), Lateral erosion ('toe-cutting') of alluvial fans by axial rivers; implications for basin analysis and architecture, *J. Geol. Soc. London*, *158*, 885–893.

Léger, L. (1909), Principes de la méthode rationnelle du peuplement des cours d'eau à salmonidés., Travaux de la Laboratoire de Pisciculture à, *Grenoble*, *1*, 533–568.

Legleiter, C. J., T. L. Phelps, and E. E. Wohl (2007), Geostatistical analysis of the effects of stage and roughness on reach- scale spatial patterns of velocity and turbulence intensity, *Geomorphology*, *83*, 322–345.

Legleiter, C. J., D. A. Roberts, and R. L. Lawrence (2009), Spectrally based remote sensing of river bathymetry, *Earth Surf. Processes Landforms*, *34*, 1039–1059.

Lehning, M., I. Voelksh, D. Gustafsson, A. N. Tuan, M. Staehli, and M. Zappa (2006), ALPINE3D: A detailed model of mountain surface processes and its application to snow hydrology, *Hydrol. Processes*, *20*, 2111–2128.

Lehre, A. K. (1981), Sediment budget of a small California Coast Range drainage basin near San Francisco, in *Erosion and Sediment Transport in Pacific Rim Steeplands*, edited by T. R. H. Davies and A. J. Pierce, *IAHS Publ.*, *132*, 123–139.

Leibundgut, C. (1998), Tracer-based assessment of vulnerability in mountainous headwaters, in *Hydrology, Water Resources and Ecology in Headwaters*, edited by K. Kovar et al., *IAHS Publ.*, *248*, 317–325.

Leier, A. L., P. G. DeCelles, and J. D. Pelletier (2005), Mountains, monsoons, and megafans, *Geology*, *33*, 289–292.

Leigh, D. S., and P. A. Webb (2006), Holocene erosion, sedimentation, and stratigraphy at Raven Fork, southern Blue Ridge Mountains, USA, *Geomorphology*, *78*, 161–177.

Leithold, E. L., D. W. Perkey, N. E. Blair, and T. N. Creamer (2005), Sedimentation and carbon burial on the northern California continental shelf: The signatures of land-use change, *Cont. Shelf Res.*, *25*, 349–371.

Leiva, J. C., G. A. Cabrera, and L. E. Lenzano (2007), 20 years of mass balances on the Piloto Glacier, Las Cuevas River basin, Mendoza, Argentina, *Global Planet. Change*, *59*, 10–16.

Lekach, J., and A. P. Schick (1982), Suspended sediment in desert floods in small catchments, *Isr. J. Earth Sci.*, *31*, 144–156.

Lekach, J., and A. P. Schick (1983), Evidence for transport of bedload in waves: Analysis of fluvial sediment samples in a small upland stream channel, *Catena*, *10*, 267–279.

Leland, H. V., and J. L. Carter (1985), Effects of copper on production of periphyton, nitrogen fixation and processing of leaf litter in a Sierra Nevada, California, stream, *Freshwater Biol.*, *15*, 155–173.

Lenhardt, M., P. Cakić, and J. Kolarevic (2004), Influences of the HEPS Djerdap I and Djerdap II dam construction on catch of economically important fish species in the Danube River, *Ecohydrol. Hydrobiol.*, *4*, 499–502.

Lenzi, M. A. (2001), Step-pool evolution in the Rio Cordon, northeastern Italy, *Earth Surf. Processes Landforms*, *26*, 991–1008.

Lenzi, M. A. (2002), Stream bed stabilization using boulder check dams that mimic step-pool morphology features in northern Italy, *Geomorphology*, *45*, 243–260.

Lenzi, M. A. (2004), Displacement and transport of marked pebbles, cobbles and boulders during floods in a steep mountain stream, *Hydrol. Processes*, *18*, 1899–1914.

Lenzi, M. A., and F. Comiti (2003), Local scouring and morphological adjustments in steep channels with check-dam sequences, *Geomorphology*, *55*, 97–109.

Lenzi, M. A., L. Marchi, and G. R. Scussel (1990), Measurement of coarse sediment transport in a small Alpine stream, in *Hydrology in Mountainous Regions I. Hydrological Measurements: The Water Cycle*, edited by H. Lang and A. Musy, *IAHS Publ.*, *193*, 283–290.

Lenzi, M. A., V. D'Agostino, C. Gregoretti, and D. Sonda (2003a), A simplified numerical model for debris-flow hazard assessment: DEFLIMO, in *Debris-Flow Hazards Mitigation: Mechanics, Prediction, and Assessment*, edited by D. Rickenmann and C. Chen, pp. 611–622, Millpress, Rotterdam.

Lenzi, M. A., L. Mao, and F. Comiti (2003b), Interannual variation of suspended sediment load and sediment yield in an alpine catchment, *Hydrol. Sci. J.*, *48*, 899–915.

Lenzi, M. A., A. Marion, and F. Comiti (2003c), Interference processes on scouring at bed sills, *Earth Surf. Processes Landforms*, *28*, 99–110.

Lenzi, M. A., A. Marion, and F. Comiti (2003d), Local scouring at grade-control structures in alluvial mountain rivers, *Water Resour. Res.*, *39*(7), 1176, doi:10.1029/2002WR001815.

Lenzi, M. A., F. Comiti, and A. Marion (2004a), Local scouring at bed sills in a mountain river: Plima River, Italian Alps, *J. Hydraul. Eng.*, *130*, 267–269.

Lenzi, M. A., L. Mao, and F. Comiti (2004b), Magnitude-frequency analysis of bed load data in an Alpine boulder bed stream, *Water Resour. Res.*, *40*, W07201, doi:10.1029/2003WR002961.

Lenzi, M. A., L. Mao, and F. Comiti (2006), When does bedload transport begin in steep boulder bed streams?, *Hydrol. Processes*, *20*, 3517–3533.

Leonard, S., G. Staidl, J. Fogg, K. Gebhardt, W. Hagenbruck, and D. Pritchard (1992), Riparian Area Management: Procedures for Ecological Site Inventory – with Special Reference to Riparian-Wetland Sites, *Bureau of Land Management TR-1737-7*, 135 pp, U.S. Dep. of the Interior, Denver, Colo.

Leopold, L. B. (1951), Rainfall frequency: An aspect of climatic variation, *Eos Trans. AGU*, *32*, 347–357.

Leopold, L. B. (1970), An improved method for size distribution of stream bed gravel, *Water Resour. Res.*, *6*, 1357–1366.

Leopold, L. B. (1976), Reversal of erosion cycle and climatic change, *Quat. Res.*, *6*, 557–562.

Leopold, L. B. (1992), Sediment size that determines channel morphology, in *Dynamics of Gravel-Bed Rivers*, edited by P. Billi et al., pp. 297–311, John Wiley, Chichester, U. K.

Leopold, L. B., and W. B. Bull (1979), Base level, aggradation, and grade, *Proc. Am. Philos. Soc.*, *123*, 168–202.

Leopold, L. B., and W. W. Emmett (1976), Bedload measurements, East Fork River, Wyoming, *Proc. Natl. Acad. Sci. U. S. A.*, *73*, 1000–1004.

Leopold, L. B., and W. B. Langbein (1962), The concept of entropy in landscape evolution, *U.S. Geol. Surv. Prof. Pap.*, *500-A*.

Leopold, L. B., and T. Maddock (1953), The hydraulic geometry of stream channels and some physiographic implications, *U.S. Geol. Surv. Prof. Pap.*, *252*, 57 pp.

Leopold, L. B., and J. P. Miller (1956), Ephemeral streams - hydraulic factors and their relation to the drainage net, *U.S. Geol. Surv. Prof. Pap.*, *282-A*, pp. 1–37.

Leopold, L. B., and M. G. Wolman (1957), River channel patterns - braided, meandering and straight, *U.S. Geol. Surv. Prof. Pap.*, *282B*, pp. 39–85.

Leopold, L. B., M. G. Wolman, and J. P. Miller (1964), *Fluvial Processes in Geomorphology*, 522 pp., W. H. Freeman, San Francisco, Calif.

Leopold, L. B., W. W. Emmett, and R. M. Myrick (1966), Channel and hillslope processes in a semiarid area, New Mexico, *U.S. Geol. Surv. Prof. Pap.*, *352-G*, pp. 193–243.

Leopold, L. B., R. Huppman, and A. Miller (2005), Geomorphic effects of urbanization in forty-one years of observation, *Proc. Am. Philos. Soc.*, *149*, 349–371.

Lepistö, A., P. Seuna, and L. Bengtsson (1994), The environmental tracer approach in storm runoff studies in forested catchments, in *FRIEND: Flow Regimes from International Experimental and Network Data*, edited by P. Seuna et al., *IAHS Publ.*, *221*, 369–379.

Lepori, F., and N. Hjerdt (2006), Disturbance and aquatic biodiversity: Reconciling contrasting views, *BioScience*, *56*, 809–818.

Lepori, F., D. Palm, E. Brännäs, and B. Malmqvist (2005), Does restoration of structural heterogeneity in streams enhance fish and macroinvertebrate diversity?, *Ecol. Appl.*, *15*, 2060–2071.

Le Roux, J. P. (2004), An integrated law of the wall for hydrodynamically transitional flow over plane beds, *Sediment. Geol.*, *163*, 311–321.

Le Roux, J. P., and M. Brodalka (2004), An Excel™-VBA programme for the analysis of current velocity profiles, *Comput. Geosci.*, *30*, 867–879.

Lespez, L. (2003), Geomorphic responses to long-term land use changes in eastern Macedonia (Greece), *Catena*, *51*, 181–208.

Lesschen, J. P., J. M. Schoorl, and L. H. Cammeraat (2009), Modelling runoff and erosion for a semi-arid catchment using a multi-scale approach based on hydrological connectivity, *Geomorphology*, *109*, 174–183.

Lester, R. E., and A. J. Boulton (2008), Rehabilitating agricultural streams in Australia with wood: A review, *Environ. Manage.*, *42*, 310–326.

Lettenmaier, D. P. (2005), Some perspectives on hydrological prediction and the role of PUB, in *Predictions in Ungauged Basins: International Perspectives on the State of the Art and Pathways Forward*, edited by S. Franks et al., *IAHS Publ.*, *301*, 46–56.

Levish, D. R. (2002), Paleohydrologic bounds: Non-exceedance information for flood hazard assessment, in *Ancient Floods, Modern Hazards: Principles and Applications of Paleoflood Hydrology*, edited by P. K. House et al., pp. 175–190, AGU, Washington, D. C.

Lewin, J. (1976), Initiation of bed forms and meanders in coarse-grained sediment, *Geol. Soc. Am. Bull.*, *87*, 281–285.

Lewin, J., and P. A. Brewer (2001), Predicting channel patterns, *Geomorphology*, *40*, 329–339.

Lewin, J., B. E. Davies, and P. J. Wolfenden (1977), Interactions between channel change and historic mining sediments, in *River Channel Changes*, edited by K. J. Gregory, pp. 353–367, John Wiley, Chichester, U. K.

Lewis, J. (1998), Evaluating the impacts of logging activities on erosion and suspended sediment transport in the Caspar Creek watersheds, in *Proceedings of the Conference on Coastal Watersheds: The Caspar Creek Story*, edited by R. R. Ziemer, pp. 55–69, Pacific Southwest Res. Sta., USDA, Albany, Calif.

Lewis, J. (2003), Turbidity-controlled sampling for suspended sediment load estimation, in *Erosion and Sediment Transport Measurement in Rivers: Technological and Methodological Advances*, edited by J. Bogen, T. Fergus, and D. E. Walling, *IAHS Publ.*, *283*, 13–20.

Lewis, S. G., and J. F. Birnie (2001), Little Ice Age alluvial fan development in Langedalen, western Norway, *Geogr. Ann.*, *83A*, 179–190.

Leyland, J., and S. E. Darby (2008), An empirical-conceptual gully evolution model for channelled sea cliffs, *Geomorphology*, *102*, 419–434.

Li, R. M. (1979), Water and sediment routing from watersheds, in *Modeling of Rivers*, edited by H. W. Shen, pp. 9-1–9-88, John Wiley, New York.

Li, S. S., R. G. Millar, and S. Islam (2008), Modelling gravel transport and morphology for the Fraser River Gravel Reach, British Columbia, *Geomorphology*, *95*, 206–222.

Li, X. Y., L. Y. Liu, S. Y. Gao, Y. J. Ma, and Z. P. Zang (2008), Stemflow in three shrubs and its effect on soil water enhancement in semiarid loess region of China, *Agric. For. Meteorol.*, *148*, 1501–1507.

Li, Y., J. Harbor, A. P. Stroeven, D. Fabel, J. Kleman, D. Fink, M. W. Caffee, and D. Elmore (2005), Ice sheet erosion patterns in valley systems in northern Sweden investigated using cosmogenic nuclides, *Earth Surf. Processes Landforms*, *30*, 1039–1049.

Li, Z., and P. D. Komar (1986), Laboratory measurements of pivoting angles for applications to selective entrainment of gravel in a current, *Sedimentology*, *33*, 413–423.

Liébault, F., and H. Piégay (2001), Assessment of channel changes due to long-term bedload supply decrease, Roubion River, France, *Geomorphology*, *36*, 167–186.

Liébault, F., and H. Piégay (2002), Causes of 20th century channel narrowing in mountain and piedmont rivers of southeastern France, *Earth Surf. Processes Landforms*, *27*, 425–444.

Liébault, F., P. Clement, H. Piégay, C. F. Rogers, G. M. Kondolf, and N. Landon (2002), Contemporary channel changes in the Eygues Basin, southern French Prealps: The relationship of subbasin variability to watershed characteristics, *Geomorphology*, *45*, 53–66.

Liebe, J. R., N. van de Giesen, M. Andreini, M. T. Walter, and T. S. Steenhuis (2009), Determining watershed response in data poor environments with remotely sensed small reservoirs as runoff gauges, *Water Resour. Res.*, *45*, W07410, doi:10.1029/2008WR007369.

Lienkaemper, G. W., and F. J. Swanson (1987), Dynamics of large woody debris in streams in old-growth Douglas-fir forests, *Can. J. For. Res.*, *17*, 150–156.

Lifton, Z. M., G. D. Thackray, R. Van Kirk, and N. F. Glenn (2009), Influence of rock strength on the valley morphometry of Big Creek, central Idaho, USA, *Geomorphology*, *111*, 173–181.

Ligon, F. K., W. E. Dietrich, and W. J. Trush (1995), Downstream ecological effects of dams: A geomorphic perspective, *BioScience*, *45*, 183–192.

Likens, G. E., F. H. Bormann, R. S. Pierce, J. S. Eaton, and N. M. Johnson (1977), *Biogeochemistry of a Forested Ecosystem*, 146 pp., Springer, New York.

Lilbaek, G., and J. W. Pomeroy (2007), Modelling enhanced infiltration of snowmelt ions into frozen soil, *Hydrol. Processes*, *21*, 2641–2649.

Lillquist, K., and K. Walker (2006), Historical glacier and climate fluctuations at Mount Hood, Oregon, *Arct. Antarct. Alp. Res.*, *38*, 399–412.

Limbrey, S. (1983), Archaeology and palaeohydrology, in *Background to Palaeohydrology: A Perspective*, edited by K. J. Gregory, pp. 189–212, John Wiley, Chichester, U. K.

Limerinos, J. T. (1970), Determination of the Manning coefficient from measured bed roughness in natural channels, *U.S. Geol. Surv. Water Supply Pap.*, *1898-B*, 47 pp.

Limousin, J. M., S. Rambal, J. M. Ourcival, and R. Joffre (2008), Modelling rainfall interception in a Mediterranean Quercus ilex ecosystem: Lessons from a throughfall exclusion experiment, *J. Hydrol.*, *357*, 57–66.

Lin, C. A., L. Wen, D. Chaumont, and M. Béland (2005), The use of coupled meteorological and hydrological models for flash flood simulation, in *Climate and Hydrology in Mountain Areas*, pp. 221–232, John Wiley, Chichester, U. K.

Lin, G.-F., L.-H. Chen, and J.-N. Lai (2006a), Assessment of risk due to debris flow events: A case study in central Taiwan, *Nat. Hazards*, *39*, 1–14.

Lin, G.-W., H. Chen, N. Hovius, M.-J. Horng, S. Dadson, P. Meunier, and M. Lines (2008), Effects of earthquake and cyclone sequencing on landsliding and fluvial sediment transfer in a mountain catchment, *Earth Surf. Processes Landforms*, *33*, 1354–1373.

Lin, H. S., W. Kogelmann, C. Walker, and M. A. Bruns (2006b), Soil moisture patterns in a forested catchment; a hydropedological perspective, *Geoderma*, *131*, 345–368.

Lin, W.-T., W.-C. Chou, C.-Y. Lin, P.-H. Huang, and J.-S. Tsai (2006c), Automated suitable drainage network extraction from digital elevation models in Taiwan's upstream watersheds, *Hydrol. Processes*, *20*, 289–306.

Lin, Z., and T. Oguchi (2006), DEM analysis on longitudinal and transverse profiles of steep mountainous watersheds, *Geomorphology*, *78*, 77–89.

Lin, Z., and T. Oguchi (2009), Longitudinal and transverse profiles of hilly and mountainous watersheds in Japan, *Geomorphology*, *111*, 17–26.

Lin, Z., T. Oguchi, Y.-G. Chen, and K. Saito (2009), Constant-slope alluvial fans and source basins in Taiwan, *Geology*, *37*, 787–790.

Liniger, H. (1992), Water and soil resource conservation and utilization on the northwest side of Mount Kenya, *Mt. Res. Dev.*, *12*, 363–373.

Link, T. E., G. N. Flerchinger, M. Unsworth, and D. Marks (2005), Water relations in an old-growth Douglas fir stand, in *Climate and Hydrology in Mountain Areas*, edited by C. de Jong, D. Collins, and R. Ranzi, pp. 147–159, John Wiley, Chichester, U. K.

Linstead, C. (2001), The effects of large woody debris accumulations on river hydraulics and implications for physical habitat, in *Hydro-ecology: Linking Hydrology and Aquatic Ecology*, edited by M. C. Acreman, *IAHS Publ., 266*, 91–99.

Lisle, T. E. (1979), A sorting mechanism for a riffle-pool sequence: Summary, *Geol. Soc. Am. Bull., 90*, 616–617.

Lisle, T. E. (1982), Effects of aggradation and degradation on riffle-pool morphology in natural gravel channels, northwestern California, *Water Resour. Res., 18*, 1643–1651.

Lisle, T. E. (1986), Stabilization of a gravel channel by large streamside obstructions and bedrock bends, Jacoby Creek, northwestern California, *Geol. Soc. Am. Bull., 97*, 999–1011.

Lisle, T. E. (1987), Overview: Channel morphology and sediment transport in steepland streams, in Erosion and Sedimentation in the Pacific Rim, in *Proceedings of the Corvallis Symp, Aug. 1987*, edited by R. L. Beschta et al., *IAHS Publ., 165*, 287–297.

Lisle, T. E. (1995), Particle size variations between bed load and bed material in natural gravel bed channels, *Water Resour. Res., 31*, 1107–1118.

Lisle, T. E. (2008), The evolution of sediment waves influenced by varying transport capacity in heterogeneous rivers, in *Gravel-Bed Rivers VI: From Process Understanding to River Restoration*, edited by H. Habersack, H. Piégay, and M. Rinaldi, pp. 443–472, Elsevier, Amsterdam.

Lisle, T. E., and M. Church (2002), Sediment transport-storage relations for degrading, gravel bed channels, *Water Resour. Res., 38*(11), 1219, doi:10.1029/2001WR001086.

Lisle, T. E., and S. Hilton (1992), The volume of fine sediment in pools: An index of sediment supply in gravel-bed streams, *Water Resour. Bull., 28*, 371–383.

Lisle, T. E., and S. Hilton (1999), Fine bed material in pools of natural gravel bed channels, *Water Resour. Res., 35*, 1291–1304.

Lisle, T. E., and M. A. Madej (1992), Spatial variation in armouring in a channel with high sediment supply, in *Dynamics of Gravel-Bed Rivers*, edited by P. Billi et al., pp. 277–293, John Wiley, Chichester, U. K.

Lisle, T. E., and M. B. Napolitano (1998), Effects of recent logging on the main channel of North Fork Caspar Creek, in *Proceedings of the Conference on Coastal Watersheds: The Caspar Creek Story*, edited by R. R. Ziemer, pp. 81–85, Pacific Southwest Res. Sta., USDA, Albany, Calif.

Lisle, T. E., H. Ikeda, and F. Iseya (1991), Formation of stationary alternate bars in a steep channel with mixed-size sediment: A flume experiment, *Earth Surf. Processes Landforms, 16*, 463–469.

Lisle, T. E., J. E. Pizzuto, H. Ikeda, F. Iseya, and Y. Kodama (1997), Evolution of a sediment wave in an experimental channel, *Water Resour. Res., 33*, 1971–1981.

Litchfield, N. J., and K. R. Berryman (2005), Correlation of fluvial terraces within the Hikurangi Margin, New Zealand: Implications for climate and baselevel controls, *Geomorphology, 68*, 291–313.

Litchfield, N., and K. Berryman (2006), Relations between postglacial fluvial incision rates and uplift rates in the North Island, New Zealand, *J. Geophys. Res., 111*, F02007, doi:10.1029/2005JF000374.

Liu, C.-M. (1992), The effectivness of check dams in controlling upstream channel stability in northeastern Taiwan, in *Erosion, Debris Flows and Environment in Mountain Regions*, edited by D. E. Walling, T. R. Davies, and B. Hasholt, *IAHS Publ., 209*, 423–428.

Liu, F., M. W. Williams, and N. Caine (2004a), Source waters and flow paths in an alpine catchment, Colorado Front Range, United States, *Water Resour. Res., 40*, W09401, doi:10.1029/2004WR003076.

Liu, J. G., P. J. Mason, N. Clerici, S. Chen, A. Davis, F. Miao, H. Deng, and L. Liang (2004b), Landslide hazard assessment in the Three Gorges area of the Yangtze River using ASTER imagery: Zigui-Badong, *Geomorphology, 61*, 171–187.

Liu, J. J., Y. Li, P. C. Su, and Z. L. Cheng (2008), Magnitude-frequency relations in debris flow, *Environ. Geol., 55*, 1345–1354.

Liu, R. Z., J. R. Ni, A. G. L. Borthwick, Z. S. Li, and O. W. H. Wai (2006), Rapid zonation of abrupt mass movement hazard; Part II, Applications, *Geomorphology, 80*, 226–235.

Liu, W., Q. Guo, and Y. Wang (2008), Temporal-spatial climate change in the last 35 years in Tibet and its geo-environmental consequences, *Environ. Geol.*, *54*, 1747–1754.

Lizarralde, M., J. Escobar, and G. Deferrari (2004), Invader species in Argentina: A review about the beaver (Castor canadensis) population situation on Tierra del Fuego ecosystem, *Interciencia*, *29*, 352–356.

Lliboutry, L., B. M. Arnao, A. Pautre, and B. Schneider (1977), Glaciological problems set by the control of dangerous lakes in Cordillera Blanca, Peru: I. Historical failures of morainic dams, their causes and prevention, *J. Glaciol.*, *18*, 239–254.

Lock, M. A., R. R. Wallace, J. W. Costerton, R. M. Ventullo, and S. E. Charton (1984), River epilithon: Toward a structural-functional model, *Oikos*, *42*, 10–22.

Löffler, J., and O. Rößler (2005), Climatologic and hydrologic coupling in the ecology of Norwegian high mountain catchments, in *Climate and Hydrology in Mountain Areas*, edited by C. de Jong, D. Collins, and R. Ranzi, pp. 185–214, John Wiley, Chichester, U. K.

Lofthouse, C., and A. Robert (2008), Riffle-pool sequences and meander morphology, *Geomorphology*, *99*, 214–223.

Loget, N., and J. Van Den Driessche (2009), Wave train model for knickpoint migration, *Geomorphology*, *106*, 376–382.

Loheide, S. P., and S. M. Gorelick (2007), Riparian hydroecology: A coupled model of the observed interactions between groundwater flow and meadow vegetation patterning, *Water Resour. Res.*, *43*, W07414, doi:10.1029/2006WR005233.

Loheide, S. P., and J. D. Lundquist (2009), Snowmelt-induced diel fluxes through the hyporheic zone, *Water Resour. Res.*, *45*, W07404, doi:10.1029/2008WR007329.

Long, J. W., A. Tecle, and B. M. Burnette (2003), Marsh development at restoration sites on the White Mountain Apache Reservation, Arizona, *J. Am. Water Resour. Assoc.*, *39*, 1345–1359.

Lopez-Moreno, J. I., M. Beniston, and J. M. Garcia-Ruiz (2008), Environmental change and water management in the Pyrenees; facts and future perspectives for Mediterranean mountains, *Global Planet. Change*, *61*, 300–312.

López-Moreno, J. I., S. Goyette, and M. Beniston (2009), Impact of climate change on snowpack in the Pyrenees: Horizontal spatial variability and vertical gradients, *J. Hydrol.*, *374*, 384–396.

Lorang, M. S., and G. Aggett (2005), Potential sedimentation impacts related to dam removal: Icicle Creek, Washington, USA, *Geomorphology*, *71*, 182–201.

Lorch, B. (1998), Transport and aquatic impacts of highway traction sand and salt near Vail Pass, Colorado, unpublished M.S. thesis, Colo. State Univ., Ft. Collins.

Loughran, R. J., and B. L. Campbell (1995), The identification of catchment sediment sources, in *Sediment and Water Quality in River Catchments*, edited by I. Foster, A. Gurnell, and B. Webb, pp. 189–205, John Wiley, Chichester, U. K.

Loukas, A., L. Vasiliades, and N. R. Dalezios (2000), Flood producing mechanisms identification in southern British Columbia, Canada, *J. Hydrol.*, *227*, 218–235.

Love, D. W. (1979), Quaternary fluvial geomorphic adjustments in Chaco Canyon, New Mexico, in *Adjustments of the Fluvial System*, edited by D. D. Rhodes and G. P. Williams, pp. 277–308, Kendall/ Hunt, Dubuque, Iowa.

Lowrance, R., R. Todd, J. Fail, O. Hendrickson, and R. Leonard (1984), Riparian forests as nutrient filters in agricultural watersheds, *BioScience*, *34*, 374–377.

Lu, H., C. J. Moran, and M. Sivapalan (2005), A theoretical exploration of catchment-scale sediment delivery, *Water Resour. Res.*, *41*, W09415, doi:10.1029/2005WR004018.

Lu, X. X., S. Zhang, and J. Xu (2010), Climate change and sediment flux from the Roof of the World, *Earth Surf. Processes Landforms*, *35*, 732–735.

Luce, C. H., and T. A. Black (1999), Sediment production from forest roads in western Oregon, *Water Resour. Res.*, *35*, 2561–2570.

Luchi, R., W. Bertoldi, G. Zolezzi, and M. Tubino (2007), Monitoring and predicting channel change in a free-evolving, small Alpine river: Ridanna Creek (north east Italy, *Earth Surf. Processes Landforms, 32*, 2104–2119.

Ludwig, R., et al. (2003), Web-based modelling of energy, water and matter fluxes to support decision making in mesoscale catchments: The integrative perspective of GLOWA-Danube, *Phys. Chem. Earth, 28*, 621–634.

Lugo, A. E., S. L. Brown, R. Dodson, T. S. Smith, and H. H. Shugart (1999), The Holdridge life zones of the conterminous United States in relation to ecosystem mapping, *J. Biogeogr., 26*, 1025–1038.

Lugt, H. J. (1983), *Vortex Flow in Nature and Technology*, 297 pp., Wiley, New York.

Luino, F. (2005), Sequence of instability processes triggered by heavy rainfall in the northern Italy, *Geomorphology, 66*, 13–39.

Lundberg, N., and R. J. Dorsey (1990), Rapid Quaternary emergence, uplift, and denudation of the Coastal Range, eastern Taiwan, *Geology, 18*, 638–641.

Lundquist, J. D., M. D. Dettinger, and D. R. Cayan (2005), Snow-fed streamflow timing at different basin scales: Case study of the Tuolumne River above Hetch Hetchy, Yosemite, California, *Water Resour. Res., 41*, W07005, doi:10.1029/2004WR003933.

Lunetta, R. S., B. L. Cosentino, D. R. Montgomery, E. M. Beamer, and T. J. Beechie (1997), GIS-based evaluation of salmon habitat in the Pacific Northwest, *Photogramm. Eng. Remote Sens., 63*, 1219–1229.

Luo, W., B. P. Grudzinski, and D. Pederson (2010), Estimating hydraulic conductivity from drainage patterns – a case study in the Oregon Cascades, *Geology, 38*, 335–338.

Lusby, G. C., and T. J. Toy (1976), An evaluation of surface-mine spoils area restoration in Wyoming using rainfall simulation, *Earth Surf. Processes, 1*, 375–386.

Lustig, L. K. (1965), Sediment yield of the Castaic watershed, western Los Angeles County, California - a quantitative geomorphic approach, *U.S. Geol. Surv. Prof. Pap., 422-F*, 23 pp.

Lvovich, M. I., and G. M. Chernogaeva (1977), The water balance of Moscow, in *Effects of Urbanization and Industrialization on the Hydrological Regime and on Water Quality, IAHS Publ.123*, pp. 48–51.

Lyons, W. B., C. A. Nezat, A. E. Carey, and D. M. Hicks (2002), Organic carbon fluxes to the ocean from high-standing islands, *Geology, 30*, 443–446.

Maas, G. S., and M. G. Macklin (2002), The impact of recent climate change on flooding and sediment supply within a Mediterranean mountain catchment, southwestern Crete, Greece, *Earth Surf. Processes Landforms, 27*, 1087–1105.

MacArthur, R. C., M. D. Harvey, and E. F. Sing (1990), Estimating sediment delivery and yield on alluvial fans, in *Hydraulics/Hydrology of Arid Lands*, edited by R. H. French, pp. 700–705, Am. Soc. of Civ. Eng., New York.

MacDonald, L. H. (2000), Evaluating and managing cumulative effects: Process and constraints, *Environ. Manage., 26*, 299–315.

MacDonald, L. H., and J. A. Hoffman (1995), Causes of peak flows in northwestern Montana and northeastern Idaho, *Water Resour. Bull., 31*, 79–86.

MacDonald, L. H., D. M. Anderson, and W. E. Dietrich (1997), Paradise threatened: Land use and erosion on St. John, U. S. Virgin Islands, *Environ. Manage., 21*, 851–863.

MacDonald, L. H., R. Sampson, D. Brady, L. Juarros, and D. Martin (2000), Predicting post-fire erosion and sedimentation risk on a landscape scale: A case study from Colorado, *J. Sustainable For., 11*, 57–87.

MacDonald, L. H., R. W. Sampson, and D. M. Anderson (2001), Runoff and road erosion at the plot and road segment scales, St. John, US Virgin Islands, *Earth Surf. Processes Landforms, 26*, 251–272.

MacFarlane, W. A., and E. Wohl (2003), Influence of step composition on step geometry and flow resistance in step-pool streams of the Washington Cascades, *Water Resour. Res., 39*(2), 1037, doi:10.1029/2001WR001238.

MacGregor, K. C., R. S. Anderson, S. P. Anderson, and E. D. Waddington (1998), Glacially driven evolution of long valley profiles and implications for alpine landscape evolution, *Eos Trans. AGU*, *79*, F337.

Mack, G. H., M. R. Leeder, and M. Carothers-Durr (2008), Modern flood deposition, erosion, and fan-channel avulsion on the semiarid Red Canyon and Palomas Canyon alluvial fans in the southern Rio Grande Rift, New Mexico, USA, *J. Sediment. Res.*, *78*, 432–442.

Mackay, D. (1991), *Working the Kauri: A social and photographic history of New Zealand's pioneer Kauri bushmen*, Random Century, Auckland, New Zealand.

Mackenthun, K. M., and W. M. Ingram (1966), Pollution and life in the water, in *Organism-Substrate Relationships in Streams*, edited by K. W. Cummins, C. A. Tryon, and R. T. Hartman, *Spec. Publ.*, *4*, 136–145.

Mackey, B. H., J. J. Roering, and J. A. McKean (2009), Long-term kinematics and sediment flux of an active earthflow, Eel River, California, *Geology*, *37*, 803–806.

Mackin, J. H. (1948), Concept of the graded river, *Geol. Soc. Am. Bull.*, *59*, 463–512.

Macklin, M. G., B. T. Rumsby, and T. Heap (1992), Flood alluviation and entrenchment – Holocene valley-floor development and transformation in the British uplands, *Geol. Soc. Am. Bull.*, *104*, 631–643.

Macklin, M. G., P. A. Brewer, K. A. Hudson-Edwards, G. Bird, T. J. Coulthard, I. A. Dennis, P. J. Lechler, J. R. Miller, and J. N. Turner (2006), A geomorphological approach to the management of rivers contaminated by metal mining, *Geomorphology*, *79*, 423–447.

MacWilliams, M. L., J. M. Wheaton, G. B. Pasternack, R. L. Street, and P. K. Kitanidis (2006), Flow convergence routing hypothesis for pool-riffle maintenance in alluvial rivers, *Water Resour. Res.*, *42*, W10427, doi:10.1029/2005WR004391.

Maddy, D., T. Demir, D. R. Bridgland, A. Veldkamp, C. Stemerdink, T. van der Schriek, and R. Westaway (2005), An obliquity-controlled early Pleistocene river terrace record from western Turkey?, *Quat. Res.*, *63*, 339–346.

Madej, M. A. (2001), Development of channel organization and roughness following sediment pulses in single-thread, gravel bed rivers, *Water Resour. Res.*, *37*, 2259–2272.

Madej, M. A. (2005), The role of organic matter in sediment budgets in forested terrain, in *Sediment Budgets 2*, edited by A. J. Horowitz and D. E. Walling, *IAHS Publ.*, *292*, 9–15.

Madej, M. A. (2010), Redwoods, restoration, and implications for carbon budgets, *Geomorphology*, *116*, 264–273.

Madej, M. A., and V. Ozaki (1996), Channel response to sediment wave propagation and movement, Redwood Creek, California, USA, *Earth Surf. Processes Landforms*, *21*, 911–927.

Madej, M. A., and V. Ozaki (2009), Persistence of effects of high sediment loading in a salmon-bearing river, northern California, in *Management and Restoration of Fluvial Systems with Broad Historical Changes and Human Impacts*, edited by L. A. James, S. L. Rathburn, and G. R. Whittecar, pp. 43–55, Geol. Soc. of Am., Boulder, Colo.

Madej, M. A., E. A. Eschenbach, C. Diaz, R. Teasley, and K. Baker (2006), Optimization strategies for sediment reduction practices on roads in steep, forested terrain, *Earth Surf. Processes Landforms*, *31*, 1643–1656.

Madole, R. F., D. P. Van Sistine, and J. A. Michael (1998), Pleistocene Glaciation in the Upper Platte River Drainage Basin, Colorado, 1:500,000 scale map, U. S. Geol. Surv., Denver, Colo.

Madsen, A. T., and A. S. Murray (2009), Optically stimulated luminescence dating of young sediments: A review, *Geomorphology*, *109*, 3–16.

Madsen, S. W. (1995), Channel Response Associated with Predicted Water and Sediment Yield Increases in Northwestern Montana, unpublished MS thesis, 230 pp., Colo. State Univ., Ft. Collins

Magilligan, F. J. (1992), Thresholds and the spatial variability of flood power during extreme floods, *Geomorphology*, *5*, 373–390.

Magilligan, F. J., and P. F. McDowell (1997), Stream channel adjustments following elimination of cattle grazing, *J. Am. Water Resour. Assoc.*, *33*, 867–878.

Magilligan, F. J., and K. H. Nislow (2001), Long-term changes in regional hydrologic regime following impoundment in a humid-climate watershed, *J. Am. Water Resour. Assoc.*, *37*, 1551–1569.

Magilligan, F. J., and M. L. Stamp (1997), Historical land-cover changes and hydrogeomorphic adjustment in a small Georgia watershed, *Ann. Assoc. Am. Geogr.*, *87*, 614–635.

Magilligan, F. J., B. Gomez, L. A. K. Mertes, L. C. Smith, N. D. Smith, D. Finnegan, and J. B. Garvin (2002), Geomorphic effectiveness, sandur development, and the pattern of landscape response during jökulhlaups: Skeiäarársandur, southeastern Iceland, *Geomorphology*, *44*, 95–113.

Magirl, C. S., R. H. Webb, and P. G. Griffiths (2005), Changes in the water surface profile of the Colorado River in Grand Canyon, Arizona, between 1923 and 2000, *Water Resour. Res.*, *41*, W05021, doi:10.1029/2003WR002519.

Magirl, C. S., J. W. Gartner, G. M. Smart, and R. H. Webb (2009), Water velocity and the nature of critical flow in large rapids on the Colorado River, Utah, *Water Resour. Res.*, *45*, W05427, doi:10.1029/2009WR007731.

Magirl, C. S., P. G. Griffiths, and R. H. Webb (2010), Analyzing debris flows with the statistically calibrated empirical model LAHARZ in southeastern Arizona, USA, *Geomorphology*, *119*, 111–124.

Magner, J. A., B. Vondracek, and K. N. Brooks (2008), Grazed riparian management and stream channel response in southeastern Minnesota (USA) streams, *Environ. Manage.*, *42*, 377–390.

Maher, E., A. M. Harvey, and D. France (2007), The impact of a major Quaternary river capture on the alluvial sediments of a beheaded river system, the Rio Alias SE Spain, *Geomorphology*, *84*, 344–356.

Maitre, V., A.-C. Cosandey, E. Desagher, and A. Parriaux (2003), Effectiveness of groundwater nitrate removal in a river riparian area; the importance of hydrogeological conditions, *J. Hydrol.*, *278*, 76–93.

Maizels, J. K. (1983), Palaeovelocity and palaeodischarge determination for coarse gravel deposits, in *Background to Palaeohydrology*, edited by K. J. Gregory, pp. 101–139, John Wiley, Chichester, U. K.

Maizels, J. K. (1989), Sedimentology and palaeohydrology of Holocene flood deposits in front of a jökulhlaup glacier, South Iceland, in *Floods: Hydrological, Sedimentological and Geomorphological Implications*, edited by K. Beven and P. Carling, pp. 239–251, John Wiley, Chichester, U. K.

Maizels, J. K. (1991), The origin and evolution of Holocene sandur deposits in areas of jökulhlaup drainage, Iceland, in *Environmental Change in Iceland: Past and Present*, edited by J. K. Maizels and C. Caseldine, pp. 267–302, Kluwer Acad., Dordrecht.

Maizels, J. K. (1997), Jökulhlaup deposits in proglacial areas, *Quat. Sci. Rev.*, *16*, 793–819.

Maizels, J. K., and A. Russell (1992), Quaternary perspectives on jökulhlaup prediction, *Quat. Proc.*, *2*, 133–152.

Major, J. J., and L. E. Mark (2006), Peak flow responses to landscape disturbances caused by the cataclysmic 1980 eruption of Mount St. Helens, Washington, *Geol. Soc. Am. Bull.*, *118*, 938–958.

Major, J. J., T. C. Pierson, and R. L. Dinehart (1999), Sediment yield following severe volcanic disturbance – two-decade perspective from Mount St. Helens, *Geol. Soc. Am. Abstr. Programs*, *31*, A-200.

Major, J. J., T. C. Pierson, R. L. Dinehart, and J. E. Costa (2000), Sediment yield following severe volcanic disturbance – a two-decade perspective from Mount St. Helens, *Geology*, *28*, 819–822.

Major, J. J., J. E. O'Connor, G. E. Grant, K. R. Spicer, H. M. Bragg, A. Rhode, D. Q. Tanner, C. W. Anderson, and J. R. Wallick (2008a), Initial fluvial response to the removal of Oregon's Marmot Dam, *Eos Trans. AGU*, *89*(27), 241–242.

Major, J. J., K. R. Spicer, A. Rhode, J. E. O'Connor, H. M. Bragg, D. Q. Tanner, C. W. Anderson, J. R. Wallick, and G. E. Grant (2008b), Initial fluvial response to the removal of Oregon's Marmot Dam, *Eos Trans. AGU*, *89*(27), 241–242.

Malanson, G. P., and D. R. Butler (1990), Woody debris, sediment, and riparian vegetation of a subalpine river, Montana, USA, *Arct. Alp. Res.*, *22*, 183–194.

Malanson, G. P., and D. R. Butler (1991), Floristic variation among gravel bars in a subalpine river, Montana, USA, *Arct. Alp. Res.*, *23*, 273–278.

Malard, F., K. Tockner, M.-J. Dole-Olivier, and J. V. Ward (2002), A landscape perspective of surface-subsurface hydrological exchanges in river corridors, *Freshwater Biol.*, *47*, 621–640.

Malarz, R. (2005), Effects of flood abrasion of the Carpathian alluvial gravels, *Catena*, *64*, 1–26.

Malet, J.-P., D. Laigle, A. Remaitre, and O. Maquaire (2005), Triggering conditions and mobility of debris flows associated to complex earthflows, *Geomorphology*, *66*, 215–235.

Malik, I., and M. Matyja (2008), Bank erosion history of a mountain stream determined by means of anatomical changes in exposed tree roots over the last 100 years (Bílá Opava River – Czech Republic), *Geomorphology*, *98*, 126–142.

Malmaeus, J. M., and M. A. Hassan (2002), Simulation of individual particle movement in a gravel streambed, *Earth Surf. Processes Landforms*, *27*, 81–97.

Malmon, D. V., S. L. Reneau, T. Dunne, D. Katzman, and P. G. Drakos (2005), Influence of sediment storage on downstream delivery of contaminated sediment, *Water Resour. Res.*, *41*, W05008, doi:10.1029/2004WR003288.

Manga, M., and J. W. Kirchner (2000), Stress partitioning in streams by large woody debris, *Water Resour. Res.*, *36*, 2373–2379.

Manners, R. B., M. W. Doyle, and M. J. Small (2007), Structure and hydraulics of natural woody debris jams, *Water Resour. Res.*, *43*, W06432, doi:10.1029/2006WR004910.

Mao, L., and M. A. Lenzi (2007), Sediment mobility and bedload transport conditions in an alpine stream, *Hydrol. Processes*, *21*, 1882–1891.

Mao, L., and N. Surian (2010), Observations on sediment mobility in a large gravel-bed river, *Geomorphology*, *114*, 326–337.

Mao, L., F. Comiti, A. Andreoli, M. A. Lenzi, and G. R. Scussel (2005), Bankfull and bed load effective discharge in a steep boulder-bed channel, in *Sediment Budgets I*, edited by D. E. Walling and A. J. Horowitz, *IAHS Publ.*, *291*, 189–195.

Mao, L., A. Andreoli, F. Comiti, and M. A. Lenzi (2008a), Geomorphic effects of large wood jams on a sub-Antarctic mountain stream, *River Res. Appl.*, *24*, 249–266.

Mao, L., F. Comiti, A. Andreoli, L. Picco, M. A. Lenzi, A. Urciulo, R. Iturraspe, and A. Iroumè (2008b), Role and management of in-channel wood in relation to flood events in Southern Andes basins, in *Monitoring, Simulation, Prevention and Remediation of Dense and Debris Flow II*, edited by D. de Wrachien, C. A. Brebbia, and M. A. Lenzi, *WIT Trans. Eng. Sci.*, *60*, 207–216.

Mao, L., G. P. Uyttendaele, A. Iroumé, and M. A. Lenzi (2008c), Field based analysis of sediment entrainment in two high gradient streams located in Alpine and Andine environments, *Geomorphology*, *93*, 368–383.

Mao, L., M. Cavalli, F. Comiti, L. Marchi, M. A. Lenzi, and M. Arattano (2009), Sediment transfer processes in two Alpine catchments of contrasting morphological settings, *J. Hydrol.*, *364*, 88–98.

Marc, V., J.-F. Didon-Lescot, and C. Michael (2001), Investigation of the hydrological processes using chemical and isotopic tracers in a small Mediterranean forested catchment during autumn recharge, *J. Hydrol.*, *247*, 215–229.

Marchand, J. P., R. D. Jarrett, and L. L. Jones (1984), *Velocity profile, water-surface slope, and bed-material size for selected streams in Colorado*, U.S. Geol. Surv. Open File Rep. 84-733.

Marchenko, S. S., A. P. Gorbunov, and V. E. Romanovsky (2007), Permafrost warming in the Tien Shan Mountains, central Asia, *Global Planet. Change*, *56*, 311–327.

Marchetti, D. W., and T. E. Cerling (2005), Cosmogenic ^{3}He exposure ages of Pleistocene debris flows and desert pavements in Capitol Reef National Park, Utah, *Geomorphology*, *67*, 423–435.

Marchetti, D. W., T. E. Cerling, J. C. Dohrenwend, and W. Gallin (2007), Ages and significance of glacial and mass movement deposits on the west side of Boulder Mountain, Utah, USA, *Palaeogeogr. Palaeoclimatol. Palaeoecol.*, *252*, 503–513.

Marchi, L., and M. Cavalli (2007), Procedures for the documentation of historical debris flows: Application to the Chieppena Torrent (Italian Alps), *Environ. Manage.*, *40*, 493–503.

Marchi, L., M. Arattano, and A. M. Deganutti (2002), Ten years of debris-flow monitoring in the Moscardo Torrent (Italian Alps), *Geomorphology*, *46*, 1–17.

Marcus, W. A., K. Roberts, L. Harvey, and G. Tackman (1992), An evaluation of methods for estimating Manning's n in small mountain streams, *Mt. Res. Dev.*, *12*, 227–239.

Marcus, W. A., S. C. Ladd, J. A. Stoughton, and J. W. Stock (1995), Pebble counts and the role of user-dependent bias in documenting sediment size distribution, *Water Resour. Res.*, *31*, 2625–2631.

Marcus, W. A., G. A. Meyer, and D. R. Nimmo (2001), Geomorphic control of persistent mine impacts in a Yellowstone Park stream and implications for the recovery of fluvial systems, *Geology*, *29*, 355–358.

Marcus, W. A., R. A. Marston, C. R. Colvard, and R. D. Gray (2002), Mapping the spatial and temporal distributions of woody debris in streams of the Greater Yellowstone Ecosystem, USA, *Geomorphology*, *44*, 323–335.

Marcus, W. A., C. J. Legleiter, R. J. Aspinall, J. W. Boardman, and R. L. Crabtree (2003), High spatial resolution hyperspectral mapping of in-stream habitats, depths, and woody debris in mountain streams, *Geomorphology*, *55*, 363–380.

Marcus, W. A., R. J. Aspinall, and R. A. Marston (2004), Geographic information systems and surface hydrology in mountains, in *Geographic Information Science and Mountain Geomorphology*, edited by M. P. Bishop and J. F. Shroder, pp. 343–379, Praxis, Chichester, U. K.

Marden, M., C. Mazengarb, A. Palmer, K. Berryman, and D. Rowan (2008), Last glacial aggradation and postglacial sediment production from the non-glacial Waipaoa and Waimata catchments, Hikurangi Margin, North Island, New Zealand, *Geomorphology*, *99*, 404–419.

Margalef, R. (1960), Ideas for a synthetic approach to the ecology of running waters, *Int. Rev. Gesamten Hydrobiol.*, *45*, 133–153.

Marinucci, M. R., F. Giorgi, M. Benitson, M. Wild, P. Tschuck, and A. Bernasconi (1995), High resolution simulation of January and July climate over the western Alpine region with a nested regional modeling system, *Theor. Appl. Climatol.*, *51*, 119–138.

Marion, A., M. Bellinello, I. Guymer, and A. Packman (2002), Effect of bed form geometry on the penetration of nonreactive solutes into a streambed, *Water Resour. Res.*, *38*(10), 1209, doi:10.1029/2001WR000264.

Marion, A., S. J. Tait, and I. K. McEwan (2003), Analysis of small-scale gravel bed topography during armoring, *Water Resour. Res.*, *39*(12), 1334, doi:10.1029/2003WR002367.

Marion, A., A. I. Packman, M. Zaramella, and A. Bottacin-Busolin (2008a), Hyporheic flows in stratified beds, *Water Resour. Res.*, *44*, W09433, doi:10.1029/2007WR006079.

Marion, A., M. Zaramella, and A. Bottacin-Busolin (2008b), Solute transport in rivers with multiple storage zones: The STIR model, *Water Resour. Res.*, *44*, W10406, doi:10.1029/2008WR007037.

Marion, D. A., and F. Weirich (2003), Equal-mobility bed load transport in a small, step-pool channel in the Ouachita Mountains, *Geomorphology*, *55*, 139–154.

Mark, B. G., and G. O. Seltzer (2005), Evaluation of recent glacier recession in the Cordillera Blanca, Peru (AD 1962–1999); spatial distribution of mass loss and climatic forcing, *Quat. Sci. Rev.*, *24*, 2265–2280.

Mark, R. K. (1992), Map of debris-flow probability, San Mateo County, California, *U.S. Geol. Surv. Misc. Invest. Ser., Map I-1257-M*.

Markham, A., and G. Day (1994), Sediment transport in the Fly River basin, Papua New Guinea, in *Variability in Stream Erosion and Sediment Transport*, edited by L. J. Olive, R. J. Loughran, and J. A. Kesby, *IAHS Publ.*, *224*, 233–239.

Marks, J. C., R. Parnell, C. Carter, E. C. Dinger, and G. A. Haden (2006), Interactions between geomorphology and ecosystem processes in travertine streams: Implications for decommissioning a dam on Fossil Creek, Arizona, *Geomorphology*, *77*, 299–307.

Marks, K., and P. Bates (2000), Integration of high-resolution topographic data with floodplain flow models, *Hydrol. Processes*, *14*, 2109–2122.

Marlow, C. B., and T. M. Pogacnik (1985), Time of grazing and cattle-induced damage to streambanks, in *Riparian Ecosystems and Their Management: Reconciling Conflicting Uses*, USDA Forest Service General Tech. Rep. RM-120, pp. 279–284, First North American Riparian Conference, Tucson, Arizona.

Marron, D. C. (1985), Colluvium in bedrock hollows on steep slopes, Redwood Creek drainage basin, northwestern California, in *Soils and Geomorphology*, edited by P. D. Jungerius, *Catena Suppl., 6*, 59–68.

Marston, R. A. (1982), The geomorphic significance of log steps in forest streams, *Ann. Assoc. Am. Geogr., 72*, 99–108.

Marston, R. A. (2008), Land, life, and environmental change in mountains, *Ann. Assoc. Am. Geogr., 98*, 507–520.

Marston, R. A. (2010), Geomorphology and vegetation on hillslopes: Interactions, dependencies, and feedback loops, *Geomorphology, 116*, 206–217.

Marston, R. A., and D. M. Furin (2004), Reclamation of surface coal mined lands in northwest Colorado, in *WorldMinds: Geographical Perspectives on 100 Problems*, edited by D. G. Janelle, B. Warf, and K. Hansen, pp. 515–519, Kluwer Acad., Dordrecht.

Marston, R. A., J. Girel, G. Pautou, H. Piégay, J.-P. Bravard, and C. Arneson (1995), Channel metamorphosis, floodplain disturbance, and vegetation development: Ain River, France, *Geomorphology, 13*, 121–131.

Marston, R. A., M. M. Miller, and L. Devkota (1998), Geoecology and mass movement in the Manaslu-Ganesh and Langtang-Jugal himals, Nepal, *Geomorphology, 26*, 139–150.

Marston, R. A., J.-P. Bravard, and T. Green (2003), Impacts of reforestation and gravel mining on the Malnant River, Haute-Savoie, French Alps, *Geomorphology, 55*, 65–74.

Marston, R. A., J. D. Mills, D. R. Wrazien, B. Bassett, and D. K. Splinter (2005), Effects of Jackson Lake Dam on the Snake River and its floodplain, Grand Teton National Park, Wyoming, USA, *Geomorphology, 71*, 79–98.

Marti, C., and G. R. Bezzola (2006), Bed load transport in braided gravel-bed rivers, in *Braided Rivers: Process, Deposits, Ecology and Management*, edited by G. H. Sambrook Smith et al., *Spec. Publ. Int. Assoc. Sedimentol., 36*, 199–215.

Martin, D. J., and L. E. Benda (2001), Patterns of instream wood recruitment and transport at the watershed scale, *Trans. Am. Fish. Soc., 130*, 940–958.

Martin, Y. (2003), Evaluation of bed load transport formulae using field evidence from the Vedder River, British Columbia, *Geomorphology, 53*, 75–95.

Martin, Y., and D. Ham (2005), Testing bedload transport formulae using morphologic transport estimates and field data: Lower Fraser River, British Columbia, *Earth Surf. Processes Landforms, 30*, 1265–1282.

Martín-Duque, J. F., M. A. Sanz, J. M. Bodoque, A. Lucía, and C. Martín-Moreno (2010), Restoring earth surface processes through landform design. A 13-year monitoring of a geomorphic reclamation model for quarries on slopes, *Earth Surf. Processes Landforms, 35*, 531–548.

Martín-Vide, J. P., and A. Andreatta (2009), Channel degradation and slope adjustment in steep streams controlled through bed sills, *Earth Surf. Processes Landforms, 34*, 38–47.

Martinez-Carreras, N., M. Soler, E. Hernandez, and F. Gallart (2007), Simulating badland erosion with KINEROS2 in a small Mediterranean mountain basin (Vallcebre, eastern Pyrenees, *Catena, 71*, 145–154.

Martinez-Castroviejo, R. (1988), Advances in fluvial geomorphology of mountain environments, *Pirineos, 132*, 65–88.

Marui, H., and H. Yoshimatsu (2007), Landslide dams formed by the 2004 mid-Niigata Prefecture earthquake in Japan, in *Progress in Landslide Science*, edited by K. Sassa et al., pp. 285–293, Springer, Berlin.

Marutani, T., M. Kasai, L. M. Reid, and N. A. Trustrum (1999), Influence of storm-related sediment storage on the sediment delivery from tributary catchments in the Upper Waipaoa River, New Zealand, *Earth Surf. Processes Landforms, 24*, 881–896.

Marzadri, A., D. Tonina, A. Bellin, G. Vignoli, and M. Tubino (2010), Semianalytical analysis of hyporheic flow induced by alternate bars, *Water Resour. Res., 46*, W07531, doi:10.1029/2009WR008285.

Masioka, M. H., R. Villalba, B. H. Luckman, M. E. Lascano, S. Delgado, and P. Stepanek (2008), 20th-century glacier recession and regional hydroclimatic changes in northwestern Patagonia, *Global Planet. Change*, *60*, 85–100.

Mason, D. P. M., T. A. Little, and R. J. Van Dissen (2006), Rates of active faulting during late Quaternary fluvial terrace formation at Saxton River, Awatere Fault, New Zealand, *Geol. Soc. Am. Bull.*, *118*, 1431–1446.

Mast, M. A., C. Kendall, D. H. Campbell, D. W. Clow, and J. Back (1995), Determination of hydrologic pathways in an alpine-subalpine basin using isotopic and chemical tracers, Loch Vale watershed, Colorado, USA, in *Biogeochemistry of Seasonally Snow-Covered Catchments*, edited by K. A. Tonnessen, M. W. Williams, and M. Tranter, *IAHS Publ.*, *228*, 263–270.

Mather, A. E., and A. J. Hartley (2005), Flow events on a hyper-arid alluvial fan: Quebrada Tambores, Salar de Atacama, northern Chile, in *Alluvial Fans: Geomorphology, Sedimentology, Dynamics*, edited by A. M. Harvey, A. E. Mather, and M. Stokes, *Geol. Soc. London Spec. Publ.*, *251*, 9–24.

Mather, A. E., and A. J. Hartley (2006), The application of drainage system analysis in constraining spatial patterns of uplift in the Coastal Cordillera of northern Chile, in *Tectonics, Climate, and Landscape Evolution*, edited by S. D. Willett et al., *Geol. Soc. Am. Spec. Pap.*, *398*, 87–99.

Mather, A. E., A. M. Harvey, and M. Stokes (2000), Quantifying long-term catchment changes of alluvial fan systems, *Geol. Soc. Am. Bull.*, *112*, 1825–1833.

Matmon, A., P. R. Bierman, J. Larsen, S. Southworth, M. J. Pavich, and M. W. Caffee (2003), Temporally and spatially uniform rates of erosion in the southern Appalachian Great Smoky Mountains, *Geology*, *31*, 155–158.

Matsui, K., and T. Ohta (2003), Estimating the snow distribution in a subalpine region using a distributed snowmelt model, in *Water Resources Systems – Water Availability and Global Change*, edited by S. Franks et al., *IAHS Publ.*, *280*, 282–291.

Matthes, G. H. (1947), Macroturbulence in natural stream flow, *Eos Trans. AGU*, *28*, 255–265.

Matyjasik, M., and N. Keate (2003), Montane wetland hydrochemistry, *Hydrol. Sci. Technol.*, *19*, 251–264.

Maurakis, E. G., W. S. Woolcott, and M. H. Sabaj (1992), Water currents in spawning areas of pebble nests of Nocomis leptocephalus (Pisces: Cyprinidae), *Southeast. Fishes Counc. Proc.*, *25*, 1–3.

Maxwell, A. R., and A. N. Papanicolaou (2001), Step-pool morphology in high-gradient streams, *Int. J. Sediment Res.*, *16*, 380–390.

May, C. L. (2002), Debris flows through different forest age classes in the central Oregon Coast Range, *J. Am. Water Resour. Assoc.*, *38*, 1097–1113.

May, C. L. (2007), Sediment and wood routing in steep headwater streams: An overview of geomorphic processes and their topographic signatures, *For. Sci.*, *53*, 119–130.

May, C. L., and R. E. Gresswell (2003a), Large wood recruitment and redistribution in headwater streams in the southern Oregon Coast Range, USA, *Can. J. For. Res.*, *33*, 1352–1362.

May, C. L., and R. E. Gresswell (2003b), Processes and rates of sediment and wood accumulation in headwater streams of the Oregon Coast Range, USA, *Earth Surf. Processes Landforms*, *28*, 409–424.

May, C. L., and R. E. Gresswell (2004), Spatial and temporal patterns of debris-flow deposition in the Oregon Coast Range, USA, *Geomorphology*, *57*, 135–149.

Mayer, B., M. Stoffel, M. Bollschweiler, J. Hübl, and F. Rudolf-Miklau (2010), Frequency and spread of debris floods on fans: A dendrogeomorphic case study from a dolomite catchment in the Austrian Alps, *Geomorphology*, *118*, 199–206.

McAllister, L. S. (2008), Reconstructing historical riparian conditions of two river basins in eastern Oregon, USA, *Environ. Manage.*, *42*, 412–425.

McArdell, B. W., and R. Faeh (2001), A computational investigation of river braiding, in *Gravel-Bed Rivers V*, edited by M. P. Mosley, pp. 73–86, N. Z. Hydrol. Soc., Wellington.

McAuliffe, J. R., L. A. Scuderi, and L. D. McFadden (2006), Tree-ring record of hillslope erosion and valley floor dynamics: Landscape responses to climate variation during the last 400 yr in the Colorado Plateau, northeastern Arizona, *Global Planet. Change*, *50*, 184–201.

McBride, M., W. C. Hession, and D. M. Rizzo (2010), Riparian reforestation and channel change: How long does it take?, *Geomorphology, 116*, 330–340.

McCabe, G. J., Jr., and L. E. Hay (1995), Hydrological effects of hypothetical climate change in the East River basin, Colorado, USA, *Hydrol. Sci. J., 40*, 303–318.

McCarthy, J. M. (2008), Factors Influencing Ecological Recovery Downstream of Diversion Dams in Southern Rocky Moutain Streams, unpublished M.S. thesis, Colo. State Univ., Ft. Collins.

McCartney, S. E., S. K. Carey, and J. W. Pomeroy (2006), Intra-basin variability of snowmelt water balance calculations in a subarctic catchment, *Hydrol. Processes, 20*, 1001–1016.

McClain, M. E., and R. J. Naiman (2008), Andean influences on the biogeochemistry and ecology of the Amazon River, *BioScience, 58*, 325–338.

McClain, M. E., et al. (2003), Biogeochemical hot spots and hot moments at the interface of terrestrial and aquatic ecosystems, *Ecosystems, 6*, 301–312.

McClain, M. E., G. Pinay, and R. M. Holmes (1999), Contrasting biogeochemical cycles of riparian forests in temperate, wet tropical, and arid regions, *1999 Annual Meeting Abstracts*, pp. 26, Ecological Society of America.

McCormick, B. C., K. N. Eshleman, J. L. Griffith, and P. A. Townsend (2009), Detection of flooding responses at the river basin scale enhanced by land use change, *Water Resour. Res., 45*, W08401, doi:10.1029/2008WR007594.

McDonald, A., S. N. Lane, N. E. Haycock, and E. A. Chalk (2004), Rivers of dreams: On the gulf between theoretical and practical aspects of an upland river restoration, *Trans. Inst. Br. Geogr., 29*, 257–281.

McDonnell, J. J. (1990), A rationale for old water discharge through macropores in a steep, humid catchment, *Water Resour. Res., 26*, 2821–2832.

McDonnell, J. J. (2003), Where does water go when it rains? Moving beyond the variable source area concept of rainfall-runoff response, *Hydrol. Processes, 17*, 1869–1875.

McDonnell, J. J., M. K. Stewart, and I. F. Owens (1991), Effect of catchment-scale subsurface mixing on stream isotopic response, *Water Resour. Res., 27*, 3065–3073.

McDonnell, J. J., B. McGlynn, K. Kendall, J. Shanley, and C. Kendall (1998), The role of near-stream riparian zones in the hydrology of steep upland catchments, in *Hydrology, Water Resources and Ecology in Headwaters*, edited by K. Kovar et al., *IAHS Publ., 248*, 173–180.

McDonnell, J. J., B. McGlynn, K. Vache, and I. Tromp-Van Meerveld (2005), A perspective on hillslope hydrology in the context of PUB, in *Predictions in Ungauged Basins: International Perspectives on the State of the Art and Pathways Forward*, edited by S. Franks et al., *IAHS Publ., 301*, 204–212.

McDonnell, J. J., et al. (2007), Moving beyond heterogeneity and process complexity: A new vision for watershed hydrology, *Water Resour. Res., 43*, W07031, doi:10.1029/2006WR005467.

McDowell, P. F. (2001), Spatial variations in channel morphology at segment and reach scales, Middle Fork John Day River, northeastern Oregon, in *Geomorphic Processes and Riverine Habitat*, edited by J. M. Dorava et al., pp. 159–172, AGU, Washington, D. C.

McDowell, P. F., and F. J. Magilligan (1997), Response of stream channels to removal of cattle grazing disturbance: Overview of western U. S. exclosure studies, in *Management of Landscapes Disturbed by Channel Incision*, edited by S. S. Y. Wang, E. J. Langendoen, and F. D. Shields, pp. 469–475, Univ. of Mississippi, Oxford.

McEwen, L. J., J. A. Matthews, R. A. Shakesby, and M. S. Berrisford (2002), Holocene gorge excavation linked to boulder fan formation and frost weathering in a Norwegian alpine periglaciofluvial system, *Arct. Antarct. Alp. Res., 34*, 345–357.

McGinness, H. M., M. C. Thoms, and M. R. Southwell (2002), Connectivity and fragmentation of flood plain-river exchanges in a semiarid, anabranching river system, in *The Structure, Function and Management Implications of Fluvial Sedimentary Systems*, edited by F. J. Dyer, M. C. Thoms, and J. M. Olley, *IAHS Publ., 276*, 19–26.

McGlynn, B. L., and J. Seibert (2003), Distributed assessment of contributing area and riparian buffering along stream networks, *Water Resour. Res.*, *39*(4), 1082, doi:10.1029/2002WR001521.

McGlynn, B. L., J. J. McDonnell, and D. D. Brammer (2002), A review of the evolving perceptual model of hillslope flowpaths at the Maimai catchments, New Zealand, *J. Hydrol.*, *257*, 1–26.

McIlroy, S. K., C. Montagne, C. A. Jones, and B. L. McGlynn (2008), Identifying linkages between land use, geomorphology, and aquatic habitat in a mixed-use watershed, *Environ. Manage.*, *42*, 867–876.

McKean, J. A., W. E. Dietrich, R. C. Finkel, J. R. Southon, and M. W. Caffee (1993), Quantification of soil production and downslope creep rates from cosmogenic ^{10}Be accumulations on a hillslope profile, *Geology*, *21*, 343–346.

McKean, J., and J. Roering (2004), Objective landslide detection and surface morphology mapping using high-resolution airborne laser altimetry, *Geomorphology*, *57*, 331–351.

McKenney, R. (2001), Channel changes and habitat diversity in a warm-water, gravel-bed stream, in *Geomorphic Processes and Riverine Habitat*, edited by J. M. Dorava et al., pp. 57–71, AGU, Washington, D. C.

McLean, S. R., and V. I. Nikora (2006), Characteristics of turbulent unidirectional flow over rough beds: Double-averaging perspective with particular focus on sand dunes and gravel beds, *Water Resour. Res.*, *42*, W10409, doi:10.1029/2005WR004708.

McLean, S. R., and J. D. Smith (1979), Turbulence measurements in the boundary layer over a sand wave field, *J. Geophys. Res.*, *84*, 7791–7808.

McLeay, D. J., I. K. Birtwell, G. F. Hartman, and G. L. Ennis (1987), Responses of Arctic grayling (Thymallus arcticus) to acute and prolonged exposure to Yukon placer mining sediment, *Can. J. Fish. Aquat. Sci.*, *44*, 658–673.

McNamara, J. P., and C. Borden (2004), Observations on the movement of coarse gravel using implanted motion-sensing radio transmitters, *Hydrol. Processes*, *18*, 1871–1884.

McNeil, W. J. (1966), Effect of the spawning bed environment on reproduction of pink and chum salmon, *U.S. Fish Wildl. Serv. Bull.*, *65*, 495–523.

McPherson, H. J. (1971), Downstream changes in sediment character in a high energy mountain stream channel, *Arct. Alp. Res.*, *3*, 65–79.

Meade, R. H., T. Dunne, J. E. Richey, U. M. Santos, and E. Salati (1985), Storage and remobilization of suspended sediment in the lower Amazon River of Brazil, *Science*, *228*, 488–490.

Measures, R., and S. Tait (2008), Quantifying the role of bed surface topography in controlling sediment stability in water- worked gravel deposits, *Water Resour. Res.*, *44*, W04413, doi:10.1029/2006WR005794.

Megahan, W. F., and C. C. Bohn (1989), Progressive, long-term slope failure following road construction and logging on noncohesive, granitic soils of the Idaho Batholith, in *Headwaters Hydrology*, pp. 501–510, Am. Water Resour. Assoc., Bethesda, Md.

Megahan, W. F., and W. J. Kidd (1972), Effects of logging and logging roads on erosion and sediment deposition from steep terrain, *J. For.*, *70*, 136–141.

Mei-e, R., and Z. Xianmo (1994), Anthropogenic influences on changes in the sediment load of the Yellow River, China, during the Holocene, *The Holocene*, *4*, 314–320.

Meigs, A. (1998), Bedrock landsliding accompanying deglaciation: Three possible examples from the Chugach/St. Elias Range, Alaska, *Eos Trans. AGU*, *79*, F337.

Meixner, T., R. C. Bales, M. W. Williams, D. H. Campbell, and J. S. Baron (2000), Stream chemistry modeling of two watersheds in the Front Range, Colorado, *Water Resour. Res.*, *36*, 77–87.

Mejia-Navarro, M., and E. E. Wohl (1994), Geological hazard and risk evaluation using GIS: Methodology and model applied to Medellín, Colombia, *Bull. Assoc. Eng. Geol.*, *31*, 459–481.

Mejia-Navarro, M., E. E. Wohl, and S. D. Oaks (1994), Geological hazards, vulnerability, and risk assessment using GIS: Model for Glenwood Springs, Colorado, *Geomorphology*, *10*, 331–354.

Meko, D. M. (1990), Inferences from tree rings on low frequency variations in runoff in the Interior Western United States, *Proceedings Sixth Annual Pacific Climate Workshop, Asilomar, California - March 5–8*,

1989, edited by J. L. Betancourt and A. M. MacKay, *Tech. Rep. 23*123–127, Calif. Dep. of Water Resour., Interagency Ecol. Studies Program.

Melack, J. M., and J. O. Sickman (1995), Snowmelt induced chemical changes in seven streams in the Sierra Nevada, California, in *Biogeochemistry of Seasonally Snow-Covered Catchments*, edited by K. A. Tonnessen, M. W. Williams, and M. Tranter, *IAHS Publ.*, *228*, 221–234.

Melchiorre, C., M. Matteuci, A. Azzoni, and A. Zanchi (2008), Artificial neural networks and cluster analysis in landslide susceptibility zonation, *Geomorphology*, *94*, 379–400.

Meleason, M. A., S. V. Gregory, and J. P. Bolte (1999), Simulation of large-wood dynamics in small streams of the Pacific Northwest, *1999 Annual Meeting Abstracts*, pp. 147, Ecological Society of America.

Meleason, M. A., R. Davies-Colley, A. Wright-Stow, J. Horrox, and K. Costley (2005), Characteristics and geomorphic effect of wood in New Zealand's native forest streams, *Int. Rev. Hydrobiol.*, *90*, 466–485.

Melelli, L., and A. Taramelli (2004), An example of debris-flow hazard modelling using GIS, *Nat. Hazards Earth Syst. Sci.*, *4*, 347–358.

Melis, T. S., R. H. Webb, P. G. Griffiths, and T. W. Wise (1995), *Magnitude and frequency data for historic debris flows in Grand Canyon National Park and vicinity, Arizona*, U.S. Geol. Surv. Water Resour. Invest. Rep. 94-4214, 285 pp.

Melis, T. S., D. J. Topping, and D. M. Rubin (2003), Testing laser-based sensors for continuous in situ monitoring of suspended sediment in the Colorado River, Arizona, in *Erosion and Sediment Transport Measurement in Rivers: Technological and Methodological Advances*, edited by J. Bogen, T. Fergus, and D. E. Walling, *IAHS Publ.*, *283*, 21–27.

Melosh, H. J. (1987), The mechanics of large rock avalanches, in *Debris Flows/Avalanches: Process, Recognition, and Mitigation*, edited by J. E. Costa and G. F. Wieczorek, *Rev. Eng. Geol.*, *7*, 41–49.

Melton, M. A. (1960), Intravalley variation in slope angles related to microclimate and erosional environment, *Geol. Soc. Am. Bull.*, *71*, 133–144.

Mendoza, J. A., P. Ulriksen, F. Picado, and T. Dahlin (2008), Aquifer interactions with a polluted mountain river of Nicaragua, *Hydrol. Processes*, *22*, 2264–2273.

Menges, E. S., and D. M. Waller (1983), Plant strategies in relation to elevation and light in floodplain herbs, *Am. Nat.*, *122*, 454–473.

Menounos, B. (2000), A Holocene debris-flow chronology for an alpine catchment, Colorado Front Range, in *Geomorphology, Human Activity and Global Environmental Change*, edited by O. Slaymaker, pp. 117–149, John Wiley, Chichester, U. K.

Menzel, L., and H. Lang (1998), Spatial variation in evapotranspiration in Swiss Alpine regions, in *Hydrology, Water Resources and Ecology in Headwaters*, edited by K. Kovar et al., *IAHS Publ.*, *248*, 115–121.

Merefield, J. R. (1995), Sediment mineralogy and the environmental impact of mining, in *Sediment and Water Quality in River Catchments*, edited by I. Foster, A. Gurnell, and B. Webb, pp. 145–160, John Wiley, Chichester, U. K.

Mernild, S. H., G. E. Liston, and B. Hasholt (2007), Snow-distribution and melt modelling for glaciers in Zackenberg River drainage basin, north-eastern Greenland, *Hydrol. Processes*, *21*, 3249–3263.

Mernild, S. H., B. Hasholt, and G. E. Liston (2008), Climatic control on river discharge simulations, Zackenberg River drainage basin, northeast Greenland, *Hydrol. Processes*, *22*, 1932–1948.

Merot, P., B. Ezzehar, C. Walter, and P. Aurousseau (1995), Mapping waterlogging of soils using digital terrain models, *Hydrol. Processes*, *9*, 27–34.

Merriam, J. L., W. H. McDowell, J. L. Tank, W. M. Wollheim, C. L. Crenshaw, and S. L. Johnson (2002), Characterizing nitrogen dynamics, retention and transport in a tropical rainforest stream using an in situ ^{15}N addition, *Freshwater Biol.*, *47*, 143–160.

Merritt, D. M., and E. E. Wohl (2002), Processes governing hydrochory along rivers: Hydraulics, hydrology, and dispersal phenology, *Ecol. Appl.*, *12*, 1071–1087.

Merritt, D. M., and E. E. Wohl (2006), Plant dispersal along rivers fragmented by dams, *River Res. Appl.*, *22*, 1–26.

Merritts, D. J., and M. Ellis (1994), Introduction to special section on tectonics and topography, *J. Geophys. Res.*, *99*(B6), 12,135–12,141.

Merritts, D. J., and K. R. Vincent (1989), Geomorphic response of coastal streams to low, intermediate, and high rates of uplift, Mendocino triple junction region, northern California, *Geol. Soc. Am. Bull.*, *100*, 1373–1388.

Merritts, D. J., K. R. Vincent, and E. E. Wohl (1994), Long river profiles, tectonisim, and eustasy: A guide to interpreting fluvial terraces, *J. Geophys. Res.*, *99*(B7), 14,031–14,050.

Mertes, L. A. K. (1997), Documentation and significance of the perirheic zone on inundated floodplains, *Water Resour. Res.*, *33*, 1749–1762.

Mertes, L. A. K. (2000), Inundation hydrology, in *Inland Flood Hazards: Human, Riparian, and Aquatic Communities*, edited by E. E. Wohl, pp. 145–166, Cambridge Univ. Press, Cambridge.

Mertes, L. A. K. (2002), Remote sensing of riverine landscapes, *Freshwater Biol.*, *47*, 799–816.

Merz, J. E., G. B. Pasternack, and J. M. Wheaton (2006), Sediment budget for salmonid spawning habitat rehabilitation in a regulated river, *Geomorphology*, *76*, 207–228.

Messerli, B., and J. D. Ives (1984), *Mountain Ecosystems: Stability and Instability*, Int. Mtn. Soc., Boulder, Colo.

Messerli, B., and J. D. Ives (Eds.) (1997), *Mountains of the World: A Global Priority*, The Parthenon, London.

Meyer, G. A. (2001), Recent large-magnitude floods and their impact on valley-floor environments of northeastern Yellowstone, *Geomorphology*, *40*, 271–290.

Meyer, G. A., and P. M. Watt (1999), Hydrological controls on the geomorphic and ecological impacts of a tailings dam-break flood on Soda Butte Creek, Montana and Yellowstone National Park, *Geol. Soc. Am. Abstr. Programs*, *31*, A-253.

Meyer, G. A., S. G. Wells, R. C. Balling, and A. J. T. Jull (1992), Response of alluvial systems to fire and climate change in Yellowstone National Park, *Nature*, *357*, 147–150.

Meyer, G. A., S. G. Wells, and A. J. T. Jull (1995), Fire and alluvial chronology in Yellowstone National Park; climatic and intrinsic controls on Holocene geomorphic processes, *Geol. Soc. Am. Bull.*, *107*, 1211–1230.

Meyer, G. A., J. L. Pierce, S. H. Wood, and A. J. T. Jull (2001), Fire, storms, and erosional events in the Idaho batholith, *Hydrol. Processes*, *15*, 3025–3028.

Meyer, J. L., D. L. Strayer, J. B. Wallace, S. L. Eggert, G. S. Helfman, and N. E. Leonard (2007), The contribution of headwater streams to biodiversity in river networks, *J. Am. Water Resour. Assoc.*, *43*, 86–103.

Meyer-Peter, E., and R. Müller (1948), Formulas for bed load transport, in *Report on Second Meeting of the International Association of Hydraulic Structures Research*, pp. 39–64, Stockholm, Sweden.

Miall, A. D. (1977), A review of the braided-river depositional environment, *Earth Sci. Rev.*, *13*, 1–62.

Michalik, A., and W. Bartnik (1994), An attempt at determination of incipient bed load motion in mountain streams, in *Dynamics and Geomorphology of Mountain Rivers*, edited by P. Ergenzinger and K.-H. Schmidt, pp. 289–299, Springer, Berlin.

Michel, R. L., D. Campbell, D. Clow, and J. T. Turk (2000), Timescales for migration of atmospherically derived sulphate through an alpine/subalpine watershed, Loch Vale, Colorado, *Water Resour. Res.*, *36*, 27–36.

Micheli, E. R., and J. W. Kirchner (2002a), Effects of wet meadow riparian vegetation on streambank erosion. 1. Remote sensing measurements of streambank migration and erodibility, *Earth Surf. Processes Landforms*, *27*, 627–639.

Micheli, E. R., and J. W. Kirchner (2002b), Effects of wet meadow riparian vegetation on streambank erosion. 2. Measurements of vegetated bank strength and consequences for failure mechanics, *Earth Surf. Processes Landforms*, *27*, 687–697.

Middleton, G. V., and J. B. Southard (1984), *Mechanics of Sediment Movement (SEPM Short Course No. 3)*, 401 pp., SEPM Soc. for Sediment, Providence, R. I.

Migon, P., M. Hradek, and K. Parzoch (2002), Extreme events in the Sudetes Mountains; their long-term geomorphic impact and possible controlling factors, *Stud. Geomorphol. Carpatho-Balcanica*, *36*, 29–48.

Mihalcea, C., C. Mayer, G. Diolaiuti, C. D'Agata, C. Smiraglia, A. Lambrecht, E. Vuillermoz, and G. Tartari (2008), Spatial distribution of debris thickness and melting from remote-sensing and meteorological data, at debris-covered Baltoro Glacier, Karakoram, Pakistan, *Ann. Glaciol.*, *48*, 49–57.

Mika, S., A. Boulton, D. Ryder, and D. Keating (2008), Ecological function in rivers: Insights from crossdisciplinary science, in *River Futures: An Integrative Scientific Approach to River Repair*, edited by G. J. Brierley and K. A. Fryirs, pp. 85–99, Island Press, Washington, D. C.

Mikos, M. (1994), The downstream fining of gravel-bed sediments in the Alpine Rhine River, in *Dynamics and Geomorphology of Mountain Rivers*, edited by P. Ergenzinger and K.-H. Schmidt, pp. 93–108, Springer, Berlin.

Milan, D. J., G. L. Heritage, A. R. G. Large, and M. E. Charlton (2001), Stage dependent variability in tractive force distribution through a riffle-pool sequence, *Catena*, *44*, 85–109.

Milan, D. J., G. L. Heritage, and D. Hetherington (2007), Application of a 3D laser scanner in the assessment of erosion and deposition volumes and channel change in a proglacial river, *Earth Surf. Processes Landforms*, *32*, 1657–1674.

Milhous, R. T. (1973), Sediment transport in a gravel-bottomed stream, unpublished Ph.D. dissertation, 232 pp., Oreg. State Univ., Corvallis.

Milhous, R. T. (1982), Effect of sediment transport and flow regulation on the ecology of gravel-bed rivers, in *Gravel-Bed Rivers*, edited by R. D. Hey, C. R. Thorne, and J. C. Bathurst, pp. 819–841, John Wiley, Chichester, U. K.

Millar, R. G. (1999), Grain and form resistance in gravel-bed rivers, *J. Hydraul. Res.*, *37*, 303–312.

Millar, R. G. (2005), Theoretical regime equations for mobile gravel-bed rivers with stable banks, *Geomorphology*, *64*, 207–220.

Millar, R. G., and M. C. Quick (1993), Effect of bank stability on geometry of gravel rivers, *J. Hydraul. Eng.*, *119*, 1343–1363.

Millar, R. G., and M. C. Quick (1994), Flow resistance of high-gradient gravel channels, in *Proceedings, ASCE Hydraulic Engineering '94 Conference, August 1–5, Buffalo, New York*, edited by G. V. Cotroneo and R. R. Rumer, pp. 717–721, Am. Soc. of Civ. Eng., New York.

Miller, A. J. (1990a), Flood hydrology and geomorphic effectiveness in the central Appalachians, *Earth Surf. Processes Landforms*, *15*, 119–134.

Miller, A. J. (1990b), Fluvial response to debris associated with mass wasting during extreme floods, *Geology*, *18*, 599–602.

Miller, A. J. (1994), Debris-fan constrictions and flood hydraulics in river canyons: Some implications from two-dimensional flow modelling, *Earth Surf. Processes Landforms*, *19*, 681–697.

Miller, A. J. (1995), Valley morphology and boundary conditions influencing spatial variations of flood flow, in *Natural and Anthropogenic Influences in Fluvial Geomorphology*, Geophys. Monogr. Ser., vol. 89, edited by J. E. Costa et al., pp. 57–81, AGU, Washington, D. C.

Miller, A. J., and B. L. Cluer (1998), Modeling considerations for simulation of flow in bedrock channels, in *Rivers Over Rock: Fluvial Processes in Bedrock Channels*, Geophys. Monogr. Ser., vol. 107, edited by K. J. Tinkler and E. E. Wohl, pp. 61–104, AGU, Washington, D. C.

Miller, A. J., and D. J. Parkinson (1993), Flood hydrology and geomorphic effects on river channels and floodplains: The flood of November 4–5, 1985, in the South Branch Potomac River basin of West Virginia, in *Geomorphic studies of the storm and flood of November 3–4, 1985, in the Upper Potomac and Cheat River Basins in West Virginia and Virginia*, edited by R. B. Jacobson, U.S. Geol. Surv. Bull., *1981*, E1–E96.

Miller, D. J., and L. Benda (2000), Effects of punctuated sediment supply on valley-floor landforms and sediment transport, *Geol. Soc. Am. Bull.*, *112*, 1814–1824.

Miller, D. J., and K. M. Burnett (2008), A probabilistic model of debris-flow delivery to stream channels, demonstrated for the Coast Range of Oregon, USA, *Geomorphology*, *94*, 184–205.

Miller, E. K., C. D. Carson, A. J. Friedland, and J. D. Blum (1995), Chemical and isotopic tracers of snowmelt flowpaths in a subalpine watershed, in *Biogeochemistry of Seasonally Snow-Covered Catchments*, edited by K. A. Tonnessen, M. W. Williams, and M. Tranter, *IAHS Publ.*, *228*, 349–353.

Miller, J., R. Barr, D. Grow, P. Lechler, D. Richardson, K. Waltman, and J. Warwick (1999), Effects of the 1997 flood on the transport and storage of sediment and mercury within the Carson River Valley, west-central Nevada, *J. Geol.*, *107*, 313–327.

Miller, J., D. Germanoski, K. Waltman, R. Tausch, and J. Chambers (2001), Influence of late Holocene hillslope processes and landforms on modern channel dynamics in upland watersheds of central Nevada, *Geomorphology*, *38*, 373–391.

Miller, J. P. (1958), *High Mountain Streams: Effects of Geology on Channel Characteristics and Bed Material, Memoir*, vol. 4, State Bur. of Mines and Miner. Resour., Socorro, N. M.

Miller, J. R. (1991), The influence of bedrock geology on knickpoint development and channel-bed degradation along downcutting streams in south-central Indiana, *J. Geol.*, *99*, 591–605.

Miller, J. R., and J. B. Ritter (1996), An examination of the Rosgen classification of natural rivers: Discussion, *Catena*, *27*, 295–299.

Miller, R. A., J. Troxell, and L. B. Leopold (1971), Hydrology of two small river basins in Pennsylvania before urbanization, *U.S. Geol. Surv. Prof. Pap.*, *701-A*, 57 pp.

Miller, T. E. (1996), Geologic and hydrologic controls on karst and cave development in Belize, *J. Cave Karst Stud.*, *58*, 100–120.

Miller, W. R., and J. I. Drever (1977), Chemical weathering and related controls on surface water chemistry in the Absaroka Mountains, Wyoming, *Geochim. Cosmochim. Acta*, *41*, 1693–1702.

Milliman, J. D., and R. H. Meade (1983), World-wide delivery of river sediment to oceans, *J. Geol.*, *91*, 1–21.

Milliman, J. D., and J. P. M. Syvitski (1992), Geomorphic/tectonic control of sediment discharge to the ocean: The importance of small mountainous rivers, *J. Geol.*, *100*, 525–544.

Mills, H. H. (1989), Hollow form as a function of boulder size in the Valley and Ridge province, southwestern Virginia, *Geology*, *17*, 595–598.

Mills, H. H. (2000), Apparent increasing rates of stream incision in the eastern United States during the late Cenozoic, *Geology*, *28*, 955–957.

Milly, P. C. D., J. Betancourt, M. Falkenmark, R. M. Hirsch, Z. W. Kundzewicz, D. P. Lettenmaier, and R. J. Stouffer (2008), Stationarity is dead: Whither water management?, *Science*, *319*, 573–574.

Milner, A. M. (1994), System recovery, in *The Rivers Handbook*, vol. 2, edited by P. Calow and G. E. Petts, pp. 76–97, Blackwell Sci., Oxford, U. K.

Milner, N. J., J. Scullion, P. A. Carling, and D. T. Crisp (1991), The effects of discharge on sediment dynamics and consequent effects on invertebrates and salmonids in upland rivers, in *Advances in Applied Biology*, vol. 6, edited by T. H. Coaker, pp. 153–220, Academic Press, London.

Milzow, C., P. Molnar, B. W. McArdell, and P. Burlando (2006), Spatial organization in the step-pool structure of a steep mountain stream (Vogelbach, Switzerland), *Water Resour. Res.*, *42*, W04418, doi:10.1029/2004WR003870.

Mimikou, M. A., P. S. Hadjisavva, and Y. S. Kouvopoulos (1991), Regional effects of climate change on water resources systems, in *Hydrology for the Water Management of Large River Basins*, edited by F. H. M. Van de Ven et al., *IAHS Publ.*, *201*, 173–182.

Minor, H.-E., W. H. Hager, and S. Canepa (2002), Does an aerated water jet reduce plunge pool scour?, in *Rock Scour due to Falling High-Velocity Jets*, edited by A. J. Schleiss and E. Bollaert, pp. 117–124, Swets and Zietlinger, Lisse.

Minshall, G. W. (1988), Stream ecosystem theory: A global perspective, *J. North Am. Benthol. Soc.*, *7*, 263–288.

Minshall, G. W., R. C. Peterson, K. W. Cummins, T. L. Bott, J. R. Sedell, C. E. Cushing, and R. L. Vannote (1983), Interbiome comparison of stream ecosystem dynamics, *Ecol. Monogr.*, *53*, 1–25.

Minshall, G. W., K. W. Cummins, R. C. Petersen, C. E. Cushing, D. A. Bruns, J. R. Sedell, and R. L. Vannote (1985), Developments in stream ecosystem theory, *Can. J. Fish. Aquat. Sci.*, *42*, 1045–1055.

Miori, S., R. Repetto, and M. Tubino (2006), A one-dimensional model of bifurcations in gravel bed channels with erodible banks, *Water Resour. Res.*, *42*, W11413, doi:10.1029/2006WR004863.

Misukoshi, H., and M. Aniya (2002), Use of contour-based DEMs for deriving and mapping topographic attributes, *Photogramm. Eng. Remote Sens.*, *68*, 83–93.

Mitchell, D. K., and F. J. Pazzaglia (1999), A field-based test of the stream power incision law, *Geol. Soc. Am. Abstr. Programs*, *31*, A-255.

Mitchell, S. G., and D. R. Montgomery (2006), Influence of a glacial buzzsaw on the height and morphology of the Cascade Range in central Washington State, USA, *Quat. Res.*, *65*, 96–107.

Mitchell, S. G., D. Montgomery, and H. Greenberg (2009), Erosional unloading, hillslope geometry, and the height of the Cascade Range, Washington State, USA, *Earth Surf. Processes Landforms*, *34*, 1108–1120.

Miyabuchi, Y., and F. Nakamura (1991), Seasonal variation of erosion processes at the headwater basin of Oboppu River in Tarumae Volcano, Hokkaido, *Trans., Jpn. Geomorphol. Union*, *12*, 367–377.

Mizutani, T. (1987), Drainage basin characteristics affecting sediment discharge from steep mountain basins, in *Erosion and Sedimentation in the Pacific Rim*, edited by R. L. Beschta et al., *IAHS Publ.165*, pp. 397–398, Proceedings of the Corvallis Symposium, Aug. 1987.

Mizutani, T. (1998), Laboratory experiment and digital simulation of multiple fill-cut terrace formation, *Geomorphology*, *24*, 353–361.

Mizuyama, T. (1991), Sediment yield and river bed change in mountain rivers, in *Fluvial Hydraulics of Mountain Regions*, edited by A. Armanini and G. DiSilvio, pp. 147–161, Springer, Berlin.

Mizuyama, T. (1993), Structural and non-structural debris-flow countermeasures, in *Hydraulic Engineering '93*, edited by H. W. Shen, S. T. Su, and F. Wen, pp. 1914–1919, Am. Soc. of Civ. Eng., New York.

Mizuyama, T., A. Yazawa, and K. Ido (1987), Computer simulation of debris flow deposit processes, in *Erosion and Sedimentation in the Pacific Rim*, edited by R. L. Beschta et al., *IAHS Publ.*, *165*, 179–190.

Mizuyama, T., S. Kobashi, and G. Ou (1992), Prediction of debris flow peak discharge, *International Symposium, Interpraevent 1992-Bern*, pp. 99–108.

Mizuyama, T., K. Kosugi, I. Sato, and S. Kobashi (1994), Runoff through underground pipes in hollows, *Proceedings of the International Symposium on Forest Hydrology, Tokyo, Japan, Oct. 1994*, pp. 233–240, Univ. of Tokyo, Tokyo.

Mizuyama, T., M. Fujita, and M. Nonaka (2003), Measurement of bed load with the use of hydrophones in mountain torrents, in *Erosion and Sediment Transport Measurement in Rivers: Technological and Methodological Advances*, edited by J. Bogen, T. Fergus, and D. E. Walling, *IAHS Publ.*, *283*, 222–227.

Modaressi, H. (2006), Climate change and ground movements, *Geosciences*, *3*, 44–49.

Moeyersons, J., M. van den Eeckhaut, J. Nyssen, T. Gebreyohannes, J. van de Wauw, J. Hofmeister, J. Poesen, J. A. Deckers, and M. Haile (2008), Mass movement mapping for geomorphological understanding and sustainable development; Tigray, Ethiopia, *Catena*, *75*, 45–54.

Moir, H. J., C. N. Gibbins, C. Soulsby, and J. Webb (2004), Linking channel geomorphic characteristics to spatial patterns of spawning activity and discharge use by Atlantic salmon (Salmo salar L.), *Geomorphology*, *60*, 21–35.

Moldan, B., and J. Cerny (1994), Small catchment research, in *Biogeochemistry of Small Catchments: A Tool for Environmental Research*, edited by B. Moldan and J. Cerny, pp. 1–29, John Wiley, Chichester, U. K.

Molina, A., G. Govers, F. Cisneros, and V. Vanacker (2009), Vegetation and topographic controls on sediment deposition and storage on gully beds in a degraded mountain area, *Earth Surf. Processes Landforms*, *34*, 755–767.

Mollo-Christensen, E. (1971), Physics of turbulent flow, *AIAA J.*, *9*, 1217–1228.

Molnar, L. (Ed.) (1990), *Hydrology of Mountain Areas*, *IAHS Publ.*, *190*, 452 pp.

Molnar, P., and P. England (1990), Late Cenozoic uplift of mountain ranges and global climate change: Chicken or egg?, *Nature*, *346*, 29–34.

Molnar, P., and J. A. Ramirez (2002), On downstream hydraulic geometry and optimal energy expenditure: Case study of the Ashley and Taieri Rivers, *J. Hydrol.*, *259*, 105–115.

Molnar, P., et al. (1994), Quaternary climate change and the formation of river terraces across growing anticlines on the north flank of the Tien Shan, China, *J. Geol.*, *102*, 583–602.

Molnar, P., P. Burlando, and W. Ruf (2002), Integrated catchment assessment of riverine landscape dynamics, *Aquat Sci.*, *64*, 129–140.

Molnar, P., R. S. Anderson, G. Kier, and J. Rose (2006a), Relationships among probability distributions of stream discharges in floods, climate, bed load transport, and river incision, *J. Geophys. Res.*, *111*, F02001, doi:10.1029/2005JF000310.

Molnar, P., P. Burlando, J. Kirsch, and E. Hinz (2006b), Model investigations of the effects of land-use changes and forest damage on erosion in mountainous environments, in *Sediment Dynamics and the Hydromorphology of Fluvial Systems*, edited by J. S. Rowan, R. W. Duck, and A. Werritty, *IAHS Publ.*, *306*, 589–600.

Monecke, K., J. Winsemann, and J. Hanisch (2001), Climatic response of Quaternary alluvial deposits in the upper Kali Gandaki Valley (West Nepal), *Global Planet. Change*, *28*, 293–302.

Monteith, H., and G. Pender (2005), Flume investigations into the influence of shear stress history on a graded sediment bed, *Water Resour. Res.*, *41*, W12401, doi:10.1029/2005WR004297.

Montgomery, D. R. (1994a), Road surface drainage, channel initiation, and slope instability, *Water Resour. Res.*, *30*, 1925–1932.

Montgomery, D. R. (1994b), Valley incision and the uplift of mountain peaks, *J. Geophys. Res.*, *99*(B7), 13,913–13,921.

Montgomery, D. R. (1999), Process domains and the river continuum, *J. Am. Water Resour. Assoc.*, *35*, 397–410.

Montgomery, D. R. (2001a), Geomorphology, river ecology, and ecosystem management, in *Geomorphic Processes and Riverine Habitat*, edited by J. M. Dorava et al., pp. 247–253, AGU, Washington, D. C.

Montgomery, D. R. (2001b), Slope distributions, threshold hillslopes, and steady-state topography, *Am. J. Sci.*, *301*, 432–454.

Montgomery, D. R. (2004a), GIS in tectonic geomorphology and landscape evolution, in *Geographic Information Science and Mountain Geomorphology*, edited by M. P. Bishop and J. F. Shroder, pp. 425–460, Praxis, Chichester, U. K.

Montgomery, D. R. (2004b), Observations on the role of lithology in strath terrace formation and bedrock channel width, *Am. J. Sci.*, *304*, 454–476.

Montgomery, D. R. (2007), Is agriculture eroding civilization's foundation?, *GSA Today*, *17*(10), 4–9.

Montgomery, D. R., and T. B. Abbe (2006), Influence of logjam-formed hard points on the formation of valley-bottom landforms in an old-growth forest valley, Queets River, Washington, USA, *Quat. Res.*, *65*, 147–155.

Montgomery, D. R., and J. M. Buffington (1997), Channel-reach morphology in mountain drainage basins, *Geol. Soc. Am. Bull.*, *109*, 596–611.

Montgomery, D. R., and J. M. Buffington (1998), Channel processes, classification, and response, in *River Ecology and Management*, edited by R. Naiman and R. Bilby, pp. 13–42, Springer, New York.

Montgomery, D. R., and W. E. Dietrich (1988), Where do channels begin?, *Nature*, *336*, 232–234.

Montgomery, D. R., and W. E. Dietrich (1989), Source areas, drainage density, and channel initiation, *Water Resour. Res.*, *25*, 1907–1918.

Montgomery, D. R., and W. E. Dietrich (1992), Channel initiation and the problem of landscape scale, *Science*, *255*, 826–830.

Montgomery, D. R., and W. E. Dietrich (1994a), A physically based model for the topographic control on shallow landsliding, *Water Resour. Res.*, *30*, 1153–1171.

Montgomery, D. R., and W. E. Dietrich (1994b), Landscape dissection and drainage area-slope thresholds, in *Process Models and Theoretical Geomorphology*, edited by M. J. Kirkby, pp. 221–246, John Wiley, Chichester, U. K.

Montgomery, D. R., and W. E. Dietrich (2002), Runoff generation in a steep, soil-mantled landscape, *Water Resour. Res.*, *38*(9), 1168, doi:10.1029/2001WR000822.

Montgomery, D. R., and E. Foufoula-Georgiou (1993), Channel network source representation using digital elevation models, *Water Resour. Res.*, *29*, 3925–3934.

Montgomery, D. R., and K. B. Gran (2001), Downstream variations in the width of bedrock channels, *Water Resour. Res.*, *37*, 1841–1846.

Montgomery, D. R., and J. López-Blanco (2003), Post-Oligocene river incision, southern Sierra Madre Occidental, Mexico, *Geomorphology*, *55*, 235–247.

Montgomery, D. R., and MacDonald (2002), Diagnostic approach to stream channel assessment and monitoring, *J. Am. Water Resour. Assoc.*, *38*, 1–16.

Montgomery, D. R., and D. B. Stolar (2006), Reconsidering Himalayan river anticlines, *Geomorphology*, *82*, 4–15.

Montgomery, D. R., R. H. Wright, and T. Booth (1991), Debris flow hazard mitigation for colluvium-filled swales, *Bull. Assoc. Eng. Geol.*, *28*, 303–323.

Montgomery, D. R., J. M. Buffington, R. D. Smith, K. M. Schmidt, and G. Pess (1995a), Pool spacing in forest channels, *Water Resour. Res.*, *31*, 1097–1105.

Montgomery, D. R., G. E. Grant, and K. Sullivan (1995b), Watershed analysis as a framework for implementing ecosystem management, *Water Resour. Bull.*, *31*, 369–386.

Montgomery, D. R., T. B. Abbe, J. M. Buffington, N. P. Peterson, K. M. Schmidt, and J. D. Stock (1996a), Distribution of bedrock and alluvial channels in forested mountain drainage basins, *Nature*, *381*, 587–589.

Montgomery, D. R., J. M. Buffington, N. P. Peterson, D. Schuett-Hames, and T. P. Quinn (1996b), Stream-bed scour, egg burial depths, and the influence of salmonid spawning on bed surface mobility and embryo survival, *Can. J. Fish. Aquat. Sci.*, *53*, 1061–1070.

Montgomery, D. R., W. E. Dietrich, and K. Sullivan (1998), The role of GIS in watershed analysis, in *Landform Monitoring, Modelling and Analysis*, edited by S. N. Lane, K. S. Richards, and J. H. Chandler, pp. 241–261, John Wiley, Chichester, U. K.

Montgomery, D. R., M. S. Panfil, and S. K. Hayes (1999), Channel-bed mobility response to extreme sediment loading at Mount Pinatubo, *Geology*, *27*, 271–274.

Montgomery, D. R., G. Balco, and S. D. Willett (2001), Climate, tectonics, and the morphology of the Andes, *Geology*, *29*, 579–582.

Montgomery, D. R., B. D. Collins, J. M. Buffington, and T. B. Abbe (2003a), Geomorphic effects of wood in rivers, in *The Ecology and Management of Wood in World Rivers*, edited by S. V. Gregory, K. L. Boyer, and A. M. Gurnell, pp. 21–47, Am. Fish. Soc., Bethesda, Md.

Montgomery, D. R., T. M. Massong, and S. C. S. Hawley (2003b), Influence of debris flows and log jams on the location of pools and alluvial channel reaches, Oregon Coast Range, *Geol. Soc. Am. Bull.*, *115*, 78–88.

Montgomery, D. R., K. M. Schmidt, W. E. Dietrich, and J. McKean (2009), Instrumental record of debris flow initiation during natural rainfall: Implications for modeling slope stability, *J. Geophys. Res.*, *114*, F01031, doi:10.1029/2008JF001078.

Moody, J. A., and D. A. Kinner (2006), Spatial structures of stream and hillslope drainage networks following gully erosion after wildfire, *Earth Surf. Processes Landforms*, *31*, 319–337.

Moody, J. A., and D. A. Martin (2001), Initial hydrologic and geomorphic response following a wildfire in the Colorado Front Range, *Earth Surf. Processes Landforms, 26*, 1049–1070.

Moody, J. A., and R. H. Meade (2008), Terrace aggradation during the 1978 flood on Powder River, Montana, USA, *Geomorphology, 99*, 387–403.

Moody, J. A., D. A. Kinner, and X. íbeda (2009), Linking hydraulic properties of fire-affected soils to infiltration and water repellency, *J. Hydrol., 379*, 291–303.

Moog, D. B., and P. J. Whiting (1998), Annual hysteresis in bed load rating curves, *Water Resour. Res., 34*, 2393–2399.

Moon, B. P., and M. J. Selby (1983), Rock mass strength and scarp forms in southern Africa, *Geogr. Ann., 65A*, 135–145.

Moore, J. R., J. W. Sanders, W. E. Dietrich, and S. D. Glaser (2009), Influence of rock mass strength on the erosion rate of alpine cliffs, *Earth Surf. Processes Landforms, 34*, 1339–1352.

Moore, N., and P. Diplas (1994), Effects of particle shape on bedload transport, in *Proceedings, ASCE Hydraulic Engineering '94 Conference, August 1–5, Buffalo, New York*, edited by G. V. Cotroneo and R. R. Rumer, pp. 800–804, Am. Soc. of Civ. Eng., New York.

Morche, D., C. Katterfeld, S. Fuchs, and K.-H. Schmidt (2006), The life-span of a small high mountain lake, the Vordere Blaue Gumpe in the Bavarian Alps, in *Sediment Dynamics and the Hydromorphology of Fluvial Systems*, edited by J. S. Rowan, R. W. Duck, and A. Werritty, *IAHS Publ., 306*, 72–81.

Moreiras, S. M. (2005), Landslide susceptibility zonation in the Rio Mendoza valley, Argentina, *Geomorphology, 66*, 345–357.

Moreiras, S. M. (2006), Frequency of debris flows and rockfall along the Mendoza River valley (central Andes), Argentina: Associated risk and future scenario, *Quat. Int., 158*, 110–121.

Morisawa, M. E. (1962), Quantitative geomorphology of some watersheds in the Appalachian Plateau, *Geol. Soc. Am. Bull., 73*, 1025–1046.

Morisawa, M. E. (1985), *Rivers: Form and Processes*, 222 pp., Longman, London.

Morisawa, M. E., and E. LaFlure (1979), Hydraulic geometry, stream equilibrium, and urbanization, in *Adjustments of the Fluvial System*, edited by D. D. Rhodes and G. P. Williams, pp. 333–350, George Allen and Unwin, London.

Moritsuna, T., U. Matsubayashi, and F. Takagi (1998), On the spatial heterogeneity of soil permeability in a mountainous forest area, *J. Hydrosci. Hydraul. Eng., 16*, 11–18.

Morrice, J. A., H. M. Valett, C. N. Dahm, and M. E. Campana (1997), Alluvial characteristics, groundwater-surface water exchange and hydrological retention in headwater streams, *Hydrol. Processes, 11*, 253–267.

Morris, A. E. L., P. C. Goebel, and B. J. Palik (2007), Geomorphic and riparian forest influences on characteristics of large wood and large-wood jams in old-growth and second-growth forests in northern Michigan, USA, *Earth Surf. Processes Landforms, 32*, 1131–1153.

Morris, A. E. L., P. C. Goebel, and B. J. Palik (2010), Spatial distribution of large wood jams in streams related to stream-valley geomorphology and forest age in northern Michigan, *River Res. Appl., 26*(7), 835–847.

Morris, S. E. (1986), The significance of rainsplash in the surficial debris cascade of the Colorado Front Range foothills, *Earth Surf. Processes Landforms, 11*, 11–22.

Morrissey, M. M., G. F. Wieczorek, and B. A. Morgan (2008), A comparative analysis of simulated and observed landslide locations triggered by Hurricane Camille in Nelson County, Virginia, *Hydrol. Processes, 22*, 524–531.

Morton, D. M., R. M. Alvarez, K. R. Ruppert, and B. Goforth (2008), Contrasting rainfall generated debris flows from adjacent watersheds at Forest Falls, southern California, USA, *Geomorphology, 96*, 322–338.

Moschen, H. (1990), Overflow weirs as gauging stations in mountain brooks, in *Hydrology in Mountainous Regions I. Hydrological Measurements: The Water Cycle*, edited by H. Lang and A. Musy, *IAHS Publ., 193*, 291–298.

Moshe, L. B., I. Haviv, Y. Enzel, E. Zilberman, and A. Matmon (2008), Incision of alluvial channels in response to continuous base level fall: Field characterization, modeling, and validation along the Dead Sea, *Geomorphology*, *93*, 524–536.

Mosley, M. P. (1973), Rainsplash and the convexity of badland divides, *Z. Geomorphol.*, *18*, 10–25.

Mosley, M. P. (1978), Erosion in the southeastern Ruahine Range; its implications for downstream river control, *N. Z. J. For.*, *23*, 21–48.

Mosley, M. P. (1982), The effect of a New Zealand beech forest canopy on the kinetic energy of water drops and on surface erosion, *Earth Surf. Processes Landforms*, *7*, 103–107.

Mosley, M. P., and D. S. Tindale (1985), Sediment variability and bed material sampling in gravel-bed rivers, *Earth Surf. Processes Landforms*, *10*, 465–482.

Moss, A. J. (1963), The physical nature of the common sandy and pebbly deposits. Part II, *Am. J. Sci.*, *261*, 297–343.

Moss, A. J. (1972), Bed-load sediments, *Sedimentology*, *18*, 159–219.

Moss, J. H. (1974), The relation of river terrace formation to glaciation in the Shoshone River basin, western Wyoming, in *Glacial Geomorphology*, edited by D. R. Coates, pp. 293–314, State Univ. of New York, Binghamton.

Moss, J. H., and W. E. Bonini (1961), Seismic evidence supporting a new interpretation of the Cody Terrace near Cody, Wyoming, *Geol. Soc. Am. Bull.*, *72*, 547–556.

Motha, J. A., P. J. Wallbrink, P. B. Hairsine, and R. B. Grayson (2003), Determining the sources of suspended sediment in a forested catchment in southeastern Australia, *Water Resour. Res.*, *39*(3), 1056, doi:10.1029/2001WR000794.

Mount, N. J., G. H. Sambrook Smith, and T. A. Stott (2005), An assessment of the impact of upland afforestation on lowland river reaches: The Afon Trannon, mid-Wales, *Geomorphology*, *64*, 255–269.

Mountain Agenda (1997), *Mountains of the World: Challenges for the 21st Century, A Contribution to Chapter 13, Agenda*, vol. 21, 36 pp., Paul Haupt Verlag, Switzerland.

Moussa, R. (2003), On morphometric properties of basins, scale effects and hydrological response, *Hydrol. Processes*, *17*, 33–58.

Moussa, R. (2008), Effect of channel network topology, basin segmentation and rainfall spatial distribution on the geomorphologic instantaneous unit hydrograph transfer function, *Hydrol. Processes*, *22*, 395–419.

Moussa, R. (2009), Definition of new equivalent indices of Horton-Strahler ratios for the derivation of the Geomorphological Instantaneous Unit Hydrograph, *Water Resour. Res.*, *45*, W09406, doi:10.1029/2008WR007330.

Moussavi-Harami, R., A. Mahboubi, and M. Khanebad (2004), Analysis of controls on downstream fining along three gravel-bed rivers in the Band-e-Golestan drainage basin NE Iran, *Geomorphology*, *61*, 143–153.

Moya, J., J. Corominas, J. P. Arcas, and C. Baeza (2010), Tree-ring based assessment of rockfall frequency on talus slopes at Solà d'Andorra, eastern Pyrenees, *Geomorphology*, *118*, 393–408.

Moyle, P. B., and J. F. Mount (2007), Homogenous rivers, homogenous faunas, *Proc. Natl. Acad. Sci. U. S. A.*, *104*, 5711–5712.

Moyle, P. B., and P. J. Randall (1998), Evaluating the biotic integrity of watersheds in the Sierra Nevada, California, *Conserv. Biol.*, *12*, 1318–1326.

Mueller, E. R., and J. Pitlick (2005), Morphologically based model of bed load transport capacity in a headwater stream, *J. Geophys. Res.*, *110*, F02016, doi:10.1029/2003JF000117.

Mueller, E. R., J. Pitlick, and J. M. Nelson (2005), Variation in the reference Shields stress for bed load transport in gravel- bed streams and rivers, *Water Resour. Res.*, *41*, W04006, doi:10.1029/2004WR003692.

Mugnier, J.-L., D. Becel, and D. Granjeon (2006), Active tectonics in the Subandean belt inferred from the morphology of the Rio Pilcomayo (Bolivia), in *Tectonics, Climate, and Landscape Evolution*, edited by S. D. Willett et al., *Spec. Pap. Geol. Soc. Am.*, *398*, 353–369.

Muhar, S., M. Jungwirth, G. Unfer, C. Wiesner, M. Poppe, S. Schmutz, S. Hohensinner, and H. Habersack (2008), Restoring riverine landscapes at the Drau River: Successes and deficits in the context of ecological integrity, in *Gravel-Bed Rivers VI: From Process Understanding to River Restoration*, edited by H. Habersack, H. Piégay, and M. Rinaldi, pp. 779–807, Elsevier, Amsterdam.

Mukhopadhyay, B., J. Cornelius, and W. Zehner (2003), Application of kinematic wave theory for predicting flash flood hazards on coupled alluvial fan-piedmont plain landforms, *Hydrol. Processes, 17*, 839–868.

Mul, M. L., R. K. Mutiibwa, J. W. A. Foppen, S. Uhlenbrook, and H. H. G. Savenije (2007), Identification of groundwater flow systems using geological mapping and chemical spring analysis in south Pare Mountains, Tanzania, *Phys. Chem. Earth, 32*, 1015–1022.

Mulder, T., and J. P. M. Syvitski (1996), Climatic and morphologic relationships of rivers: Implications of sea level fluctuations on river loads, *J. Geol., 103*, 285–299.

Mulholland, P. J., et al. (2001), Inter-biome comparison of factors controlling stream metabolism, *Freshwater Biol., 46*, 1503–1517.

Mummey, D. L., P. D. Stahl, and J. S. Buyer (2002), Soil microbiological properties 20 years after surface mine reclamation: Spatial analysis of reclaimed and undisturbed sites, *Soil Biol. Biochem., 34*, 1717–1725.

Mundie, J. H., and D. G. Crabtree (1997), Effects on sediments and biota of cleaning a salmonid spawning channel, *Fish. Manage. Ecol., 4*, 111–126.

Muneepeerakul, R., A. Rinaldo, S. A. Levin, and I. Rodriguez-Iturbe (2008), Signatures of vegetational functional diversity in river basins, *Water Resour. Res., 44*, W01431, doi:10.1029/2007WR006153.

Murphy, M. L., and J. D. Hall (1981), Varied effects of clear-cut logging on predators and their habitat in small streams of the Cascade Mountains, Oregon, *Can. J. Fish. Aquat. Sci., 38*, 137–145.

Murray, A. B., and C. Paola (1994), A cellular model of braided rivers, *Nature, 371*, 54–57.

Murray, A. B., and C. Paola (1997), Properties of a cellular braided-stream model, *Earth Surf. Processes Landforms, 22*, 1001–1025.

Murray, A. B., M. A. F. Knaapen, M. Tal, and M. L. Kirwan (2008), Biomorphodynamics: Physical-biological feedbacks that shape landscapes, *Water Resour. Res., 44*, W11301, doi:10.1029/2007WR006410.

Muskatirovic, J. (2008), Analysis of bedload transport characteristics of Idaho streams and rivers, *Earth Surf. Processes Landforms, 33*, 1757–1768.

Mussetter, R. A., M. D. Harvey, and R. D. Tenney (1999), Geologic and geomorphic associations with Colorado pikeminnow spawning, lower Yampa River, Colorado, *Geol. Soc. Am. Abstr. Programs, 31*, A-483.

Muste, M., F. Braileanu, and R. Ettema (2000), Flow and sediment transport measurements in a simulated ice-covered channel, *Water Resour. Res., 36*, 2711–2720.

Mutz, M. (2003), Hydraulic effects of wood in streams and rivers, in *The Ecology and Management of Wood in World Rivers*, edited by S. V. Gregory, K. L. Boyer, and A. M. Gurnell, pp. 93–107, Am. Fish. Soc., Bethesda, Md.

Muzik, I. (2002), A first-order analysis of the climate change effect on flood frequencies in a subalpine watershed by means of a hydrological rainfall-runoff model, *J. Hydrol., 267*, 65–73.

Mürle, U., J. Ortlepp, and M. Zahner (2003), Effects of experimental flooding on riverine morphology, structure and riparian vegetation: The River Spöl, Swiss National Park, *Aquat Sci., 65*, 191–198.

Myers, L. H. (1989), Riparian Area Management: Inventory and Monitoring Riparian Areas, *Bureau of Land Management TR-1737-7*, 79 pp, U.S. Dep. of the Interior, Denver, Colo.

Myers, T. J., and S. Swanson (1991), Aquatic habitat condition index, stream type, and livestock bank damage in northern Nevada, *Water Resour. Bull., 27*, 667–677.

Myers, T. J., and S. Swanson (1992), Variation of stream stability with stream type and livestock bank damage in northern Nevada, *Water Resour. Bull., 28*, 743–754.

Myers, T. J., and S. Swanson (1996a), Long-term aquatic habitat restoration: Mahogany Creek, Nevada, as a case study, *Water Resour. Bull.*, *32*, 241–252.

Myers, T. J., and S. Swanson (1996b), Temporal and geomorphic variations of stream stability and morphology: Mahogany Creek, Nevada, *Water Resour. Bull.*, *32*, 253–265.

Nadeau, T.-L., and M. C. Rains (2007), Hydrological connectivity between headwater streams and downstream waters: How science can inform policy, *Hydrol. Processes*, *43*, 118–133.

Naden, P. (1987), Modelling gravel-bed topography from sediment transport, *Earth Surf. Processes Landforms*, *12*, 353–367.

Nadler, C. T., and S. A. Schumm (1981), Metamorphosis of South Platte and Arkansas Rivers, eastern Colorado, *Phys. Geogr.*, *2*, 95–115.

Nador, A., E. Thamo-Boszo, A. Magyari, and E. Babinszki (2007), Fluvial responses to tectonics and climate change during the late Weichselian in the eastern part of the Pannonian Basin (Hungary), *Sediment. Geol.*, *202*, 174–192.

Naef, F., and G. R. Bezzola (1990), Hydrology and morphological consequences of the 1987 flood event in the upper Reuss valley, in *Hydrology in Mountainous Regions II. Artificial Reservoirs, Water and Slopes*, edited by R. O. Sinniger and M. Monbaron, *IAHS Publ.*, *194*, 339–346.

Naef, F., P. Horat, A. G. Milnes, and E. Hoehn (1990), Anomalous hydrological behaviour of an Alpine stream (Varuna, Poschiavo, southern Switzerland) and its interpretation in terms of the geology of the catchment, in *Hydrology in Mountainous Regions II. Artificial Reservoirs, Water and Slopes*, edited by R. O. Sinniger and M. Monbaron, *IAHS Publ.*, *194*, 347–354.

Naeth, M. A., and D. S. Chanasyk (1996), Runoff and sediment yield under grazing in foothills fescue grasslands of Alberta, *Water Resour. Bull.*, *32*, 89–95.

Nagle, G. (2007), Evaluating 'natural channel design' stream projects, *Hydrol. Processes*, *21*, 2539–2545.

Nagle, G. N., T. J. Fahey, J. C. Ritchie, and P. B. Woodbury (2007), Variations in sediment sources and yields in the Finger Lakes and Catskills regions of New York, *Hydrol. Processes*, *21*, 828–838.

Naiman, R. J., and H. DéCamps (1997), The ecology of interfaces: Riparian zones, *Annu. Rev. Ecol. Syst.*, *28*, 621–658.

Naiman, R. J., and J. M. Melillo (1984), Nitrogen budget of a subarctic stream altered by beaver (Castor canadensis), *Oecologia*, *62*, 150–155.

Naiman, R. J., and J. R. Sedell (1979), Relationships between metabolic parameters and stream order in Oregon, *Can. J. Fish. Aquat. Sci.*, *37*, 834–847.

Naiman, R. J., J. M. Melillo, and J. E. Hobbie (1986), Ecosystem alteration of boreal forest streams by beaver (Castor canadensis), *Ecology*, *67*, 1254–1269.

Naiman, R. J., C. A. Johnston, and J. C. Kelley (1988), Alteration of North American streams by beaver, *BioScience*, *38*, 753–762.

Naiman, R. J., D. G. Lonzarich, T. J. Beechie, and S. C. Ralph (1992), General principles of classification and the assessment of conservation potential in rivers, in *River Conservation and Management*, edited by P. J. Boon, P. Calow, and G. E. Petts, pp. 93–123, John Wiley, Chichester, U. K.

Naiman, R. J., K. L. Fetherston, S. McKay, and J. Chen (1998), Riparian forests, in *River Ecology and Management: Lessons from the Pacific Coastal Ecoregion*, edited by R. J. Naiman and R. E. Bilby, pp. 289–323, Springer, New York.

Naiman, R. J., H. Décamps, and M. E. McClain (2005), *Riparia: Ecology, Conservation, and Management of Streamside Communities*, Elsevier, Amsterdam.

Nakagawa, H., T. Tsujimoto, and Y. Shimizu (1991), Turbulent flow with small relative submergence, in *Fluvial Hydraulics of Mountain Regions*, edited by A. Armanini and G. DiSilvio, pp. 33–44, Springer, Berlin.

Nakamura, F., and N. Shin (2001), The downstream effects of dams on the regeneration of riparian tree species in northern Japan, in *Geomorphic Processes and Riverine Habitat*, edited by J. M. Dorava et al., pp. 173–181, AGU, Washington, D. C.

Nakamura, F., and F. J. Swanson (1993), Effects of coarse woody debris on morphology and sediment storage of a mountain stream system in western Oregon, *Earth Surf. Processes Landforms*, *18*, 43–61.

Nakamura, F., and F. J. Swanson (2003), Dynamics of wood in rivers in the context of ecological disturbance, in *The Ecology and Management of Wood in World Rivers*, edited by S. V. Gregory, K. L. Boyer, and A. M. Gurnell, pp. 279–297, Am. Fish. Soc., Bethesda, Md.

Nakamura, F., T. Araya, and S. Higashi (1987), Influence of river channel morphology and sediment production on residence time and transport distance, in *Erosion and Sedimentation in the Pacific Rim*, edited by R. L. Beschta et al., *IAHS Publ.*, *165*, 355–364.

Nakamura, F., H. Maita, and T. Araya (1995), Sediment routing analyses based on chronological changes in hillslope and riverbed morphologies, *Earth Surf. Processes Landforms*, *20*, 333–346.

Nakamura, F., Y. Kawaguchi, D. Nakano, and H. Yamada (2008), Ecological responses to anthropogenic alterations of gravel-bed rivers in Japan, from floodplain river segments to the microhabitat scale: A review, in *Gravel-Bed Rivers VI: From Process Understanding to River Restoration*, edited by H. Habersack, H. Piégay, and M. Rinaldi, pp. 501–523, Elsevier, Amsterdam.

Nakano, S., K. Maekawa, and S. Yamamoto (1990), Change of the life cycle of Japanese charr following artificial lake construction by damming, *Nippon Suisan Gakkaishi*, *56*, 1901–1905.

Nakatani, K., T. Wada, Y. Satofuka, and T. Mizuyama (2008), Development of "Kanako", a wide use 1-D and 2-D debris flow simulator equipped with GUI, *WIT Trans. Eng. Sci.*, *60*, 49–58.

Nanninga, P. M., and R. J. Wasson (1985), Calculation of the volume of an alluvial fan, *J. Int. Assoc. Math. Geol.*, *17*, 53–65.

Nanson, G. C. (1974), Bedload and suspended-load transport in a small, steep mountain stream, *Am. J. Sci.*, *274*, 471–486.

Nanson, G. C., and H. F. Beach (1977), Forest succession and sedimentation on a meandering-river floodplain, northeastern British Columbia, Canada, *J. Biogeogr.*, *4*, 229–251.

Nanson, G. C., and H. Q. Huang (1999), Anabranching rivers: Divided efficiency leading to fluvial diversity, in *Varieties of Fluvial Form*, edited by A. J. Miller and A. Gupta, pp. 477–494, John Wiley, Chichester, U. K.

Nanson, G. C., and A. D. Knighton (1996), Anabranching rivers: Their cause, character and classification, *Earth Surf. Processes Landforms*, *21*, 217–239.

Nanson, G. C., and R. W. Young (1981), Downstream reduction of rural channel size with contrasting urban effects in small coastal streams of southeastern Australia, *J. Hydrol.*, *52*, 239–255.

Napolitano, M. B. (1998), Persistence of historical logging impacts on channel form in mainstem North Fork Caspar Creek, in *Proceedings of the Conference on Coastal Watersheds: The Caspar Creek Story*, edited by R. R. Ziemer, pp. 97–101, Pacific Southwest Res. Sta., USDA, Albany, Calif.

Nardi, F., E. R. Vivoni, and S. Grimaldi (2006), Investigating a floodplain scaling relation using a hydrogeomorphic delineation method, *Water Resour. Res.*, *42*, W09409, doi:10.1029/2005WR004155.

Nash, D. B., and J. S. Beaujon (2006), Modeling degradation of terrace scarps in Grand Teton National Park, USA, *Geomorphology*, *75*, 400–407.

Nash, J. E. (1957), The form of the instantaneous unit hydrograph, in *Comptes Rendus et Rapports Toronto*, pp. 114–118, IAHS Press, Wallingford, U. K.

Nash, L. L., and P. H. Gleick (1991), Sensitivity of streamflow in the Colorado basin to climatic changes, *J. Hydrol.*, *125*, 221–241.

National Academy (1992), *Restoration of Aquatic Ecosystems: Science, Technology, and Public Policy*, 552 pp., National Academy Press, Washington, D. C.

National Research Council (1991), *Opportunities in the Hydrologic Sciences*, 348 pp., National Academy Press, Washington, D. C.

Navratil, O., and M.-B. Albert (2010), Non-linearity of reach hydraulic geometry relations, *J. Hydrol.*, *388*, 280–290.

Navratil, O., M.-B. Albert, E. Hérouin, and J.-M. Gresillon (2006), Determination of bankfull discharge magnitude and frequency: Comparison of methods on 16 gravel-bed river reaches, *Earth Surf. Processes Landforms*, *31*, 1345–1363.

Nayak, A., D. Marks, D. G. Chandler, and M. Seyfried (2010), Long-term snow, climate, and streamflow trends at the Reynolds Creek Experimental Watershed, Owyhee Mountains, Idaho, United States, *Water Resour. Res.*, *46*, W06519, doi:10.1029/2008WR007525.

Naylor, S., and E. J. Gabet (2007), Valley asymmetry and glacial verus nonglacial erosion in the Bitterroot Range, Montana, USA, *Geology*, *35*, 375–378.

Neal, C., A. Avila, and F. Roda (1995), Modelling the long-term impacts of atmospheric pollution deposition and repeated forestry cycles on stream water chemistry for a holm oak forest in northeastern Spain, *J. Hydrol.*, *168*, 51–71.

Neal, C., T. Hill, S. Alexander, B. Reynolds, S. Hill, A. J. Dixon, M. Harrow, M. Neal, and C. J. Smith (1997), Stream water quality in acid sensitive UK upland areas: An example of potential water quality remediation based on groundwater manipulation, *Hydrol. Earth Syst. Sci.*, *1*, 185–196.

Neal, E. G., M. T. Waler, and C. Coffeen (2002), Linking the Pacific Decadal Oscillation to seasonal stream discharge patterns in southeast Alaska, *J. Hydrol.*, *263*, 188–197.

Neary, D. G., and L. W. Swift, Jr. (1987), Rainfall thresholds for triggering a debris avalanching event in the southern Appalachian Moutains, in *Debris Flows/Avalanches: Process, Recognition, and Mitigation*, edited by J. E. Costa and G. F. Wieczorek, *Rev. Eng. Geol.*, *7*, 81–92.

Necea, D., W. Fielitz, and L. Matenco (2005), Late Pliocene-Quaternary tectonics in the frontal part of the SE Carpathians: Insights from tectonic geomorphology, *Tectonophysics*, *410*, 137–156.

Nedeltcheva, T. H., C. Piedallu, J. C. Gegout, J. M. Stussi, J. P. Boudot, N. Angeli, and E. Dambrine (2006), Influence of granite mineralogy, rainfall, vegetation and relief on stream water chemistry (Vosges Mountains, north-eastern France), *Chem. Geol.*, *231*, 1–15.

Needham, D. J., and R. D. Hey (1992), Dynamic modelling of bed waves, in *Dynamics of Gravel-Bed Rivers*, edited by P. Billi et al., pp. 401–414, John Wiley, Chichester, U. K.

Needham, J. G., and J. T. Lloyd (1930), *The Life of Inland Waters*, 438 pp., Charles C. Thomas, Springfield, Ill.

Nelson, J. M., M. W. Schmeeckle, and R. L. Shreve (2001), Turbulence and particle entrainment, in *Gravel-Bed Rivers V*, edited by M. P. Mosley, pp. 221–240, N. Z. Hydrol. Soc., Wellington, New Zealand.

Nelson, J. M., R. L. Shreve, S. R. McLean, and T. G. Drake (1995), Role of near-bed turbulence structure in bed transport and bed form mechanics, *Water Resour. Res.*, *31*, 2071–2086.

Nelson, J. M., J. P. Bennett, and S. M. Wiele (2003), Flow and sediment-transport modeling, in *Tools in Fluvial Geomorphology*, edited by G. M. Kondolf and H. Piégay, pp. 539–576, John Wiley, Chichester, U. K.

Nepf, H. M. (1999), Drag, turbulence, and diffusion in flow through emergent vegetation, *Water Resour. Res.*, *35*, 479–489.

Newbury, R., and M. Gaboury (1993), Exploration and rehabilitation of hydraulic habitats in streams using principles of fluvial behaviour, *Freshwater Biol.*, *29*, 195–210.

Newman, B. D., E. R. Vivoni, and A. R. Groffman (2006), Surface water-groundwater interactions in semiarid drainages of the American southwest, *Hydrol. Processes*, *20*, 3371–3394.

Newson, M. D., and A. R. G. Large (2006), 'Natural' rivers, 'hydromorphological quality' and river restoration: A challenging new agenda for applied fluvial geomorphology, *Earth Surf. Processes Land-forms*, *31*, 1606–1624.

Newson, M. D., and G. J. Leeks (1987), Transport processes at the catchment scale, in *Sediment Transport in Gravel-Bed Rivers*, edited by C. R. Thorne, J. C. Bathurst, and R. D. Hey, pp. 187–223, John Wiley, Chichester, U. K.

Newson, M. D., and M. Robinson (1983), Effects of agricultural drainage on upland streamflow: Case study in mid-Wales, *J. Environ. Manage.*, *17*, 333–348.

Nezat, C. A., E. Y. Graham, N. L. Green, W. B. Lyons, K. Neumann, A. E. Carey, and M. Hicks (1999), The physical and chemical controls on rare earth element concentrations in dissolved load, suspended load, and river channel sediments of the Hokitika River, New Zealand, *Geol. Soc. Am. Abstr. Programs*, *31*, A–256.

Ni, J. R., R. Z. Liu, O. W. H. Wai, A. G. L. Borthwick, and X. D. Ge (2006), Rapid zonation of abrupt mass movement hazard; Part I, General principles, *Geomorphology*, *80*, 214–225.

Nicholas, A. P. (2000), Modelling bedload yield in braided gravel bed rivers, *Geomorphology*, *36*, 89–106.

Nicholas, A. P. (2001), Computational fluid dynamics modelling of boundary roughness in gravel-bed rivers: An investigation of the effects of random variability in bed elevation, *Earth Surf. Processes Landforms*, *26*, 345–362.

Nicholas, A. P., and T. A. Quine (2007a), Crossing the divide: Representation of channels and processes in reduced-complexity river models at reach and landscape scales, *Geomorphology*, *90*, 318–339.

Nicholas, A. P., and T. A. Quine (2007b), Modeling alluvial landform change in the absence of external environmental forcing, *Geology*, *35*, 527–530.

Nicholas, A. P., and G. H. Sambrook Smith (1999), Numerical simulation of three-dimensional flow hydraulics in a braided channel, *Hydrol. Processes*, *13*, 913–929.

Nicholas, A. P., R. Thomas, and T. A. Quine (2006), Cellular modelling of braided river form and process, in *Braided Rivers: Process, Deposits, Ecology and Management*, edited by G. H. Sambrook Smith et al., *Spec. Publ. Int. Assoc. Sedimentol.*, *36*, 137–151.

Nichols, K. K., P. R. Bierman, M. Caffee, R. Finkel, and J. Larsen (2005a), Cosmogenically enabled sediment budgeting, *Geology*, *33*, 133–136.

Nichols, K. K., P. R. Bierman, R. Finkel, and J. Larsen (2005b), Long-term sediment generation rates for the Upper Rio Chagres basin: Evidence from cosmogenic ^{10}Be, in *The Rio Chagres, Panama: A Multidisciplinary Profile of a Tropical Watershed*, edited by R. S. Harmon, pp. 297–313, Springer, Dordrecht, The Netherlands.

Nichols, K. K., P. R. Bierman, W. R. Foniri, A. R. Gillespie, M. Caffee, and R. Finkel (2006), Dates and rates of arid region geomorphic processes, *GSA Today*, *16*(8), 4–11.

Nickolotsky, A., and R. T. Pavlowsky (2007), Morphology of step-pools in a wilderness headwater stream: The importance of standardizing geomorphic measurements, *Geomorphology*, *83*, 294–306.

Niedzialek, J. M., and F. L. Ogden (2005), Runoff production in the Upper Rio Chagres watershed, Panama, in *The Rio Chagres, Panama: A Multidisciplinary Profile of a Tropical Watershed*, edited by R. S. Harmon, pp. 149–168, Springer, Dordrecht, The Netherlands.

Niemann, J. D., N. M. Gasparini, G. E. Tucker, and R. L. Bras (2001), A quantitative evaluation of Playfair's Law and its use in testing long-term stream erosion models, *Earth Surf. Processes Landforms*, *26, 1317–1332*.

Niemi, N. A., M. Oskin, D. W. Burbank, A. M. Heimsath, and E. J. Gabet (2005), Effects of bedrock landslides on cosmogenically determined erosion rates, *Earth Planet. Sci. Lett.*, *237*, 480–498.

Nienow, P., M. Sharp, and I. Willis (1998), Seasonal changes in the morphology of the subglacial drainage system, Haut Glacier d'Arolla, Switzerland, *Earth Surf. Processes Landforms*, *23*, 825–843.

Nihlgard, B. J., W. T. Swank, and M. J. Mitchell (1994), Biological processes and catchment studies, in *Biogeochemistry of Small Catchments: A Tool for Environmental Research*, edited by B. Moldan and J. Cerny, pp. 133–161, John Wiley, Chichester, U. K.

Nik, A. R. (1988), Water yield changes after forest conversion to agricultural landuse in Peninsular Malaysia, *J. Trop. For. Sci.*, *1*, 67–84.

Nikora, V. (2008), Hydrodynamics of gravel-bed rivers: Scale issues, in *Gravel-Bed Rivers VI: From Process Understanding to River Restoration*, edited by H. Habersack, H. Piégay, and M. Rinaldi, pp. 61–81, Elsevier, Amsterdam.

Nikora, V., and J. Walsh (2004), Water-worked gravel surfaces: High-order structure functions at the particle scale, *Water Resour. Res.*, *40*, W12601, doi:10.1029/2004WR003346.

Nikora, V. I., D. G. Goring, and B. J. F. Biggs (1998), On gravel-bed roughness characterization, *Water Resour. Res.*, *34*, 517–527.

Nikora, V., H. Habersack, T. Huber, and I. McEwan (2002), On bed particle diffusion in gravel bed flows under weak bed load transport, *Water Resour. Res.*, *38*(6), 1081, doi:10.1029/2001WR000513.

Nikora, V. I., A. Sukhodolov, and P. M. Rowinski (1997), Statistical sand wave dynamics in one-directional water flows, *J. Fluid Mech.*, *351*, 17–39.

Nilsson, C., and K. Berggren (2000), Alterations of riparian ecosystems caused by river regulation, *BioScience*, *50*, 783–792.

Nilsson, C., and R. Jansson (1995), Floristic differences between riparian corridors of regulated and free-flowing boreal rivers, *Reg. Rivers Res. Manage.*, *11*, 55–66.

Nilsson, C., and M. Svedmark (2002), Basic principles and ecological consequences of changing water regimes: Riparian plant communities, *Environ. Manage.*, *30*, 468–480.

Nilsson, C., A. Ekblad, M. Gardfjell, and B. Carlberg (1991a), Long-term effects of river regulation on river margin vegetation, *J. Appl. Ecol.*, *28*, 963–987.

Nilsson, C., M. Gardfjell, and G. Grelsson (1991b), Importance of hydrochory in structuring plant communities along rivers, *Can. J. Bot.*, *69*, 2631–2633.

Nilsson, C., E. Nilsson, M. E. Johansson, M. Dynesius, G. Grelsson, S. Xiong, R. Jansson, and M. Danvind (1993), Processes structuring riparian vegetation, *Curr. Topics Bot. Res.*, *1*, 419–431.

Nilsson, C., A. Ekblad, M. Dynesius, S. Backe, M. Gardfjell, B. Carlberg, S. Hellqvist, and R. Jansson (1994), A comparison of species richness and traits of riparian plants between a main river channel and its tributaries, *J. Ecol.*, *82*, 281–295.

Nilsson, C., R. Jansson, and U. Zinko (1997), Long-term responses of river-margin vegetation to water-level regulation, *Science*, *276*, 798–800.

Nilsson, C., J. E. Pizzuto, G. E. Moglen, M. A. Palmer, E. H. Stanley, N. E. Bockstael, and L. C. Thompson (2003), Ecological forecasting and the urbanization of stream ecosystems: Challenges for economists, hydrologists, geomorphologists, and ecologists, *Ecosystems*, *6*, 659–674.

Nilsson, C., et al. (2005a), Forecasting environmental responses to restoration of rivers used as log floatways: An interdisciplinary challenge, *Ecosystems*, *8*, 779–800.

Nilsson, C., C. A. Reidy, M. Dynesius, and C. Revenga (2005b), Fragmentation and flow regulation of the world's large river systems, *Science*, *308*, 405–408.

Niswonger, R. G., D. E. Prudic, G. Pohll, and J. Constantz (2005), Incorporating seepage losses into the unsteady streamflow equations for simulating intermittent flow along mountain front streams, *Water Resour. Res.*, *41*, W06006, doi:10.1029/2004WR003677.

Nolan, K. M., and D. C. Marron (1985), Contrast in stream-channel response to major storms in two mountainous areas of California, *Geology*, *13*, 135–138.

Nolan, K. M., D. C. Marron, and L. M. Collins (1984), *Stream channel response to the January 3–5, 1982, storm in the Santa Cruz mountains, West Central California*, U.S. Geol. Surv. Open File Rep. 84-248.

Nordstrom, D. K. (2009), Acid rock drainage and climate change, *J. Geochem. Explor.*, *100*, 97–104.

North, C. P., and G. L. Warwick (2007), Fluvial fans: Myths, misconceptions, and the end of the terminal-fan model, *J. Sediment. Res.*, *77*, 693–701.

Norton, J. B., F. Bowannie, P. Peynetsa, W. Quandelacy, and S. F. Siebert (2002), Native American methods for conservation and restoration of semiarid ephemeral streams, *J. Soil Water Conserv.*, *57*, 250–258.

Norton, S. A., R. Wagai, T. Navratil, J. M. Kaste, and F. A. Rissberger (2000), Response of a first-order stream in Maine to short-term in-stream acidification, *Hydrol. Earth Syst. Sci.*, *4*, 383–391.

Nott, J. F., and D. M. Price (1994), Plunge pools and paleoprecipitation, *Geology*, *22*, 1047–1050.

Nott, J. F., R. Young, and I. McDougall (1996), Wearing down, wearing back, and gorge extension in the long-term denudation of a highland mass: Quantitative evidence from the Shoalhaven Catchment, southeastern Australia, *J. Geol.*, *104*, 224–232.

Nott, J. F., M. F. Thomas, and D. M. Price (2001), Alluvial fans, landslides and late Quaternary climate change in the wet tropics of northeast Queensland, *Aust. J. Earth Sci.*, *48*, 875–882.

Nowakowski, A. L., and E. Wohl (2008), Influences on wood load in mountain streams of the Bighorn National Forest, Wyoming, USA, *Environ. Manage.*, *42*, 557–571.

Nowell, A. R. M., and M. Church (1979), Turbulent flow in a depth-limited boundary layer, *J. Geophys. Res.*, *84*, 4816–4824.

Nunnally, N. R. (1978), Improving channel efficiency without sacrificing fish and wildlife habitat: The case for stream restoration, in *Strategies for Protection and Management of Floodplain Wetlands and Other Riparian Ecosystems*, USDA Forest Service General Technical Report WO-12, pp. 394–399.

Nunokawa, M., T. Gomi, J. N. Negishi, and O. Nakahura (2008), A new method to measure substrate coherent strength of Stenopsyche marmorata, *Landscape Ecol. Eng.*, *4*(2), 125–131.

Nyssen, J., et al. (2009), How soil conservation affects the catchment sediment budget – a comprehensive study in the north Ethiopian highlands, *Earth Surf. Processes Landforms*, *34*, 1216–1233.

Obled, C., I. D. Cluckie, and C. G. Collier (1991), Reflections on rainfall information requirements for operational rainfall-runoff modelling, in *Hydrological Applications of Weather Radar*, edited by I. D. Cluckie and C. G. Collier, pp. 469–482, Ellis Horwood, London.

O'Brien, A. K., K. C. Rice, O. P. Bricker, M. M. Kennedy, and R. T. Anderson (1997), Use of geochemical mass balance modelling to evaluate the role of weathering in determining stream chemistry in five mid-Atlantic watersheds on different lithologies, *Hydrol. Processes*, *11*, 719–744.

O'Brien, J. S. (1987), A case study of minimum streamflow for fishery habitat in the Yampa River, in *Sediment Transport in Gravel-Bed Rivers*, edited by C. R. Thorne, J. C. Bathurst, and R. D. Hey, pp. 921–946, John Wiley, Chichester, U. K.

O'Brien, J. S. (1993), Hydraulic modeling and mapping of mud and debris flows, in *Hydraulic Engineering '93*, vol. 2, edited by H. W. Shen, S. T. Su, and F. Wen, pp. 1762–1767, Am. Soc. of Civ. Eng., New York.

O'Brien, J. S., and P. Y. Julien (1988), Laboratory analysis of mudflow properties, *J. Hydraul. Eng.*, *114*, 877–887.

O'Connor, J. E. (1993), *Hydrology, hydraulics, and geomorphology of the Bonneville Flood, Spec. Pap.* Geol. Soc. Am., *274*, 83 pp.

O'Connor, J. E., and S. F. Burns (2009), Cataclysms and controversy – aspects of the geomorphology of the Columbia River Gorge, in *Volcanoes to Vineyards: Geologic Field Trips Through the Dynamic Landscape of the Pacific Northwest*, edited by J. E. O'Connor, R. J. Dorsey, and I. P. Madin, pp. 237–251, Geol. Soc. of Am., Boulder, Colo.

O'Connor, J. E., and J. E. Costa (1993), Geologic and hydrologic hazards in glacierized basins in North America resulting from 19th and 20th century global warming, *Nat. Hazards*, *8*, 121–140.

O'Connor, J. E., and J. E. Costa (2004), Spatial distribution of the largest rainfall-runoff floods from basins between 2.6 and 26,000 km^2 in the United States and Puerto Rico, *Water Resour. Res.*, *40*, W01107, doi:10.1029/2003WR002247.

O'Connor, J. E., and R. H. Webb (1988), Hydraulic modeling for paleoflood analysis, in *Flood Geomorphology*, edited by V. R. Baker, R. C. Kochel, and P. C. Patton, pp. 393–402, John Wiley, New York.

O'Connor, J. E., R. H. Webb, and V. R. Baker (1986), Paleohydrology of pool-and-riffle pattern development: Boulder Creek, Utah, *Geol. Soc. Am. Bull.*, *97*, 410–420.

O'Connor, J. E., L. L. Ely, E. E. Wohl, L. E. Stevens, T. S. Melis, V. S. Kale, and V. R. Baker (1994), A 4500-year record of large floods on the Colorado River in the Grand Canyon, Arizona, *J. Geol.*, *102*, 1–9.

O'Connor, J. E., J. H. Hardison, and J. E. Costa (2001), *Debris flows from failures of Neoglacial-age moraine dams in the Three Sisters and Mount Jefferson Wilderness Areas, Oregon, U.S. Geol. Surv. Prof. Pap.*, *1606*, 93 pp.

O'Connor, J. E., M. A. Jones, and T. L. Haluska (2003), Flood plain and channel dynamics of the Quinault and Queets Rivers, Washington, USA, *Geomorphology*, *51*, 31–59.

O'Connor, J. E., J. Major, and G. Grant (2008), Down with the dams: Unchaining U. S. rivers, *Geotimes*, March.

Odum, H. T. (1957), Trophic structure and productivity of Silver Springs, Florida, *Ecol. Monogr.*, *27*, 55–112.

Oerlemans, J., and E. J. L. Klok (2004), Effect of summer snowfall on glacier mass balance, *Ann. Glaciol.*, *38*, 97–100.

O'Farrell, C. R., A. M. Heimsath, and J. M. Kaste (2007), Quantifying hillslope erosion rates and processes for a coastal California landscape over varying timescales, *Earth Surf. Processes Landforms*, *32, 544–560.*

OFDA (US Office of Foreign Disaster Assistance) (1988), *Disaster History: Significant Data on Major Disasters Worldwide, 1900-Present*, Agency for International Development, Washington, D. C.

Oguchi, T. (1994a),), Average erosional conditions of Japanese mountains estimated from the frequency and magnitude of landslides, paper presented at International Symposium on Forest Hydrology Tokyo.

Oguchi, T. (1994b), Late Quaternary geomorphic development of alluvial fan-source basin systems: The Yamagata region, Japan, *Geogr. Rev. Jpn.*, *67B*, 81–100.

Oguchi, T. (1994c), Late Quaternary geomorphic development of mountain river basins based on landform classification: The Kitakami region, northeast Japan, *Bull. Dep. Geogr. Univ. Tokyo*, *26*, 15–32.

Oguchi, T. (1996a), Factors affecting the magnitude of post-glacial hillslope incision in Japanese mountains, *Catena*, *26*, 171–186.

Oguchi, T. (1996b), Late Quaternary hillslope erosion rates in Japanese mountains estimated from landform classification and morphometry, *Z. Geomorphol.*, *106*, 169–181.

Oguchi, T. (1997a), Channel incision and sediment production in Japanese mountains in relation to past and future climatic change, in *Proceedings of the Conference on Management of Landscapes Disturbed by Channel Incision*, edited by S. S. Y. Wang, E. J. Langendoen, and F. D. Shields, Jr., pp. 867–872.

Oguchi, T. (1997b), Drainage density and relative relief in humid steep mountains with frequent slope failure, *Earth Surf. Processes Landforms*, *22*, 107–120.

Oguchi, T. (1997c), Late Quaternary sediment budget in alluvial-fan-source-basin systems in Japan, *J. Quat. Sci.*, *12*, 381–390.

Oguchi, T., and C. T. Oguchi (2004), Late Quaternary rapid talus dissection and debris flow deposition on an alluvial fan in Syria, *Catena*, *55*, 125–140.

Oguchi, T., and H. Ohmori (1994), Analysis of relationships among alluvial fan area, source basin area, basin slope, and sediment yield, *Z. Geomorphol.*, *38*, 405–420.

Ohmori, H. (1991), Change in the mathematical function type describing the longitudinal profile of a river through an evoluationary process, *J. Geol.*, *99*, 97–110.

Ohmori, H., and H. Shimazu (1994), Distribution of hazard types in a drainage basin and its relation to geomorphological setting, *Geomorphology*, *10*, 95–106.

Ohmura, A. (2006), Changes in mountain glaciers and ice caps during the 20th century, *Ann. Glaciol.*, *43*, 361–368.

Ohsumi Works Office (1995), *Debris Flow at Sakurajima 2*, 81 pp., Kyushu Reg. Construct. Bur., Min. of Construct., Japan.

Okuda, S. (1978), *Observation on the Motion of Debris Flow and Its Geomorphological Effects*, 24 pp., Int. Geogr. Union, Commun. Field Exp. Geomorphol., Paris, France.

Okunishi, K., S. Okuda, and H. Suwa (1987), A large-scale debris avalanche as an episode in slope-channel processes, in *Erosion and Sedimentation in the Pacific Rim*, edited by R. L. Beschta et al., *IAHS Publ.*, *165*, 225–232.

Oldmeadow, D. F., and M. Church (2006), A field experiment on streambed stabilization by gravel structures, *Geomorphology*, *78*, 335–350.

Ollesch, G., I. Kistner, R. Meissner, and K.-E. Lindenschmidt (2006), Modelling of snowmelt erosion and sediment yield in a small low-mountain catchment in Germany, *Catena*, *68*, 161–176.

Ollesch, G., Y. Sukhanovski, I. Kistner, M. Rode, and R. Meissner (2005), Characterization and modelling of the spatial heterogeneity of snowmelt erosion, *Earth Surf. Processes Landforms*, *30*, 197–211.

Olley, J. M., and A. S. Murray (1994), Origins of variability in the ^{230}Th/^{232}Th ratio in sediments, in *Variability in Stream Erosion and Sediment Transport*, edited by L. J. Olive, R. J. Loughran, and A. J. Kesby, *IAHS Publ.*, *224*, 65–70.

O'Loughlin, E. M. (1986), Prediction of surface saturation zones in natural catchments by topographic analysis, *Water Resour. Res.*, *22*, 794–804.

Olson, R., and W. A. Hubert (1994), *Beaver: Water Resources and Riparian Habitat Manager*, 48 pp., Univ. of Wyoming, Laramie.

Onda, Y., Y. Komatsu, M. Tsujimura, and J. Fujihara (2001), The role of subsurface runoff through bedrock on storm flow generation, *Hydrol. Processes*, *15*, 1693–1706.

O'Neal, J. S., and T. H. Sibley (1999), A biological evaluation of stream rehabilitation: Comparison of the effects of large woody debris and engineered alternative, *1999 Annual Meeting Abstracts*, pp. 160, Ecological Society of America.

O'Neill, M. P., and A. D. Abrahams (1984), Objective identification of pools and riffles, *Water Resour. Res.*, *20*, 921–926.

Ono, Y. (1990), Alluvial fans in Japan and South Korea, in *Alluvial Fans: A Field Approach*, edited by A. H. Rachocki and M. Church, pp. 91–107, John Wiley, New York.

Onodera, S.-I., M. Sawano, M. Saito, and H. Takahashi (2007), Effect of frequent storms on nutrient discharge in a mountainous coastal catchment, western Japan, in *Water Quality and Sediment Behaviour of the Future: Predictions for the 21st Century*, edited by B. W. Webb and D. de Boer, *IAHS Publ.*, *314*, 108–116.

Orlandini, S. (2002), On the spatial variation of resistance to flow in upland channel networks, *Water Resour. Res.*, *38*(10), 1197, doi:10.1029/2001WR001187.

Orsborn, J. F., and J. W. Anderson (1986), Stream improvements and fish response: A bio-engineering assessment, *Water Resour. Bull.*, *22*, 381–388.

Ortega, J. A., and G. G. Heydt (2009), Geomorphological and sedimentological analysis of flash-flood deposits: The case of the 1997 Rivillas flood (Spain), *Geomorphology*, *112*, 1–14.

Ortlepp, J., and U. Mürle (2003), Effects of experimental flooding on brown trout (Salmo trutta fario L.): The River Spöl, Swiss National Park, *Aquat. Sci.*, *65*, 232–238.

Orwin, J. F., and C. C. Smart (2005), An inexpensive turbidimeter for monitoring suspended sediment, *Geomorphology*, *68*, 3–15.

Osborne, L. L., and M. J. Wiley (1992), Influence of tributary spatial position on the structure of warmwater fish communities, *Can. J. Fish. Aquat. Sci.*, *49*, 671–681.

Osman, A. M., and C. R. Thorne (1988), Riverbank stability analysis I: Theory, *J. Hydraul. Eng.*, *114*, 134–150.

Osmundson, D. B., R. J. Ryel, V. L. Lamarra, and J. Pitlick (2002), Flow-sediment-biota relations: Implications for river regulation effects on native fish abundance, *Ecol. Appl.*, *12*, 1719–1739.

Osterkamp, T. E. (2005), The recent warming of permafrost in Alaska, *Global Planet. Change*, *49*, 187–202.

Osterkamp, W. R., and E. R. Hedman (1977), Variation of width and discharge for natural high-gradient stream channels, *Water Resour. Res.*, *13*, 256–258.

Osterkamp, W. R., and C. R. Hupp (1984), Geomorphic and vegetative characteristics along three northern Virginia streams, *Geol. Soc. Am. Bull.*, *95*, 1093–1101.

Osterkamp, W. R., and C. R. Hupp (1987), Dating and interpretation of debris flows by geologic and botanical methods at Whitney Creek Gorge, Mount Shasta, California, in *Debris Flows/Avalanches: Process, Recognition, and Mitigation*, edited by J. E. Costa and G. F. Wieczorek, *Rev. Eng. Geol.*, *7*, 157–163.

Osterkamp, W. R., and T. J. Toy (1994), The healing of disturbed hillslopes by gully gravure, *Geol. Soc. Am. Bull.*, *106*, 1233–1241.

Osterkamp, W. R., C. R. Hupp, and J. C. Blodgett (1986), Magnitude and frequency of debris flows, and areas of hazard on Mount Shasta, northern California, *U.S. Geol. Surv. Prof. Pap., 1396-C*, 21 pp.

Oswald, E. B., and E. Wohl (2008), Wood-mediated geomorphic effects of a jökulhlaup in the Wind River Mountains, Wyoming, *Geomorphology, 100*, 549–562.

Oswood, M. W., A. M. Milner, and J. G. Irons, III (1992), Climate change and Alaskan rivers and streams, in *Global Climate Change and Freshwater Ecosystems*, edited by P. Firth and S. G. Fisher, pp. 192–210, Springer, New York.

Otto, J.-C., L. Schrott, M. Jaboyedoff, and R. Dikau (2009), Quantifying sediment storage in a high alpine valley (Turtmanntal, Switzerland), *Earth Surf. Processes Landforms, 34*, 1726–1742.

Ouchi, S. (1985), Response of alluvial rivers to slow active tectonic movement, *Geol. Soc. Am. Bull., 96*, 504–515.

Ouchi, S. (2004), Flume experiments on the horizontal stream offset by strike-slip faults, *Earth Surf. Processes Landforms, 29*, 161–173.

Ouchi, S., and T. Mizuyama (1989), Volume and movement of Tombi Landslide in 1858, Japan, *Trans. Jpn. Geomorphol. Union, 10*, 27–51.

Ouimet, W. B., K. X. Whipple, and D. E. Granger (2009), Beyond threshold hillslopes: Channel adjustment to base-level fall lin tectonically active mountain ranges, *Geology, 37*, 579–582.

Overton, C. K., M. A. Radko, and R. L. Nelson (1993), *Fish habitat conditions: Using the Northern/intermountain region's inventory procedures for detecting differences on two differently managed watersheds*, USDA For. Serv. General Tech. Rep. INT-300, 14 pp.

Owen, L. A., and M. C. Sharma (1998), Rates and magnitudes of paraglacial fan formation in the Garhwal Himalaya: Implications for landscape evolution, *Geomorphology, 26*, 171–184.

Owens, P., and O. Slaymaker (1992), Late Holocene sediment yields in small alpine and subalpine drainage basins, British Columbia, in *Erosion, Debris Flows and Environment in Mountain Regions*, edited by D. E. Walling, T. R. Davies, and B. Hasholt, *IAHS Publ., 209*, 147–154.

Pack, R. T. (2005), Application of airborne and spaceborne remote sensing methods, in *Debris-Flow Hazards and Related Phenomena*, edited by M. Jakob and O. Hungr, pp. 275–289, Springer, Berlin.

Packman, A. I., and N. H. Brooks (2001), Hyporheic exchange of solutes and colloids with moving bed forms, *Water Resour. Res., 37*, 2591–2605.

Page, M., M. Marden, M. Kasai, B. Gomez, D. Peacock, H. Betts, T. Parkner, T. Marutani, and N. Trustrum (2008), Changes in basin-scale sediment supply and transfer in a rapidly transformed New Zealand landscape, in *Gravel-Bed Rivers VI: From Process Understanding to River Restoration*, edited by H. Habersack, H. Piégay, and M. Rinaldi, pp. 337–358, Elsevier, Amsterdam.

Palacios, D., J. J. Zamorano, and A. Gómez (2001), The impact of present lahars on the geomorphologic evolution of proglacial gorges: Popocatepetl, Mexico, *Geomorphology, 37*, 15–42.

Palmer, M. A., R. F. Ambrose, and N. L. Poff (1997), Ecological theory and community restoration ecology, *Restor. Ecol., 5*, 291–300.

Palmer, M. A., A. E. Bely, and K. E. Berg (1992), Response of invertebrates to lotic disturbance: A test of the hyporheic refuge hypothesis, *Oecologia, 89*, 182–194.

Palmer, M. A., et al. (2005), Standards for ecologically successful river restoration, *J. Appl. Ecol., 42*, 208–217.

Palmer, M. A., et al. (2010), Mountaintop mining consequences, *Science, 327*, 148–149.

Palmer, T. (1994), *Lifelines: The Case for River Conservation*, 254 pp., Island Press, Washington, D. C.

Pan, B., D. Burbank, Y. Wang, G. Wu, J. Li, and Q. Guan (2003), A 900 k.y. record of strath terrace formation during glacial-interglacial transitions in Northwest China, *Geology, 31*, 957–960.

Panek, T., V. Smolkova, J. Hradecky, and K. Kirchner (2007), Landslide dams in the northern part of the Czech Flysch Carpathians: Geomorphic evidence and imprints, *Stud. Geomorphol. Carpatho-Balcanica, 41*, 77–96.

Paola, C. (1996), Incoherent structure: Turbulence as a metaphor for stream braiding, in *Coherent Flow Structures in Open Channels*, edited by P. J. Ashworth et al., pp. 705–723, Wiley, Chichester, U. K.

Paola, C. (2001), Modelling stream braiding over a range of scales, in *Gravel-Bed Rivers V*, edited by M. P. Mosley, pp. 11–46, N. Z. Hydrol. Soc., Wellington.

Paola, C., G. Parker, R. Seal, S. K. Sinha, J. B. Southard, and P. R. Wilcock (1992), Downstream fining by selective deposition in a laboratory flume, *Science*, *258*, 1757–1760.

Papa, M., S. Egashira, and T. Itoh (2004), Critical conditions of bed sediment entrainment due to debris flow, *Nat. Hazards Earth Syst. Sci.*, *4*, 469–474.

Papanicolaou, A. N., and A. Schuyler (2003), Cluster evolution and flow-frictional characteristics under different sediment availabilities and specific gravity, *J. Eng. Mech.*, *129*, 1206–1219.

Papanicolaou, A. N., K. Strom, A. Schuyler, and N. Talebbeydokhti (2003), The role of sediment specific gravity and availability on cluster evolution, *Earth Surf. Processes Landforms*, *28*, 69–86.

Papanicolaou, A. N., A. Bdour, and E. Wicklein (2004), One-dimensional hydrodynamic/sediment transport model applicable to steep mountain streams, *J. Hydraul. Res.*, *42*(4), 357–375.

Park, C. C. (1977a), Man-induced changes in stream channel capacity, in *River Channel Changes*, edited by K. J. Gregory, pp. 121–144, John Wiley, Chichester, U. K.

Park, C. C. (1977b), World-wide variations in hydraulic geometry exponents of stream channels: An analysis and some observations, *J. Hydrol.*, *33*, 133–146.

Parker, C., A. Simon, and C. R. Thorne (2008), The effects of variability in bank material properties on riverbank stability: Goodwin Creek, Mississippi, *Geomorphology*, *101*, 533–543.

Parker, G. (1979), Hydraulic geometry of active gravel rivers, *J. Hydraul. Div. Am. Soc. Civ. Eng.*, *105*, 1185–1201.

Parker, G. (1990), Surface-based bedload transport relation for gravel rivers, *J. Hydraul. Res.*, *28*, 417–436.

Parker, G. (1991), Downstream variation of grain size in gravel rivers: Abrasion versus selective sorting, in *Fluvial Hydraulics of Mountain Regions*, edited by A. Armanini and G. DiSilvio, pp. 347–360, Springer, Berlin.

Parker, G. (1996), Some speculations on the relation between channel morphology and channel-scale flow structures, in *Coherent Flow Structures in Open Channels*, edited by P. J. Ashworth et al., pp. 423–458, John Wiley, Chichester, U. K.

Parker, G., and P. C. Klingeman (1982), On why gravel bed streams are paved, *Water Resour. Res.*, *18*, 1409–1423.

Parker, G., and L. A. Perg (2005), Probabilistic formulation of conservation of cosmogenic nuclides: Effect of surface elevation fluctuations on approach to steady state, *Earth Surf. Processes Landforms*, *30*, 1127–1144.

Parker, G., and A. W. Peterson (1980), Bar resistance of gravel-bed streams, *J. Hydraul. Div. Am. Soc. Civ. Eng.*, *106*, 1559–1575.

Parker, G., and C. M. Toro-Escobar (2002), Equal mobility of gravel in streams: The remains of the day, *Water Resour. Res.*, *38*(11), 1264, doi:10.1029/2001WR000669.

Parker, G., P. C. Klingeman, and D. G. McLean (1982), Bedload and size distribution in gravel-bed streams, *J. Hydraul. Div. Am. Soc. Civ. Eng.*, *108*, 544–571.

Parker, G., Y. Cui, J. Imran, and W. E. Dietrich (1997), Flooding in the lower Ok Tedi, Papua New Guinea due to the disposal of mine tailings and its amelioration, in *International Seminar on Recent Trends of Floods and their Preventive Measures, Sapporo, Japan, 20–21 June 1996*, pp. 21–48.

Parker, G., C. M. Toro-Escobar, M. Ramey, and S. Beck (2003), Effect of floodwater extraction on mountain stream morphology, *J. Hydraul. Eng.*, *129*, 885–895.

Parker, G., P. R. Wilcock, C. Paola, W. E. Dietrich, and J. Pitlick (2007), Physical basis for quasi-universal relations describing bankfull hydraulic geometry of single-thread gravel bed rivers, *J. Geophys. Res.*, *112*, F04005, doi:10.1029/2006JF000549.

Parker, G., M. A. Hassan, and P. Wilcock (2008a), Adjustment of the bed surface size distribution of gravel-bed rivers in response to cycled hydrographs, in *Gravel-Bed Rivers VI: From Process Understanding to River Restoration*, edited by H. Habersack, H. Piégay, and M. Rinaldi, pp. 241–289, Elsevier, Amsterdam.

Parker, G., T. Muto, Y. Akamatsu, W. E. Dietrich, and J. W. Lauer (2008b), Unravelling the conundrum of river response to rising sea-level from laboratory to field: Part II, The Fly-Strickland River system, Papua New Guinea, *Sedimentology*, *55*, 1657–1686.

Parker, M. (1986), Beaver, Water Quality, and Riparian Systems. Wyoming Water and Streamside Zone Conferences, Wyoming Water Res. Cent., Univ. of Wyoming Laramie.

Parker, R. S. (1977), *Experimental study of drainage basin evolution and its hydrologic implications*, Colo. State Univ. Hydrol. Pap., vol. 90, 58 pp.

Parrett, C. (1987), *Fire-related debris flows in the Beaver Creek drainage, Lewis and Clark County, Montana*, U.S. Geol. Surv. Water Supply Pap., *2330*, pp. 57–67.

Parrett, C., and S. R. Holnbeck (1994), Relation between largest known flood discharge and elevation in Montana, in *Proceedings, ASCE Hydraulic Engineering '94 Conference, August 1–5, Buffalo, New York*, edited by G. V. Cotroneo and R. R. Rumer, pp. 870–874, Am. Soc. of Civ. Eng., New York.

Parriaux, A., and G. F. Nicoud (1990), Hydrological behavior of glacial deposits in mountainous areas, in *Hydrology of Mountainous Areas*, edited by L. Molnar, *IAHS Publ.*, *190*, 291–312.

Parsons, M., C. A. McLoughlin, K. A. Kotschy, K. H. Rogers, and M. W. Rountree (2005), The effects of extreme floods on the biophysical heterogeneity of river landscapes, *Front. Ecol. Environ.*, *3*, 487–494.

Passalacqua, P., T. D. Trung, E. Foufoula-Georgiou, G. Sapiro, and W. E. Dietrich (2010), A geometric framework for channel network extraction from lidar: Nonlinear diffusion and geodesic paths, *J. Geophys. Res.*, *115*, F01002, doi:10.1029/2009JF001254.

Pasternack, G. B., C. R. Ellis, K. A. Leier, B. L. Vallé, and J. D. Marr (2006), Convergent hydraulics at horseshoe steps in bedrock rivers, *Geomorphology*, *82*, 126–145.

Pasternack, G. B., C. R. Ellis, and J. D. Marr (2007), Jet and hydraulic jump near-bed stresses below a horseshoe waterfall, *Water Resour. Res.*, *43*, W07449, doi:10.1029/2006WR005774.

Patrick, R. (1995), Rivers of the United States, vol. 2, *Chemical and Physical Characteristics*, Chap. 7, pp. 195–228, Wiley, New York.

Patton, P. C. (1988a), Drainage basin morphometry and floods, in *Flood Geomorphology*, edited by V. R. Baker, R. C. Kochel, and P. C. Patton, pp. 51–64, John Wiley, New York.

Patton, P. C. (1988b), Geomorphic response of streams to floods in the glaciated terrain of southern New England, in *Flood Geomorphology*, edited by V. R. Baker, R. C. Kochel, and P. C. Patton, pp. 261–277, John Wiley, New York.

Patton, P. C., and V. R. Baker (1976), Morphometry and floods in small drainage basins subject to diverse hydrogeomorphic controls, *Water Resour. Res.*, *12*, 941–952.

Patton, P. C., and P. J. Boison (1986), Processes and rates of formation of Holocene alluvial terraces in Harris Wash, Escalante River basin, south-central Utah, *Geol. Soc. Am. Bull.*, *97*, 369–378.

Patton, P. C., and S. A. Schumm (1975), Gully erosion, northern Colorado: A threshold phenomenon, *Geology*, *3*, 88–90.

Paul, S. K., S. K. Bartarya, P. Rautela, and A. K. Mahajan (2000), Catastrophic mass movement of 1998 monsoons at Malpa in Kali Valley, Kumaun Himalaya (India), *Geomorphology*, *35*, 169–180.

Payn, R. A., M. N. Gooseff, B. L. McGlynn, K. E. Bencala, and S. M. Wondzell (2009), Channel water balance and exchange with subsurface flow along a mountain headwater stream in Montana, United States, *Water Resour. Res.*, *45*, W11427, doi:10.1029/2008WR007644.

Payne, B. A., and M. F. Lapointe (1997), Channel morphology and lateral stability: Effects on distribution of spawning and rearing habitat for Atlantic salmon in a wandering cobble-bed river, *Can. J. Fish. Aquat. Sci.*, *54*, 2627–2736.

Pazzaglia, F. J., and M. T. Brandon (2001), A fluvial record of long-term steady-state uplift and erosion across the Cascadia forearc high, western Washington State, *Am. J. Sci.*, *301*, 385–431.

Pazzaglia, F. J., and T. W. Gardner (1993), Fluvial terraces of the lower Susquehanna River, *Geomorphology*, *8*, 83–113.

Pazzaglia, F. J., T. W. Gardner, and D. J. Merritts (1998), Bedrock fluvial incision and longitudinal profile development over geologic time scales determined by fluvial terraces, in *Rivers Over Rock: Fluvial Processes in Bedrock Channels*, Geophys. Monogr. Ser., vol. 107, edited by K. J. Tinkler and E. E. Wohl, pp. 207–235, AGU, Washington, D. C.

Pearce, A. J., and A. I. McKerchar (1979), Upstream generation of storm runoff, in *Physical Hydrology: New Zealand Experience*, edited by D. L. Murray and P. Ackroyd, pp. 165–192, N. Z. Hydrol. Soc., Wellington North.

Pearce, J. T., F. J. Pazzaglia, E. B. Evenson, D. E. Lawson, R. B. Alley, D. Germanoski, and J. D. Denner (2003), Bedload component of glacially discharged sediment: Insights from the Matanuska Glacier, Alaska, *Geology*, 31, 7–10.

Pearce, S. A., F. J. Pazzaglia, and M. C. Eppes (2004), Ephemeral stream response to growing folds, *Geol. Soc. Am. Bull.*, 116, 1223–1239.

Pearlstine, L., H. McKellar, and W. Kitchens (1985), Modelling the impacts of a river diversion on bottomland forest communities in the Santee River floodplain, South Carolina, *Ecol. Modell.*, 29, 283–302.

Peart, M. R., and A. W. Jayawardena (1994), Some observations on bedload movement in a small stream in Hong Kong, in *Variability in Stream Erosion and Sediment Transport*, edited by L. J. Olive, R. J. Loughran, and J. A. Kesby, *IAHS Publ.*, 224, 71–76.

Pederson, J. L., F. Pazzaglia, and G. Smith (2000), Ancient hillslope deposits: Missing links in the study of climate controls on sedimentation, *Geology*, 28, 27–30.

Pederson, J. L., G. Smith, and F. Pazzaglia (2001), Comparing the modern, Quaternary, and Neogene records of climate-controlled hillslope sedimentation in southeast Nevada, *Geol. Soc. Am. Bull.*, 113, 305–319.

Pederson, J. L., M. D. Anders, T. M. Rittenour, W. D. Sharp, J. C. Gosse, and K. E. Karlstrom (2006), Using fill terraces to understand incision rates and evolution of the Colorado River in eastern Grand Canyon, Arizona, *J. Geophys. Res.*, 111, F02003, doi:10.1029/2004JF000201.

Pelletier, J. D., and C. Rasmussen (2009), Geomorphically based predictive mapping of soil thickness in upland watersheds, *Water Resour. Res.*, 45, W09417, doi:10.1029/2008WR007319.

Pelletier, J. D., L. Mayer, P. A. Pearthree, P. K. House, K. A. Demsey, J. E. Klawon, and K. R. Vincent (2005), An integrated approach to flood hazard assessment on alluvial fans using numerical modeling, field mapping, and remote sensing, *Geol. Soc. Am. Bull.*, 117, 1167–1180.

Pelletier, J. D., M. Cline, and S. B. DeLong (2007), Desert pavement dynamics: Numerical modeling and field-based calibration, *Earth Surf. Processes Landforms*, 32, 1913–1927.

Pelletier, J. D., S. B. DeLong, M. L. Cline, C. D. Harrington, and G. N. Keating (2008), Dispersion of channel-sediment contaminants in distributary fluvial systems: Application to fluvial tephra and radionuclide redistribution following a potential volcanic eruption at Yucca Mountain, *Geomorphology*, 94, 226–246.

Pelletier, J. D., P. G. DeCelles, and G. Zandt (2010), Relationships among climate, erosion, topography, and delamination in the Andes: A numerical modeling investigation, *Geology*, 38, 259–262.

Pellicciotti, F., J. Helbing, A. Rivera, V. Favier, J. Corripio, J. Araos, J.-E. Sicart, and M. Carenzo (2008), A study of hte energy balance and melt regime on Juncal Norte Glacier, semi-arid Andes of central Chile, using melt models of different complexity, *Hydrol. Processes*, 22, 3980–3997.

Pelpola, C. P., and E. J. Hickin (2004), Long-term bed load transport rate based on aerial-photo and ground penetrating radar surveys of fan-delta growth, Coast Mountains, British Columbia, *Geomorphology*, 57, 169–181.

Penck, W. (1924), *Die morphologische Analyse: Ein Kapital der Physikalischen Geologie*, Geographische Abhandlungen, 2 Reihe, Heft 2, Engelhorn, Stuttgart.

Penck, W. (1953), *Morphological Analysis of Land Forms: A Contribution to Physical Geology*, 429 pp., translated by H. Czech and K. C. Boswell, Macmillan, London.

Pennak, R. W. (1971), Towards a classification of lotic habitats, *Hydrobiologia*, 38, 321–334.

Pereira, H. C. (1989), *Policy and Practice in the Management of Tropical Watersheds*, 237 pp., Westview Press, Boulder, Colo.

Pérez-Peña, J. V., J. M. Azañón, G. Booth-Rea, A. Azor, and J. Delgado (2009), Differentiating geology and tectonics using a spatial autocorrelation technique for the hypsometric integral, *J. Geophys. Res.*, *114*, F02018, doi:10.1029/2008JF001092.

Pérez-Peña, J. V., A. Azor, J. M. Azañón, and E. A. Keller (2010), Active tectonics in the Sierra Nevada (Betic Cordillera, SE Spain): Insights from geomorphic indexes and drainage pattern analysis, *Geomorphology*, *119*, 74–87.

Perona, P., P. Molnar, M. Savina, and P. Burlando (2009), An observation-based stochastic model for sediment and vegetation dynamics in the floodplain of an Alpine braided river, *Water Resour. Res.*, *45*, W09418, doi:10.1029/2008WR007550.

Persico, L., and G. Meyer (2009), Holocene beaver damming, fluvial geomorphology, and climate in Yellowstone National Park, Wyoming, *Quat. Res.*, *71*, 340–353.

Personius, S. F., H. M. Kelsey, and P. C. Graben (1993), Evidence for regional stream aggradation in the central Oregon Coast Range during the Pleistocene-Holocene transition, *Quat. Res.*, *40*, 297–308.

Pess, G. R., T. B. Abbe, T. A. Drury, and D. R. Montgomery (1998), Biological evaluation of engineered log jams: North Fork Stillaguamish River, Washington, *Eos Trans. AGU*, *79*, F346.

Peters, N. E. (1994), Hydrologic studies, in *Biogeochemistry of Small Catchments: A Tool for Environmental Research*, edited by B. Moldan and J. Cerny, pp. 207–228, John Wiley, Chichester, U. K.

Peterson, B. J., et al. (2001), Control of nitrogen export from watersheds by headwater streams, *Science*, *292*, 86–90.

Peterson, D., R. Smith, S. Hager, J. Hicke, M. Dettinger, and K. Huber (2005), River chemistry as a monitor of Yosemite Park mountain hydroclimates, *Eos Trans. AGU*, *86*, 285–288.

Petit, F. (1987), The relationship between shear stress and the shaping of the bed of a pebble-loaded river, La Rulles-Ardenne, *Catena*, *14*, 453–468.

Petit, F. (1989), The evaluation of grain shear stress from experiments in a pebble-bedded flume, *Earth Surf. Processes Landforms*, *14*, 499–508.

Petit, F. (1990), Evaluation of grain shear stresses required to initiate movement of particles in natural rivers, *Earth Surf. Processes Landforms*, *15*, 135–148.

Petit, F., F. Gob, G. Houbrechts, and A. A. Assani (2005), Critical specific stream power in gravel-bed rivers, *Geomorphology*, *69*, 92–101.

Petley, D. N., F. Mantovani, M. H. Bulmer, and A. Zannoni (2005), The use of surface monitoring data for the interpretation of landslide movement patterns, *Geomorphology*, *66*, 133–147.

Petrie, J., and P. Diplas (2000), Statistical approach to sediment sampling accuracy, *Water Resour. Res.*, *36*, 597–605.

Pettit, N. E., and R. J. Naiman (2006), Flood-deposited wood creates regeneration niches for riparian vegetation on a semi-arid South African river, *J. Veg. Sci.*, *17*, 615–624.

Petts, G. E. (1977), Channel response to flow regulation: The case of the River Derwent, Derbyshire, in *River Channel Changes*, edited by K. J. Gregory, pp. 145–164, John Wiley, Chichester, U. K.

Petts, G. E. (1984), *Impounded Rivers: Perspectives for Ecological Management*, 326 pp., John Wiley, Chichester, U. K.

Petts, G. E. (1989), Historical analysis of fluvial hydrosystems, in *Historical Change of Large Alluvial Rivers: Western Europe*, edited by G. E. Petts, pp. 1–18, John Wiley, Chichester, U. K.

Phanikumar, M. S., I. Aslam, C. Shen, D. T. Long, and T. C. Voice (2007), Separating surface storage from hyporheic retention in natural streams using wavelet decomposition of acoustic Doppler current profiles, *Water Resour. Res.*, *43*, W05406, doi:10.1029/2006WR005104.

Philbrick, S. S. (1970), Horizontal configuration and the rate of erosion of Niagara Falls, *Geol. Soc. Am. Bull.*, *81*, 3723–3732.

Phillips, J. D. (2002), Geomorphic impacts of flash flooding in a forested headwater basin, *J. Hydrol.*, *269*, 236–250.

Phillips, J. D. (2003), Sources of nonlinear complexity in geomorphic systems, *Prog. Phys. Geogr.*, *26*, 339–361.

Phillips, J. D. (2010), The job of the river, *Earth Surf. Processes Landforms*, *35*, 305–313.

Phillips, J. D., and L. Park (2009), Forest blowdown impacts of Hurricane Rita on fluvial systems, *Earth Surf. Processes Landforms*, *34*, 1069–1081.

Phillips, J. D., M. Marden, and B. Gomez (2007), Residence time of alluvium in an aggrading fluvial system, *Earth Surf. Processes Landforms*, *32*, 307–316.

Phillips, J. D., S. McCormack, J. Duan, J. P. Russo, A. M. Schumacher, G. N. Tripathi, R. B. Brockman, A. B. Mays, and S. Pulugurtha (2010), Origin and interpretation of knickpoints in the Big South Fork River basin, Kentucky-Tennessee, *Geomorphology*, *114*, 188–198.

Phillips, L. F., and S. A. Schumm (1987), Effect of regional slope on drainage networks, *Geology*, *15*, 813–816.

Phillips, P. J., and J. M. Harlin (1984), Spatial dependency of hydraulic geometry exponents in a subalpine stream, *J. Hydrol.*, *71*, 277–283.

Phippen, S. J., and E. Wohl (2003), An assessment of land use and other factors affecting sediment loads in the Rio Puerco watershed, New Mexico, *Geomorphology*, *52*, 269–287.

Phipps, R. L. (1985), *Collecting, preparing, crossdating, and measuring tree increment cores*, U.S. Geol. Surv. Water Resour. Invest. Rep. 85-4148, 48 pp.

Pickup, G. (1991), Event frequency and landscape stability on the floodplain systems of arid central Australia, *Quat. Sci. Rev.*, *10*, 463–473.

Pickup, G., and R. F. Warner (1976), Effects of hydrologic regime on magnitude and frequency of dominant discharge, *J. Hydrol.*, *29*, 51–75.

Pickup, G., R. J. Higgins, and I. Grant (1983), Modelling sediment transport as a moving wave - the transfer and deposition of mining waste, *J. Hydrol.*, *60*, 281–301.

Piégay, H., and A. M. Gurnell (1997), Large woody debris and river geomorphological pattern: Examples from S. E. France and S. England, *Geomorphology*, *19*, 99–116.

Piégay, H., and S. A. Schumm (2003), System approaches in fluvial geomorphology, in *Tools in Fluvial Geomorphology*, edited by G. M. Kondolf and H. Piégay, pp. 105–134, John Wiley, Chichester, U. K.

Piégay, H., D. E. Walling, N. Landon, Q. He, F. Liébault, and R. Petiot (2004), Contemporary changes in sediment yield in an alpine mountain basin due to afforestation (the upper Drome in France), *Catena*, *55*, 183–212.

Piégay, H., et al. (2005), Public perception as a barrier to introducing wood in rivers for restoration purposes, *Environ. Manage.*, *36*, 665–674.

Piégay, H., G. Grant, F. Nakamura, and N. Trustrum (2006), Braided river management: From assessment of river behaviour to improved sustainable development, in *Braided Rivers: Process, Deposits, Ecology and Management*, edited by G. H. Sambrook Smith et al., *Spec. Publ. Int. Assoc. Sedimentol.*, *36*, 257–275.

Pierce, J. L., and G. A. Meyer (2008), Long-term fire history from alluvial fan sediments: The role of drought and climate variability, and implications for management of Rocky Mountain forests, *Int. J. Wildl. Fire*, *17*, 84–95.

Pierce, J. L., G. A. Meyer, and A. J. T. Jull (2004), Fire-induced erosion and millennial-scale climate change in northern ponderosa pine forests, *Nature*, *432*, 87–90.

Pierson, F. B., P. R. Robichaud, C. A. Moffet, K. E. Spaeth, C. J. Williams, S. P. Hardegree, and P. E. Clark (2008), Soil water repellency and infiltration in coarse-textured soils of burned and unburned sagebrush ecosystems, *Catena*, *74*, 98–108.

Pierson, T. C. (1980), Erosion and deposition by debris flows at Mt. Thomas, North Canterbury, New Zealand, *Earth Surf. Processes*, *5*, 227–247.

Pierson, T. C. (2005), Hyperconcentrated flow – transitional process between water flow and debris flow, in *Debris-Flow Hazards and Related Phenomena*, edited by M. Jakob and O. Hungr, pp. 159–202, Springer, Berlin.

Pierson, T. C. (2007), Dating young geomorphic surfaces using age of colonizing Douglas fir in southwestern Washington and northwestern Oregon, USA, *Earth Surf. Processes Landforms*, *32*, 811–831.

Pike, A. S. (2006), Application of digital terrain analysis to estimate hydrological variables in the Luquillo Mountains of Puerto Rico, in *Climate Variability and Change—Hydrological Impacts, IAHS Publ.308*, pp. 81–85.

Pike, A. S. (2010), Longitudinal patterns in stream channel geomorphology and aquatic habitat in the Luquillo Mountains of Puerto Rico, Ph.D. dissertation, 260 pp., Univ. of Pa., Philadelphia.

Pike, A. S., and F. N. Scatena (2010), Riparian indicators of flow frequency in a tropical montane stream network, *J. Hydrol.*, *382*, 72–87.

Pike, A. S., F. N. Scatena, and E. E. Wohl (2010), Longitudinal patterns in stream channel geomorphology in the tropical mountain streams of the Luquillo Mountains, Puerto Rico, *Earth Surf. Processes Landforms*.

Pike, R. J., and S. Sobieszczyk (2008), Soil slip/debris flow localized by site attributes and wind-driven rain in the San Francisco Bay region storm of January 1982, *Geomorphology*, *94*, 290–313.

Pike, R. J., et al. (1998), Slope failure and shoreline retreat during northern California's latest El Nino, *GSA Today*, *8*, 1–6.

Pinay, G., H. DéCamps, E. Chauvet, and E. Fustec (1990), Functions of ecotones in fluvial systems, in *Ecology and Management of Aquatic-terrestrial Ecotones*, edited by R. J. Naiman and H. DéCamps, UNESCO, Paris and The Parthenon, Carnforth, U. K.

Pircher, W. (1990), The contribution of hydropower reservoirs to flood control in the Austrian Alps, in *Hydrology in Mountainous Regions II. Artificial Reservoirs, Water and Slopes*, edited by R. O. Sinniger and M. Monbaron, *IAHS Publ.*, *194*, 3–10.

Pitlick, J. (1988), The response of coarse-bed rivers to large floods in California and Colorado, unpublished Ph.D. dissertation, 137 pp., Colo. State Univ., Ft. Collins.

Pitlick, J. (1994a), Coarse sediment transport and the maintenance of fish habitat in the upper Colorado River, in *Proceedings, ASCE Hydraulic Engineering '94 Conference, August 1–5, Buffalo, New York*, edited by G. V. Cotroneo and R. R. Rumer, pp. 855–859, Am. Soc. of Civ. Eng., New York.

Pitlick, J. (1994b), Relation between peak flows, precipitation, and physiography for five mountainous regions in the western USA, *J. Hydrol.*, *158*, 219–240.

Pitlick, J., and M. M. Van Steeter (1998), Geomorphology and endangered fish habitats of the upper Colorado River. 2. Linking sediment transport to habitat maintenance, *Water Resour. Res.*, *34*, 303–316.

Pitlick, J., and P. R. Wilcock (2001), Relations between streamflow, sediment transport, and aquatic habitat in regulated rivers, in *Geomorphic Processes and Riverine Habitat*, edited by J. M. Dorava et al., pp. 185–198, AGU, Washington, D. C.

Pitlick, J. C., and C. R. Thorne (1987), Sediment supply, movement and storage in an unstable gravel-bed river, in *Sediment Transport in Gravel-Bed Rivers*, edited by C. R. Thorne, J. C. Bathurst, and R. D. Hey, pp. 151–183, John Wiley, Chichester, U. K.

Pizzuto, J. E. (2002), Effects of dam removal on river form and process, *BioScience*, *52*, 683–691.

Pizzuto, J. E. (2003), Numerical modeling of alluvial landforms, in *Tools in Fluvial Geomorphology*, edited by G. M. Kondolf and H. Piégay, pp. 577–595, John Wiley, Chichester, U. K.

Pizzuto, J. E., W. C. Hession, and M. McBride (2000), Comparing gravel-bed rivers in paired urban and rural catchments of southeastern Pennsylvania, *Geology*, *28*, 79–82.

Pizzuto, J. E., G. Moglen, M. Palmer, and K. Nelson (2008), Two model scenarios illustrating the effects of land use and climate change on gravel riverbeds of suburban Maryland, USA, in *Gravel-Bed Rivers VI: From process understanding to river restoration*, edited by H. Habersack, H. Piégay, and M. Rinaldi, pp. 359–386, Elsevier, Amsterdam.

Pizzuto, J. E., M. O'Neal, and S. Stotts (2010), On the retreat of forested, cohesive riverbanks, *Geomorphology*, *116*, 341–352.

Planchon, O., N. Silvera, R. Gimenez, D. Favis-Mortlock, J. Wainwright, Y. Le Bissonais, and G. Govers (2005), An automated salt-tracing gauge for flow-velocity measurement, *Earth Surf. Processes Landforms*, *30*, 833–844.

Platts, W. S. (1981), *Effects of livestock grazing*, USDA For. Serv. General Tech. Rep. PNW-124, 25 pp.

Platts, W. S., W. F. Megahan, and G. W. Minshall (1983), *Methods for evaluating stream, riparian, and biotic conditions*, USDA For. Serv. General Tech. Rep. INT-138, 70 pp.

Platts, W. S., et al. (1987), Methods for evaluating riparian habitats with applications to management, *General Tech. Rep. INT-221*, 177 pp, U.S. Dep. of Agric., For. Serv., Ogden, Utah.

Platts, W. S., R. J. Torquemada, M. L. McHenry, and C. K. Graham (1989), Changes in salmon spawning and rearing habitat from increased delivery of fine sediment to the South Fork Salmon River, Idaho, *Trans. Am. Fish. Soc.*, *118*, 274–283.

Poff, N. L. (1992), Regional hydrologic response to climate change: An ecological perspective, in *Global Climate Change and Freshwater Ecosystems*, edited by P. Firth and S. G. Fisher, pp. 88–115, Springer, New York.

Poff, N. L. (1996), A hydrogeography of unregulated streams in the United States and an examination of scale-dependence in some hydrological descriptors, *Freshwater Biol.*, *36*, 71–91.

Poff, N. L. (1997), Landscape filters and species traits: Towards mechanistic understanding and prediction in stream ecology, *J. North Am. Benthol. Soc.*, *16*, 391–409.

Poff, N. L., and J. D. Allan (1995), Functional organization of stream fish assemblages in relation to hydrological variability, *Ecology*, *76*, 606–627.

Poff, N. L., and D. D. Hart (2002), How dams vary and why it matters for the emerging science of dam removal, *BioScience*, *52*, 659–668.

Poff, N. L., and J. V. Ward (1989), Implications of streamflow variability and predictability for lotic community structure: A regional analysis of streamflow patterns, *Can. J. Fish. Aquat. Sci.*, *46*, 1805–1818.

Poff, N. L., and J. V. Ward (1990), Physical habitat template of lotic ecosystems: Recovery in the context of historical pattern of spatiotemporal heterogeneity, *Environ. Manage.*, *14*, 629–645.

Poff, N. L., and E. Wohl (1999), An integrated hydrologic-geomorphic aquatic classification for fluvial ecosystems, *Ecological Society of America, 1999 Annual Meeting Abstracts*, p. 31.

Poff, N. L., J. D. Allan, M. B. Bain, J. R. Karr, K. L. Prestegaard, B. D. Richter, R. E. Sparks, and J. C. Stromberg (1997), The natural flow regime: A paradigm for river conservation and restoration, *BioScience*, *47*, 769–784.

Poff, N. L., B. P. Bledsoe, and C. O. Cuhaciyan (2006), Hydrologic variation with land use across the contiguous United States: Geomorphic and ecological consequences for stream ecosystems, *Geomorphology*, *79*, 264–285.

Poff, N. L., J. D. Olden, D. M. Merritt, and D. M. Pepin (2007), Homogenization of regional river dynamics by dams and global biodiversity implications, *Proc. Natl. Acad. Sci. U. S. A.*, *104*, 5732–5737.

Poisson, B., and J.-P. Avouac (2004), Holocene hydrological changes inferred from alluvial stream entrenchment in north Tian Shan (northwestern China), *J. Geol.*, *112*, 231–249.

Pollen, N. (2006), Temporal and spatial variability of root reinforcement of streambanks: Accounting for soil shear strength and moisture, *Catena*, *69*, 197–205.

Pollen, N., and A. Simon (2005), Estimating the mechanical effects of riparian vegetation on streambank stability using a fiber bundle model, *Water Resour. Res.*, *41*, W07025, doi:10.1029/2004WR003801.

Pollen, N., A. Simon, and A. Collison (2004), Advances in assessing the mechanical and hydrologic effects of riparian vegetation on streambank stability, in *Riparian Vegetation and Fluvial Geomorphology*, edited by S. J. Bennett and A. Simon, pp. 125–139, AGU, Washington, D. C.

Pollen-Bankhead, N., and A. Simon (2010), Hydrologic and hydraulic effects of riparian root networks on streambank stability: Is mechanical root-reinforcement the whole story?, *Geomorphology, 116*, 353–362.

Pollock, M. M., T. J. Beechie, and C. E. Jordan (2007), Geomorphic changes upstream of beaver dams in Bridge Creek, an incised stream channel in the interior Columbia River basin, eastern Oregon, *Earth Surf. Processes Landforms, 32*, 1174–1185.

Polvi, L. E. (2009), Characterization of riparian zones in mountain valleys of the Colorado Front Range, M.S. thesis, Colo. State Univ., Fort Collins.

Pomeroy, J. W., D. M. Gray, T. Brown, N. R. Hedstrom, W. L. Quinton, R. J. Granger, and S. K. Carey (2007), The Cold Regions Hydrological Model: A platform for basing process representation and model structure on physical evidence, *Hydrol. Processes, 21*, 2650–2667.

Ponton, J. R. (1972), Hydraulic geometry in the Green and Birkenhead basins, British Columbia, in *Mountain Geomorphology: Geomorphological Processes in the Canadian Cordillera*, edited by H. O. Slaymaker and H. J. McPherson, pp. 151–160, Tantalus Res. Ltd., Vancouver, Canada.

Poole, G. C. (2002), Fluvial landscape ecology: Addressing uniqueness within the river discontinuum, *Freshwater Biol., 47*, 641–660.

Poole, G. C., C. A. Frissell, and S. C. Ralph (1997), In-stream habitat unit classification: Inadequacies for monitoring and some consequences for management, *J. Am. Water Resour. Assoc., 33*, 879–896.

Pope, R. J. J., and K. N. Wilkinson (2005), Reconciling the roles of climate and tectonics in late Quaternary fan development on the Spartan Piedmont, Greece, in *Alluvial Fans: Geomorphology, Sedimentology, Dynamics*, edited by A. M. Harvey, A. E. Mather, and M. Stokes, *Geol. Soc. Spec. Publ., 251*, 133–152.

Pornprommin, A., and N. Izumi (2010), Inception of stream incision by seepage erosion, *J. Geophys. Res., 115*, F02022, doi:10.1029/2009JF001369.

Post, A., and L. R. Mayo (1971), *Glacier dammed lakes and outburst floods in Alaska*, U.S. Geol. Surv. Hydrol. Invest. Atlas HA-455, 3 sheets.

Post, D. A., and M. G. Hartcher (2006), Evaluating uncertainty in modelled sediment delivery in data-sparse environments: Application to the Mae Chaem Catchment, Thailand, in *Predictions in Ungauged Basins: Promise and Progress*, edited by M. Sivapalan et al., *IAHS Publ., 303*, 80–89.

Post, D. A., I. G. Littlewood, and B. F. Croke (2005), New directions for top-down modelling: Introducing the PUB Top-Down Modelling Working Group, in *Predictions in Ungauged Basins: International Perspectives on the State of the Art and Pathways Forward*, edited by S. Franks et al., *IAHS Publ., 301*, 125–133.

Powell, D. M. (1998), Patterns and processes of sediment sorting in gravel-bed rivers, *Prog. Phys. Geogr., 22*, 1–32.

Powell, D. M., I. Reid, J. B. Laronne, and L. Frostick (1996), Bed load as a component of sediment yield from a semiarid watershed of the northern Negev, in *Erosion and Sediment Yield: Global and Regional Perspectives*, edited by D. E. Walling and B. W. Webb, *IAHS Publ., 236*, 389–397.

Powell, D. M., I. Reid, and J. B. Laronne (2001), Evolution of bed load grain size distribution with increasing flow strength and the effect of flow duration on the caliber of bed load sediment yield in ephemeral gravel bed rivers, *Water Resour. Res., 37*, 1463–1474.

Powell, J. W. (1875), *Exploration of the Colorado River of the West (1869–72)*, U.S. Gov. Print. Off., Washington, D. C.

Powell, J. W. (1876), *Report on the Geology of the Eastern Portion of the Uinta Mountains*, U.S. Gov. Print. Off., Washington, D. C.

Power, M. E., W. E. Dietrich, and J. C. Finlay (1996), Dams and downstream aquatic biodiversity: Potential food web consequences of hydrologic and geomorphic change, *Environ. Manage., 20*, 887–895.

Power, M. E., A. Sun, M. Parker, W. E. Dietrich, and J. T. Wootton (1995), Hydraulic food-chain models: An approach to the study of food-web dynamics in large rivers, *BioScience, 45*, 159–167.

Pratt, B., D. W. Burbank, A. Heimsath, and T. Ojha (2002), Impulsive alluviation during early Holocene strengthened monsoons, central Nepal Himalaya, *Geology, 30*, 911–914.

Pratt-Sitaula, B., D. W. Burbank, A. M. Heimsath, and T. Ojha (2004), Landscape disequilibrium on 1000-10,000 year scales Marsyandi River, Nepal, central Himalaya, *Geomorphology*, *58*, 233–241.

Pratt-Sitaula, B., M. Garde, D. W. Burbank, M. Oskin, A. Heimsath, and E. Gabet (2007), Bedload-to-suspended load ratio and rapid bedrock incision from Himalayan landslide-dam lake record, *Quat. Res.*, *68*, 111–120.

Prent, M. T. H., and E. J. Hickin (2001), Annual regime of bedforms, roughness and flow resistance, Lillooet River, British Columbia, BC, *Geomorphology*, *41*, 369–390.

Press, F., and R. Siever (1986), *Earth*, 4th ed., 656 pp., W. H. Freeman, San Francisco, Calif.

Prestegaard, K. L. (1983a), Bar resistance in gravel bed streams at bankfull stage, *Water Resour. Res.*, *19*, 472–476.

Prestegaard, K. L. (1983b), Variables influencing water-surface slopes in gravel-bed streams at bankfull stage, *Geol. Soc. Am. Bull.*, *94*, 673–678.

Price, K., and D. S. Leigh (2006), Morphological and sedimentological responses of streams to human impact in the southern Blue Ridge Mountains, USA, *Geomorphology*, *78*, 142–160.

Price, L. W. (1981), *Mountains and Man: A Study of Process and Environment*, 506 pp., Univ. of Calif. Press, Berkeley.

Price, L. W. (2002), *Mountains*, 72 pp., Voyageur Press, Stillwater, Minn.

Price, M. F., and R. G. Barry (1997), Climate change, in *Mountains of the World: A Global Priority*, edited by B. Messerli and J. D. Ives, pp. 409–445, The Parthenon, London.

Priesnitz, K., and E. Schunke (2002), The fluvial morphodynamics of two small permafrost drainage basins, Richardson Mountains, northwestern Canada, *Permafrost Periglacial Processes*, *13*, 207–217.

Pringle, C. M. (2001), Hydrologic connectivity and the management of biological reserves: A global perspective, *Ecol. Appl.*, *11*, 981–998.

Pringle, C. M. (2003), What is hydrologic connectivity and why is it ecologically important?, *Hydrol. Processes*, *17*, 2685–2689.

Pristachová, G. (1990), Quantitative geomorphology, stream networks, and instantaneous unit hydrograph, in *Hydrology of Mountainous Areas*, edited by L. Molnar, *IAHS Publ.*, *190*, 369–375.

Pritchard, D., G. G. Roberts, N. J. White, and C. N. Richardson (2009), Uplift histories from river profiles, *Geophys. Res. Lett.*, *36*, L24301, doi:10.1029/2009GL040928.

Prochaska, A. B., P. M. Santi, and J. D. Higgins (2008), Debris basin and deflection berm design for fire-related debris-flow mitigation, *Environ. Eng. Geosci.*, *14*, 297–313.

Procter, J., S. J. Cronin, I. C. Fuller, G. Lube, and V. Manville (2010), Quantifying the geomorphic impacts of a lake-breakout lahar, Mount Ruapehu, New Zealand, *Geology*, *38*, 67–70.

Prokopovich, N. P. (1984), Occurrence of mercury in dredge tailings near Folsom South Canal, California, *Bull. Assoc. Eng. Geol.*, *XXI*, 531–543.

Prosser, I. P., and B. Abernethy (1996), Predicting the topographic limits to a gully network using a digital terrain model and process thresholds, *Water Resour. Res.*, *32*, 2289–2298.

Prosser, I. P., and W. E. Dietrich (1995), Field experiments on erosion by overland flow and their implications for a digital terrain model of channel initiation, *Water Resour. Res.*, *31*, 2867–2876.

Prosser, I. P., and P. Rustomji (2000), Sediment transport capacity relations for overland flow, *Prog. Phys. Geogr.*, *24*, 179–193.

Prosser, I. P., W. E. Dietrich, and J. Stevenson (1995), Flow resistance and sediment transport by concentrated overland flow in a grassland valley, *Geomorphology*, *13*, 71–86.

Prosser, I. P., A. O. Hughes, and I. D. Rutherfurd (2000), Bank erosion of an incised upland channel by subaerial processes: Tasmania, Australia, *Earth Surf. Processes Landforms*, *25*, 1085–1101.

Pruess, J., E. E. Wohl, and R. D. Jarrett (1998), Methodology and implications of maximum paleodischarge estimates for mountain channels, upper Animas River basin, Colorado, USA, *Arct. Alp. Res.*, *30*, 40–50.

Pyrce, R. S. (1995), A field investigation of planimetric knickpoint morphology from rock-bed sections of Niagara Escarpment fluvial systems, M.A. thesis, 119 pp., Wilfrid Laurier Univ., Waterloo, Ont., Canada.

Pyrce, R. S., and P. E. Ashmore (2003a), Particle path length distributions in meandering gravel-bed streams: Results from physical models, *Earth Surf. Processes Landforms*, *28*, 951–966.

Pyrce, R. S., and P. E. Ashmore (2003b), The relation between particle path length distributions and channel morphology in gravel-bed streams: A synthesis, *Geomorphology*, *56*, 167–187.

Qiu, Z. (2009), Assessing critical source areas in watersheds for conservation buffer planning and riparian restoration, *Environ. Manage.*, *44*, 968–980.

Quinn, J. M., N. R. Phillips, and S. M. Parkyn (2007), Factors influencing retention of coarse particulate organic matter in streams, *Earth Surf. Processes Landforms*, *32*, 1186–1203.

Rabeni, C. F., and G. S. Wallace (1998), The influence of flow variation on the ability to evaluate the biological health of headwater streams, in *Hydrology, Water Resources and Ecology in Headwaters*, edited by K. Kovar et al., *IAHS Publ.*, *248*, 411–417.

Radecki-Pawlik, A. (2002), Bankfull discharge in mountain streams: Theory and practice, *Earth Surf. Processes Landforms*, *27*, 115–123.

Rader, R. B., and T. A. Belish (1999), Influence of mild to severe flow alterations on invertebrates in three mountain streams, *Reg. Rivers Res. Manage.*, *15*, 353–363.

Rădoane, M., and N. Rădoane (2005), Dams, sediment sources and reservoir silting in Romania, *Geomorphology*, *71*, 112–125.

Rădoane, M., N. Rădoane, and D. Dumitriu (2003), Geomorphological evolution of longitudinal river profiles in the Carpathians, *Geomorphology*, *50*, 293–206.

Raff, D. A., J. A. Ramírez, and J. L. Smith (2004), Hillslope drainage development with time: A physical experiment, *Geomorphology*, *62*, 169–180.

Raffaelli, D. G., A. G. Hildrew, and P. S. Giller (1994), Scale, pattern and process in aquatic systems: Concluding remarks, in *Aquatic Ecology: Scale, Pattern and Process*, edited by P. S. Giller, A. G. Hildrew, and D. G. Raffaelli, pp. 601–606, Blackwell Sci., Oxford, U. K.

Rahaman, W., S. K. Singh, R. Sinha, and S. K. Tandon (2009), Climate control on erosion distribution over the Himalaya during the past ~100 ka, *Geology*, *37*, 559–562.

Ralph, S. C., G. C. Poole, L. L. Conquest, and R. J. Naiman (1994), Stream channel morphology and woody debris in logged and unlogged basins of western Washington, *Can. J. Fish. Aquat. Sci.*, *51*, 37–51.

Ralston, D. R., and A. G. Morilla (1974), Ground-water movement through an abandoned tailings pile, in *Water Resources Problems Related to Mining*, edited by R. F. Hadley and D. T. Snow, pp. 174–183, Am. Water Resour. Assoc., Minneapolis, Minn.

Ramanathan, R., A. Guin, R. W. Ritzi, D. F. Dominic, V. L. Freedman, T. D. Scheibe, and I. A. Lunt (2010), Simulating the heterogeneity in braided channel belt deposits: 1. A geometric-based methodology and code, *Water Resour. Res.*, *46*, W04515, doi:10.1029/2009WR008111.

Ramírez, E., B. Francou, P. Ribstein, M. Descloitres, R. Guérin, J. Mendoza, R. Gallaire, B. Pouyaud, and E. Jordan (2001), Small glaciers disappearing in the tropical Andes: A case study in Bolivia: Glaciar Chacaltaya (16°S), *J. Glaciol.*, *47*, 187–194.

Ranalli, A. J., and D. L. Macalady (2010), The importance of the riparian zone and in-stream processes in nitrate attenuation in undisturbed and agricultural watersheds – a review of the scientific literature, *J. Hydrol.*, *389*, 406–415.

Rango, A., and J. Martinec (2000), Hydrological effects of a changed climate in humid and arid mountain region, *World Resour. Rev.*, *12*, 493–508.

Rango, A., E. Gomez-Landesa, M. Bleiweiss, K. Havstad, and K. Tanksley (2003), Improved satellite snow mapping, snowmelt runoff forecasting, and climate change simulations in the upper Rio Grande Basin, *World Resour. Rev.*, *15*, 25–41.

Raper, S. C. B., and R. J. Braithwaite (2006), Low sea level rise projections from mountain glaciers and icecaps under global warming, *Nature*, *439*, 311–313.

Rapp, A. (1960), Recent development of mountain slopes in Karkevagge and surroundings, northern Sweden, *Geogr. Ann.*, *41*, 65–200.

Rapp, A., and R. Nyberg (1981), Alpine debris flows in northern Scandinavia, *Geogr. Ann.*, *63A*, 183–196.

Rapp, A., and L. Strömquist (1976), Slope erosion due to extreme rainfall in the Scandinavian Mountains, *Geogr. Ann.*, *58A*, 193–200.

Rasemann, S., J. Schmidt, L. Schrott, and R. Dikau (2004), Geomorphmetry in mountain terrain, in *Geographic Information Science and Mountain Geomorphology*, edited by M. P. Bishop and J. F. Shroder, pp. 101–145, Praxis, Chichester, U. K.

Rathburn, S. L., and E. E. Wohl (2001), One-dimensional sediment transport modeling of pool recovery along a mountain channel after a reservoir sediment release, *Reg. Rivers Res. Manage.*, *17*, 251–273.

Rathburn, S. L., and E. Wohl (2003), Predicting fine sediment dynamics along a pool-riffle mountain channel, *Geomorphology*, *55*, 111–124.

Rathburn, S. L., D. M. Merritt, E. E. Wohl, J. S. Sanderson, and H. A. L. Knight (2009), Characterizing environmental flows for maintenance of river ecosystems: North Fork Cache la Poudre River, Colorado, in *Management and Restoration of Fluvial Systems with Broad Historical Changes and Human Impacts*, edited by L. A. James, S. L. Rathburn, and G. R. Whittecar, *Spec. Pap. Geol. Soc. Am.*, *451*, 143–157.

Raven, E. K., S. N. Lane, and R. I. Ferguson (2010), Using sediment impact sensors to improve the morphological sediment budget approach for estimating bedload transport rates, *Geomorphology*, *119*, 125–134.

Raven, E. K., S. N. Lane, R. I. Ferguson, and L. J. Bracken (2009), The spatial and temporal patterns of aggradation in a temperate, upland, gravel-bed river, *Earth Surf. Processes Landforms*, *34*, 1181–1197.

Rawat, J. S., and M. S. Rawat (1994), Accelerated erosion and denudation in the Nana Kosi watershed, Central Himalaya, India. Part I: Sediment load, *Mt. Res. Dev.*, *14*, 25–38.

Raymo, M. E., and W. F. Ruddiman (1992), Tectonic forcing of the late Cenozoic climate, *Nature*, *359*, 117–122.

Raymo, M. E., W. F. Ruddiman, and P. N. Froelich (1988), Influence of late Cenozoic mountain building on geochemical cycles, *Geology*, *16*, 649–653.

Recking, A. (2010), A comparison between flume and field bed load transport data and consequences for surface-based bed load transport prediction, *Water Resour. Res.*, *46*, W03518, doi:10.1029/2009WR008007.

Recking, A., P. Frey, A. Paquier, P. Belleudy, and J. Y. Champagne (2008), Feedback between bed load transport and flow resistance in gravel and cobble bed rivers, *Water Resour. Res.*, *44*, W05412, doi:10.1029/2007WR006219.

Recking, A., P. Frey, A. Paquier, and P. Belleudy (2009), An experimental investigation of mechanisms involved in bed load sheet production and migration, *J. Geophys. Res.*, *114*, F03010, doi:10.1029/2008JF000990.

Redmond, K. T., Y. Enzel, P. K. House, and F. Biondi (2002), Climate variability and flood frequency at decadal to millennial time scales, in *Ancient Floods, Modern Hazards: Principles and Applications of Paleoflood Hydrology*, edited by P. K. House et al., pp. 21–45, AGU, Washington, D. C.

Rees, D. E. (1994), Indirect effects of heavy metals observed in macroinvertebrate availability, brown trout, (Salmo trutta) diet composition, and bioaccumulation in the Arkansas River, Colorado, unpublished M.S. thesis, 47 pp., Colo. State Univ., Ft. Collins.

Rees, H. G., and D. N. Collins (2006a), An assessment of the potential impacts of climatic warming on glacier-fed river flow in the Himalaya, in *Climate Variability and Change – Hydrological Impacts*, edited by S. Demuth et al., *IAHS Publ.*, *308*, 473–478.

Rees, H. G., and D. N. Collins (2006b), Regional differences in response of flow in glacier-fed Himalayan rivers to climatic warming, *Hydrol. Processes*, *20*, 2157–2169.

Reeves, G. H., L. E. Benda, K. M. Burnett, P. A. Bisson, and J. R. Sedell (1995), A disturbance-based ecosystem approach to maintaining and restoring freshwater habitats of evolutionarily significant units of anadromous salmonids in the Pacific Northwest, *Am. Fish. Soc. Symp.*, *17*, 334–349.

Reeves, G. H., K. M. Burnett, and E. V. McGarry (2003), Sources of large wood in the main stem of a fourth-order watershed in coastal Oregon, *Can. J. For. Res.*, *33*, 1363–1370.

Refsgard, J. C., and B. Storm (1995), MIKE SHE, in *Computer Models of Watershed Hydrology*, edited by V. P. Singh, pp. 809–846, Water Resour. Publ., Highland Ranch, Colo.

Reid, D. E., and E. J. Hickin (2008), Flow resistance in steep mountain streams, *Earth Surf. Processes Landforms*, *33*, 2211–2240.

Reid, D. E., E. J. Hickin, and S. C. Babakaiff (2010), Low-flow hydraulic geometry of small, steep mountain streams in southwest British Columbia, *Geomorphology*, *122*, 39–55.

Reid, I., and L. E. Frostick (1986), Dynamics of bedload transport in Turkey Brook, a coarse-grained alluvial channel, *Earth Surf. Processes Landforms*, *11*, 143–155.

Reid, I., J. T. Layman, and L. E. Frostick (1980), The continuous measurement of bedload discharge, *J. Hydraul. Res.*, *18*, 243–249.

Reid, I., L. E. Frostick, and J. L. Layman (1985), The incidence and nature of bedload transport during flood flows in coarse-grained alluvial channels, *Earth Surf. Processes Landforms*, *10*, 33–44.

Reid, I., L. E. Frostick, and A. C. Brayshaw (1992), Microform roughness elements and the selective entrainment and entrapment of particles in gravel-bed rivers, in *Dynamics of Gravel-Bed Rivers*, edited by P. Billi et al., pp. 253–275, John Wiley, Chichester, U. K.

Reid, L. M., and T. Dunne (1984), Sediment production from forest road surfaces, *Water Resour. Res.*, *20*, 1753–1761.

Reid, L. M., and T. Dunne (2003), Sediment budgets as an organizing framework in fluvial geomorphology, in *Tools in Fluvial Geomorphology*, edited by G. M. Kondolf and H. Piégay, pp. 463–500, John Wiley, Chichester, U. K.

Reid, L. M., N. J. Dewey, T. E. Lisle, and S. Hilton (2010), The incidence and role of gullies after logging in a coastal redwood forest, *Geomorphology*, *117*, 155–169.

Reid, S. C., S. N. Lanc, J. M. Berney, and J. Holden (2007a), The timing and magnitude of coarse sediment transport events within an upland, temperate gravel-bed river, *Geomorphology*, *83*, 152–182.

Reid, S. C., S. N. Lane, D. R. Montgomery, and C. J. Brookes (2007b), Does hydrological connectivity improve modelling of coarse sediment delivery in upland environments?, *Geomorphology*, *90*, 263–282.

Reidenbach, M. A., M. Limm, M. Hondzo, and M. T. Stacey (2010), Effects of bed roughness on boundary layer mixing and mass flux across the sediment-water interface, *Water Resour. Res.*, *46*, W07530, doi:10.1029/2009WR008248.

Reinfelds, I., T. Cohen, P. Batten, and G. Brierley (2004), Assessment of downstream trends in channel gradient, total and specific stream power: A GIS approach, *Geomorphology*, *60*, 403–416.

Reinhardt, L., D. Jerolmack, B. J. Cardinale, V. Vanacker, and J. Wright (2010), Dynamics interactions of life and its landscape: Feedbacks at the interface of geomorphology and ecology, *Earth Surf. Processes Landforms*, *35*, 78–101.

Reiser, D. W., and T. C. Bjornn (1979), *Influence of forest and rangeland management on anadromous fish habitat in the western U. S. and Canada. Part I: Habitat requirements of anadromous salmonids*, USDA For. Serv. General Tech. Rep. PNW-96, 54 pp.

Remondo, J., J. Soto, A. González-Díez, J. R. Díaz de Terán, and A. Cendrero (2005), Human impact on geomorphic processes and hazards in mountain areas in northern Spain, *Geomorphology*, *66*, 69–84.

Ren, D., D. J. Karoly, and L. M. Leslie (2007), Temperate mountain glacier-melting rates for the period 2001-30: Estimates from three coupled GCM simulations for the Greater Himalayas, *J. Appl. Meteorol. Climatol.*, *46*, 890–899.

Renard, K. G., G. R. Foster, G. A. Weesies, and J. P. Porter (1991), RUSLE- revised universal soil loss equation, *J. Soil Water Conserv.*, *46*, 30–33.

Reneau, S. L. (2000), Stream incision and terrace development in Frijoles Canyon, Bandelier National Monument, New Mexico, and the influence of lithology and climate, *Geomorphology*, *32*, 171–193.

Reneau, S. L., and D. P. Dethier (1996), Late Pleistocene landslide dammed-lakes along the Rio Grande, White Rock Canyon, New Mexico, *Geol. Soc. Am. Bull.*, *108*, 1492–1507.

Reneau, S. L., and W. E. Dietrich (1987), The implications of hollows in debris flow studies; examples from Marin County, California, in *Debris Flows/Avalanches: Process, Recognition, and Mitigation*, edited by J. E. Costa and G. F. Wieczorek, *Rev. Eng. Geol.*, *7*, 165–180.

Reneau, S. L., D. Katzman, G. A. Kuyumjian, A. Lavine, and D. V. Malmon (2007), Sediment delivery after a wildfire, *Geology*, *35*, 151–154.

Rengers, F., and E. Wohl (2007), Trends of grain sizes on gravel bars in the Rio Chagres, Panama, *Geomorphology*, *83*, 282–293.

Renwick, W. H. (1977), Erosion caused by intense rainfall in a small catchment in New York State, *Geology*, *5*, 361–364.

Resh, V. H., A. V. Brown, A. P. Covich, M. E. Gurtz, H. W. Li, G. W. Minshall, S. R. Reice, A. L. Sheldon, J. B. Wallace, and R. Wissmar (1988), The role of disturbance in stream ecology, *J. North Am. Benthol. Soc.*, *7*, 433–455.

Reusser, L., and P. Bierman (2007), Accuracy assessment of LiDAR-derived DEMs of bedrock river channels: Holtwood Gorge, Susquehanna River, *Geophys. Res. Lett.*, *34*, L23S06, doi:10.1029/2007GL031329.

Reusser, L. J., and P. R. Bierman (2010), Using meteoric ^{10}Be to track fluvial sand through the Waipaoa River basin, New Zealand, *Geology*, *38*, 47–50.

Reusser, L. J., P. R. Bierman, M. J. Pavich, E. Zen, J. Larsen, and R. C. Finkel (2004), Rapid late Pleistocene incision of Atlantic passive-margin river gorges, *Science*, *305*, 499–502.

Reusser, L. J., P. R. Bierman, M. J. Pavich, J. Larsen, and R. C. Finkel (2006), An episode of rapid bedrock channel incision during the last glacial cycle, measured with ^{10}Be, *Am. J. Sci.*, *306*, 69–102.

Reynolds, B., P. J. Chapman, M. C. French, A. Jenkins, and H. S. Wheater (1995a), Major, minor, and trace element chemistry of surface waters in the Everest region of Nepal, in *Biogeochemistry of Seasonally Snow-Covered Catchments*, edited by K. A. Tonnessen, M. W. Williams, and M. Tranter, *IAHS Publ.*, *228*, 405–412.

Reynolds, B., W. H. Robertson, M. Hornung, and P. A. Stevens (1995b), Forest manipulation and solute production: Modelling the nitrogen response to clearcutting, in *Solute Modelling in Catchment Systems*, edited by S. T. Trudgill, pp. 211–233, John Wiley, Chichester, U. K.

Reynolds, J. B., R. C. Simmons, and A. R. Burkholder (1989), Effects of placer mining discharge on health and food of Arctic grayling, *Water Resour. Bull.*, *25*, 625–635.

Rhea, J. O., and L. O. Grant ,(1974), Topographic influences on snowfall patterns in mountainous terrain, in *Adv. Concepts and Techniques in the Study of Snow and Ice Resources*, pp. 182–192, Natl. Acad. of Sci., Washington, D. C

Rhoades, R., X. Zapata, and J. Arangundy (2006), Climate change in Cotacachi, in *Development with Identity: Community, Culture and Sustainability in the Andes*, edited by R. E. Rhoades, pp. 64–74, CAB Int., Wallingford, U. K.

Rhoads, B. L., and A. N. Sukhodolov (2001), Field investigation of three-dimensional flow structure at stream confluences: 1. Thermal mixing and time-averaged velocities, *Water Resour. Res.*, *37*, 2393–2410.

Rhoads, B. L., and A. N. Sukhodolov (2004), Spatial and temporal structure of shear layer turbulence at a stream confluence, *Water Resour. Res.*, *40*, W06304, doi:10.1029/2003WR002811.

Rice, K. C., and O. P. Bricker (1995), Seasonal cycles of dissolved constituents in streamwater in two forested catchments in the mid-Atlantic region of the eastern USA, *J. Hydrol.*, *170*, 137–158.

Rice, S. P. (1998), Which tributaries disrupt downstream fining along gravel-bed rivers?, *Geomorphology*, *22*, 39–56.

Rice, S. P. (1999), The nature and controls on downstream fining within sedimentary links, *J. Sediment. Res.*, *69*, 32–39.

Rice, S. P., M. T. Greenwood, and C. B. Joyce (2001), Macroinvertebrate community changes at coarse sediment recruitment points along two gravel bed rivers, *Water Resour. Res.*, *37*, 2793–2803.

Rice, S. P., T. Buffin-Bélanger, J. Lancaster, and I. Reid (2008), Movements of a macroinvertebrate, (Potamophylax latipennis) across a gravel-bed substrate: Effects of local hydraulics and micro-topography under increasing discharge, in *Gravel-Bed Rivers VI: From Process Understanding to River Restoration*, edited by H. Habersack, H. Piégay, and M. Rinaldi, pp. 637–660, Elsevier, Amsterdam.

Rice, S. P., J. Lancaster, and P. Kemp (2010), Experimentation at the interface of fluvial geomorphology, stream ecology and hydraulic engineering and the development of an effective, interdisciplinary river science, *Earth Surf. Processes Landforms*, *35*, 64–77.

Richards, K. S. (1973), Hydraulic geometry and channel roughness – a non-linear system, *Am. J. Sci.*, *273*, 877–896.

Richards, K. S. (1976), The morphology of riffle-pool sequences, *Earth Surf. Processes*, *1*, 71–88.

Richards, K. S. (1978), Simulation of flow geometry in a riffle-pool stream, *Earth Surf. Processes*, *3*, 345–354.

Richards, K. S. (1987), *River Channels: Environment and Process*, 391 pp., Blackwell, Oxford, U. K.

Richards, K. S. (1990), Fluvial geomorphology: Initial motion of bed material in gravel-bed rivers, *Prog. Phys. Geogr.*, *14*, 395–415.

Richards, K. S. (1993), Sediment delivery and the drainage network, in *Channel Network Hydrology*, edited by K. Beven and M. J. Kirkby, pp. 221–254, John Wiley, Chichester, U. K.

Richards, K. S., and L. M. Milne (1979), Problems in the calibration of an acoustic device for the observation of bedload transport, *Earth Surf. Processes*, *4*, 335–346.

Richards, K. S., and R. Wood (1977), Urbanization, water redistribution, and their effect on channel processes, in *River Channel Changes*, edited by K. J. Gregory, pp. 369–388, John Wiley, New York.

Richardson, K., and P. A. Carling (2005), *A Typology of Sculpted Forms in Open Bedrock Channels, Spec. Pap. Geol. Soc. Am.*, *392*.

Richardson, K., and P. A. Carling (2006), The hydraulics of a straight bedrock channel: Insights from solute dispersion studies, *Geomorphology*, *82*, 98–125.

Richardson, K., I. Benson, and P. A. Carling (2003), An instrument to record sediment movement in bedrock channels, in *Erosion and Sediment Transport Measurement in Rivers: Technological and Methodological Advances*, edited by J. Bogen, T. Fergus, and D. E. Walling, *IAHS Publ.*, *283*, 228–235.

Richardson, M. C., M.-J. Fortin, and B. A. Branfireun (2009), Hydrogeomorphic edge detection and delineation of landscape functional units from LiDAR digital elevation models, *Water Resour. Res.*, *45*, W10441, doi:10.1029/2008WR007518.

Richmond, A. D., and K. D. Fausch (1995), Characteristics and function of large woody debris in subalpine Rocky Mountain streams in northern Colorado, *Can. J. Fish. Aquat. Sci.*, *52*, 1789–1802.

Richter, B. D. (2010), Re-thinking environmental flows: From allocations and reserves to sustainability boundaries, *River Res. Appl.*, *26*(8), 1052–1063.

Richter, B. D., J. Baumgartner, J. Powell, and D. Braun (1996), A method for assessing hydrologic alteration within ecosystems, *Conserv. Biol.*, *10*, 1163–1174.

Richter, B. D., J. Baumgartner, W. Robert, and D. Braun (1997), How much water does a river need?, *Freshwater Biol.*, *37*, 231–249.

Richter, B. D., J. Baumgartner, R. Wigington, D. Braun, and J. Powell (1998), A spatial assessment of hydrologic alteration within a river network, *Regul. Rivers*, *14*, 329–340.

Richter, B. D., R. Mathews, D. L. Harrison, and R. Wigington (2003), Ecologically sustainable water management: Managing river flows for ecological integrity, *Ecol. Appl.*, *13*, 206–224.

Rickenmann, D. (1990), Debris flows 1987 in Switzerland: Modelling and fluvial sediment transport, in *Hydrology in Mountainous Regions II. Artificial Reservoirs, Water and Slopes*, edited by R. O. Sinniger and M. Monbaron, *IAHS Publ.*, *194*, 371–378.

Rickenmann, D. (1991a), Bed load transport and hyperconcentrated flow at steep slopes, in *Fluvial Hydraulics of Mountain Regions*, edited by A. Armanini and G. DiSilvio, pp. 429–441, Springer, Berlin.

Rickenmann, D. (1991b), Hyperconcentrated flow and sediment transport at steep slopes, *J. Hydraul. Eng.*, *117*, 1419–1439.

Rickenmann, D. (1994a), An alternative equation for the mean velocity in gravel-bed rivers and mountain torrents, *Proceedings, ASCE Hydraulic Engineering '94 Conference, August 1–5, Buffalo, New York*, edited by G. V. Cotroneo and R. R. Rumer, pp. 672–676, Am. Soc. of Civ. Eng., New York.

Rickenmann, D. (1994b), Bedload transport and discharge in the Erlenbach stream, in *Dynamics and Geomorphology of Mountain Rivers*, edited by P. Ergenzinger and K.-H. Schmidt, pp. 53–66, Springer, Berlin.

Rickenmann, D. (1997), Sediment transport in Swiss torrents, *Earth Surf. Processes Landforms*, *22*, 937–951.

Rickenmann, D. (2001), Comparison of bed load transport in torrents and gravel bed streams, *Water Resour. Res.*, *37*, 3295–3305.

Rickenmann, D. (2005), Runout prediction methods, in *Debris-Flow Hazards and Related Phenomena*, edited by M. Jakob and O. Hungr, pp. 305–324, Springer, Berlin.

Rickenmann, D., and T. Koch (1997), Comparison of debris flow modelling approaches, in *Debris-Flow Hazards Mitigation: Mechanics, Prediction, and Assessment*, edited by C.-L. Chen, pp. 576–585, Am. Soc. of Civ. Eng., New York.

Rickenmann, D., V. D'Agostino, G. Dalla Fontana, and L. Marchi (1998), New results from sediment transport measurements in two Alpine torrents, in *Hydrology, Water Resources and Ecology in Headwaters*, edited by K. Kovar et al., *IAHS Publ.*, *248*, 283–289.

Ricklefs, R. E. (1987), Community diversity: Relative roles of local and regional processes, *Science*, *235*, 167–171.

Rico, M., G. Benito, and A. Barnolas (2001), Combined palaeoflood and rainfall-runoff assessment of mountain floods ,(Spanish Pyrenees), *J. Hydrol.*, *245*, 59–72.

Riedel, J. L., R. A. Haugerud, and J. J. Clague (2007), Geomorphology of a Cordilleran Ice Sheet drainage network through breached divides in the North Cascade Mountains of Washington and British Columbia, *Geomorphology*, *91*, 1–18.

Righter, K., M. Caffee, J. Rosas-Elguera, and V. Valencia (2010), Channel incision in the Rio Atenguillo, Jalisco, Mexico, defined by [36]Cl measurements of bedrock, *Geomorphology*, *120*, 279–292.

Rigon, E., F. Comiti, L. Mao, and M. A. Lenzi (2008), Relationships among basin area, sediment transport mechanisms and wood storage in mountain basins of the Dolomites ,(Italian Alps), *WIT Trans. Eng. Sci.*, *60*, 163–172.

Riihimaki, C. A., R. S. Anderson, E. B. Safran, D. P. Dethier, R. C. Finkel, and P. R. Bierman (2006), Longevity and progressive abandonment of the Rocky Flats surface, Front Range, Colorado, *Geomorphology*, *78*, 265–278.

Rijskijk, A., and L. A. Bruijnzeel (1991), *Erosion, sediment yield and land-use patterns in the Upper Konto watershed, East Java, Indonesia, Part III, results of the 1989–1990 measuring campaign*, Konto River Project Communication Ser., vol. 18, DHV Consultants, The Netherlands.

Riley, S. J. (1972), Comparison of morphometric measures of bankfull, *J. Hydrol.*, *17*, 23–31.

Rinaldi, M., and N. Casagli (1999), Stability of streambanks formed in partially saturated soils and effects of negative pore water pressures: The Sieve River, (Italy), *Geomorphology*, *26*, 253–277.

Rinaldi, M., and S. E. Darby (2008), Modelling river-bank erosion processes and mass failure mechanisms: Progress towards fully coupled simulations, in *Gravel-Bed Rivers VI: From Process Understanding to River Restoration*, edited by H. Habersack, H. Piégay, and M. Rinaldi, pp. 213–239, Elsevier, Amsterdam.

Rinaldi, M., and A. Simon (1998), Bed-level adjustments in the Arno River, central Italy, *Geomorphology*, *22*, 57–71.

Ritsema, C. J., H. Kuipers, L. Kleiboer, E. van den Elsen, K. Oostindie, J. G. Wesseling, J.-W. Wolthuis, and P. Havinga (2009), A new wireless underground network system for continuous monitoring of soil water contents, *Water Resour. Res.*, *45*, W00D36, doi:10.1029/2008WR007071.

Ritter, D. F. (1967), Rates of denudation, *J. Geol. Educ.*, *15*, 154–159.

Ritter, D. F. (1988), Floodplain erosion and deposition during the December 1982 floods in southeast Missouri, in *Flood Geomorphology*, edited by V. R. Baker, R. C. Kochel, and P. C. Patton, pp. 243–259, John Wiley, New York.

Ritter, D. F., R. C. Kochel, and J. R. Miller (1995), *Process Geomorphology*, 3rd ed., 546 pp., Wm. C. Brown, Dubuque, Iowa.

Ritter, D. F., R. C. Kochel, and J. R. Miller (1999), The disruption of Grassy Creek: Implications concerning catastrophic events and thresholds, *Geomorphology*, *29*, 323–338.

Ritter, J. B., and T. W. Gardner (1993), Hydrologic evolution of drainage basins disturbed by surface mining, central Pennsylvania, *Geol. Soc. Am. Bull.*, *105*, 101–115.

Ritter, J. B., J. R. Miller, and J. Husek-Wulforst (2000), Environmental controls on the evolution of alluvial fans in Buena Vista Valley, north central Nevada, during late Quaternary time, *Geomorphology*, *36*, 63–87.

Rivenbark, B. L., and C. R. Jackson (2004), Average discharge, perennial flow initiation, and channel initiation – small southern Appalachian basins, *J. Am. Water Resour. Assoc.*, *40*, 639–646.

Roberson, J. A., and C. T. Crowe (1993), *Engineering Fluid Mechanics*, 823 pp., John Wiley, New York.

Robert, A. (1998), Characteristics of velocity profiles along riffle-pool sequences and estimates of bed shear stress, *Geomorphology*, *19*, 89–98.

Robert, A., A. G. Roy, and B. De Serres (1993), Space-time correlations of velocity measurements at a roughness transition in a gravel-bed river, in *Turbulence: Perspectives on Flow and Sediment Transport*, edited by N. J. Clifford, J. R. French, and J. Hardisty, pp. 165–183, John Wiley, Chichester, U. K.

Roberts, G., J. Hudson, G. Leeks, and C. Neal (1994), The hydrological effects of clear-felling established coniferous forestry in an upland area of mid-Wales, in *Integrated River Basin Development*, edited by C. Kirby and W. R. White, pp. 187–199, John Wiley, Chichester, U. K.

Roberts, G. G., and N. White (2010), Estimating uplift rate histories from river profiles using African examples, *J. Geophys. Res.*, *115*, B02406, doi:10.1029/2009JB006692.

Roberts, M. D., M. P. O'Neill, J. P. Dobrowolski, J. C. Schmidt, and P. G. Wolf (1998), Hydro-geomorphic influences on clonal recruitment of cottonwood in a narrow, steep gradient, mountain valley, *Eos Trans. AGU*, *79*, F346.

Roberts, M. J., A. J. Russell, F. S. Tweed, and O. Knudsen (2000), Ice fracturing during jökulhlaups: Implications for englacial floodwater routing and outlet development, *Earth Surf. Processes Landforms*, *25*, 1429–1446.

Roberts, M. J., A. J. Russell, F. S. Tweed, and O. Knudsen (2001), Controls on englacial sediment deposition during November 1996 jökulhlaup, Skeiäarárjökull, Iceland, *Earth Surf. Processes Landforms*, *26*, 935–952.

Roberts, R. G., and M. Church (1986), The sediment budget in severely disturbed watersheds, Queen Charlotte Ranges, British Columbia, *Can. J. For. Res.*, *16*, 1092–1106.

Robinson, C. T., and S. Matthaei (2007), Hydrological heterogeneity of an alpine stream-lake network in Switzerland, *Hydrol. Processes*, *21*, 3146–3154.

Robinson, D. A., C. S. Campbell, J. W. Hopmans, B. K. Hornbuckle, S. B. Jones, R. Knight, F. Ogden, J. Selker, and O. Wendroth (2008), Soil moisture measurement for ecological and hydrological watershed-scale observatories: A review, *Vadose Zone J.*, *7*, 358–389.

Robinson, R. A. J., J. Q. G. Spencer, M. R. Strecker, A. Richter, and R. N. Alonso (2005), Luminescence dating of alluvial fans in intramontane basins of NW Argentina, in *Alluvial Fans: Geomorphology, Sedimentology, Dynamics*, edited by A. M. Harvey, A. E. Mather, and M. Stokes, *Geol. Soc. Spec. Publ.*, *251*, 153–168.

Robinson, S. K. (1990), Coherent motions in the turbulent boundary layer, *Annu. Rev. Fluid Mech.*, *104*, 387–405.

Robison, E. G., and R. L. Beschta (1990), Coarse woody debris and channel morphology interactions for undisturbed streams in southeastern Alaska, USA, *Earth Surf. Processes Landforms*, *15*, 149–156.

Rockwell, T. K., E. A. Keller, M. N. Clark, and D. L. Johnson (1984), Chronology and rates of faulting of Ventura River terraces, California, *Geol. Soc. Am. Bull.*, *95*, 1466–1474.

Rodriguez-Iturbe, I. (1993), The geomorphological unit hydrograph, in *Channel Network Hydrology*, edited by K. Beven and M. J. Kirkby, pp. 43–68, John Wiley, Chichester, U. K.

Rodriguez-Iturbe, I., and A. Rinaldo (1997), *Fractal River Basins: Chance and Self-Organization*, 547 pp., Cambridge Univ. Press, New York.

Rodriguez-Iturbe, I., and J. B. Valdès (1979), The geomorphologic structure of hydrologic response, *Water Resour. Res.*, *15*, 1409–1420.

Rodriguez-Iturbe, I., E. Ijjasz-Vasquez, R. L. Bras, and D. G. Tarboton (1992a), Power law distributions of discharge mass and energy in river basins, *Water Resour. Res.*, *28*, 1089–1093.

Rodriguez-Iturbe, I., A. Rinaldo, R. Rigon, R. L. Bras, A. Marani, and E. Ijjasz-Vasquez (1992b), Energy dissipation, runoff prediction, and the three-dimensional structure of river basins, *Water Resour. Res.*, *28*, 1095–1103.

Roe, G. H., D. R. Montgomery, and B. Hallett (2002), Effects of orographic precipitation variations on the concavity of steady-state river-state river profiles, *Geology*, *50*, 143–146.

Roering, J. J. (2004), Soil creep and convex-upward velocity profiles: Theoretical and experimental investigation of disturbance-driven sediment transport on hillslopes, *Earth Surf. Processes Landforms*, *29*, 1597–1612.

Roering, J. J. (2008), How well can hillslope evolution models explain topography? Simulating soil transport and production with high-resolution topographic data, *Geol. Soc. Am. Bull.*, *120*, 1248–1262.

Roering, J. J., and M. Gerber (2005), Fire and the evolution of steep, soil-mantled landscapes, *Geology*, *33*, 349–352.

Roering, J. J., J. W. Kirchner, and W. E. Dietrich (1999), Evidence for nonlinear, diffusive sediment transport on hillslopes and implications for landscape morphology, *Water Resour. Res.*, *35*, 853–870.

Roering, J. J., J. W. Kirchner, and W. E. Dietrich (2001a), Hillslope evolution by nonlinear, slope-dependent transport: Steady state morphology and equilibrium adjustment timescales, *J. Geophys. Res.*, *106*(B8), 16,499–16,513.

Roering, J. J., J. W. Kirchner, L. S. Sklar, and W. E. Dietrich (2001b), Hillslope evolution by nonlinear creep and landsliding: An experimental study, *Geology*, *29*, 143–146.

Roering, J. J., P. Almond, P. Tonkin, and J. McKean (2002), Soil transport driven by biological processes over millennial time scales, *Geology*, *30*, 1115–1118.

Roering, J. J., K. M. Schmidt, J. D. Stock, W. E. Dietrich, and D. R. Montgomery (2003), Shallow landsliding, root reinforcement, and the spatial distribution of trees in the Oregon Coast Range, *Can. Geotech. J.*, *40*, 237–253.

Roering, J. J., P. Almond, P. Tonkin, and J. McKean (2004), Constraining climatic controls on hillslope dynamics using a coupled model for the transport of soil and tracers: Application to loess-mantled hilslopes, South Island, New Zealand, *J. Geophys. Res.*, *109*, F01010, doi:10.1029/2003JF000034.

Roering, J. J., J. W. Kirchner, and W. E. Dietrich (2005), Characterizing structural and lithologic controls on deep-seated landsliding: Implications for topographic relief and landscape evolution in the Oregon Coast Range, USA, *Geol. Soc. Am. Bull.*, *117*, 654–668.

Roering, J. J., J. T. Perron, and J. W. Kirchner (2007), Functional relationships between denudation and hillslope form and relief, *Earth Planet. Sci. Lett.*, *264*, 245–258.

Roesli, U., and C. Schindler (1990), Debris flows 1987 in Switzerland: Geological and hydrogeological aspects, in *Hydrology in Mountainous Regions II. Artificial Reservoirs, Water and Slopes*, edited by R. O. Sinniger and M. Monbaron, *IAHS Publ.*, *194*, 379–386.

Rolauffs, P., D. Herring, and S. Lohse (2001), Composition, invertebrate community and productivity of a beaver dam in comparison to other stream habitat types, *Hydrobiologia*, *459*, 201–212.

Roline, R. A., and J. R. Boehmke (1981), Heavy metals pollution of the upper Arkansas River, Colorado, and Its Effects on the Distribution of the Aquatic Macrofauna, *REC-ERC-81-15*, 71 pp, U. S. D. I. Bureau of Reclamation.

Rood, S. B., G. M. Samuelson, J. K. Weber, and K. A. Wyrwot (2005), Twentieth-century decline in streamflows from the hydrographic apex of North America, *J. Hydrol.*, *306*, 215–233.

Roper, B. B., J. M. Buffington, E. Archer, C. Moyer, and M. Ward (2008), The role of observer variation in determining Rosgen stream types in northeastern Oregon mountain streams, *J. Am. Water Resour. Assoc.*, *44*, 417–427.

Rose, J. (1984), Alluvial terraces of an equatorial river, Melinau drainage basin, Sarawak, *Z. Geomorphol.*, *28*, 155–177.

Rosenbloom, N., and R. S. Anderson (1994), Hillslope and channel evolution in a marine terraced landscape, Santa Cruz, California, *J. Geophys. Res.*, *99*, 14,013–14,029.

Rosenfeld, C. L. (1998), Storm induced mass wasting in the Oregon Coast Range, USA, in *Geomorphological Hazards in High Mountain Areas*, edited by J. Kalvoda and C. L. Rosenfeld, pp. 167–176, Kluwer Acad., The Netherlands.

Rosgen, D. L. (1994), A classification of natural rivers, *Catena*, *22*, 169–199.

Rosgen, D. L., and H. L. Silvey (1996), *Applied River Morphology*, 390 pp., Wildland Hydrology, Pagosa Springs, Colo.

Ross, J., and R. Gilbert (1999), Lacustrine sedimentation in a monsoon environment: The record from Phewa Tal, middle mountain region of Nepal, *Geomorphology*, *27*, 307–323.

Rossi, M., G. Guzetti, P. Reichenbach, A. C. Mondini, and S. Peruccacci (2010), Optimal landslide susceptibility zonation based on multiple forecasts, *Geomorphology*, *114*, 129–142.

Rot, B. W., R. J. Naiman, and R. E. Bilby (2000), Stream channel configuration, landform, and riparian forest structure in the Cascade Mountains, Washington, *Can. J. Fish. Aquat. Sci.*, *57*, 699–707.

Roth, G., and P. La Barbera (1997), Morphological characterization of channel initiation, *Phys. Chem. Earth*, *22*, 329–332.

Roth, G., P. La Barbera, and M. Greco (1996), On the description of the basin effective drainage structure, *J. Hydrol.*, *187*, 119–135.

Röthlisberger, H., and H. Lang (1987), Glacial hydrology, in *Glacio-Fluvial Sediment Transfer*, edited by A. M. Gurnell and M. J. Clark, pp. 207–284, John Wiley, Chichester, U. K.

Rowbotham, D. N., and D. Dudycha (1998), GIS modelling of slope stability in Phewa Tal watershed, Nepal, *Geomorphology*, *26*, 151–170.

Rowe, L. K., and A. J. Pearce (1994), Hydrology and related changes after harvesting native forest catchments and establishing Pinus radiata plantations. Part 2. The native forest water balance and changes in streamflow after harvesting, *Hydrol. Processes*, *8*, 291–297.

Roy, A. G., and A. D. Abrahams (1980), Rhythmic spacing and origin of pools and riffles: Discussion, *Geol. Soc. Am. Bull.*, *91*, 248–250.

Roy, A. G., T. Buffin-Bélanger, and S. Del (1996), Scales of turbulent coherent flow structures in a gravel-bed river, in *Coherent Flow Structures in Open Channels*, edited by P. J. Ashworth et al., pp. 147–164, John Wiley, Chichester, U. K.

Roy, P. K., and A. Mazumdar (2005), Hydrological impacts of climatic variability on water resources of the Damodar River basin, India, in *Regional Hydrological Impacts of Climatic Change: Impact Assessment and Decision Making; Part 1*, edited by T. Wagener et al., *IAHS Publ.*, *295*, 147–156.

Rubin, Y., I. A. Lunt, and J. S. Bridge (2006), Spatial variability in river sediments and its link with river channel geometry, *Water Resour. Res.*, *42*, W06D16, doi:10.1029/2005WR004853.

Ruff, M., and K. Czurda (2008), Landslide susceptibility analysis with a heuristic approach in the eastern Alps (Vorarlberg, Austria), *Geomorphology*, *94*, 314–324.

Ruhe, R. V. (1952), Topographic discontinuities of the Des Moines lobe, *Am. J. Sci.*, *250*, 46–56.

Ruiz-Villanueva, V., A. Díez-Herrero, M. Stoffel, M. Bollschweiler, J. M. Bodoque, and J. A. Ballesteros (2010), Dendrogeomorphic analysis of flash floods in a small ungauged mountain catchment (central Spain), *Geomorphology*, *118*, 383–392.

Rumsby, B. T., J. Brasington, J. A. Langham, S. J. McLelland, R. Middleton, and G. Rollinson (2008), Monitoring and modelling particle and reach-scale morphological change in gravel-bed rivers: Applications and challenges, *Geomorphology*, *93*, 40–54.

Runkel, R. L., and S. C. Chapra (1993), An efficient numerical solution of the transient storage equations for solute transport in small streams, *Water Resour. Res.*, *29*, 211–215.

Runkel, R. L., D. M. McKnight, and H. Rajaram (2003), Modeling hyporheic zone processes, *Adv. Water Resour.*, *26*, 901–905.

Runkel, R. L., B. A. Kimball, K. Walton-Day, and P. L. Verplanck (2007), A simulation-based approach for estimating premining water quality; Red Mountain Creek, Colorado, *Appl. Geochem.*, *22*, 1899–1918.

Rushmer, E. L. (2007), Physical-scale modelling of jökulhlaups (glacial outburst floods) with contrasting hydrograph shapes, *Earth Surf. Processes Landforms*, *32*, 954–963.

Rushmer, E. L., A. J. Russell, F. S. Tweed, O. Knudsen, and P. M. Marren (2002), The role of hydrograph shape in controlling glacier outburst flood (jökulhlaup) sedimentation, in *The Structure, Function and Management Implications of Fluvial Sedimentary Systems*, edited by F. J. Dyer, M. C. Thoms, and J. M. Olley, *IAHS Publ.*, *276*, 305–313.

Ruslan, I. W. (1995), Impact of urbanisation and uphill land clearances on the sediment yield of an urbanising catchment of Pulau Pinang, Malaysia, in *Postgraduate Research in Geomorphology: Selected Papers from the 17th BGRG Postgraduate Symposium*, edited by S. J. McLelland, A. R. Skellern, and P. R. Porter, pp. 28–33, Br. Geomorphol. Res. Grp., Sch. of Geogr., Univ. of Leeds, Leeds, U. K.

Russell, A. J., F. G. M. Van Tatenhove, and R. S. W. Van de Wal (1995), Effects of ice-front collapse and flood generation on a proglacial river channel near Kangerlussuaq (Sondre Stromfjord), West Greenland, *Hydrol. Processes*, *9*, 213–226.

Russell, A. J., F. S. Tweed, and O. Knudsen (2000), Flash flood at Sólheimajökull heralds the reawakening of an Icelandic subglacial volcano, *Geol. Today*, *16*(3), 102–106.

Russell, S. O. (1972), Behavior of steep creeks in a large flood, in *Mountain Geomorphology: Geomorphic Processes in the Canadian Cordillera*, edited by H. O. Slaymaker and H. J. McPherson, pp. 223–227, Tantalus Res., Vancouver, Canada.

Rustomji, P., and I. Prosser (2001), Spatial patterns of sediment delivery to valley floors: Sensitivity to sediment transport capacity and hillslope hydrology relations, *Hydrol. Processes*, *15*, 1003–1018.

Rutherfurd, I. D., and J. R. Grove (2004), The influence of trees on stream bank erosion: Evidence from root-plate abutments, in *Riparian Vegetation and Fluvial Geomorphology*, edited by S. J. Bennett and A. Simon, pp. 141–152, AGU, Washington, D. C.

Rutherfurd, I. D., P. Bishop, and T. Loffler (1994), Debris flows in northeastern Victoria, Australia: Occurrence and effects on the fluvial system, in *Variability in Stream Erosion and Sediment Transport*, edited by L. J. Olive, R. J. Loughran, and J. A. Kesby, *IAHS Publ.*, *224*, 359–369.

Ryan, S. E., Bedload transport patterns in pool-riffle and step-pool stream systems, in *Effects of Human-Induced Changes on Hydrologic Systems*, pp. 669–678, Am. Water Resources Assoc., 1994a.

Ryan, S. E. ,(1994b,), Effects of transbasin diversion on flow regime, bedload transport, and channel morphology in Colorado mountain streams, unpublished Ph.D. dissertation, 236 pp., Univ. of Colorado, Boulder.

Ryan, S. E. (1997), Morphologic response of subalpine streams to transbasin flow diversion, *J. Am. Water Resour. Assoc.*, *33*, 839–854.

Ryan, S. E., and M. K. Dixon (2008), Spatial and temporal variability in stream sediment loads using examples from the Gros Ventre Range, Wyoming, USA, in *Gravel-Bed Rivers VI: From Process Understanding to River Restoration*, edited by H. Habersack, H. Piégay, and M. Rinaldi, pp. 387–407, Elsevier, Amsterdam.

Ryan, S. E., and G. E. Grant (1991), Downstream effects of timber harvesting on channel morphology in Elk River basin, Oregon, *J. Environ. Qual.*, *20*, 60–72.

Ryan, S. E. and C. A. Troendle, Measuring bedload in coarse-grained mountain channels: procedures, problems, and recommendations, in *Water Resources Education, Training, and Practice: Opportunities for the Next Century*, pp. 949–958, American Water Resources Association, 1997.

Ryan, S. E., L. S. Porth, and C. A. Troendle (2002), Defining phases of bedload transport using piecewise regression, *Earth Surf. Processes Landforms*, *27*, 971–990.

Ryan, S. E., L. S. Porth, and C. A. Troendle (2005), Coarse sediment transport in mountain streams in Colorado and Wyoming, USA, *Earth Surf. Processes Landforms*, *30*, 269–288.

Ryder, D., G. J. Brierley, R. Hobbs, G. Kyle, and M. Leishman (2008), Vision generation: What do we seek to achieve in river rehabilitation?, in *River Futures: An Integrative Scientific Approach to River Repair*, edited by G. J. Brierley and K. A. Fryirs, pp. 16–27, Island Press, Washington, D. C.

Ryder, J. M. (1971a), The stratigraphy and morphology of paraglacial alluvial fans in south-central British Columbia, *Can. J. Earth Sci.*, *8*, 279–298.

Ryder, J. M. (1971b), Some aspects of the morphometry of paraglacial alluvial fans in south-central British Columbia, *Can. J. Earth Sci.*, *8*, 1252–1264.

Ryder, J. M., and M. Church (1986), The Lillooet terraces of Fraser River: A palaeoenvironmental enquiry, *Can. J. Earth Sci.*, *23*, 869–884.

Sabaj, M. H., E. G. Maurakis, and W. S. Woolcott (2000), Spawning behaviors in the bluehead chub, Nocomis leptocephalus, N. micropogon and Central stoneroller, Campostoma anomalum, *Am. Midl. Nat.*, *144*, 187–201.

Sable, K. A., and E. Wohl (2006), The relationship of lithology and watershed characteristics to fine sediment deposition in streams of the Oregon Coast Range, *Environ. Manage.*, *37*, 659–670.

Sabo, J. L., et al. (2005), Riparian zones increase regional species richness by harboring different, not more, species, *Ecology*, *86*, 56–62.

Saco, P. M., and P. Kumar (2002), Kinematic dispersion in stream networks 1. Coupling hydraulic and network geometry, *Water Resour. Res.*, *38*(11), 1244, doi:10.1029/2001WR000695.

Safran, E. B. (2003), Geomorphic interpretation of low-temperature thermochronologic data: Insights from two-dimensional thermal modeling, *J. Geophys. Res.*, *108*(B4), 2189, doi:10.1029/2002JB001870.

Safran, E. B., P. R. Bierman, R. Aalto, T. Dunne, K. X. Whipple, and M. W. Caffee (2005), Erosion rates driven by channel network incision in the Bolivian Andes, *Earth Surf. Processes Landforms*, *30*, 1007–1024.

Sah, M. P., and R. K. Mazari (1998), Anthropogenically accelerated mass movement, Kulu Valley, Himachal Pradesh, India, *Geomorphology*, *26*, 123–138.

Saito, H., D. Nakayama, and H. Matsuyana (2009), Comparison of landslide susceptibility based on a decision-tree model and actual landslide occurrence: The Akaishi Mountains, Japan, *Geomorphology*, *109*, 108–121.

Saito, H., D. Nakayama, and H. Matsuyana (2010), Relationship between the initiation of a shallow landslide and rainfall intensity-duration thresholds in Japan, *Geomorphology*, *118*, 161–175.

Saito, K., and T. Oguchi (2005), Slope of alluvial fans in humid regions of Japan, Taiwan and the Philippines, *Geomorphology*, *70*, 147–162.

Salant, N. L., C. E. Renshaw, and F. J. Magilligan (2006), Short and long-term changes to bed mobility and bed composition under altered sediment regimes, *Geomorphology*, *76*, 43–53.

Salant, N. L., C. E. Renshaw, F. J. Magilligan, J. M. Kaste, K. H. Nislow, and A. M. Heimsath (2007), The use of short-lived radionuclides to quantify transitional bed material transport in a regulated river, *Earth Surf. Processes Landforms*, *32*, 509–524.

Salas, J. D., E. E. Wohl, and R. D. Jarrett (1994), Determination of flood characteristics using systematic, historical and paleoflood data, in *Coping with Floods*, edited by G. Rossi, N. Harmancioglu, and V. Yevjevich, pp. 111–134, Kluwer, Dordrecht.

Salcher, B. C., R. Faber, and M. Wagreich (2010), Climate as main factor controlling the sequence development of two Pleistocene alluvial fans in the Vienna Basin (eastern Austria) – a numerical modelling approach, *Geomorphology*, *115*, 215–227.

Salciarini, D., J. W. Godt, W. Z. Savage, P. Conversini, R. L. Baum, and J. A. Michael (2006), Modeling regional initiation of rainfall-induced shallow landslides in the eastern Umbria region of central Italy, *Landslides*, *3*, 181–194.

Salehin, M., A. I. Packman, and M. Paradis (2004), Hyporheic exchange with heterogeneous streambeds: Laboratory experiments and modeling, *Water Resour. Res.*, *40*, W11504, doi:10.1029/2003WR002567.

Salo, E. O., and T. W. Cundy (Eds.) (1987), *Streamside Management: Forestry and Fishery Implications*, Univ. of Washington, Institute of Forest Resources, Seattle.

Salo, J., R. Kalliola, I. Häkkinen, Y. Mäkinen, P. Niemelä, M. Puhakka, and P. D. Coley (1986), River dynamics and the diversity of Amazon lowland forest, *Nature*, *322*, 254–258.

Samadi, A., E. Amiri-Tokaldany, and S. E. Darby (2009), Identifying the effects of parameter uncertainty on the reliability of riverbank stability modeling, *Geomorphology*, *106*, 219–230.

Sambrook Smith, G. H., and A. P. Nicholas (2005), Effect on flow structure of sand deposition on a gravel bed: Results from a two-dimensional flume experiment, *Water Resour. Res.*, *41*, W10405, doi:10.1029/2004WR003817.

Sambrook Smith, G. H., J. L. Best, C. S. Bristow, and G. E. Petts (2006), Braided rivers: Where have we come in 10 years? Progress and future needs, in *Braided Rivers: Process, Deposits, Ecology and Management*, edited by G. H. Sambrook Smith et al., *Spec. Publ. Int. Assoc. Sedimentol.*, *36*, 1–10.

Sanborn, P., M. Geertsema, A. J. T. Jull, and B. Hawkes (2006), Soil and sedimentary charcoal evidence from Holocene forest fires in an inland temperate rainforest, east-central British Columbia, Canada, *Holocene*, *16*, 415–427.

Sanborn, S. C., and B. P. Bledsoe (2006), Predicting streamflow regime metrics for ungauged streams in Colorado, Washington, and Oregon, *J. Hydrol.*, *325*, 241–261.

Sanchez, G., Y. Rolland, M. Corsini, R. Braucher, D. Bourlès, M. Arnold, and G. Aumaître (2010), Relationships between tectonics, slope instability and climate change: Cosmic ray exposure dating of active faults, landslides and glacial surfaces in the SW Alps, *Geomorphology*, *117*, 1–13.

Sander, G. C., P. B. Hairsine, L. Beuselinck, and G. Govers (2002), Steady state sediment transport through an area of net deposition: Multisize class solutions, *Water Resour. Res.*, *38*(6), 1087, doi:10.1029/2001WR000323.

Sandoz, M. (1964), *The Beaver Men: Spearheads of Empire*, 335 pp., Univ. of Nebraska Press, Lincoln.

Santi, P. M., V. G. De Wolfe, J. D. Higgins, S. H. Cannon, and J. E. Gartner (2008), Sources of debris flow material in burned areas, *Geomorphology*, *95*, 310–321.

Santini, M., S. Grimaldi, F. Nardi, A. Petroselli, and M. C. Rulli (2009), Pre-processing algorithms and landslide modelling on remotely sensed DEMs, *Geomorphology*, *113*, 110–125.

Sapozhnikov, V. B., and E. Foufoula-Georgiou (1996), Self-affinity in braided rivers, *Water Resour. Res.*, *32*, 1429–1439.

Sapozhnikov, V. B., and E. Foufoula-Georgiou (1999), Horizontal and vertical self-organization of braided rivers toward a critical state, *Water Resour. Res.*, *35*, 843–851.

Sarah, P. (2004), Nonlinearity of ecogeomorphic processes along Mediterranean-arid transect, *Geomorphology*, *60*, 303–317.

Sasahara, K., M. Kaneko, M. Takeuchi, N. Minami, and A. Subarkah (2005), Erosion of deposits from the pyroclastic flow that occurred on Mt. Merapi, Indonesia in July 1998, in *Sediment Budgets I*, edited by D. E. Walling and A. J. Horowitz, *IAHS Publ.*, *291*, 332–338.

Sass, O. (2006), Determination of the internal structure of alpine talus deposits using different geophysical methods (Lechtaler Alps, Austria), *Geomorphology*, *80*, 45–58.

Sass, O., and M. Krautblatter (2007), Debris flow-dominated and rockfall-dominated talus slopes: Genetic models derived from GPR measurements, *Geomorphology*, *86*, 176–192.

Sassa, K., and G. H. Wang (2005), Mechanism of landslide-triggered debris flows: Liquefaction phenomena due to the undrained loading of torrent deposits, in *Debris-Flow Hazards and Related Phenomena*, edited by M. Jakob and O. Hungr, pp. 81–104, Springer, Berlin.

Sato, H. P., T. Sekiguchi, R. Kojiroi, Y. Suzuki, and M. Iida (2005), Overlaying landslide distribution on the earthquake source, geological and topographical data: The Mid Niigata prefecture earthquake in 2004, Japan, *Landslides*, *2*, 143–152.

Sato, T., Y. Kurashige, and K. Hirakawa (1997), Slow mass movement in the Taisetsu Mountains, Hokkaido, Japan, *Permafrost Periglacial Processes*, *8*, 347–357.

Saucedo, R., J. L. Macias, D. Sarocchi, M. Bursik, and B. Rupp (2008), The rain-triggered Atenquique volcaniclastic debris flows of October 16, 1955 at Nevado de Colima Volcano, Mexico, *J. Volcanol. Geotherm. Res.*, *173*, 69–83.

Sauer, V. B., R. E. Curtis, L. Santiago-Rivera, and R. Gonzalez (1985), Quantifying flood discharges in mountainous tropical streams, *International Symposium on Tropical Hydrology and Second Caribbean Islands Water Resources Congress*, edited by F. Quinones and A. V. Sanchez, pp. 104–108, Am. Water Resour. Assoc., Bethesda, Md.

Saunders, I., and A. Young (1983), Rates of surface processes on slopes, slope retreat and denudation, *Earth Surf. Processes Landforms*, *8*, 473–501.

Savage, W., and R. Baum (2005), Instability of steep slopes, in *Debris-Flow Hazards and Related Phenomena*, edited by M. Jakob and O. Hungr, pp. 53–79, Springer, Berlin.

Sawada, T., and T. Takahashi (1994), Sediment yield on bare slopes, paper presented at International Symposium on Forest Hydrology, Univ. of Tokyo, Tokyo.

Sawada, T., K. Ashida, and T. Takahashi (1983), Relationship between channel pattern and sediment transport in a steep gravel bed river, *Z. Geomorphol.*, *46*, 55–66.

Sawyer, A. H., and M. B. Cardenas (2009), Hyporheic flow and residence time distributions in heterogeneous cross-bedded sediment, *Water Resour. Res.*, *45*, W08406, doi:10.1029/2008WR007632.

Sawyer, A. M., G. B. Pasternack, H. J. Moir, and A. A. Fulton (2010), Riffle-pool maintenance and flow convergence routing observed on a large gravel-bed river, *Geomorphology*, *114*, 143–160.

Saxena, P. B., and S. Prakash (1982), A study of the morphometric determinants of the stage of cycle of erosion in the Nayar Basin (Garhwal Himalayas), Perspectives in Geomorphology, vol. IV, in *Essays on Indian Geomorphology*, edited by H. S. Sharma, pp. 77–92, Concept, New Delhi.

Sayama, T., and J. J. McDonnell (2009), A new time-space accounting scheme to predict stream water residence time and hydrograph source components at the watershed scale, *Water Resour. Res.*, *45*, W07401, doi:10.1029/2008WR007549.

Saynor, M. J., R. J. Loughran, W. D. Erskine, and P. F. Scott (1994), Sediment movement on hillslopes measured by caesium-137 and erosion pins, in *Variability in Stream Erosion and Sediment Transport*, edited by L. J. Olive, R. J. Loughran, and J. A. Kesby, *IAHS Publ.*, *224*, 87–93.

Scanlon, T. M., S. M. Ingram, and A. L. Riscassi (2010), Terrestrial and in-stream influences on the spatial variability of nitrate in a forested headwater catchment, *J. Geophys. Res.*, *115*, G02022, doi:10.1029/2009JG001091.

Scarpino, R. V. (1985), *Great River: An Environmental History of the Upper Mississippi, 1850–1950*, Univ. of Mo. Press, Columbia.

Scatena, F. N., and M. C. Larsen (1991), Physical aspects of Hurricane Hugo in Puerto Rico, *Biotropica*, *23*, 317–323.

Schaller, M., F. von Blanckenburg, N. Hovius, A. Veldkamp, M. W. van den Berg, and P. W. Kubik (2004), Paleoerosion rates from cosmogenic [10]Be in a 1.3 Ma terrace sequence; response of the River Meuse to changes in climate and rock uplift, *J. Geol.*, *112*, 127–144.

Schaller, M., N. Hovius, S. D. Willett, S. Ivy-Ochs, H. A. Synal, and M. C. Chen (2005), Fluvial bedrock incision in the active mountain belt of Taiwan from in situ-produced cosmogenic nuclides, *Earth Surf. Processes Landforms*, *30*, 955–971.

Schaller, M., and T. A. Ehlers (2006), Limits to quantifying climate driven changes in denudation rates with cosmogenic radionuclides, *Earth Planet. Sci. Lett.*, *248*, 153–167.

Scharer, K. M., D. W. Burbank, J. Chen, and R. J. Weldon (2006), Kinematic models of fluvial terraces over active detachment folds: Constraints on the growth mechanism of the Kashi-Atushi fold system, Chinese Tian Shan, *Geol. Soc. Am. Bull.*, *118*, 1006–1021.

Scheidegger, A. E. (1995), Geojoints and geostresses, in *Mechanics of Jointed and Faulted Rock*, edited by H.-P. Rossmanith, pp. 3–35, A. A. Balkema, Rotterdam.

Scheidegger, A. E., and R. Hantke (1994), On the genesis of river gorges, *Trans. Jpn. Geomorphol. Union*, *15*, 91–110.

Scheidl, C., and D. Rickenmann (2010), Empirical prediction of debris-flow mobility and deposition on fans, *Earth Surf. Processes Landforms*, *35*, 157–173.

Schick, A. P. (1970), Desert floods: Interim results of observations in the Nahal Yael Research Watershed, southern Israel, 1965–1970, *Symposium on the Results of Research on Representative and Experimental Basins*, pp. 478–493, IASH-UNESCO, Wellington, New Zealand.

Schick, A. P. (1974), Formation and obliteration of desert stream terraces - a conceptual analysis, *Z. Geomorphol.*, *21*, 88–105.

Schick, A. P., and J. Lekach (1987), A high magnitude flood in the Sinai Desert, in *Catastrophic Flooding*, edited by L. Mayer and D. Nash, pp. 381–410, Allen and Unwin, Boston, Mass.

Schick, A. P., and J. Lekach (1993), An evaluation of two 10-year sediment budgets, Nahal Yael, Israel, *Phys. Geogr.*, *14*, 225–238.

Schick, A. P., and D. Magid (1978), Terraces in arid stream valleys: A probability model, *Catena*, *5*, 237–250.

Schick, A. P., J. Lekach, and M. A. Hassan (1982), Bed load transport in desert floods: Observations in the Negev, in *Sediment Transport in Gravel-Bed Rivers*, edited by C. R. Thorne, J. C. Bathurst, and R. D. Hey, pp. 617–642, John Wiley, Chichester, U. K.

Schick, A. P., M. A. Hassan, and J. Lekach (1987a), A vertical exchange model for coarse bedload movement: Numerical considerations, in *Geomorphological Models: Theoretical and Empirical Aspects*, edited by F. Ahnert, *Catena Suppl.*, *10*, 73–83.

Schick, A. P., J. Lekach, and M. A. Hassan (1987b), Vertical exchange of coarse bedload in desert streams, in *Desert Sediments: Ancient and Modern*, edited by L. Frostick and I. Reid, *Geol. Soc. Spec. Publ.*, *35*, 7–16.

Schick, A. P., T. Grodek, and M. G. Wolman (1999), Hydrologic processes and geomorphic constraints on urbanization of alluvial fan slopes, *Geomorphology*, *31*, 325–335.

Schiefer, E., B. Menounos, and O. Slaymaker (2006), Extreme sediment delivery events recorded in the contemporary sediment record of a montane lake, southern Coast Mountains, British Columbia, *Can. J. Earth Sci.*, *43*, 1777–1790.

Schild, A., P. Singh, and J. Hübl (1998), Application of GIS for hydrological modelling in high mountain areas of the Austrian Alps, in *Hydrology, Water Resources and Ecology in Headwaters*, edited by K. Kovar et al., *IAHS Publ.*, *248*, 569–576.

Schildgen, T., D. P. Dethier, P. Bierman, and M. Caffee (2002), [26]Al and [10]Be dating of late Pleistocene and Holocene fill terraces; a record of fluvial deposition and incision, Colorado Front Range, *Earth Surf. Processes Landforms*, *27*, 773–787.

Schildgen, T. F., T. A. Ehlers, D. M. Whipp, M. C. van Soest, K. X. Whipple, and K. V. Hodges (2009), Quantifying canyon incision and Andean Plateau surface uplift, southwest Peru: A thermochronometer and numerical modeling approach, *J. Geophys. Res.*, *114*, F04014, doi:10.1029/2009JF001305.

Schiefer, E., M. A. Hassan, B. Menounos, C. P. Pelpola, and O. Slaymaker (2010), Interdecadal patterns of total sediment yield from a montane catchment, southern Coast Mountains, British Columbia, Canada, *Geomorphology*, *118*, 207–212.

Schindl, G., M. Studnicka, A. Eckelhart, and W. Summer (2005), Hydrological and instrumentation aspects of monitoring and analysing suspended sediment transport crossing international borders, in *Sediment Budgets I*, edited by D. E. Walling and A. J. Horowitz, *IAHS Publ.*, *291*, 227–240.

Schlichting, H. (1968), *Boundary-Layer Theory*, 747 pp., McGraw-Hill, New York.

Schimel, D. S., T. G. F. Kittel, S. Running, R. Monson, A. Turnipseed, and D. Anderson (2002), Carbon sequestration studied in western U. S. mountains, *Eos Trans. AGU*, *83*, 445–449.

Schleusener, R. A., G. L. Smith, and M. C. Chen (1962), Effect of flow diversion for irrigation on peak rates of runoff from watersheds in and near the Rocky Mountain foothills of Colorado, *Int. Assoc. Hydrol. Bull.*, *7*, 53–61.

Schlosser, I. J. (1991), Stream fish ecology: A landscape perspective, *BioScience*, *41*, 704–712.

Schmal, R., and T. Wesche (1989), Historical implications of the railroad crosstie industry on current riparian and stream habitat management in the central Rocky Mountains, in *Practical Approaches to Riparian Resource Management*, edited by R. E. Gresswell, B. A. Barton, and J. L. Kershner, pp. 189, U.S. Bur. of Land Manage., Billings, Mont.

Schmeeckle, M. W., J. M. Nelson, J. Pitlick, and J. P. Bennett (2001), Interparticle collision of natural sediment grains in water, *Water Resour. Res.*, *37*, 2377–2391.

Schmeeckle, M. W., J. M. Nelson, and R. L. Shreve (2007), Forces on stationary particles in near-bed turbulent flows, *J. Geophys. Res.*, *112*, F02003, doi:10.1029/2006JF000536.

Schmidt, J. C. (1990), Recirculating flow and sedimentation in the Colorado River in Grand Canyon, Arizona, *J. Geol.*, *98*, 709–724.

Schmidt, J. C., D. M. Rubin, and H. Ikeda (1993), Flume simulation of recirculating flow and sedimentation, *J. Geol.*, *29*, 2925–2939.

Schmidt, K.-H. (1994), River channel adjustment and sediment budget in response to a catastrophic flood event (Lainbach catchment, southern Bavaria), in *Dynamics and Geomorphology of Mountain Rivers*, edited by P. Ergenzinger and K.-H. Schmidt, pp. 109–127, Springer, Berlin.

Schmidt, K.-H., and P. Ergenzinger (1992), Bedload entrainment, travel lengths, step lengths, rest periods - studied with passive (iron, magnetic) and active (radio) tracer techniques, *Earth Surf. Processes Landforms*, *17*, 147–165.

Schmidt, K.-H., and D. Gintz (1995), Results of bedload tracer experiments in a mountain river, in *River Geomorphology*, edited by E. J. Hickin, pp. 37–54, John Wiley, Chichester, U. K.

Schmidt, K.-H., and D. Morche (2006), Sediment output and effective discharge in two small high mountain catchments in the Bavarian Alps, Germany, *Geomorphology*, *80*, 131–145.

Schmidt, K. M., J. J. Roering, J. D. Stock, W. E. Dietrich, D. R. Montgomery, and T. Schaub (2001), The variability of root cohesion as an influence on shallow landslide susceptibility in the Oregon Coast Range, *Can. Geotech. J.*, *38*, 995–1024.

Schmitt, L., G. Maire, P. Nobelis, and J. Humbert (2007), Quantitative morphodynamic typology of rivers: A methodological study based on the French Upper Rhine basin, *Earth Surf. Processes Landforms*, *32*, 1726–1746.

Schmocker-Fackel, P., and F. Naef (2010), More frequent flooding? Changes in flood frequency in Switzerland since 1850, *J. Hydrol.*, *381*, 1–8.

Schmutz, S., A. Zitek, S. Zobl, M. Jungwirth, N. Knopf, E. Kraus, T. Bauer, and T. Kaufmann (2002), Integrated approach to the conservation and restoration of Danube salmon, Hucho hucho, populations in Austria, in *Conservation of Freshwater Fishes: Options for the Future*, edited by M. L. Collares-Pereira, M. M. Coelho, and I. G. Cowx, pp. 157–173, Blackwell Sci., Oxford, U. K.

Schnackenberg, E. S., and L. H. MacDonald (1998), Detecting cumulative effects on headwater streams in the Routt National Forest, Colorado, *J. Am. Water Resour. Assoc.*, *34*, 1163–1177.

Schneeberger, C., H. Blatter, A. Abe-Ouchi, and M. Wild (2003), Modelling changes in the mass balance of glaciers of the Northern Hemisphere for a transient $2XCO_2$ scenario, *J. Hydrol.*, *282*, 145–163.

Schneider, H., M. Schwab, and F. Schlunegger (2008), Channelized and hillslope sediment transport and the geomorphology of mountain belts, *Int. J. Earth Sci.*, *97*, 179–192.

Schnorbus, M., and Y. Alila (2004), Forest harvesting impacts on the peak flow regime in the Columbia Mountains of southeastern British Columbia: An investigation using long-term numerical modeling, *Water Resour. Res.*, *40*, W05205, doi:10.1029/2003WR002918.

Schöberl, F. (1991), Continuous simulation of sediment transport in the case of glacierized watershed, in *Fluvial Hydraulics of Mountain Regions*, edited by A. Armanini and G. DiSilvio, pp. 71–81, Springer, Berlin.

Schoellhamer, D. H., and S. A. Wright (2003), Continuous measurement of suspended-sediment discharge in rivers by use of optical backscatterance sensors, in *Erosion and Sediment Transport Measurement in Rivers: Technological and Methodological Advances*, edited by J. Bogen, T. Fergus, and D. E. Walling, *IAHS Publ.*, *283*, 28–36.

Schoklitsch, A. (1962), *Handbuch des Wasserbaus*, 3rd ed., Springer, Vienna.

Schrott, L., G. Hufschmidt, M. Hankammer, T. Hoffmann, and R. Dikau (2003), Spatial distribution of sediment storage types and quantification of valley fill deposits in an alpine basin, Reintal, Bavarian Alps, Germany, *Geomorphology*, *55*, 45–63.

Schuerch, P., A. L. Densmore, B. W. McArdell, and P. Molnar (2006), The influence of landsliding on sediment supply and channel change in a steep mountain catchment, *Geomorphology*, *78*, 222–235.

Schulze, O., R. Roth, and O. Pieper (1994), Probable maximum precipitation in the Upper Harz Mountains, in *FRIEND: Flow Regimes from International Experimental and Network Data*, edited by P. Senna et al., *IAHS Publ.*, *221*, 315–321.

Schumm, S. A. (1956), The role of creep and rainwash on the retreat of badland slopes, *Am. J. Sci.*, *254*, 693–706.

Schumm, S. A. (1960), *The Shape of Alluvial Channels in Relation to Sediment Type*, U.S. Geol. Surv. Prof. Pap., *352B*, pp. 17–30.

Schumm, S. A. (1963a), *A Tentative Classification of Alluvial River Channels*, U.S. Geol. Surv. Circ., *477*, 10 pp.

Schumm, S. A. (1963b), *The Disparity Between Present Rates of Denudation and Orogeny*, U.S. Geol. Surv. Prof. Pap., *454-H*, 13 pp.

Schumm, S. A. (1964), Seasonal variations of erosion rates and processes on hillslopes in western Colorado, *Z. Geomorphol. Suppl.*, *5*, 215–238.

Schumm, S. A. (1966), The development and evolution of hillslopes, *J. Geol. Educ.*, *14*, 98–104.

Schumm, S. A. (1973), Geomorphic thresholds and the complex response of drainage systems, in *Fluvial Geomorphology*, edited by M. Morisawa, pp. 299–310, State Univ. of New York, Binghamton.

Schumm, S. A. (1977), *The Fluvial System*, 338 pp., John Wiley, New York.

Schumm, S. A. (1979), Geomorphic thresholds: The concept and its applications, *Trans. Inst. Br. Geogr.*, *4*, 485–515.

Schumm, S. A. (1980), Some applications of the concept of geomorphic thresholds, in *Thresholds in Geomorphology*, edited by D. R. Coates and J. D. Vitek, pp. 473–485, George Allen and Unwin, London.

Schumm, S. A. (1981), Evolution and response of the fluvial system, sedimentologic implications, *Spec. Publ. Soc. Econ. Paleontol. Mineral.*, *31*, 19–29.

Schumm, S. A. (1985), Patterns of alluvial rivers, *Annu. Rev. Earth Planet. Sci.*, *13*, 5–27.

Schumm, S. A. (1986), Alluvial river response to active tectonics, in *Active Tectonics*, pp. 80–94, Natl. Acad. Press, Washington, D. C.

Schumm, S. A. (1997), Drainage density: Problems of prediction and application, in *Process and Form in Geomorphology*, edited by D. R. Stoddart, pp. 15–45, Routledge, London.

Schumm, S. A. (1999), Causes and controls of channel incision, in *Incised River Channels: Processes, Forms, Engineering and Management*, edited by S. E. Darby and A. Simon, pp. 19–33, John Wiley, Chichester, U. K.

Schumm, S. A. (2005), *River Variability and Complexity*, 234 pp., Cambridge Univ. Press, Cambridge, U. K.

Schumm, S. A., and R. F. Hadley (1957), Arroyos and the semiarid cycle of erosion, *Am. J. Sci.*, *25*, 161–174.

Schumm, S. A., and R. F. Hadley (1961), *Progress in the Application of Landform Analysis in Studies of Semiarid Erosion*, U.S. Geol. Surv. Circ., *437*, 14 pp.

Schumm, S. A., and H. R. Khan (1972), Experimental study of channel patterns, *Geol. Soc. Am. Bull.*, *83*, 1755–1770.

Schumm, S. A., and R. W. Lichty (1965), Time, space and causality in geomorphology, *Am. J. Sci.*, *263*, 110–119.

Schumm, S. A., and G. C. Lusby (1963), Seasonal variation of infiltration capacity and runoff on hillslopes in western Colorado, *J. Geophys. Res.*, *68*, 3655–3666.

Schumm, S. A., and R. S. Parker (1973), Implications of complex response of drainage systems for Quaternary alluvial stratigraphy, *Nature*, *243*, 99–100.

Schumm, S. A., and M. A. Stevens (1973), Abrasion in place: A mechanism for rounding and size reduction of coarse sediments in rivers, *Geology*, *1*, 37–40.

Schumm, S. A., M. D. Harvey, and C. C. Watson (1984), *Incised Channels: Morphology, Dynamics and Control*, 200 pp., Water Resour. Publ., Loveland, Colo.

Schumm, S. A., M. P. Mosley, and W. E. Weaver (1987), *Experimental Fluvial Geomorphology*, 413 pp., John Wiley, New York.

Schumm, S. A., W. D. Erskine, and J. W. Tilleard (1996), Morphology, hydrology, and evolution of the anastomosing Owens and King Rivers, Victoria, Australia, *Geol. Soc. Am. Bull.*, *108*, 1212–1224.

Schumm, S. A., J. F. Dumont, and J. M. Holbrook (2000), *Active Tectonics and Alluvial Rivers*, 276 pp., Cambridge Univ. Press, Cambridge, U. K.

Schuster, P. F., D. P. Krabbenhoft, D. L. Naftz, L. D. Cecil, M. L. Olson, J. F. DeWild, and J. R. Green (2002), A 270-year ice core record of atmospheric mercury deposition to western North America, *Environ. Sci. Technol.*, *36*, 2303–2310.

Schwarz, M., P. Lehmann, and D. Or (2010), Quantifying lateral root reinforcement in steep slopes – from a bundle of roots to tree stands, *Earth Surf. Processes Landforms*, *35*, 354–367.

Schwindt, A. R., J. W. Fournie, D. H. Landers, C. B. Schreck, and M. L. Kent (2008), Mercury concentrations in salmonids from western U. S. national parks and relationships with age and macrophage aggregates, *Environ. Sci. Technol.*, *42*, 1365–1370.

Scott, G. R. (1975), Cenozoic surfaces and deposits in the southern Rocky Mountains, in *Cenozoic History of the Southern Rocky Mountains*, edited by B. F. Curtis, *Mem. Geol. Soc. Am.*, *144*, 227–248.

Scott, K. M. (1971), *Origin and Sedimentology of 1969 Debris Flows Near Glendora, California*, U.S. Geol. Surv. Prof. Pap., *750-C*, pp. C242–C247.

Scott, K. M., and G. C. Gravlee, Jr. (1968), *Flood Surge on the Rubicon River, California - Hydrology, Hydraulics, and Boulder Transport*, U.S. Geol. Surv. Prof. Pap., *422-M*, pp. M1–M38.

Scott, M. L., G. T. Auble, and J. M. Friedman (1997), Flood dependency of cottonwood establishment along the Missouri River, Montana, USA, *Ecol. Appl.*, *7*, 677–690.

Scott, M. L., J. M. Friedman, and G. T. Auble (1996), Fluvial process and the establishment of bottomland trees, *Geomorphology*, *14*, 327–339.

Scott, P. F., and W. D. Erskine (1994), Geomorphic effects of a large flood on fluvial fans, *Earth Surf. Processes Landforms*, *19*, 95–108.

Seal, R., and C. Paola (1995), Observations of downstream fining on the North Fork Toutle River near Mount St. Helens, Washington, *Water Resour. Res.*, *31*, 1409–1419.

Seal, R., C. Toro-Escobar, Y. Cui, C. Paola, G. Parker, J. B. Southard, and P. R. Wilcock (1998), Downstream fining by selective deposition; theory, laboratory, and field observations, in *Gravel-Bed Rivers in the Environment*, edited by P. C. Klingeman et al., pp. 61–84, Water Resour. Publ., Highlands Ranch, Colo.

Sear, D. A. (1996), Sediment transport processes in pool-riffle sequences, *Earth Surf. Processes Landforms*, *21*, 241–262.

Sear, D. A. (2003), Event bed load yield measurement with load cell bed load traps and prediction of bed load yield from hydrograph shape, in *Erosion and Sediment Transport Measurement in Rivers: Technological and Methodological Advances*, edited by J. Bogen, T. Fergus, and D. E. Walling, *IAHS Publ.*, *283*, 146–153.

Sear, D. A., and N. W. Arnell (2006), The application of palaeohydrology in river management, *Catena*, *66*, 169–183.

Sear, D. A., M. W. E. Lee, P. A. Carling, R. J. Oakey, and M. B. Collins (2003), An assessment of the accuracy of the Spatial Integration Method (SIM) for estimating coarse bedload transport in gravel-bedded streams using tracers, in *Erosion and Sediment Transport Measurement in Rivers: Technological and Methodological Advances*, edited by J. Bogen, T. Fergus, and D. E. Walling, *IAHS Publ.*, *283*, 164–171.

Sear, D. A., J. M. Wheaton, and S. E. Darby (2008), Uncertain restoration of gravel-bed rivers and the role of geomorphology, in *Gravel-Bed Rivers VI: From Process Understanding to River Restoration*, edited by H. Habersack, H. Piégay, and M. Rinaldi, pp. 739–761, Elsevier, Amsterdam.

Sedell, J. R., F. H. Everest, and F. J. Swanson (1982), Fish habitat and streamside management: Past and present, paper presented at Annual Meeting, Soc. of Am. For.

Sedell, J. R., and J. L. Froggatt (1984), Importance of streamside forests to large rivers: The isolation of the Willamette River, Oregon, USA, from its floodplain by snagging and streamside forest removal, *Verh. Int. Verein. Limnol.*, *22*, 1828–1834.

Sedell, J. R., J. E. Richey, and F. J. Swanson (1989), The river continuum concept: A basis for expected ecosystem behavior of very large rivers?, *Can. Spec. Publ. Fish. Aquat. Sci.*, *106*, 110–127.

Segura, C., and J. Pitlick (2010), Scaling frequency of channel-forming flows in snowmelt-dominated streams, *Water Resour. Res.*, *46*, W06524, doi:10.1029/2009WR008336.

Segura, R., V. Arancibia, M. C. Zuniga, and P. Pasten (2006), Distribution of copper, zinc, lead and cadmium concentrations in stream sediments from the Mapocho River in Santiago, Chile, *J. Geochem. Explor.*, *91*, 71–80.

Seibert, J., K. Bishop, A. Rodhe, and J. J. McDonnell (2003), Groundwater dynamics along a hillslope: A test of the steady-state hypothesis, *Water Resour. Res.*, *39*(1), 1014, doi:10.1029/2002WR001404.

Seidl, M. A., and W. E. Dietrich (1992), The problem of channel erosion into bedrock, in *Functional Geomorphology*, edited by K.-H. Schmidt and J. de Ploey, *Catena Suppl.*, *23*, 101–124.

Seidl, M. A., W. E. Dietrich, and J. W. Kirchner (1994), Longitudinal profile development into bedrock: An analysis of Hawaiian channels, *J. Geol.*, *102*, 457–474.

Seidl, M. A., R. C. Finkel, M. W. Caffee, G. B. Hudson, and W. E. Dietrich (1997), Cosmogenic isotope analyses applied to river longitudinal profile evolution: Problems and interpretations, *Earth Surf. Processes Landforms*, *22*, 195–209.

Seitzinger, S. P., R. V. Styles, E. W. Boyer, R. B. Alexander, G. Billen, R. W. Howarth, B. Mayer, and N. Van Breemen (2002), Nitrogen retention in rivers: Model development and application to watersheds in the northeastern U. S. A, *Biogeochemistry*, *57/58*, 199–237.

Selby, M. J. (1980), A rock mass strength classification for geomorphic purposes: With tests from Antarctica and New Zealand, *Z. Geomorphol.*, *24*, 31–51.

Selby, M. J. (1982), *Hillslope Materials and Processes*, 264 pp., Oxford Univ. Press, Oxford, U. K.

Selkowitz, D. J., D. B. Fagre, and B. A. Reardon (2002), Interannual variations in snowpack in the Crown of the Continent Ecosystem, *Hydrol. Processes*, *16*, 3651–3665.

Seminara, G., L. Solari, and G. Parker (2002), Bed load at low Shields stress on arbitrarily sloping beds: Failure of the Bagnold hypothesis, *Water Resour. Res.*, *38*(11), 1249, doi:10.1029/2001WR000681.

Seo, J. I., and F. Nakamura (2009), Scale-dependent controls upon the fluvial export of large wood from river catchments, *Earth Surf. Processes Landforms*, *34*, 786–800.

Seo, J. I., F. Nakamura, D. Nakano, H. Ichiyanagi, and K. W. Chun (2008), Factors controlling the fluvial export of large woody debris, and its contribution to organic carbon budgets at watershed scales, *Water Resour. Res.*, *44*, W04428, doi:10.1029/2007WR006453.

Seong, Y. B., L. A. Owen, M. P. Bishop, A. Bush, P. Clendon, L. Copland, R. C. Finkel, U. Kamp, and J. F. Shroder (2008), Rates of fluvial bedrock incision within an actively uplifting orogen: Central Karakoram Mountains, northern Pakistan, *Geomorphology*, *97*, 274–286.

Sepulveda, S. A., S. Rebolledo, and G. Vargas (2006), Recent catastrophic debris flows in Chile; geological hazard, climatic relationships and human response, *Quat. Int.*, *158*, 83–95.

SER , (Society for Ecological Restoration Science and Policy Working Group) (2002), *The SER Primer on Ecological Restoration*, Soc. for Ecol. Restor., Tucson, Ariz.

Serban, M., M. G. Macklin, P. A. Brewer, D. Balteanu, and G. Bird (2004), The impact of metal mining activities on the upper Tisa River basin, Romania and transboundary river pollution, *Stud. Geomorphol. Carpatho-Balcanica*, *38*, 97–111.

Sevruk, B., K. Matokova-Sadlonova, and L. Toskano (1998), Topography effects on small-scale precipitation variability in the Swiss pre-Alps, in *Hydrology, Water Resources and Ecology in Headwaters*, edited by K. Kovar et al., *IAHS Publ.*, *248*, 51–58.

Seydell, H., I. B. E. Wawra, and U. C. E. Zanke (2008), Evaluating vertical velocities between the stream and the hyporheic zone from temperature data, in *Gravel-Bed Rivers VI: From Process Understanding to River Restoration*, edited by H. Habersack, H. Piégay, and M. Rinaldi, pp. 109–131, Elsevier, Amsterdam.

Shaban, A., G. Faour, M. Khawlie, and C. Abdallah (2004), Remote sensing application to estimate the volume of water in the form of snow on Mount Lebanon, *Hydrol. Sci. J.*, *49*, 643–653.

Shanley, J. B., and A. Chalmers (1999), The effect of frozen soil on snowmelt runoff at Sleepers River, Vermont, *Hydrol. Processes*, *13*, 1843–1857.

Shanley, J. B., C. Kendall, M. R. Albert, and J. P. Hardy (1995), Chemical and isotopic evolution of a layered eastern U. S. snowpack and its relation to stream-water composition, in *Biogeochemistry of Seasonally Snow-Covered Catchments*, edited by K. A. Tonnessen, M. W. Williams, and M. Tranter, *IAHS Publ.*, *228*, 329–338.

Shao, S., Y.-M. E. Lo, and G. Wang (2002), Simulation of fan formation using a debris mass model, *J. Hydraul. Res.*, *40*, 425–433.

Sharma, A., L. Marshall, and D. Nott (2005), A Bayesian view of rainfall-runoff modelling: Alternatives for parameter estimate, model comparison and hierarchical model development, in *Predictions in Ungauged Basins: International Perspectives on the State of the Art and Pathways Forward*, edited by S. Franks et al., *IAHS Publ.*, *301*, 299–311.

Sharma, U. C., and V. Sharma (2003), Mathematical model for predicting soil erosion by flowing water in ungauged watersheds, in *Erosion Prediction in Ungauged Basins: Integrating Methods and Techniques*, edited by D. de Boer et al., *IAHS Publ.*, *279*, 79–83.

Sharp, R. P. (1942), Mudflow levees, *J. Geomorphol.*, *5*, 222–227.

Sharp, W. D., K. R. Ludwig, O. A. Chadwick, R. Amundson, and L. L. Glaser (2003), Dating fluvial terraces by [230]Th/U on pedogenic carbonate, Wind River Basin, Wyoming, *Quat. Res.*, *59*, 139–150.

Shaw, J. R., and D. J. Cooper (2008), Linkages among watersheds, stream reaches, and riparian vegetation in dryland ephemeral stream networks, *J. Hydrol.*, *350*, 68–82.

Sheffer, N. A., M. Rico, Y. Enzel, G. Benito, and T. Grodek (2008), The palaeoflood record of the Gardon River, France: A comparison with the extreme 2002 flood event, *Geomorphology*, *98*, 71–83.

Shelford, V. E. (1911), Ecological succession. I. Stream fishes and the method of physiographic analysis, *Biol. Bull.*, *21*, 9–35.

Sheridan, G. J., and P. J. Noske (2007), A quantitative study of sediment delivery and stream pollution from different forest road types, *Hydrol. Processes*, *21*, 387–398.

Shieh, C.-L., Y.-R. Guh, and S.-Q. Wang (2007), The application of range of variability approach to the assessment of a check dam on riverine habitat alteration, *Environ. Geol.*, *52*, 427–435.

Shields, F. D., A. Brookes, and J. Haltiner (1999), Geomorphological approaches to incised stream channel restoration in the United States and Europe, in *Incised River Channels: Processes, Forms, Engineering and Management*, edited by S. E. Darby and A. Simon, pp. 371–394, John Wiley, Chichester, U. K.

Shields, F. D., and C. J. Gippel (1995), Prediction of effects of woody debris removal on flow resistance, *J. Hydraul. Eng.*, *121*, 341–354.

Shih, S.-M., and P. D. Komar (1990a), Differential bedload transport rates in a gravel-bed stream: A grain-size distribution approach, *Earth Surf. Processes Landforms*, *15*, 539–552.

Shih, S.-M., and P. D. Komar (1990b), Hydraulic controls of grain-size distributions of bedload gravels in Oak Creek, Oregon, USA, *Sedimentology*, *37*, 367–376.

Shimazu, H. (1994), Segmentation of Japanese mountain rivers and its causes based on gravel transport processes, *Trans. Jpn. Geomorphol. Union*, *15*, 111–128.

Shimazu, H., and T. Oguchi (1996), River processes after rapid valley-filling due to large landslides, *GeoJournal*, *38*, 339–344.

Shlemon, R. J., R. H. Wright, and D. R. Montgomery (1987), Anatomy of a debris flow, Pacifica, California, in *Debris Flows/Avalanches: Process, Recognition, and Mitigation*, edited by J. E. Costa and G. F. Wieczorek, *Rev. Eng. Geol.*, *7*, 181–199.

Shrestha, D. P., J. A. Zinck, and E. van Ranst (2004), Modelling land degradation in the Nepalese Himalaya, *Catena*, *57*, 135–156.

Shroba, R. R., P. W. Schmidt, E. J. Crosby, W. R. Hansen, and J. M. Soule (1979), *Geologic and geomorphic effects in the Big Thompson Canyon area, Larimer County, U.S. Geol. Surv. Prof. Pap.*, *1115B*, pp. 87–152.

Shroder, J. F., and M. P. Bishop (2004), Mountain geomorphic systems, in *Geographic Information Science and Mountain Geomorphology*, edited by M. P. Bishop and J. F. Shroder, pp. 33–73, Praxis, Chichester, U. K.

Shroder, J. F., Jr., M. P. Bishop, and R. Scheppy (1998), Catastrophic flood flushing of sediment, western Himalaya, Pakistan, in *Geomorphological Hazards in High Mountain Areas*, edited by J. Kalvoda and C. L. Rosenfeld, pp. 27–48, Kluwer Acad., The Netherlands.

Shucksmith, J. D., J. B. Boxall, and I. Guymer (2010), Effects of emergent and submerged natural vegetation on longitudinal mixing in open channel flow, *Water Resour. Res.*, *46*, W04504, doi:10.1029/2008WR007657.

Shvidchenko, A. B., G. Pender, and T. B. Hoey (2001), Critical shear stress for incipient motion of sand/gravel streambeds, *Water Resour. Res.*, *37*, 2273–2283.

Shyu, J. B. H., K. Sieh, J.-P. Avouac, W.-S. Chen, and Y.-G. Chen (2006), Millennial slip rate of the Longitudinal Valley fault from river terraces: Implications for convergence across the active suture of eastern Taiwan, *J. Geophys. Res.*, *111*, B08403, doi:10.1029/2005JB003971.

Sickman, J. O., J. M. Melack, and J. L. Stoddard (2002), Regional analysis of inorganic nitrogen yield and retention in high-elevation ecosystems of the Sierra Nevada and Rocky Mountains, *Biogeochemistry*, *57–58*, 341–374.

Sidle, R. C. (1988), Bed load transport regime of a small forest stream, *Water Resour. Res.*, *24*, 201–218.

Sidle, R. C. (2005), Influence of forest harvesting activities on debris avalanches and flows, in *Debris-Flow Hazards and Related Phenomena*, edited by M. Jakob and O. Hungr, pp. 387–409, Springer, Berlin.

Sidle, R. C., S. Noguchi, Y. Tsuboyama, and K. Laursen (2001), A conceptual model of preferential flow systems in forested hillslopes: Evidence of self-organization, *Hydrol. Processes*, *15*, 1675–1692.

Sieben, J. (1997), Modelling of hydraulics and morphology in mountain rivers, Ph.D. dissertation, 223 pp., Technical Univ. of Delft, The Netherlands.

Sigafoos, R. S. (1961), Vegetation in relation to flood frequency near Washington, D. C., *U.S. Geol. Surv. Prof. Pap.*, *424-C*, pp. C-248–C-250.

Sigafoos, R. S. (1964), Botanical evidence of floods and flood-plain deposition, *U.S. Geol. Surv. Prof. Pap.*, *485-A*, 35 pp.

Silsby, D. G., and G. L. Larson (1983), A comparison of streams in logged and unlogged areas of Great Smoky Mountains National Park, *Hydrobiologia*, *102*, 99–111.

Sime, L. C., R. I. Ferguson, and M. Church (2007), Estimating shear stress from moving boat acoustic Doppler velocity measurements in a large gravel bed river, *Water Resour. Res.*, *43*, W03418, doi:10.1029/2006WR005069.

Simon, A., and J. Castro (2003), Measurement and analysis of alluvial channel form, in *Tools in Fluvial Geomorphology*, edited by G. M. Kondolf and H. Piégay, pp. 291–322, John Wiley, Chichester, U. K.

Simon, A., and A. J. C. Collison (2001), Pore-water pressure effects on the detachment of cohesive streambeds: Seepage forces and matric suction, *Earth Surf. Processes Landforms*, 26, 1421–1442.

Simon, A., and A. J. C. Collison (2002), Quantifying the mechanical and hydrological effects of riparian vegetation on streambank stability, *Earth Surf. Processes Landforms*, 27, 527–546.

Simon, A., and C. R. Hupp (1986), Channel evolution in modified Tennessee channels, paper presented at Fourth Federal Interagency Sedimentation Conference, U.S. Dep. of Agric. Res. Serv., Las Vegas, Nev.

Simon, A., and M. Rinaldi (2006), Disturbance, stream incision, and channel evolution: The roles of excess transport capacity and boundary materials in controlling channel response, *Geomorphology*, 79, 361–383.

Simon, A., and R. E. Thomas (2002), Processes and forms of an unstable alluvial system with resistant, cohesive streambeds, *Earth Surf. Processes Landforms*, 27, 699–718.

Simon, A., M. C. Larsen, and C. R. Hupp (1990), The role of soil processes in determining mechanisms of slope failure and hillslope development in a humid-tropical forest, eastern Puerto Rico, *Geomorphology*, 3, 263–286.

Simon, A., A. Curini, S. Darby, and E. J. Langendoen (1999), Streambank mechanics and the role of bank and near-bank processes in incised channels, in *Incised River Channels: Processes, Forms, Engineering and Management*, edited by S. E. Darby and A. Simon, pp. 123–152, John Wiley, Chichester, U. K.

Simon, A., A. Curini, S. E. Darby, and E. Langendoen (2000), Bank and near bank processes in an incised channel, *Geomorphology*, 35, 193–217.

Simon, A., W. Dickerson, and A. Heins (2004), Suspended-sediment transport rates at the 1.5-year recurrence interval for ecoregions of the United States: Transport conditions at the bankfull and effective discharge?, *Geomorphology*, 58, 243–262.

Simon, A., M. Doyle, M. Kondolf, F. D. Shields, B. Rhoads, and M. McPhillips (2007), Critical evaluation of how the Rosgen classification and associated "natural channel design" methods fail to integrate and quantify fluvial processes and channel response, *J. Am. Water Resour. Assoc.*, 43, 1117–1131.

Simoni, S., F. Zanotti, G. Bertoldi, and R. Rigon (2008), Modelling the probability of occurrence of shallow landslides and channelized debris flows using GEOtop-FS, *Hydrol. Processes*, 22, 532–545.

Simons, D. B., and R. M. Li (1976), *Procedure for Estimating Model Parameters of a Mathematical Model*, USDA Forest Serv., Rocky Mtn For. and Range Exp. Sta., Flagstaff, Ariz.

Simons, D. B., and E. V. Richardson (1966), *Resistance to flow in alluvial channels*, U.S. Geol. Surv. Prof. Pap., 422J, 61 pp.

Simons, D. B., R. M. Li, and M. A. Stevens (1975), *Developments of Models for Predicting Water and Yield from Storms on Small Watersheds*, USDA Forest Serv., Rocky Mtn For and Range Exp. Sta., Flagstaff, Ariz.

Simons, Li and Associates, Inc. (1982), Final Report: Debris and Flood Control Plan for Portland and Cascade Creeks at Ouray, Colorado, Colo. Water Board, Denver.

Simpson, C. J., and D. G. Smith (2001), The braided Milk River, northern Montana, fails the Leopold-Wolman discharge-gradient test, *Geomorphology*, 41, 337–353.

Simpson, G., and S. Castelltort (2006), Coupled model of surface water flow, sediment transport and morphological evolution, *Comput. Geosci.*, 32, 1600–1614.

Singer, M. B. (2008), A new sampler for extracting bed material sediment from sand and gravel beds in navigable rivers, *Earth Surf. Processes Landforms*, 33, 2277–2284.

Singer, M. B., and T. Dunne (2006), Modeling the influence of river rehabilitation scenarios on bed material sediment flux in a large river over decadal timescales, *Water Resour. Res.*, 42, W12415, doi:10.1029/2006WR004894.

Singh, A., K. Fienberg, D. J. Jerolmack, J. Marr, and E. Foufoula-Georgiou (2009), Experimental evidence for statistical scaling and intemittency in sediment transport rates, *J. Geophys. Res.*, *114*, F01025, doi:10.1029/2007JF000963.

Singh, A., F. Porté-Angel, and E. Foufoula-Georgiou (2010), On the influence of gravel bed dynamics on velocity power spectra, *Water Resour. Res.*, *46*, W04509, doi:10.1029/2009WR008190.

Singh, M., I. B. Singh, and G. Mueller (2007), Sediment characteristics and transportation dynamics of the Ganga River, *Geomorphology*, *86*, 144–175.

Singh, P. (2006), Estimates and analysis of suspended sediment from a glacierized basin in the Himalayas, in *Sediment Dynamics and the Hydromorphology of Fluvial Systems*, edited by J. S. Rowan, R. W. Duck, and A. Werritty, *IAHS Publ.*, *306*, 21–27.

Singh, P., and L. Bengtsson (2004), Hydrological sensitivity of a large Himalayan basin to climate change, *Hydrol. Processes*, *18*, 2363–2385.

Singh, P., and L. Bengtsson (2005), Impact of warmer climate on melt and evaporation for the rainfed, snowfed and glacierfed basins in the Himalayan region, *J. Hydrol.*, *300*, 140–154.

Singh, P., G. Spitzbart, H. Hübl, and H. W. Weinmeister (1998), The role of snowpack in producing floods under heavy rainfall, in *Hydrology, Water Resources and Ecology in Headwaters*, edited by K. Kovar et al., *IAHS Publ.*, *248*, 89–95.

Singh, P., K. S. Ramasastri, N. Kumar, and M. Arora (2000), Correlations between discharge and meteorological parameters and runoff forecasting from a highly glacierized Himalayan Basin, *Hydrol. Sci. J.*, *45*, 637–652.

Singh, P., M. Arora, and N. K. Goel (2006), Effect of climate change on runoff of a glacierized Himalayan basin, *Hydrol. Processes*, *20*, 1979–1992.

Singh, T., and J. Kaur (Eds.) (1985), *Integrated Mountain Research*, Himalayan Books, New Delhi.

Singh, V. (2002), Is hydrology kinematic?, *Hydrol. Processes*, *16*, 667–716.

Sinniger, R. O., and M. Monbaron (Eds.) (1990), *Hydrology in Mountainous Regions II, artificial reservoirs, water and slopes*, *IAHS Publ.*, *194*, 446 pp.

Sivapalan, M., T. Wagener, S. Uhlenbrook, E. Zehe, V. Lakshmi, X. Liang, Y. Tachikawa, and P. Kumar (2006), *Predictions in ungauged basins: Promise and progress, IAHS Publ.*, *303*.

Skalak, K., and J. Pizzuto (2010), The distribution and residence time of suspended sediment stored within the channel margins of a gravel-bed bedrock river, *Earth Surf. Processes Landforms*, *35*, 435–446.

Skermer, N. A., and D. F. VanDine (2005), Debris flows in history, in *Debris-Flow Hazards and Related Phenomena*, edited by M. Jakob and O. Hungr, pp. 25–51, Springer, Berlin.

Skewes, O., F. Gonzalez, R. Olave, A. Ávila, V. Vargas, P. Paulsen, and H. E. König (2006), Abundance and distribution of American beaver, Castor canadensis, (Kuhl 1820) in Tierra del Fuego and Navarino islands, Chile, *Eur. J. Wildl. Res.*, *52*, 292–296.

Sklar, L. S., and W. E. Dietrich (1998), River longitudinal profiles and bedrock incision models: Stream power and the influence of sediment supply, in *Rivers Over Rock: Fluvial Processes in Bedrock Channels, Geophys. Monogr. Ser.*, vol. 107, edited by K. J. Tinkler and E. E. Wohl, pp. 237–260, AGU, Washington, D. C.

Sklar, L. S., and W. E. Dietrich (2001), Sediment and rock strength controls on river incision into bedrock, *Geology*, *29*, 1087–1090.

Sklar, L. S., and W. E. Dietrich (2004), A mechanistic model for river incision into bedrock by saltating bed load, *Water Resour. Res.*, *40*, W06301, doi:10.1029/2003WR002496.

Sklar, L. S., and W. E. Dietrich (2006), The role of sediment in controlling steady-state bedrock channel slope: Implications of the saltation-abrasion incision model, *Geomorphology*, *82*, 58–83.

Sklar, L. S., W. E. Dietrich, E. Foufoula-Georgiou, B. Lashermes, and D. Bellugi (2006), Do gravel bed river size distributions record channel network structure?, *Water Resour. Res.*, *42*, W06D18, doi:10.1029/2006WR005035.

Sklar, L. S., J. Fadde, J. G. Venditti, P. Nelson, M. A. Wydzga, Y. Cui, and W. E. Dietrich (2009), Translation and dispersion of sediment pulses in flume experiments simulating gravel augmentation below dams, *Water Resour. Res.*, *45*, W08439, doi:10.1029/2008WR007346.

Slaney, P. A., T. G. Halsey, and H. A. Smith (1977), Some effects of forest harvesting on salmonid rearing habitat in two streams in the central interior of British Columbia, *Fish. Manage. Rep. 71*, 26 pp, B. C. Min. of Recreation and Conserv.

Slaymaker, H. O. (1972), Sediment yield and sediment control in the Canadian Cordillera, in *Mountain Geomorphology: Geomorphological Processes in the Canadian Cordillera*, edited by H. O. Slaymaker and H. J. McPherson, pp. 235–245, Tantalus Res. Ltd., Vancouver, B. C., Canada.

Slaymaker, H. O. (1974), Alpine hydrology, in *Arctic and Alpine Environments*, edited by J. D. Ives and R. G. Barry, pp. 134–155, Methuen, London.

Slaymaker, H. O., and H. J. McPherson (Eds.) (1972), *Mountain Geomorphology*, Tantalus Res. Ltd., Vancouver, B. C., Canada.

Slaymaker, O. (2006), Towards the identification of scaling relations in drainage basin sediment budgets, *Geomorphology, 80*, 8–19.

Slaymaker, O., C. Souch, B. Menounos, and G. Filipelli (2003), Advances in Holocene mountain geomorphology inspired by sediment budget methodology, *Geomorphology, 55*, 305–316.

Sletten, K., and L. H. Blikra (2007), Holocene colluvial (debris-flow and water-flow) processes in eastern Norway; stratigraphy, chronology, and palaeoenvironmental implications, *J. Quat. Sci., 22*, 619–635.

Sloan, J., J. R. Miller, and N. Lancaster (2001), Response and recovery of the Eel River, California, and its tributaries to floods in 1955, 1964, and 1997, *Geomorphology, 36*, 129–154.

Smart, C. C. (1988), A deductive model of karst evolution based on hydrological probability, *Earth Surf. Processes Landforms, 13*, 271–288.

Smart, G. M. (1984), Sediment transport formula for steep channels, *J. Hydraul. Eng., 110*, 267–276.

Smart, G. M. (1994), Turbulent velocities in a mountain river, *Proceedings, ASCE Hydraulic Engineering '94 Conference, August 1–5, Buffalo, New York*, edited by G. V. Cotroneo and R. R. Rumer, pp. 844–848, Am. Soc. of Civ. Eng., New York.

Smart, G. M., and M. N. R. Jaeggi (1983), *Sediment Transport on Steep Slopes, Versuchsanstalt für Wasserbau, Hydrologie und Glaziologie, Mitteilungen*, vol. 64, Eidgenössische Technische Hochschule Zürich, Switzerland.

Smith, A. M., and P. K. Zawada (1988), The role of the geologist in flood contingency planning, *S. Afr. Geol. Surv. Pap.*, vol. 7.3, 9 pp.

Smith, C., and B. Croke (2005), Sources of uncertainty in estimating suspended sediment load, in *Sediment Budgets 2*, edited by A. J. Horowitz and D. E. Walling, *IAHS Publ., 292*, 136–143.

Smith, D. G., and C. M. Pearce (2002), Ice jam-caused fluvial gullies and scour holes on northern river flood plains, *Geomorphology, 42*, 85–95.

Smith, H. G., and D. Dragovich (2008), Sediment budget analysis of slope-channel coupling and in-channel sediment storage in an upland catchment, southeastern Australia, *Geomorphology, 101*, 643–654.

Smith, J. A., M. L. Baeck, K. L. Meierdiercks, P. A. Nelson, A. J. Miller, and E. J. Holland (2005a), Field studies of the storm event hydrologic response in an urbanizing watershed, *Water Resour. Res., 41*, W10413, doi:10.1029/2004WR003712.

Smith, J. A., P. Sturdevant-Rees, M. L. Baeck, and M. C. Larsen (2005b), Tropical cyclones and the flood hydrology of Puerto Rico, *Water Resour. Res., 41*, W06020, doi:10.1029/2004WR003530.

Smith, J. D. (2004), The role of riparian shrubs in preventing floodplain unraveling along the Clark Fork of the Columbia River in the Deer Lodge Valley, Montana, in *Riparian Vegetation and Fluvial Geomorphology*, edited by S. J. Bennett and A. Simon, pp. 71–85, AGU, Washington, D. C.

Smith, N. (1971), *A History of Dams*, Peter Davies, London.

Smith, R. B. (2006), Progress on the theory of orographic precipitation, in *Tectonics, Climate, and Landscape Evolution*, edited by S. D. Willett et al., *Spec. Pap. Geol. Soc. Am., 398*, 1–16.

Smith, R. D., and R. L. Beschta (1994), A mechanism of pool formation and maintenance in forest streams, *Proceedings, ASCE Hydraulic Engineering '94 Conference, August 1–5, Buffalo, New York*, edited by G. V. Cotroneo and R. R. Rumer, pp. 824–828, Am. Soc. of Civ. Eng., New York.

Smith, R. D., R. C. Sidle, P. E. Porter, and J. R. Noel (1993), Effects of experimental removal of woody debris on the channel morphology of a forest, gravel-bed stream, *J. Hydrol.*, *52*, 153–178.

Smith, R. E. (2002), *Infiltration Theory for Hydrologic Applications*, AGU, Washington, D. C.

Smith, R. E., D. C. Goodrich, and C. L. Unkrich (1999), Simulation of selected events on the Catsop catchment by KINEROS2: A report for the GCTE conference on catchment scale erosion models, *Catena*, *37*, 457–475.

Smith, S. M., and K. L. Prestegaard (2005), Hydraulic performance of a morphology-based stream channel design, *Water Resour. Res.*, *41*, W11413, doi:10.1029/2004WR003926.

Smith, T. R. (2010), A theory for the emergence of channelized drainage, *J. Geophys. Res.*, *115*, F02023, doi:10.1029/2008JF001114.

Snelder, T. H., and B. J. F. Biggs (2002), Multiscale River Environment Classification for water resources management, *J. Am. Water Resour. Assoc.*, *38*, 1225–1239.

Snelder, T. H., P. Cattanéo, A. M. Suren, and B. J. F. Biggs (2004), Is the River Environment Classification an improved landscape-scale classification of rivers?, *J. North Am. Benthol. Soc.*, *23*, 580–598.

Snelder, T. H., B. J. F. Biggs, and R. A. Woods (2005), Improved eco-hydrological classification of rivers, *River Res. Appl.*, *21*, 609–628.

Snelder, T. H., H. Pella, J.-G. Wasson, and N. Lamoroux (2008), Definition procedures have little effect on performance of environmental classifications of streams and rivers, *Environ. Manage.*, *42*, 771–788.

Snow, D. T. (1964), Landslide of Cerro Condor-Sencca, Department of Ayacucho, Peru, in *Engineering Geology Case Histories*, edited by G. A. Kiersch, pp. 1–6, Geol. Soc. of Am., Boulder, Colo.

Snyder, N. P. (2009), Studying stream morphology with airborne laser elevation data, *Eos Trans. AGU*, *90*(6), 45–46.

Snyder, N. P., and L. L. Kammer (2008), Dynamic adjustments in channel width in response to a forced diversion: Gower Gulch, Death Valley National Park, California, *Geology*, *36*, 187–190.

Snyder, N. P., K. X. Whipple, G. E. Tucker, and D. J. Merritts (2000), Landscape response to tectonic forcing: Digital elevation model analysis of stream profiles in the Mendocino triple junction region, northern California, *Geol. Soc. Am. Bull.*, *112*, 1250–1263.

Snyder, N. P., K. X. Whipple, G. E. Tucker, and D. J. Merritts (2003), Channel response to tectonic forcing: Field analysis of stream morphology and hydrology in the Mendocino triple junction region, northern California, *Geomorphology*, *53*, 97–127.

Solbrig, O. T. (1992), From Genes to Ecosystems: A Research Agenda for Biodiversity, *Report of a IUBS-SCOPE-UNESCO Workshop*, 123 pp, Int. Union for Biol. Sci., (IUBS), Orsay, France.

Sommerfeld, R. A., R. C. Musselman, and G. L. Wooldridge (1990), Comparison of estimates of snow input with a small alpine catchment, *J. Hydrol.*, *120*, 295–307.

Sorriso-Valvo, M., L. Antronico, and E. Le Pera (1998), Controls on modern fan morphology in Calabria, southern Italy, *Geomorphology*, *24*, 169–187.

Sosio, R., G. B. Crosta, and P. Frattini (2007), Field observations, rheological testing and numerical modelling of a debris-flow event, *Earth Surf. Processes Landforms*, *32*, 290–306.

Soulliere, E. J., and T. J. Toy (1986), Rilling of hillslopes reclaimed before 1977 surface mining law, Dave Johnston Mine, Wyoming, *Earth Surf. Processes Landforms*, *11*, 293–205.

Soulsby, C., R. Malcolm, R. Helliwell, and R. C. Ferrier (1999), Hydrogeochemistry of montane springs and their influence on streams in the Cairngorm Mountains, Scotland, *Hydrol. Earth Syst. Sci.*, *3*, 409–419.

Sousa, W. P. (1984), The role of disturbance in natural communities, *Annu. Rev. Ecol. Syst.*, *15*, 353–391.

Spaliviero, M. (2003), Historic fluvial development of the Alpine-foreland Tagliamento River, Italy, and consequences for floodplain management, *Geomorphology*, *52*, 317–333.

Spencer, C. N., B. R. McClelland, and J. A. Stanford (1991), Shrimp stocking, salmon collapse, and eagle displacement, *BioScience*, *41*, 14–21.

Splinter, D. K., D. C. Dauwalter, R. A. Marston, and W. L. Fisher (2010), Ecoregions and stream morphology in eastern Oklahoma, *Geomorphology*, *122*, 117–128.

Spreafico, M. (2006), Flash floods in mountain areas, in *Climate Variability and Change – Hydrological Impacts*, edited by S. Demuth et al., *IAHS Publ., 308*, 232–238.

Spreafico, M., and C. Lehmann (1994), Sediment transport observations in Switzerland, in *Variability in Stream Erosion and Sediment Transport*, edited by L. J. Olive, R. J. Loughran, and J. A. Kesby, *IAHS Publ., 224*, 259–268.

Springer, G. S. (2002), Caves and their potential use in paleoflood studies, in *Ancient Floods, Modern Hazards: Principles and Applications of Paleoflood Hydrology*, edited by P. K. House et al., pp. 329–343, AGU, Washington, D. C.

Springer, G. S., and J. S. Kite (1997), River-derived slackwater sediments in caves along Cheat River, West Virginia, *Geomorphology, 18*, 91–100.

Springer, G. S., H. S. Dowdy, and L. S. Eaton (2001), Sediment budgets for two mountainous basins affected by a catastrophic storm: Blue Ridge Mountains, Virginia, *Geomorphology, 37*, 135–148.

Springer, G. S., E. E. Wohl, J. A. Foster, and D. G. Boyer (2003), Testing for reach-scale adjustments of hydraulic variables to soluble and insoluble strata: Buckeye Creek and Greenbrier River, West Virginia, *Geomorphology, 56*, 201–217.

Springer, G. S., S. Tooth, and E. E. Wohl (2006), Theoretical modeling of stream potholes based upon empirical observations from the Orange River, Republic of South Africa, *Geomorphology, 82*, 160–176.

Sridhar, V., and A. Nayak (2010), Implications of climate-driven variability and trends for the hydrologic assessment of the Reynolds Creek Experimental Watershed, Idaho, *J. Hydrol., 385*, 183–202.

Stahl, K., R. D. Moore, J. M. Shea, D. Hutchinson, and A. J. Cannon (2008), Coupled modelling of glacier and streamflow response to future climate scenarios, *Water Resour. Res., 44*, W02422, doi:10.1029/2007WR005956.

Staley, D. M., T. A. Wasklewicz, and J. S. Blaszczynski (2006), Surficial patterns of debris flow deposition on alluvial fans in Death Valley, CA, using airborne laser SWATH mapping data, *Geomorphology, 74*, 152–163.

Stallard, R. F., and J. M. Edmond (1983), Geochemistry of the Amazon 2: The influence of the geologic and weathering environment on the dissolved load, *J. Geophys. Res., 88*, 9671–9688.

Stam, A. C., M. J. Mitchell, H. R. Krouse, and J. S. Kahl (1992), Stable sulfur isotopes of sulfate in precipitation and stream solutions in a northern hardwood watershed, *Water Resour. Res., 28*, 231–236.

Stanford, J. A. (2006), Landscapes and riverscapes, in *Methods in Stream Ecology*, edited by R. F. Hauer and G. A. Lamberti, pp. 3–21, Elsevier, Amsterdam.

Stanford, J. A., and F. R. Hauer (1992), Mitigating the impacts of stream and lake regulation in the Flathead River catchment, Montana, USA: An ecosystem perspective, *Aquat. Conserv. Mar. Freshwater Ecosyst., 2*, 35–63.

Stanford, J. A., and J. V. Ward (1988), The hyporheic habitat of river ecosystems, *Nature, 335*, 64–66.

Stanford, J. A., and J. V. Ward (1993), An ecosystem perspective of alluvial rivers: Connectivity and the hyporheic corridor, *J. North Am. Benthol. Soc., 12*, 48–60.

Stanford, J. A., F. R. Hauer, and J. V. Ward (1988), Serial discontinuity in a large river system, *Verh. Int. Verein. Limnol., 23*, 1114–1118.

Stanford, J. A., J. V. Ward, W. J. Liss, C. A. Frissell, R. N. Williams, J. A. Lichatowich, and C. C. Coutant (1996), A general protocol for restoration of regulated rivers, *Reg. Rivers Res. Manage., 12*, 391–413.

Stanley, E. H., and A. J. Boulton (1995), Hyporheic processes during flooding and drying in a Sonoran Desert stream. I. Hydrologic and chemical dynamics, *Arch. Hydrobiol., 134*, 1–26.

Stanley, E. H., and M. W. Doyle (2003), Trading off: The ecological effects of dam removal, *Front. Ecol. Environ., 1*, 15–22.

Stanley, E. H., S. G. Fisher, and N. B. Grimm (1997), Ecosystem expansion and contraction in streams, *BioScience, 47*, 427–435.

Stanley, E. H., M. A. Luebke, M. W. Doyle, and D. W. Marshall (2002), Short-term changes in channel form and macroinvertebrate communities following low-head dam removal, *J. North Am. Benthol. Soc., 21*, 172–187.

Stark, C. P. (2006), A self-regulating model of bedrock river channel geometry, *Geophys. Res. Lett.*, *33*, L04402, doi:10.1029/2005GL023193.

Stark, C. P., and F. Guzzetti (2009), Landslide rupture and the probability distribution of mobilized debris volumes, *J. Geophys. Res.*, *114*, F00A02, doi:10.1029/2008JF001008.

Stark, C. P., E. Foufoula-Georgiou, and V. Ganti (2009), A nonlocal theory of sediment buffering and bedrock channel evolution, *J. Geophys. Res.*, *114*, F01029, doi:10.1029/2008JF000981.

Stark, C. P., J. R. Barbour, Y. S. Hayakawa, T. Hattanji, N. Hovius, H. Chen, C.-W. Lin, M.-J. Horng, K.-Q. Xu, and Y. Fukahata (2010), The climatic signature of incised river meanders, *Science*, *327*, 1497–1501.

Starkel, L. (1972), The role of catastrophic rainfall in the shaping of the relief of the Lower Himalaya (Darjeeling Hills), *Geogr. Pol.*, *21*, 103–147.

Starkel, L. (1988), Tectonic, anthropogenic and climatic factors in the history of the Vistula River valley downstream of Cracow, in *Lake, Mire and River Environments During the Last 15,000 Years*, edited by G. Lang and C. Schluchter, pp. 161–170, A. A. Balkema, Rotterdam.

Starkel, L. (2003), Climatically controlled terraces in uplifting mountain areas, *Quat. Sci. Rev.*, *22*, 2189–2198.

Starkel, L., R. Soja, and D. J. Michczynska (2006), Past hydrological events reflected in Holocene history of Polish rivers, *Catena*, *66*, 24–33.

Starnes, L. B., and D. C. Gasper (1995), Effects of surface mining on aquatic resources in North America, *Fisheries*, *20*, 20–23.

Statzner, B., and D. Borchardt (1994), Longitudinal patterns and processes along streams: Modelling ecological responses to physical gradients, in *Aquatic Ecology: Scale, Pattern and Process*, edited by P. S. Giller, A. G. Hildrew, and D. G. Raffaelli, pp. 113–140, Blackwell Sci., Oxford, U. K.

Statzner, B., and O. Peltret (2006), Assessing potential abiotic and biotic complications of crayfish-induced gravel transport in experimental streams, *Geomorphology*, *74*, 245–256.

Statzner, B., and P. Sagnes (2008), Crayfish and fish as bioturbators of streambed sediments: Assessing joint effects of species with different mechanistic abilities, *Geomorphology*, *93*, 267–287.

Statzner, B., U. Fuchs, and L. W. G. Higler (1996), Sand erosion by mobile predaceous stream insects: Implications for ecology and hydrology, *Water Resour. Res.*, *32*, 2279–2287.

Stedinger, J. R., and V. R. Baker (1987), Surface water hydrology: Historical and paleoflood information, *Rev. Geophys.*, *25*, 119–124.

Stedinger, J. R., and T. A. Cohn (1986), Flood frequency analysis with historical and paleoflood information, *Water Resour. Res.*, *22*, 785–793.

Stein, O. R., and P. Y. Julien (1993), Criterion delineating the mode of headcut migration, *J. Hydraul. Eng.*, *119*, 37–50.

Sterling, S., and O. Slaymaker (2007), Lithologic control of debris torrent occurrence, *Geomorphology*, *86*, 307–319.

Sterling, S. M., and M. Church (2002), Sediment trapping characteristics of a pit trap and the Helley-Smith sampler in a cobble gravel bed river, *Water Resour. Res.*, *38*(8), 1144, doi:10.1029/2000WR000052.

Sternberg, H. (1875), Untersuchungen über Längen- und Querprofile geschiebeführender Flüsse, *Zeitschrift für Bauwesen*, *25*, 483–506.

Stewardson, M. (2005), Hydraulic geometry of stream reaches, *J. Hydrol.*, *306*, 97–111.

Stewart, G., R. Anderson, and E. Wohl (2005), Two-dimensional modeling of habitat suitability as a function of discharge on two Colorado rivers, *River Res. Appl.*, *21*, 1061–1074.

Stewart, I. T., D. R. Cayan, and M. D. Dettinger (2004), Changes in snowmelt runoff timing in western North America under a 'business as usual' climate change scenario, *Clim. Change*, *62*, 217–232.

Stewart, J. H., and V. C. LaMarche, Jr. (1968), *Erosion and deposition produced by the flood of December 1964 on Coffee Creek, Trinity County, California*, U.S. Geol. Surv. Prof. Pap., 422-K, pp. K1–K22.

Stewart, M. D., P. D. Bates, M. G. Anderson, D. A. Price, and T. P. Burt (1999), Modelling floods in hydrologically complex lowland river reaches, *J. Hydrol.*, *223*, 85–106.

Stock, J. D., and W. E. Dietrich (2003), Valley incision by debris flows: Evidence of a topographic signature, *Water Resour. Res.*, *39*(4), 1089, doi:10.1029/2001WR001057.

Stock, J. D., and W. E. Dietrich (2006), Erosion of steepland valleys by debris flows, *Geol. Soc. Am. Bull.*, *118*, 1125–1148.

Stock, J. D., and D. R. Montgomery (1999), Geologic constraints on bedrock river incision using the stream power law, *J. Geophys. Res.*, *14*(B3), 4983–4993.

Stock, J. D., D. R. Montgomery, B. D. Collins, W. E. Dietrich, and L. Sklar (2005), Field measurements of incision rates following bedrock exposure: Implications for process controls on the long profiles of valleys cut by rivers and debris flows, *Geol. Soc. Am. Bull.*, *117*, 174–194.

Stock, J. D., K. M. Schmidt, and D. M. Miller (2007), Controls on alluvial fan long-profiles, *Geol. Soc. Am. Bull.*, *120*, 619–640.

Stocker-Mittaz, C., M. Hoelzle, and W. Haeberli (2002), Modelling alpine permafrost distribution based on energy-balance data; a first step, *Permafrost Periglacial Processes*, *13*, 271–282.

Stoddard, M. A., and J. P. Hayes (2005), The influence of forest management on headwater stream amphibians at multiple spatial scales, *Ecol. Appl.*, *15*, 811–823.

Stoffel, M., D. Conus, M. A. Grichting, I. Lievre, and G. Maitre (2008), Unraveling the patterns of late Holocene debris-flow activity on a cone in the Swiss Alps: Chronology, environment and implications for the future, *Global Planet. Change*, *60*, 222–234.

Stoffel, M., I. Lievre, D. Conus, M. A. Grichting, H. Raetzo, H. W. Gaertner, and M. Monbaron (2005), 400 years of debris-flow activity and triggering weather conditions: Ritigraben, Valais, Switzerland, *Arct. Antarct. Alp. Res.*, *37*, 387–395.

Stokes, M., and A. E. Mather (2003), Tectonic origin and evolution of a transverse drainage: The Rio Almanzora, Betic Cordillera, southeast Spain, *Geomorphology*, *50*, 59–81.

Stokes, S., and D. Walling (2003), Radiogenic and isotopic methods for the direct dating of fluvial sediments, in *Tools in Fluvial Geomorphology*, edited by G. M. Kondolf and H. Piégay, pp. 233–267, John Wiley, Chichester, U. K.

Stonestrom, D. A., and J. Constantz (Eds.) (2003), Heat as a tool for studying the movement of ground water near streams, *U.S. Geol. Surv. Circ.*, *1260*, 45 pp.

Storey, R. G., K. W. F. Howard, and D. D. Williams (2003), Factors affecting riffle-scale hyporheic exchange flows and their seasonal changes in a gaining stream: A three-dimensional groundwater flow model, *Water Resour. Res.*, *39*(2), 1034, doi:10.1029/2002WR001367.

Stott, T. (1997), A comparison of stream bank erosion processes on forested and moorland streams in the Balquhidder catchments, central Scotland, *Earth Surf. Processes Landforms*, *22*, 383–399.

Stottlemyer, R. (2001), Processes regulating watershed chemical export during snowmelt, Fraser experimental forest, Colorado, *J. Hydrol.*, *245*, 177–195.

Stoughton, J. A., and W. A. Marcus (2000), Persistent impacts of trace metals from mining on floodplain grass communities along Soda Butte Creek, Yellowstone National Park, *Environ. Manage.*, *25*, 305–320.

Stover, S. C., and D. R. Montgomery (2001), Channel change and flooding, Skokomish River, Washington, *J. Hydrol.*, *243*, 272–286.

Strahler, A. N. (1950), Equilibrium theory of erosional slopes approached by frequency distribution analysis, Part II, *Am. J. Sci.*, *248*, 673–696.

Strahler, A. N. (1952a), Dynamic basis of geomorphology, *Geol. Soc. Am. Bull.*, *63*, 923–938.

Strahler, A. N. (1952b), Hypsometric (area altitude) analysis of erosional topography, *Geol. Soc. Am. Bull.*, *63*, 1117–1142.

Strahler, A. N. (1957), Quantitative analysis of watershed geomorphology, *Eos Trans. AGU*, *38*, 913–920.

Strahler, A. N. (1964), Quantitative geomorphology of drainage basins and channel networks, in *Handbook of Applied Hydrology*, edited by V. T. Chow, pp. 4-40–4-74, McGraw Hill, New York.

Strand, R. I. (1975), *Bureau of Reclamation Procedures for Predicting Sediment Yield, Agricultural Research Service*, vol. ARS-S-40, pp. 10–15.

Strasser, U., and P. Ethcevers (2005), Using subgrid parameterisation and a forest canopy climate model for improving forecasts of snowmelt runoff, in *Climate and Hydrology in Mountain Areas*, edited by C. de Jong, D. Collins, and R. Ranzi, pp. 29–44, John Wiley, Chichester, U. K.

Straumann, R. K., and O. Korup (2009), Quantifying postglacial sediment storage at the mountain-belt scale, *Geology*, *37*, 1079–1082.

Strayer, D. L., M. E. Power, W. F. Fagan, S. T. A. Pickett, and J. Belnap (2003), A classification of ecological boundaries, *BioScience*, *53*, 723–729.

Strecker, M. R., G. E. Hilley, J. R. Arrowsmith, and I. Cout (2003), Differential structural and geomorphic mountain-front evolution in an active continental collision zone: The Northwest Pamir, southern Kyrgyzstan, *Geol. Soc. Am. Bull.*, *115*, 166–181.

Strom, K. B., and A. N. Papanicolaou (2009), Occurrence of cluster microforms in mountain rivers, *Earth Surf. Processes Landforms*, *34*, 88–98.

Strom, K., A. N. Papanicolaou, N. Evangelopoulos, and M. Odeh (2004), Microforms in gravel bed rivers: Formation, disintegration, and effects on bedload transport, *J. Hydraul. Eng.*, *130*, 1–14.

Stromberg, J. C., D. T. Patten, and B. D. Richter (1991), Flood flows and dynamics of Sonoran riparian forests, *Rivers*, *2*, 221–235.

Stromberg, J. C., B. D. Richter, D. T. Patten, and L. G. Wolden (1993), Response of a Sonoran riparian forest to a 10-year return flood, *Great Basin Nat.*, *53*, 118–130.

Strozzi, T., R. Delaloye, A. Kääb, C. Ambrosi, E. Perruchoud, and U. Wegmüller (2010), Combined observations of rock mass movements using satellite SAR interferometry, differential GPS, airborne digital photogrammetry, and airborne photography interpretation, *J. Geophys. Res.*, *115*, F01014, doi:10.1029/2009JF001311.

Strunk, H. (1992), Reconstructing debris flow frequency in the southern Alps back to AD 1500 using dendrogeomorphical analysis, in *Erosion, Debris Flows and Environment in Mountain Regions*, edited by D. E. Walling, T. R. Davies, and B. Hasholt, *IAHS Publ.*, *209*, 299–306.

Sturdevant-Rees, P., J. A. Smith, J. Morrison, and M. L. Baeck (2001), Tropical storms and the flood hydrology of the central Appalachians, *Water Resour. Res.*, *37*, 2143–2168.

Sturm, M., J. Beget, and C. Benson (1987), Observations of jökulhlaups from ice-dammed Strandline Lake, Alaska: Implications for paleohydrology, in *Catastrophic Flooding*, edited by L. Mayer and D. Nash, pp. 79–94, Allen and Unwin, Boston.

Stüve, P. E. (1990), Spatial and temporal variation of flow resistance in an Alpine river, in *Hydrology in Mountainous Regions I. Hydrological Measurements: The Water Cycle*, edited by H. Lang and A. Musy, *IAHS Publ.*, *193*, 307–314.

Sueker, J. K. (1995), Chemical hydrograph separation during snowmelt for three headwater basins in Rocky Mountain National Park, Colorado, in *Biogeochemistry of Seasonally Snow-Covered Catchments*, edited by K. A. Tonnessen, M. W. Williams, and M. Tranter, *IAHS Publ.*, *228*, 271–281.

Sueker, J. K., J. N. Ryan, C. Kendall, and R. D. Jarrett (2000), Determination of hydrologic pathways during snowmelt for alpine/subalpine basins, Rocky Mountain National Park, Colorado, *Water Resour. Res.*, *36*, 63–75.

Sugai, T. (1993), River terrace development by concurrent fluvial processes and climatic changes, *Geomorphology*, *6*, 243–252.

Sukhodolov, A. N., and B. L. Rhoads (2001), Field investigation of three-dimensional flow structure at stream confluences: 2. Turbulence, *Water Resour. Res.*, *37*, 2411–2424.

Sullivan, A. B., and J. I. Drever (2001), Spatiotemporal variability in stream chemistry in a high-elevation catchment affected by mine drainage, *J. Hydrol.*, *252*, 237–250.

Sullivan, K., T. E. Lisle, C. A. Dolloff, G. E. Grant, and L. M. Reid (1987), Stream channels: The link between forests and fishes, in *Streamside Management: Forestry and Fishery Implications*, edited by E. O. Salo and T. W. Cundy, pp. 39–97, Univ. of Washington, Institute of Forest Resources, Seattle.

Sullivan, T. J., B. J. Cosby, A. T. Herlihy, J. R. Webb, A. J. Bulger, K. U. Snyder, P. F. Brewer, E. H. Gilbert, and D. L. Moore (2004), Regional model projections of future effects of sulfur and nitrogen deposition on streams in the Southern Appalachian Mountains, *Water Resour. Res.*, *40*, W02101, doi:10.1029/2003WR001998.

Summerfield, M. A., and N. J. Hulton (1994), Natural controls of fluvial denudation rates in major world drainage basins, *J. Geophys. Res.*, *99*(B7), 13,871–13,883.

Sun, S., and H. Deng (2003), A catchment surface runoff simulation for land surface model study, in *Weather Radar Information and Distributed Hydrological Modelling*, edited by Y. Tachikawa et al., *IAHS Publ.*, *282*, 308–314.

Surian, N. (1999), Channel changes due to river regulation: The case of the Piave River, Italy, *Earth Surf. Processes Landforms*, *24*, 1135–1151.

Surian, N. (2002), Downstream variation in grain size along an Alpine river: Analysis of controls and processes, *Geomorphology*, *43*, 137–149.

Surian, N., and E. D. Andrews (1999), Estimation of geomorphically significant flows in alpine streams of the Rocky Mountains, Colorado (USA), *Reg. Rivers Res. Manage.*, *15*, 273–288.

Surian, N., and A. Cisotto (2007), Channel adjustments, bedload transport and sediment sources in a gravel-bed river, Brenta River, Italy, *Earth Surf. Processes Landforms*, *32*, 1641–1656.

Surian, N., L. Mao, M. Giacomin, and L. Ziliani (2009), Morphological effects of different channel-forming discharges in a gravel-bed river, *Earth Surf. Processes Landforms*, *34*, 1093–1107.

Suszka, L. (1991), Modification of the transport rate formula for steep channels, in *Fluvial Hydraulics of Mountain Regions*, edited by A. Armanini and G. DiSilvio, pp. 59–70, Springer, Berlin.

Sutherland, A. J. (1987), Static armour layers by selective erosion, in *Sediment Transport in Gravel-Bed Rivers*, edited by C. R. Thorne, J. C. Bathurst, and R. D. Hey, pp. 243–268, John Wiley, Chichester, U. K.

Sutherland, D. G., M. H. Ball, S. J. Hilton, and T. E. Lisle (2002), Evolution of a landslide-induced sediment wave in the Navarro River, California, *Geol. Soc. Am. Bull.*, *114*, 1036–1048.

Sutherland, R. A. (1991), Caesium-137 and sediment budgeting within a partially closed drainage basin, *Z. Geomorphol.*, *35*, 47–63.

Suzuki, K., and K. Kato (1991), Mobile armouring of bed surface in steep slope river with gravel and sand mixture, in *Fluvial Hydraulics of Mountain Regions*, edited by A. Armanini and G. DiSilvio, pp. 393–404, Springer, Berlin.

Swank, W. T. (1986), Biological control of solute losses from forest ecosystems, in *Solute Processes*, edited by S. T. Trudgill, pp. 85–139, John Wiley, Chichester, U. K.

Swank, W. T., and C. E. Johnson (1994), Small catchment research in the evaluation and development of forest management practices, in *Biogeochemistry of Small Catchments: A Tool for Environmental Research*, edited by B. Moldan and J. Cerny, pp. 383–408, John Wiley, Chichester, U. K.

Swanson, F. J., G. W. Lienkaemper, and J. R. Sedell (1976), *History, physical effects, and management implications of large organic debris in western Oregon streams*, USDA For. Serv. General Tech. Rep. PNW-56

Swanson, F. J., L. E. Benda, S. H. Duncan, G. E. Grant, W. F. Megahan, L. M. Reid, and R. R. Ziemer (1987), Mass failures and other processes of sediment production in Pacific Northwest forest landscapes, in *Streamside Management: Forestry and Fishery Implications*, edited by E. O. Salo and T. W. Cundy, pp. 9–38, Univ. of Washington, Institute of Forest Resources, Seattle.

Swanson, F. J., T. K. Kratz, N. Caine, and R. G. Woodmansee (1988), Landform effects on ecosystem patterns and processes, *BioScience*, *38*, 92–98.

Swanson, F. J., S. M. Wondzell, and G. E. Grant (1992), Landforms, disturbance, and ecotones, in *Landscape Boundaries: Consequences for Biotic Diversity and Ecological Flows*, edited by A. J. Hansen and F. di Castri, pp. 304–323, Springer, New York.

Swanson, F. J., S. L. Johnson, S. V. Gregory, and S. A. Acker (1998), Flood disturbance in a forested mountain landscape, *BioScience*, *48*, 681–689.

Swanston, D. N., and F. J. Swanson (1976), Timber harvesting, mass erosion, and steepland forest geomorphology in the Pacific Northwest, in *Geomorphology and Engineering*, edited by D. R. Coates, pp. 199–221, Dowden, Hutchinson, and Ross, Stroudsburg, Pa.

Sweeney, B. W. (1992), Streamside forests and the physical, chemical, and trophic characteristics of piedmont streams in eastern North America, *Water Sci. Technol.*, 26, 2653–2673.

Sweeney, B. W., T. L. Bott, J. K. Jackson, L. A. Kaplan, J. D. Newbold, L. J. Standley, W. C. Hession, and R. J. Horwitz (2004), Riparian deforestation, stream narrowing, and loss of stream ecosystem services, *Proc. Natl. Acad. Sci.*, 101, 14,132–14,137.

Szupiany, R. N., M. L. Amsler, D. R. Parsons, and J. L. Best (2009), Morphology, flow structure, and suspended bed sediment transport at two large braid-bar confluences, *Water Resour. Res.*, 45, W05415, doi:10.1029/2008WR007428.

Tabacchi, E., A. M. Planty-Tabacchi, M. J. Salinas, and H. Decamps (1996), Landscape structure and diversity in riparian plant communities: A longitudinal comparative study, *Reg. Rivers Res. Manage.*, 12, 367–390.

Taboada, A., and N. Estrada (2009), Rock-and-soil avalanches: Theory and simulation, *J. Geophys. Res.*, 114, F03004, doi:10.1029/2008JF001072.

Tagata, S., T. Yamakoshi, Y. Doi, K. Sasahara, H. Nishimoto, and H. Nagura (2005), Post-eruption sediment budget of a small catchment on the Miyakejima volcano, Japan, in *Sediment Budgets I*, edited by D. E. Walling and A. J. Horowitz, *IAHS Publ.*, 291, 37–45.

Tague, C. (2009), Assessing climate change impacts on alpine stream-flow and vegetation water use: Mining the linkages with subsurface hydrologic processes, *Hydrol. Processes*, 23(12), 1815–1819, doi:10.1002/hyp.7288.

Tague, C., and L. Band (2001), Simulating the impact of road construction and forest harvesting on hydrologic response, *Earth Surf. Processes Landforms*, 26, 135–151.

Tague, C., and G. E. Grant (2004), A geological framework for interpreting the low-flow regimes of Cascade streams, Willamette River Basin, Oregon, *Water Resour. Res.*, 40, W04303, doi:10.1029/2003WR002629.

Tague, C., and G. E. Grant (2009), Groundwater dynamics mediate low-flow response to global warming in snow-dominated alpine regions, *Water Resour. Res.*, 45, W07421, doi:10.1029/2008WR007179.

Tague, C., M. Farrell, G. Grant, J. Choate, and A. Jefferson (2008a), Deep groundwater mediates streamflow response to climate warming in the Oregon Cascades, *Clim. Change*, 86, 1–2.

Tague, C., S. Valentine, and M. Kotchen (2008b), Effect of geomorphic channel restoration on streamflow and groundwater in a snowmelt-dominated watershed, *Water Resour. Res.*, 44, W10415, doi:10.1029/2007WR006418.

Taillefumier, F., and H. Piégay (2003), Contemporary land use changes in prealpine Mediterranean mountains; a multivariate GIS-based approach applied to two municipalities in the southern French Prealps, *Catena*, 51, 267–296.

Takahashi, G. (1990), A study on the riffle-pool concept, *Trans. Jpn. Geomorphol. Union*, 11, 319–336.

Takahashi, T. (1991a), *Debris Flows*, A. A. Balkema, Rotterdam.

Takahashi, T. (1991b), Mechanics and the existence criteria of various types of flows during massive sediment transport, in *Fluvial Hydraulics of Mountain Regions*, edited by A. Armanini and G. DiSilvio, pp. 267–278, Springer, Berlin.

Takahashi, T. (2007), Progress in debris flow modeling, in *Progress in Landslide Science*, edited by K. Sassa et al., pp. 59–77, Springer, Berlin.

Takahashi, T., H. Nakagawa, and S. Kuang (1987), Estimation of debris flow hydrograph on varied slope bed, in *Erosion and Sedimentation in the Pacific Rim*, edited by R. L. Beschta et al., *IAHS Publ.*, 165, 167–177.

Takahashi, T., H. Nakagawa, Y. Satofuka, and K. Kawaike (2001), Flood and sediment disasters triggered by 1999 rainfall in Venezuela; a river restoration plan for an alluvial fan, *J. Nat. Disaster Sci.*, 23, 65–82.

Takao, A., J. N. Negishi, M. Nunokawa, T. Gomi, and O. Nakahara (2006), Potential influences of a net-spinning caddisfly (Trichoptera: Stenopsyche marmorata) on stream substratum stability in heterogeneous field environments, *J. North Am. Benthol. Soc.*, *25*, 545–555.

Tal, M., and C. Paola (2007), Dynamic single-thread channels maintained by the interaction of flow and vegetation, *Geology*, *35*, 347–350.

Tal, M., and C. Paola (2010), Effects of vegetation on channel morphodynamics: Results and insights from laboratory experiments, *Earth Surf. Processes Landforms*, *35*, 1014–1028.

Tal, M., K. Gran, A. B. Murray, C. Paola, and D. M. Hicks (2004), Riparian vegetation as a primary control on channel characteristics in multi-thread rivers, in *Riparian Vegetation and Fluvial Geomorphology*, edited by S. J. Bennett and A. Simon, pp. 43–58, AGU, Washington, D. C.

Talbot, T., and M. Lapointe (2002), Numerical modeling of gravel bed river response to meander straightening: The coupling between the evolution of bed pavement and long profile, *Water Resour. Res.*, *38*(6), 1074, doi:10.1029/2001WR000330.

Tanaka, Y., Y. Onda, and Y. Agata (1993), Effect of rock properties on the longitudinal profiles of river beds: Comparison of the mountain rivers in granite and Paleozoic sedimentary rock basins, *Geogr. Rev. Jpn.*, *66A*, 203–216.

Tarboton, D. G., R. L. Bras, and I. Rodriguez-Iturbe (1991), On the extraction of channel networks from digital elevation data, *Hydrol. Processes*, *5*, 81–100.

Tarboton, D. G., R. L. Bras, and I. Rodriguez-Iturbe (1992), A physical basis for drainage density, *Geomorphology*, *5*, 59–76.

Tarolli, P., and G. Dalla Fontana (2009), Hillslope-to-valley transition morphology: New opportunities from high resolution DTMs, *Geomorphology*, *113*, 47–56.

Taylor, M. P., and R. G. H. Kesterton (2002), Heavy metal contamination of an arid river environment: Gruben River, Namibia, *Geomorphology*, *42*, 311–327.

Taylor, S., X. Feng, J. W. Kirchner, R. Osterhuber, B. Klaue, and C. E. Renshaw (2001), Isotopic evolution of a seasonal snowpack and its melt, *Water Resour. Res.*, *37*, 759–769.

Taylor, S. B., and J. S. Kite (2006), Cmoparative geomorphic analysis of surficial deposits at three central Appalachian watersheds: Implications for controls on sediment transport efficiency, *Geomorphology*, *78*, 22–43.

Teleki, P. G. (1972), Areal sorting of bed-load material: The hypothesis of velocity reversal: Discussion, *Geol. Soc. Am. Bull.*, *83*, 911–914.

Tennekes, H., and J. L. Lumley (1994), *A First Course in Turbulence*, 300 pp., MIT Press, Cambridge, Mass.

Terblanche, D. E., G. G. S. Pegram, and M. P. Mittermaier (2001), The development of weather radar as a research and operational tool for hydrology in South Africa, *J. Hydrol.*, *241*, 3–25.

Tesfa, T. K., D. G. Tarboton, D. G. Chandler, and J. P. McNamara (2009), Modeling soil depth from topographic and land cover attributes, *Water Resour. Res.*, *45*, W10438, doi:10.1029/2008WR007474.

Tetzlaff, D., C. Soulsby, P. J. Bacon, A. F. Youngson, C. Gibbins, and I. A. Malcolm (2007a), Connectivity between landscapes and riverscapes – a unifying theme in integrating hydrology and ecology in catchment science?, *Hydrol. Processes*, *21*, 1385–1389.

Tetzlaff, D., C. Soulsby, S. Waldron, I. A. Malcolm, P. J. Bacon, S. M. Dunn, A. Lilly, and A. F. Youngson (2007b), Conceptualization of runoff processes using a geographical information system and tracers in a nested mesoscale catchment, *Hydrol. Processes*, *21*, 1289–1307.

Teixeira, E. C., and P. C. Caliari (2005), Estimation of the concentration of suspended solids in rivers from turbidity measurement: Error assessment, in *Sediment Budgets 1*, edited by D. E. Walling and A. J. Horowitz, *IAHS Publ.*, *291*, 151–160.

Thamo-Bozso, E., Z. Kercsmar, and A. Nador (2002), Tectonic control on changes in sediment supply: Quaternary alluvial systems, Körös sub-basin, SE Hungary, *Geol. Soc. Spec. Publ.*, *191*, 37–53.

Thang, H. C., and N. A. Chappell (2005), Minimising the hydrological impact of forest harvesting in Malaysia's rainforests, in *Forests, Water and People in the Humid Tropics*, edited by M. Bonell and L. A. Bruijnzeel, pp. 852–865, Cambridge Univ. Press, Cambridge, U. K.

Thibodeaux, L. J., and J. D. Boyle (1987), Bedform-generated convective transport in bottom sediment, *Nature*, *325*, 341–343.

Thom, B. A., P. S. Kavanagh, and K. K. Jones (2001), Reference site selection and survey results, 2000, *Monitoring Program Rep. OPSW-ODFW-2001-6*, Oreg. Dep. of Fish and Wildl., Portland.

Thomas, M. F., J. Nott, and D. M. Price (2001), Late Quaternary stream sedimentation in the humid tropics; a review with new data from NE Queensland, Australia, *Geomorphology*, *39*, 53–68.

Thomas, M. F., J. Nott, A. S. Murray, and D. M. Price (2007), Fluvial response to late Quaternary climate change in NE Queensland, Australia, *Palaeogeogr. Palaeoclimatol. Palaeoecol.*, *251*, 119–136.

Thomas, R., and A. P. Nicholas (2002), Simulation of braided river flow using a new cellular routing scheme, *Geomorphology*, *43*, 179–195.

Thomas, R., A. P. Nicholas, and T. A. Quine (2007), Cellular modelling as a tool for interpreting historic braided river evolution, *Geomorphology*, *90*, 302–317.

Thompson, A. (1986), Secondary flows and the pool-riffle unit: A case study of the processes of meander development, *Earth Surf. Processes Landforms*, *11*, 631–642.

Thompson, A. (1987), Channel response to flood events in a divided upland stream, in *International Geomorphology 1986, Part I*, edited by V. Gardiner, pp. 691–709, John Wiley, Chichester, U. K.

Thomspon, C., and J. Croke (2008), Channel flow competence and sediment transport in upland streams in southeast Australia, *Earth Surf. Processes Landforms*, *33*, 329–352.

Thompson, C., J. Croke, R. Ogden, and P. J. Wallbrink (2006), A morpho-statistical classification of mountain river reach types in southeastern Australia, *Geomorphology*, *81*, 43–65.

Thompson, C., E. Rhodes, and J. Croke (2007), The storage of bed material in mountain stream channels as assessed using Optically Stimulated Luminescence dating, *Geomorphology*, *83*, 307–321.

Thompson, C., J. Croke, and I. Takken (2008), A catchment-scale model of mountain stream channel morphologies in southeast Australia, *Geomorphology*, *95*, 119–144.

Thompson, D. M. (1994), Hydraulics and sediment transport processes in a pool-riffle Rocky Mountain stream, unpublished M.S. thesis, 288 pp., Colo. State Univ., Ft. Collins.

Thompson, D. M. (1995), The effects of large organic debris on sediment processes and stream morphology in Vermont, *Geomorphology*, *11*, 235–244.

Thompson, D. M. , (1997), Hydraulics and pool geometry, unpublished Ph.D. dissertation, 260 pp., Colo. State Univ., Ft. Collins.

Thompson, D. M. (2001), Random controls on semi-rhythmic spacing of pools and riffles in constriction-dominated rivers, *Earth Surf. Processes Landforms*, *26*, 1195–1212.

Thompson, D. M. (2002a), Geometric adjustment of pools to changes in slope and discharge: A flume experiment, *Geomorphology*, *46*, 257–265.

Thompson, D. M. (2002b), Long-term effect of instream habitat-improvement structures on channel morphology along the Blackledge and Salmon Rivers, Connecticut, USA, *Environ. Manage.*, *29*, 250–265.

Thompson, D. M. (2003), A geomorphic explanation for a meander cutoff following channel relocation of a coarse-bedded river, *Environ. Manage.*, *31*, 385–400.

Thompson, D. M. (2004), The influence of pool length on local turbulence production and energy slope: A flume experiment, *Earth Surf. Processes Landforms*, *29*, 1341–1358.

Thompson, D. M. (2006a), Changes in pool size in response to a reduction in discharge: A flume experiment, *River Res. Appl.*, *22*, 343–351.

Thompson, D. M. (2006b), Did the pre-1980 use of in-stream structures improve streams? A reanalysis of historical data, *Ecol. Appl.*, *16*, 784–796.

Thompson, D. M. (2006c), The role of vortex shedding in the scour of pools, *Adv. Water Resour.*, *29*, 121–129.

Thompson, D. M. (2007), The characteristics of turbulence in a shear zone downstream of a channel constriction in a coarse-grained forced pool, *Geomorphology*, *83*, 199–214.

Thompson, D. M. (2008), The influence of lee sediment behind large bed elements on bedload transport rates in supply-limited channels, *Geomorphology*, *99*, 420–432.

Thompson, D. M., and K. S. Hoffman (2001), Equilibrium pool dimensions and sediment-sorting patterns in coarse-grained, New England channels, *Geomorphology, 38*, 301–316.

Thompson, D. M., and G. N. Stull (2002), The development and historic use of habitat structures in channel restoration in the United States: The grand experiment in fisheries management, *Geogr. Phys. Quat., 56*, 45–60.

Thompson, D. M., and E. E. Wohl (2009), The linkage between velocity patterns and sediment entrainment in a forced-pool and riffle unit, *Earth Surf. Processes Landforms, 34*, 177–192.

Thompson, D. M., E. E. Wohl, and R. D. Jarrett (1996), A revised velocity-reversal and sediment-sorting model for a high-gradient, pool-riffle stream, *Phys. Geogr., 17*, 142–156.

Thompson, D. M., J. M. Nelson, and E. E. Wohl (1998), Interactions between pool geometry and hydraulics, *Water Resour. Res., 34*, 3673–3681.

Thompson, D. M., E. E. Wohl, and R. D. Jarrett (1999), Velocity reversals and sediment sorting in pools and riffles controlled by channel constrictions, *Geomorphology, 27*, 229–241.

Thompson, S. M., and J. E. Adams (1979), Suspended load in some major rivers of New Zealand, in *Physical Hydrology: New Zealand Experience*, edited by D. L. Murray and P. Ackroyd, pp. 213–229, N. Z. Hydrol. Soc. Inc., Wellington North.

Thonon, I., and M. Van Der Perk (2003), Measuring suspended sediment characteristics using a LISST-ST in an embanked flood plain of the River Rhine, in *Erosion and Sediment Transport Measurement in Rivers: Technological and Methodological Advances*, edited by J. Bogen, T. Fergus, and D. E. Walling, *IAHS Publ., 283*, 37–44.

Thorndycraft, V. R., G. Benito, M. Barriendos, and M. C. Llasat (2002), *Palaeofloods, Historical Data and Climatic Variability: Applications in Flood Risk Assessment*, Centro de Ciencias Meioambientales, Madrid, Spain.

Thorne, C. R. (1990), Effect of vegetation on riverbank erosion and stability, in *Vegetation and Erosion*, edited by J. B. Thornes, pp. 125–114, Wiley, Chichester, U. K.

Thorne, C. R., J. C. Bathurst, and R. D. Hey (1987), *Sediment Transport in Gravel-Bed Rivers*, John Wiley, Chichester, U. K.

Thorne, S. D. , (1997), Stable bed features and their influence on flow structure, sediment transport and channel evolution in high mountain streams, unpublished Ph.D. dissertation, 301 pp., Florida State Univ., Tallahassee.

Thornes, J. B., and I. Alcantara-Ayala (1998), Modelling mass failure in a Mediterranean mountain environment: Climatic, geological, topographical and erosional controls, *Geomorphology, 24*, 87–100.

Thouvenot-Korppoo, M., G. Billen, and J. Garnier (2009), Modelling benthic denitrification processes over a whole drainage network, *J. Hydrol., 379*, 239–250.

Thurman, E. M. (1985), *Organic Geochemistry of Natural Waters*, 497 pp., Martinus Nijhoff/Dr. W. Junk, Dordrecht.

Tianche, L., R. L. Schuster, and J. Wu (1986), Landslide dams in south-central China, in *Landslide Dams: Processes, Risk, and Mitigation*, pp. 146–162, Am. Soc. of Civ. Eng., New York.

Ticehurst, J. L., H. P. Cresswell, N. J. McKenzie, and M. R. Glover (2007), Interpreting soil and topographic properties to conceptualise hillslope hydrology, *Geoderma, 137*, 279–292.

Tinkler, K. J. (1993), Fluvially sculpted rock bedforms in Twenty Mile Creek, Niagara Peninsula, Ontario, *Can. J. Earth Sci., 30*, 945–953.

Tinkler, K. J. (1997a), Critical flow in rockbed streams with estimated values for Manning's n, *Geomorphology, 20*, 147–164.

Tinkler, K. J. (1997b), Indirect velocity measurement from standing waves in rockbed rivers, *J. Hydraul. Eng., 123*, 918–921.

Tinkler, K. J., and E. E. Wohl (1998), A primer on bedrock channels, in *Rivers Over Rock: Fluvial Processes in Bedrock Channels, Geophys. Monogr. Ser.*, vol. 107, edited by K. J. Tinkler and E. E. Wohl, pp. 1–18, AGU, Washington, D. C.

Tinkler, K. J., J. W. Pengelly, W. G. Parkins, and G. Asselin (1994), Postglacial recession of Niagara Falls in relation to the Great Lakes, *Quat. Res.*, *42*, 20–29.

Tockner, K., F. Schiemer, C. Baumgartner, G. Kum, E. Weigand, I. Zweimüller, and J. V. Ward (1999), The Danube restoration project: Species diversity patterns across connectivity gradients in the floodplain system, *Reg. Rivers Res. Manage.*, *15*, 245–258.

Tockner, K., C. Baumgartner, F. Schiemer, and J. V. Ward (2000a), Biodiversity of a Danubian floodplain: Structural, functional and compositional aspects, in *Biodiversity in Wetlands: Assessment, Function and Conservation*, vol. 1, edited by B. Gopal, W. J. Junk, and J. A. Davis, pp. 141–159, Backhuys, Leiden, The Netherlands.

Tockner, K., F. Malard, and J. V. Ward (2000b), An extension of the flood pulse concept, *Hydrol. Processes*, *14*, 2861–2883.

Todd, A., D. McKnight, and L. Wyatt (2003), Abandoned mines, mountain sports, and climate variability: Implications for the Colorado tourism economy, *Eos Trans. AGU*, *84*(38), 377–380.

Tomkin, J. H. (2007), Coupling glacial erosion and tectonics at active orogens: A numerical modeling study, *J. Geophys. Res.*, *112*, F02015, doi:10.1029/2005JF000332.

Tomkin, J. H., M. T. Brandon, F. J. Pazzaglia, J. R. Barbour, and S. D. Willett (2003), Quantitative testing of bedrock incision models for the Clearwater River, NW Washington state, *J. Geophys. Res.*, *108*(B6), 2308, doi:10.1029/2001JB000862.

Tomkin, J. H., and J. Braun (2002), The influence of alpine glaciation on the relief of tectonically active mountain belts, *Am. J. Sci.*, *302*, 169–190.

Tomkins, J. D., and S. F. Lamoureux (2005), Multiple hydroclimatic controls over Recent sedimentation in proglacial Mirror Lake, southern Selwyn Mountains, Northwest Territories, *Can. J. Earth Sci.*, *42*, 1589–1599.

Tonina, D., and J. M. Buffington (2007), Hyporheic exchange in gravel bed rivers with pool-riffle morphology: Laboratory experiments and three-dimensional modeling, *Water Resour. Res.*, *43*, W01421, doi:10.1029/2005WR004328.

Tonina, D., C. H. Luce, B. Rieman, J. M. Buffington, P. Goodwin, S. R. Clayton, S. M. Alì, J. J. Barry, and C. Berenbrock (2008), Hydrological response to timber harvest in northern Idaho: Implications for channel scour and persistence of salmonids, *Hydrol. Processes*, *22*, 3223–3235.

Tooth, S. (1999), Floodouts in central Australia, in *Varieties of Fluvial Form*, edited by A. J. Miller and A. Gupta, pp. 219–247, John Wiley, Chichester, U. K.

Tooth, S., and T. S. McCarthy (2004), Anabranching in mixed bedrock-alluvial rivers: The example of the Orange River above Augrabies Falls, Northern Cape Province, South Africa, *Geomorphology*, *57*, 235–262.

Tooth, S., H. Rodnight, G. A. T. Duller, T. S. McCarthy, P. M. Marren, and D. Brandt (2007), Chronology and controls of avulsion along a mixed bedrock-alluvial river, *Geol. Soc. Am. Bull.*, *119*, 452–461.

Törnlund, E., and L. Östlund (2002), Floating timber in northern Sweden: The construction of floatways and transformation of rivers, *Environ. Hist.*, *8*, 85–106.

Torres, R., P. Mouginis-Mark, S. Self, H. Garbeil, K. Kallianpur, and R. Quiambao (2004), Monitoring the evolution of the Pasig-Potrero alluvial fan, Pinatubo Volcano, using a decade of remote sensing data, *J. Volcanol. Geotherm. Res.*, *138*, 371–392.

Toth, E., A. Brath, and A. Montanari (2000), Comparison of short-term rainfall prediction models for real-time flood forecasting, *J. Hydrol.*, *239*, 132–147.

Townsend, C. R., and A. G. Hildrew (1994), Species traits in relation to a habitat template for river systems, *Freshwater Biol.*, *31*, 265–275.

Toy, T. J. (1984), Geomorphology of surface-mined lands in the western United States, in *Development and Applications of Geomorphology*, edited by J. E. Costa and P. J. Fleisher, pp. 133–170, Springer, Berlin.

Toy, T. J. (1989), Geomorphic design and management of disturbed lands, in *Sediment and the environment*, edited by R. F. Hadley and E. D. Ongley, *IAHS Publ.*, *184*, 145–153.

Toy, T. J., and W. R. Osterkamp (1995), The applicability of RUSLE to geomorphic studies, *J. Soil Water Conserv.*, *50*, 498–503.

Toy, T. J., G. R. Foster, and K. G. Renard (1999), RUSLE for mining, construction and reclaimed lands, *J. Soil Water Conserv.*, *54*, 462–467.

Toyos, G., R. Gunasekera, G. Zanchetta, C. Oppenheimer, R. Sulpizio, M. Favalli, and M. T. Pareschi (2008), GIS-assisted modelling for debris flow hazard assessment based on the events of May 1998 in the area of Sarno, southern Italy; II, Velocity and dynamic pressure, *Earth Surf. Processes Landforms*, *33*, 1693–1708.

Trainor, K., and M. Church (2003), Quantifying variability in stream channel morphology, *Water Resour. Res.*, *39*(9), 1248, doi:10.1029/2003WR001971.

Tranter, M., P. Brimblecombe, T. D. Davies, C. E. Vincent, P. W. Abrahams, and I. Blackwood (1986), The composition of snowfall, snowpack and meltwater in the Scottish Highlands - evidence for preferential elution, *Atmos. Environ.*, *20*, 517–525.

Trayler, C. R. (1997), Spatial and temporal variability in sediment movement, and the role of woody debris in a sub-alpine stream, Colorado, unpublished M.S. thesis, 155 pp., Colo. State Univ., Ft. Collins.

Trayler, C. R., and E. E. Wohl (2000), Seasonal changes in bed elevation in a step-pool channel, Rocky Mountains, Colorado, USA, *Arct. Antarct. Alp. Res.*, *32*, 95–103.

Tribe, S., and M. Church (1999), Simulations of cobble structure on a gravel streambed, *Water Resour. Res.*, *35*, 311–318.

Trieste, D. J. (1994), Supercritical flows versus subcritical flows in natural channels, *Proceedings, ASCE Hydraulic Engineering '94 Conference, August 1–5, Buffalo, New York*, edited by G. V. Cotroneo and R. R. Rumer, pp. 732–736, Am. Soc. of Civ. Eng., New York.

Trieste, D. J., and R. D. Jarrett (1987), Roughness coefficients of large floods, *Proceed., Conf. On Irrig. Systems for the 21st Century*, edited by L. G. James and M. J. English, pp. 32–40, Am. Soc. of Civ. Eng., Portland, Oreg.

Trimble, S. W. (1994), Erosional effects of cattle on streambanks in Tennessee, USA, *Earth Surf. Processes Landforms*, *19*, 451–464.

Trimble, S. W. (1997a), Contribution of stream channel erosion to sediment yield from an urbanizing watershed, *Science*, *278*, 1442–1444.

Trimble, S. W. (1997b), Stream channel erosion and change resulting from riparian forests, *Geology*, *25*, 467–469.

Trimble, S. W. (2004), Effects of riparian vegetation on stream channel stability and sediment budgets, in *Riparian Vegetation and Fluvial Geomorphology*, edited by S. J. Bennett and A. Simon, pp. 153–169, AGU, Washington, D. C.

Trimble, S. W., and A. C. Mendel (1995), The cow as a geomorphic agent - a critical review, *Geomorphology*, *13*, 233–253.

Triska, F. J., V. C. Kennedy, R. J. Avanzio, G. W. Zellweger, and K. E. Bencala (1989), Retention and transport of nutrients in a third-order stream in northwestern California: Hyporheic processes, *Ecology*, *70*, 1893–1905.

Tritton, D. J. (1988), *Physical Fluid Dynamics*, 519 pp., Clarendon Press, Oxford.

Troendle, C. A. (1993), Sediment transport for instream flow/channel maintenance, *Proceedings of the Technical Workshop on Sediments, 3–7 Feb. 1992*, pp. 1–4, For. Serv. Environ. Prot. Agency, Washington, D. C.

Troendle, C. A., J. M. Nankervis, and S. E. Ryan , (1996), Sediment transport from small, steep-gradient watersheds in Colorado and Wyoming, in *Sedimentation Technologies for Management of Natural Resources in the 21st Century*, pp. IX-39–IX-45, Sixth Federal Interagency Sedimentation Conference, March 1996, Las Vegas, Nev.

Troll, C. (1954), Über das Wesen der Hochgebirgsnatur, *Jahrbuch Deutsch. Alpenvereins*, *80*, 142–157.

Tropeano, D., and L. Turconi (2004), Using historical documents for landslide, debris flow and stream flood prevention: Applications in northern Italy, *Nat. Hazards*, *31*, 663–679.

Trustrum, N. A., B. Gomez, M. J. Page, L. M. Reid, and D. M. Hicks (1999), Sediment production, storage and output: The relative role of large magnitude events in steepland catchment, *Z. Geomorphol.*, *115*, 71–86.

Tsai, H., and Q. C. Sung (2003), Geomorphic evidence for an active pop-up zone associated with the Chelungpu Fault in central Taiwan, *Geomorphology*, *56*, 31–47.

Tsai, H., Z.-Y. Hseu, W.-S. Huang, and Z.-S. Chen (2007), Pedogenic approach to resolving the geomorphic evolution of the Pakua River terraces in central Taiwan, *Geomorphology*, *83*, 14–28.

Tsujimoto, T. (1991), Bed-load transport in steep channels, in *Fluvial Hydraulics of Mountain Regions*, edited by A. Armanini and G. DiSilvio, pp. 89–102, Springer, Berlin.

Tsujimura, M., Y. Onda, and J. Ito (2001), Stream water chemistry in a steep headwater basin with high relief, *Hydrol. Processes*, *15*, 1847–1858.

Tubino, M., and W. Bertoldi (2008), Bifurcations in gravel-bed streams, in *Gravel-Bed Rivers VI: From Process Understanding to River Restoration*, edited by H. Habersack, H. Piégay, and M. Rinaldi, pp. 133–160, Elsevier, Oxford, U. K.

Tucker, G. E. (2009), Natural experiments in landscape evolution, *Earth Surf. Processes Landforms*, *34*, 1450–1460.

Tucker, G. E., and D. N. Bradley (2010), Trouble with diffusion: Reassessing hillslope erosion laws with a particle-based model, *J. Geophys. Res.*, *115*, F00A10, doi:10.1029/2009JF001264.

Tucker, G. E., and G. R. Hancock (2010), Modelling landscape evolution, *Earth Surf. Processes Landforms*, *35*, 28–50.

Tucker, G. E., L. Arnold, R. L. Bras, H. Flores, E. Istanbulluoglu, and P. Sólyom (2006), Headwater channel dynamics in semiarid rangelands, Colorado high plains, USA, *Geol. Soc. Am. Bull.*, *118*, 959–974.

Tunç, M., U. Çamdali, and C. Parmaksizoğlu (2006), Comparison of Turkey's electrical energy consumption and production with some European countries and optimization of future electrical power supply investments in Turkey, *Energy Policy*, *34*, 50–59.

Tunnicliffe, J., A. S. Gottesfeld, and M. Mohamed (2000), High resolution measurement of bedload transport, *Hydrol. Processes*, *14*, 2631–2643.

Tunusluoglu, M. C., C. Gokceoglu, H. A. Nefeslioglu, and H. Sonmez (2008), Extraction of potential debris source areas by logistic regression technique: A case study from Barla, Besparmak and Kapi mountains , (NW Taurids, Turkey), *Environ. Geol.*, *54*, 9–22.

Turconi, L., S. K. De, D. Tropeano, and G. Savio (2010), Slope failure and related processes in the Mt. Rocciamelone area , (Cenischia Valley, Western Italian Alps), *Geomorphology*, *114*, 115–128.

Turk, J. K., B. R. Goforth, R. C. Graham, and K. J. Kendrick (2008), Soil morphology of a debris flow chronosequence in an coniferous forest, southern California, USA, *Geoderma*, *146*, 157–165.

Turkelboom, F., J. Poesen, and G. Trebuil (2008), The multiple land degradation effects caused by land-use intensification in tropical steeplands: A catchment study from northern Thailand, *Catena*, *75*, 102–116.

Turowski, J. M., and D. Rickenmann (2009), Tools and cover effects in bedload transport observations in the Pitzbach, Austria, *Earth Surf. Processes Landforms*, *34*, 26–37.

Turowski, J. M., D. Lague, A. Crave, and N. Hovius (2006), Experimental channel response to tectonic uplift, *J. Geophys. Res.*, *111*, F03008, doi:10.1029/2005JF000306.

Turowski, J. M., D. Lague, and N. Hovius (2007), Cover effect in bedrock abrasion: A new derivation and its implications for the modeling of bedrock channel morphology, *J. Geophys. Res.*, *112*, F04006, doi:10.1029/2006JF000697.

Turoswki, J. M., N. Hovius, H. Meng-Long, D. Lague, and C. Men-Chiang (2008), Distribution of erosion across bedrock channels, *Earth Surf. Processes Landforms*, *33*, 353–363.

Turowski, J. M., D. Lague, and N. Hovius (2009a), Response of bedrock channel width to tectonic forcing: Insights from a numerical model, theoretical considerations, and comparison with field data, *J. Geophys. Res.*, *114*, F03016, doi:10.1029/2008JF001133.

Turowski, J. M., E. M. Yager, A. Badoux, D. Rickenmann, and P. Molnar (2009b), The impact of exceptional events on erosion, bedload transport and channel stability in a step-pool channel, *Earth Surf. Processes Landforms*, *34*, 1661–1673.

Tweed, F. S., and A. J. Russell (1999), Controls on the formation and sudden drainage of glacier-impounded lakes: Implications for jökulhlaup characteristics, *Prog. Phys. Geogr.*, *23*, 79–110.

Tweto, O. (1979), Geologic map of Colorado, 1:500,000 scale map, U.S. Geol. Surv., Denver, Colo.

Uchida, T., N. Ohte, A. Kimoto, T. Mizuyama, and L. Changhua (2000), Sediment yield on a devastated hill in southern China: Effects of microbiotic crust on surface erosion process, *Geomorphology*, *32*, 129–145.

Uchida, T., K. Kosugi, and T. Mizuyama (2002), Effects of pipe flow and bedrock groundwater on runoff generation in a steep headwater catchment in Ashiu, central Japan, *Water Resour. Res.*, *38*(7), 1119, doi:10.1029/2001WR000261.

Uchida, T., Y. Asano, N. Ohte, and T. Mizuyama (2003), Analysis of flowpath dynamics in a steep unchanneled hollow in the Tanakami Mountains of Japan, *Hydrol. Processes*, *17*, 417–430.

Uehlinger, U., R. Zah, and H. Bürgi (1998), The Val Roseg project: Temporal and spatial patterns of benthic algae in an Alpine stream ecosystem influenced by glacier runoff, in *Hydrology, Water Resources and Ecology in Headwaters*, edited by K. Kovar et al., *IAHS Publ.*, *248*, 419–424.

Ugarte, A., and M. Madrid (1994), Roughness coefficient in mountain rivers, *Proceedings, ASCE Hydraulic Engineering '94 Conference, August 1–5, Buffalo, New York*, edited by G. V. Cotroneo and R. R. Rumer, pp. 652–656, Am. Soc. of Civ. Eng., New York.

Uhlenbrook, S., and S. Hoeg (2003), Quantifying uncertainties in tracer-based hydrograph separations: A case study for two-, three- and five-component hydrograph separations in a mountainous catchment, *Hydrol. Processes*, *17*, 431–453.

Uhlenbrook, S., and D. Tetzlaff (2005), Operational weather radar assessment of convective precipitation as an input to flood modelling in mountainous basins, in *Climate and Hydrology in Mountain Areas*, edited by C. de Jong, D. Collins, and R. Ranzi, pp. 233–246, John Wiley, Chichester, U. K.

Uhlenbrook, S., and J. Wenninger (2006), Identification of flow pathways along hillslopes using electrical resistivity tomography , (ERT), in *Predictions in Ungauged Basins: Promise and Progress*, edited by M. Sivapalan et al., *IAHS Publ.*, *303*, 15–20.

Uhlenbrook, S., M. Frey, C. Leibundgut, and P. Maloszewski (2002), Hydrograph separations in a mesoscale mountainous basin at event and seasonal time scales, *Water Resour. Res.*, *38*(6), 1096, doi:10.1029/2001WR000938.

Uhlenbrook, S., J. J. McDonnell, and C. Leibundgut (2003), Runoff generation and implications for river basin modelling, *Hydrol. Processes*, *17*, 2.

Urbonas, B., and B. Benik (1995), Stream stability under a changing environment, in *Stormwater Runoff and Receiving Systems: Impact, Monitoring, and Assessment*, edited by E. E. Herricks, pp. 77–101, Lewis, Boca Raton, Fla.

U.S. Army Corps of Engineers (1998), *HEC-6 Scour and Deposition in Rivers and Reservoirs User's Manual, version 4.1*, Hydrol. Eng. Cent., Davis, Calif.

U.S. Department of Transportation (1979), *Restoration of Fish Habitat in Relocated Streams*, 63 pp., Fed. Highway Admin., Washington, D. C.

USDI , (U.S. Department of the Interior, Bureau of Reclamation) , (1993), *Riparian area management: Process for assessing proper functioning condition*, Technical Ref. 1737-9, 51 pp.

USFS , (U.S. Forest Service) (1992), *R1-WATSED-PC Handbook*, USDA For. Serv. Reg. 1, Missoula, Mont.

U.S. National Academy of Sciences (1983), *Changing Climate*, Natl. Acad. of Sci., Natl. Acad. Press, Washington, D. C.

Valdiya, K. S. (1987), *Environmental Geology: Indian Context*, Tata-McGraw Hill, New Delhi.

Valdiya, K. S. (1997), High dams in central Himalaya in context of active faults, seismicity, and societal problems, *J. Geol. Soc. India*, *49*, 479–494.

Valero-Garcés, B. L., A. Navas, J. Machín, and D. Walling (1999), Sediment sources and siltation in mountain reservoirs: A case study from the Central Spanish Pyrenees, *Geomorphology*, *28*, 23–41.

Valett, H. M. (1993), Surface-hyporheic interactions in a Sonoran Desert stream: Hydrologic exchange and diel periodicity, *Hydrobiologia*, *259*, 133–144.

Valett, H. M., S. G. Fisher, and E. H. Stanley (1990), Physical and chemical characteristics of the hyporheic zone of a Sonoran Desert stream, *J. North Am. Benthol. Soc.*, *9*, 201–215.

Valett, H. M., S. F. Fisher, N. B. Grimm, E. H. Stanley, and A. J. Boulton (1992), Hyporheic-surface water exchange: Implications for the structure and functioning of desert stream ecosystems, *Proceedings, First International Conference on Groundwater Ecology*, edited by J. A. Stanford and J. J. Simon, pp. 395–405, Am. Water Resour. Assoc., Bethesda, Md.

Valett, H. M., S. F. Fisher, N. B. Grimm, and P. Camill (1994), Vertical hydrologic exchange and ecological stability of a desert stream ecosystem, *Ecology*, *75*, 548–560.

Valla, P. G., P. A. van der Beek, and D. Lague (2010), Fluvial incision into bedrock: Insights from morphometric analysis and numerical modeling of gorges incising glacial hanging valleys, (Western Alps, France), *J. Geophys. Res.*, *115*, F02010, doi:10.1029/2008JF001079.

Vallance, J. W. (2005), Volcanic debris flows, in *Debris-Flow Hazards and Related Phenomena*, edited by M. Jakob and O. Hungr, pp. 247–274, Springer, Berlin.

Vallé, B. L., and G. B. Pasternack (2002), TDR measurements of hydraulic jump aeration in the South Fork of the American River, California, *Geomorphology*, *42*, 153–165.

Vallé, B. L., and G. B. Pasternack (2006), Submerged and unsubmerged natural hydraulic jumps in a bedrock step-pool mountain channel, *Geomorphology*, *82*, 146–159.

Valyrakis, M., P. Diplas, C. L. Dancey, K. Greer, and A. O. Celik (2010), Role of instantaneous force magnitude and duration on particle entrainment, *J. Geophys. Res.*, *115*, F02006, doi:10.1029/2008JF001247.

Vanacker, V., F. von Blanckenburg, G. Govers, A. Molina, J. Poesen, J. Deckers, and P. W. Kubik (2007), Restoring dense vegetation can slow mountain erosion to near natural benchmark levels, *Geology*, *35*, 303–306.

Van Breemen, N., et al. (2002), Where did all the nitrogen go? Fate of nitrogen inputs to large watersheds in the northeastern U. S. A, *Biogeochemistry*, *57/58*, 267–293.

Vance, D., M. Bickle, S. Ivy-Ochs, and P. W. Kubik (2003), Erosion and exhumation in the Himalaya from cosmogenic isotope inventories of river sediments, *Earth Planet. Sci. Lett.*, *206*, 273–288.

Van Cleve, K., C. T. Dyrness, G. M. Marion, and R. Erickson (1993), Control of soil development on the Tanana River floodplain, interior Alaska, *Can. J. For. Res.*, *23*, 941–955.

Van De Wiel, M. J., and T. J. Coulthard (2010), Self-organized criticality in river basins: Challenging sedimentary records of environmental change, *Geology*, *38*, 87–90.

Van De Wiel, M. J., and S. E. Darby (2007), A new model to analyse the impact of woody riparian vegetation on the geotechnical stability of riverbanks, *Earth Surf. Processes Landforms*, *32*, 2185–2198.

Van De Wiel, M. J., T. J. Coulthard, M. G. Macklin, and J. Lewin (2007), Embedding reach-scale fluvial dynamics within the CAESAR cellular automaton landscape evolution model, *Geomorphology*, *90*, 283–301.

Van Der Beek, P., and P. Bishop (2003), Cenozoic river profile development in the Upper Lachlan catchment (SE Australia) as a test of quantitative fluvial incision models, *J. Geophys. Res.*, *108*(B6), 2309, doi:10.1029/2002JB002125.

Vanderbilt, K. L., K. Lajtha, and F. J. Swanson (2003), Biogeochemistry of unpolluted forested watersheds in the Oregon Cascades: Temporal patterns of precipitation and stream nitrogen fluxes, *Biogeochemistry*, *62*, 87–117.

Van Der Schrier, G., D. Efthymiadis, K. R. Briffa, and P. D. Jones (2007), European Alpine moisture variability for 1800–2003, *Int. J. Climatol.*, *27*, 415–427.

Van Haveren, B. P. (1991), Placer mining and sediment problems in interior Alaska, in *Proceedings of the 5th Federal Interagency Sedimentation Conference*, vol. 2, edited by S.-S. Fan and Y.-H. Kuo, pp. 10-69–10-74, Fed. Energy Regul. Comm., Washington, D. C.

Van Niekerk, A. W., G. L. Heritage, L. J. Broadhurst, and B. P. Moon (1999), Bedrock anastomosing channel systems: Morphology and dynamics in the Sabie River, Mpumalanga Province, South Africa, in *Varieties of Fluvial Form*, edited by A. J. Miller and A. Gupta, pp. 33–51, John Wiley, Chichester, U. K.

Van Nieuwenhuyse, E. E., and J. D. LaPerriere (1986), Effects of placer gold mining on primary production in subarctic streams of Alaska, *Water Resour. Bull.*, *22*, 91–99.

Vannote, R. L., G. W. Minshall, K. W. Cummins, J. R. Sedell, and C. E. Cushing (1980), The river continuum concept, *Can. J. Fish. Aquat. Sci.*, *37*, 130–137.

Van Rompaey, A., P. Bazzolfi, R. J. A. Jones, and L. Montanarella (2005), Modeling sediment yields in Italian catchments, *Geomorphology*, *65*, 157–169.

Van Schaik, N. L. M. B., S. Schnabel, and V. G. Jetten (2008), The influence of preferential flow on hillslope hydrology in a semi-arid watershed , (in the Spanish Dehesas), *Hydrol. Processes*, *22*, 3844–3855.

Van Steeter, M. M., and J. Pitlick (1998), Geomorphology and endangered fish habitats of the upper Colorado River: 1. Historic changes in streamflow, sediment load, and channel morphology, *Water Resour. Res.*, *34*, 287–302.

Van Steijn, H., J. De Ruig, and F. Hoozemans (1988), Morphological and mechanical aspects of debris flows in parts of the French Alps, *Z. Geomorphol.*, *32*, 143–161.

Varnes, D. J. (1958), Landslide types and processes, in *Landslides and Engineering Practice*, edited by E. Eckel, *Highway Res. Board Spec. Rep.*, *29*, 20–47.

Vasquez, D. A., D. H. Smith, and B. F. Edwards (2002), Influence of terrain on scaling laws for river networks, *Water Resour. Res.*, *38*(11), 1260, doi:10.1029/2000WR000152.

Vaux, W. G. (1968), Interchange of stream and intragravel water in a salmon spawning riffle, *Spec. Sci. Rep. 405*, 11 pp, U.S. Fish and Wildl. Serv., Washington, D. C.

Veatch, A. C. (1906), Geology and underground water resources of northern Louisiana and southern Arkansas, *U.S. Geol. Surv. Prof. Pap.*, *46*, 422 pp.

Veblen, T. T., and J. A. Donnegan (2005), Historical range of variability for forest vegetation of the national forests of the Colorado Front Range, *Final report, USDA For. Serv. Agreement 1102-0001-99-033*, 151 pp, Rocky Mtn. Reg., Golden, Colo.

Velbel, M. A. (1992), Geochemical mass balances and weathering rates in forested watersheds of the southern Blue Ridge. III. Cation budgets and the weathering rate of amphibole, *Am. J. Sci.*, *292*, 58–78.

Velbel, M. A. (1993), Weathering and pedogenesis at the watershed scale: Some recent lessons from studies of acid-deposition effects, *Chem. Geol.*, *107*, 337–339.

Velbel, M. A. (1995), Interaction of ecosystem processes and weathering processes, in *Solute Modelling in Catchment Systems*, edited by S. T. Trudgill, pp. 193–209, John Wiley, Chichester, U. K.

Veldkamp, A., and L. A. Tebbens (2001), Registration of abrupt climate changes within fluvial systems: insights from numerical modelling experiments, *Global Planet. Change*, *28*, 129–144.

Venditti, J. G., W. E. Dietrich, P. A. Nelson, M. A. Wydzga, J. Fadde, and L. Sklar (2010), Mobilization of coarse surface layers in gravel-bedded rivers by finer gravel bed load, *Water Resour. Res.*, *46*, W07506, doi:10.1029/2009WR008329.

Verbunt, M., J. Gurtz, K. Jasper, H. Lang, P. Warmerdam, and M. Zappa (2003), The hydrological role of snow and glaciers in alpine river basins and their distributed modeling, *J. Hydrol.*, *282*, 36–55.

Verdú, J. M., R. J. Batalla, and J. A. Martínez-Casasnovas (2005), High-resolution grain-size characterization of gravel bars using imagery analysis and geo-statistics, *Geomorphology*, *72*, 73–93.

Vergara, W., A. M. Deeb, A. M. Valencia, R. S. Bradley, B. Francou, A. Zarzar, A. Grünwaldt, and S. M. Haeussling (2007), Economic impacts of rapid glacier retreat in the Andes, *Eos Trans. AGU*, *88*(25), 261–264.

Verhaar, P. M., P. M. Biron, R. I. Ferguson, and T. B. Hoey (2008), A modified morphodynamic model for investigating the response of rivers to short-term climate change, *Geomorphology*, *101*, 674–682.

Vericat, D., R. J. Batalla, and C. Garcia (2006), Breakup and reestablishment of the armour layer in a large gravel-bed river below dams: The lower Ebro, *Geomorphology*, *76*, 122–136.

Vericat, D., R. J. Batalla, and C. N. Gibbins (2008), Sediment entrainment and depletion from patches of fine material in a gravel-bed river, *Water Resour. Res.*, *44*, W11415, doi:10.1029/2008WR007028.

Vericat, D., M. Church, and R. J. Batalla (2006), Bed load bias: Comparison of measurements obtained using two (76 and 152 mm) Helley-Smith samplers in a gravel bed river, *Water Resour. Res.*, *42*, W01402, doi:10.1029/2005WR004025.

Verneaux, J. (1973), Cours d'eau de Franche-Comté (massif du Jura), Recherches écologiques sur le réseau hydrographique du Doubs – Essai de biotypologie, *Ann. Sci. Univ. Besancon Zool.*, *3*, 79–90.

Verstraeten, G., and J. Poesen (1999), The nature of small-scale flooding, muddy floods and retention pond sedimentation in central Belgium, *Geomorphology*, *29*, 275–292.

Verstraeten, G., J. Poesen, J. de Vente, and X. Koninckx (2003), Sediment yield variability in Spain: A quantitative and semiqualitative analysis using reservoir sedimentation rates, *Geomorphology*, *50*, 327–348.

Veyrat-Charvillon, S., and M. Memier (2006), Stereophotogrammetry of archive data and topographic approaches to debris-flow torrent measurements: Calculation of channel-sediment states and a partial sediment budget for Manival torrent (Isere, France), *Earth Surf. Processes Landforms*, *31*, 201–219.

Vezzoli, G. (2004), Erosion in the western Alps (Dora Baltea Basin); 2, Quantifying sediment yield, *Sediment. Geol.*, *171*, 247–259.

Vianello, A., and D. D'Agostino (2007), Bankfull width and morphological units in an alpine stream of the dolomites (northern Italy), *Geomorphology*, *83*, 266–281.

Villa, F., A. Tamburini, M. Deamicis, S. Sironi, V. Maggi, and G. Rossi (2008), Volume decrease of Rutor Glacier (western Italian Alps) since Little Ice Age; a quantitative approach combining GPR, GPS and cartography, *Geogr. Fis. Dinam. Quat.*, *31*, 63–70.

Vincent, K. R., and J. G. Elliott (1999), Effect of ore milling on the Animas River channels and flood plain near Eureka, Colorado, *Geol. Soc. Am. Abstr. Programs*, *31*, A–253.

Vischer, D. (1989), Impact of 18th and 19th century river training works: Three case studies from Switzerland, in *Historical Change of Large Alluvial Rivers: Western Europe*, edited by G. E. Petts, H. Moller, and A. L. Roux, pp. 19–40, John Wiley, Chichester, U. K.

Viseras, C., M. L. Calvache, J. M. Soria, and J. Fernandez (2003), Differential features of alluvial fans controlled by tectonic or eustatic accommodation space; examples from the Betic Cordillera, Spain, *Geomorphology*, *50*, 181–202.

Vitek, J. D., and D. F. Ritter (1993), Geomorphology in the USA, in *The Evolution of Geomorphology*, edited by H. J. Walker and W. E. Grabau, pp. 469–481, John Wiley, New York.

Viviroli, D., H. H. Dürr, B. Messerli, M. Meybeck, and R. Weingartner (2007), Mountains of the world, water towers for humanity: Typology, mapping, and global significance, *Water Resour. Res.*, *43*, W07447, doi:10.1029/2006WR005653.

Viviroli, D., R. Weingartner, and B. Messerli (2003), Assessing the hydrological significance of the world's mountains, *Mt. Res. Dev.*, *23*, 32–40.

Vivoni, E. R., R. S. Bowman, R. L. Wyckoff, R. T. Jakubowski, and K. E. Richards (2006), Analysis of a monsoon flood event in an ephemeral tributary and its downstream hydrologic effects, *Water Resour. Res.*, *42*, W03404, doi:10.1029/2005WR004036.

Vivoni, E. R., J. C. Rodríguez, and C. J. Watts (2010), On the spatiotemporal variability of soil moisture and evapotranspiration in a mountainous basin within the North American monsoon region, *Water Resour. Res.*, *46*, W02509, doi:10.1029/2009WR008240.

Vogt, J. V., R. Colombo, and F. Bertolo (2003), Deriving drainage networks and catchment boundaries: A new methodology combining digital elevation data and environmental characteristics, *Geomorphology*, *51*, 281–298.

Vogt, K. A., et al. (1996), Litter dynamics along stream, riparian and upslope areas following Hurricane Hugo, Luquillo Experimental Forest, Puerto Rico, *Biotropica*, *28*, 458–470.

Volker, H. X., T. A. Wasklewicz, and M. A. Ellis (2007), A topographic fingerprint to distinguish alluvial fan formative processes, *Geomorphology*, *88*, 34–45.

Vollmer, S., and M. G. Kleinhans (2007), Predicting incipient motion, including the effect of turbulent pressure fluctuations in the bed, *Water Resour. Res.*, *43*, W05410, doi:10.1029/2006WR004919.

Von Blanckenburg (2005), The control mechanisms of erosion and weathering at basin scale from cosmogenic nuclides in river sediment, *Earth Planet. Sci. Lett.*, *237*, 462–479.

Von Humboldt, A. (1852), *Cosmos: A Sketch of a Physical Description of the Universe*, translated by E. C. Otté, Harper and Brothers, New York.

Von Storch, H., E. Zorita, and U. Cubasch (1993), Downscaling of climate changes to regional scales: An application to winter rainfall in the Iberian Peninsula, *J. Clim.*, *6*, 1161–1171.

Vuichard, D., and M. Zimmermann (1987), The 1985 catastrophic drainage of a moraine-dammed lake, Khumbu Himal, Nepal: Cause and consequences, *Mt. Res. Dev.*, *7*, 91–110.

Vuille, M., G. Kaser, and I. Juen (2008), Glacier mass balance variability in the Cordillera Blanca, Peru and its relationship with climate and the large-scale circulation, *Global Planet. Change*, *62*, 14–28.

Wagener, S. M., and J. D. LaPerriere (1985), Effects of placer mining on the invertebrate communities of an interior Alaska stream, *Freshwater Invert. Biol.*, *4*, 208–214.

Wagner, K. (2007), Mental models of flash floods and landslides, *Risk Anal.*, *27*, 671–782.

Wahl, K. L. (1994), Bias in regression estimates of Manning's n, *Proceedings, ASCE Hydraulic Engineering '94 Conference, August 1–5, Buffalo, New York*, edited by G. V. Cotroneo and R. R. Rumer, pp. 727–731, Am. Soc. of Civ. Eng., New York.

Waide, J. B., W. H. Caskey, R. L. Todd, and L. R. Boring (1998), Changes in soil nitrogen pools and transformations following forest clearcutting, in *Forest Hydrology and Ecology at Coweeta*, edited by W. T. Swank and D. A. Crossley, Jr., *Ecol. Stud.*, *66*, 221 232.

Wainwright, J., and J. D. A. Millington (2010), Mind, the gap in landscape-evolution modelling, *Earth Surf. Processes Landforms*, *35*, 842–855.

Walcott, R. C., and M. A. Summerfield (2008), Scale dependence of hypsometric integrals: An analysis of southeast African basins, *Geomorphology*, *96*, 174–186.

Walder, J. S., and J. E. Costa (1996), Outburst floods from glacier-dammed lakes: The effect of mode of lake drainage on flood magnitude, *Earth Surf. Processes Landforms*, *21*, 701–723.

Walder, J. S., and J. E. O'Connor (1997), Methods for predicting peak discharge of floods caused by failure of natural and constructed earthen dams, *Water Resour. Res.*, *33*, 2237–2348.

Walford, G. M., and W. L. Baker (1995), Classification of the riparian vegetation along a 6-km reach of the Animas River, southwestern Colorado, *Great Basin Nat.*, *55*, 287–303.

Walker, L. R., and F. S. Chapin (1986), Physiological controls over seedling growth in primary succession on an Alaskan floodplain, *Ecology*, *67*, 1508–1523.

Walker, L. R., J. C. Zasada, and F. S. Chapin (1986), The role of life history processes in primary succession on an Alaskan floodplain, *Ecology*, *67*, 1243–1253.

Wallace, J. B., and J. R. Webster (1996), The role of macroinvertebrates in stream ecosystem function, *Annu. Rev. Entomol.*, *41*, 115–139.

Wallace, J. B., S. L. Eggerton, J. L. Meyer, and J. R. Webster (1997), Multiple trophic levels of a forest stream linked to terrestrial litter inputs, *Science*, *277*, 102–104.

Wallbrink, P., W. Blake, S. Doerr, R. Shakesby, G. Humphreys, and P. English (2005), Using tracer based sediment budgets to assess redistribution of soil and organic material after severe bush fires, in *Sediment Budgets 2*, edited by A. J. Horowitz and D. E. Walling, *IAHS Publ.*, *292*, 223–230.

Wallbrink, P. J., J. M. Olley, and A. S. Murray (1994), Measuring soil movement using [137]Cs: Implications of reference site variability, in *Variability in Stream Erosion and Sediment Transport*, edited by L. J. Olive, R. J. Loughran, and J. A. Kesby, *IAHS Publ.*, *224*, 95–102.

Wallerstein, N. P. (2003), Dynamic model for constriction scour caused by large woody debris, *Earth Surf. Processes Landforms*, *28*, 49–68.

Walling, D. E. (1983), The sediment delivery problem, *J. Hydrol.*, *65*, 209–237.

Walling, D. E. (2003), Using environmental radionuclides as tracers in sediment budget investigations, in *Erosion and Sediment Transport Measurement in Rivers: Technological and Methodological Advances*, edited by J. Bogen, T. Fergus, and D. E. Walling, *IAHS Publ.*, *283*, 57–78.

Walling, D. E., A. L. Collins, H. M. Sichingabula, and G. J. L. Leeks (2003), Use of reconnaissance measurements to establish catchment sediment budgets: A Zambian example, in *Erosion Prediction in Ungauged Basins: Integrating Methods and Techniques*, edited by D. de Boer et al., *IAHS Publ.*, *279*, 3–12.

Walling, D. E., and D. Fang (2003), Recent trends in the suspended sediment loads of the world's rivers, *Global Planet. Change*, *39*, 111–126.

Walling, D. E., and A. H. A. Kleo (1979), Sediment yield of rivers in areas of low precipitation: A global view, in *The Hydrology of Mountainous Areas*, *IAHS Publ.128*, pp. 479–493.

Walling, D. E., and B. W. Webb (1983), Patterns of sediment yield, in *Background to Palaeohydrology*, edited by K. J. Gregory, John Wiley, Chichester, U. K.

Walling, D. E., and B. W. Webb (1987), Suspended load in gravel-bed rivers; UK experience, in *Sediment Transport in Gravel-Bed Rivers*, edited by C. R. Thorne, J. C. Bathurst, and R. D. Hey, pp. 691–732, John Wiley, Chichester, U. K.

Walling, D. E., and B. W. Webb (1996), Erosion and sediment yield: A global overview, in *Erosion and Sediment Yield: Global and Regional Perspectives*, edited by D. E. Walling and B. W. Webb, *IAHS Publ.*, *236*, 3–19.

Wallis, E., R. MacNally, and P. S. Lake (2008), A Bayesian analysis of physical habitat changes at tributary confluences in cobble-bed upland streams of the Acheron River basin, Australia, *Water Resour. Res.*, *44*, W11421, doi:10.1029/2008WR006831.

Wallis, P. M., H. B. N. Hynes, and S. A. Telang (1981), The importance of groundwater in the transportation of allochthonous dissolved organic matter to the streams draining a small mountain basin, *Hydrobiologia*, *79*, 77–90.

Walsh, J., and D. M. Hicks (2002), Braided channels: Self-similar or self-affine?, *Water Resour. Res.*, *38*(6), 1082, doi:10.1029/2001WR000749.

Walter, R. C., and D. J. Merritts (2008), Natural streams and the legacy of water-powered mills, *Science*, *319*, 299–304.

Wang, G., H. Ma, J. Qian, and J. Chang (2004), Impact of land use changes on soil carbon, nitrogen and phosphorus and water pollution in an arid region of northwest China, *Soil Use Manage.*, *20*, 32–39.

Wang, G., Y. Wang, Y. Li, and H. Cheng (2007), Influences of alpine ecosystem responses to climatic change on soil properties on the Qinghai-Tibet Plateau, China, *Catena*, *70*, 506–514.

Wang, G., A. Suemine, and W. H. Schulz (2010), Shear-rate-dependent strength control on dynamics of rainfall-triggered landslides, Tokushima Prefecture, Japan, *Earth Surf. Processes Landforms*, *35*, 407–416.

Wang, J., M. Ma, and P. Federicis (2001a), Simulating snowmelt runoff in mountainous watersheds of Italy using GIS and remote sensing data, *J. Glaciol. Geocryol.*, *23*, 436–441.

Wang, J., Y. Shen, A. Lu, L. Wang, and Z. Shi (2001b), Impact of climate change on snow meltwater runoff in the mountain regions of Northwest China, *J. Glaciol. Geocryol.*, *23*, 28–33.

Wang, N. L., and M. K. Han (1993), Geomorphology in China (The People's Republic of), in *The Evolution of Geomorphology*, edited by H. J. Walker and W. E. Grabau, pp. 93–105, John Wiley, New York.

Wang, X., X. Zhong, S. Liu, and M. Li (2008), A non-linear technique based on fractal methodology for describing gully-head changes associated with land-use in an arid environment in China, *Catena*, *72*, 106–112.

Wang, Y., and Y.-K. Tung (2006), Stochastic generation of geomorphological instantaneous unit hydrograph-based flow hydrograph, *Int. J. River Basin Manage.*, *4*, 49–56.

Wang, Z.-Y., P. Qi, and C. S. Melching (2009), Fluvial hydraulics of hyperconcentrated floods in Chinese rivers, *Earth Surf. Processes Landforms*, *34*, 981–993.

Wang, Z.-Y., J. Xu, and C. Li (2004), Development of step-pool sequences and its effects in resistance and stream bed stability, *Int. J. Sediment Res.*, *19*, 161–171.

Warburton, J. (1990), Comparison of bed load yield estimates for a glacial meltwater stream, in *Hydrology in Mountainous Regions I. Hydrological Measurements: The Water Cycle*, edited by H. Lang and A. Musy, *IAHS Publ.*, *193*, 315–323.

Warburton, J. (1994), Channel change in relation to meltwater flooding, Bas Glacier d'Arolla, Switzerland, *Geomorphology*, *11*, 141–149.

Warburton, J., and T. Davies (1994), Variability of bedload transport and channel morphology in a braided river hydraulic model, *Earth Surf. Processes Landforms*, *19*, 403–421.

Ward, A. S., M. N. Gooseff, and K. Singha (2010a), Characterizing hyporheic transport processes – interpretation of electrical geophysical data in coupled stream-hyporheic zone systems during solute tracer studies, *Adv. Water Resour.*, in press.

Ward, A. S., M. N. Gooseff, and K. Singha (2010b), Imaging hyporheic zone solute transport using electrical resistivity, *Hydrol. Processes*, *24*, 948–953.

Ward, D. J., J. A. Spotila, G. S. Hancock, and J. M. Galbraith (2005), New constraints on the late Cenozoic incision history of the New River, Virginia, *Geomorphology*, *72*, 54–72.

Ward, J. V. (1989), The four-dimensional nature of lotic ecosystems, *J. North Am. Benthol. Soc.*, *8*, 2–8.

Ward, J. V. (1992), A mountain river, in *The Rivers Handbook*, edited by P. Calow and G. E. Petts, pp. 493–510, Blackwell Sci., Oxford, U. K.

Ward, J. V., P. Burgherr, M. O. Gessner, F. Malard, C. T. Robinson, K. Tockner, U. Uehlinger, and R. Zah (1998), The Val Roseg project: Habitat heterogeneity and connectivity gradients in a glacial flood-plain system, in *Hydrology, Water Resources and Ecology in Headwaters*, edited by K. Kovar et al., *IAHS Publ.*, *248*, 425–432.

Ward, J. V., and B. C. Kondratieff (1992), *An Illustrated Guide to the Mountain Stream Insects of Colorado*, 191 pp., Univ. Press of Colo., Niwot.

Ward, J. V., and J. A. Stanford (1983), The serial discontinuity concept of lotic ecosystems, in *Dynamics of Lotic Ecosystems*, edited by T. D. Fontaine and S. M. Bartell, pp. 29–42, Ann Arbor Sci., Ann Arbor, Mich.

Ward, J. V., and J. A. Stanford (1995), The serial discontinuity concept: Extending the model to floodplain rivers, *Reg. Rivers Res. Manage.*, *10*, 159–168.

Ward, J. V., K. Tockner, U. Uehlinger, and F. Malard (2001), Understanding natural patterns and processes in river corridors as the basis for effective river restoration, *Reg. Rivers Res. Manage.*, *17*, 311–323.

Ward, J. V., K. Tockner, D. B. Arscott, and C. Claret (2002), Riverine landscape diversity, *Freshwater Biol.*, *47*, 517–539.

Warren, D. R., E. S. Bernhardt, R. O. Hall, and G. E. Likens (2007), Forest age, wood and nutrient dynamics in headwater streams of the Hubbard Brook Experimental Forest, NH, *Earth Surf. Processes Landforms*, *32*, 1154–1163.

Warrick, J. A., and L. A. K. Mertes (2009), Sediment yield from tectonically active semiarid Western Transverse Ranges of California, *Geol. Soc. Am. Bull.*, *121*, 1054–1070.

Warrick, J. A., D. M. Rubin, P. Ruggiero, J. N. Harney, A. E. Draut, and D. Buscombe (2009), Cobble cam: Grain-size measurements of sand to boulder from digital photographs and autocorrelation analyses, *Earth Surf. Processes Landforms*, *34*, 1811–1821.

Wasson, R. J. (1978), A debris flow at Reshun, Pakistan, Hindu Kush, *Geogr. Ann.*, *60A*, 151–159.

Wasson, R. J., R. K. Mazari, B. Starr, and G. Clifton (1998), The recent history of erosion and sedimentation on the Southern Tablelands of southeastern Australia: Sediment flux dominated by channel incision, *Geomorphology*, *24*, 291–308.

Waters, J. V., S. J. Jones, and H. A. Armstrong (2010), Climatic controls on late Pleistocene alluvial fans, Cyprus, *Geomorphology*, *115*, 228–251.

Wathen, S. J., and T. B. Hoey (1998), Morphologic controls on the downstream passage of a sediment wave in a gravel-bed stream, *Earth Surf. Processes Landforms*, *23*, 715–730.

Wathen, S. J., T. B. Hoey, and A. Werritty (1997), Quantitative determination of the activity of within-reach sediment storage in a small gravel-bed river using transit time and response time, *Geomorphology*, *20*, 113–134.

Wathen, S. J., R. I. Ferguson, T. B. Hoey, and A. Werritty (1995), Unequal mobility of gravel and sand in weakly bimodal river sediments, *Water Resour. Res.*, *31*, 2087–2096.

Wathne, B. M., A. Henriksen, and S. Norton (1990), Buffering capacity of river substrates during acid episodes, in *Acidic Deposition: Its Nature and Impacts*, edited by G. D. Holmes and F. T. Last, pp. 405, Royal Soc. of Edinburgh, Edinburgh, U. K.

Watterson, N. A., and J. A. Jones (2006), Flood and debris flow interactions with roads promote the invasion of exotic plants along steep mountain streams, western Oregon, *Geomorphology*, *78*, 107–123.

Waylen, M. J. (1979), Chemical weathering in a drainage basin underlain by Old Red Sandstone, *Earth Surf. Processes*, *4*, 167–178.

Waythomas, C. F. (2001), Formation and failure of volcanic debris dams in the Chakachatna River valley associated with eruptions of the Spurr volcanic complex, Alaska, *Geomorphology*, *39*, 111–129.

Waythomas, C. F., and R. D. Jarrett (1994), Flood geomorphology of Arthurs Rock Gulch, Colorado: Paleoflood history, *Geomorphology*, *11*, 15–40.

Waythomas, C. F., J. S. Walder, R. G. McGimsey, and C. A. Neal (1996), A catastrophic flood caused by drainage of a caldera lake at Aniakchak Volcano, Alaska, and its implications for volcanic hazards assessment, *Geol. Soc. Am. Bull.*, *108*, 861–871.

Webb, A. A., and W. D. Erskine (2003), Distribution, recruitment, and geomorphic significance of large woody debris in an alluvial forest stream: Tonghi Creek, southeastern Australia, *Geomorphology*, *51*, 109–126.

Webb, J. A., and C. R. Fielding (1999), Debris flow and sheetflood fans of the Northern Prince Charles Mountains, East Antarctica, in *Varieties of Fluvial Form*, edited by A. J. Miller and A. Gupta, pp. 317–341, John Wiley, Chichester, U. K.

Webb, R. H. (1987), *Debris flows from tributaries of the Colorado River, Grand Canyon National Park, Arizona: Executive summary*, U.S. Geol. Surv. Open File Rep. 87-117, 7 pp.

Webb, R. H. (1996), *Grand Canyon, a century of change: rephotography of the 1889–1890 Stanton Expedition*, Univ. of Ariz. Press, Tucson.

Webb, R. H., and V. R. Baker (1987), Changes in hydrologic conditions related to large floods on the Escalante River, south-central Utah, in *Regional Flood Frequency Analysis*, edited by V. P. Singh, pp. 309–323, D. Reidel, Dordrecht, The Netherlands.

Webb, R. H., and J. L. Betancourt (1990), *Climatic variability and flood frequency of the Santa Cruz River, Pima County, Arizona*, U.S. Geol. Surv. Open File Rep. 90-553, 69 pp.

Webb, R. H., and S. L. Rathburn (1988), Paleoflood hydrology research in the southwestern United States, *Transp. Res. Rec.*, *1201*, 9–21.

Webb, R. H., P. T. Pringle, and G. R. Rink (1987), Debris flows from tributaries of the Colorado River, Grand Canyon National Park, Arizona, *U.S. Geol. Surv. Open File Rep.*, 87-118, 64 pp.

Webb, R. H., P. T. Pringle, S. L. Reneau, and G. R. Rink (1988), Monument Creek debris flow, 1984: Implications for formation of rapids on the Colorado River in Grand Canyon National Park, *Geology*, *16*, 50–54.

Webb, R. H., S. S. Smith, and V. A. McCord (1991), *Historic Channel Change of Kanab Creek, Southern Utah and Northern Arizona*, Grand Canyon Nat. Hist. Assoc. Monogr., vol. 9, Grand Canyon, Ariz.

Webb, R. H., T. S. Melis, P. G. Griffiths, and J. G. Elliott (1999), Reworking of aggraded debris fans, in *The Controlled Flood in Grand Canyon*, edited by R. H. Webb et al., pp. 37–51, AGU, Washington, D. C.

Webb, R. H., J. Belnap, and J. S. Weisheit (2004), *Cataract Canyon: A Human and Environmental History of the Rivers in Canyonlands*, The Univ. of Utah Press, Salt Lake City.

Webb, R. H., P. G. Griffiths, and L. P. Rudd (2008), Holocene debris flows on the Colorado Plateau: The influence of clay mineralogy and chemistry, *Geol. Soc. Am. Bull.*, *120*, 1010–1020.

Weber, W. A. (1965), Plant geography in the southern Rocky Mountains, in *The Quaternary in the United States*, edited by H. E. Wright and D. G. Frey, pp. 453–468, Princeton Univ. Press, Princeton, N. J.

Webster, J. R., and B. C. Patten (1979), Effects of watershed perturbation on stream potassium and calcium dynamics, *Ecol. Monogr.*, *49*, 51–72.

Webster, J. R., and W. T. Swank (1985), Within-stream factors affecting nutrient transport from forested and logged watersheds, in *Proceedings of Forestry and Water Quality: A Mid-South Symposium*, edited by B. G. Blackmon, pp. 18–40, Univ. of Ark., Little Rock.

Webster, J. R., et al. (2003), Factors affecting ammonium uptake in streams – an inter-biome perspective, *Freshwater Biol.*, *48*, 1329–1352.

Wegmann, K. W., and F. J. Pazzaglia (2002), Holocene strath terraces, climate change, and active tectonics; the Clearwater River basin, Olympic Peninsula, Washington State, *Geol. Soc. Am. Bull.*, *114*, 731–744.

Weigert, A., and J. Schmidt (2005), Water transport under winter conditions, *Catena*, *64*, 193–208.

Weichert, R. B., G. R. Bezzola, and H.-E. Minor (2008), Bed morphology and generation of step-pool channels, *Earth Surf. Processes Landforms*, *33*, 1678–1692.

Weiler, M., and J. McDonnell (2004), Virtual experiments; a new approach for improving process conceptualization in hillslope hydrology, *J. Hydrol.*, *285*, 3–18.

Weingartner, R., M. Barben, and M. Spreafico (2003), Floods in mountain areas: An overview based on examples from Switzerland, *J. Hydrol.*, *282*, 10–24.

Weirich, F. H. (1987), Sediment transport and deposition by fire-related debris flows in southern California, in *Erosion and Sedimentation in the Pacific Rim*, edited by R. L. Beschta et al., *IAHS Publ.*, *165*, 283–283.

Weiss, F. H. (1994), Luminophor experiments in the Saalach and Salzach rivers, in *Dynamics and Geomorphology of Mountain Rivers*, edited by P. Ergenzinger and K.-H. Schmidt, pp. 83–91, Springer, Berlin.

Weissel, J. K., and M. A. Seidl (1998), Inland propagation of erosional escarpments and river profile evolution across the southeast Australian passive continental margin, in *Rivers Over Rock: Fluvial Processes in Bedrock Channels*, Geophys. Monogr. Ser., vol. 107, edited by K. J. Tinkler and E. E. Wohl, pp. 189–206, AGU, Washington, D. C.

Weissmann, G. S., J. F. Mount, and G. E. Fogg (2002), Glacially driven cycles in accumulation space and sequence stratigraphy of a stream-dominated alluvial fan, San Joaquin Valley, California, USA, *J. Sediment. Res.*, *72*, 240–251.

Weissmann, G. S., G. L. Bennett, and A. L. Lansdale (2005), Factors controlling sequqence development on Quaternary fluvial fans, San Joaquin basin, California, USA, in *Alluvial Fans: Geomorphology, Sedimentology, Dynamics*, edited by A. M. Harvey, A. E. Mather, and M. Stokes, *Geol. Soc. Spec. Publ.*, *251*, 169–186.

Welcomme, R. L., R. A. Ryder, and J. R. Sedell (1989), Dynamics of fish assemblages in river systems - a synthesis, *Can. Spec. Publ. Fish. Aquat. Sci.*, *106*, 569–577.

Wellnitz, T. A., N. L. Poff, G. Cosyleon, and B. Steury (2001), Current velocity and spatial scale as determinants of the distribution and abundance of two rheophilic herbivorous insects, *Landscape Ecol.*, *16*, 111–120.

Wells, W. G., II (1987), The effects of fire generation on the generation of debris flows in southern California, in *Debris Flows/Avalanches: Process, Recognition, and Mitigation*, edited by J. E. Costa and G. F. Wieczorek, *Rev. Eng. Geol.*, *7*, 105–114.

Wells, W. G., P. M. Wohlgemuth, A. G. Campbell, and F. H. Weirich (1987), Postfire sediment movement by debris flows in the Santa Ynez Mountains, California, in *Erosion and Sedimentation in the Pacific Rim*, edited by R. L. Beschta et al., *IAHS Publ.*, *165*, 275–276.

Welsch, D. L., C. N. Kroll, J. J. McDonnell, and D. A. Burns (2001), Topographic controls on the chemistry of subsurface stormflow, *Hydrol. Processes*, *15*, 1925–1938.

Welty, J. J., T. Beechie, K. Sullivan, D. M. Hyink, R. E. Bilby, C. Andrus, and G. Pess (2002), Riparian aquatic interaction simulator (RAIS): A model of riparian forest dynamics for the generation of large woody debris and shade, *For. Ecol. Manage.*, *162*, 299–318.

Wemple, B. C., J. A. Jones, and G. E. Grant (1996), Channel network extension by logging roads in two basins, western Cascades, Oregon, *Water Resour. Bull.*, *32*, 1195–1207.

Wemple, B. C., F. J. Swanson, and J. A. Jones (2001), Forest roads and geomorphic process interactions, Cascade Range, Oregon, *Earth Surf. Processes Landforms*, *26*, 191–204.

Wende, R. (1999), Boulder bedforms in jointed-bedrock channels, in *Varieties of Fluvial Form*, edited by A. J. Miller and A. Gupta, pp. 189–216, John Wiley, Chichester, U. K.

Werritty, A. (1992), Downstream fining in a gravel-bed river in southern Poland: Lithologic controls and the role of abrasion, in *Dynamics of Gravel-Bed Rivers*, edited by P. Billi et al., pp. 333–350, John Wiley, Chichester, U. K.

West, A. J., A. Galy, and M. Bickle (2005), Tectonic and climatic controls on silicate weathering, *Earth Planet. Sci. Lett.*, *235*, 211–228.

Westaway, R., M. Pringle, S. Yurtmen, T. Demir, D. Bridgland, G. Rowbotham, and D. Maddy (2004), Pliocene and Quaternary regional uplift in western Turkey: The Gediz River terrace staircase and the volcanism at Kula, *Tectonophysics*, *391*, 121–169.

Westbrook, C. J., D. J. Cooper, and B. W. Baker (2006), Beaver dams and overbank floods influence groundwater-surface water interactions of a Rocky Mountain riparian area, *Water Resour. Res.*, *42*, W06404, doi:10.1029/2005WR004560.

Westbrook, C. J., D. J. Cooper, and B. W. Baker (2010), Beaver assisted river valley formation, *River Res. Appl.*, doi: 10.1002/rra.1359.

Westerling, A. L., H. G. Hidalgo, D. R. Cayan, and T. W. Swetnam (2006), Warming and earlier spring increase western U. S. forest wildfire activity, *Science*, *313*(5789), 940–943.

Wetzel, K. (1994), The significance of fluvial erosion, channel storage and gravitational processes in sediment production in a small mountainous catchment area, in *Dynamics and Geomorphology of Mountain Rivers*, edited by P. Ergenzinger and K.-H. Schmidt, pp. 141–160, Springer, Berlin.

Wheaton, J. M., G. B. Pasternack, and J. E. Merz (2004a), Spawning habitat rehabilitation – I. Conceptual approach and methods, *Int. J. River Basin Manage.*, *2*, 3–20.

Wheaton, J. M., G. B. Pasternack, and J. E. Merz (2004b), Spawning habitat rehabilitation – II. Using hypothesis development and testing in design, Mokelumne River, California, USA, *Int. J. River Basin Manage.*, *2*, 3–20.

Wheaton, J. M., J. Brasington, S. E. Darby, and D. A. Sear (2010), Accounting for uncertainty in DEMs from repeat topographic surveys: Improved sediment budgets, *Earth Surf. Processes Landforms*, *35*, 136–156.

Whetton, P. H., M. R. Haylock, and R. Galloway (1996), Climate change and snow cover duration in the Australian Alps, *Clim. Change*, *32*, 447–449.

Whipple, K. X. (1992), Predicting debris-flow runout and deposition on fans: The importance of the flow hydrograph, in *Erosion, Debris Flows and Environment in Mountain Regions*, edited by D. E. Walling, T. R. Davies, and B. Hasholt, *IAHS Publ.*, *209*, 337–345.

Whipple, K. X. (1993), Interpreting debris-flow hazard from study of fan morphology, in *Hydraulic Engineering '93*, vol. 2, edited by H. W. Shen, S. T. Su, and F. Wen, pp. 1302–1307, Am. Soc. Civ. Eng., New York.

Whipple, K. X. (2004), Bedrock rivers and the geomorphology of active orogens, *Annu. Rev. Earth Planet. Sci.*, *32*, 151–185.

Whipple, K. X., and T. Dunne (1992), The influence of debris-flow rheology on fan morphology, Owens Valley, California, *Geol. Soc. Am. Bull.*, *104*, 887–900.

Whipple, K. X., and C. R. Trayler (1996), Tectonic control of fan size: The importance of spatially variable subsidence rates, *Basin Res.*, *8*, 351–366.

Whipple, K. X., and G. E. Tucker (1999), Dynamics of stream-power river incision model: Implications for height limits of mountain ranges, landscape response timescales, and research needs, *J. Geophys. Res.*, *104*(B8), 17,661–17,774.

Whipple, K. X., E. Kirby, and S. H. Brocklehurst (1999), Geomorphic limits to climate-induced increases in topographic relief, *Nature*, *401*, 39–43.

Whipple, K. X., N. P. Snyder, and K. Dollenmayer (2000), Rates and processes of bedrock incision by the Upper Ukak River since the 1912 Novarupta ash flow in the Valley of Ten Thousand Smokes, Alaska, *Geology*, *28*, 835–838.

Whitaker, A., Y. Alila, J. Beckers, and D. Toews (2002), Evaluating peak flow sensitivity to clear-cutting in different elevation bands of a snowmelt-dominated mountainous catchment, *Water Resour. Res.*, *38*(9), 1172, doi:10.1029/2001WR000514.

White, A. F., A. E. Blum, M. S. Schulz, D. V. Vivit, D. A. Stonestrom, M. Larsen, S. F. Murphy, and D. Eberl (1998), Chemical weathering in a tropical watershed, Luquillo Mountains, Puerto Rico: I. Long-term versus short-term weathering fluxes, *Geochim. Cosmochim. Acta*, *62*, 209–226.

White, J. Q., G. B. Pasternack, and H. J. Moir (2010), Valley width variation influences riffle-pool location and persistence on a rapidly incising gravel-bed river, *Geomorphology*, *121*, 206–221.

White, P. S. (1979), Pattern, process, and natural disturbance in vegetation, *Bot. Rev.*, *45*, 229–299.

White, S. (1995), Soil erosion and sediment yield in the Philippines, in *Sediment and Water Quality in River Catchments*, edited by I. Foster, A. Gurnell, and B. Webb, pp. 391–406, John Wiley, Chichester, U. K.

White, W. B. (1988), *Geomorphology and Hydrology of Karst Terrains*, 464 pp., Oxford Univ. Press, New York.

White, W. D., and S. G. Wells (1979), Forest-fire devegetation and drainage basin adjustments in mountainous terrain, in *Adjustments of the Fluvial System*, edited by D. D. Rhodes and G. P. Williams, pp. 199–223, George Allen and Unwin, London.

White, W. R., R. Bettess, and E. Paris (1982), Analytical approach to river regime, *J. Hydraul. Div. Am. Soc. Civ. Eng.*, *108*, 1179–1193.

Whiteman, C. D. (2000), *Mountain Meteorology: Fundamentals and Applications*, Oxford Univ. Press, New York.

Whiting, P. J. (1997), The effect of stage on flow and components of the local force balance, *Earth Surf. Processes Landforms*, *22*, 517–530.

Whiting, P. J. (2003), Flow measurement and characterization, in *Tools in Fluvial Geomorphology*, edited by G. M. Kondolf and H. Piégay, pp. 323–346, John Wiley, Chichester, U. K.

Whiting, P. J., and J. B. Bradley (1993), A process-based classification system for headwater streams, *Earth Surf. Processes Landforms*, *18*, 603–612.

Whiting, P. J., and W. E. Dietrich (1985), The role of bedload sheets in the transport of heterogeneous sediment (abstract), *Eos Trans. AGU*, *66*, 910.

Whiting, P. J., and W. E. Dietrich (1991), Convective accelerations and boundary shear stress over a channel bar, *Water Resour. Res.*, *27*, 783–796.

Whiting, P. J., and J. G. King (2003), Surface particle sizes on armoured gravel streambeds: Effects of supply and hydraulics, *Earth Surf. Processes Landforms*, *28*, 1459–1471.

Whiting, P. J., W. E. Dietrich, L. B. Leopold, T. G. Drake, and R. L. Shreve (1988), Bedload sheets in heterogeneous sediment, *Geology*, *16*, 105–108.

Whiting, P. J., J. F. Stamm, D. B. Moog, and R. L. Orndorff (1999), Sediment-transporting flows in headwater streams, *Geol. Soc. Am. Bull.*, *111*, 450–466.

Whiting, P. J., G. Matisoff, W. Fornes, and F. M. Soster (2005), Suspended sediment sources and transport distances in the Yellowstone River basin, *Geol. Soc. Am. Bull.*, *117*, 515–529.

Whitley, J. R., and R. S. Campbell (1974), Some aspects of water quality and biology of the Missouri River, *Trans. Miss. Acad. Sci.*, *8*, 60–72.

Whittaker, A. C., P. A. Cowie, M. Attal, G. E. Tucker, and G. P. Roberts (2007a), Bedrock channel adjustment to tectonic forcing: Implications for predicting river incision rates, *Geology*, *35*, 103–106.

Whittaker, A. C., P. A. Cowie, M. Attal, G. E. Tucker, and G. P. Roberts (2007b), Contrasting transient and steady-state rivers crossing active normal faults: New field observations from the Central Apennines, Italy, *Basin Res.*, *19*, 529–556.

Whittaker, A. C., M. Attal, P. A. Cowie, G. E. Tucker, and G. Roberts (2008), Decoding temporal and spatial patterns of fault uplift using transient river long profiles, *Geomorphology*, *100*, 506–526.

Whittaker, J. G. (1987a), Modelling bed-load transport in steep mountain streams, in *Erosion and Sedimentation in the Pacific Rim*, edited by R. L. Beschta et al., *IAHS Publ.165*, pp. 319–332, Proceed of the Corvallis Symposium, Aug 1987.

Whittaker, J. G. (1987b), Sediment transport in step-pool systems, in *Sediment Transport in Gravel-Bed Rivers*, edited by C. R. Thorne, J. C. Bathurst, and R. D. Hey, pp. 545–579, John Wiley, Chichester, U. K.

Whittaker, J. G., and M. N. R. Jaeggi (1982), Origins of step-pool systems in mountain streams, *J. Hydraul. Div. Am. Soc. Civ. Eng.*, *108*, 758–773.

Wiberg, P. L., and J. D. Smith (1987), Calculations of the critical shear stress for motion of uniform and heterogeneous sediments, *Water Resour. Res.*, *23*, 1471–1480.

Wiberg, P. L., and J. D. Smith (1991), Velocity distribution and bed roughness in high-gradient streams, *Water Resour. Res.*, *27*, 825–838.

Wichmann, V., T. Heckmann, F. Haas, and M. Becht (2009), A new modelling approach to delineate the spatial extent of alpine sediment cascades, *Geomorphology*, *111*, 70–78.

Wickham, M. G. (1967), Physical microhabitat of trout, unpublished M.S. thesis, 42 pp., Colo. State Univ., Ft. Collins.

Wicks, J. M., and J. C. Bathurst (1996), SHESED: A physically based, distributed erosion and sediment yield component for the SHE hydrological modeling system, *J. Hydrol.*, *175*, 213–238.

Wieczorek, G. F. (1987), Effect of rainfall intensity and duration on debris flows in central Santa Cruz Mountains, California, in *Debris Flows/Avalanches: Process, Recognition, and Mitigation*, edited by J. E. Costa and G. F. Wieczorek, *Rev. Eng. Geol.*, *7*, 93–104.

Wieczorek, G. F. (1993), Assessment and prediction of debris-flow hazards, in *Hydraulic engineering '93*, edited by H. W. Shen, S. T. Su, and F. Wen, pp. 1272–1283, Am. Soc. of Civ. Eng., New York.

Wieczorek, G. F., and T. Glade (2005), Climatic factors influencing the occurrence of debris flows, in *Debris-Flow Hazards and Related Phenomena*, edited by M. Jakob and O. Hungr, pp. 325–362, Springer, Berlin.

Wieczorek, G. F., and P. P. Leahy (2008), Landslide hazard mitigation in North America, *Environ. Eng. Geosci.*, *14*, 133–144.

Wieczorek, G. F., E. W. Lips, and S. D. Ellen (1989), Debris flows and hyperconcentrated floods along the Wasatch Front, Utah, 1983 and 1984, *Bull. Assoc. Eng. Geol.*, *26*, 191–208.

Wieczorek, G. F., G. Mandrone, and L. Decola (1997), The influence of hillslope shape on debris-flow initiation, *Debris-Flow Hazards Mitigation: Mechanics, Prediction, and Assessment: Proceedings of the First International Conference*, pp. 21–31, Am. Soc. of Civ. Eng., Reston, Va.

Wiedmer, M., D. R. Montgomery, A. R. Gillespie, and H. Greenberg (2010), Late Quaternary megafloods from Glacial Lake Altna, southcentral Alaska, USA, *Quat. Res.*, *73*, 413–424.

Wiener, J. G., D. P. Krabbenhoft, G. H. Heinz, and A. M. Scheuhammer , (2002), Ecotoxicology of mercury, in *Handbook of Ecotoxicology*, 2nd ed., edited by D. J. Hoffman et al., pp. 407–461, CRC Press, Boca Raton, Fla.

Wiens, J. A. (2002), Riverine landscapes: Taking landscape ecology into the water, *Freshwater Biol.*, *47*, 501–515.

Wiesmann, U. (1992), Socioeconomic viewpoints on highland-lowland systems: A case study on the northwest side of Mount Kenya, *Mt. Res. Dev.*, *12*, 375–381.

Wilcock, P. R. (1992a), Experimental investigation of the effect of mixture properties on transport dynamics, in *Dynamics of Gravel-Bed Rivers*, edited by P. Billi et al., pp. 109–139, John Wiley, Chichester, U. K.

Wilcock, P. R. (1992b), Flow competence: A criticism of a classic concept, *Earth Surf. Processes Landforms*, *17*, 289–298.

Wilcock, P. R. (1993), Critical shear stress of natural sediments, *J. Hydraul. Eng.*, *119*, 491–505.

Wilcock, P. R. (1996), Estimating local bed shear stress from velocity observations, *Water Resour. Res.*, *32*, 3361–3366.

Wilcock, P. R. (1997), Friction between science and practice: The case of river restoration, *Eos Trans. AGU*, *78*(41), 454.

Wilcock, P. R. (2001), The flow, the bed, and the transport; interaction in flume and field, in *Gravel-Bed Rivers V*, edited by M. P. Mosley, pp. 183–219, N. Z. Hydrol. Soc., Wellington, New Zealand.

Wilcock, P. R., and J. C. Crowe (2003), Surface-based transport model for mixed-size sediment, *J. Hydraul. Eng.*, *129*, 120–128.

Wilcock, P. R., and B. T. DeTemple (2005), Persistence of armor layers in gravel-bed streams, *Geophys. Res. Lett.*, *32*, L08402, doi:10.1029/2004GL021772.

Wilcock, P. R., and R. M. Iverson (Eds.) (2003), *Prediction in Geomorphology*, AGU, Washington, D. C.

Wilcock, P. R., and S. T. Kenworthy (2002), A two-fraction model for the transport of sand/gravel mixtures, *Water Resour. Res.*, *38*(10), 1194, doi:10.1029/2001WR000684.

Wilcock, P. R., A. F. Barta, and C. C. C. Shea (1994), Estimating local bed shear stress in large gravel-bed rivers, *Proceedings, ASCE Hydraulic Engineering '94 Conference, August 1–5, Buffalo, New York*, edited by G. V. Cotroneo and R. R. Rumer, pp. 834–838, Am. Soc. of Civ. Eng., New York.

Wilcock, P. R., A. F. Barta, C. C. Shea, G. M. Kondolf, W. V. G. Matthews, and J. Pitlick (1996a), Observations of flow and sediment entrainment on a large gravel-bed river, *Water Resour. Res.*, *32*, 2897–2909.

Wilcock, P. R., G. M. Kondolf, W. V. G. Matthews, and A. F. Barta (1996b), Specification of sediment maintenance flows for a large gravel-bed river, *Water Resour. Res.*, *32*, 2911–2921.

Wilcock, P. R., S. T. Kenworthy, and J. C. Crowe (2001), Experimental study of the transport of mixed sand and gravel, *Water Resour. Res.*, *37*, 3349–3358.

Wilcock, P. R., et al. (2003), When models meet managers: Examples from geomorphology, in *Prediction in Geomorphology*, edited by P. R. Wilcock and R. M. Iverson, pp. 27–40, AGU, Washington, D. C.

Wilcox, A. C., and E. E. Wohl (2006), Flow resistance dynamics in step-pool stream channels: 1. Large woody debris and controls on total resistance, *Water Resour. Res.*, *42*, W05418, doi:10.1029/2005WR004277.

Wilcox, A. C., and E. E. Wohl (2007), Field measurements of three-dimensional hydraulics in a step-pool channel, *Geomorphology*, *83*, 215–231.

Wilcox, A. C., J. M. Nelson, and E. E. Wohl (2006), Flow resistance dynamics in step-pool channels: 2. Partitioning between grain, spill, and woody debris resistance, *Water Resour. Res.*, *42*, W05419, doi:10.1029/2005WR004278.

Wiley, M. J., L. L. Osborne, and R. W. Larimore (1990), Longitudinal structure of an agricultural prairie river system and its relationship to current stream ecosystem theory, *Can. J. Fish. Aquat. Sci.*, *47*, 373–384.

Wilford, D. J., M. E. Sakals, J. L. Innes, and R. C. Sidle (2005), Fans with forests: Contemporary hydrogeomorphic processes on fans with forest in west central British Columbia, Canada, in *Alluvial Fans: Geomorphology, Sedimentology, Dynamics*, edited by A. M. Harvey, A. E. Mather, and M. Stokes, *Geol. Soc. Spec. Publ.*, *251*, 25–40.

Wilhelm, F. (1994), Human impact and exploitation of water resources in the Northern Alps (Tyrol and Bavaria), in *Dynamics and Geomorphology of Mountain Rivers*, edited by P. Ergenzinger and K.-H. Schmidt, pp. 15–35, Springer, Berlin.

Wilkerson, F. D., and G. L. Schmid (2003), Debris flows in Glacier National Park, Montana: Geomorphology and hazards, *Geomorphology*, *55*, 317–328.

Wilkinson, S. N., R. J. Keller, and I. D. Rutherfurd (2004), Phase-shifts in shear stress as an explanation for the maintenance of pool-riffle sequences, *Earth Surf. Processes Landforms*, *29*, 737–753.

Wilkinson, S. N., J. M. Olley, A. M. Read, and R. D. Derose (2005), Targeting erosion control using spatially distributed sediment budgets, in *Sediment Budgets 2*, edited by A. J. Horowitz and D. E. Walling, *IAHS Publ.*, *292*, 65–72.

Wilkinson, S. N., I. P. Prosser, and A. O. Hughes (2006), Predicting the distribution of bed material accumulation using river network sediment budgets, *Water Resour. Res.*, *42*, W10419, doi:10.1029/2006WR004958.

Wilkinson, S. N., I. D. Rutherfurd, and R. J. Keller (2008), An experimental test of whether bar instability contributes to the formation, periodicity and maintenance of pool-riffle sequences, *Earth Surf. Processes Landforms*, *33*, 1742–1756.

Willett, S. D. (1999), Orogeny and orography: The effects of erosion on the structure of mountain belts, *J. Geophys. Res.*, *104*(B12), 28,957–28,981.

Willett, S. D., and M. T. Brandon (2002), On steady states in mountain belts, *Geology*, *30*, 175–178.

Willett, S. D., C. Beaumont, and P. Fullsack (1993), Mechanical model for the tectonics of doubly vergent compressional orogens, *Geology*, *21*, 371–374.

Willett, S. D., R. Slingerland, and N. Hovius (2001), Uplift, shortening, and steady state topography in active mountain belts, *Am. J. Sci.*, *301*, 445–485.

Willett, S. D., N. Hovius, M. T. Brandon, and D. M. Fisher (2006), *Tectonics, Climate, and Landscape Evolution*, Spec. Pap. Geol. Soc. Am., *398*.

Willgoose, G., and G. Hancock (1998), Revising the hypsometric curve as an indicator of form and process in transport-limited catchments, *Earth Surf. Processes Landforms*, *23*, 611–623.

Willgoose, G., R. L. Bras, and I. Rodriguez-Iturbe (1991), Results from a new model of river basin evolution, *Earth Surf. Processes Landforms*, *16*, 237–254.

Willi, H. P. (1991), Review of mountain river training procedures in Switzerland, in *Fluvial Hydraulics of Mountain Regions*, edited by A. Armanini and G. Di Silvio, pp. 317–329, Springer, Braunschweig, Germany.

Williams, C. A., and D. J. Cooper (2005), Mechanisms of riparian cottonwood decline along regulated rivers, *Ecosystems*, *8*, 382–395.

Williams, G. P. (1978a), Bank-full discharge of rivers, *Water Resour. Res.*, *14*, 1141–1154.

Williams, G. P. (1978b), *The case of the shrinking channels – the North Platte and Platte Rivers in Nebraska*, U.S. Geol. Surv. Circ., *781*, 48 pp.

Williams, G. P. (1983), Paleohydrological methods and some examples from Swedish fluvial environments. I. Cobble and boulder deposits, *Geogr. Ann.*, *65A*, 227–243.

Williams, G. P., and H. P. Guy (1971), Debris avalanches - a geomorphic hazard, in *Environmental Geomorphology*, edited by D. R. Coates, pp. 25–46, State Univ. of New York, Binghamton.

Williams, G. P., and H. P. Guy (1973), Erosional and depositional aspects of Hurricane Camille in Virginia, 1969, U.S. Geol. Surv. Prof. Pap., *804*, 80 pp.

Williams, G. P., and D. L. Rosgen (1989), Measured total sediment loads (suspended and bedloads) for 93 United States streams, U.S. Geol. Surv. Open File Rep., *89-67*, 128 pp.

Williams, G. P., and M. G. Wolman (1984), Effects of dams and reservoirs on surface-water hydrology; changes in rivers downstream from dams, U.S. Geol. Surv. Prof. Pap., *1286*, 83 pp.

Williams, G. P., and M. G. Wolman (1985), Effects of dams and reservoirs on surface-water hydrology - changes in rivers downstream from dams, U.S. Geol. Surv. Natl. Water Summary 1985, pp. 83–88.

Williams, M. W., and T. Platts-Mills (1998), Selectivity of chemical weathering in high elevation catchments of the Colorado Front Range, *Eos Trans. AGU*, *79*, F337.

Willis, I. C., N. S. Arnold, and B. W. Brock (2002), Effect of snowpack removal on energy balance, melt and runoff in a small supraglacial catchment, *Hydrol. Processes*, *16*, 2721–2749.

Willis, K. J., and H. J. B. Birks (2006), What is natural? The need for a long-term perspective in biodiversity conservation, *Science*, *314*, 1261–1265.

Wilson, R. C., J. D. Torikai, and S. D. Ellen (1992), Development of rainfall warning thresholds for debris flows in the Honolulu district, Oahu, U.S. Geol. Surv. Open File Rep. 92-521, 35 pp.

Winiger, M., M. Gumpert, and H. Yamout (2005), Karakorum-Hindukush-western Himalaya: Assessing high-altitude water resources, *Hydrol. Processes*, *19*, 2329–2338.

Winter, T. C. (2007), The role of ground water in generating streamflow in headwater areas and in maintaining base flow, *J. Am. Water Resour. Assoc.*, *43*, 15–25.

Winterbourn, M. J., and C. R. Townsend (1991), Streams and rivers: One-way flow systems, in *Fundamentals of Aquatic Ecology*, 2nd ed., edited by R. S. K. Barnes and K. H. Mann, pp. 230–242, Blackwell Sci., Oxford, U. K.

Wipf, S., C. Rixen, M. Fischer, B. Schmid, and V. Stoeckli (2005), Effects of ski piste preparation on alpine vegetation, *J. Appl. Ecol.*, *42*, 306–316.

Wischmeier, W. H., and D. D. Smith (1978), *Predicting rainfall-erosion losses - a guide to conservation planning*, *Agric. Handb.*, vol. 537.

Wishart, D., J. Warburton, and L. Bracken (2008), Gravel extraction and planform change in a wandering gravel-bed river: The River Wear, northern England, *Geomorphology*, *94*, 131–152.

Wisniewski, P. A., and F. J. Pazzaglia (2002), Epeirogenic controls on Canadian River incision and landscape evolution, Great Plains of northeastern New Mexico, *J. Geol.*, *110*, 437–456.

Wissmar, R. C., and F. J. Swanson (1990), Landscape disturbances and lotic ecotones, in *Ecology and Management of Aquatic-Terrestrial Ecotones*, edited by R. J. Naiman and H. DéCamps.

Witt, A., B. D. Malamud, M. Rossi, F. Guzzetti, and S. Peruccacci (2010), Temporal correlations and clustering of landslides, *Earth Surf. Processes Landforms*, *35*, 1138–1156.

Wittenberg, L., and M. D. Newson (2005), Particle clusters in gravel-bed rivers: An experimental morphological approach to bed material transport and stability concepts, *Earth Surf. Processes Landforms*, *30*, 1351–1368.

Wobus, C. W., G. E. Tucker, and R. S. Anderson (2006a), Self-formed bedrock channels, *Geophys. Res. Lett.*, *33*, L18408, doi:10.1029/2006GL027182.

Wobus, C. W., B. T. Crosby, and K. X. Whipple (2006b), Hanging valleys in fluvial systems: Controls on occurrence and implications for landscape evolution, *J. Geophys. Res.*, *111*, F02017, doi:10.1029/2005JF000406.

Wobus, C. W., K. X. Whipple, and K. V. Hodges (2006c), Neotectonics of the central Nepalese Himalaya: Constraints from geomorphology, detrital $^{40}Ar/^{39}Ar$ thermochronology, and themal modeling, *Tectonics*, *25*, TC4011, doi:10.1029/2005TC001935.

Wobus, C., K. X. Whipple, E. Kirby, N. Snyder, J. Johnson, K. Spyropolou, B. Crosby, and D. Sheehan (2006d), Tectonics from topography: Procedures, promise, and pitfalls, in *Tectonics, Climate, and Landscape Evolution*, edited by S. D. Willett et al., *Spec. Pap. Geol. Soc. Am.*, *398*, 55–74.

Wohl, E. E. (1992a), Bedrock benches and boulder bars: Floods in the Burdekin Gorge of Australia, *Geol. Soc. Am. Bull.*, *104*, 770–778.

Wohl, E. E. (1992b), Gradient irregularity in the Herbert Gorge of northeastern Australia, *Earth Surf. Processes Landforms*, *17*, 69–84.

Wohl, E. E. (1993), Bedrock channel incision along Piccaninny Creek, Australia, *J. Geol.*, *101*, 749–761.

Wohl, E. E. (1995), Estimating flood magnitude in ungauged mountain channels, Nepal, *Mt. Res. Dev.*, *15*, 69–76.

Wohl, E. E. (1998), Bedrock channel morphology in relation to erosional processes, in *Rivers Over Rock: Fluvial Processes in Bedrock Channels*, *Geophys. Monogr. Ser.*, vol. 107, edited by K. J. Tinkler and E. E. Wohl, pp. 133–151, AGU, Washington, D. C.

Wohl, E. E. (1999), Incised bedrock channels, in *Incised River Channels: Processes, Forms, Engineering and Management*, edited by S. Darby and A. Simon, pp. 187–218, John Wiley, Chichester, U. K.

Wohl, E. E. (2000a), Geomorphic effects of floods, in *Inland Flood Hazards: Human, Riparian, and Aquatic Communities*, edited by E. E. Wohl, pp. 167–193, Cambridge Univ. Press, Cambridge, U. K.

Wohl, E. E. (2000b), Substrate influences on step-pool sequences in the Christopher Creek drainage, Arizona, *J. Geol.*, *108*, 121–129.

Wohl, E. E. (2001), *Virtual rivers: Lessons from the mountain rivers of the Colorado Front Range*, Yale Univ. Press, New Haven, Conn.

Wohl, E. E. (2004), Limits of downstream hydraulic geometry, *Geology*, *32*, 897–900.

Wohl, E. E. (Ed.) (2005), *Downstream hydraulic geometry along a tropical mountain river The Rio Chagres, Panama: A Multidisciplinary Profile of a Tropical Watershed*, edited by R. S. Harmon, pp. 169–188, Springer, Dordrecht, The Netherlands.

Wohl, E. E. (2006), Human impacts to mountain streams, *Geomorphology*, *79*, 217–248.

Wohl, E. E. (2007), Channel-unit hydraulics on a pool-riffle channel, *Phys. Geogr.*, *28*, 233–248.

Wohl, E. E. (2008a), The effect of bedrock jointing on the formation of straths in the Cache la Poudre River drainage, Colorado Front Range, *J. Geophys. Res.*, *113*, F01007, doi:10.1029/2007JF000817.

Wohl, E. E. (2008b), Review of effects of large floods in resistant-boundary channels, in *Gravel-Bed Rivers VI: From Process Understanding to River Restoration*, edited by H. Habersack, H. Piégay, and M. Rinaldi, pp. 181–212, Elsevier, Amsterdam.

Wohl, E. E., and D. Cadol (2010), Patterns and controls on wood distribution in old-growth forest streams of the Colorado Front Range, USA, *Geomorphology*, in press.

Wohl, E. E., and D. A. Cenderelli (1998), Flooding in the Himalaya Mountains, in *Flood Studies in India*, edited by V. S. Kale, *Mem. Geol. Soc. India*, *41*, 77–99.

Wohl, E. E., and D. A. Cenderelli (2000), Sediment deposition and transport patterns following a reservoir sediment release, *Water Resour. Res.*, *36*, 319–333.

Wohl, E. E., and G. C. L. David (2008), Consistency of scaling relations among bedrock and alluvial channels, *J. Geophys. Res.*, *113*, F04013, doi:10.1029/2008JF000989.

Wohl, E. E., and Y. Enzel (1995), Data for palaeohydrology, in *Global Continental Palaeohydrology*, edited by K. J. Gregory, L. Starkel, and V. R. Baker, pp. 23–59, John Wiley, Chichester, U. K.

Wohl, E. E., and J. R. Goode (2008), Wood dynamics in headwater streams of the Colorado Rocky Mountains, *Water Resour. Res.*, *44*, W09429, doi:10.1029/2007WR006522.

Wohl, E. E., and T. Grodek (1994), Channel bed-steps along Nahal Yael, Negev desert, Israel, *Geomorphology*, *9*, 117–126.

Wohl, E. E., and H. Ikeda (1997), Experimental simulation of channel incision into a cohesive substrate at varying gradient, *Geology*, *25*, 295–298.

Wohl, E. E., and H. Ikeda (1998), The effect of roughness configuration on velocity profiles in an artificial channel, *Earth Surf. Processes Landforms*, *23*, 159–169.

Wohl, E. E., and K. Jaeger (2009a), A conceptual model for the longitudinal distribution of wood in mountain streams, *Earth Surf. Processes Landforms*, *34*, 329–344.

Wohl, E. E., and K. Jaeger (2009b), Geomorphic implications of hydroclimatic differences among step-pool channels, *J. Hydrol.*, *374*, 148–161.

Wohl, E. E., and C. J. Legleiter (2003), Controls on pool characteristics along a resistant-boundary channel, *J. Geol.*, *111*, 103–114.

Wohl, E. E., and D. M. Merritt (2001), Bedrock channel morphology, *Geol. Soc. Am. Bull.*, *113*, 1205–1212.

Wohl, E. E., and D. M. Merritt (2005), Prediction of mountain stream morphology, *Water Resour. Res.*, *41*, W08419, doi:10.1029/2004WR003779.

Wohl, E. E., and D. M. Merritt (2008), Reach-scale channel geometry of mountain streams, *Geomorphology*, *93*, 168–185.

Wohl, E. E., and D. J. Merritts (2007), What is a natural river?, *Geogr. Compass*, *1*(4), 871–900.

Wohl, E. E., and T. Oguchi (2004), Geographic information systems and mountain hazards, in *Geographic Information Science and Mountain Geomorphology*, edited by M. P. Bishop and J. F. Shroder, pp. 309–341, Praxis, Chichester, U. K.

Wohl, E. E., and P. A. Pearthree (1991), Debris flows as geomorphic agents in the Huachuca Mountains of southeastern Arizona, *Geomorphology*, *4*, 273–292.

Wohl, E. E., and G. Springer (2005), Bedrock channel incision along the Upper Rio Chagres basin, Panama, in *The Rio Chagres, Panama: A Multidisciplinary Profile of a Tropical Watershed*, edited by R. S. Harmon, pp. 189–209, Springer, Dordrecht, The Netherlands.

Wohl, E. E., and D. M. Thompson (2000), Velocity characteristics along a small step-pool channel, *Earth Surf. Processes Landforms*, *25*, 353–367.

Wohl, E. E., and A. C. Wilcox (2005), Channel geometry of mountain streams in New Zealand, *J. Hydrol.*, *300*, 252–266.

Wohl, E. E., K. R. Vincent, and D. J. Merritts (1993), Pool and riffle characteristics in relation to channel gradient, *Geomorphology*, *6*, 99–110.

Wohl, E. E., N. Greenbaum, A. P. Schick, and V. R. Baker (1994a), Controls on bedrock channel incision along Nahal Paran, Israel, *Earth Surf. Processes Landforms*, *19*, 1–13.

Wohl, E. E., D. J. Anthony, S. W. Madsen, and D. M. Thompson (1996), A comparison of surface sampling methods for coarse fluvial sediments, *Water Resour. Res.*, *32*, 3219–3226.

Wohl, E. E., S. Madsen, and L. MacDonald (1997), Characteristics of log and clast bed-steps in step-pool streams of northwestern Montana, USA, *Geomorphology*, *20*, 1–10.

Wohl, E. E., D. M. Thompson, and A. J. Miller (1999), Canyons with undulating walls, *Geol. Soc. Am. Bull.*, *111*, 949–959.

Wohl, E. E., D. A. Cenderelli, and M. Mejia-Navarro (2001), Channel change from extreme floods in canyon rivers, in *Applying Geomorphology to Environmental Management*, edited by D. J. Anthony et al., pp. 149–174, Water Resour. Publ., Littleton, Colo.

Wohl, E. E., J. N. Kuzma, and N. E. Brown (2004), Reach-scale channel geometry of a mountain river, *Earth Surf. Processes Landforms*, *29*, 969–981.

Wohl, E. E., P. L. Angermeier, B. Bledsoe, G. M. Kondolf, L. MacDonnell, D. M. Merritt, M. A. Palmer, N. L. Poff, and D. Tarboton (2005), River restoration, *Water Resour. Res.*, *41*, W10301, doi:10.1029/2005WR003985.

Wohl, E. E., D. Cooper, L. Poff, F. Rahel, D. Staley, and D. Winters (2007), Assessment of stream ecosystem function and sensitivity in the Bighorn National Forest, Wyoming, *Environ. Manage.*, *40*, 284–302.

Wohl, E. E., M. Palmer, and G. M. Kondolf (2008), River management in the United States, in *River Futures: An Integrative Scientific Approach to River Repair*, edited by G. J. Brierley and K. A. Fryirs, pp. 174–200, Island Press, Washington, D. C.

Wohl, E. E., F. Ogden, and J. Goode (2009), Episodic wood loading in a mountainous neotropical watershed, *Geomorphology*, *111*, 149–159.

Wohl, E. E., R. H. Webb, V. R. Baker, and G. Pickup (1994b), *Sedimentary flood records in the bedrock canyons of rivers in the monsoonal region of Australia*, *Water Resour. Pap.*, vol. 107, 102 pp., Colo. State Univ., Ft. Collins.

Wolcott, J., and M. Church (1991), Strategies for sampling spatially heterogeneous phenomena: The example of river gravels, *J. Sediment. Petrol.*, *61*, 534–543.

Wolf, E. C., D. J. Cooper, and N. T. Hobbs (2007), Hydrologic regime and herbivory stabilize an alternative state in Yellowstone National Park, *Ecol. Appl.*, *17*, 1572–1587.

Wolkowinsky, A. J., and D. E. Granger (2004), Early Pleistocene incision of the San Juan River, Utah, dated with ^{26}Al and ^{10}Be, *Geology*, *32*, 749–752.

Wolman, M. G. (1954), A method of sampling coarse river-bed material, *Eos Trans. AGU*, *35*, 951–956.

Wolman, M. G. (1955), *The natural channel of Brandywine Creek, Pennsylvania*, *U.S. Geol. Surv. Prof. Pap.*, *271*.

Wolman, M. G. (1959), Factors influencing erosion of a cohesive river bank, *Am. J. Sci.*, *257*, 204–216.

Wolman, M. G. (1967), A cycle of sedimentation and erosion in urban river channels, *Geogr. Ann.*, *49A*, 385–395.

Wolman, M. G., and R. Gerson (1978), Relative scales of time and effectiveness of climate in watershed geomorphology, *Earth Surf. Processes*, *3*, 189–208.

Wolman, M. G., and L. B. Leopold (1957), *River flood plains: Some observations on their formation*, *U.S. Geol. Surv. Prof. Pap.*, *282-C*.

Wolman, M. G., and J. P. Miller (1960), Magnitude and frequency of forces in geomorphic processes, *J. Geol.*, *68*, 54–74.

Wolman, M. G., and A. P. Schick (1967), Effects of construction on fluvial sediment, urban and suburban areas of Maryland, *Water Resour. Res.*, *3*, 451–464.

Wolter, A., B. Ward, and T. Millard (2010), Instability in eight sub-basins of the Chilliwack River valley, British Columbia, Canada: A comparison of natural and logging-related landslides, *Geomorphology*, *120*, 123–132.

Womack, W. R., and S. A. Schumm (1977), Terraces of Douglas Creek, northwestern Colorado: An example of episodic erosion, *Geology*, *5*, 72–76.

Wondzell, S. M., J. LaNier, R. Haggerty, R. D. Woodsmith, and R. T. Edwards (2009), Changes in hyporheic exchange flow following experimental wood removal in a small, low-gradient stream, *Water Resour. Res.*, *45*, W05406, doi:10.1029/2008WR007214.

Wondzell, S. M., and F. J. Swanson (1996a), Seasonal and storm dynamics of the hyporheic zone of a fourth-order mountain stream. I. Hydrologic processes, *J. North Am. Benthol. Soc.*, *15*, 3–19.

Wondzell, S. M., and F. J. Swanson (1996b), Seasonal and storm dynamics of the hyporheic zone of a fourth-order mountain stream. II: Nitrogen cycling, *J. North Am. Benthol. Soc.*, *15*, 20–34.

Wondzell, S. M., and F. J. Swanson (1999), Floods, channel change, and the hyporheic zone, *Water Resour. Res.*, *35*, 555–567.

Wong, M., G. Parker, P. DeVries, T. M. Brown, and S. J. Burges (2007), Experiments on the dispersion of tracer stones under lower-regime plane-bed equilibrium bed load transport, *Water Resour. Res.*, *43*, W03440, doi:10.1029/2006WR005172.

Woo, M.-K., and R. Thorne (2003), Streamflow in the Mackenzie Basin, Canada, *Arctic*, *56*, 328–340.

Woo, M.-K., and R. Thorne (2006), Snowmelt contribution to discharge from a large mountainous catchment in subarctic Canada, *Hydrol. Processes*, *20*, 2129–2139.

Woo, M.-K., D. L. Kane, S. K. Carey, and D. Yang (2008), Progress in permafrost hydrology in the new millennium, *Permafrost Periglacial Processes*, *19*, 237–254.

Woodhouse, C. A., S. T. Gray, and D. M. Meko (2006), Updated streamflow reconstructions for the Upper Colorado River basin, *Water Resour. Res.*, *42*, W05415, doi:10.1029/2005WR004455.

Woods, S. W., A. Birkas, and R. Ahl (2007), Spatial variability of soil hydrophobicity after wildfires in Montana and Colorado, *Geomorphology*, *86*, 465–479.

Woods, S. W., L. H. MacDonald, and C. J. Westbrook (2006), Hydrologic interactions between an alluvial fan and a slope wetland in the central Rocky Mountains, USA, *Wetlands*, *26*, 230–243.

Woodward, J. C. (1995), Patterns of erosion and suspended sediment yield in Mediterranean river basins, in *Sediment and Water Quality in River Catchments*, edited by I. Foster, A. Gurnell, and B. Webb, pp. 365–389, John Wiley, Chichester, U. K.

Wooldridge, C. L., and E. J. Hickin (2002), Step-pool and cascade morphology, Mosquito Creek, British Columbia: A test of four analytical techniques, *Can. J. Earth Sci.*, *39*, 493–503.

Wooster, J. K., S. R. Dusterhoff, Y. Cui, L. S. Sklar, W. E. Dietrich, and M. Malko (2008), Sediment supply and relative size distribution effects on fine sediment infiltration into immobile gravels, *Water Resour. Res.*, *44*, W03424, doi:10.1029/2006WR005815.

World Meteorological Organization (1987), Concept of the global energy and water cycle experiment, *WMO/TD 215*, World Meteorol. Organ., Geneva, Switzerland.

Wotton, R. S. (1986), The use of silk life-lines by larvae of Simulium noelleri (Diptera), *Aquat. Insects*, *8*, 255–261.

Wright, A., W. A. Marcus, and R. Aspinall (2000), Evaluation of multispectral, fine scale digital imagery as a tool for mapping stream morphology, *Geomorphology*, *33*, 107–120.

Wright, J. P. (1999), Effects of an ecosystem engineer, the beaver, on regional patterns of species richness of the streamside herbaceous plant community, *1999 Annual Meeting Abstracts*, pp. 326., Ecological Society of America.

Wright, S. A., J. C. Schmidt, T. S. Melis, D. J. Topping, and D. M. Rubin (2008), Is there enough sand? Evaluating the fate of Grand Canyon sandbars, *GSA Today*, *18*(8), 4–10.

Wroblicky, G. J., M. E. Campana, H. M. Valett, and C. N. Dahm (1998), Seasonal variation in surface-subsurface water exchange and lateral hyporheic area of two stream-aquifer systems, *Water Resour. Res.*, *34*, 317–328.

Wroten, W. H. , (1956), The railroad tie industry in the central Rocky Mountain region: 1867–1900, unpublished Ph.D. dissertation, 287 pp., Colo. State Univ., Ft. Collins.

Wu, C.-H., and S.-C. Chen (2009), Determining landslide susceptibility in central Taiwan from rainfall and six site factors using the analytical hierarchy process method, *Geomorphology*, *112*, 190–204.

Wu, T. H. (2003), Assessment of landslide hazard under combined loading, *Can. Geotech. J.*, *40*, 821–829.

Wu, T. H., W. P. McKinnell, and D. N. Swanton (1979), Strength of tree roots and landslides on Prince of Wales Island, Alaska, *Can. Geotech. J.*, *16*, 19–33.

Wu, W., F. D. Shields, S. J. Bennett, and S. S. Y. Wang (2005), A depth-averaged two-dimensional model for flow, sediment transport, and bed topography in curved channels with riparian vegetation, *Water Resour. Res.*, *41*, W03015, doi:10.1029/2004WR003730.

Wu, W., and S. S. Y. Wang (2004), A depth-averaged two-dimensional numerical model of flow and sediment transport in open channels with vegetation, in *Riparian Vegetation and Fluvial Geomorphology*, edited by S. J. Bennett and A. Simon, pp. 253–265, AGU, Washington, D. C.

Wulf, H., B. Bookhagen, and D. Scherler (2010), Seasonal precipitation gradients and their impact on fluvial sediment flux in the northwest Himalaya, *Geomorphology*, *118*, 13–21.

Wyatt, A. M., and S. W. Franks (2006), The Multi-Model approach to rainfall-runoff modelling, in *Predictions in Ungauged Basins: Promise and Progress*, edited by M. Sivapalan et al., *IAHS Publ.*, *303*, 134–144.

Wynn, T. M., M. B. Henderson, and D. H. Vaughan (2008), Changes in streambank erodibility and critical shear stress due to subaerial processes along a headwater stream, southwestern Virginia, USA, *Geomorphology*, *97*, 260–273.

Wyżga, B. (1991), Present-day downcutting of the Raba River channel (Western Carpathians, Poland) and its environmental effects, *Catena*, *18*, 551–556.

Wyżga, B. (1996), Changes in the magnitude and transformation of flood waves subsequent to the channelization of the Raba River, Polish Carpathians, *Earth Surf. Processes Landforms*, *21*, 749–763.

Wyżga, B. (2008), A review on channel incision in the Polish Carpathian rivers during the 20th century, in *Gravel-Bed Rivers VI: From Process Understanding to River Restoration*, edited by H. Habersack, H. Piégay, and M. Rinaldi, pp. 525–555, Elsevier, Amsterdam.

Wyżga, B., and J. Zawiejska (2005), Wood storage in a wide mountain river: Case study of the Czarny Dunajec, Polish Carpathians, *Earth Surf. Processes Landforms*, *30*, 1475–1494.

Xiao, C., et al. (2007), Observed changes of cryosphere in China over the second half of the 20th century: An overview, *Ann. Glaciol.*, *46*, 382–390.

Xie, X., S. Norra, Z. A. Berner, and D. Stueben (2005), A GIS-supported multivariate statistical analysis of relationships among stream water chemistry, geology and land use in Baden-Wuerttemberg, Germany, *Water Air Soil Pollut.*, *167*, 39–57.

Yager, E. M., J. W. Kirchner, and W. E. Dietrich (2007), Calculating bed load transport in steep boulder bed channels, *Water Resour. Res.*, *43*, W07418, doi:10.1029/2006WR005432.

Yair, A., and J. De Ploey (1979), Field observations and laboratory experiments concerning the creep process of rock blocks in an arid environment, *Catena*, *6*, 245–258.

Yair, A., and H. Lavee (1985), Runoff generation in arid and semi-arid zones, in *Hydrological Forecasting*, edited by M. G. Anderson and T. P. Burt, pp. 183–220, John Wiley, New York.

Yalin, M. S. (1963), An expression for bedload transportation, *J. Hydraul. Div. Am. Soc. Civ. Eng.*, *89*, 221–250.

Yamada, S. (1999a), Mountain ordering: A method for classifying mountains based on their morphometry, *Earth Surf. Processes Landforms*, *24*, 653–660.

Yamada, S. (1999b), The role of soil creep and slope failure in the landscape evolution of a head water basin: Field measurements in a zero order basin of northern Japan, *Geomorphology, 28,* 329–344.

Yamazaki, G., Y. Ishii, D. Kobayashi, N. Ishikawa, and H. Shibata (2005), Cl⁻ budget and subsurface runoff process during the snowmelt season in a nival mountainous watershed, *J. Jpn. Soc. Snow Ice, 67,* 477–491.

Yang, C. T. (1971), Formation of riffles and pools, *Water Resour. Res., 7,* 1567–1574.

Yang, C. T. (1976), Minimum unit stream power and fluvial hydraulics, *J. Hydraul. Div. Am. Soc. Civ. Eng., 102,* 919–934.

Yang, C. T., and C. C. S. Song (1979), Theory of minimum rate of energy dissipation, *J. Hydraul. Div. Am. Soc. Civ. Eng., 105,* 769–784.

Yang, C. T., C. C. S. Song, and M. J. Woldenberg (1981), Hydraulic geometry and minimum rate of energy dissipation, *Water Resour. Res., 17,* 1014–1018.

Yang, C. T., M. A. Trevino, and F. J. Simoes (1998), *User's manual for generalized stream tube model for alluvial river simulation, ver. 2.0,* U.S. Bur. of Reclam. Tech. Serv. Group, Denver, Colo.

Yang, D., T. Koike, and H. Tanizawa (2003), Effect of precipitation spatial distribution on the hydrological response in the upper Tone River of Japan, in *Weather Radar Information and Distributed Hydrological Modelling,* edited by Y. Tachikawa et al., *IAHS Publ., 282,* 194–202.

Yanites, B. J., R. H. Webb, P. G. Griffiths, and C. S. Magirl (2006), Debris flow deposition and reworking by the Colorado River in Grand Canyon, Arizona, *Water Resour. Res., 42,* W11411, doi:10.1029/2005WR004847.

Yanites, B. J., G. E. Tucker, K. J. Mueller, and Y.-G. Chen (2010a), How rivers react to large earthquakes: Evidence from central Taiwan, *Geology, 38,* 639–642.

Yanites, B. J., G. E. Tucker, K. J. Mueller, Y.-G. Chen, T. Wilcox, S.-Y. Huang, and K.-W. Shi (2010b), Incision and channel morphology across active structures along the Peikang River, central Taiwan: Implications for the importance of channel width, *Geol. Soc. Am. Bull., 122,* 1192–1208.

Yanosky, T. M. (1982a), *Effects of flooding upon woody vegetation along parts of the Potomac River flood plain,* U.S. Geol. Surv. Prof. Pap., *1206,* 21 pp.

Yanosky, T. M. (1982b), Hydrologic inferences from ring widths of flood-damaged trees, Potomac River, Maryland, *Environ. Geol., 4,* 43–52.

Yanosky, T. M. (1983), *Evidence of floods on the Potomac River from anatomical abnormalities in the wood of flood-plain trees,* U.S. Geol. Surv. Prof. Pap., *1296,* 42 pp.

Yanosky, T. M. (1984), Documentation of high summer flows on the Potomac River from the wood anatomy of ash trees, *Water Resour. Bull., 20,* 241–250.

Yanosky, T. M., and R. D. Jarrett (2002), Dendrochronologic evidence for the frequency and magnitude of paleofloods, in *Ancient Floods, Modern Hazards: Principles and Applications of Paleoflood Hydrology,* edited by P. K. House et al., pp. 77–89, AGU, Washington, D. C.

Yarnell, S. M., J. F. Mount, and E. W. Larsen (2006), The influence of relative sediment supply on riverine habitat heterogeneity, *Geomorphology, 80,* 310–324.

Yasuda, Y., and I. Ohtsu (1999), Flow resistance of skimming flow in stepped channels, in Proceedings of the 28th IAHR Congress [CD-ROM], Int. Assoc. of Hydraul. Eng. and Res., Madrid.

Yatsu, E. (1955), On the longitudinal profile of the graded river, *Eos Trans. AGU, 36,* 655–663.

Yeats, R. S., and V. C. Thakur (2008), Active faulting south of the Himalayan front; establishing a new plate boundary, *Tectonophysics, 453,* 63–73.

Yesenov, U. Y., and A. S. Degovets (1979), Catastrophic mudflow on the Bolshaya Almatinka River in 1977, *Soviet Hydrol., 18,* 158–160.

Yetemen, O., E. Istanbulluoglu, and E. R. Vivoni (2010), The implications of geology, soils, and vegetation on landscape morphology: Inferences from semi-arid basins with complex vegetation patterns in central New Mexico, USA, *Geomorphology, 116,* 246–263.

Yochum, S. E. (2010), Flow resistance prediction in high-gradient streams, Ph.D. dissertation, Colo. State Univ., Fort Collins.

Yokoyama, S. (1999), Rapid formation of river terraces in non-welded ignimbrite along the Hishida River, Kyushu, Japan, *Geomorphology*, *30*, 291–304.

Yoo, K., R. Amundson, A. M. Heimsath, and W. E. Dietrich (2005), Process-based model linking pocket gopher , (Thomomys bottae) activity to sediment transport and soil thickness, *Geology*, *33*, 917–920.

Yoo, K., R. Amundson, A. M. Heimsath, W. E. Dietrich, and G. H. Brimhall (2007), Integration of geochemical mass balance with sediment transport to calculate rates of soil chemical weathering and transport on hillslopes, *J. Geophys. Res.*, *112*, F02013, doi:10.1029/2005JF000402.

Young, A. (1972), *Slopes*, Oliver and Boyd, Edinburgh, U. K.

Young, D. R., I. C. Burke, and D. H. Knight (1984), Water relations of high-elevation phreatophytes in Wyoming, *Am. Midl. Nat.*, *114*, 384–392.

Young, G. J. (1980), Monitoring glacier outburst floods, *Nord. Hydrol.*, *11*, 285–300.

Young, M. K. (1994), Movement and characteristics of stream-borne coarse woody debris in adjacent burned and undisturbed watersheds in Wyoming, *Can. J. For. Res.*, *24*, 1933–1938.

Young, M. K., D. Haire, and M. A. Bozek (1994), The effect and extent of railroad tie drives in streams of southeastern Wyoming, *West, J. Appl. For.*, *9*, 125–130.

Young, M. K., E. A. Mace, E. T. Ziegler, and E. K. Sutherl (2006), Characterizing and contrasting instream and riparian coarse wood in western Montana basins, *For. Ecol. Manage.*, *226*, 26–40.

Young, M. K., R. N. Schmal, and C. M. Sobczak , (1990), Railroad tie drives and stream channel complexity: Past impacts, current status, and future prospects, in *Proceedings of the Annual Meeting of Society of American Foresters, Spokane, Washington, Publ.*vol. 89-02, pp. 126–130, Soc. of Am. For., Bethesda, Md

Young, R. W. (1985), Waterfalls: Form and process, *Z. Geomorphol.*, *55*, 81–95.

Young, R. W. (1989), Crustal constraints on the evolution of the continental divide of eastern Australia, *Geology*, *17*, 528–530.

Young, R. W., and I. McDougall (1993), Long-term landscape evolution: Early Miocene and Modern rivers in southern New South Wales, Australia, *J. Geol.*, *101*, 35–49.

Young, R. W., and C. R. Twidale (1993), Geomorphology in Australia, in *The Evolution of Geomorphology*, edited by H. J. Walker and W. E. Grabau, pp. 29–43, John Wiley, New York.

Young, W. J., and T. R. H. Davies (1991), Bedload transport processes in a braided gravel-bed river model, *Earth Surf. Processes Landforms*, *16*, 499–511.

Young, W. J., J. M. Olley, I. P. Prosser, and R. F. Warner (2001), Relative changes in sediment supply and sediment transport capacity in a bedrock-controlled river, *Water Resour. Res.*, *37*, 3307–3320.

Yu, B. (1995), Contribution of heavy rainfall to rainfall erosivity, runoff, and sediment transport in the wet tropics of Australia, in *Natural and Anthropogenic Influences in Fluvial Geomorphology, Geophys. Monogr. Ser.*, vol. 89, edited by J. E. Costa et al., pp. 113–123, AGU, Washington, D. C.

Yu, B. (2005), Process-based erosion modelling: Promises and progress, in *Forests, Water and People in the Humid Tropics*, edited by M. Bonell and L. A. Bruijnzeel, pp. 790–810, Cambridge Univ. Press, Cambridge, U. K.

Yu, B., and D. Neil (1994), Temporal and spatial variation of sediment yield in the Snowy Mountains region, Australia, in *Variability in Stream Erosion and Sediment Transport*, edited by L. J. Olive, R. J. Loughran, and J. A. Kesby, *IAHS Publ.*, *224*, 281–289.

Yu, S., C. Zhu, and F. Wang (2003), Radiocarbon constraints on the Holocene flood deposits of the Ning-Zhen Mountains, lower Yangtze River area of China, *J. Quat. Sci.*, *18*, 521–525.

Yuan, Y., X. Shao, W. Wei, S. Yu, Y. Gong, and V. Trouet (2007), The potential to reconstruct Manasi River streamflow in the northern Tien Shan Mountains (NW China), *Tree Ring Res.*, *63*, 81–93.

Yumoto, M., T. Ogata, N. Matsuoka, and E. Matsumoto (2006), Riverbank freeze-thaw erosion along a small mountain stream, Nikko volcanic area, central Japan, *Permafrost Periglacial Processes*, *17*, 325–339.

Zalasiewicz, J., et al. (2008), Are we now living in the Anthropocene?, *GSA Today*, *18*(2), 4–8.

Zanuttigh, B., and A. Lamberti (2007), Instability and surge development in debris flows, *Rev. Geophys.*, *45*, RG3006, doi:10.1029/2005RG000175.

Zaprowski, B. J., E. B. Evenson, F. J. Pazzaglia, and J. B. Epstein (2001), Knickzone propagation in the Black Hills and northern High Plains: A different perspective on the late Cenozoic exhumation of the Laramide Rocky Mountains, *Geology*, *29*, 547–550.

Zehe, E., H. Lee, and M. Sivapalan (2005), Derivation of closure relations and commensurate state variables for meso-scale models using the REW approach, in *Predictions in Ungauged Basins: International Perspectives on the State of the Art and Pathways Forward*, edited by S. Franks et al., *IAHS Publ.*, *301*, 134–158.

Zeitler, P. K., et al. (2001), Erosion, Himalayan geodynamics, and the geomorphology of metamorphism, *GSA Today*, *11*(1), 4–8.

Zelt, R. B., and E. E. Wohl (2004), Channel and woody debris characteristics in adjacent burned and unburned wtersheds a decade after wildfire, Park County, Wyoming, *Geomorphology*, *57*, 217–233.

Zen, E., and K. L. Prestegaard (1994), Possible hydraulic significance of two kinds of potholes: Examples from the paleo-Potomac River, *Geology*, *22*, 47–50.

Zêzere, J. L., R. A. C. Garcia, S. C. Oliveira, and E. Reis (2008), Probabilistic landslide risk analysis considering direct costs in the area north of Lisbon (Portugal), *Geomorphology*, *94*, 467–495.

Zgheib, P. W. (1990), Large bed element channels in steep mountain streams, in *Hydrology in Mountainous Regions II - Artificial Reservoirs, Water and Slopes*, edited by R. O. Sinniger and M. Monbaron, *IAHS Publ.*, *194*, 277–283.

Zhang, B., and R. S. Govindaraju (2003), Geomorphology-based artificial neural networks , (GANNs) for estimation of direct runoff over watersheds, *J. Hydrol.*, *273*, 18–34.

Zhang, K., K. Liu, and J. Yang (2004), Asymmetrical valleys created by the geomorphic response of rivers to strike-slip fault, *Quat. Res.*, *62*, 310–315.

Zhang, T., R. G. Barry, K. Knowles, F. Ling, and R. L. Armstrong (2003a), Distribution of seasonally and perennially frozen ground in the Northern Hemisphere, *Permafrost: Proceedings of the 8th International Conference on Permafrost, Zurich, Switzerland, 21–25 July 2003*, pp. 1289–1294, A. A. Balkema, Lisse, Netherlands.

Zhang, T., M. Cen, G. Zhou, and X. Wu (2006), A new method for debris flow detection using multi-temporal DEMs without ground control points, *Int. J. Remote Sens.*, *27*, 4911–4921.

Zhang, Y., J. N. Negishi, J. S. Richardson, and R. Kolodziejczyk (2003b), Impacts of marine-derived nutrients on stream ecosystem functioning, *Proc. R. Soc. London, Ser. B*, *270*, 2117–2123.

Zhao, Y., S. Wang, L. Duan, Y. Lei, P. Cao, and J. Hao (2008), Primary air pollutant emissions of coal-fired power plants in China: Current status and future predictions, *Atmos. Environ.*, *42*, 8442–8452.

Zicheng, K., and L. Jing (1987), Erosion processes and effects of debris flows, in *Erosion and Sedimentation in the Pacific Rim*, edited by R. L. Beschta et al., *IAHS Publ.*, *165*, 233–242.

Ziegler, A. D., R. A. Sutherland, and T. W. Giambelluca (2001), Acceleration of Horton overland flow and erosion by footpaths in an upland agricultural watershed in northern Thailand, *Geomorphology*, *41*, 249–262.

Zielinski, T. (2003), Catastrophic flood effects in alpine/foothill fluvial systems (a case study from the Sudetes Mts, SW Poland), *Geomorphology*, *54*, 293–306.

Ziemer, R. R. (1981), Storm flow response to road building and partial cutting in small streams of northern California, *Water Resour. Res.*, *17*, 907–917.

Ziemer, R. R. (1992), Effect of logging on subsurface pipeflow and erosion: Coastal northern California, USA, in *Erosion, Debris Flows and Environment in Mountain Regions*, edited by D. E. Walling, T. R. Davies, and B. Hasholt, *IAHS Publ.*, *209*, 187–197.

Ziemer, R. R. (1998), Monitoring watersheds and streams, *Proceedings of the Conference on Coastal Watersheds: The Caspar Creek Story*, edited by R. R. Ziemer, pp. 129–134, Pacific Southwest Res. Sta., USDA, Albany, Calif.

Zierholz, C., I. P. Prosser, P. J. Fogarty, and P. Rustomji (2001), In-stream wetlands and their significance for channel filling and the catchment sediment budget, Jugiong Creek, New South Wales, *Geomorphology*, *38*, 221–235.

Zimmermann, A., and M. Church (2001), Channel morphology, gradient profiles and bed stresses during flood in a step-pool channel, *Geomorphology*, *40*, 311–327.

Zimmermann, A., M. Church, and M. A. Hassan (2008), Video-based gravel transport measurements with a flume mounted light table, *Earth Surf. Processes Landforms*, *33*, 2285–2296.

Zimmermann, A., M. Church, and M. A. Hassan (2010), Step-pool stability: Testing the jammed state hypothesis, *J. Geophys. Res.*, *115*, F02008, doi:10.1029/2009JF001365.

Zimmermann, M. (2004), Managing debris flow risks: Security measures for a hazard-prone resort in Switzerland, *Mt. Res. Dev.*, *24*, 19–23.

Zimmerman, R. C., J. C. Goodlett, and G. H. Comer (1967), The influence of vegetation on channel form of small streams, in *Symposium on River Morphology*, *Int. Assoc. Sci. Hydrol. Publ.*vol. 75, pp. 255–275.

Zisu, N. S., N. Greenbaum, M. Inbar, and A. Flexer (2003), Morphometric analysis of the Naftali Mountain front, *Isr. J. Earth Sci.*, *52*, 191–202.

Zolezzi, G., A. Bellin, M. C. Bruno, B. Maiolini, and A. Siviglia (2009), Assessing hydrological alterations at multiple temporal scales: Adige River, Italy, *Water Resour. Res.*, *45*, W12421, doi:10.1029/2008WR007266.

Zong, L., and H. Nepf (2010), Flow and deposition in and around a finite patch of vegetation, *Geomorphology*, *116*, 363–372.

Zorn, M., K. Natek, and B. Komac (2006), Mass movements and flash-floods in Slovene Alps and surrounding mountains, *Stud. Geomorphol. Carpatho-Balcanica*, *40*, 127–145.

Zuhal, A., S. Aynur, S. Arda, and S. A. Unal (2007), Calibration/validation of satellite derived snow products with in situ data over the mountainous eastern part of Turkey, in *Remote Sensing for Environmental Monitoring and Change Detection*, edited by M. Owe and C. M. Neale, *IAHS Publ.*, *316*, 157–168.

SUBJECT INDEX

—bar, 92, 119, 151, 161, 171–172, 207, 231–232, 243, 274, 277, 366
—flow, 2–3, 7, 11, 18, 32, 59, 72, 75, 80, 90, 114, 121, 141–154, 160–161, 171–172, 174, 189, 193, 195–196, 207, 212, 215–216, 224, 226–229, 231–233, 238–240, 242–243, 246, 258, 261, 277, 284, 292, 302, 322, 325, 327, 336–337, 345, 349, 350–351, 354
—see also Manning's n
—form, 3, 7, 11, 20, 32, 72, 74–75, 90, 92, 114, 119, 121, 141–147, 149–154, 157, 160–161, 166, 193, 195–196, 207, 215–216, 224, 226–227, 242–243, 246, 249, 258, 261, 277, 284, 292, 323, 327, 336–337, 349–350, 354
—grain, 2,7, 12, 38, 74–75, 80, 141–142, 145–152, 154, 157, 166, 171, 173–174, 189, 193, 195, 226–227, 231, 233, 238, 240, 243, 258
—partitioning, 83, 149–153, 240
—see also roughness height
—spill, 149–150, 152, 154
—substrate/boundary, 7, 41, 75, 80, 83, 114, 119, 174, 189, 224, 226–229, 231–233, 238–242, 258, 292, 302, 322–323, 351
Resistance coefficient, 141–143, 145–146, 148–149, 227, 349
—see also Manning's n
Response reaches, 242–243, 247, 249, 258
Reynolds number, 109, 156–158
Rills, 37, 47, 59
Riparian ecosystems/biota, 1, 293, 311, 319, 342, 2, 3, 226, 258, 285–286, 319
River biological integrity, 341–342
River chemistry, 4–5, 128–130, 132–133, 141, 350
—see also contaminant
—see also dissolved organic carbon
—see also ionic pulse
—see also nitrogen
—see also nutrient
—see also total dissolved solids
—see also trace metals
River continuum concept, 277–279, 282–283, 293
River health, 342
River management, 114, 254, 259, 339–340, 352
River physical integrity, 342
River rehabilitation, 339–341
River restoration, 114, 263, 268, 292, 311, 321, 343–345, 350–352
River training, 321–323
Roads, 221, 231, 298–299, 301, 305, 344

Rock-mass strength, 23, 229
Roughness height, 75, 145–146, 150, 153–154, 162, 238, 240
—see also resistance
Runoff (see overland flow)

Sackung, 29
Salmonid, 132, 216–217, 222, 264, 287–288, 291, 300, 327, 341–342
Sandur, 253
Sapping, 50, 59–60
Sediment budget, 202, 204
Sediment connectivity, 3, 184, 350
Sediment fingerprinting, 210, 213
Sediment sampling, 18
—bulk, 30, 167, 227
—DH-48 sampler, 211
—Helley-Smith sampler, 179–182
—in situ clast measurements, 167
Sediment transport, 7, 13, 21, 23, 31–32, 35, 37–38, 40, 47, 58, 62, 65, 80, 83–86, 92–94, 99, 102, 104–106, 111–112, 114–115, 117–118, 120, 122, 142, 158–159, 162, 170, 172, 176–177, 182–184, 186–187, 191, 193, 195–196, 198, 200–201, 205, 209, 211–213, 224, 226, 229, 233–235, 237, 239, 241, 246, 248–249, 251, 254–255, 263, 301, 310, 314, 318, 321, 323, 325, 336, 346, 350, 351, 353
—bedload, 2, 78, 83, 85–86, 120, 123, 151, 164, 170, 172–173, 175, 177, 179–185, 187–189, 194, 200, 202, 204, 208–211, 241, 253, 274, 287, 306, 336, 350, 353
—bedload discharge, 179, 182, 190, 194–195, 241, 350
—bedload pulse, 187, 206
—bedload sheets, 161, 182, 183
—bedload transport distances, 200, 210, 219, 241
—bedload transport equations, 191, 195, 227
—bedload transport rates, 183, 184, 188, 200, 210
—bedload yield, 179, 197
—capacity-limited, 176, 191, 237
—compound Poisson model, 188
—see also downstream fining
—see also entrainment
—episodic bedload movement, 185
—equal mobility, 176, 186,
—hysteresis, 184–185, 208, 209
—selective entrainment, 177, 179, 227